Handbook of Experimental Pharmacology

Volume 108/I

GTPases in Biology I

Contributors

K. Aktories, D.L. Altschuler, M.W. Anderson, C. Barlowe, E. Bergmann, J. Bian, L. Birnbaumer, G.M. Bokoch, H.R. Bourne, P. Brennwald, C.C. Burgess, E.S. Burstein, S.L. Campbell-Burk, M.-F. Carlier, L. Carnell, J. Cavallius, R.A. Cerione, P. Chardin, P.S. Charifson, R.A. Chavez, Y.-T. Chen, G.J. Clark, J. Coburn, T.A. Darden, M.A. De Matteis, C.J. Der, J. Downward, P. Dupree, T. Evans, L.A. Feig, G. Fischer von Mollard, C.K. Foley, D. Gallwitz, T.K. Ghosh, J.B. Gibbs, D.L. Gill, R. Gilmore, B.D. Gomperts, R.S. Goody, A. Hall, M.J. Hart, J.S. Herskovits, L.A. Huber, A. Hwang, H. Itoh, R. Jahn, F. Jurnak, I. Just, R.A. Kahn, K. Kaibuchi, Y. Kaziro, A. Kikuchi, S.-H. Kim, E.G. Lapetina, D. Leonard, T.H.W. Lillie, A. Lütcke, A. Luini, I.G. Macara, K. Matsumoto, F. McCormick, W.C. Merrick, M.V. Milburn, S.G. Miller, H.-P.H. Moore, J. Moss, P. Novick, V.M. Olkkonen, E.F. Pai, D. Pantaloni, L.G. Pedersen, S.R. Pfeffer, G.G. Privé, P.J. Rapiejko, A.J. Ridley, T. Sasaki, R. Schekman, K. Shinjo, A.D. Short, K. Simons, D.W. Stacey, M. Strom, T.C. Südhof, Y. Takai, K. Tanaka, A. Toh-e, M. Torti, R.B. Vallee, T.E. Van Aken, M. Vaughan, R.T. Waldron, R.A. Weinberg, M. Wessling-Resnick, A. Wittinghofer, Y.-A. Yoon, H. Yu, M. Zerial, Y. Zheng

Editors

Burton F. Dickey and Lutz Birnbaumer

Springer-Verlag
Berlin Heidelberg New York London Paris
Tokyo Hong Kong Barcelona Budapest

Professor BURTON F. DICKEY, M.D.
Baylor College of Medicine
Department of Medicine, Cell Biology and
Experimental Therapeutics, Room 3C-385
Houston VA Medical Center
2002 Holcombe Boulevard
Houston, TX 77030, USA

Professor LUTZ BIRNBAUMER, Ph.D.
Baylor College of Medicine
Department of Cell Biology, Molecular Physiology and Biophysics,
Medicine and Neurosciences
One Baylor Plaza
Houston, TX 77030, USA

With 117 Figures, Some in Colour and 29 Tables

Library of Congress Cataloging-in-Publication Data. GTPases in Biology / contributors, K. Aktories . . . [et al.] : editors, Burton F. Dickey and Lutz Birnbaumer. p cm. – (Handbook of experimental pharmacology : v. 108) Includes bibliographical references and index
ISBN-13: 978-3-642-78269-5 e-ISBN-13: 978-3-642-78267-1
DOI: 10.1007/ 978-3-642-78267-1

1 Guanosine triphosphatase I. Aktories, K II. Dickey, Burton F , 1953– III. Birnbaumer, Lutz. IV Series. QP905.H3 vol. 108 [QP609.G83] 615' 1 s – dc 20 574.19'253] 93-21636

Softcover reprint of the hardcover 1st edition 1993

Typesetting: Best-set Typesetter Ltd , Hong Kong
27/3130/SPS-5 4 3 2 1 0 – Printed on acid-free paper

List of Contributors

AKTORIES, K., Institut für Pharmakologie und Toxikologie der Universität des Saarlandes, D-66421 Homburg/Saar, Germany

ALTSCHULER, D.L., Division of Cell Biology, Burroughs Wellcome Co., 3030 Cornwallis Road, Research Triangle Park, NC 27709, USA

ANDERSON, M.W., National Institute of Environmental Health Sciences, Research Triangle Park, NC 27709, USA

BARLOWE, C., Department of Molecular and Cell Biology and Howard Hughes Medical Institute, University of California, Berkeley, CA 94720, USA

BERGMANN, E., Department of Biochemistry, University of California at Riverside, Riverside, CA 92521, USA

BIAN, J., Department of Biological Chemistry, University of Maryland School of Medicine, Baltimore, MD 21201, USA

BIRNBAUMER, L., Departments of Cell Biology and Molecular Physiology and Biophysics and Division of Neurosciences, Baylor College of Medicine, Room 149-E, One Baylor Plaza, Houston, TX 77030, USA

BOKOCH, G.M., Departments of Immunology and Cell Biology, IMM-14, R221, The Scripps Research Institute, 10666 N. Torrey Pines Road, La Jolla, CA 92037, USA

BOURNE, H.R., Departments of Pharmacology and Medicine and the Cardiovascular Research Institute, University of California, Box 0450, S-1210, San Francisco, CA 94143, USA

BRENNWALD, P., Department of Cell Biology, Yale University School of Medicine, 333 Cedar Street, New Haven, CT 06510, USA

BURGESS, C.C., Present address: Ciba-Corning Diagnostics, 333 Coney St., E. Walpole, MA 02032, USA

BURSTEIN, E.S., Department of Pathology, Medical Alumni Building, University of Vermont, Burlington, VT 05405-0068, USA

Campbell-Burk, S.L., Department of Biochemistry and Biophysics, CB#7260 University of North Carolina, Chapel Hill, NC 27599-7260, USA

Carlier, M.-F., Laboratoire d'Enzymologie, C.N.R.S., F-91198 Gif-sur-Yvette, France

Carnell, L., Department of Molecular and Cell Biology, Division of Cell and Developmental Biology 571 LSA, University of California, Berkeley, CA 94720, USA

Cavallius, J., Department of Biochemistry, School of Medicine, Case Western Reserve University, 10900 Euclid Avenue, Cleveland, OH 44106-4935, USA

Cerione, R.A., Department of Pharmacology, Schurman Hall, Cornell University, Ithaca, NY 14853, USA

Chardin, P., Institut de Pharmacologie Moléculaire et Cellulaire du CNRS, UPR 411, 660 route des Lucioles, F-06560 Valbonne (Sophia Antipolis), France

Charifson, P.S., Glaxo Inc., 5 Moore Drive, Research Triangle Park, NC 27709, USA

Chavez, R.A., Department of Molecular and Cell Biology, Division of Cell and Developmental Biology 571 LSA, University of California, Berkeley, CA 94720, USA

Chen, Y.-T., Department of Molecular and Cell Biology, Division of Cell and Developmental Biology 571 LSA, University of California, Berkeley, CA 94720, USA

Clark, G.J., University of North Carolina at Chapel Hill, Department of Pharmacology, Chapel Hill, NC 27599-7365, USA

Coburn, J., Division of Rheumatology, Box 406, New England Medical Center, 750 Washington St., Boston, MA 02111, USA

Darden, T.A., National Institute of Environmental Health Sciences, Research Triangle Park, NC 27709, USA

De Matteis, M.A., Unit of Physiopathology of Secretion, Istituto Mario Negri, Consorzio Mario Negri Sud, I-66030 Santa Maria Imbaro (CH), Italy

Der, C.J., University of North Carolina at Chapel Hill, Department of Pharmacology, Chapel Hill, NC 27599, USA

Downward, J., Imperial Cancer Research Fund Laboratories, Lincoln's Inn Fields, London WC2A 3PX, Great Britain

Dupree, P., European Molecular Biology Laboratory, Cell Biology Programme, P.O. Box 10.2209, D-69012 Heidelberg, Germany

Evans, T., Department of Cell Biology, Genentech, Inc., 469 Pt. San Bruno Boulevard, South San Francisco, CA 94080, USA

Feig, L.A., Department of Biochemistry, Tufts University School of Medicine, 136 Harrison Ave., Boston, MA 02111, USA

Fischer von Mollard, G., Howard Hughes Medical Institute and Department of Pharmacology, Boyer Center for Molecular Medicine, Yale University School of Medicine, New Haven, CT 06510, USA

Foley, C.K., Cray Research Inc., P.O. Box 12746, Research Triangle Park, NC 27709, USA

Gallwitz, D., Department of Molecular Genetics, Max-Planck-Institute for Biophysical Chemistry, P.O. Box 2841, D-37018 Göttingen, Germany

Ghosh, T.K., Department of Biological Chemistry, University of Maryland School of Medicine, Baltimore, MD 21201, USA

Gibbs, J.B., Department of Cancer Research, Merck Research Laboratories, West Point, PA 19486, USA

Gill, D.L., Department of Biological Chemistry, University of Maryland School of Medicine, Baltimore, MD 21201, USA

Gilmore, R., Department of Biochemistry and Molecular Biology, University of Massachusetts Medical School, 55 Lake Avenue North, Worcester, MA 01655, USA

Gomperts, B.D., Department of Physiology, University College London, University Street, London WC1E 6JJ, Great Britain

Goody, R.S., Max-Planck-Institut für Molekulare Physiologie, Abteilung Physikalische Biochemie, Postfash 10 26 64, D-44026 Dortmund, Germany

Hall, A., MRC Laboratory for Molecular Cell Biology, University College London, Gower Street, London WCIE 6BT, Great Britain

Hart, M.J., Onyx Pharmaceuticals, 3031 Research Drive Richmond, CA 94806, USA

Herskovits, J.S., Cell Biology Group, Worcester Foundation for Experimental Biology, 222 Maple Avenue, Shrewsbury, MA 01545, USA

Huber, L.A., European Molecular Biology Laboratory, Cell Biology Programme, P.O. Box 10.2209, D-69012 Heidelberg, Germany

HWANG, A., Chiron Corp. 4560, Horton St., Emeryville, CA 54608-2916, USA

ITOH, H., Tokyo Institute of Technology, Department of Biological Sciences, Faculty of Bioscience and Biotechnology, 4259 Nagatsuda, Midoriku, Yokohama 227, Japan

JAHN, R., Howard Hughes Medical Institute and Department of Pharmacology, Boyer Center for Molecular Medicine, Yale University School of Medicine, New Haven, CT 06510, USA

JURNAK, F., Department of Biochemistry, University of California at Riverside, Riverside, CA 92521, USA

JUST, I., Institut für Pharmakologie und Toxikologie der Universität des Saarlandes, D-66421 Homburg/Saar, Germany

KAHN, R.A., Laboratory of Biological Chemistry, Developmental Therapeutics Program, Division of Cancer Treatment, National Cancer Institute, Bldg. 37, Room 5D-02, Bethesda, MD 20892, USA

KAIBUCHI, K., Department of Biochemistry, Kobe University School of Medicine, Kobe 650, Japan

KAZIRO, Y., Department of Molecular Biology, DNAX Research Institute of Molecular and Cellular Biology, 901 California Avenue, Palo Alto, CA 94304-1104, USA

KIKUCHI, A., Department of Biochemistry, Kobe University School of Medicine, Kobe 650, Japan

KIM, S.-H., University of California, Melvin Calvin Laboratory, Berkeley, CA 94720, USA

LAPETINA, E.G., Division of Cell Biology, Burroughs Wellcome Co., 3030 Cornwallis Road, Research Triangle Park, NC 27709, USA

LEONARD, D., Department of Biochemistry, Cellular, and Molecular Biology, Department of Pharmacology, Schurman Hall, Cornell University, Ithaca, NY 14853, USA

LILLIE, T.H.W., Department of Physiology, University College London, University Street, London WC1E 6JJ, Great Britain

LÜTCKE, A., European Molecular Biology Laboratory, Cell Biology Programme, P.O. Box 10.2209, D-69012 Heidelberg, Germany

LUINI, A., Laboratory of Molecular Neurobiology, Istituto Mario Negri, Consorzio Mario Negri Sud, I-66030 Santa Maria Imbaro (CH), Italy

MACARA, I.G., Department of Pathology, Medical Alumni Building, University of Vermont, Burlington, VT 05405-0068, USA

MATSUMOTO, K., Department of Molecular Biology, Faculty of Science, Nagoya University, Chikusa-Ku, Nagoya 464-01, Japan

MCCORMICK, F., Onyx Pharmaceuticals, 3031 Research Drive, Building A, Richmond, CA 94806, USA

MERRICK, W.C., Department of Biochemistry, School of Medicine, Case Western Reserve University, 10900 Euclid Avenue, Cleveland, OH 44106-4935, USA

MILBURN, M.V., Department of Biochemistry and Molecular Biology, Harvard University, Cambridge, MA 02138, USA

MILLER, S.G., Department of Molecular and Cell Biology, Parnassus Pharmaceuticals, Inc. Alameda, CA 94502, USA

MOORE, H.-P.H., Department of Molecular and Cell Biology, Division of Cell and Developmental Biology 571 LSA, University of California, Berkeley, CA 94720, USA

MOSS, J., Laboratory of Cellular Metabolism, Bldg. 10, Room 5N-307, National Heart, Lung, and Blood Institute, National Institutes of Health, Bethesda, MD 20892, USA

NOVICK, P., Department of Cell Biology, Yale University School of Medicine, 333 Cedar Street, New Haven, CT 06510, USA

OLKKONEN, V.M., European Molecular Biology Laboratory, Cell Biology Programme, P.O. Box 10.2209, D-69012 Heidelberg, Germany

PAI, E.F., Department of Biochemistry, University of Toronto, Toronto, Ontario, Canada M5S 1A8

PANTALONI, D., Laboratoire d'Enzymologie, C.N.R.S., F-91198 Gif-sur-Yvette, France

PEDERSEN, L.G., Department of Chemistry, University of North Carolina at Chapel Hill, Chapel Hill, NC 27514, USA

PFEFFER, S.R., Department of Biochemistry, Stanford University School of Medicine, Stanford, CA 94305, USA

PRIVÉ, G.G., Institute of Molecular Biology, University of California, Los Angeles, CA 90024, USA

RAPIEJKO, P.J., Department of Biochemistry and Molecular Biology, University of Massachusetts Medical School, 55 Lake Avenue North, Worcester, MA 01655, USA

RIDLEY, A.J., Chester Beatty Laboratories, Institute of Cancer Research, Fulham Road, London SW3 6JB, Great Britain

SASAKI, T., Department of Biochemistry, Kobe University School of Medicine, Kobe 650, Japan

SCHEKMAN, R., Department of Molecular and Cell Biology and Howard Hughes Medical Institute, University of California, Berkeley, CA 94720, USA

SHINJO, K., Pfizer Central Research, Nagoya, Japan

SHORT, A.D., Department of Biological Chemistry, University of Maryland School of Medicine, Baltimore, MD 21201, USA

SIMONS, K., European Molecular Biology Laboratory, Cell Biology Programme, P.O. Box 10.2209, D-69012 Heidelberg, Germany

STACEY, D.W., The Department of Molecular Biology, The Cleveland Clinic Foundation, NC2-151, 9500 Euclid Avenue, Cleveland, OH 44106, USA

STROM, M., Department of Molecular Genetics, Max-Planck-Institute for Biophysical Chemistry, P.O. Box 2841, D-37018 Göttingen, Germany

SÜDHOF, T.C., Howard Hughes Medical Institute and Department of Molecular Genetics, Southwestern Medical Center, University of Texas, Dallas, TX 75235, USA

TAKAI, Y., Department of Biochemistry, Kobe University School of Medicine, Kobe 650, Japan

TANAKA, K., Department of Biochemistry, Kobe University School of Medicine, Kobe 650, Japan

TOH-E, A., Department of Biology, Faculty of Science, University of Tokyo, Hongo, Tokyo 113, Japan

TORTI, M., Division of Cell Biology, Burroughs Wellcome Co., 3030 Cornwallis Road, Research Triangle Park, NC 27709, USA

VALLEE, R.B., Cell Biology Group, Worcester Foundation for Experimental Biology, 222 Maple Avenue, Shrewsbury, MA 01545, USA

VAN AKEN, T.E., Du Pont Merck Pharmaceutical Company, Department of Structural Biology, P.O. Box 80328, Wilmington, DE 19880-0328, USA

VAUGHAN, M., Laboratory of Cellular Metabolism, Bldg. 10, Room 5N-307, National Heart, Lung, and Blood Institute, National Institutes of Health, Bethesda, MD 20892, USA

WALDRON, R.T., Department of Biological Chemistry, University of Maryland School of Medicine, Baltimore, MD 21201, USA

Weinberg, R.A., Whitehead Institute for Biomedical Research, Massachusetts Institute of Technology, Nine Cambridge Center, Cambridge, MA 02142, USA

Wessling-Resnick, M., Department of Nutrition, Division of Biological Sciences, Harvard School of Public Health, 665 Huntington Avenue, Boston, MA 02115, USA

Wittinghofer, A., Max-Planck-Institut für molekulare Physiologie Abteilung Strukturelle Biologie, Postfash 10 26 64, D-44026 Dortmund, Germany

Yoon, Y.-A., Department of Molecular and Cell Biology, Division of Cell and Developmental Biology 571 LSA, University of California, Berkeley, CA 94720, USA

Yu, H., Department of Molecular and Cell Biology, Division of Cell and Developmental Biology 571 LSA, University of California, Berkeley, CA 94720, USA

Zerial, M., European Molecular Biology Laboratory, Cell Biology Programme, P.O. Box 10.2209, D-69012 Heidelberg, Germany

Zheng, Y., Department of Pharmacology, Schurman Hall, Cornell University, Ithaca, NY 14853, USA

Preface

The question that may be raised is why a book on GTPases, especially as so much has been written lately in so many primary journals and review publications. Moreover, I received Pedro Cuatrecasas' invitation to edit this book after having just completed the editing, with Ravi Iyengar, of *G Proteins* for Academic Press. Understandably, my initial reaction was negative. Prodded by Pedro, and witnessing the breakneck pace at which new information was appearing in journals, I finally agreed. But even so, I decided to delay until I could better define how to organize and focus the book. A review of *G Proteins* by Peter Gierschik, which he titled "Balanced but unpackaged," proved helpful for the selection of topics and details of execution. *G Proteins* had been on trimeric G proteins only. Gierschik asked for "something on the small GTPases" and also more about structure. He also suggested more cross referencing, more serious editing by the editors, and faster publication. The arrival in Houston, from Boston, of my co-editor Burt Dickey, with his interest in small GTPases and GTPase structure, and his willingness to aid me, gave me – us – the impetus of requesting chapters.

It soon proved that Burt was doing all the work, as some of the authors know. Burt would not take a "no" for an answer, and insisted, again and again. The consequence of the broadened scope was that a 30–40 chapter book became what it is now – a two volume compendium. The original deadline for receipt of the manuscripts was July 1992. The last chapters arrived around New Year. The original manuscripts represented the state-of-the-art in mid–late 1992, but thankfully many of the authors have taken advantage of an opportunity to add remarks on important recent developments.

I am not sure that we have been able to cope with all of Peter Gierschik's (valid) criticisms. We hope that errors have been kept to a minimum. But he also wished that authors would not contradict each other, or at least that the editors would edit out the contradictions. Yet, for the most part contradictions are based on the working hypotheses of the individual investigators and represent their legitimate points of view. They are in fact, in exposition of the evolution of thoughts and arise because the experiments are not yet clear. We have thus left them, as far as they exist. Authors also change their mind. For example, I am sure that if one looked carefully, Henry Bourne no longer supports the idea that $\beta\gamma$ dimers promote GDP release. And I have been swayed to recognize $\beta\gamma$ dimers as regulators of effectors.

We have attempted to standardize nomenclature throughout. However, given the lack of agreement on usage even among workers within a field such as the Ras family, we have not been rigid in imposing terminology. Also, the nomenclature can be expected to evolve. Certain usages, however, were suggested. The term "GTPases" has been used to denote the superfamily because it is informative and short; "G protein" is reserved for members of the heterotrimeric signal-transducing family out of deference to the pleas of workers in this field. "Ras-related" or "small" is applied to the extended family of GTPases of molecular masses 19 to 26 kDa, including ARFs, in accordance with recommendations from a FASEB-sponsored conference, although it is recognized that ARFs possess several structural features which could warrant their designation as a distinct family. Proteins that modulate guanine nucleotide exchange are designated GEPs (guanine nucleotide exchange proteins), with a GDS (guanine nucleotide dissociation stimulator) accelerating exchange, and a GDI (guanine nucleotide dissociation inhibitor) retarding it. A protein that accelerates intrinsic GTPase activity is designated a GAP (GTPase activating protein), and one that retards it a GIP (GTPase inhibitory protein).

To finally answer the question raised in the first line of this preface, we expect this book to serve as a compendium which investigators and advanced students alike can go to and find much of what they need if they are interested in regulatory GTPases, be they large or small, be it structural or functional. This must be coupled with the realization that the information covers knowledge and concepts as they stood at mid to late 1992.

Some of the most recent developments could not be included. For example we do not treat the role of molecular diversity of adenylyl cyclases in single cells. Only hints are presented about the complexity of PLCβ's, which at this time number 4 in the mammalian genome, each responding to a different extent to α's of the α_q family and to $\beta\gamma$ dimers. The specific roles of the many Rab's in the ever-expanding intracellular membrane compartments needs updating the role of trimeric G proteins in the regulation of vesicle budding is absent, as is an extensive treatment of vesicle coating and uncoating. Heidi Hamm's and Paul Seglar's crystal of transducin α is absent. Maybe it is time to begin editing a follow up?

Houston, TX, USA — LUTZ BIRNBAUMER
November 1993

The obvious answer to the question raised by Lutz Birnbaumer, "Why a book on GTPases?", is that it has been four years since the last book on this still evolving field was published. More significantly, the discovery of entirely new members of the GTPase superfamily and considerable progress in elucidating the functions and molecular structures of individual GTPases makes this an apt time to update the subject. Given the rate at which the superfamily continues to grow, it will become less and less reasonable to consider all its branches in a single book. The size to which this volume grew should be a warning to any future editors!

The unique perspective of the variety and the shared features of GTPases afforded by an inclusive volume such as this makes it almost inevitable to begin to ask fundamental questions about why cells contain so many GTPases, and what the origins are of the basic GTPase mechanism. Answers to such questions can at present only be speculative, but studies of molecular evolution allow us to make some educated guesses. The ribosomal machinery, in which RNA functions both in information storage and in catalysis, may be thought of as an archaic remnant of life's origins in an RNA world. As such, the molecular species which make up this machine appear to be some of the most ancient and highly conserved among all contemporary cellular organisms. Included among the components of the translation apparatus are a number of related GTPase initiation and elongation factors. Several of these GTPases appear to function as proofreaders which neatly reconcile the inherently conflicting demands for speed and for accuracy in protein synthesis. Similarly, the uniquely successful organism which has given rise to all currently known cellular organisms must have been the one that was best able to satisfy the competitive needs for both reproductive speed and accuracy. One can imagine the elegant molecular mechanisms of the cellular synthetic machinery (including the GTPase proofreaders) being forged on the anvil of intense primordial competition.

If the initiation and elongation factors are indeed the ancestral GTPases, then the basic mechanism of GTP binding and hydrolysis has subsequently been appropriated for the regulation of a surprisingly wide array of biochemical processes. Other avenues for speculation include the coincidence that the two great families of intracellular regulatory switches – the protein kinases and the GTPases – both depend on interaction with a purine for their regulatory function. Are their shared, albeit limited, structural motifs evidence of a common origin, or of convergent evolution? Study of the superfamily of GTPases, in addition to providing us a panorama of the extraordinary breadth of the regulatory biology of cells, may yet afford us some insight into the very nature and origins of life.

Houston, TX, USA
November 1993

BURTON F. DICKEY

Contents

CHAPTER 6

Dynamin, A Microtubule-Activated GTPase Involved in Endocytosis

CHAPTER 7

Transmembrane Protein Translocation: Signal Recognition Particle and Its Receptor in the Endoplasmic Reticulum

CHAPTER 8

GTPases and Actin as Targets for Bacterial Toxins

CHAPTER 13

Conformational Switch and Structural Basis for Oncogenic Mutations of *Ras* Proteins

CHAPTER 14

Structural and Mechanistic Aspects of the GTPase Reaction of H-ras p21

CHAPTER 34

The Biology of ADP-Ribosylation Factors

CHAPTER 35

Molecular Characterization of ADP-Ribosylation Factors

CHAPTER 43

ADP-Ribosylation of Small GTPases by *Clostridium botulinum* Exoenzyme C3 and *Pseudomonas aeruginosa* Exoenzyme S

Contents of Companion Volume 108/II

Section I
Biological Importance of GTPase-Driven Switches

CHAPTER 1

GTPases Everywhere!

H.R. Bourne

A. Introduction

Everywhere we look in biology, GTPases abound. Although the first examples appeared in apparently unrelated contexts – protein synthesis, cancer, hormone action, vision – we now recognize GTPases as members of a large superfamily, encoded by genes apparently descended from a single progenitor. The ability of a primordial GTPase to oscillate between GTP- and GDP-bound conformations furnished an opportunity for evolution to elaborate a diverse and versatile panoply of regulatory switches.

To introduce the GTPases, this chapter will begin by sketching the GTPase cycle and switch mechanism used by all these proteins. We will then consider how these switches are used in regulating other proteins in the cell, outlining several well-defined mechanisms and pointing to mysteries not yet resolved. Finally, we will briefly describe examples of GTPases that are organized in regulatory cascades – some well understood, others just now emerging into view.

Before we begin, a warning to readers: This chapter may pique your curiosity or challenge you to disagree with its generalizations. Other chapters will amply satisfy your hunger for large numbers of nourishing facts.

B. The GTPase Cycle and the Molecular Switch

Allosteric regulatory proteins switch between conformations that are determined by the presence or absence of a bound ligand. Key features of the GTPase cycle (Fig. 1) distinguish all members of the GTPase superfamily from simpler allosteric switches. Different ligands, GTP and GDP, respectively, stabilize the active and inactive conformations of GTPases, and transitions between the conformations occur in discrete, separately regulated steps.

The transition from G·GDP to G·GTP begins with the release of bound GDP. The rate of dissociation is determined by the structure of the binding pocket of the GTPase protein, and is also (in most cases) subject to regulation by guanine nucleotide exchange proteins (GEPs). GEPs fall into one of two categories (Fig. 1): guanine nucleotide dissociation stimulator (GDS) proteins stimulate and guanine nucleotide dissociation inhibitor (GDI)

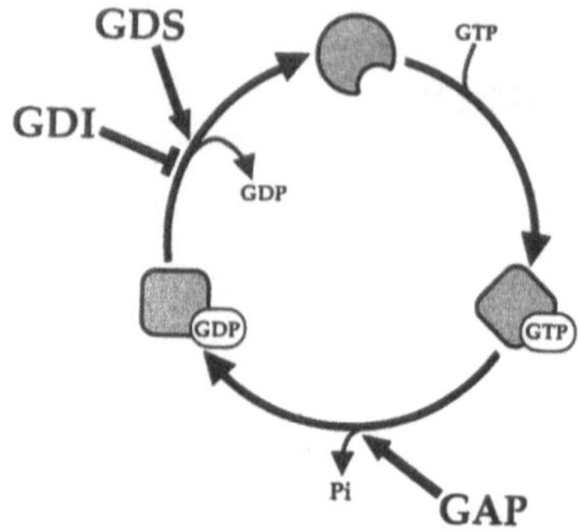

Fig. 1. The basic GTPase cycle. The first step, dissociation of GDP from the G–GDP complex, is accelerated and slowed, respectively, by guanine nucleotide dissociation stimulators (*GDSs*) and inhibitors (*GDIs*). GTP binds to the transient "empty" state of the GTPase, and an intrinsic enzymatic activity of G–GTP then converts it to G–GDP

proteins inhibit dissociation of GDP. Following release of GDP, the "empty site" form of the GTPase rapidly binds GTP. The usually short duration of the empty site conformation has obscured the possibility that GTP binding may be regulated, in some cases, separately from GDP dissociation (see below).

The GTPase cycle allows the duration of the active G·GTP conformation to be precisely controlled. In the absence of other proteins, the intrinsic GTPase activity of G·GTP determines its rate of inactivation. Individual GTPases hydrolyze GTP at widely varying rates, with intrinsic $k_{cat \cdot GTP}$ values ranging from $\sim 10^{-3}$ to $\sim 10\,min^{-1}$. Many GTPases are also regulated by specific GTPase activating proteins (GAPs; Fig. 1), which can increase $k_{cat \cdot GTP}$ by factors of 10^2–10^5. With or without a GAP, however, the critical advantage of inactivation by GTP hydrolysis is that the duration of each G·GTP complex – and therefore the concentration of G·GTP relative to G·GDP – can be regulated in ways *independent* from the influences that determine how rapidly G·GTP is formed. This feature makes the GTPase switch more versatile than switches that rely simply upon binding and dissociation of a regulatory ligand.

C. Structure of the GTPase Switch

The common architectural design shared by GTPases, above and beyond the fact that they perform the same biochemical reaction, indicates their probable origin from a single progenitor. Most of the GTPases discussed in this book contain stretches of amino acid sequence that are conserved across the entire GTPase superfamily; in the primary structure of each GTPase, these stretches appear in the same order, although spacing between them may differ. Figure 2 depicts the distributions of five such regions in proteins that belong to three different GTPase subfamilies: Gα proteins, the α subunits of the heterotrimeric signaling G proteins; p21^{ras}, the most extensively studied example of a large family of monomeric GTPases; and elongation factor EF-Tu, a bacterial protein involved in ribosomal protein synthesis.

These stretches of amino acid sequence are conserved because each region forms a part of the guanine nucleotide binding pocket of the GTPase, as shown by the two presently available three-dimensional structures, those of p21^{ras} and EF-Tu (see chapters in Part II). The three-dimensional structures of the guanine nucleotide binding domains of p21^{ras} and EF-Tu are almost superimposable, although their amino acid sequences are only 16% identical (VALENCIA et al. 1991). Consequently, we infer a closely similar three-dimensional structure for Gα – and, by extension, for all GTPases that share the key sequence motifs.

Investigators in many laboratories are trying to determine how these conserved structures perform the biochemical steps of the GTPase cycle, and the degree to which these mechanisms vary in different members of the superfamily (see chapters in Part II). Three-dimensional structures of the GDP- and GTP-bound forms of p21^{ras} identify two regions of GTP-induced conformational change, often called 'switch regions" (designated in Fig. 2 by open boxes, numbered 1 and 2). Limited evidence indicates that GTP also changes conformation of cognate regions in other GTPases. The available three-dimensional structures place constraints on ideas about the enzymatic mechanism responsible for hydrolyzing GTP, although that mechanism is still not well understood (see Chaps. 13, 14, 16). Although different GTPases probably perform this reaction in similar ways, wide variations in intrinsic $k_{cat \cdot GTP}$ values indicate differences in detail. The molecular mechanism of GDP–GTP exchange will remain mysterious, at least until a

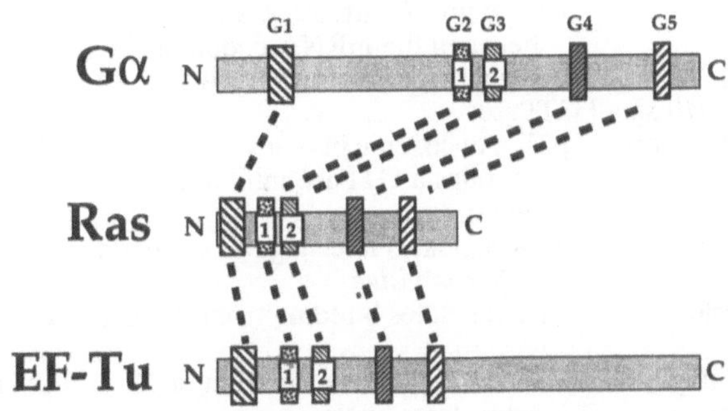

Fig. 2. Similar primary structures of GTPases in three families. Boxes *G1–G5* represent stretches of sequence that are conserved among members of the GTPase superfamily and that form the guanine nucleotide binding pocket of these proteins. The *open boxes* represent "switch regions" 1 and 2, which undergo GTP-induced conformational change. *EF-Tu*, elongation factor Tu

GTPase structure is determined in the "empty guanine nucleotide site" conformation.

Although structural features of the basic switch are highly conserved, individual members of the superfamily must communicate with specific proteins, which may regulate the cycle and/or respond to regulation by different conformations of the GTPase itself. Many laboratories are using mutated and chimeric proteins, antibodies, and peptides in attempts to identify and locate (in one dimension and three dimensions) the specific knobs and whistles responsible for such communication. So far these attempts have been most successful with Gα proteins (see Chaps. 10, 48, 49) in part because many of the proteins that communicate with Gα are well characterized and in part because Gα proteins are more amenable to experimentation, in that (unlike p21^{ras} or EF-Tu, for instance) they are not essential for cells to survive and proliferate.

D. Primary Structures Identify GTPases with Related Functions

Table 1 lists families of GTPases covered in other chapters of this book. Amino acid sequence motifs in regions G1–G5 (Fig. 2) identify proteins in these families as unequivocal members of the GTPase superfamily. (The

Table 1. GTPase families

Family	Example(s)	See
Heterotrimertic G-proteins (hormonal and sensory signaling)	G_s, the stimulatory regulator of adenylyl cyclase G_t (transducin) mediates retinal phototransduction	Chap. III and
Initiation and elongation factors (ribosomal protein synthesis)	EF-Tu proofreads association between the mRNA codon and the anticodon of aa-tRNA	Chaps. II and IX
Monomeric (ras-related) small GTPases		
Ras family (proliferation, differentiation)	p21^{ras}, oncogenic in many cells if it contains a GTPase-inhibiting mutation	Part IIIA
Rab family (vesicle traffic)	Sec4 regulates late stage of secretion in *S. cerevisiae*	Part IIIB
ARF family (vesicle traffic)	Sar1 regulates budding from the ER	Part IIIB
Rho family (assembly and function of actin cytoskeleton)	Rho-A stimulates formation of actin stress fibers	Chap. XXXV
SRP/SR family (translocates polypeptides into the ER)	SRP54, a GTPase component of the SRP	Chap. VII
Tubulins and cytoskeletal motor GTPases	α and β tubulin polymerize to form microtubules	Chaps. V and VI

tubulins and dynamin-related GTPases constitute exceptions; related more distantly to the GTPase superfamily and even to one another, this group is not further discussed in this chapter.) In general, regions G1, G3, and G4 contain sequences that are highly conserved throughout the superfamily, while sequences in G2, G4, and elsewhere are usually conserved within an individual GTPase family and serve to identify its members.

Within a family defined by similarities in amino acid sequence, the individual GTPases turn out to perform similar or closely related functions. This generalization is especially striking with respect to the Gα subunits of heterotrimeric G-proteins and the GTPases of ribosomal protein synthesis. Structurally, members of each family resemble one another much more closely than they resemble members of other families. In their GTP-bound active forms, the well-characterized Gα proteins (Chap. 3) transduce signals between ligand-regulated GEPs (receptors for hormones and sensory stimuli) and effectors, some of which act as GAPs. The elongation and initiation factors (EFs and IFs) of ribosomal protein synthesis use their GTPase cycle in a quite different way, to catalyze and monitor assembly of complex macromolecular structures (Chaps. 2, 9). The parallel similarities of structure and function suggest that each of these families arose by duplication of a gene that encoded the first member of the family; in each case, such a progenitor may have performed a function roughly similar to those of its descendants. This idea probably applies also to the signal recognition particle (SRP)/signal recognition particle receptor (SR) family, which includes SRP54, the 54-kDa subunit of the SRP, and SRα and SRβ, the α and SRβ subunits of the SR. These GTPases, which resemble one another in primary structure, are components of the machinery that translocates polypeptides into the endoplasmic reticulum (ER); their functions are not understood in molecular detail (see Chap. 7).

Proteins in the large and rapidly expanding group of monomeric GTPases are all structurally related to proteins encoded by Ras proto-oncogenes. This group is further subdivided into at least four families – Ras, Rab, ADP-ribosylation factor (ARF), and Rho – each of which exhibits distinctive features of primary structure, again suggesting derivation from specialized progenitors (see Chap. 12 and Chaps. in Part III). Within each of these four structurally defined families individual proteins tend to perform related functions – although, as with the SRP/SR family, molecular details of these functions are often still obscure.

E. Uses of the GTPase Switch: Stoichiometric Activation

The different conformations induced by binding GDP vs. GTP determine the affinities of GTPase machines for other macromolecules. While this general principle applies to all regulatory functions of GTPases, the associations and dissociations of different GTPases are choreographed in different ways.

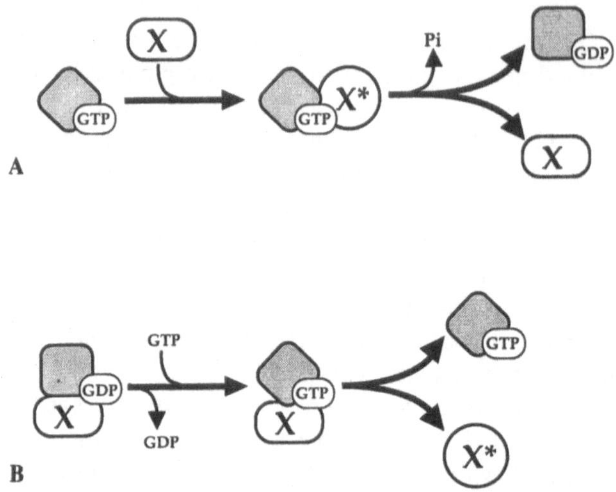

Fig. 3A,B. Stoichiometric activation by GTPases by binding effector (**A**) and releasing efforter (**B**). **A** G–GTP binds to *X*, activating it (G–GTP–X*). Hydrolysis of GTP inactivates *X*, which dissociates from G–GDP. *X** may (or may not) act as a GAP for G·GDP binds and inactivates *X*. Replacement of GDP by GTP causes dissociation of free, active *X** from G–GTP

This choreography is best understood for the heterotrimeric signaling G-proteins (Fig. 3). Here the GEP is the receptor that detects extracellular hormones and sensory signals; in its ligand-activated from, this GEP catalyzes replacement of GDP by GTP in the guanine nucleotide binding pocket of Gα, and Gα·GTP dissociates from the GEP and from the G-protein's $\beta\gamma$ subunit complex. In several well-documented cases, Gα·GTP transmits its signal by binding to an effector protein (X in Fig. 3A) and stoichiometrically regulating its activity (X*). Such an effector may be an adenylyl cyclase, a phosphodiesterase (PDE), a phospholipase, or an ion channel (see Chaps. 55–58). Although effector activation is not catalytic, the GTPase cycle ensures amplification of the signal that was originally detected by the receptor GEP. Amplification takes place because an activated receptor requires only a few milliseconds to convert Gα·GDP to Gα.· GTP remains able to stimulate the effector for a much longer time, determined by its $k_{cat \cdot GTP}$. This means both that the second messenger signal can persist for a relatively protracted period, longer than the original signal, and also that a single activated receptor can activate multiple Gα molecules and, therefore, multiple effectors.

If it depends only on the intrinsic $k_{cat \cdot GTP}$ of Gα· GTP, the effector signal may last for tens of seconds. Alternatively, a GAP activity in the effector itself may limit the persistence of the second messenger signal by accelerating conversion of Gα·GTP to α·GDP. Such GAP activities were recently reported for two effectors, phospholipase C-β1 and a retinal cGMP-PDE (ARSHAVSKY and BOWNDS 1992; BERSTEIN et al. 1992).

Certain heterotrimeric G-proteins illustrate a second strategy for harnessing the GTP-induced change in the conformation of Gα to generate a signal. In this case (Fig. 3B), Gα·GDP binds and sequesters the effector molecule X in an inactive form; replacement of GDP by GTP causes Gα·GTP to release a free, active effector, X*. This effector molecule is the $\beta\gamma$ subunit complex of the G-protein, which mediates responses to mating pheromones in *S. cerevisiae*, and, in mammalian cells, regulates certain isozymes of adenylyl cyclase and phospholipase C (see Chaps. 55, 58, 62).

F. Uses of the GTPase Switch: Assembling a Complex

GTPases are also used to catalyze and monitor the assembly of macromolecular complexes. Figure 4A illustrates how GTPase-catalyzed association can "proofread" formation of a complex between two macromolecules, X and Y: The GTP-bound form of the GTPase binds first to X and then to Y, which interacts both with the GTPase and with X. A GAP activity intrinsic to Y then triggers hydrolysis of GTP, and the GDP-bound form of the GTPase dissociates; X*·Y*, the key reaction product, has been assembled.

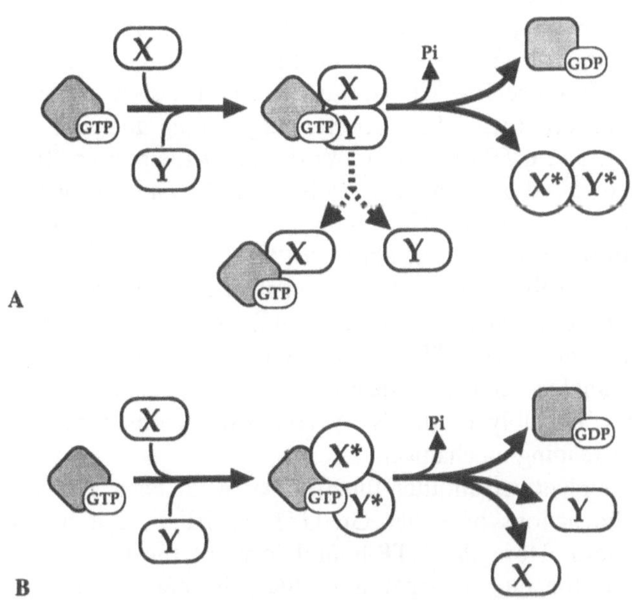

Fig. 4A,B. Assembly of multimeric complexes by GTPases. G–GTP binds sequentially to *X* and *Y*. Thereafter, as described in the text, different mechanisms are employed for proofreading (**A**) and for association of co-effectors (**B**)

The proofreading process uses this GTPase-dependent assembly process in a subtle way. Proofreading works by comparing the rate at which GTP is hydrolyzed in the ternary complex of X, Y, and GTPase vs. the rate at which Y dissociates from the complex of X and GTPase (dashed arrows in Fig. 4A). The first rate, determined by the $k_{cat} \cdot _{GTP}$ of GTP hydrolysis under the influence of a GAP (molecule Y), serves as a standard for monitoring the second. An imperfect fit between X and Y causes them to dissociate more rapidly (dashed arrows) than would a perfectly matched pair. To take advantage of this difference, the rate of GTP hydrolysis can be timed to determine which pairs of X and Y molecules pass muster. The proofreading process allows "correct" X·Y complexes to assemble after GTP is hydrolyzed; in contrast, an X·Y complex that fails to persist beyond a preset time period, dictated by the $k_{cat} \cdot _{GTP}$, is discarded.

In the best understood example of proofreading by a GTPase, the rate of GTP hydrolysis by EF-Tu is used to monitor the accuracy of codon–anticodon pairing in ribosomal protein synthesis (see Chap. 2). Here proofreading is necessary because differences in affinity between correct and incorrect codon–anticodon pairs are often not great enough to assure the required high degree of fidelity. To increase fidelity, the programmed ribosome (a GAP for EF-Tu) presents a specific mRNA codon, corresponding to Y, for possible pairing with X, the anticodon of an aminoacyl-tRNA; the relative rates of codon–anticodon dissociation vs. Y-triggered GTP hydrolysis by EF-Tu determine whether a particular amino acyl-tRNA is used or rejected. At an earlier step in translation, IF-2 uses its GTPase activity in a similar way, to monitor assembly of an initiation complex.

Like stoichiometric activation by a GTPase such as Gα·GTP (Fig. 3A), proofreading GTPases such as EF-Tu use GTP hydrolysis as a timing device – but in a very different way. For proofreading, timing does not begin until the complex of X and GTPase binds to Y; the timer then "fires" (hydrolyzes GTP) at a rate dictated by Y, a GAP, rather than by an intrinsic $k_{cat} \cdot _{GTP}$ of the GTPase. In contrast, the Gα timer is "set" as soon as GTP binds, and GTP hydrolysis terminates its ability to "fire" (to activate the effector). GAP-regulated hydrolysis of GTP is the central reaction in proofreading, while it performs only an ancillary "turn-off" role in stoichiometric activation of an effector by Gα·GTP. Consequently, binding of an hydrolysis-resistant GTP analog characteristically enhances effector activation but drastically slows assembly of the X · Y complexes formed by a GTPase-dependent proofreading mechanism.

Figure 4B depicts yet another function for a GTPase, less well documented than the actions of either Gα·GTP or EF-Tu but incorporating features from both. Here the GTP-bound form of the GTPase promotes assembly of two (or more) target molecules, X and Y, into a trimeric GTPase·X*·Y* complex that acts as an effector. As in the proofreading case, the reaction brings X and Y together, and Y can act as a GAP. As in the case of Gα·GTP, however, the resulting hydrolysis of GTP serves to terminate activation of the co-effectors – and their association as well. As with $G\alpha \cdot GTP$, moreover, inhibition of GTP hydrolysis or binding of an hydrolysis resistant GTP analog would enhance activation of the coeffectors.

A number of recently discovered GTPases are candidates for roles as promoters of assembly reactions in one or the other of the scenarios depicted in Fig. 4. Parallel to the proofreading paradigm (Fig. 4A), some monomeric GTPases may catalyze vectorial movement of membrane vesicles, monitoring correct association or fusion of specific vesicles with target organelles. This notion (BOURNE 1988) may apply to some members of the Rab family, which are often associated with a specific vesicle or organelle where they appear to regulate vesicle movements (or fusion events) that can be inhibited by hydrolysis resistant GTP analogs (see Chapters in Part IIIB). For one of these proteins, a product of the *Sec4* gene in *S. cerevisiae*, genetic evidence bears out a prediction of the proofreading analogy – that is, a late stage of the secretory pathway is blocked in much the same way by two kinds of *Sec4* mutations, those that inhibit the GTPase activity of the Sec4 protein and those that simply inactivate it (see Chap. 4).

To date, no biochemical evidence unequivocally identifies a monomeric GTPase that acts by assembling coeffector molecules (Fig. 4B). We have known for many years that GTPase-inhibiting mutations create Ras proteins that contribute to neoplastic transformation and that stimulate proliferation, differentiation, and other responses in cultured mammalian cells, or activate adenylyl cyclase in *S. cerevisiae* (see chapters in Part IIIA). More recently, Rac and Rho proteins carrying such mutations were found to produce specific changes in morphology and actin assembly in fibroblasts (RIDLEY and HALL 1992; RIDLEY et al. 1992). These observations are consistent either with direct activation of a single effector molecule, by analogy to $G\alpha \cdot GTP$ (Fig. 3A), or with the coeffector model (Fig. 4B). In the latter case, specific GAP molecules (e.g., ras-GAP, neurofibromin, etc.) could serve as one element of the coeffector complex (see Chap. 23).

Recent biochemical studies (ABO et al. 1991; KNAUS et al. 1991; MIZUNO et al. 1992) strongly suggest that the coeffector model may apply to at least one effect of an activated monomeric GTPase. The respiratory burst of neutrophils, triggered by a number of extracellular stimuli, serves to produce superoxide anion, which plays a key role in host defense against many microorganisms (CLARK 1990). $NADPH^+$-oxidase, an enzyme in the neutrophil membrane, can produce superoxide only when it forms a complex with a pair of essential proteins, which are found in the cytoplasm of unstimulated neutrophils. In vitro, these cytoplasmic proteins can associate with membranes and activate $NADPH^+$-oxidase, but *only* in the presence of a Rac protein that is in its GTP-bound active state (ABO et al. 1991; KNAUS et al. 1991; MIZUNO et al. 1992). Although many biochemical details of this process remain to be established, a likely interpretation is that $Rac \cdot GTP$ promotes assembly of a coeffector complex that includes key cytoplasmic proteins and $NADPH^+$-oxidase.

G. Other Potential Uses of the GTPase Switch

We have focused almost exclusively on the GTP-bound form of the GTPase switch, because very little firm information points to important regulatory roles of the switch in its other states, GDP-bound or "empty" (see Fig. 1). The potential stability of these other two states suggests nonetheless that we should be looking for circumstances in which evolution, a relentless opportunist, has found a use for one or both of them.

The binding and sequestration of $\beta\gamma$ subunits by $G\alpha \cdot GDP$, described above and depicted abstractly in Fig. 3B, represent a well-documented regulatory function for G·GDP. In addition, the GDP-bound α subunit (α_t) of retinal transducin can associate with and inactivate the catalytic component of an effector enzyme, a cGMP-PDE, in vitro (KROLL et al. 1989). As suggested elsewhere (BOURNE and STRYER 1992), inactivation of this PDE by $\alpha_t \cdot GDP$ may play an essential role in turning off the response of a retinal rod cell to a light pulse.

Because empty site forms of GTPases seem to play a transient role in the GTPase cycle, their regulatory potential has been neglected. Yet it is clear that the empty forms of GTPases – stabilized in vitro by removing guanine nucleotides or in vivo by mutations that reduce affinity for binding guanine nucleotides – can bind specifically to GEPs, with well-documented consequences. Such consequences include altered conformation of a GEP (for instance, stabilization of the MII form of rhodopsin by "empty" transducin, as described in Chap. 60) and sequestration of GEPs followed by dominant negative blocking of a regulatory pathway (for instance, mutant Ras proteins that turn off pathways regulated by normal Ras proteins, as described in Chap. 19). Moreover, man-made mutations in rhodopsin produce a GEP that responds to a stimulus (light) by binding its GTPase (transducin) and stabilizing its empty state, apparently even preventing its activation by GTP (FRANKE et al. 1990). The fact that mutations in a GEP or a GTPase can stabilize GEP·GTPase complexes raises the possibility that nature has found pathways in which stable versions of these complexes play significant regulatory roles. Such complexes might be sought among the SRP/SR protein machinery that translocates proteins into the ER (see Chap. 7).

H. Cascades of GTPases

The protein kinases have shown us how cells may use a large array of structurally similar molecular switches positioned at key points in a regulatory pathway. GTPase switches participate in similarly complex pathways. In controlling ribosomal protein synthesis, one GTPase completes an assembly and passes it on to the next GTPase. GTPases may also act directly on one another, just as one kinase can phosphorylate another: the primary structure of guanine nucleotide exchange factor (GEF) exhibits the tell-tale

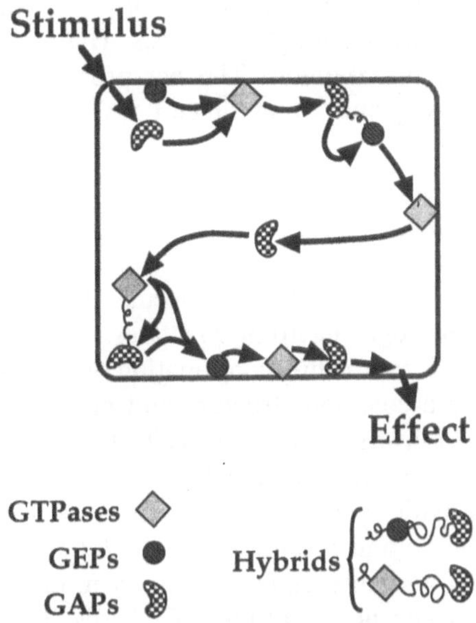

Fig. 5. Imaginary regulatory cascade of GTPases, mediating cellular response to an external stimulus. In some cases GTPases, GEPs, and GAPs may be linked together in hybrid proteins, as described in the text

features of a GTPase (see Chap. 9); in its GTP-bound form GEF is a GEP for eIF-2 (DHOLAKIA and WAHBA 1989).

Outside of ribosomal protein synthesis, regulatory cascades of GTPases have not been documented in biochemical detail. As noted earlier, however, vesicle traffic in secretory and endocytotic pathways may be controlled at several steps by sequentially acting GTPases. We infer the existence of other GTPase cascades, such as the imaginary version depicted in Fig. 5, from several kinds of observations. For instance, genetic interactions among genes for GTPases, GEPs, and GAPs have begun to reveal a regulatory pathway in which these proteins act to control the position and formation of buds in *S. cerevisiae* (CHANT et al. 1991; CHANT and HERSKOWITZ 1991). Similarly, effects of dominant active and dominant negative mutant GTPases injected into Swiss 3T3 cells have revealed a regulatory network of GTPases, including Ras, Rac, and Rho, that controls formation of stress fibers and membrane ruffles (RIDLEY and HALL 1992; RIDLEY et al. 1992).

In another example, the amino acid sequence of a protein called p190 reveals an unmistakable GTPase domain at its amino terminus and a GAP-like domain near its carboxy terminus (SETTLEMAN et al. 1992). We do not yet know either the function of the GTPase or the specific target of the GAP (it could be the GTPase domain, of course). Combined with the fact that p190 itself was discovered by virtue of its association in cells with another protein, Ras-GAP, the colocation of a GTPase and a second GAP in the same protein strongly suggests that p190 is a pivotal component in a regulatory network that involves multiple GTPases. Hybrid proteins like p190 may turn out to be quite common. The bcr protein has a Rac1-specific GAP at one end and sequences that suggest a GEP activity at the other (DIEKMANN et al. 1991).

I. Perspectives

As in many other fields, the tools of molecular biology have fuelled an explosion of new knowledge regarding GTPases. From all this work, the most exciting finding is a compelling confirmation of the unity of all biology: all these GTPases use a common structure and mechanism to perform an extraordinary variety of diverse tasks in the cell. As a practical consequence, investigators find powerful clues for tackling their own experimental problems in observations made in far distant fields. Almost anything we learn about one GTPase can help us to understand the others.

This principle will certainly endure, even as future investigations discover an overwhelming cornucopia of novel GTPases. The new proteins will use the GTPase structure and switch in all the familiar ways as well as in ways we cannot imagine. Now scattered observations suggest the existence of regulatory cascades of GTPases. These hints will lead to detailed maps of regulatory pathways and discovery of new ones. An exhilarating prospect – and somehow a slightly daunting one as well. Ah, what fun the GTPases have given us! It is hard to imagine that future GTPases will give us more.

References

Abo A, Pick E, Hall A, Totty N, Teahan CG, Segal AW (1991) Activation of the NADPH oxidase involves the small GTP-binding protein p21^{rac1}. Nature 353: 668–670

Arshavsky VY, Bownds MD (1992) Regulation of deactivation of photoreceptor G protein by its target enzyme and cGMP. Nature 357:416–417

Berstein G, Blank JL, Jhon DY, Exton JH, Rhee SG, Ross EM (1992) Phospholipase C-β1 is a GTPase activating protein (GAP) for $G_{q/11}$, its physiologic regulator, Cell 70:411–418

Bourne HR (1988) Do GTPases direct membrane traffic in secretion? Cell 53:669–671

Bourne HR, Stryer L (1992) The target set the tempo. Nature 358:541–543

Chant J, Corrado K, Pringle JR, Herskowitz I (1991) Yeast BUD5, encoding a putative GDP–GTP exchange factor, is necessary for bud site selection and interacts with bud formation gene BEM1. Cell 65:1213–1224

Chant J, Herskowitz I (1991) Genetic control of bud site selection in yeast by a set of gene products that constitute a morphogenetic pathway. Cell 65:1203–1212

Clark RA (1990) The human neutrophil respiratory burst oxidase. J Infect Dis 161:1140–1147

Dholakia JN, Wahba AJ (1989) Mechanism of the nucleotide exchange reaction in eukaryotic polypeptide chain initiation. Characterization of the guanine nucleotide exchange factor as a GTP-binding protein. J Biol Chem 264:546–550

Diekmann D, Brill S, Garrett MD, Totty N, Hsuan J, Monfries C, Hall C, Lim L, Hall A (1991) Bcr encodes a GTPase-activating protein for p21^{rac}. Nature 351:400–402

Franke RR, König B, Sakmar TP, Khorana HG, Hofmann KP (1990) Rhodopsin mutants that bind but fail to activate transducin. Science 250:123–125

Knaus UG, Heyworth PG, Evans T, Curnutte JT, Bokoch GM (1991) Regulation of phagocyte oxygen radical production by the GTP-binding protein Rac 2. Science 254:1512–1515

Kroll S, Phillips WJ, Cerione RA (1989) The regulation of the cyclic GMR phosphodiesterase by the GDP-bound form of the α subunit of transducin. J Biol Chem 264:4490–4497

Mizuno T, Kaibuchi K, Ando S, Musha T, Hiraoka K, Takaishi K, Asada M, Nunoi H, Matsuda I, Takai Y (1992) Regulation of the superoxide-generating NADPH oxidase by a small GTP-binding protein and its stimulatory and inhibitory GDP/GTP exchange proteins. J Biol Chem 267:10215–10218

Ridley AJ, Hall A (1992) The small GTP-binding protein rho regulates the assembly of focal adhesions and actin stress fibers in response to growth factors. Cell 70:389–399

Ridley AJ, Paterson HF, Johnston CL, Diekmann D, Hall A (1992) The small GTP-binding protein rac regulates growth factor-induced membrane ruffling. Cell 70:401–410

Settleman J, Narasimhan V, Foster LC, Weinberg RA (1992) Molecular cloning of cDNAs encoding the GAP-associated protein p190: implications for a signaling pathway from ras to the nucleus. Cell 69:539–549

Valencia A, Kjeldgaard M, Pai EF, Sander C (1991) GTPase domains of *ras* p21 oncogene protein and elongation factor Tu: analysis of three-dimensional structures, sequence families, and functional sites. Proc Natl Acad Sci USA 88: 5443–5447

CHAPTER 2

Proofreading in the Elongation Cycle of Protein Synthesis

E. BERGMANN and F. JURNAK

A. Introduction

The objective of the chapter is to review the role of GTPases in the elongation cycle of bacterial ribosomal biosynthesis illustrated in Fig. 1. In addition to the ribosome and aminoacyl-tRNA, three cytoplasmic protein factors participate in the elongation cycle in vivo. Two of these, elongation factor (EF-)Tu and EF-G, are GTPases. In its active form, EF-Tu-GTP forms a ternary complex with aminoacyl-tRNA and positions it in the A site of the ribosome. The ribosome stimulates the hydrolysis of EF-Tu-GTP, causing the release of EF-Tu from the ribosome in its inactive GDP form (Fig. 1a). Ribosomal proteins catalyze the next step, the peptidyl transfer reaction in which the peptide bond between the aminoacyl group of the A-site tRNA and the peptidyl chain on the P-site tRNA is formed. Translocation then occurs, with EF-G-GTP catalyzing the transfer of the peptidyl-tRNA from the A site to the P site of the ribosome (Fig. 1b). Translocation is also accompanied by GTP hydrolysis, with EF-G-GDP being released from the ribosome. The third elongation factor, EF-Ts, catalyzes the exchange of GTP for GDP on EF-Tu.

It is well established that a minimum of two GTP molecules are hydrolyzed to GDP for each addition of an amino acid to a polypeptide chain during the elongation step. Because it is also known that protein synthesis can occur in vitro in the absence of EF-Tu and EF-G and because the energy of the cleaved triphosphates does not appear in the form of a chemical bond, the role of the GTPase reactions was not readily apparent when the elongation cycle was first established approximately 25–30 years ago. HOPFIELD (1974) and NINIO (1975) are credited for introducing the concept of kinetic proofreading to ribosomal biosynthesis and for suggesting that the EF-Tu-GTPase reaction provides the energy to allow the preferential flow of incorrect substrate along a discard branch. Many experiments have subsequently been performed to test the validity of the original kinetic proofreading hypothesis and to elaborate upon the details relevant to protein synthesis. This review will summarize the current theories and supporting data.

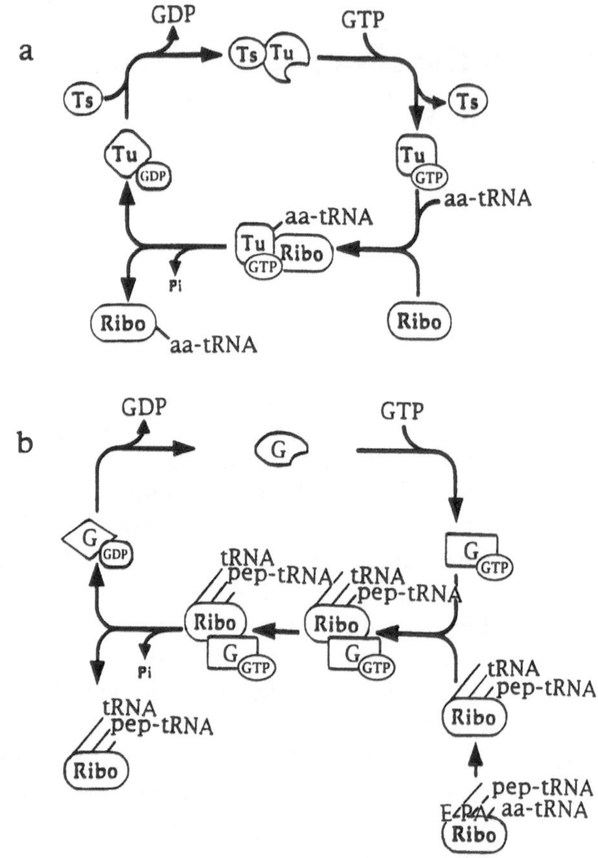

Fig. 1. a Scheme of the GTPase cycle of EF-Tu and the EF-Tu-dependent binding of aminoacyl-tRNA to the ribosomal A site. The three different conformational states of EF-Tu are indicated, as is the sequence of the reactions. **b** The sequence of reactions in the GTPase cycle of elongation factor G and its relation to the peptidyl transfer and translocation reactions of the ribosome. Ribo is the mRNA-programmed ribosome with the three tRNA binding sites, E, P and A. Pep-tRNA, aa-tRNA, and tRNA are peptidyl-tRNA, aminoacyl-tRNA and deacylated tRNA respectively

B. General Concepts

Mechanisms of kinetic proofreading, their requirements, theoretical justifications, and some practical considerations for NTPases have recently been reviewed (YARUS 1992a,b). For the sake of clarity in reviewing the application of kinetic proofreading to the elongation cycle of protein synthesis, some simplified concepts are briefly presented herein.

I. Specificity

The metabolism of every living cell depends upon the necessary accuracy of metabolic pathways and important reactions. The emphasis is on the *necessary*, rather than *excessive* accuracy. Reactions which are catalyzed by enzymes or involve recognition by biological macromolecules can be extremely accurate and substrate specific. DNA replication is an example of a very low error-rate reaction, with one mutation occurring per 10^8–10^{12} incorporated nucleotides (ENGLISCH et al. 1985). The theoretical limit of the specificity of biochemical reactions is given by a ratio of rate constants for the noncognate (k^{nc}) and cognate (k^c) substrate:

$$\text{specificity} = k^{nc}/k^c \qquad (1)$$

In a simple reaction, the ratio of the rate constants approximates the difference in free energies of activation between the cognate and noncognate species:

$$\text{specificity} = (k_{cat}/K_M)^{nc}/(k_{cat}/K_M)^c = \exp(-\Delta\Delta G^{\#}/RT) \qquad (2)$$

Equation 2 gives the maximum specificity attainable in a single step or, for that matter, in a linear steady-state pathway. The specificity can be quite large if the difference in free energies of activation, $\Delta\Delta G^{\#}$, is due to a large number of interactions at the molecular level. If, however, $\Delta\Delta G^{\#}$ between cognate and noncognate species is only due to one hydrogen bond or the few interactions of an additional methyl group, the maximum specificity is severely limited.

Another consideration is the negative consequence of implementing a large specificity in a single step. Reactions with a large specificity require that the energy barrier separating the enzyme–substrate complex and the enzyme–product complex is large for noncognate and small for cognate substrates. Such a condition can only be achieved when many interactions govern the formation of the complex. The formation and transformation of many such interactions during binding and catalysis would require time, effectively slowing the overall reaction rate and decreasing the yield of a pathway. The effect becomes clearer if one considers competing reactions with a large excess of noncognate substrate. If competing with a cognate substrate, the large excess of slowly reacting noncognate substrate will occupy significant amounts of enzyme, suppressing large yields of the correct product and effectively slowing down the overall rate. The inverse relationship between specificity and speed is an important consideration in many biological systems.

To avoid the large differences in potential free energy and the inevitable slow rate constant, it is preferable to distribute the discrimination between cognate and noncognate substrate over multiple specific steps in a pathway. However, the effective discrimination cannot be enhanced by distributing specificity over multiple steps in a *linear steady-state* pathway. As a consequence of the steady-state condition,

$$d[I_n]/dt = 0 \qquad (3)$$

no intermediate (I) can accumulate in a linear steady-state pathway, regardless of the number (n) of multiple discriminatory steps. In steps with high specificity, the discriminated noncognate intermediate accumulates only until its increased concentration compensates for the reduced rate constant. In other words, as long as the steady-state condition is fulfilled, any substrate, cognate or noncognate, entering the linear pathway *will exit as product* at the end, rather than be discarded in an unreacted form. The rate of the overall pathway is determined by the rate constant of the rate-limiting step and the substrate concentration. Similarly, the maximum achievable accuracy is determined by the specificity of the most specific step according to the following relationship:

$$v^c/v^{nc} = k^c[S^c]/k^{nc}[S^{nc}] = ([S^c]/S^{nc}])/\text{specificity} \tag{4}$$

where v^c and v^{nc} are the steady-state rates for the cognate and noncognate substrates. When the cognate and noncognate concentrations, $[S^c]$ and $[S^{nc}]$, are equivalent, the accuracy depends upon only the rate constants of one step, the most discriminatory one. The inherent specificity of all other steps is lost in a linear steady-state pathway.

II. Proofreading

A kinetic mechanism which could improve the specificity of a pathway beyond that possible in a single step is proofreading. In its simplest form, proofreading provides an "emergency exit" for the noncognate intermediate from a steady-state pathway. Any intermediate that might accumulate could then exit through the "proofreading" branch before its conversion into product. The following is an example of a simple branched pathway which has the potential to proofread:

$$\begin{array}{ccccc} A & \xrightarrow{k_{AB}} & B & \xrightarrow{k_{BC}} & C \\ & & \downarrow k_{BA^*} & & \\ & & A^* & & \end{array} \tag{5}$$

To proofread, discrimination by differential rate constants, ${k_{BC}}^c$ and ${k_{BC}}^{nc}$, must follow the nonspecific branch BA* which has equivalent rate constants for cognate and noncognate substrates. Secondly, the "off" rate of the proofreading branch must be faster than the rate for the discriminatory step. Such an arrangement will allow noncognate substrate to dissociate quickly rather than accumulate. An important consequence of the branch step is that greater accuracy can be achieved at the same speed than in a similar linear steady-state pathway with the same set of rate constants for k_{AB} and k_{BC}. Thus, the proofreading branch dampens the inverse effect of accuracy upon speed and facilitates a compromise.

Another vital component of any successful proofreading mechanism is the consumption of energy. In order to make the discard reaction BA* a distinct pathway from the entrance reaction AB, some form of energy is necessary. The minimum energy required must compensate for the potential energy difference between A and A*. The more frequently the discard branch is used, the greater is the energy consumption. In his original model, Hopfield postulated that the required energy for proofreading in protein elongation is derived from the hydrolysis of GTP. The observation was pivotal for testing the kinetic proofreading model because energy consumption, such as GTP hydrolysis, could be measured experimentally.

The simple proofreading mechanism described is very inefficient because the proofreading branch is fast and nonspecific. Consequently, significant amounts of cognate substrate will also exist through the proofreading branch. The following is an example of a more sophisticated and powerful mechanism:

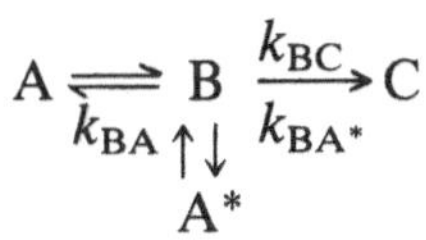

In this mechanism, the rate-limiting step, k_{BC}, is nonspecific and the dissociation rate constants, k_{BA} and k_{BA*}, are the specific ones, that is, faster for noncognate substrates. Further improvement in efficiency is achieved by placing the following constraints on the rate constants: $k_{BA*}{}^{nc} > k_{BC} > k_{BA*}{}^{c}$. With $k_{BA*}{}^{nc} > k_{BC}$, noncognate substrates preferentially dissociate through the proofreading branch rather than form product. Similarly, cognate substrates preferentially form product, and cognate substrate loss through the branch is minimized. The example illustrates the proper use of specificity, which is an essential feature of all efficient proofreading mechanisms. Other proofreading strategies are discussed thoroughly by YARUS (1992a,b).

The maintenance of an appropriate relationship among rate constants is also important for balancing accuracy, speed, and energy consumption. In the second proofreading mechanism described, the overall rate of the reaction could be increased by increasing the rate-limiting step, assumed to be k_{BC} in the example. However, if $k_{BC} > k_{BA*}{}^{nc} > k_{BA*}{}^{c}$, then most cognate and noncognate substrate would be accepted and accuracy would be decreased. On the other hand, if k_{BC} were slowed and the relative rate relationship became $k_{BA*}{}^{nc} > k_{BA*}{}^{c} > k_{BC}$, more cognate substrate would be rejected through the proofreading branch and accuracy would be increased. The price of high accuracy would be a slower overall reaction rate and a large energy consumption through excessive use of the discard branch. Obviously, the example is a simple one; nevertheless, it correctly illustrates the parameter compromise required for a successful proofreading mechanism.

Accuracy and speed are inversely related but, inverse as previously noted, the presence of a proofreading branch dampens the effect of accuracy upon speed. Further attenuation of the inverse relationship between accuracy and speed can be achieved by increasing the number of properly selected specific reactions in a branched pathway. Specificity, which occurs as a result of the energy barrier separating the enzyme–substrate complex and the enzyme–product complex, need not be as large in each reaction of a multiple-step branched pathway as that required for a single step reaction. Thus, as long as specificity is properly applied, the same accuracy can be achieved at faster speeds by distributing the specificity over an increased number of reactions in a branched pathway.

In summary, a pathway with proofreading branches can achieve a higher substrate specificity than a single reaction or a linear pathway. Three additional conditions, including energy consumption, "constrained" use of specificity, and a compromise between yield and accuracy, are also essential for any successful proofreading mechanism (YARUS 1992a,b).

C. Parameters of Protein Biosynthesis

In vivo, the overall accuracy of prokaryotic protein biosynthesis is estimated to lie between 10^{-3} and 10^{-4}. As reviewed by KURLAND and GALLANT (1986), the fidelity of translation of any given codon is not constant because it is dependent upon a heterogeneous collection of factors, such as cell type and method of error detection. Similar arguments can be made for translational efficiency. The in vivo rate of elongation in *Escherichia coli* for several proteins is estimated to be between 10 and 20 peptide bonds/s for each ribosome (PEDERSON 1984). Over the years, it has become apparent that the ribosome does not fully exploit its available specificity (KURLAND and GALLANT 1986). Rather, it employs sophisticated mechanisms, including proofreading, to balance the necessary accuracy with speed at reasonable energy costs. Studies with mutants and antibiotics have been particularly informative in elucidating the details of the relationship between accuracy and speed (KURLAND and GALLANT 1986 and references therein; TAPIO and KURLAND 1986; HORNIG et al. 1987; FAXEN et al. 1988; TAPIO and ISAKSSON 1988; BILGIN et al. 1988, 1992; RICHTER DAHLFORS and KURLAND 1990a; DIAZ and EHRENBERG 1991).

Given the complexity and the number of required macromolecules, it is not surprising that in vitro ribosomal biosynthesis assays are experimentally difficult. Two of the major problems, that of achieving the observed accuracy and speed of in vivo translation, have both been overcome. In vitro accuracy was improved by the rapid preparation of more active ribosomes (JELENC 1980) and by the addition of an optimized mix of inorganic ions and polyamines (JELENC and KURLAND 1979). The best in vitro systems today typically use ribosomes which are 60%–80% active in binding Phe-tRNA and which translate with an error frequency of $2–6 \times 10^{-4}$. In vitro elongation rates of 8–12 peptides/s at 37°C were achieved in the accuracy-optimized ribosomal assay system by including a preliminary incubation of the poly(U)-primed ribosomes with N-acetylphenylalanyl-tRNA to eliminate the rate-limiting step of polyphenylalanine initiation (WAGNER et al. 1982). Despite these achievements, the variability of ribosome and factor preparations often makes it difficult to reproduce measurements in the same laboratory.

Many studies have focused on testing and refining the concept of kinetic proofreading in protein biosynthesis since its introduction in the mid 1970s. Consequently, many experiments have been carried out to measure various kinetic parameters of the elongation reactions in order to identify the specific subset that are primarily responsible for the translational accuracy observed in vivo. In order to measure single reaction rates or to elucidate other details of the elongation cycle, the in vitro ribosomal biosynthesis assay has been cleverly manipulated in a variety of ways to decouple the individual reactions or to shift the rate-limiting step. For example, the omission of EF-G blocks translocation and permits the study of a single turnover of the EF-Tu-dependent reactions during elongation (THOMPSON and STONE 1977). The omission of EF-Ts slows polypeptide synthesis by making the dissociation of GDP from EF-Tu the rate-limiting step (RUUSALA et al. 1982). The substitution of the slowly hydrolyzing GTPγS for GTP forces the ribosome-catalyzed EF-Tu-GTPase reaction to be the rate-limiting step (ECCLESTON et al. 1985). Although each assay modification has yielded new insight into the proofreading process during the elongation cycle, each is also fraught with its own set of pitfalls. The study of isolated reactions under pre-steady-state conditions will only yield information about the attainable specificity or rate of a particular reaction under specific in vitro conditions, but will not provide the actual rate or accuracy of the same reaction under steady-state or in vivo conditions. Similarly, slowing down a particular step for experimental expediency raises concerns about missing the faster details of the altered reaction as well as about changing the details of the normal steady state. The objective of the review is not to critique the individual studies, but to present the consensus view of translational proofreading and to indicate the areas which require further investigation.

D. EF-Tu-Dependent Kinetic Proofreading

It has long been established that in vitro ribosomal biosynthesis can be carried out in the absence of EF-Tu, but at the cost of decreased accuracy. Because Hopfield's original hypothesis proposed a plausible role for EF-Tu in the elongation cycle, the experiments to test the concept of kinetic proofreading in protein biosynthesis focused on the EF-Tu-dependent reactions.

The simplest mechanism for the role of EF-Tu in proofreading is the minimal kinetic model lucidly reviewed by THOMPSON (1988). According to the model, translational accuracy is achieved in three steps during each elongation cycle, as illustrated in Fig. 2. The first step involves the initial selection of aminoacyl-tRNA. EF-Tu-GTP complexes with either cognate

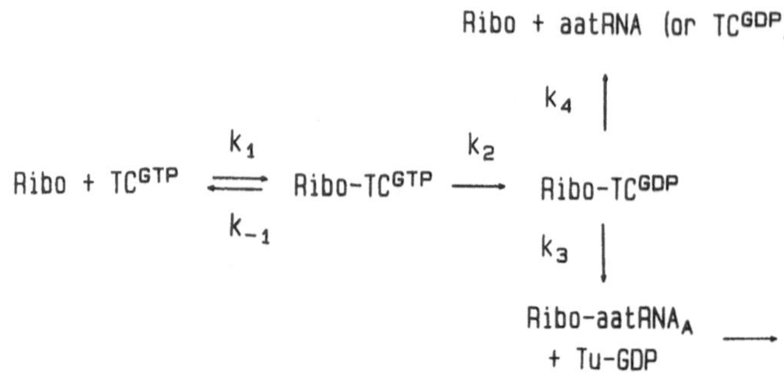

Fig. 2. Scheme for the reactions involved in the selection and proofreading of aminoacyl-tRNA. Reaction steps are labelled with the appropriate rate constants. Ribo indicates the poly(U)-programmed ribosome with an empty A site, and TC is an abbreviation for the ternary complex, EF-Tu-GTP–aminoacyl-tRNA

or noncognate aminoacyl-tRNA, and the resulting EF-Tu-GTP–aminoacyl-tRNA complex then binds to the A site of the ribosome. The "on" rate, k_1, is very fast and close to the theoretical diffusion limit. The "off" rate, k_{-1}, is much slower and depends upon multiple molecular interactions including, but not necessarily limited to, those between the codon of the mRNA and the anticodon of the aminoacyl-tRNA (DIAZ and EHRENBERG 1991). Because some noncognate tRNAs are near-cognates, differing from cognate tRNAs only by a few weak interactions, the specificity of the initial selection step is generally thought to be insufficient to account for the observed translational accuracy. In the second step of the kinetic mechanism, irreversible hydrolysis of GTP occurs. The rate of hydrolysis of EF-Tu-GTP in the presence of the ribosome, k_2, is faster than $k_{-1}{}^{c}$ for cognate tRNA but slower than $k_{-1}{}^{nc}$ for noncognate tRNA. Consequently, most noncognate tRNAs dissociate from the ribosome before GTP hydrolysis occurs. A few incorrect, probably near-cognate tRNAs remain associated with the ribosome even after GTP hydrolysis. Therefore, an additional step with a branch is essential to improve the accuracy of translation. It is this particular EF-Tu-mediated branch to which the term "proofreading" in protein synthesis generally alludes. In the proofreading step, either EF-Tu-GDP or aminoacyl-tRNA (uncomplexed or complexed to hydrolyzed EF-Tu-GDP) dissociates from the ribosome, with rate constants of k_3 and k_4 respectively. If k_3 is faster than k_4, then EF-Tu-GDP dissociates first, allowing the remaining A-site aminoacyl-tRNA to assume a proper position for the peptidyl transfer reaction (MOAZED and NOLLER 1989). If k_3 is slower than k_4, as in the case of near-cognate tRNAs, then the incorrect aminoacyl-tRNA dissociates first, leaving an empty A site available for another round of EF-Tu-GTP–aminoacyl-tRNA association. The minimal kinetic model is consistent with the original Hopfield scheme but does not attempt to ascertain the impact of other EF-Tu-independent elongation cycle reactions, such as peptidyl transfer or translocation, upon translational accuracy.

Considerable experimental data exist which support the basic tenets of the minimal kinetic model (for review, see KURLAND and GALLANT 1986; THOMPSON 1988). Because in vitro ribosomal assays are difficult, absolute values for the rate constants vary tremendously from laboratory to laboratory. Yet most investigators are in general agreement that $k_{-1}^{c} \ll k_2 < k_{-1}^{nc}$, with a specificity of 1000–4000 for initial selection of cognate over noncognate tRNA, and that $k_4^{c} < k_3 < k_4^{nc}$, with a specificity of approximately 100 for the proofreading step. A particularly important confirmation of the model was the measurement of the accompanying energy consumption, that is, EF-Tu-GTPs hydrolyzed per peptide bond formed (THOMPSON and STONE 1977; RUUSALA et al. 1982; EHRENBERG et al. 1990). Most agree that the number is close to one for cognate tRNAs and much higher for near-cognate tRNAs. Controversy remains, however, regarding k_2. Based on experimental data, Thompson suggests that the rates of EF-Tu-mediated reactions act as an internal kinetic standard in the proofreading mechanism (THOMPSON 1988). As an internal standard, the ribosome-catalyzed EF-Tu-GTP hydrolysis rate, k_2, and the rate of EF-Tu-GDP dissociation from the ribosome, k_3, are independent of the type of A-site tRNA, i.e., k_2 and k_3 are the same for cognate, near-cognate, and noncognate tRNAs. THOMPSON and DIX (1982) actually measure a 6–12-fold difference in the k_2s with Phe-tRNA and Leu$_2$-tRNA in the A site of poly(U)-programmed ribosomes under single turnover conditions. They attribute the difference to statistical errors of the experiment and consider it to be too small to have much impact upon overall accuracy. PINGOUD et al. (1990) believe that the ribosome behaves like an allosteric enzyme in the EF-Tu-dependent cycle, altering relevant K_Ms and k_{cat}s, depending upon the presence of cognate or noncognate tRNA in the A site. Their computer simulation of the elongation cycle under steady-state conditions specifically indicates that the EF-Tu-GTPase reaction is under allosteric control. The investigators also demonstrate that a 25-fold difference in k_2 has a more favorable impact upon the overall accuracy, speed, and energy costs of elongation than does a 100-fold discrimination factor in k_4.

In summary, k_2 is dependent upon the type of A-site tRNA in the allosteric model, and independent in the internal kinetic standard model. Both models are compatible with the acceptable proofreading strategies for the proper use of specificity presented by Yarus (1992a,b), but only the internal kinetic standard model could achieve the maximal specificity theoretically possible for two specific reactions. The allosteric model is similar to Yarus's third strategy, which must be rescued by an additional specific reaction. However, the allosteric model has the potential of achieving a greater overall specificity than the internal kinetic standard model, although it cannot achieve the maximal specificity theoretically possible for three specific reactions. Because maximal accuracy does not appear to be important in ribosomal biosynthesis, what is the significance of k_2? If the internal kinetic standard model is correct, then discrimination occurs in two reactions. In the allosteric model, specificity occurs in three reactions and has a greater potential to facilitate the necessary compromise between accuracy and speed at a reasonable energy cost. Given the experimental difficulties, the k_2 controversy is not likely to be resolved soon.

E. EF-Tu-Independent Error Correction Mechanisms

No reasonable set of rate constants for the EF-Tu-dependent elongation reactions correctly predicts the observed translational accuracy and speed in vivo. Therefore, the more recent investigations have focused on the role and kinetic parameters of other EF-Tu-independent elongation reactions.

I. Peptidyl Transfer

PINGOUD et al. (1990) implicate the peptidyl transfer rate, suggesting that the rate is allosterically controlled by the type of A-site tRNA. Their hypothesis is based upon the same computer simulations which predicted an allosterically controlled EF-Tu-GTPase reaction. The computations indicate that in vivo speeds could only be achieved by varying k_{-1}, k_2, and k_4. To predict accuracy as well as speed, a discrimination factor of approximately 100 in the peptidyl transfer rate is necessary. Theoretical considerations have also lead NINIO (1986) to independently predict that the peptidyl transfer rate plays a role in the fine tuning of ribosomal accuracy. He postulates that the ribosome can switch between two states: a "low" accuracy one with a fast peptidyl transfer rate when cognate tRNAs are abundant, and a "high" accuracy one with a slower peptidyl transfer rate in the absence of cognate tRNAs. The slower the peptidyl transfer rate, the more time the ribosome has to switch to the high accuracy state which may be needed when there is a shortage of cognate tRNAs. Both hypotheses share the view that peptidyl transfer is another potential reaction contributing to translational accuracy, but not the primary source of specificity. At present, there is insufficient experimental data to support or refute either hypothesis.

II. EF-G-Dependent Translocation

In view of the proofreading role that GTP plays in EF-Tu-dependent elongation reactions, it seems only natural to ascribe a similar role to the GTP hydrolysis associated with EF-G-mediated translocation. In such a model, EF-G-GTP hydrolysis would provide the energy for the preferential discard of peptidyl-tRNAs from the ribosome if an incorrect amino acid had been erroneously incorporated into the polypeptide chain. It must be kept in mind, however, that proofreading has a strong disadvantage at the translocation step: the longer the polypeptide chain, the more cellular resources would be wasted by the discard branch. Interestingly, there is data to support the presence of an "editing" mechanism in which peptidyl-tRNAs are discarded during translocation, but there is no data to suggest that the discard is accompanied by excessive EF-G-GTP hydrolysis. Currently, it is believed that EF-G plays only a catalytic role, increasing the translocation rate by a factor of approximately 500 (GAST et al. 1987). It is also believed that the energy derived from EF-G-GTP hydrolysis is only used to maintain the proper conformation of EF-G required for complexation with the ribosome (PARMEGGIANI and SANDER 1981).

The first person to suggest "editing" at the level of translocation was MENNINGER (1978), who observed dissociation of peptidyl-tRNAs from elongating ribosomes in vivo by inhibiting the activity of peptidyl transfer RNA hydrolase of temperature-sensitive *E. coli* cells. However, he was unable to distinguish between cognate and noncognate tRNAs to demonstrate a preferential dissociation of inappropriate peptidyl-tRNAs. Several in vitro translocation studies under pre-steady-state conditions have shown that translocation is a slow and, perhaps, the rate-limiting step of protein synthesis, at least in the early stages of translation (ROBERTSON et al. 1986a; GAST et al. 1987). The initial translocation rate was also shown to be dependent upon the type of tRNA bound to the ribosome, suggesting an allosteric role for the ribosome. The presence of Leu-tRNA in either the A or P site of poly(U)-programmed ribosomes reduced the initial translocation rate constant for Phe-tRNA by a factor of 10 (GAST et al. 1987). The investigators speculated that retardation of the translocation rate in the presence of ribosome-bound noncognate tRNAs may function as a form of editing by increasing the probability of dissociation of incorrect short peptidyl-tRNAs. Because the overall translocation rate appears to be much faster in the later stages of translation (WAGNER et al. 1982; ANDERSSON et al. 1984; BILIGIN et al. 1988), the editing mechanism at the level of translocation may be switched off with longer peptidyl-tRNAs.

EF-G-GTP hydrolysis is known to occur after translocation of the A-site peptidyl-tRNAs into the P site but before release of EF-G from the ribosome (PARMEGGIANI and SANDER 1981; ROBERTSON et al. 1986b). Experiments to determine the total number of GTPs hydrolyzed per peptide bond formed in poly(U)-primed ribosomal assays in the presence of EF-G are complicated by several factors. Particularly troublesome is the fast ribosome-dependent rate of EF-G-GTP hydrolysis, which is uncoupled from translocation (CHINALI and PARMEGGIANI 1980). RICHTER DAHLFORS and KURLAND (1990b) recently circumvented the GTPase problems by a clever manipulation of the poly(U)-primed ribosomal assay, and measured a stoichiometric ratio close to 1 for the number of hydrolyzed EF-G-GTPs coupled to translocation per peptide bond in polyphenylalanine. The investigators also found that the ratio did not change upon substitution of the wild type EF-G with a hyperaccurate variant of EF-G, suggesting that the increase in accuracy was not accompanied by the excessive energy consumption normally expected in maximally efficient kinetic proofreading mechanisms. However, no one has yet attempted to measure the difference in the coupled EF-G-GTPs consumed per translocation or per peptidyl-tRNA dissociation between correct peptidyl-tRNAs and those containing errors. Such an experiment would constitute an alternate test of whether some form of kinetic proofreading exists at the level of translocation.

III. Allosteric Linkage Between A and E Sites

In recent years, the discovery of a third tRNA ribosomal binding site, termed the exit or E site, has led some investigators to re-evaluate the necessity for any kinetic proofreading mechanism during the elongation cycle. According to a recent review by NIERHAUS (1990), an error correction mechanism which includes ribosomal allostery, but not kinetic proofreading, is sufficient to explain a translational accuracy of 1:1000. In the three-site allosteric model, when the E site is occupied with a deacylated tRNA from the previous elongation cycle, the A site has a low affinity for aminoacyl-tRNA and can only accept the codon–anticodon interaction. A correct match triggers a slow conformational change in the A site from a state of low affinity to a state of high affinity for aminoacyl-tRNA. The "low to high" affinity switch at the A site is accompanied by a corresponding, but opposite switch from "high to low" affinity at the E site and the subsequent dissociation of the E-site tRNA. Noncognate tRNAs dissociate before the slow "low to high" affinity switch in the A site. Thus, translational accuracy is a function of only the specificity of the initial aminoacyl-tRNA selection step. Hydrolysis of EF-Tu-GTP occurs after the initial A-site occupation, but is not coupled to the allosteric transition between the A and E sites. The GTP serves only to maintain the proper conformation of EF-Tu required for the unfavorable binding of aminoacyl-tRNA to the low affinity conformation of the A site. GTP hydrolysis does not provide the energy for the discard branch in the three-site allosteric model. Whether allostery or kinetic proofreading is the predominant mechanism that accompanies the specific selection of cognate tRNAs under the experimental conditions cited by Nierhaus is not clear.

F. Summary

Translational accuracy in bacterial protein synthesis is generally attributed to the initial selection of cognate aminoacyl-tRNAs for A-site association and to the proofreading branch following ribosome-catalyzed EF-Tu-GTP hydrolysis. Collectively, the experimental data support a prominent kinetic proofreading role for EF-Tu in ribosomal biosynthesis under normal cellular growth conditions. However, the failure of models to predict the observed in vivo accuracy and speed using only kinetic parameters of EF-Tu-dependent reactions suggests that the question of translational accuracy has not been completely resolved. Some other major issues which must be investigated further include: the allosteric behavior of the ribosome and its comparative significance with kinetic proofreading in translational accuracy; and the involvement of other reactions in error correction, such as editing in translocation and their relative importance during various stages of translation. Finally, of great interest to guanine nucleotide aficionados is the role, if any, of the hydrolyzed GTPs in translocation, initiation, and termination in some type of kinetic proofreading mechanism. Conceivably, the ribosomes may have stored reserves of specificity into every reaction and may utilize more than one type of error correction mechanism to facilitate the compromise among accuracy, speed, and energy consumption in response to changes in environmental or genetic factors. The answers must await further experimentation.

Acknowledgements. The authors thank Alfred Pingoud of Medizinische Hochschule, Hannover, Germany, and Peter Brzovic of the University of California, Riverside, for helpful discussions. The authors also acknowledge the support of the United States Public Health Service (GM#26895) during preparation of the review.

References

Andersson SGE, Buckingham RH, Kurland CG (1984) Does codon composition influence ribosome function? EMBO J 3/1:91–94

Bilgin N, Kirsebom LA, Ehrenberg M, Kurland CG (1988) Mutations in ribosomal proteins L7/L12 perturb EF-G and EF-Tu functions. Biochimie 70:611–618

Bilgin N, Claesens F, Pahverk H, Ehrenberg M (1992) Kinetic properties of Escherichia coli ribosomes with altered forms of S12. J Mol Biol 224:1011–1027

Chinali G, Parmeggiani A (1980) The coupling with polypeptide synthesis of the GTPase activity dependent on elongation factor G. J Biol Chem 255:7455–7459

Diaz I, Ehrenberg M (1991) ms^2i^6A deficiency enhances proofreading in translation. J Mol Biol 222:1161–1171

Eccleston JF, Dix DB, Thompson RC (1985) The rate of cleavage of GTP on the binding of phe-tRNA-elongation factor Tu GTP to poly(U)-programmed ribosomes of Escherichia coli. J Biol Chem 260/30:16237–16241

Ehrenberg M, Rojas AM, Weiser J, Kurland CG (1990) How many EF-Tu molecules participate in aminoacyl-tRNA binding and peptide bond formation in Escherichia coli translation? J Mol Biol 211:739–749

Englisch U, Gauss D, Friest W, Englisch S, Sternbach H, von der Haar F (1985) Error rates of the replication and expression of genetic information. Angew Chem Int Ed Engl 24:1015–1025

Faxen M, Kirsebom LA, Isaksson LA (1988) Is efficiency of suppressor tRNAs controlled at the level of ribosomal proofreading in vivo? J Bacteriol 170/8: 3756–3760

Gast FU, Peters F, Pingoud A (1987) The role of translocation in ribosomal accuracy: translocation rates for cognate and noncognate aminoacyl- and peptidyl-tRNAs on Escherichia coli ribosomes. J Biol Chem 262/25:11920–11926

Hopfield JJ (1974) Kinetic proofreading: a new mechanism for reducing errors in biosynthetic processes requiring high specificity. Proc Natl Acad Sci USA 71/10:4135–4139

Hornig H, Woolley P, Luhrmann R (1987) Decoding at the ribosomal A site: antibiotics, misreading and energy of aminoacyl-tRNA binding. Biochimie 69:803–813

Jelenc PC (1980) Rapid purification of highly active ribosomes from Escherichia coli. Anal Biochem 105:369–374
Jelenc PC, Kurland CG (1979) Nucleoside triphosphate regeneration decreases the frequency of translation errors. Proc Natl Acad Sci USA 76:3174–3178
Kurland CG, Gallant JA (1986) The secret life of the ribosome. In: Kirkwood TBL, Rosenberger RF, Galas DJ (eds) Accuracy in molecular processes. Chapman and Hall, Cambridge, pp 127–157
Menninger JR (1978) The accumulation as peptidyl-transfer RNA of isoaccepting transfer RNA families in Escherichia coli with temperature-sensitive peptidyl-transfer RNA hydrolase. J Biol Chem 251/19:6808–6813
Moazed D, Noller HF (1989) Interaction of tRNA with 23S rRNA in the ribosomal A, P, and E sites. Cell 57:585–597
Nierhaus KH (1990) The allosteric three-site model for the ribosomal elongation cycle: features and future. Biochemistry 29/21:4997–5008
Ninio J (1975) Kinetic amplification of enzyme discrimination. Biochimie 57:587–595
Ninio J (1986) Fine tuning ribosomal accuracy. FEBS Lett 196/1:1–4
Parmeggiani A, Sander G (1981) Properties and regulation of the GTPase activities of elongation factors Tu and G, and of initiation factor 2. Mol Cell Biochem 35:129–158
Pedersen S (1984) Escherichia coli ribosomes translate in vivo with variable rate. EMBO J 3:2895–2890
Pingoud A, Gast FU, Peters F (1990) The influence of the concentrations of elongation factors and tRNAs on the dynamics and accuracy of protein biosynthesis. Biochim Biophys Acta 1050:252–258
Richter Dahlfors AA, Kurland CG (1990a) Novel mutants of elongation factor G. J Mol Biol 215:549–557
Richter Dahlfors AA, Kurland CG (1990b) Stoichiometry of elongation factor G function in translation. J Mol Biol 216:311–314
Robertson JM, Paulsen H, Wintermeyer W (1986a) Pre-steady-state kinetics of ribosomal translocation J Mol Biol 192:351–360
Robertson JM, Urbanke C, Chinali G, Wintermeyer W, Parmeggiani A (1986b) Mechanism of ribosomal translocation: translocation limits the rate of Escherichia coli elongation factor G-promoted GTP hydrolysis. J Mol Biol 189:653–662
Ruusala T, Ehrenberg M, Kurland CG (1982) Is there proofreading during polypeptide synthesis? EMBO J 1/6:741–745
Tapio S, Kurland CG (1986) Mutant EF-Tu increases missense error in vitro. Mol Gen Genet 205:186–188
Tapio S, Isaksson LA (1988) Antagonistic effects of mutant elongation factor Tu and ribosomal protein S12 on control of translational accuracy, suppression and cellular growth. Biochimie 70:273–281
Thompson RC (1988) EFTu provides an internal kinetic standard for translational accuracy. Trends Biochem Sci 13:91–93
Thompson RC, Dix DB (1982) Accuracy of protein synthesis: a kinetic study of the reaction of poly(U)-programmed ribosomes with a leucyl-tRNA elongation factor Tu-GTP complex. J Biol Chem 257/12:6677–6682
Thompson RC, Stone PJ (1977) Proofreading of the codon-anticodon interaction on ribosomes. Proc Natl Acad Sci USA 74:198–202
Wagner EGH, Jelenc PC, Ehrenberg M, Kurland CG (1982) Rate of elongation of polyphenylalanine in vitro. Eur J Biochem 122:193–197
Yarus M (1992a) Proofreading, NTPases and translation: constraints on accurate biochemistry. Trends Biochem Sci 17:130–133
Yarus M (1992b) Proofreading, NTPases and translation: successful increase in specificity. Trends Biochem Sci 17:171–174

CHAPTER 3

A New Look at Receptor-Mediated Activation of a G-protein

L. BIRNBAUMER

The intimate kinetic processes of hormonal activation of a G-protein-mediated pathway are complex and involve the interplay of a receptor with the G-protein, the activation of the G-proteins by GTP, the G-protein subunit dissociation reaction, the turnoff of the activated α subunit by its endogenous GTPase, the reassociation of the G-protein subunits, and the reversible and presumably cyclical oscillation of the receptor between a minimum of two states – one with low and the other with high affinity for the hormone. Work by Lefkowitz's group in the late 1970s provided solid evidence that for the β-adrenergic receptor (βAR), the activation-triggering state of the receptor is the state with high agonist affinity or R_H (DELEAN et al. 1980; KENT et al. 1980). Work by CASSEL and SELINGER (1978) and later that by MICHEL and LEFKOWITZ (1982) showed that agonist stimulation accelerates an exchange reaction in which bound nucleotide, be it GDP- or GTP-like, is rapidly released from the system in an agonist-specific manner. Facilitation of guanine nucleotide binding by a receptor was confirmed with purified components in the mid 1980s (CERIONE et al. 1985, 1986). This led to the now widely held concept that one role, and very likely the sole role, of the receptor is to promote nucleotide exchange at the GDP–GTP binding site, with the implicit assumption that the automatic fate of a G-protein with GTP bound to it is its activation ("self-activation"), and that consequently the receptor is not required for the activation reaction proper. However, work by IYENGAR and myself in 1980 indicated that G-protein activation is stimulated by the agonist-activated receptor (IYENGAR et al. 1980a). This result predicted that a stable complex should form, composed of the receptor with a ligand bound to it and activated G-protein with GTP bound to it – a quaternary complex. In the absence of further considerations, such an activated complex would have to "turn off," either through the GTPase reaction or through dissociation of the activating ligand from the receptor. Since G-protein activation depended on the high affinity form of the receptor, the stable quaternary complex formed was predicted to have high ligand affinity. Yet experimental data, including those stemming from the very original discovery of nucleotide regulation of hormone-binding, did not appear to support this prediction, as addition of GTP, or nonhydrolyzable GTP analogues, promoted the formation of the low affinity rather than the high affinity state (e.g., ROJAS and BIRNBAUMER 1985). Furthermore, studies by TOLKOVSKY and LEVITZKI (1978) on the dependence of the rate of activation of adenylyl cyclase by GTPase-resistant GMP–P(NH)P showed that under activating conditions, where it was predicted that a stable receptor–G-protein complex should be formed, the receptor – rather than being locked to a single G-protein molecule – jumped from one G-protein to another, acting as a catalyst of activation rather than as an allosteric activator working in a stoichiometric mode.

The discovery by GILMAN's group that the G_s-protein dissociates upon activation (HOWLETT and GILMAN 1978; NORTHUP et al. 1983a,b) provided a thermodynamic explanation for the apparent impasse. The impasse was given by the rules of microscopic reversibility, which establish that if receptors are to stabilize an activated form of the G-protein, they could only do so by having an intrinsic affinity for the activated form that is higher than that for the unactivated form. Because through break-up the G-protein ceases to exist as such, the subunit dissociation reaction relieved the system from the need of maintaining a stable receptor–G-protein complex.

In 1985 we proposed model with a most likely reaction sequence in the activation–deactivation cycle of a G-protein that would accommodate the GTPase reaction as well as the low affinity–high affinity transitions of the receptor. In this model, the $\beta\gamma$ dimer dissociated from a HR·G* complex (hormone, receptor in its high affinity state, GTP, and trimeric G-protein in its activated conformation) to give a HR·α^* plus a free $\beta\gamma$. This was followed by dissociation of the HR·α^* complex into HR plus α^* (reviewed in BIRNBAUMER et al. 1985, 1990). Since purified receptors exhibit low agonist affinity that can be converted by addition of nucleotide-free G-protein into a state with high affinity, the receptor in free HR had to be in the low affinity state. At which point the transition to low affinity occurred could not be determined (for details, see BIRNBAUMER et al. 1990).

The choice of freeing the receptor from the activated trimer by dissociation of $\beta\gamma$, rather than by dissociation of α^*, was arbitrary and at the time the proposal was made, there was no reason to sway one way or the other. The discovery that $\beta\gamma$ dimers are elements of signaling in mammalian cells (CAMPS et al. 1992), together with the need to explain why the dose–response curve for agonist-mediated activation of the phospholipase C (PLC) is right-shifted with respect to that for activation of adenylyl cyclase for many receptors that exhibit dual signaling (TAGUCHI and FIELD 1988; VANSANDE et al. 1990; GUDERMANN et al. 1992; CHABRE et al. 1992), led me to reconsider to the previous choice of the path by which the receptor-heterotrimer breaks down. The search for a mechanism that would allow for activation of adenylyl cyclase with the α subunit of G_s and of PLC with the $\beta\gamma$ of G_s led to the development of the scheme presented in Fig. 1. This scheme drawn up in 1992, differs from the 1985 scheme (e.g., BIRNBAUMER et al. 1985) in that the HR·G* complex resolves into free-activated α-GTP (α^*) plus HR·$\beta\gamma$, instead of free $\beta\gamma$ plus HR·α^*. In so doing, it is possible to assign distinct signaling potential(s) to α and $\beta\gamma$, and to account for the experimental data.

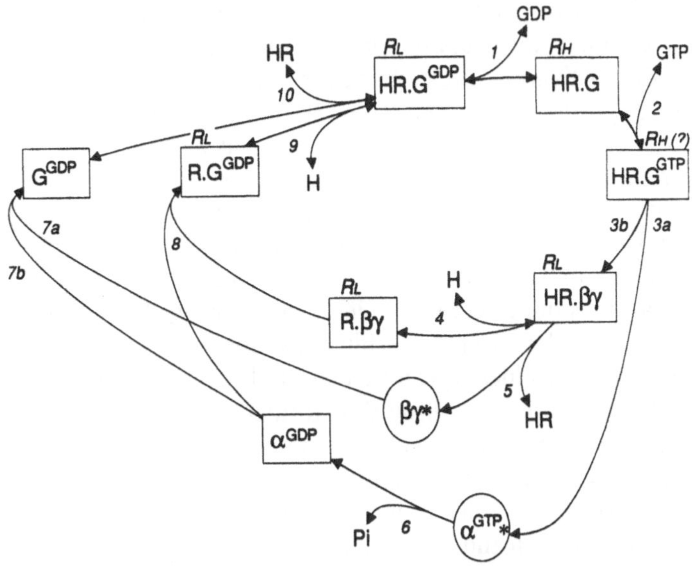

Fig. 1. Model of G-protein activation by receptors allowing for signaling through the α subunit and/or the $\beta\gamma$ dimer. Species enclosed by *circles* are signaling-competent; species enclosed by *squares* are signaling incompetent. Reactions *1* and *2* constitute the nucleotide exchange reaction plus the receptor-induced "activation" reaction of the holo-G-protein. Reaction *3* leads to formation of free α^* (*3b*) and HR$\cdot\beta\gamma$, the precursor of $\beta\gamma$ (*3a*). This reaction is required for the cycle to move forward, which in turn is required to obtained signal amplification. Amplification may occur because, during the time span during which the α is active, any of the $\beta\gamma$ species (i.e., HR$\cdot\beta\gamma$, R$\cdot\beta\gamma$ or $\beta\gamma$) is able to combine with the deactivated α-GDP subunit from another G-protein molecule, thus leading to activation of a second, or even more G-protein(s) during the life span of the first. Although not shown, direct reassociation of α-GDP with HR$\cdot\beta\gamma$ is also allowed. Reaction *5* leads to activation of $\beta\gamma$. Reaction *4* leads to loss of $\beta\gamma$ precursor, but does not interfere with signaling through α because of the ability of R$\cdot\beta\gamma$ to recycle through reactions *8*, forming R$\cdot$G(GDP), and *9*, forming HR$\cdot$G(GDP). Reaction *7* serves two purposes: firstly, *7a*, to deactivate $\beta\gamma$, secondly, *7b*, to recycle α-GDP (and $\beta\gamma$) for renewed activation through reaction *10*

Evidence supporting an interaction of the receptor with $\beta\gamma$ in the absence α – and hence for the plausibility of α dissociating to give receptor–$\beta\gamma$ complexes – was provided by Cerione's laboratory, where biochemical approaches were used to measure the formation of interactive complexes between transducin $\beta\gamma$ and rhodopsin (Phillips and Cerione 1992).

Figure 1 shows a series of self-explanatory complexes involving a free receptor (R), hormone or agonist-occupied receptor (HR), unactivated forms of a trimeric G-protein with either GDP or GTP bound to it, an intermediary hormone-receptor-G-protein complex devoid of nucleotide, and two active signaling species, α^* – mediating α subunit-dependent signaling – and $\beta\gamma$ – mediating $\beta\gamma$-dependent signaling. It is specifically assumed that the HR$\cdot\beta\gamma$ and R$\cdot\beta\gamma$ complexes are unable to mediate effector regulation and that their sole fate is to eventually reassociate with GTPase-deactivated α-GDP.

Reassociation of α-GDP with R·$\beta\gamma$ (reaction 8) results in a receptor–G-protein complex that, upon hormone binding, becomes competent to reinitiate the activation cycle (reaction 9). Reassociation of signaling-competent $\beta\gamma$ with α-GDP (reaction 7) forms a trimeric G-protein with GDP that needs to interact with a receptor and hormone (reaction 10) to, likewise, reinitiate the G-protein activation cycle.

The main feature of the reaction sequences shown in Fig. 1 is that only HR·$\beta\gamma$ (and not R·$\beta\gamma$) serves as a source of active $\beta\gamma$. As a consequence, the dissociation of H is effectively a turnoff reaction for the $\beta\gamma$-signaling pathway that does not affect the α-signaling pathway. With receptor ligands that interact with the receptor at high reaction rates, such as neurotransmitters and small peptides, it can be expected that reaction 4 is much faster than reaction 5 and that very little $\beta\gamma$ is formed, even though the α signaling pathway may be proceeding at close to maximal. It is easy to see that, at 90% receptor occupancy, the rate-limiting step in α signaling has become the GTPase turnoff reaction, so that most of the G-protein is in the α^* form. However, since the ligand dissociation rate is rapid, only a very small proportion of the G-protein's $\beta\gamma$-signaling pathway is operative. In fact, most of the $\beta\gamma$ remains in the R·$\beta\gamma$ form, unable to generate active $\beta\gamma$, but poised to reassociate with the GTPase turnoff product α-GDP, to drive the reactivation aspect of the cycle (reactions 8 → 9 → 1 → 2). Under these conditions, if the rate of reaction 4 is much faster than that of reaction 5, and that of reaction 3 (HR·G* decay) much faster than that of reaction 6 (GTPase), the steady-state level of $\beta\gamma$ could easily be only 1% or less of that of α^*. However, since HR·$\beta\gamma$ can form not only from HR·G*, but also from R·$\beta\gamma$ by reversal of reaction 4, the decay of HR·$\beta\gamma$ to R·$\beta\gamma$ can be prevented by driving the reaction backward with increasing concentrations of H. This predicts that the dose–response curve for hormone-stimulated activation of the α- and $\beta\gamma$-signaling pathways differ, and that the difference will be proportional to the k_{off} rate constant for H dissociation from HR (forward motion of reaction 4) and inversely proportional to the rate constant of HR dissociation from $\beta\gamma$ (forward motion of reaction 5).

The issue of which of the receptor species has high affinity for agonists (R_H state) and which has low affinity (R_L state) has also been addressed in Fig. 1, albeit in an only partial manner. Of the six receptor species shown, four can confidently be classified as having the receptor in the R_L state (HR·$\beta\gamma$, R·$\beta\gamma$, R·G(GDP), and HR·G(GDP)). Thus, it has been shown that purified and G-protein-free receptors only exhibit low affinity for agonists (e.g., CERIONE et al. 1984) and that reconstitution of receptors with $\beta\gamma$ dimers does not lead to the appearance of high agonist affinity (FLORIO and STERNWEIS 1989). This indicates that the receptor in HR·$\beta\gamma$ and R·$\beta\gamma$ is in the R_L state. Addition of GDP or GDPβS to membranes generates R_L, indicating that the receptor in R·G(GDP) and HR·G(GDP) is also in the R_L state (e.g., RODBELL et al. 1971; IYENGAR et al. 1980b; MATTERA et al. 1985; ROJAS and BIRNBAUMER 1985). On the other hand, for the β-adrenergic and most neurotransmitter receptors, as well as for several peptide hormone receptors such as the glucagon and chemoattractant (fMLP) receptors, the R_H state of the receptor has been trapped in in vitro assays by mere removal of guanine nucleotides or upon reconstitution of receptor into phospholipid vesicles together with nucleotide-free holo-G-protein to form HR·G (e.g., RODBELL 1971; DELEAN et al. 1980; IYENGAR et al. 1980b; CERIONE et al. 1984). This is the intermediate that needs to form between reactions 1 and 2, i.e., after GDP has dissociated and before the site is occupied by GTP. There are no direct measurements of the affinity state of a receptor in HR·G(GTP), in which G(GTP) is presumably primed to dissociate into its α^* and $\beta\gamma$ components. In solution, with purified G-protein and in the absence of a receptor, the dissociation reaction is driven by nonhydrolyzable GTP analogues and relatively high levels (5–10 m*M*) of Mg^{2+}, but this has not been observed when GTP is used (J. CODINA, T. SUNYER, and L. BIRNBAUMER, unpublished). The failure to obtain subunit dissociation with GTP has been ascribed to the GTPase activity of the G-protein, followed by extremely slow dissociation of the GDP so that accumulation of the dissociated GTP-liganded state could not be seen. However, it could equally well be that GTP is not able to induce by itself the same conformational change that nonhydrolyzable GTP analogues do, and that the role of the HR complex in this part of the cycle is to enable the activated α-GTP complex to undergo a full conformational change that dissociates into α^* and $\beta\gamma$. In support of this, conformational changes of G_0 due to GTP, measured as changes in intrinsic fluorescence, were found to be much less pronounced than those obtained with GTPγS, and a single cycle GTPase assay, while proceeding at a fast rate, engaged only a small fraction of the G-protein to which GTP was added (HIGASHIJIMA et al. 1987).

Although it is not germane to the overall arguments of signaling through α as compared to $\beta\gamma$, one plausible working hypothesis is that the transition to the G(GTP) state is associated with a HR_H to HR_L transition. On the other hand, the simple two-state model for the receptor implicit in this hypothesis is not appropriate either. Agonist-dependent phosphorylation of the receptor by receptor kinase, such as shown by LEFKOWITZ's group for phosphorylation of β-adrenergic receptor (βAR) by β-adrenergic receptor kinase (βARK), is an example of a G-protein-independent conformational change that is different from the R_L to R_H change induced by the G-protein. What is germane to the model, however, is that α signaling is driven by the HR_H complex and that $\beta\gamma$ signaling is driven by the HR_L complex. JAKOBS and GIERSCHIK and their collaborators have presented kinetic evidence to the effect that the low-affinity form of the fMLP receptor is active in signaling (GIERSCHIK et al. 1989) and, more recently, that fMLP receptors in neutrophils and HL-60 cells activate PLCβ primarily, if not exclusively, through the $\beta\gamma$ pathway (CAMPS et al. 1992).

The scheme of Fig. 1 also presents the plausible stage of the receptor-mediated G-protein activation cycle at which homologous desensitization by receptor kinase can occur. Like $\beta\gamma$ signaling, this phenomenon requires GTP, agonist, and high receptor occupation, indicating that the target of βARK is one of the HR complexes. Furthermore, targeting of βARK to the receptor was shown to be $\beta\gamma$ dependent (PITCHER et al. 1992). This makes the HR·$\beta\gamma$ complex that results as an obligatory by-product of receptor-mediated activation of the G-protein the most likely substrate for βARK. Since in all likelihood dissociation rates of H from HR·$\beta\gamma$ are faster than dissociation rates of $\beta\gamma$ from HR·$\beta\gamma$, and since in general ligand affinities are correlated with decreasing rates of dissociation, the model predicts that at equal overall receptor occupancy, desensitization should be more pronounced with high affinity ligands than with low affinity ligands. It is tempting to suggest further that the role of molecules such as arrestin and β-arrestin is to stabilize $\beta\gamma$ on R, thus impeding the participation of the receptor in the activation of new G-protein molecules, and hence clinching the uncoupling reaction initiated by the phosphorylation of the receptor.

Like all models, the one discussed here presents a working hypothesis that should lead to new experiments. The results will then impose changes on the model. This will increase our understanding, pose new questions, and lead to new hypotheses, which will, in turn, be tested. At present, however, it seems clear that the data on α and $\beta\gamma$ signaling, together with thermodynamic constraints arising from microscopic reversibility rules, have de facto ruled out the possibility that receptors act as mere catalyzers of nucleotide exchange. Nucleotide exchange is but the initial step of a series of reactions in which the receptor participates by shepherding the G-protein through its subunit dissociation/reassociation cycle.

References

Birnbaumer L, Hildebrandt JD, Codina J, Mattera R, Cerione RA, Hildebrandt JD, Sunyer T, Rojas FJ, Caron MG, Lefkowitz RJ, Iyengar R (1985) Structural basis of adenylyl cyclase stimulation and inhibition by distinct guanine nucleotide regulatory proteins. In: Cohen P, Houslay MD (eds) Molecular mechanisms of signal transduction. Elsevier/North Holland, Amsterdam, pp 131–182

Birnbaumer L, Abramowitz J, Brown AM (1990) Signal transduction by G proteins. Biochim Biophys Acta 1031:163–224

Camps M, Hou C, Sidiropoulos D, Stock JB, Jakobs KH, Gierschik P (1992) Stimulation of phospholipase C by guanine-nucleotide-binding protein $\beta\gamma$ subunits. Eur J Biochem 206:821–831

Cassel D, Selinger Z (1978) Mechanism of adenylate cyclase activation through the beta=adrenergic receptor: catecholamine-induced displacement of bound GDP by GTP. Proc Natl Acad Sci USA 75:4155–4159

Chabre O, Conklin BR, Lin HY, Lodish HF, Wilson E, Ives HE, Catanzariti L, Hemmings BA, Bourne HR (1992) A recombinant calcitonin receptor independently stimulates 3′,5′-cyclic adenosine monophosphate and Ca^{2+}/inositol phosphate signaling pathways. Mol Endocrinol 6:551–556

Cerione RA, Codina J, Benovic JL, Lefkowitz RJ, Birnbaumer L, Caron MG (1984) The mammalian $beta_2$-adrenergic receptor: reconstitution of the pure receptor with the pure stimulatory nucleotide binding protein (N_s) of the adenylate cyclase system. Biochemistry 23:4519–4525

Cerione RA, Staniszewski C, Benovic JL, Lefkowitz RJ, Caron MC, Gierschick P, Somers R, Spiegel AL, Codina J, Birnbaumer L (1985) Specificity of the functional interactions of the beta-adrenergic receptor and rhodopsin with guanine nucleotide regulatory proteins reconstituted in phospholipid vesicles. J Biol Chem 260:1493–1500

Cerione RA, Regan JW, Nakata H, Codina J, Benovic JL, Gierschick P, Somers RL, Spiegel AM, Birnbaumer L, Lefkowitz RJ, Caron MG (1986) Functional reconstitution of the alpha$_2$-adrenergic receptor with guanine nucleotide regulatory proteins in phospholipid vesicles. J Biol Chem 261:3901–3909

DeLean A, Stadel JM, Lefkowitz RJ (1980) A ternary complex model explains the agonist-specific binding properties of the adenylate cyclase-coupled beta-adrenergic receptor. J Biol Chem 255:7108–7117

Florio VA, Sternweis PC (1989) Mechanism of muscarinic receptor action on G_0 in reconstituted phospholipid vesicles. J Biol Chem 264:3909–3915

Gierschik P, Stisslinger M, Sidiropolous D, Herrmann E, Jakobs KH (1989) Dual Mg^{2+} control of formyl peptide receptor – G protein interaction in HL-60 cells. Evidence that the low agonist affinity receptor interacts with and activates the G protein. Eur J Biochem 183:97–105

Gudermann T, Birnbaumer M, Birnbaumer L (1992) Evidence for dual coupling of the murine LH receptor to adenylyl cyclase and phosphoinositide breakdown/Ca^{2+} mobilization. Studies with the cloned murine LH receptor expressed in L cells. J Biol Chem 267:4479–4488

Higashijima T, Ferguson KM, Smigel MD, Gilman AG (1987) The effect of GTP and Mg^{2+} on the GTPase activity and the fluorescent properties of G_0. J Biol Chem 262:757–761

Howlett AC, Gilman AG (1980) Hydrodynamic properties of the regulatory component of adenylate cyclase. J Biol Chem 260:2861–2866

Iyengar R, Abramowitz J, Bordelon-Riser ME, Birnbaumer L (1980a) Hormone receptor-mediated stimulation of adenylyl cyclase systems: nucleotide effects and analysis in terms of a two-state model for the basic receptor-affected enzyme. J Biol Chem 255:3558–3564

Iyengar R, Abramowitz J, Bordelon-Riser ME, Blume AJ, Birnbaumer L (1980b) Regulation of hormone-receptor coupling to adenylyl cyclases: effects of GTP and GDP. J Biol Chem 255:10312–10321

Kent RS, DeLean A, Lefkowitz RJ (1980) A quantitative analysis of beta-adrenergic receptor interactions: resolution of high and low affinity states of the receptor by computer modeling of ligand binding data. Mol Pharmacol 17:14–23

Mattera R, Pitts BJR, Entman MS, Birnbaumer L (1985) Guanine nucleotide regulation of a mammalian myocardial receptor system. Evidence for homo- and heterotropic cooperativity in ligand binding analyzed by computer assisted curve fitting. J Biol Chem 260:7410–7421

Michel T, Lefkowitz RJ (1982) Hormonal inhibition of adenylate cyclase. Alpha$_2$-adrenergic receptors promote release of [^{3}H]Guanylylimidodiphosphate from platelet membrane. J Biol Chem 257:13557–13563

Northup JK, Sternweis PC, Gilman AG (1983a) The subunits of the stimulatory regulatory component of adenylate cyclase. Resolution, activity and properties of the 35,000 dalton (beta) subunit. J Biol Chem 258:11361–11368

Northup JK, Smigel MD, Sternweis PC, Gilman AG (1983b) The subunits of the stimulatory regulatory component of adenylate cyclase. Resolution of the activated 45,000-dalton (alpha) subunit. J Biol Chem 258:11369–11376

Phillips WJ, Cerione RA (1992) Rhodopsin/transducin interactions: I. Characterization of the binding of the transducin-$\beta\gamma$ subunit complex to rhodopsin using fluorescence spectroscopy. J Biol Chem 267:17032–17039

Pitcher JA, Inglese J, Higgins JB, Arriza JL, Casey PJ, Kim C, Benovic JL, Kwatra MM, Caron MG, Lefkowitz RJ (1992) Role of $\beta\gamma$ subunits of G proteins in targeting of the β-adrenergic receptor kinase to membrane-bound receptors. Science 257:1264–1267

Rodbell M, Krans HMJ, Pohl SL, Birnbaumer L (1971) The glucagon-sensitive adenyl cyclase system in plasma membranes of rat liver: IV. Binding of glucagon: effect of guanyl nucleotides. J Biol Chem 246:1872–1876

Rojas FJ, Birnbaumer L (1985) Regulation of glucagon receptor binding. Lack of effect of Mg and preferential role for GDP. J Biol Chem 260:7829–7835

Taguchi M, Field JB (1988) Effects of thyroid-stimulating hormone, carbachol, norepinephrine, and adenosine 3′,5′-monophosphate on polyphosphatidylinositol phosphate hydrolysis in dog thyroid slices. Endocrinology 123:2019–2026

Tolkovsky AM, Levitzki A (1978) Mode of coupling between the ß-adrenergic receptor and adenylate cyclase in turkey erythrocytes. Biochemistry 17:3795–3810

VanSande J, Raspe E, Perret J, Lejeune C, Manhaut C, Vassart G, Dumont JE (1990) Thyrotropin activates both the cAMP and the PIP2 cascade in CHO cells expressing the human cDNA of the TSH receptor. Mol Cell Endocrinol 74:R1–R6

CHAPTER 4

Small GTPases and Vesicle Trafficking: Sec4p and its Interaction with Up- and Downstream Elements

P. NOVICK and P. BRENNWALD

A. Introduction

Eukaryotic cells are characterized by the presence of many intracellular membrane bounded organelles. Transfer of material from one organelle to another is generally mediated by vesicular transport. By this mechanism, a vesicle buds from a donor organelle and migrates through the cytoplasm until it recognizes the appropriate target organelle (PALADE 1975; for a recent review see ROTHMAN and ORCI 1992). The fusion of the vesicle membrane with that of the target organelle completes the transfer of both membrane constituents and lumenal contents. The wide variety of potential donors and potential targets necessitates an efficient mechanism for maintaining the specificity of the transport process. This can be separated into two distinct steps. First, the appropriate proteins must be incorporated into a transport vesicle. This process is known as sorting and involves a collection of membrane-bound receptors and cytoplasmically disposed coat proteins that serve to define the vesicle constituents. The second, and equally important step, is the process by which each vesicle is brought into contact with the appropriate target organelle. While cytoskeletal elements are involved in the transport of vesicles to the appropriate region of the cell, a membrane–membrane recognition mechanism also must be involved since different target organelles can exist in very close proximity. Over the last 5 years, much attention has been focused on a particular class of ras-related GTP binding proteins as possible elements of the molecular machinery that generates the high degree of specificity required for this vesicle-target recognition process.

The first member of this class of proteins to be implicated in secretion was the product of the *SEC4* gene. *SEC4* was identified in a screen for temperature-sensitive yeast mutants that fail to secrete cell wall proteins following a shift to the restrictive temperature (NOVICK et al. 1980). The temperature-sensitive *sec4-8* allele results in a block in protein secretion at the post-Golgi stage of the secretory pathway within 5 min of a shift to 37°C (SALMINEN and NOVICK 1987). As a result of the block in secretion, secretory vesicles accumulate in the mutant cells. Sequence analysis of *SEC4* indicated that the encoded protein shares 32% identity with *ras* (SALMINEN and NOVICK 1987). More extensive sequence identity (45%) was seen with the yeast

Ypt1 protein. While there had been considerable speculation regarding the function of Ypt1, in time it was shown that Ypt1, like Sec4, plays a role in protein secretion, though at an earlier stage of the pathway (BACON et al. 1989; BAKER et al. 1990; SEGEV 1991; SEGEV et al. 1988).

Subsequent to the identification of *SEC4* and *YPT1*, a large number of mammalian homologs, principally termed *rab* genes, were isolated by low stringency hybridization and polymerase chain reaction (PCR) using sequences derived from conserved domains (CHAVRIER et al. 1990a,b, 1992; ELFERINK et al. 1992; HAUBRUCK et al. 1987; MATSUI et al. 1988; TOUCHOT et al. 1987; ZAHRAOUI et al. 1989). These genes encode proteins that are significantly more similar to Sec4 and Ypt1 than to *ras* itself. Thus, Sec4, Ypt1, and the rab proteins form a distinct subfamily of the *ras* superfamily. Several lines of evidence suggest that many, if not all rab proteins play roles analogous to those of Sec4 and Ypt1, each regulating a distinct vesicular transport event.

Subcellular localization of these proteins has demonstrated that each is associated with a particular stage of the exocytic or endocytic pathways. Sec4 was shown to be associated with the cytoplasmic surface of secretory vesicles and the plasma membrane (GOUD et al. 1988). Ypt1 and the closely related mammalian homolog, rab1, are associated with the endoplasmic reticulum (ER), the Golgi, and the intermediate vesicles that carry material from the ER to the Golgi (PLUTNER et al. 1991; SEGEV 1991; SEGEV et al. 1988). rab2 is also found on the intermediate compartment between the ER and Golgi (CHAVRIER et al. 1990a). rab6 is located on the medial and *trans*-Golgi (GOUD et al. 1990). rab4 and 5 are on early endosomes, while rab7 is on late endosomes (CHAVRIER et al. 1990a; VAN DER SLUIJS et al. 1991). The neuronally expressed protein, rab3A, is associated with the surface of synaptic vesicles (FISCHER VON MOLLARD et al. 1990; MIZOGUCHI et al. 1990), and related proteins are associated with chromaffin granules (DARCHEN et al. 1990). In total, the localization data is consistent with the hypothesis that each member of the family controls a unique vesicular transport event.

While direct evidence shows that Sec4 and Ypt1 control vesicle traffic, such evidence regarding the function of the rab proteins has been more limited. Antibody directed against rab1 blocks transport of protein from the ER to the Golgi in vitro (PLUTNER et al. 1991). Rab5 has been shown by in vivo and in vitro experiments to be required for fusion of endosomes (GORVEL et al. 1991). A peptide designed to mimic a region of rab3A will trigger exocytosis of zymogen granules in permeabilized pancreatic acinar cells (PADFIELD et al. 1992), yet inhibit transport from the ER to the Golgi (PLUTNER et al. 1990).

All GTP-binding proteins undergo a nucleotide cycle in which GTP is first bound, then hydrolyzed, and the resulting GDP dissociates to allow a new round of GTP binding (see BOURNE et al. 1990 for review). The GDP-bound protein is different in conformation from the GTP-bound form. Since both the intrinsic rates of GTP hydrolysis and GDP dissociation are generally low for members of the ras superfamily, these proteins can exist stably in either conformation. They can, therefore, act as switches. It is thought that the GTP-bound forms of these proteins can interact with a "downstream" effector protein to initiate a specific biochemical process, while the GDP forms are inactive in this regard. The rates of GTP hydrolysis and the rates of nucleotide exchange are often controlled through interactions with accessory proteins. Many members of the ras superfamily interact with specific GTPase activating proteins (GAPs) to stimulate their very low intrinsic rates of GTP hydrolysis (ADARI et al. 1988; BECKER et al. 1991; BURSTEIN et al. 1991; GARRETT et al. 1989; GIBBS et al. 1988). Exchange proteins have been identified for a number of the members of the ras superfamily that stimulate the low intrinsic rates of nucleotide dissociation and, thus, hasten the binding of GTP (BURSTEIN and MACARA 1992; DOWNWARD et al. 1990; HUANG et al. 1990; KAIBUCHI et al. 1991). In the cases of the rab and rho branches of the ras superfamily, additional elements have been found that can both slow the dissociation of GDP and solubilize the protein when it is in its GDP-bound form (MATSUI et al. 1990; SASAKI et al. 1990; ARAKI et al. 1990; UEDA et al. 1990; ISOMURA et al. 1991). These are known as GDP dissociation inhibitors (GDIs). GDIs may play an important role in coupling the nucleotide state of these proteins to their association with the membrane.

Just as a switch in an electric circuit can be used in a variety of ways, this basic biochemical switching mechanism can be used to control diverse cellular processes. In the case of the yeast Ras proteins, the nucleotide-based switch is used to control the synthesis of cAMP (TODA et al. 1985). Regulated by the nutritional state of the cell, Cdc25 is thought to catalyze the release of pre-bound GDP from Ras allowing binding of GTP (ROBINSON et al. 1987). Ras in its GTP-bound state stimulates the CYR1 gene product, adenylyl cyclase (FIELD et al. 1988). The level of cAMP synthesis reflects the fraction of Ras that is in the GTP-bound state. Ras can, therefore, be considered a signal transducer. The *IRA1* and *IRA2* gene products encode GAPs which stimulate the hydrolysis of GTP by Ras and down-regulate the pathway (TANAKA et al. 1990). The mammalian ras protein also acts in a signal transduction mode, but adenylyl cyclase is not a downstream effector in this case. Inhibiting GTP hydrolysis either through a mutation in ras or through an inhibition of GAP function allows ras to be maintained in the GTP-bound state for prolonged periods (BARBACID 1987; TRAHEY and MCCORMICK 1987). This, in turn, is a signal for cell proliferation. The mode of action of the other members of the ras superfamily is not as clearly understood.

The current model regarding Sec4 function, and by analogy the function of the other members of the Sec4/Ypt1/rab subfamily, is that the switch is used not in a signal transduction mode, but in a cyclical mode. By this we mean that Sec4 must cycle between its GDP- and GTP-bound forms to fulfill its function. The level of function is determined not by the fraction of the protein in the GTP-bound state, but by the rate of productive GTP binding, hydrolysis, and exchange. The function of this cycle is to ensure that the appropriate vesicle is brought to the appropriate target organelle. An analogy can be made with the function of another GTP-binding protein, EF-Tu (KAZIRO 1978). In this case, the cycle of nucleotide binding and hydrolysis is used to ensure the correct delivery of an aminoacyl-tRNA to the ribosome. In the case of Sec4 and the many related rab proteins, the nucleotide cycle would serve to maintain the fidelity of vesicular transport. This model is consistent with the observations from many laboratories that different members of the rab family of proteins are associated with different stages of

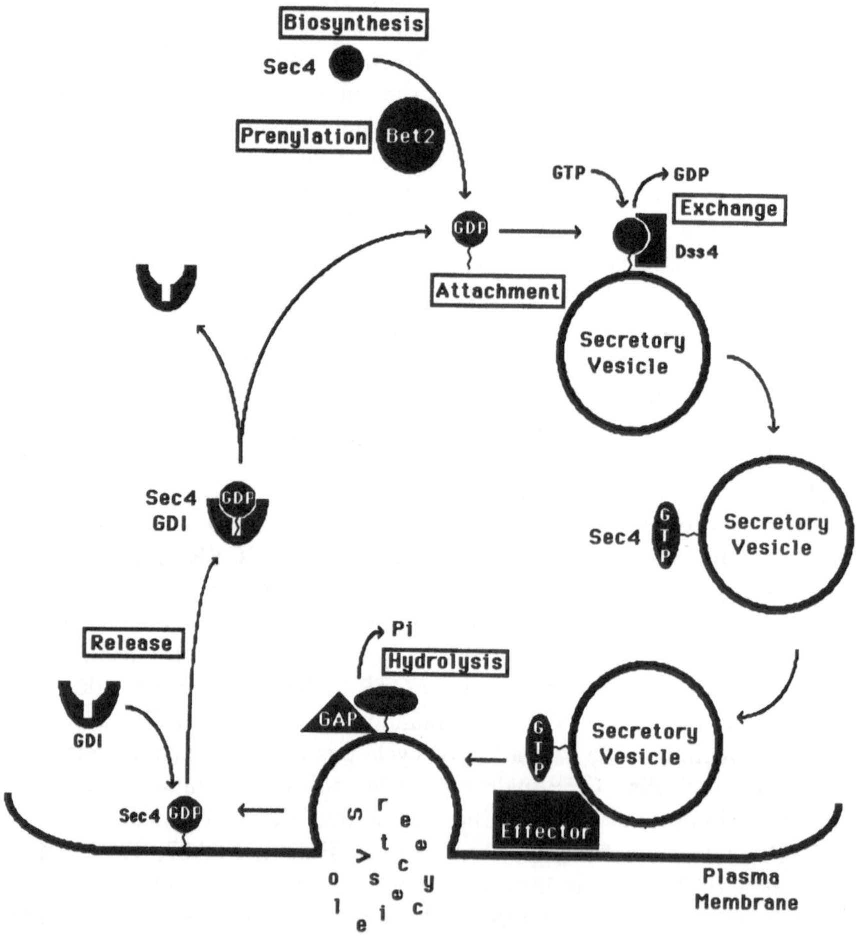

Fig. 1. The Sec4 cycle. Newly synthesized Sec4p is shown as the *small black circle* at the *top*. Sec4p is the prenylated (*squiggly line*) by a Bet2p-containing geranylgeranyl transferase. After this, Sec4p enters its functional cycle of nucleotide exchange (catalyzed by the guanine nucleotide dissociation stimulator (*GDS*) DSS4) and hydrolysis (presumably catalyzed by a GTPase activating protein, *GAP*), as well as a cycle of membrane attachment and release mediated by GDP dissociation inhibitor (*GDI*). The GDP-bound Sec4p is depicted *rounded*, and the GTP-bound Sec4p is depicted *ovoid* to reflect two presumed distinct conformational states

the exocytotic or endocytic pathways and serve to regulate distinct vesicular transport events.

With these ideas in mind, similar models were proposed by BOURNE (1988) and by us (NOVICK et al. 1988). Our current version is presented in Fig 1. A central feature of the model is the proposed coupling of the cycle of nucleotide binding and hydrolysis with the observed cycle of subcellular localization. We will consider the start of the cycle to be Sec4 in its GDP-bound state. Since the GDP-bound form can be solubilized by interaction with GDI, the cycle begins with Sec4 as a complex with GDI in the cytosol. Interaction with an exchange protein removes the GDP, allowing a GTP to bind. This causes dissociation of GDI, allowing membrane attachment. The site of membrane association is dictated by the location of the exchange protein. This we propose to be the site of secretory vesicle budding in the *trans* Golgi. Sec4 remains on the secretory vesicle in its GTP-bound state until the vesicle interacts with an effector complex on the target organelle, the plasma membrane. This interaction triggers a series of events leading to exocytosis. GAP may be a component of the effector complex which would allow coupling of GTP hydrolysis by Sec4 to fusion of the secretory vesicle. The actual point of hydrolysis of GTP by Sec4 in this model may occur either just prior to or soon after fusion with the plasma membrane. This leaves Sec4 in its GDP-bound state, which allows solubilization by GDI and initiation of a new cycle.

Another mutation constructed was a deletion of the carboxy-terminal cysteines. This led to the production of Sec4 protein which bound nucleotides normally, but which failed to attach to membranes and behaved as a recessive loss of function allele in yeast. Attachment was also blocked by mutations in the mevalonate biosynthetic pathway (P.J. NOVICK, unpublished observation) and by mutations in the BET2 gene, which has been shown to encode a novel geranylgeranyl transferase (ROSSI et al. 1991). Recently, Bet2-dependent addition of geranylgeranyl to Sec4 has been demonstrated in vitro in the laboratory of Dr. Ferro-Novick (Y. JIANG, G. ROSSI and S. FERRO-NOVICK, submitted). Together these data indicate that Sec4 is modified by the addition of geranylgeranyl moieties to the carboxy-terminal cysteines, thereby facilitating membrane attachment.

Deletion of the carboxy-terminal cysteines from the Sec4-I^{133} allele prevented its dominant lethal effects. This can be explained if membrane attachment of Sec4-I^{133} is necessary for its interfering interaction with an accessory protein. This predicts that the target accessory protein is membrane-associated.

B. The Sec4 Cycle

I. A Cycle of Sec4 Localization

Through cell fractionation studies Sec4 was localized to the cytoplasmic face of secretory vesicles and the plasma membrane. A small, soluble pool was consistently detected as well (GOUD et al. 1988). Blocking exocytosis by incubating a *sec6-4* mutant strain at the restrictive temperature led to a time-dependent shift of Sec4 from the plasma membrane to the enlarged vesicular pool. By performing the experiment as a pulse-chase study we were able to show that the pre-existing Sec4 protein associated with the plasma membrane at permissive temperature redistributes to the vesicular pool during a subsequent chase at the restrictive temperature. From this data we suggested that Sec4 may normally cycle through the cell. We further suggested that this recycling pathway utilizes a soluble intermediate.

II. Intrinsic Properties of Sec4

As a first step towards understanding Sec4 function, we determined its intrinsic nucleotide binding and hydrolysis properties (KABCENELL et al. 1990). Sec4 was purified from the soluble pool of an overproducing yeast strain. The purified protein was monomeric and bound both GTP and GDP with high affinity. GTP was hydrolyzed to GDP by Sec4, but at the rate of one mole of GTP hydrolyzed by one mole of Sec4 every 14h. GDP dissociated from Sec4 with a half-time of 4 min, quite fast for a member of the fras superfamily. In contrast, GTP appeared to dissociate with the same kinetics as GTP hydrolysis, suggesting that the nucleotide was first hydrolyzed, and then the resulting GDP rapidly dissociated.

III. GTP Binding and Membrane Attachment Are Essential for Sec4 Function

A number of mutations were made in Sec4 by in vitro mutagenesis, and these were introduced into yeast to test their function (WALWORTH et al. 1989). One mutation changed an asparagine to an isoleucine at position 133. The analogous mutation in *ras* is oncogenic and encodes a protein with a decreased affinity for guanine nucleotides and, therefore, an increased exchange rate. The Sec4-I^{133} protein, like the analogous *ras* mutant, fails to bind GTP to a significant extent. In vivo, this allele behaves as a dose-dependent, dominant lethal mutation. At levels of expression close to normal, it interferes with the function of the wild-type Sec4 protein. This suggests that it binds to one of the accessory proteins, but fails to function, thereby titrating out an essential element of the secretory machinery. A good candidate for this target accessory protein is a component which catalyzes nucleotide exchange. Since Sec4-I^{133} has a low affinity for nucleotide, it may tie up an exchange protein and thereby interfere with the function of normal Sec4.

IV. GTP Hydrolysis Is Important for Sec4 Function

One feature of our model is the implication that hydrolysis of GTP is an obligatory step in Sec4 function. Without hydrolysis Sec4 could not recycle.

This is in striking contrast to what is known regarding the function of the related protein, ras. In the case of ras, a block in the hydrolysis step leads to the build up of GTP-bound protein. This leads to constitutive activation of the downstream signal transduction pathway and to oncogenesis. Thus, GTP hydrolysis is not necessary for ras function, but rather for appropriate down-regulation. As proposed for Sec4, a block in GTP hydrolysis would prevent recycling and, thus, lead to a loss of Sec4 function. We have tested this prediction by constructing an allele, $Q{\rightarrow}L^{79}$, of Sec4 that binds GTP normally, but is blocked in hydrolysis (WALWORTH et al. 1992). This glutamine is thought, by reference to the ras crystal structure, to hydrogen bond a water molecule which is positioned to attack the phosphoryl bond of GTP (KRENGEL et al. 1990; MILBURN et al. 1990; PAI et al. 1989). Leucine cannot participate in such an interaction, thereby blocking hydrolysis.

As predicted, the purified Sec4-L^{79} protein has an undetectably low intrinsic GTP hydrolysis rate. Furthermore, the rate of hydrolysis is stimulated by a GAP activity (see below), yet only to 30% of the GAP-stimulated rate of the wild-type protein. As a result, the Sec4-L^{79} protein accumulates in yeast in its GTP-bound form, while the wild-type protein is predominantly in the GDP-bound form. Sec4-L^{79} can function as the sole source of Sec4 in vivo. The cells, however, exhibit recessive cold-sensitivity for growth, and the rate of protein secretion is slowed, leading to the accumulation of secretory vesicles at 14°C. By genetic criteria, as well, the L^{79} mutation represents a loss of function allele. Thus, contrary to the function of *ras*, slowing GTP hydrolysis leads to a loss of Sec4 function rather than a gain of function (WALWORTH et al. 1992). Our finding that hydrolysis of GTP is a necessary step in the cycle of Sec4 function supports the proposed model.

As attractive as the model may be, many more questions were raised than were answered. Is the basic premise of an obligatory nucleotide cycle in fact the case? If so, how is membrane attachment coupled to the nucleotide form of the protein? How is Sec4 reversibly attached to membranes? What controls nucleotide exchange? What controls nucleotide hydrolysis? At what point does GTP hydrolysis occur? What is the effector stimulated by GTP-bound Sec4? Many of these questions have now been addressed and are, at least, partially answered. Since the rab proteins are highly conserved in structure, it is likely that the information regarding Sec4 function in yeast exocytosis will be directly relevant to the function of the rab proteins in the many vesicular transport events in mammalian cells.

C. Accessory Proteins in the Sec4 Cycle

I. A Specific Sec4 GAP Is Present in Yeast and in Mammalian Cells

Having established that the nucleotide cycle of Sec4 is critical for its function, we have turned to the identification of components which function together with Sec4 through its cycle. Since hydrolysis of GTP by Sec4 is essential for function and the intrinsic GTP turnover time (14 h) is seven times the yeast-doubling time (2 h), we have sought evidence for a GTPase activating protein. We have found that addition of a yeast extract will stimulate the hydrolysis of GTP prebound to Sec4. The GAP activity is predominantly membrane bound, although some activity is found in the soluble fraction as well. This GAP is specific for Sec4 as demonstrated by competition studies: excess Sec4 bound to GTPγS will inhibit the activity while excess Ypt1 bound to GTPγS will not (WALWORTH et al. 1992). Homogenates of rat pancreas also possess an activity capable of stimulating the GTPase activity of Sec4. Like yeast Sec4-GAP, the mammalian activity is predominantly membrane-bound, and competition experiments with Ypt1 and Sec4 suggest that it is also specific for Sec4 (JENA et al. 1992).

II. GDI from Bovine Brain and Yeast Solubilizes Sec4 in a Nucleotide-Specific Fashion

The intrinsic off-rate of GDP from Sec4 was found to be quite high for a member of the *ras* superfamily (KABCENELL et al. 1990). Based on this observation we speculated that nucleotide exchange might be negatively regulated. TAKAI and colleagues have described a protein isolated from bovine brain, termed rab3A GDP dissociation inhibitor (rab3A GDI) because of its ability to slow the rate of dissociation of GDP from rab3A (SASAKI et al. 1990). In fact, GDI can inhibit the dissociation of GDP from Sec4 as well (SASAKI et al. 1991). TAKAI and colleagues also found that this GDI was able to solubilize rab3A from membranes when rab3A was in its GDP-bound form, but not in its GTP-bound form (ARAKI et al. 1990). We have repeated these observations using yeast membranes and find that bovine GDI will also solubilize both Sec4 and Ypt1 in a nucleotide-specific fashion. We have gone on to show that yeast cytosol contains activities which release Sec4 or Ypt1 from membranes in a nucleotide-dependent fashion (GARRETT, KABCENELL, ZAHNER, TAKAI, CHENEY, and NOVICK, submitted). This activity may explain in part the mechanism underlying the proposed coupling of the nucleotide cycle with the cycle of Sec4 localization. Through the action of GDI, Sec4 would be released from membranes only after GTP hydrolysis. This release would allow recycling of Sec4 through the cytoplasm onto a new round of secretory vesicles.

III. Suppressors from Yeast and Rat Brain Encode Nucleotide Exchange Proteins

Using a genetic approach we recently isolated a dominant suppressor of the *sec4-8* mutant, termed *DSS4* for dominant suppressor of *sec4-8* (MOYA et al. 1993). The suppressing allele was cloned and sequenced and shown to encode a 17 kDa hydrophilic protein which did not show homology to any sequences in a search of the database. A possible function for this gene product became apparent when lysates derived from yeast or bacteria overexpressing *DSS4* were shown to possess an activity which stimulates the intrinsic rate of dissociation of prebound GDP from Sec4. Deletion of *DSS4* does not affect cell growth in a wild-type background. However, deletion of *DSS4* in combination with many of the late acting *sec* mutants dramatically reduces the growth rate. A mammalian homolog, *dss4*, has been identified in a study done in collaboration with Dr. Pietro De Camilli's laboratory (BURTON et al. 1993). A rat-brain cDNA library constructed in a yeast expression vector was screened for suppression of the *sec4-8* mutation. One of the suppressing cDNAs contained a predicted amino acid sequence that is 27% identical to that of Dss4 and has been termed Mss4 for mammalian suppressor of sec4-8. Similar to Dss4, an Mss4 fusion protein expressed in bacteria was found to stimulate the dissociation of prebound GDP from Sec4. These proteins can therefore be categorized as guanine nucleotide dissociation stimulators (GDSs). Mss4, like Dss4, did not share any detectable similarity to any known GDSs or guanine nucleotide exchange factors (GEFs); therefore, Dss4 and Mss4 may comprise a new class of GDFs.

D. A Potential Downstream Effector of Sec4 Function: The Sec8/Sec15 Complex

We have, so far, discussed the cyclical nature of Sec4 and described several factors which may regulate this cycle at different points. What is missing from this picture are components which respond to this cycle to allow fusion of the vesicle with the target membrane. Several of the late acting *sec* genes show strong genetic interactions with SEC4 and therefore are possible effectors of Sec4. Double mutants combining *sec4-8* with *sec2-41*, *sec5-24*, *sec8-9*, *sec10-2*, *sec15-1* are lethal under conditions which are permissive to any of the single mutants. Moreover, the same set of mutants are suppressed by duplication of *SEC4* (SALMINEN and NOVICK 1987). We would predict that Sec4 effectors should, in addition to showing genetic interactions, be localized to the downstream compartment, in this case the plasma membrane. Recently we have shown that two of the late-acting sec gene products, the Sec8 and Sec15 proteins, are present together as part of a 20S complex (BOWSER and NOVICK 1991). This complex is found both in the cytosol and peripherally associated with the plasma membrane, but is not associated with secretory vesicles (BOWSER et al. 1992; BOWSER and NOVICK 1991). Previously we found that overexpression of Sec15p caused the accumulation of a patch of both Sec4p and Sec15p by immunofluorescence and the appearance of a cluster of vesicles by electron microscopy, demonstrating further the close interrelationship of these gene products in promoting

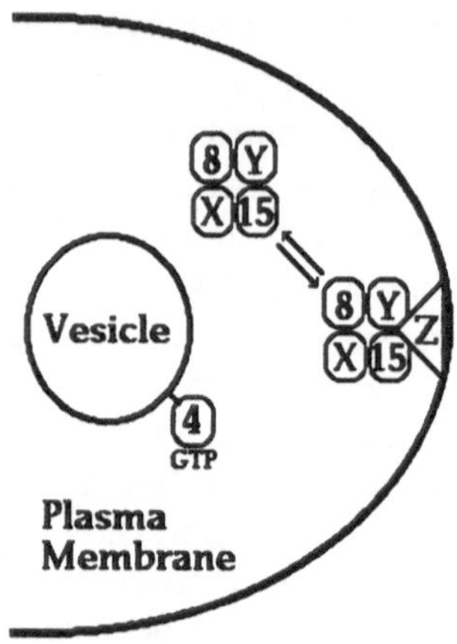

Fig. 2. The Sec8/Sec15 complex may be an effector of vesicle-bound Sec4. The Sec8p and Sec15 proteins are present together in a 20S complex along with other presently unidentified proteins (designated *X* and *Y*). This complex is present in two pools, one cytosolic and the other on the plasma membrane. The interaction with the plasma membrane may be mediated by an, as yet hypothetical, integral membrane protein(s) present on the plasma membrane (designated *Z*). Genetic and biochemical evidence suggests this complex may be functioning as an effector of vesicle-bound Sec4 to help bring about vesicle fusion with the plasma membrane

vesicle transport and fusion (SALMINEN and NOVICK 1989). In addition, the *SEC8* coding sequence shows a low, but potentially significant sequence similarity with a region of the *CYR1* gene which is involved in interactions with the *RAS* gene product (BOWSER et al. 1992). Since the adenylate cyclase encoded by *CYR1* is known to be an effector of activated Ras in yeast (FIELD et al. 1988), this region of Sec8p may be mediating an analogous interaction with Sec4p. The combination of the large number of genetic interactions between Sec4 and two members of this complex and its presence on the plasma membrane make the Sec8/Sec15 complex an excellent candidate for the Sec4 "effector complex" (see Fig. 2). We are currently exploring this idea in more detail with a combination of biochemical and genetic approaches in order to identify other components of this complex and to establish its biological role in exocytosis.

References

Adari H, Lowy DR, Willumsen BM, Der CJ, McCormick F (1988) Guanosine triphosphatase activating protein (GAP) interacts with the p21 *ras* effector binding domain. Science 240:518–521

Araki S, Kikuchi A, Hata Y, Isomura M, Takai Y (1990) Regulation of reversible binding of smg p25A, a ras p21-like GTP-binding protein, to synaptic plasma membranes and vesicles by its specific regulatory protein, GDP dissociation inhibitor. J Biol Chem 265:13007–13015

Bacon RA, Salminen A, Ruohola HNP, Ferro-Novick S (1989) The GTP-binding protein Ypt1 is required for transport in vitro: The golgi apparatus is defective in ypt1 mutants. J Cell Biol 109:1015–1022
Baker D, Wuestehube L, Schekman R, Botstein D, Segev N (1990) GTP-binding Ypt1 protein and Ca^{2+} function independently in a cell-free protein transport reaction. Proc Natl Acad Sci USA 87:355–359
Barbacid M (1987) ras Genes. Annu Rev Biochem 56:779–827
Becker J, Tan TJ, Trepte HH, Galwitz D (1991) Mutational analysis of the putative effector domain of the GTP-binding Ypt1 protein suggests specific regulation by a novel GAP activity. EMBO J 10:785–792
Bourne HR (1988) Do GTPases direct membrane traffic in secretion? Cell 53:669–671
Bourne HR, Sanders DA, McCormick F (1990) The GTPase superfamily: a conserved switch for diverse cell functions. Nature 348:125–132
Bowser R, Novick P (1991) Sec15 protein, an essential component of the exocytic apparatus, is associated with the plasma membrane and with a soluble 19.5S particle. J Cell Biol 112:1117–1131
Bowser R, Muller H, Govindan B, Novick P (1992) Sec8p and Sec15p are components of a plasma membrane associated 19.5S particle that may function downstream of Sec4p to control exocytosis. J Cell Biol 118:1041–1056
Burstein ES, Macara IG (1992) Characterization of a guanine nucleotide releasing factor and a GTPase-activating protein that are specific for the ras-related protein p25 rab3A. Proc Natl Acad Sci USA 89:1154–1158
Burstein ES, Linko-Stenz K, Macara IG (1991) Regulation of the GTPase activity of the ras-like protein p25rab3A: evidence for a rab3A specific GAP. J Biol Chem 266:2689–2692
Burton J, Roberts D, Montaldi M, Novick P, De Camilli P (1993) A mammalian guanine-nucleotide-releasing protein enhances function of yeast secretory protein Sec4. Nature 361:464–467
Chavrier P, Parton RG, Hauri HP, Simons K, Zerial M (1990a) Localization of low molecular weight GTP-binding proteins to exocytic and endocytic compartments. Cell 62:317–329
Chavrier P, Vingron M, Sander C, Simons K, Zerial M (1990b) Molecular cloning of YPT1/SEC4-related cDNAs from an epithelial cell line. Mol Cell Biol 10: 6578–6585
Chavrier P, Simons K, Zerial M (1992) The complexity of the Rab and Rho GTP-binding protein subfamilies revealed by a PCR cloning approach. Gene 112: 261–264
Darchen F, Zahraoui A, Hammel F, Monteils M, Tavitian A, Scherman D (1990) Association of the GTP-binding protein rab3A with bovine adrenal chromaffin granules. Proc Natl Acad Sci USA 87:5692–5696
Downward J, Riel R, Wu L, Weinberg R (1990) Identification of a nucleotide exchange promoting activity for p21 ras. Proc Natl Acad Sci USA 87:5998–6002
Elferink LA, Anzai K, Scheller RH (1992) rab15, a novel low molecular weight GTP-binding protein specifically expressed in rat brain. J Biol Chem 267: 5768–5775
Field J, Nikawa J, Broek D, MacDonald B, Rodgers L, Wilson I, Lerner R, Wigler M (1988) Purification of a RAS-responsive adenyl cyclase complex from Saccharomyces cerevisiae by use of an epitope addition method. Mol Cell Biol 8:2159–2165
Fischer von Mollard G, Mignery GA, Baumert M, Perin MS, Hanson TJ, Burger PM, Jahn R, Sudhof TC (1990) rab3 is a small GTP-binding protein exclusively localized to synaptic vesicles. Proc Natl Acad Sci USA 87:1988–1992
Garrett MD, Self AJ, van Oers C, Hall A (1989) Identification of distinct cytoplasmic targets for ras/Rras and rho regulatory proteins. J Biol Chem 264:10–13
Gibbs JB, Schaber MD, Allard WJ, Sigal IS, Scolnick EM (1988) Purification of ras GTPase activating protein from bovine brain. Proc Natl Acad Sci USA 85:5026–5030

Gorvel JP, Chavrier P, Zerial M, Gruenberg J (1991) Rab5 controls early endosome fusion in vitro. Cell 64:915–925
Goud B, Salminen A, Walworth N, Novick PJ (1988) A GTP-binding protein required for secretion rapid associates with secretory vesicles and the plasma membrane in yeast. Cell 53:753–768
Goud B, Zahraoui A, Tavitian A, Saraste J (1990) A small GTP-binding protein associated with the Golgi. Nature 345:553–556
Haubruck H, Disela C, Wagner P, Gallwitz D (1987) The ras-related ypt protein is an ubiquitous eukaryotic protein: isolation and sequence analysis of mouse cDNA clones highly homologous to the yeast YPT1 gene. EMBO J 6:4049–4053
Huang Y, Kung H, Kamata T (1990) Purification of a factor capable of stimulating the guanine nucleotide exchange reaction of ras proteins and its effect on ras small molecular mass G proteins. Proc Natl Acad Sci USA 87:8008–8012
Isomura M, Kikuchi A, Ohga N, Takai Y (1991) Regulation of binding of rhoB p20 to membranes by its specific regulatory protein, GDP dissociation inhibitor. Oncogene 6:119–124
Jena BP, Brennwald P, Garrett M, Novick P, Jamieson JD (1992) Distinct GAP's in mammalian pancreas act on the yeast GTP-binding proteins Ypt1 and Sec4. FEBS Lett 309:5–9
Kabcenell AK, Goud B, Northup JK, Novick PJ (1990) Binding and hydrolysis of guanine nucleotides by Sec4p, a yeast protein involved in the regulation of vesicular traffic. J Biol Chem 265:9366–9372
Kaibuchi K, Mizuno T, Fujioka H, Yamamoto T, Kishi K, Fukumoto Y, Hori Y, Takai Y (1991) Molecular cloning of the cDNA for stimulatory GDP/GTP exchange protein for *smg* p21s (*ras* p21-like small GTP-binding proteins) and characterization of stimulatory GDP/GTP exchange protein. Mol Cell Biol 11:2873–2880
Kaziro Y (1978) The role of guanosine 5'-triphosphate in polypeptide chain elongation. Biochem Biophys Acta 505:95–127
Krengel U, Schlichting I, Scherer A, Schuhman R, Frech M, John J, Kabsch W, Pai E, Wittinghofer A (1990) Three-dimensional structure of H-ras mutants: molecular basis for their inability to function as signal switch molecules. Cell 62:609–617
Matsui Y, Kikuchi A, Kondo J, Hishida T, Teranishi Y, Takai Y (1988) Nucleotide and deduced amino acid sequence of GTP-binding protein family with molecular weight of 25000 from bovine brain. J Biol Chem 263:11071–11074
Matsui Y, Kikuchi A, Araki S, Hata Y, Kondo J, Teranishi Y, Takai Y (1990) Molecular cloning and characterization of a novel type of regulatory protein (GDI) for smg p25A, a ras-like GTP-binding protein. Mol Cell Biol 10:4116–4122
Milburn MV, Tong L, deVos AM, Brunger A, Yamaizumi Z, Nishimura S, Kim SH (1990) Molecular switch for signal transduction: structural differences between active and inactive forms of protooncogenic *ras* proteins. Science 247:939–945
Mizoguchi A, Kim S, Ueda T, Kikuchi A, Yorifuji H, Hirokawa N, Takai Y (1990) Localization and subcellular distribution of smg p25A, a ras p21-like GTP-binding protein, in rat brain. J Biol Chem 265:11872–11879
Moya M, Roberts D, Novick P (1993) *DSS4-1*, is a dominant suppressor or *sec4-8* that encodes a nucleotide exchange protein that aids Sec4p function. Nature 361:460–463
Novick P, Field C, Schekman R (1980) Identification of 23 complementation groups required for post-translational events in the yeast secretory pathway. Cell 21: 205–215
Novick PJ, Goud B, Salminen A, Walworth NC, Nair J, Potenza M (1988) Regulation of vesicular traffic by a GTP-binding protein on the cytoplasmic surface of secretory vesicles in yeast. Cold Spring Harbor Symp Quant. Biol 53:637–647
Padfield PJ, Balch WE, Jamieson JD (1992) A synthetic peptide of the rab3a effector domain stimulates amylase release from permeabilized pancreatic acini. Proc Natl Acad Sci USA 89:1656–1660

Pai EF, Kabash W, Krengel U, Holmes KC, John J, Wittinghofer A (1989) Structure of the guanine-nucleotide-binding domain of the Ha-ras oncogene product p21 in the triphosphate conformation. Nature 341:209–214

Palade G (1975) Intracellular aspects of the process of protein secretion. Science 189:347–358

Plutner H, Schwaninger R, Pind S, Balch WE (1990) Synthetic peptides of the Rab effector domain inhibit vesicular transport through the secretory pathway. EMBO J 9:2375–2383

Plutner H, Cox A, Pind S, Khosravi-Far R, Bourne J, Swaninger R, Der C, Balch W (1991) Rab1 regulates vesicular transport between the endoplasmic reticulum and successive Golgi compartments. J Cell Biol 115:31–43

Robinson L, Gibbs J, Marshall M, Sigal I, Tatchell K (1987) CDC25: a component of the RAS-adenylate cyclase pathway in Saccharomyces cerevisiae. Science 235:1218–1221

Rossi G, Jiang Y, Newman A, Ferro-Novick S (1991) Dependence of Ypt1 and Sec4 membrane attachment on Bet2. Nature 351:158–161

Rothman J, Orci L (1992) Molecular dissection of the secretory pathway. Nature 355:409–415

Salminen A, Novick PJ (1987) A ras-like protein is required for a post-Golgi event in yeast secretion. Cell 49:527–538

Salminen A, Novick P (1989) The Sec15 protein responds to the function of the GTP binding protein, Sec4, to control vesicular traffic in yeast. J Cell Biol 109:1023–1036

Sasaki T, Kikuchi A, Araki S, Hata Y, Isomura M, Kuroda S, Takai Y (1990) Purification and characterization from bovine brain cytosol of a protein that inhibits dissociation of the GDP form and the subsequent binding of GTP to smg p25A, a ras p21-like GTP-binding protein. J Biol Chem 265:2333–2337

Sasaki T, Kaibuchi K, Kabcenell AK, Novick PJ, Takai Y (1991) A mammalian inhibitory GDP/GTP exchange protein (GDP Dissociation Inhibitor) for smg p25A is active on the yeast SEC4 protein. Mol Cell Biol 11:2909–2912

Segev N (1991) Mediation of the attachment or fusion step in vesicular transport by the GTP-binding Ypt1 protein. Science 252:1553–1556

Segev N, Mulholland J, Bostein D (1988) The yeast GTP-binding YPT1 protein and a mamalian counterpart are associated with the secretion machinery. Cell 52:915–924

Tanaka K, Nakafuku M, Satoh T, Marshall MS, Gibbs JB, Matsumoto K, Kaziro Y, Toh-e A (1990) S. cerevisiae genes IRA1 and IRA2 encode protiens that may be functionally equivalent to mammalian ras GTPase activation protein. Cell 60:803–807

Toda T, Uno I, Ishikawa T, Powers S, Kataoka T, Broek D, Cameron S, Broach J, Matsumoto K, Wigler M (1985) In yeast, RAS proteins are controlling elements of the adenylate cyclase. Cell 40:27–36

Touchot N, Chardin P, Tavitian A (1987) Four additional membranes of the ras gene superfamily isolated by an oligonucleotide strategy: molecular cloning of YPT-related cDNAs from a rat btain library. Proc Natl Acad Sci USA 84:8210–8214

Trahey M, McCormick F (1987) A cytoplasmic protein stimulates normal N-ras p21 GTPase, but does not affect oncogenic mutants. Science 238:542–545

Ueda T, Kikuchi A, Ohga N, Yamamoto J, Takai Y (1990) Purification and characterization from bovine brain cytosol of a novel regulatory protein inhibiting the dissociation of GDP from and the subsequent binding of GTP to rhoB p20, a ras p21-like GTP-binding protein. J Biol Chem 265:9373–9380

Van der Sluijs P, Hull M, Zahraoui A, Tavitian A, Goud B, Mellman I (1991) The small GTP-binding protein rab4 is associated with early endosomes. Proc Natl Acad Sci USA 88:6313–6317

Walworth NC, Goud B, Kabcenell AK, Novick PJ (1989) Mutational analysis of SEC4 suggests a cyclical mechanism for the regulation of vesicular traffic. Cell 8:1685–1693

Walworth NC, Brennwald P, Kabcenell AK, Garrett M, Novick P (1992) Hydrolysis of GTP by Sec4 protein plays an important role in vesicular transport and is stimulated by a GTPase activating protein in yeast. Mol Cell Biol 12:2017–2028
Zahraoui A, Touchot N, Chardin P, Tavitian A (1989) The human Rab genes encode a family of GTP-binding proteins related to yeast YPT1 and SEC4 products involved in secretion. J Biol Chem 264:12394–12401

CHAPTER 5

Cytoskeletal Assembly: The Actin and Tubulin Nucleotidases

M.-F. CARLIER and D. PANTALONI

A. Introduction

Nucleotide hydrolysis is used to drive and regulate a variety of cellular processes. In particular, the assembly of cytoskeletal polymers such as actin filaments and microtubules is an energy-consuming reaction. Because many cell motility processes (shape changes, locomotion, karyokinesis, cytokinesis, etc.) are generated by the dynamic behavior of these polymers, it is important to understand how nucleotide hydrolysis, associated to actin and tubulin polymerization, is used in a regulatory fashion. The aim of this paper is to show that the basic molecular mechanism of GTPases involved in transmembrane signaling, protein translocation, and control of proliferation (BOURNE et al. 1990, 1991; SPIEGEL 1992) also accounts for the regulation of the dynamics of actin filaments and microtubules. We will try to show how subtle changes in the rate constants of some essential reactions in the nucleotidase cycle affect the dynamics of actin filaments and microtubules and could be used as a means of regulation in vivo. It is hoped that by comparing the detailed mechanisms of different GTPases and of the F-actin/microtubule systems, fruitful ideas will emerge to better understand the different biological functions in which all these proteins are involved.

B. The Nucleotidase Cycle in the Polymerization of Actin and Tubulin

Monomeric globular actin (G-actin) binds ATP very tightly ($K_A \sim 10^{10} M^{-1}$); conversely tubulin, which is a $\alpha\beta$ dimer, binds GTP tightly ($K_A \sim 10^8 M^{-1}$). Neither G-actin nor dimeric tubulin hydrolyze ATP or GTP; nucleotide hydrolysis is conditional and is triggered by protein–protein interactions taking place upon self-association of actin into F-actin filaments and of tubulin into microtubules. A complete polymerization cycle includes the following consecutive reactions pictured in Fig. 1: incorporation of an NTP subunit into the polymer; hydrolysis of NTP on the polymerized subunit; liberation of P_i into the medium, while NDP remains bound to the polymer; dissociation of an NDP subunit, which has to exchange NTP for bound NDP before being committed to another cycle of assembly. It has been shown (CARLIER and PANTALONI 1978, 1988) that following NTP hydrolysis and P_i

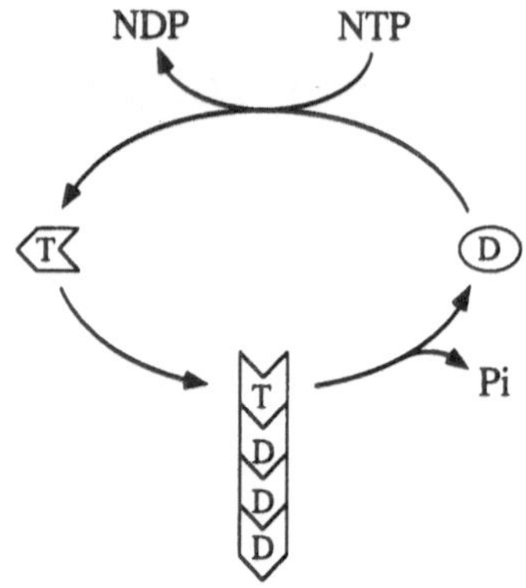

Fig. 1. Nucleotidase cycle in actin and tubulin polymerization. *T* and *D* refer to subunits having *NTP* and *NDP* bound respectively; the NTP-bound monomer is in the "activated" state, in which it self-assembles. Nucleotide hydrolysis occurs on the polymer. Depolymerization leads to the production of "inactive" NDP monomers that have a conformation different from NTP monomers

release, the interactions between subunits in the polymer are weakened; therefore, nucleotide hydrolysis regulates the dynamics of F-actin and microtubules by making the polymers more easily depolymerizable, a property which forms the basis of their highly dynamic behavior in the cell.

From the above description, a parallel can be drawn between the actin/tubulin system and the GTPase family. In both cases, the proteins exist in a NTP-bound, "active" state, in which they can interact with another "effector" macromolecule, a reaction which enhances nucleotide hydrolysis; in the "inactive" NDP-bound state, the proteins dissociate from the effector. The GTPase cycle is regulated by the effector or GAP (GTPase activating protein), which acts on the hydrolysis rate constant k_{cat}, and by the GDS (guanine nucleotide dissociation stimulator), which acts on the process of regeneration of the GTP-bound GTPase. In the case of actin/tubulin, another identical NTP monomer plays the role of GAP, and no GDS has been found so far for the polymers, i.e., nucleotide readily exchanges on G-actin and tubulin, while it is not exchangeable in the F-actin and microtubule polymers. Some G-actin-binding proteins, such as profilin, are known to accelerate the dissociation of bound nucleotide (POLLARD and COOPER 1986). In the case of tubulin, nucleoside diphosphokinase was proposed to be able to directly phosphorylate GDP bound to dimeric tubulin, thus bypassing the nucleotide exchange reaction (PENNINGROTH and KIRSCHNER 1977; HUITOREL et al. 1984), but this proposal was recently contradicted (MELKI et al. 1992).

The above characteristic features of the nucleotidase reaction in F-actin and microtubules are similar to those of the subclass of EF-Tu and *ras* p21, in the sense that NTP hydrolysis is strictly conditional. On the other hand, because of the absence of an effective GDS, actin and tubulin can be considered as analogous to EF-G, the bacterial elongation factor (KAZIRO 1978).

Despite these analogies in mechanism, neither actin nor tubulin possess the consensus sequences characteristic of the GTPase family. Further, the X-ray crystallographic structure of actin (KABSCH et al. 1990) does not show any similarity with the known three-dimensional structures of EF-Tu (JURNAK 1985) or *ras* p21 (PAI et al. 1989). On the other hand, the actin structure shows striking similarities to hexokinase and the ATPase fragment of the heat shock cognate HSC70 (HOLMES and KABSCH 1991). Interestingly, this fragment is an unregulated ATPase (FLAHERTY et al. 1990), although the ATPase activity of HSP70, under native conditions, is linked to the unfolding of clathrin vesicles via a switching mechanism similar to that of GTPases.

Tubulin structure is known at 18 Å resolution from the X-ray fiber diffraction pattern of microtubules (BEESE et al. 1987), and displays a three-domain structure, like GTPases.

C. Elementary Steps in NTP Hydrolysis on Actin Filaments and Microtubules: The Regulation of Polymer Assembly

A puzzling observation made in the early studies of tubulin polymerization into microtubules was that, while the polymer was made of GDP-bound subunits, GTP was necessary for polymer assembly, and microtubules could not be assembled from dimeric GDP-tubulin (WEISENBERG et al. 1976). In the case of actin, G-ADP-actin can polymerize into F-ADP-actin filaments, but with a critical concentration 25-fold higher than from G-ATP-actin, although in both cases the filaments are F-ADP-actin. Nucleotide hydrolysis is not necessary for polymerization, since microtubules, for example, can be polymerized from GMP-PNP-tubulin (ARAI and KAZIRO 1977) and then appear very stable (WEISENBERG and DEERY 1976). If GTP hydrolysis destabilizes microtubules, it is difficult to understand how these structures can ever be formed and maintained in solution with a low critical concentration (the critical concentration is the concentration of monomers at equilibrium with the polymer). Indeed, if GTP hydrolysis were mechanistically tightly coupled to tubulin–tubulin interaction, microtubules could never build up. The clue to this paradox was provided by a kinetic analysis of the polymerization process and associated nucleotide hydrolysis. No tight kinetic coupling can be demonstrated or between microtubule assembly and GTP hydrolysis, or between polymerization of ATP-actin and ATP hydrolysis (for review see CARLIER 1989). In both cases, cleavage of the γ-phosphate follows the incorporation of an NTP subunit into the polymer. It was later found that P_i release is slower than cleavage of the γ-phosphate (CARLIER and PANTALONI 1988; MELKI et al. 1990), so that the polymerization cycle has to be written as follows:

Scheme I

$$\begin{array}{ccccc} \text{NDP} & & & & \text{NTP} \\ \\ \text{M-NTP} & & & & \text{M-NDP} \\ \downarrow & & & & \uparrow \\ \text{P-NTP} & \rightarrow & \text{P-NDP-P}_i & \rightarrow & \text{P-NDP} + \text{P}_i \end{array}$$

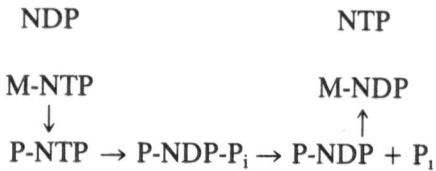

where M and P represent the monomer and polymer state respectively. Scheme I shows that the polymer can be in different states according to which nucleotide (NTP, NDP-P_i, NDP) is bound. It was shown that P_i release is the elementary step linked to the destabilization of protein–protein interactions in the polymer (for review, see Korn et al. 1987; Carlier et al. 1989). Two concepts are necessary to understand how the dynamics of actin filaments and microtubules is regulated by nucleotide hydrolysis. First, actin filaments and microtubules are fibrous polar polymers that grow endwise; only the subunits that are at the ends of the polymers can undergo dissociation from the polymer or association to monomers, leading to elongation. The internal subunits, which form the bulk of the polymer, interact with four neighbors and are comparable to molecules inside a crystal, i.e., they cannot dissociate from the polymer body. Secondly, as these polymers grow by endwise addition of NTP subunits, nucleotide is hydrolyzed and phosphate is released in a delayed fashion, which results in the existence of a "cap" of newly incorporated NTP and NDP-P_i terminal subunits at the ends of the growing polymers, while the inside consists of "order" NDP subunits. NTP and NDP-P_i subunits interact strongly with each other and dissociate slowly from the polymer ends, and therefore maintain a high stability of the edifice – that is, a low monomer–polymer equilibrium dissociation constant. The stochastic monomer–polymer exchange reactions that occur at steady state sometimes lead to loss of the NTP/NDP-P_i cap and exposure of NDP ends, which triggers the rapid depolymerization of a large number of NDP subunits. This behavior is possible because the drop in free energy linked to P_i release is not big enough for a NDP subunit to dissociate from the core of the polymer, where it is surrounded by four neighbors; however, the weakness of interactions between polymerized NDP subunits becomes fully manifest when these subunits happen to be in a terminal position in which the number of contacts with neighbors is lower. The stochastic gain and loss of the NTP cap results in what has been called the "dynamic instability" behavior of microtubules (Mitchison and Kirschner 1984), which was shown to be an essential functional property of microtubules in vivo. The fact that NTP hydrolysis is not tightly coupled to the assembly process results in a strongly nonlinear monomer concentration dependence of the rate of microtubule and actin filament growth (Carlier et al. 1984a,b). This nonlinearity in turn is linked to the production of an oscillatory regime in microtubule assembly under some conditions (Carlier et al. 1987; Melki et al. 1988). The mechanism described above allows a fine regulation of microtubule and actin filament dynamics, by modulation of the following essential rate parameters affecting the size of the NTP/NDP-P_i cap: the rate of P_i release, and the two rates of dissociation of NTP/NDP-P_i and NDP subunits from polymer ends. Quantitative comparison of the values of these parameters in the actin and tubulin systems illustrates this point: GTP hydrolysis and P_i release on microtubules are one order of magnitude faster than their counterparts on F-actin. In addition, the rate of dissociation of GDP-tubulin from microtubules is at least two orders of magnitude faster than the rate of dissociation of GTP/GDP-P_i subunits, whereas the difference between these two rates is only 5- to 10-fold in the actin system. Hence, microtubules are much more dynamic than actin filaments. The dynamic behavior of individual microtubules as well as the macroscopic properties of bulk solutions can be well described by a model within which microtubules undergo transitions between two states, "capped" and "uncapped" (Hill and Chen 1984; Hill 1985; Chen and Hill 1985; Bayley et al. 1990). An appreciation of the extent of dynamics of the polymers is provided by measuring the rate of nucleotide hydrolysis at steady state, which accounts for the amplitude of the fluctuations in length. The steady-state NTPase appears much higher in microtubule solutions than in F-actin solutions. Microtubules would be less dynamic if the rate of GTP hydrolysis was slower.

The above description of the NTPase switch in microtubules and actin filaments shows a striking similarity with the "timer" function of EF-Tu GTPase in the kinetic proofreading of the codon–anticodon match in the translation of mRNA; in this case, too, the function is regulated by the two rates of GTP hydrolysis and of GDP–EF-Tu dissociation from the mRNA-ribosome–EFTu complex.

D. Nucleotide and Metal Ion Binding to Actin and Tubulin

Actin binds tightly a β, γ-bidentate complex of ATP with a divalent metal ion that can be Ca^{2+} or Mg^{2+} (Carlier 1991 for review). Whether Ca-ATP or Mg-ATP is bound affects actin conformation and kinetic parameters for nucleotide exchange, polymerization, and ATP hydrolysis. Mg-ATP most likely is the relevant ligand bound to actin in vivo.

Following ATP hydrolysis on F-actin, the metal ion remains bound to the β-phosphate of ADP. The presence of divalent metal ion increases the affinity of both ATP and ADP to G-actin about 10^6-fold. Therefore, in its metal–nucleotide binding properties, actin is similar to EF-Tu and *ras* p21, for which Mg^{2+} ion increases the affinity of both GTP and GDP (Arai et al. 1974; De Vendittis et al. 1986).

The situation is somewhat different for tubulin, which interestingly appears to be more similar to heterotrimeric G-proteins (Gilman 1987) in its metal–nucleotide binding properties. Indeed, Mg^{2+} enhances the affinity of

GTP, but not that of GDP, for dimeric tubulin (CORREIA et al. 1987). The stereochemistry of Mg-GTP binding to tubulin has been studied using β, γ-Cr-GTP (CARLIER et al. 1991) and shows that the preferred configuration is Δ pseudo-axial (same as for *ras* p21). Both P_i and metal ion are released into the medium following GTP hydrolysis on microtubules. The difference in the metal–NDP/NTP binding properties of EF-Tu, *ras* p21, and actin on the one hand, and heterotrimeric G-proteins and tubulin on the other hand may be linked to the different binding of the divalent metal ion to protein residues in the nucleotide site, and may have some implications in the nucleotidase mechanism, but this issue must await further work.

The rate of nucleotide dissociation from actin and tubulin is at least one order of magnitude faster than the rate of GDP dissociation from G-proteins in the absence of an effector, i.e., 10^{-1} to $10\,\text{min}^{-1}$ for actin and tubulin. This rate is nonetheless slow enough for a nonnegligible amount of NDP dimer to accumulate in a transient fashion, at steady state, as a result of monomer–polymer exchange reactions, according to Scheme I. Since the NDP monomer does not polymerize, the slow rate of nucleotide exchange provides a possible modulation of the monomer/polymer mass ratio in the cell. The implication of the rate of nucleotide exchange in the dynamics of cytoskeletal polymers has been emphasized (CARLIER et al. 1989; MELKI et al. 1988). Whether this regulation mechanism effectively operates in vivo is a matter of speculation, because the [GDP tubulin]/[GTP tubulin] ratio in the cell is not known. Any putative protein that could enhance the rate of NTP regeneration on G-actin or tubulin would act as a nucleating agent, by increasing the local concentration of NTP monomers. In this respect, γ-tubulin has recently been described as a centrosomal protein involved in microtubule nucleation (JOSHI et al. 1992), but the exact mechanism by which nucleation is triggered by γ-tubulin is unknown.

E. Probing the Nucleotidase Mechanism of Actin and Tubulin using AlF_4^- and BeF_3^-, H_2O

Fluoride ions have long been known to activate the *a* subunits of heterotrimeric G-proteins, and it was demonstrated that the actual activating factor was AlF_4^- (STERNWEIS and GILMAN 1982), which bound to the GDP-bound protein exclusively (FERGUSON et al. 1986; HIGASHIJIMA et al. 1987). It was hypothesized (BIGAY et al. 1987) that AlF_4^- and BeF_3^-, H_2O could act as structural (tetrahedral) analogs of inorganic phosphate that would bind to the site of the γ-phosphate of GTP and reconstitute an active GTP-like state of the G-protein. It was actually demonstrated that AlF_4^- and BeF_3^-, H_2O bound to GDP microtubules (CARLIER et al. 1988) and F-ADP-actin filaments (COMBEAU and CARLIER 1988) in competition with P_i, mimicked all the effects of P_i in stabilizing the protein–protein interactions of the polymer, and reconstituted the very stable NDP-P_i polymer. It was further shown that while P_i binds to the NDP polymers with a relatively modest affinity ($\sim 10^3 M^{-1}$) and in rapid equilibrium, AlF_4^- and BeF_3^-, H_2O bound with an affinity three orders of magnitude higher and in slow association–dissociation equilibrium (CARLIER et al. 1989; COMBEAU and CARLIER 1988). These binding characteristics are similar to those of vanadate binding to myosin-ADP (GOODNO 1979), which, together with other information, suggested that the conformation of the NDP-AlF_4^- state was energetically closer to the NTP state or to the transition state in nucleotide hydrolysis than to the NDP-P_i state reconstituted by the binding of P_i. Consistent with this suggestion, it was found that AlF_4^-, and not AlF_3, was actually bound to F-ADP-actin and microtubules (COMBEAU and CARLIER 1989). The same stoichiometry of four fluorides bound per aluminum atom has also been found for AlF_4^- binding to the F_1-ATPase (DUPUIS et al. 1989). Recently, AlF_4^- has been shown in interact with the β–γ bridging oxygen of GDP and the Mg^{2+} ion, which is also coordinated to the β-phosphate of GDP in the Gα protein (HIGASHIJIMA et al. 1991). The exact configuration of AlF_4^- bound to the site of the γ-phosphate of the nucleotide is an important issue in understanding the molecular mechanism of nucleotidases. AlF_4^- and BeF_3^-, H_2O are indeed of general use in probing the structure of kinetic intermediates of other nucleotidases involved in a switch mechanism such as recA protein (MOREAU and CARLIER 1989) and actomyosin (PHAN and REISLER 1992).

An intriguing point concerning the binding of AlF_4^- and BeF_3^- to nucleotidases is the observation that they bind to polymerized F-ADP actin, or to GDP-microtubules, or to non-conditional GTPases such as Gα, but not to monomeric ADP-G-actin or GDP-tubulin, or to GDP-EF-Tu or to GDP-ras. The general conclusion seems to be that there are two NDP states, one in which the site of the γ-phosphate is still "open" (NDP polymer) and one in which it is occluded (NDP monomer). The transition between these two states is the irreversible conformational switch in the GTPase cycle. The open NDP state must have a sufficient lifetime for AlF_4^- to bind; this is the case for the F-ADP-actin and microtubules. It is tempting to propose that AlF_4^- could bind to the GDP-EF-Tu–ribosome complex, or to the GDP-ras–GAP complex if these had a sufficient lifetime; once dissociated from the effector, GDP-EF-Tu and GDP-ras exist essentially in the state in which the site of the γ-phosphate is occluded. The fact that Gα subunits of heterotrimeric GTPases can hydrolyze GTP in a non-conditional fashion and bind AlF_4^- in the absence of effector indicates that some of the steps of the GTPase cycle are partially reversible, e.g., GDP-Gα can exist in the open configuration of the site of the γ-phosphate.

F. Conclusions

The kinetic analysis of nucleotide hydrolysis in actin and tubulin polymerization has brought information on the crucial steps involved in the regulation of the dynamics of actin filaments and microtubules:

1. Phosphate release, and not cleavage of the γ-phosphate, is the reaction which is coupled to the destabilization of the protein–protein interactions in the polymer lattice. The same conclusion has been reached for the actomyosin ATPase, in which P_i release is linked to force development (Hibberd and Trentham 1985) and the DNA dependent recA ATPase (Moreau and Carlier 1989). Therefore, microtubules and actin filaments can be in different structural states depending on which nucleotide is bound.
2. P_i and P_i analog binding studies have pointed to a different conformation of the polymerized and unpolymerized NDP subunit; therefore, another conformation change accompanies the dissociation of an NDP-subunit from the polymer.
3. The rate of nucleotide exchange on the monomer is slow enough to play a role in the regulation of dynamics in vivo as well as in vitro.

Future research using reconstituted controlled cellular systems, site-directed mutagenesis, and high resolution structural studies will show how these elementary reactions can be regulated in vivo, and which amino acid residues of the proteins are involved in the conformational switch of actin and tubulin.

Acknowledgements. This work was supported by C.N.R.S., Ligue Nationale Française contre le Cancer, and Association pour la Recherche contre le Cancer.

References

Arai K, Kaziro Y (1977; Role of GTP in the assembly of microtubules. J Biochem (Tokyo) 82:1063–1071

Arai K, Kawakita M, Kaziro Y (1974) Studies on the polypeptide elongation feature from E. coli. V. Properties of various complexes containing EF-Tu and EF-Ts. J Biochem (Tokyo) 76:293–306

Bayley PM, Schilstra MJ, Martin SR (1990) Microthubule dynamic instability: numerical simulation of microtubule transition properties using a Lateral Cap model. J Cell Sci 95:33–48

Beese L, Stubbs G, Cohen C (1987) Microtubule structure at 18 A resolution. J Mol Biol 194:257–264

Bigay J, Deterre P, Pfister C, Chabre M (1987) Fluoride complexes of aluminium or beryllium act on G-proteins as reversibly bound analogues of the gamma phosphate of GTP. EMBO J 6:2907–2913

Bourne JR, Sanders DA, McCormick F (1990) The GTPase superfamily: a conserved switch for diverse cell functions. Nature 348:125–132

Bourne JR, Sanders DA, McCormick F (1991) The GTPase superfamily: conserved structure and molecular mechanism. Nature 349:117–127

Carlier MF (1989) Role of nucleotide hydrolysis in the dynamics of actin filament and microtubules. Int Rev Cytol 115:139–170

Carlier MF (1991) Actin: protein structure and filament dynamics. J Biol Chem 266:1–4

Carlier MF, Pantaloni D (1978) Kinetic analysis of cooperativity in tubulin polymerization in the presence of guanosine di- or triphosphate nucleotides. Biochemistry 17:1908–1915

Carlier MF, Pantaloni D (1988) Binding of phosphate to F-ADP-actin and the role of F-ADP-Pi-actin in ATP-actin polymerization. J Biol Chem 263:817–825
Carlier MF, Hill TL, Chen U (1984a) Interference of GTP hydrolysis in the mechanism of microtubule assembly: an experimental study. Proc Natl Acad Sci USA 81:772–776
Carlier MF, Pantaloni D, Korn ED (1984b) Evidence for an ATP cap at the ends of actin filaments and its regulation of the F-actin steady state. J Biol Chem 259:9983–9986
Carlier MF, Melki R, Pantaloni D, Hill TL, Chen Y (1987) Synchronous oscillations in microtubule polymerization. Proc Natl Acad Sci USA 84:5257–5261
Carlier MF, Didry D, Melki R, Chabre M, Pantaloni D (1988) Stabilization of microtubules by inorganic phosphate and its structural analogues, the fluoride complexes of aluminium and beryllium. Biochemistry 27:3555–3559
Carlier MF, Didry D, Simon C, Pantaloni D (1989) Mechanism of GTP hydrolysis in tubulin polymerization; characterization of the kinetic intermediate microtubule-GDP-Pi using phosphate analogues. Biochemistry 28:1783–1791
Carlier MF, Didry D, Valentin-Ranc C (1991) Interaction between chromium GTP and tubulin. Stereochemistry of GTP binding, GTP hydrolysis, and microtubule stabilization. J Biol Chem 266:12361–12368
Chen Y, Hill TL (1985) Theoretical treatment of microtubules disappearing in solution. Proc Natl Acad Sci USA 82:4127–4131
Combeau C, Carlier MF (1989) Probing the mechanism of ATP hydrolysis on F-actin using vanadate and the structural analogs of phosphate BeF-3 and AlF-4. J Biol Chem 263:17429–17436
Combeau C, Carlier MF (1989) Characterization of the aluminum and beryllium fluoride species bound to F-actin and microtubules at the site of the gamma-phosphate of the nucleotide. J Biol Chem 264:19017–19021
Correia JJ, Baty LT, Williams RC Jr (1987) Mg^{2+} dependence of guanine nucleotide binding to tubulin. J Biol Chem 262:17278–17284
De Vendittis E, Zahn R, Fasano O (1986) Regeneration of the GTP-bound from the GDP-bound form of human and yeast has proteins by nucleotide exchange. Stimulatory effect of organic and inorganic phosphates. Dur J Biochem 161:473–478
Dupuis A, Issartel JP, Vignais PV (1989) Direct identification of the fluoroaluminate and fluoroberyllate species responsible for inhibition of the mitochondrial F_1 ATPase. FEBS Lett 255:47–52
Ferguson KM, Higashijima T, Smigel MD, Gilman AG (1986) The influence of bound GDP on the knetics of guanine nucleotide binding to G proteins. J Biol Chem 261:7393–7399
Gilman AG (1987) G proteins: transducers of receptor-generated signals. Annu Rev Biochem 56:615–649
Goadno C (1979) Proc Natl Acad Sci USA 76:2602–2624
Hibberd MG, Dantzig JA, Trentham DR, Goldman YE (1985) Phosphate release and force generation in skeletal muscle fibers. Science 228:1317–1319
Higashijima T, Ferguson KM, Sternweis PC, Ross EM, Smigel MD, Gilman AG (1987) The effect of activating ligands on the intrinsic fluorescence of guanine nucleotide-binding regulatory proteins. J Biol Chem 262:752–756
Higashijima T, Graziano MP, Suga H, Kainosho M, Gilman AG (1991) 19F and 31P NMR spectroscopy of G protein alpha subunits. Mechanism of activation by A13 and F-. J Biol Chem 266:3396–3401
Hill TL (1985) Phase-change kinetics for a microtubule with two free ends. Proc Natl Acad Sci USA 82:431–435
Hill TL, Chen Y (1984) Phase changes at the end of a microtubule with a GTP cap. Proc Natl Acad Sci USA 81:5772–5776
Holmes KC, Kabsch W (1991) Musde proteins-actin. Curr Opin Struct Biol 1:270–280
Huitorel P, Simon C, Pantaloni D (1984) Nuclocide diphosphate kinase from brain. Purification and effect on microtubule assembly in vitro. Eur J Biochem 144:233–241

Joshi HC, Palacios MJ, McNamara L, Cleveland DW (1992) Gamma-tubulin is a centrosomal protein required for cell cycle dependent microtubule nucleation. Nature 356:80–83
Jurnak F (1985) Structure of the GDP domain of EF-Tu and location of the amino acids homologous to ras oncogene proteins. Science 230:32–36
Kabsch W, Mannherz HG, Suck D, Pai EF, Holmes KC (1990) Atomic structure of the actin: DNase I complex. Nature 347:37–44
Kaziro Y (1978) The role of guanosine 5′-triphosphate in polypeptide chain elongation. Biochim Biophys Acta 505:95–127
Korn ED, Carlier MF, Pantaloni D (1987) Actin polymerization and ATP hydrolysis. Science 238:638–644
Melki R, Carlier MF, Pantaloni D (1988) Oscillations in microtubule polymerization: the rate of GTP regeneration on tubulin controls the period. EMBO J 7:2653–2659
Melki R, Carlier MF, Pantaloni D (1990) Direct evidence for GTP and GDP-Pi intermediates in microtubule assembly. Biochemistry 29:8921–8932
Melki R, Lascu I, Carlier M-F, Verion M (1992) Nucleotide diphosphate kinase does not directly interact with tubulin nor microtubules. Biochem, Biophys Res Comm 187:65–72
Mitchison T, Kirschner MW (1984) Dynamic instability of microtubule growth. Nature 312:237–242
Moreau P, Carlier MF (1989) RecA protein- promoted cleavage of LexA repressor in the presence of ADP and structural analogues of inorganic phosphate, the fluoride complexes of aluminium and beryllium. J Biol Chem 264:2302–2306
Pai EF, Kabsch W, Krenkel U, Holmes KC, John J, Wittinghofer A (1989) Structure of the guanine nucleotide-binding domain of the Ha-ras oncogene product p21 in the triphosphate conformation. Nature 341:209–214
Phan BC, Reisler E (1992) Inhibition of Myesin ATPase by Beryllium fluoride Biochemistry 31:4787–4793
Penningroth S, Kirschner MW (1977) Nucleotide binding and phosphorylation in microtubule assembly in vitro. J Mol Biol 115:643–673
Pollard TD, Cooper JA (1986) Actin and actin-binding proteins. A critical evaluation of mechanisms and functions. Annu Rev Biochem 55:987–1035
Spiegel AM (1992) G proteins in cellular control. Curr Opin Cell Biol 4:203–211
Sternweis PC, Gilman AG (1982) Aluminium: a requirement for activation of the regulatory component of adenylate cyclase by flueride. Proc Natl Acad Sci USA 79:4888–4891
Weisenberg RC, Deery WJ (1976) Role of nucleotide hydrolysis in microtubule assembly. Nature 263:792–793
Weisenberg RC, Deery WJ, Dickinson PJ (1976) Tubulin-nucleotide interactions during the polymerization and depolymerization of microtubules. Biochemistry 15:4248–4254

CHAPTER 6

Dynamin, A Microtubule-Activated GTPase Involved in Endocytosis

R.B. VALLEE, J.S. HERSKOVITS, and C.C. BURGESS

A. Introduction

Dynamin is a 100-kDa polypeptide initially identified in brain tissue by virtue of its copurification with microtubules (SHPETNER and VALLEE 1989). Molecular cloning and biochemical analysis revealed dynamin to be a GTPase (OBAR et al. 1990; SHPETNER and VALLEE 1992). Identification of a possible homologue of dynamin in *Drosophila* together with recent transfection studies with mutant forms of rat dynamin have indicated an important role for the protein in the initial steps of the endocytic pathway.

Dynamin is unrelated to the other GTPases described in this volume, except in the consensus sequence elements correlated with GTP binding. However, a number of dynamin-related proteins have been identified (see Fig. 1). These proteins are all of very high molecular weight compared to most of the known GTPases, and together they constitute a new family. The functional relationship between these proteins is a question of considerable interest, the resolution of which promises to provide important new insight into the mechanism of endocytosis as well as other processes of fundamental importance to the cell.

This chapter will review the existing data on the properties of dynamin, with emphasis on its known functional domains. Evidence on the properties of the *Drosophila shibire* gene will be reviewed along with recent data from transfection studies in mammalian cells in an attempt to define the functional cycle of dynamin in the cell.

B. Structure and Enzymatic Properties

Molecular cloning of rat brain dynamin revealed the three well-conserved consensus sequence elements seen in most GTPases (OBAR et al. 1990). In addition, homology was observed between dynamin and two other proteins, Mx and VPSlp. Mx was originally cloned from a mouse strain resistant to myxovirus infection (STAEHELI et al. 1986). It is induced by interferon and confers resistance to infection by specific viruses. *VPS1* is a yeast gene selected in a screen for secretory mutants (ROTHMAN et al. 1990). Mutations in *VPS1* produce a defect in sorting of proteases such as carboxypeptidase Y from the Golgi apparatus to the vacuole, a degradative organelle func-

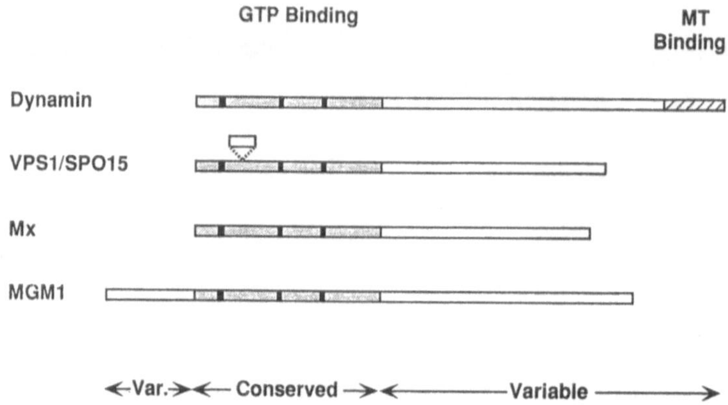

Fig. 1. Diagram of *dynamin* and related polypeptides. *Blackened bars* represent GTP-binding consensus sequence elements conserved among almost all GTPases. *Shaded* N-terminal region is highly conserved among all dynamin family members. *Cross-hatched* C-terminal region represents basic, proline-rich domain unique to dynamin. This portion of the molecule binds microtubules, and its removal abolishes microtubule-activation of the GTPase activity (see text). From VALLEE (1992)

tionally comparable to the lysosome. The *SPO15* mutation, which is defective in meiotic spindle function, was recently found to be in the same gene (YEH et al. 1991). How the meiotic phenotype relates to the membrane-sorting phenotype is uncertain. Recently, another dynamin-related gene has been identified in yeast, *MGM-1* (JONES and FANGMAN 1992). *MGM-1* mutants are defective in mitochondrial genome replication.

Sequence conservation between the members of this family resides primarily within the GTP-binding domain, a region consisting of approximately 300 amino acids. The same region shows little obvious relationship to other GTPases beyond the few amino acids comprising the GTP-binding consensus sequence elements. Thus, the amino terminal portion of the dynamin-related proteins is likely to perform some common family-specific function.

The remaining two-thirds of dynamin shows some additional homology to the comparable region of VPS1p, but little detectable homology to Mx or MGM-1. Dynamin extends further in the C-terminal direction than the other proteins when the group is aligned as in Fig. 1. The C-terminal extension of rat dynamin consisting of approximately 100 amino acids is distinctive in composition, having a pI of 12.5 and containing 32% proline.

The function of this domain has been investigated by proteolytic digestion (HERSKOVITS et al. 1991; and submitted for publication). Exposure of dynamin to papain led to the production of a relatively stable 90-kDa fragment. A 7- to 9-kDa fragment was concurrently produced which could be detected by immunoblotting using an antibody directed at the carboxy-terminal 20 amino acids of the rat sequence. In contrast to intact dynamin, which cosediments very efficiently with microtubules (SHPETNER and VALLEE

1992), the 90-kDa fragment showed no apparent binding. However, a substantial fraction of the small C-terminal fragment continued to cosediment with microtubules, indicating that it contained the microtubule-binding domain.

Purified dynamin was found to exhibit a relatively high GTPase activity compared with regulatory GTPases ($0.04–0.2\,s^{-1}$; SHPETNER and VALLEE 1992). The activity could be greatly stimulated (up to $6\,s^{-1}$) by microtubules. Papain digestion abolished this effect without detectably altering the basal level of GTPase activity (HERSKOVITS et al. 1991) consistent with the dissociation of the microtubule-binding region from the remainder of the molecule.

While the physiological significance of microtubules in dynamin function is uncertain (see below), these results suggest that the C-terminal domain represents an important regulatory site within the dynamin molecule . In addition, they suggest that the C and N termini may lie in close proximity in the folded molecule, to explain the effect of C-terminal microtubule bind-

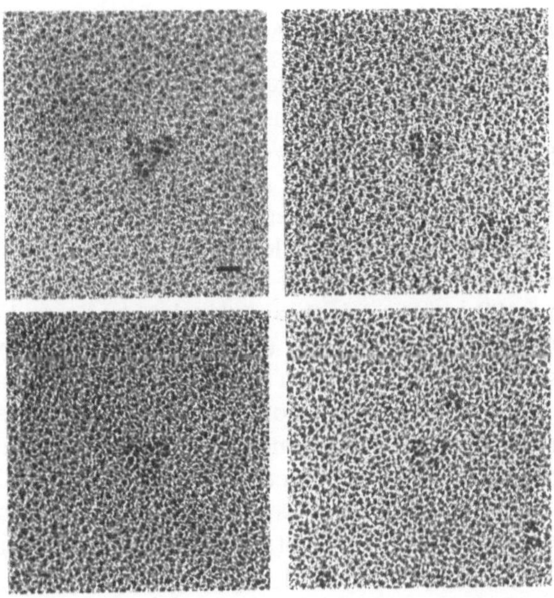

Fig. 2. Electron microscopy of dynamin. Diethylaminoethanol-purified dynamin was sprayed onto mica and rotary shadowed as previously described (TYLER and BRANTON 1980). Four individual molecules are shown, and a drawing of a dynamin molecule is provided at the right. The molecules generally appear to have two globular domains seen at the top of each image, connected to a base by two stalks. In this regard dynamin has an overall appearance similar to that of dynein (GOODENOUGH and HEUSER 1984; VALLEE et al. 1988). However dynamin is much smaller, and there is no evidence of functional similarity between the globular domains of dynamin and the force producing globular "head" domains of dynein. *Bar*: $10\,\mu$

ing on N-terminal GTPase activity. Because the GTP-binding consensus sequence elements are spatially constrained by the requirement to contact the GTP molecule, the N-terminal portion of dynamin is probably globular and conformationally related to other GTPases such as ras and EF-Tu. The organization of the large central portion of the dynamin polypeptide is uncertain, but the C-terminal domain of the molecule would lie near the globular GTP-binding region in our view.

Electron microscopy of dynamin has revealed an apparent bilaterally symmetric structure (Fig. 2). This suggests that the protein may exist as a dimer, which would in turn help to explain its ability to cross-link microtubules (SHPETNER and VALLEE 1989) despite the presence of only a single high-affinity microtubule-binding site (HERSKOVITS et al. 1991). It is tempting to speculate that the globular domains observed by electron microscopy correspond to the GTPase region, but further work will be needed to learn more fully how the dynamin polypeptide is organized within the native molecule.

C. The *Drosophila shibire* Gene

What may be the *Drosophila* homologue of dynamin is encoded by the *shibire* gene (CHEN et al. 1991; VAN DER BLIEK and MEYEROWITZ 1991). The amino acid sequences deduced from two different full-length fly cDNAs (see below) are 68% identical and 81% similar to the rat sequence, values considerably higher than those obtained in comparisons of rat dynamin with Mx, VPS1/SPO15, or MGM-1. The level of homology also remains high throughout almost the entire length of the rat and fly sequences. Within the carboxy-terminal 100 amino acids this pattern breaks down; however, both fly and rat sequences exhibit the same unusually high content of proline and high pI, suggesting that in this portion of the molecule amino acid composition is structurally and functionally more important than primary sequence. As in the case of mammalian dynamin, the fly protein was observed to cosediment with microtubules (CHEN et al. 1991).

Apparent alternative splicing was found to occur at two sites within the coding region of *shibire* (CHEN et al. 1991, 1992; VAN DER BLIEK and MEYEROWITZ 1991). At amino acid 634, insertions of zero, four, or six amino acids were identified. At amino acid 835, alternative short and long C-terminal tails can be produced. Because this region of the molecule is involved in the regulation of GTPase activity in vitro, as discussed above, it seems reasonable to speculate that alternative splicing may affect either the affinity or the specificity of the interaction of the *Drosophila* dynamin with regulatory molecules.

Seven temperature-sensitive alleles of *shibire* have been identified (POODRY 1990), and the phenotype of the *shi^{ts}* mutants has been well characterized at the behavioral and cellular level. *shi*^{ts} files become rapidly paralyzed at restrictive temperature (GRIGLIATTI et al. 1973), a condition which can be just as rapidly reversed at permissive temperature. Paralysis has been traced to a defect in synaptic transmission (KOSAKA and IKEDA 1983a; POODRY and EDGAR 1979). Exocytosis of synaptic vesicle contents is apparently unaffected (IKEDA et al. 1976; SALKOFF and KELLY 1978). However, synaptic terminals become progressively depleted of synaptic vesicles, as the number of coated and noncoated pits increase dramatically at the plasma membrane.

These data have been interpreted in terms of a specific block in the rebudding of synaptic vesicles after they have fused with the plasma membrane. Curiously, return to permissive temperature does not result in the immediate formation of synaptic vesicles. Instead, the pits expand greatly in size, and then become fragmented into smaller vesicles (KOENIG and IKEDA 1989).

Examination of other cell types, in particular nephrocytes and oocytes, has revealed a general defect in budding of vesicles from the plasma membrane (KOSAKA and IKEDA 1983b; KESSEL et al. 1989). Uptake of fluid-phase endocytic markers was abolished in these cells at restrictive temperature. A considerable increase in the number of membrane pits forming at the cell surface was also observed, while the number of membrane profiles corresponding to endocytic intermediates was greatly decreased. This again suggested a block in the formation of vesicles at the initial step in endocytosis, but normal recycling of vesicles back to the plasma membrane. Oddly, while only coated pits were seen in nephrocytes and oocytes under restrictive conditions, mostly noncoated pits were observed at the neuromuscular junction (KOENIG and IKEDA 1989).

The nature of the involvement of microtubules in *shi*^{ts} function is uncertain. Microtubules are not seen in substantial number at the synapse, nor are coated pits and vesicles normally seen in proximity to microtubules. Furthermore, cell surface receptor uptake and recycling are not affected by microtubule-disrupting drugs. Nonetheless, it is conceivable that microtubules affect the early stages of endocytosis in an indirect manner which has not previously been appreciated, or that tubulin in some cryptic form – for example, associated with the plasma membrane – is involved in *shibire* function. Tubulin is, so far, the only protein found capable of stimulating the dynamin GTPase (SHPETNER and VALLEE 1992). No appreciable effect was observed with clathrin coats, actin, vimentin, or polyglutamic acid. Maximal stimulation was observed well below physiological microtubule concentrations (~0.1 mg/ml), and was inhibited by the well-characterized microtubule-associated protein MAP 2.

Despite these considerations, a search for physiological regulatory elements and effectors other than, or in addition to, microtubules seems to be warranted, and efforts to identify such factors are under way in our laboratory.

Table 1. Diagram of mutant dynamin constructs used for transfection. From HERSKOVITS et al. (1993)

Construct		Dynamin distribution	α-Adaptin distribution	Tf uptake
D-1	1–851	diffuse	disperse	+
K44E	1–851	punctate, linear	clustered	–
S45N	1–851	punctate, linear	clustered	–
D208N	1–851	diffuse	disperse	+
K44E/C-794	1–794	punctate, linear	clustered	–
K44E/C-663	1–663	diffuse	disperse	+
C-794	1–794	diffuse	disperse	
C-663	1–663	diffuse	disperse	+
N-272	272–851	punctate	clustered	–
N-272/C-663	272–663	diffuse	disperse	+
N-456	456–851	diffuse	disperse	
N-651	651–851	diffuse	disperse	+

The conserved GTP-binding consensus sequence elements are indicated by vertical tick marks near the 5′-end of the dynamin molecule. X represents point mutations

D. Transfection of Dynamin into Cultured Mammalian Cells

Immunological analysis of the subcellular distribution of dynamin has produced conflicting results. SCAIFE and MARGOLIS (1990) reported that the rat brain protein was mostly insoluble, and saw what was described as a very weak punctate staining pattern by immunofluorescence microscopy of cultured PC12 cells. However, despite the use of confocal microscopy in that study, individual immunoreactive spots could not, in general, be resolved. Immunoblotting of controlled-pore glass-fractionated synaptosomal membranes revealed that synaptic vesicles lacked appreciable dynamin, which, instead, was found in a mixed, large-vesicle fraction. Our laboratory has seen only low level staining of cultured cells by immunofluorescence microscopy using several polyclonal antidynamin antibodies.

To gain further insight into the function of dynamin in vivo and to learn more about the specific role of GTP hydrolysis in the dynamin functional cycle, we have transfected COS-7 African green monkey kidney cells with wild-type and mutant dynamin cDNAs (Table 1; HERSKOVITS et al. 1993). The protein was subsequently visualized by immunofluorescence microscopy using a rabbit antibody generated against a glutathione-S-transferase/rat dynamin fusion protein and another rabbit antibody generated against a synthetic peptide corresponding to the C-terminal 20 amino acids of rat dynamin. Endogenous dynamin was undetectable in the COS-7 cells under the immunofluorescence conditions used, so that the distribution of the rat protein alone could be readily detected (Fig. 3).

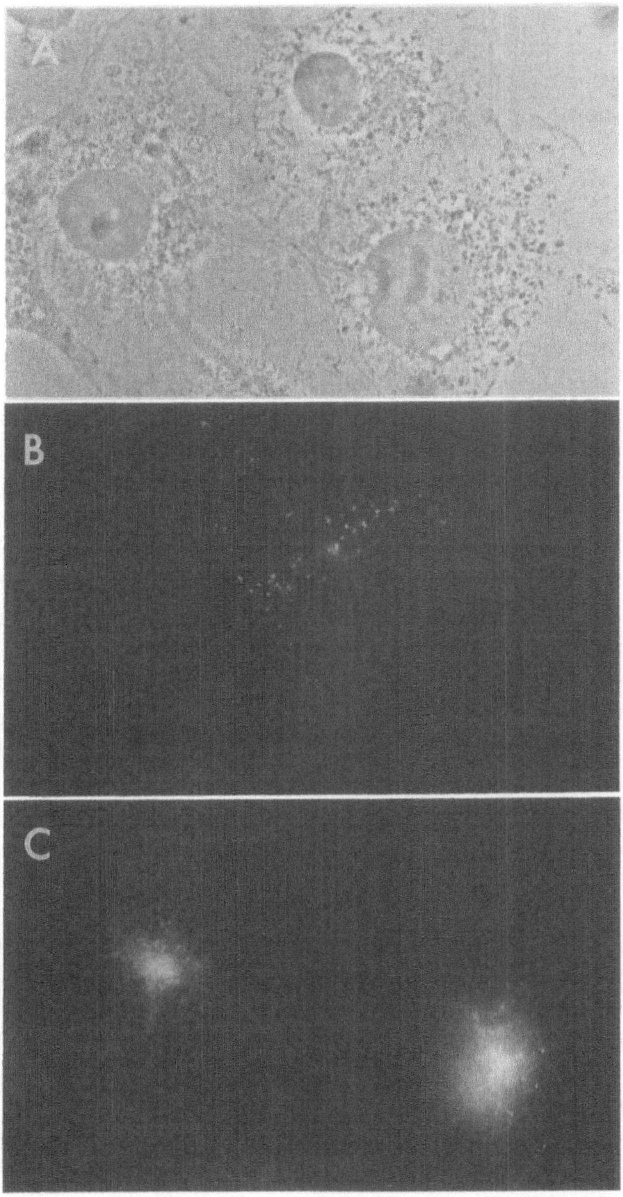

Fig. 3A–C. Effect of mutant dynamin overexpression on transferrin uptake (see Herskovits et al. 1993). COS-7 cells were transfected with the rat brain dynamin N-terminal deletion cDNA construct N-272 (see Table 1). The cells were exposed to fluorescein isothiocyanate (FITC) transferrin for 1 h and then fixed and stained with an antibody against the C terminus of rat dynamin, which does not recognize the endogenous primate protein. **A** phase contrast; **B** antidynamin immunofluorescence; **C** FITC-transferrin. Of the three cells fully in the field of view, the cell at the upper middle is positive for dynamin and shows no clear transferrin uptake, while the other two cells are negative for dynamin and show extensive transferrin uptake. Dynamin is found in a punctate pattern in the transfected cell

Transfection with wild-type dynamin yielded a diffuse pattern of immunofluorescence. Together with the large fraction of soluble dynamin in brain tissue homogenates (SHPETNER and VALLEE 1989), these results may indicate that the protein spends a substantial part of its functional cycle free in the cytoplasm. No significant colocalization with microtubules was observed in intact cells. Occasional weak microtubule staining was observed in detergent-extracted cells, but this may simply reflect the ability of the protein to bind to microtubules under cell-free conditions.

Constructs defective in GTP binding were produced to examine differences in the distribution of dynamin in different states of occupancy of the GTP-binding site. Point mutations in the first GTP-binding consensus sequence element (constructs K44E and S45N), which correspond to dominant negative mutations in ras (SIGAL et al. 1986; FEIG and COOPER 1988; FARNSWORTH and FEIG 1991), produced a mixed distribution pattern. A fraction of the protein still appeared soluble, but much of it was in a punctate state. Elongated structures were also seen which may be tubular elements. An even more dramatic change was observed if the entire GTP-binding domain was eliminated (construct N-272, encoding protein beginning at amino acid 272). In this case the mutant protein was found entirely in bright spots of variable size. A point mutation in the third GTP-binding consensus sequence element (construct D208N) which is oncogenic in ras (FEIG et al. 1986; SIGAL et al. 1986) had no effect on the distribution of dynamin.

To determine the effect of these mutations on endocytosis, transferrin uptake was monitored. K44E and S45N completely blocked transferrin uptake as did the N-terminal deletion mutant N-272. No effect was observed with construct D208N.

Removal of the C-terminal 188 amino acids of dynamin from the inhibitory point mutant constructs K44E and S45N or the inhibitory N-terminal deletion construct N-272 reversed the effects of the inhibitory mutations on the dynamin distribution and on endocytosis. Deletion of the C-terminal region from wild-type dynamin as a control had no phenotypic effect.

Examination of the distribution of tubulin, BiP, β-COP, and the 58-kDa Golgi antigen indicated no apparent effect on any of these markers of overexpression of the wild-type or mutant forms of dynamin. The normal uniform punctate distribution of clathrin heavy chain was seen in cells overexpressing the wild-type construct; however, in the inhibitory mutant constructs, the clathrin spots were found to be clustered. To determine whether all classes of clathrin-containing structure were affected, the cells were labeled with antibodies to α- and γ-adaptin, which stain plasma-membrane-derived and Golgi-derived clathrin-containing structures, respectively (ROBINSON and PEARSE 1986; AHLE et al. 1988). α-Adaptin-positive spots were clustered in a pattern similar to that seen with the clathrin-positive spots. However, γ-adaptin was found in its normal uniformly dispersed distribution, with a concentration of spots near the Golgi apparatus.

While we do not yet fully understand the significance of the alteration in clathrin and α-adaptin distribution, these results suggest that only plasma-membrane-derived coated vesicles are affected by the dynamin mutations.

While all of these results support a specific role for dynamin in endocytosis, uptake of fluid phase endocytic markers is not abolished in cells expressing the inhibitory mutant forms of dynamin, though it is not possible to say as yet whether the rate of fluid-phase uptake is descreased (HERSKOVITS et al. 1993). This result appears to be an indication that at least some aspects of endocytosis and pinocytosis occur independently of dynamin. *shi*^{ts} at restrictive temperature have been reported to be completely blocked in fluid-phase endocytosis, though receptor-mediated endocytosis has not been assayed. The basis for this apparent difference in the function of dynamin between flies and mammals is uncertain. It is possible that additional isoforms of mammalian dynamin will prove to control non-receptor-mediated endocytosis, or that mammalian cultured cells take up fluid-phase markers through routes not controlled by dynamin.

It is also of interest in this regard that one site of alternative splicing in fly dynamin is within the basic, proline-rich C-terminal domain. Conceivably, differences in the structure of this region could specify different subcellular targets for dynamin action, such as clathrin-mediated vs. non-clathrin-mediated endocytosis. In the *shi*^{ts} fly we presume that all activities directed by the *shi* gene are defective, perhaps accounting for the complete block in endocytosis. Further work will be required to resolve this issue and the detailed mechanism of action of the protein.

References

Ahle S, Mann A, Eichelsbacher U, Ungewickell E (1988) Structural relationships between clathrin assembly from the Golgi and the plasma membrane. EMBO J 7:919–929

Chen MS, Obar RA, Schroeder CC, Austin TW, Poodry CA, Wadsworth SC, Vallee RB (1991) Multiple forms of dynamin are encoded by *shibire*, a *Drosophila* gene involve in endocytosis. Nature 351:583–586

Chen MS, Burgess CC, Vallee RB, Wadsworth SC (1992) Developmental stage- and tissue-specific expression of *shibire*, a *Drosophila* gene involve in endocytosis. J Cell Sci 103:619–628

Farnsworth CL, Feig LA (1991) Dominant inhibitory mutations in the Mg^{2+}-binding site of ras^H prevent its activation by GTP. Mol Cell Biol 11:4822–4829

Feig LA, Cooper GM (1988) Inhibition of NIH 3T3 cell proliferation by a mutant ras protein with preferential affinity for GDP. Mol Cell Biol 8:3235–3243

Feig LA, Pan B-T, Roberts TM, Cooper GM (1986) Isolation of ras GTP-binding mutants using an *in situ* colony-binding assay. Proc Natl Acad Sci USA 83: 4607–4611

Goodenough U, Heuser J (1984) Structural comparison of purified dynein proteins with *in situ* dynein arms. J Mol Biol 180:1083–1118

Grigliatti TA, Hall L, Rosenbluth R, Suzuki DT (1973) Temperature-sensitive mutations in *Drosophila melanogaster*. Mol Gen Genet 120:107–114

Herskovits JS, Schroeder CC, Vallee RB (1991) Functional domains of dynamin. J Cell Biol 115:34a
Herskovits JS, Burgert CC, Obar R, Vallee RB (1993) Effects of mutant rat dynamin on endocytosis. J Cell Biol 122:565–578
Ikeda K, Ozawa S, Hagiwara S (1976) Synaptic transmission reversibly conditioned by single-gene mutation in *Drosophila melanogaster*. Nature 259:489–491
Jones B, Fangman W (1992) Mitochondrial DNA maintenance in yeast requires a protein containing a region related to the GTP-binding domain of dynamin. Genes Dev 6:380–389
Kessell I, Holst B, Roth TF (1989) Membranous intermediates in endocytosis are labile, as shown in a temperature-sensitive mutant. Proc Natl Acad Sci USA 86:4968–4972
Kim Y-T, Wu C-F (1990) Allelic interactions at the shibire locus of Drosophila: effects on behavior. J Neurogenet 7:1–14
Koenig JH, Ikeda K (1989) Disappearance and reformation of synaptic vesicle membrane upon transmitter release observed under reversible blockage of membrane retrieval. J Neurosci 9:3844–3860
Kosaka T, Ikeda K (1983a) Possible temperature-dependent blockage of synaptic vesicle recycling induced by a single gene mutation in Drosophila. J Neurobiol 14:207–225
Kosaka T, Ikeda K (1983b) Reversible blockage of membrane retrieval and endocytosis in the garland cell of the temperature-sensitive mutant of Drosophila melanogaster, shibire^{ts1}. J Cell Biol 97:499–507
Obar R, Collins CA, Hammarback JA, Shpetner HS, Vallee RB (1990) Molecular cloning of the microtubule-associated mechanochemical enzyme dynamin reveals homology with a new family of GTP-binding proteins. Nature 347:256–261
Poodry CA, Edgar L (1979) Reversible alterations in the neuromuscular junctions of Drosophila melanogaster bearing a temperature-sensitive mutation, shibire. J Cell Biol 81:520–527
Poodry CA (1990) shi: shibire. Drosophila Information Service 68:207–208
Robinson MS, Pearse BMF (1986) Immunofluorescence localization of 100K coated vesicle proteins. J Cell Biol 102:48–54
Rothman JH, Raymond CK, Gilbert T, O'Hara PJ, Stevens TH (1990) A putative GTP binding protein homologous to interferon-inducible Mx proteins performs an essential function in yeast protein sorting. Cell 61:1063–1074
Salkoff L, Kelly L (1978) Temperature-induced seizure and frequency-dependent neuromuscular block in a ts mutant of Drosophila. Nature 273:156–158
Scaife R, Margolis RL (1990) Biochemical and immunochemical analysis of rat brain dynamin interaction with microtubules and organelles in vivo and in vitro. J Cell Biol 111:3023–3033
Shpetner HS, Vallee RB (1989) Identification of dynamin, a novel mechanochemical enzyme that mediates interactions between microtubules. Cell 59:421–432
Shpetner HS, Vallee RB (1992) Dynamin is a GTPase which is stimulated to high levels of activity by microtubules. Nature 355:733–735
Sigal IS, Gibbs JB, D'Alonzo JS, Temeles GL, Wolanski BS, Socher SH Scolnick, EM (1986) Mutant ras-encoded proteins with altered nucleotide binding exert dominant biological effects. Proc Natl Acad Sci USA 83:952–956
Staeheli P, Haller O, Boll W, Lindenmann H, Weissmann, C (1986) Mx protein: constitutive expression in 3T3 cells transformed with cloned Mx cDNA confers selective resistance to influenza virus. Cell 44:147–158
Tyler JM, Branton D (1980) Rotary shadowing of extended molecules dried from glycerol. J Ultrastruct Res 71:95–102
Vallee RB, Wall JS, Paschal BM, Shpetner HS (1988) Microtubule-associated protein 1C from brain is a two-headed cytosolic dynein. Nature 332:561–563
Vallee R (1992) Dynamin: motor protein or regulatory GTPase. J Musc Res Cell Motil 13:493–496

van der Bliek AM, Meyerowitz EM (1991) Dynamin-like protein encoded by the Drosophila shibire gene associated with vesicular traffic. Nature 351:411–414
Yeh E, Driscoll R, Coltrera M, Olins A, Bloom K (1991) A dynamin-like protein encoded by the yeast sporulation gene SPO15. Nature 349:713–715

Note Added in Proof

Since the submission of this manuscript, acidic phospholipids and SH3 domains of signal transducing proteins have been found to interact with dynamin and in some cases stimulate its GTPase activity (see VALLEE and SHPETNER, 1993, Nature, News and Views, in press). In addition, inhibition of transferrin uptake by mutant recombinant dynamin has also been observed by another laboratory (VAN DER BLIEK et al. J Cell Biol 122:553–563).

CHAPTER 7

Transmembrane Protein Translocation: Signal Recognition Particle and Its Receptor in the Endoplasmic Reticulum

P.J. Rapiejko and R. Gilmore

A. Introduction

The membrane-bound ribosomes of the rough endoplasmic reticulum (RER) are engaged in the synthesis of secretory proteins, resident lumenal proteins of the exocytic and endocytic membrane systems, and the majority of cellular integral membrane proteins. The information responsible for the selective delivery of ribosomes to the RER is contained in an amino terminal signal sequence. Ribosomes synthesizing proteins with RER-specific signal sequences are subsequently targeted to a membrane-bound translocation site or "translocon." The translocon is a multicomponent protein assembly that mediates the unidirectional transport of proteins or protein domains across the RER membrane. Transport of nascent polypeptide chains across the endoplasmic reticulum (ER) membrane has been proposed to occur through a proteinaceous transport site or channel (Gilmore and Blobel 1985; Simon and Blobel 1991) that may consist of several integral membrane proteins which have been identified by cross-linking to nascent polypeptides (Krieg et al. 1989; High et al. 1991; Kellaris et al. 1991; Görlich et al. 1992). Upon entry into the RER lumen, the nascent polypeptide undergoes modifications and folding reactions that result in the assembly of a mature protein. The focus of this article will be upon the roles of signal recognition particle (SRP) and SRP receptors, two multisubunit GTP-binding proteins that mediate the initial phases of the protein translocation reaction.

B. The Signal Recognition Particle and Its Receptor

A soluble ribonucleoprotein complex known as the SRP specifically binds with high affinity to those nascent polypeptides containing an appropriate signal sequence (Walter et al. 1981). The six polypeptide subunits of SRP (72, 68, 54, 19, 14, and 9 kDa) are organized into three functional domains as defined by their binding to separate regions of the 7SL-RNA (SRP-RNA) (Walter and Blobel 1982; Siegel and Walter 1988). The 54 kDa subunit of SRP (SRP54) binds to the signal sequence of a nascent polypeptide shortly after it emerges from the large ribosomal subunit (Walter and Blobel 1981; Walter et al. 1981; Krieg et al. 1986; Kurzchalia et al.

1986). High affinity binding of SRP54 to the signal sequence arrests or reduces the rate of elongation of the nascent polypeptide in vitro (WALTER and BLOBEL 1981; WOLIN and WALTER 1989).

The SRP–ribosome complex is targeted to the ER membrane via binding of the SRP to the SRP receptor, or docking protein (GILMORE et al. 1982; MEYER et al. 1982). The SRP receptor is a heterodimeric integral membrane protein composed of a 68-kDa α subunit (SRα) (GILMORE et al. 1982; MEYER et al. 1982) and a 30-kDa β subunit (SRβ) (TAJIMA et al. 1986). Targeting of the SRP–ribosome complex to its receptor results in the dissociation of the SRP from the signal sequence (GILMORE and BLOBEL 1983), membrane insertion of the nascent polypeptide, and release of the elongation arrest of translation (WALTER and BLOBEL 1981; GILMORE et al. 1982).

C. Protein Translocation Across the Rough Endoplasmic Reticulum Requires GTP

The standard in vitro protein translocation assay contains a cell-free protein translation system, mRNA for a secretory protein, and microsomal membrane vesicles. The ribonucleotide dependence of the translocation reaction was not detected until investigators incubated preassembled SRP–ribosome–nascent polypeptide complexes with microsomal membranes in the absence of nucleotides (PERARA et al. 1986). Translocation of secretory proteins and integration of membrane proteins was found to require GTP in a capacity distinct from polypeptide elongation (CONNOLLY and GILMORE 1986; HOFFMAN and GILMORE 1988). The GTP-dependent reaction step was shown to occur after targeting of the SRP–ribosome complex to the membrane, but before membrane insertion of the nascent polypeptide (CONNOLLY and GILMORE 1986). This observation provided an impetus for experiments designed to identify the relevant GTP-binding protein(s) and define their precise function. Subsequent analysis showed that the SRP receptor-mediated dissociation of the SRP from the signal sequence required GTP (CONNOLLY and GILMORE 1989). Biochemical evidence for GTP-binding sites in SRα and SRβ was provided by photolabeling experiments, and by the finding that SRα will bind GTP after denaturing gel electrophoresis (CONNOLLY and GILMORE 1989). The nonhydrolyzable GTP analogue, guanylyl-5′-imidodiphosphate (Gpp(NH)p), can replace GTP in a transport assay that contains an excess of microsomal membranes relative to SRP–ribosome complexes (CONNOLLY and GILMORE 1986; HOFFMAN and GILMORE 1988). Further experimentation showed that the SRP dissociates from the signal sequence when SRP–ribosome complexes are incubated with microsomal membranes in the presence of Gpp(NH)p. However, under these conditions, the SRP fails to subsequently dissociate from the membrane-bound SRP receptor, thereby preventing subsequent rounds of ribosome targeting (CONNOLLY et al. 1991).

A

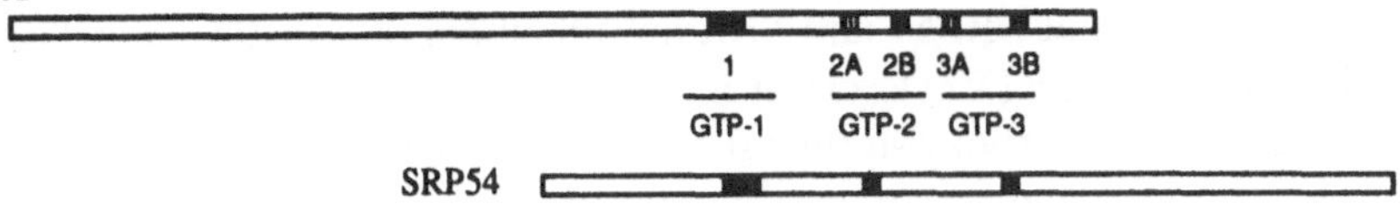

B

	GTP - 1	GTP - 2	GTP - 3
Consensus	GXXXXGK$_{S}^{T}$	DXXG	NKXD
SRα	GVNGVGKS	DTAG[a]	TKFD[a]
SRP54	GLQGSGKT	DTSG	TKLD
ftsY	GVNGVGKT	DTAG	TKLD
ffh	GLQGAGKT	DTAG	TKVD

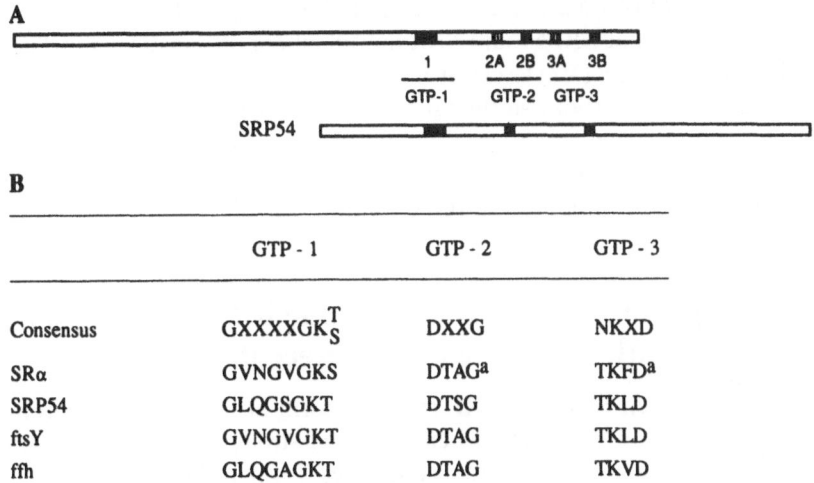

Fig. 1. A Bar diagrams for SRP54 and SRα have been aligned. The GTP-binding motifs (GTP-1, GTP-2, and GTP-3) are designated by black segments in the diagrams. **B** The consensus amino acid sequence for the GTP-binding motifs is shown above the sequences for SRα, SRP54, ftsY, and ffh. The sequence alignment is that of BERNSTEIN et al. (1989) and ROMISH et al. (1989). [a] Only the GTP-2B and GTP-3B elements are shown; the GTP-2A (DAAG) and GTP-3A (NTPD) have been omitted for clarity

Although the amino acid sequence of SRα had been reported previously (LAUFFER et al. 1985), a subsequent examination of the sequence (CONNOLLY and GILMORE 1989) revealed that it contains sequence motifs that are similar to the consensus elements found in GTP-binding proteins (DEVER et al. 1987). However, the GTP-binding consensus motifs in SRα were unusual in several respects. Although SRα contains a GTP-1 motif that matches the accepted consensus sequence (DEVER et al. 1987), two potential GTP-2 and two potential GTP-3 consensus elements were identified (Fig. 1). Neither the GTP-3A sequence (NTPD) nor the GTP-3B sequence (TKFD) perfectly matched the typical NKXD consensus sequence. This atypical feature raised several questions concerning the putative GTP-binding site in SRα. Which of the putative GTP-3 motifs in SRα participate in GTP binding? What is the significance of the apparent divergence from the NKXD consensus motif? Insight into these questions was provided by the cDNA cloning of SRP54 (BERNSTEIN et al. 1989; ROMISH et al. 1989). Examination of the SRP54 sequence revealed a GTP-binding domain (G domain) that could be aligned with the G domains in SRα and with two *Escherichia coli* proteins of unknown function, ffh and ftsY (BERNSTEIN et al. 1989; ROMISH et al. 1989). The sequence homology between SRα and SRP54 in the G domain was calculated to be 47%. The GTP-2 and GTP-3 elements in SRP54 can be aligned with the GTP-2B and GTP-3B elements in SRα, while the significance of the GTP-2A and GTP-3A elements is not clear. Conservation of threonine instead of asparagine in the GTP-3 motif as well as several other features indicate that SRP54 and SRα are members of a distinct subclass of GTP-binding proteins (Valencia et al. 1991). In addition to the G domain, SRP54 contains a carboxyl-terminal methionine-rich domain (M domain) which has been shown to contain binding sites for the signal sequence and the SRP-RNA (Zopf et al. 1990). More recently, it has been reported that SRβ contains a GTP-binding site (Ogg et al. 1992).

D. Binding and Hydrolysis of Guanine Ribonucleotides by Signal Recognition Particle and Its Receptor

Complexes between SRP and SRP receptor form at reduced ionic strength (50 m*M* KOAc) and dissociate when the ionic strength is raised (Gilmore et al. 1982). SRP–SRP receptor complexes formed in the presence of Gpp(NH)p do not dissociate when the ionic strength is raised, unlike complexes formed in the presence of GTP or GDP (Connolly et al. 1991). We hypothesize that at least one of the three GTP-binding sites is occupied with Gpp(NH)p in the ionic strength-insensitive complex. Binding of GTP to purified preparations of SRP and SRP receptor has been examined (T. Connolly and R. Gilmore, manuscript in preparation). Neither SRP nor SRP receptor bound appreciable quantities of [α-^{32}P]GTP when the purified proteins were incubated separately for 1 h. Rapid, saturable binding of labeled guanine nucleotide was detected when SRP and the SRP receptor were combined. The results were consistent with a single bound guanine nucleotide per SRP–SRP receptor complex. SRP does not hydrolyze GTP at a detectable rate (Poritz et al. 1990). SRP–SRP receptor complexes hydrolyze GTP at a rate that is approximately 40-fold greater than that shown by SRP receptor alone (T. Connolly and R. Gilmore, manuscript in preparation). The complete SRP particle is not essential for the GTP hydrolysis reaction. GTP hydrolysis occurs in reactions that contain SRP receptor plus an SRP subparticle composed of SRP54 and the *E. coli* 4.5S RNA (Poritz et al. 1990). The GTP binding and hydrolysis experiments utilized purified proteins, hence the GTPase cycle detected with this assay probably represents only one of three GTP hydrolysis cycles that function during a protein translocation reaction. A model (Fig. 2) for the GTPase cycle of the SRP–SRP receptor complex is an adaption of GTPase cycle diagrams depicted in the review by Bourne et al. (1990). Several crucial aspects of this model have not been adequately tested, so the identification of SRα as the site that exchanges GTP must be considered tentative. We speculate that both SRP54 and SRα are in the GDP-bound form when purified (Fig. 2, item a). Formation of an SRP–SRP receptor complex (item b) causes GDP to dissociate from a single GTP-binding site (item c). We suggest that SRP54 serves as a guanine nucleotide dissociation stimulator (GDS) for SRα. GTP binding to SRα (item d) stabilizes the SRP–SRP

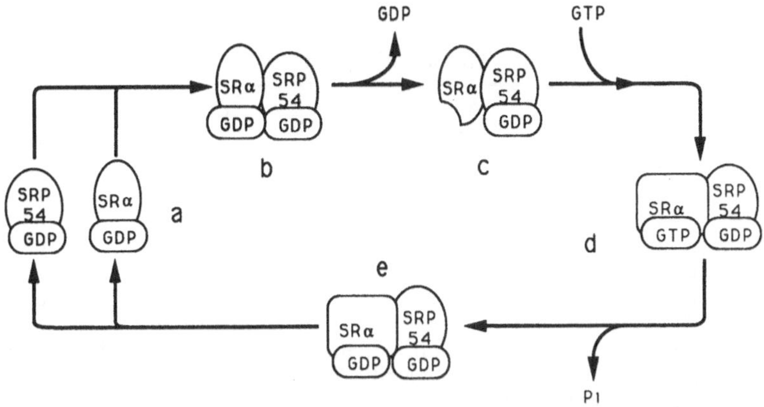

Fig. 2. A model for the GTPase cycle for SRP–SRP receptor complexes. Only SRα and SRP54 are shown. The diagrams shown here are based upon diagrams for other GTPases (Bourne et al. 1990)

receptor complex. Hydrolysis of GTP (item e) destabilizes the complex, allowing dissociation of the two proteins (item a). Hydrolysis of GTP by SRP–SRP receptor complexes was not dependent upon the addition of an exogenous GTPase activating protein (GAP).

E. Site-Directed Mutagenesis of SRα

The identification of three protein subunits in SRP and SRP receptors with GTP-binding sites indicates that our current understanding of the role of GTP in protein translocation is far from complete. To overcome the inherent complexity of this system, it was necessary to devise experimental strategies that would allow the selective analysis of each of the three putative GTP-binding protein subunits. With this goal in mind, we introduced point mutations in the GTP-binding motifs of SRα (Table 1). Several of the point mutations were analogous to those constructed previously in H-*ras* p21 that either impair or abolish GTP binding (Clanton et al. 1986; Sigal et al. 1986). The altered amino acids were in segments of SRα that should contact the triphosphate (SRα 1-7), magnesium (SRα 2-6), and guanine ring (SRα 3-1, 3-2, and 3-5) of GTP based on the structure of EF-Tu (la Cour et al. 1985) and H-*ras* p21 (Pai et al. 1990). The functional properties of the altered SRα subunits were analyzed in vitro by repopulating SRα-deficient microsomal membranes with the in vitro translated SRα. Microsomal membranes containing several of these SRα mutants were defective in protein translocation. For example, conversion of the threonine in the GTP-3B consensus motif to a lysine (SRα 3-3) eliminated translocation activity, while the conversion of the asparagine to lysine in the GTP-3A element (SRα 3-4) did not significantly impair the function of SRα (Rapiejko and Gilmore

Table 1. Summary of the mutations in SRα

Name[a]	Sequence[b]	Translocation (%)[c]	Complexes[d]	Affinity[e]
wt	Wild-type	100	yes	3.0 μM
3-1	T*E*FD	0	no	nd
3-2	*N*KFD	32	yes	3.0 mM
3-3	*K*KFD	0	no	nd
3-5	TKF*A*	0	no	nd
3A-4	*K*TPD	65	yes	nd
2-6	DTAG*Q*	66	yes	7.0 μM
1-7	GVNGVG*N*S	0	no	nd

[a] The first number for each point mutation name designates the GTP element that has been altered. Point mutation designated 3-1, 3-2, etc. are in the GTP-3B motif, while 3A-4 is in the GTP-3A motif.
[b] The altered amino acid is underlined in the second column.
[c] Translocation activity for microsomal membranes repopulated with SRα wild-type is defined as 100%. Membranes repopulated with SRα mutants have been normalized to this value.
[d] "Complexes" refers to the ability of a mutant SRP receptor to form Gpp(NH)p stabilized complexes with SRP.
[e] Affinity is the Gpp(NH)p concentration required for half maximal complex formation with SRP.

1992). In agreement with the protein alignment analysis, these data suggest that the GTP-2B and GTP-3B motifs of SRα are involved in GTP binding. Replacement of the atypical threonine in the GTP-3B element with an asparagine residue (SRα 3-2) markedly reduced the translocation activity of the repopulated membranes (Table 1). The translocation defect in SRα 3-2 is due to a 50- to 100-fold reduced affinity for GTP relative to the wild-type SRα (RAPIEJKO and GILMORE 1992). Formation of the Gpp(NH)p-stabilized SRP–SRP receptor complex was shown to require a functional GTP-binding site in SRα, suggesting that this site is occupied by Gpp(NH)p in the high affinity complex.

F. The Sorting and Targeting Functions of Signal Recognition Particle are GTP Independent

Based upon analogy to the kinetic proofreading model for EF-Tu function (THOMPSON et al. 1986), the GTP-binding site in SRP54 could control the fidelity of signal sequence recognition (BERNSTEIN et al. 1989). Alternatively, the GTP-bound form of SRP54 might have an enhanced affinity for the SRP receptor, thereby regulating the targeting reaction (RAPIEJKO and GILMORE 1992). However, data to support either hypothesis is lacking. To determine whether SRP54 requires GTP prior to contact with the SRP receptor, we have examined the nucleotide dependence of the assembly of the SRP–ribosome complexes. These experiments demonstrated that binding of

SRP to a signal sequence is neither stimulated nor inhibited by GTP or Gpp(NH)p (P.J. Rapiejko and R. Gilmore, unpublished observation). Furthermore, the presence of Gpp(NH)p during signal sequence binding did not interfere with the subsequent membrane insertion of the signal sequence. Targeting of the SRP–ribosome complex to the membrane was also found to be independent of GTP or Gpp(NH)p (P.J. Rapiejko and R. Gilmore, unpublished observation). Therefore, neither hypothesis presented above for a function of the GTP-binding site in SRP54 was validated. Taken together with the results of the SRα mutagenesis experiment, we suggest that SRP54 binds GTP after contact with SRα. Mutagenesis studies of SRP54 comparable to those described with SRα will be important in defining which reaction step requires a functional GTP-binding site in SRP54.

G. Current Models for GTP Function During Protein Translocation

A model for the function of the GTP-binding sites in SRα and SRP54 that is based in part upon the observations described above, and in part upon current models for the hydrolysis cycles of other GTP-binding proteins (Bourne et al. 1990), is shown in Fig. 3. Due to the complexity of these interlocking GTP hydrolysis cycles, several aspects of this model remain untested and should be considered speculative. First we shall consider SRα and SRP54 from the perspective of possible reaction intermediates and protein factors (GDSs and GAPs) that might control the two interlocking GTPase cycles. In analogy to other GTP-binding proteins, we propose that

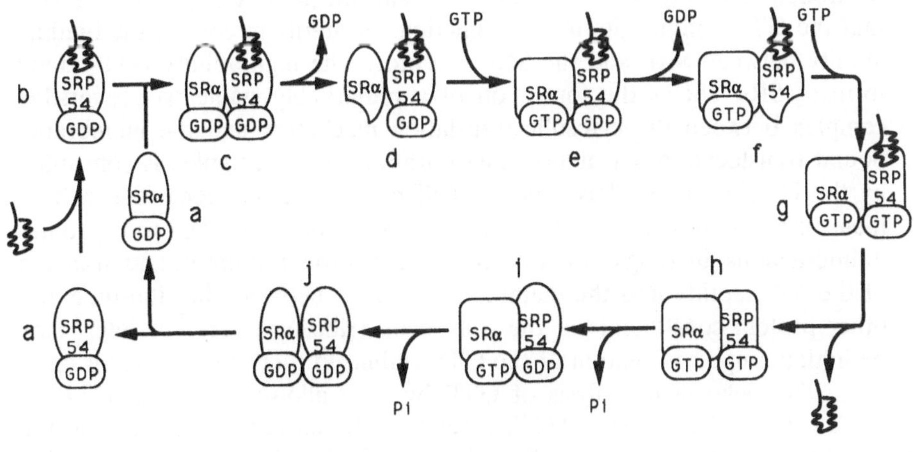

Fig. 3. A model for GTPase cycles of SRα and SRP54 in protein translocation. Only SRα and SRP54 are shown. The ribosome-bound nascent polypeptide is designated by the loop bound to SRP54 in the intermediates *b–g*

both SRP54 and SRα are in the GDP-bound form prior to initiation of a translocation reaction (Fig. 3, items a and a′). For simplicity, ribosomes bearing a nascent polypeptide are designated by the curled line adjacent to SRP54 in (item a). Binding of SRP54 to the ribosome-bound signal sequence is nucleotide independent (item b), as is the initial interaction between SRP receptor and SRP54 (item c). We suggest that SRP54 serves as a GDS for SRα (items d and e) as proposed in Fig. 2 for the GTPase cycle of the SRP–SRP receptor complex. The hypothesis that SRα exchanges nucleotide prior to SRP54 is based upon the finding that membranes repopulated with the SRα mutant 3-2 are blocked prior to SRP dissociation from the signal sequence (P.J. Rapiejko and R. Gilmore, unpublished observation). Although this reaction order is consistent with current data, an alternative model that predicts simultaneous binding of GTP to both SRα and SRP54 is not incompatible with current data (Ogg et al. 1992). We propose that the GTP-bound form of SRα is a GDS for SRP54, provided the M domain of SRP54 is bound to a signal sequence (Fig. 3, items f and g). Once both binding sites are occupied by GTP, the signal sequence dissociates from SRP54 (item h). The intermediate depicted in item h was previously trapped by inclusion of Gpp(NH)p in place of GTP (Connolly et al. 1991). Intermediates i and j (Fig. 3) are obtained by hydrolysis of GTP by SRP54 and SRα. The order of hydrolysis reactions has not been defined, and current data cannot discriminate between simultaneous and sequential hydrolysis reactions of SRα and SRP54. We suggest that hydrolysis of GTP by SRP54 is conditional, and only occurs after dissociation of the signal sequence. Likewise, hydrolysis of GTP by SRα may require that SRP54 is in the GDP-bound form.

It is likely that GTP hydrolysis in this system insures that the targeting and insertion reactions are unidirectional, by providing a mechanism for the controlled assembly and disassembly of reaction intermediates. We suggest that the GTP-binding site in SRα functions primarily to control the binding affinity between SRP and the SRP receptor. The high affinity GTP-bound form of SRα allows the formation of a remarkably stable, yet reversible complex between the cytosolic translation machinery and the membrane-bound translocation machinery. The duration of this complex is controlled by the interlocking GTPase cycles of SRα and SRP54. Membrane attachment of ribosomes engaged in translocation has been ascribed to two types of interactions: binding of the ribosome to a ribosome receptor, and insertion of the polypeptide into the membrane. Contact between the ribosome and other protein constituents of the translocon (ribosome receptor, etc.) may be initiated upon formation of the GTP-stabilized SRP–SRP receptor complex. The delayed hydrolysis of GTP by SRα allows binding of GTP by SRP54. Once SRP54 binds GTP, it releases the nascent polypeptide to the translocon. Both binding and hydrolysis of GTP by SRP54 are presumably controlled by SRα and by signal sequences. Inappropriate binding of GTP by SRP54 would result in premature dissociation of SRP from the signal sequence. Efficient insertion of the nascent polypeptide into the RER would be insured by the inherent delay in GTP hydrolysis that is a characteristic of GTP-binding proteins, so that SRP54 is unable to rebind the signal sequence prior to membrane insertion of the polypeptide. Upon completion of GTP hydrolysis by both SRP54 and SRα, dissociation of the SRP–SRP receptor complex allows both proteins to function in subsequent targeting reactions.

Although we have avoided speculation concerning the role of the GTP-binding site in SRβ due to insufficient information, it is not inconceivable that a GTPase cycle involving this subunit might control the cyclic association between the SRP receptor and the translocon. Recently, OGG et al. (1992) proposed that the GTP-binding sites in SRα might serve a similar function. OGG and colleagues proposed that SRP54 and SRα undergo simultaneous guanine nucleotide exchange reactions. Binding of GTP by these two protein subunits permits the SRP to present the nascent chain to the translocation site. In doing so, the SRP releases the nascent chain and the SRP receptor dissociates from the translocon. The subsequent simultaneous hydrolysis of GTP by SRα and SRP54 allows dissociation of the SRP and the SRP receptor. Although the two models share a number of common features, the differences between the two models indicate that a consensus view for this extraordinarily complex GTPase cycle has not yet been achieved.

References

Bernstein HD, Poritz MA, Strub K, Hoben PJ, Brenner S, Walter P (1989) The amino acid sequence of the 54 kDa subunit of the signal recognition particle suggests a model for signal sequence recognition. Nature 340:482–486

Bourne HR, Sanders DA, McCormick F (1990) The GTPase superfamily: a conserved switch for diverse cell functions. Nature 348:125–132

Clanton DJ, Hattori S, Shih TY (1986) Mutations of the ras gene product p21 that abolish guanine nucleotide binding. Proc Natl Acad Sci USA 83:5076–5080

Connolly T, Gilmore R (1986) Formation of a functional ribosome-membrane junction during protein translocation requires the participation of a GTP-binding protein. J Cell Biol 103:2253–2261

Connolly T, Gilmore R (1989) The signal recognition particle receptor mediates the GTP-dependent displacement of SRP from the signal sequence of the nascent polypeptide. Cell 57:599–610

Connolly T, Rapiejko PJ, Gilmore R (1991) Requirement of GTP hydrolysis for dissociation of the signal recognition particle from its receptor. Science 252: 1171–1173

Dever TE, Glynias MJ, Merrick WC (1987) GTP-binding proteins: three consensus sequence elements with distinct spacing. Proc Natl Acad Sci USA 84:1814–1818

Gilmore R, Blobel G (1983) Transient involvement of the signal recognition particle and its receptor in the microsomal membrane prior to protein translocation. Cell 35:677–685

Gilmore R, Blobel G (1985) Translocation of secretory proteins across the microsomal membrane occurs through an environment acessible to aqueous perturbants. Cell 42:497–505

Gilmore R, Walter P, Blobel G (1982) Protein translocation across the endoplasmic reticulum: II. Isolation and characterization of the signal recognition particle receptor. J Cell Biol 95:470–477

Görlich D, Hartmann E, Prehn S, Rapoport TA (1992) A protein of the endoplasmic reticulum involved early in polypeptide translocation. Nature 357:47–52

High S, Görlich D, Wiedman M, Rapoport TA, Dobberstein B (1991) The identification of proteins in the proximity of signal-anchor sequences during their targeting to and insertion into the membrane of the RER. J Cell Biol 113:35–44

Hoffman K, Gilmore R (1988) Guanosine triphosphate promotes the posttranslational integration of opsin into the endoplasmic reticulum. J Biol Chem 263:4381–4385

Kellaris KV, Bowen S, Gilmore R (1991) Endoplasmic reticulum translocation intermediates are adjacent to a non-glycosylated 34 kD integral membrane protein. J Cell Biol 114:21–33

Krieg UC, Johnson AE, Walter P (1989) Protein translocation across the endoplasmic reticulum: identification by photocross-linking of a 39-kD integral membrane glycoprotein as part of a putative translocation tunnel. J Cell Biol 109:2033–2043

Krieg UC, Walter P, Johnson AE (1986) Photocrosslinking of the signal sequence of nascent preprolactin to the 54 kD polypeptide of the signal recognition particle. Proc Natl Acad Sci USA 83:8604–8608

Kurzchalia TV, Wiedmann M, Girshovich AS, Bochkareva ES, Bielka H, Rapoport TA (1986) The signal sequence of nascent preprolactin interacts with the 54 kD polypeptide of the signal recognition particle. Nature 320:634–636

la Cour TFM, Nyborg J, Thirup S, Clark BFC (1985) Structural details of the binding of guanosine diphosphate to elongation factor Tu from E. coli. as studied by X-ray crystallography. EMBO J 4:2385–2388

Lauffer L, Garcia PD, Harkins RN, Coussens L, Ullrich A, Walter P (1985) Topology of signal recognition particle receptor in endoplasmic reticulum membrane. Nature 318:334–338

Meyer DI, Krause E, Dobberstein B (1982) Secretory protein translocation across membranes – the role of the "docking protein". Nature 297:647–650

Ogg SC, Nunnari JM, Miller JD, Walter P (1992) The role of GTP in protein targeting to the endoplasmic reticulum. In: Neupert W, Lill R (eds) Membrane biogenesis and protein targeting. Elsevier, Amsterdam (in press)

Pai EF, Krengel U, Petsko GA, Goody RS, Kabsch W, Wittinghofer A (1990) Refined crystal structure of the triphosphate conformation of H-ras p21 at 1.35 Å resolution: implications for the mechanism of GTP hydrolysis. EMBO J 9:2351–2359

Perara E, Rothman RE, Lingappa VR (1986) Uncoupling translocation from translation: implications for transport of proteins across membranes. Science 232: 348–352

Poritz MA, Bernstein HD, Strub K, Zopf D, Wilhelm H, Walter P (1990) An E. coli ribonucleoprotein containing 4.5S RNA resembles mammalian signal recognition particle. Science 250:1111–1117

Rapiejko PJ, Gilmore R (1992) Protein translocation across the ER requires a functional GTP binding site in the α subunit of the signal recognition particle receptor. J Cell Biol 117:493–503

Romish K, Webb J, Herz J, Prehn S, Frank R, Vingron M, Dobberstein B (1989) Homology of the 54K protein of signal recognition particle, docking protein and two E. coli proteins with putative GTP-binding domains. Nature 340:478–482

Siegel V, Walter P (1988) Each of the activities of signal recognition particle (SRP) is contained within a distinct domain: analysis of biochemical mutants of SRP. Cell 52:39–49

Sigal IS, Gibbs JB, D'Alonzo JS, Temeles GL, Wolanski BS, Socher SH, Scolnick EM (1986) Mutant ras-encoded proteins with altered nucleotide binding exert dominant biological effects. Proc Natl Acad Sci USA 83:952–956

Simon SM, Blobel G (1991) A protein-conducting channel in the endoplasmic reticulum. Cell 65:371–380

Tajima S, Lauffer L, Rath VL, Walter P (1986) The signal recognition particle receptor is a complex that contains two distinct polypeptide chains. J Cell Biol 103:1167–1178

Thompson RC, Dix DB, Karim AM (1986) The reaction of ribosomes with elongation factor Tu-GTP complexes. Aminoacyl-tRNA-independent reactions in the elongation cycle determine the accuracy of protein synthesis. J Biol Chem 261:4868–4874

Valencia A, Chardin P, Wittinghofer A, Sander C (1991) The ras protein family: evolutionary tree and role of conserved amino acids. Biochemistry 30:4637–4648

Walter P, Blobel G (1981) Translocation of proteins across the endoplasmic reticulum: III. Signal recognition protein (SRP) causes signal sequence-dependent and site-specific arrest of chain elongation that is released by microsomal membranes. J Cell Biol 91:557–561

Walter P, Blobel G (1982) Signal recognition particle contains a 7S RNA essential for protein translocation across the endoplasmic reticulum. Nature 299:691–698

Walter P, Ibrahimi I, Blobel G (1981) Translocation of proteins across the endoplasmic reticulum: I. Signal recognition protein (SRP) binds to in-vitro-assembled polysomes synthesizing secretory protein. J Cell Biol 91:545–550

Wolin SL, Walter P (1989) Signal recognition particle mediates a transient elongation arrest of preprolactin in reticulocyte lysate. J Cell Biol 109:2617–2622

Zopf D, Bernstein HD, Johnson AE, Walter P (1990) The methionine-rich domain of the 54 kD protein subunit of the signal recognition particle contains an RNA binding site and can be crosslinked to a signal sequence. EMBO J 9:4511–4517

CHAPTER 8

GTPases and Actin as Targets for Bacterial Toxins

K. AKTORIES and I. JUST

A. Introduction

It has been discussed in detail in various other chapters of this volume how multiple GTPases operate as switches and amplifiers to control signal transduction, cell motile functions, differentiation, and/or proliferation of cells. These GTPases are sensitive targets for bacterial ADP-ribosylating toxins. Studies within the last three decades have greatly increased our knowledge about the structure and functions of these ADP-ribosylating toxins and their pathogenetic role in diseases. Moreover, ADP-ribosylating toxins have proved to be invaluable tools in identifying their protein substrates (GTPases) and in studying the physiological functions of these eukaryotic regulatory proteins. The usefulness of the toxins as biochemical instruments is based on the following properties: with few exceptions, these extremely potent agents are able to enter intact cells and/or to covalently modify the target GTPase with high selectivity, thereby inhibiting or enhancing signal pathways or regulatory mechanisms which are controlled by GTPases.

The bacterial toxins can be divided into at least four groups. A first group, *Pseudomonas* exotoxin A and diphtheria toxin ADP-ribosylate elongation factor 2 (EF-2) which participates in peptide elongation in eukaryotes. A second group of toxins affect heterotrimeric G-proteins; among these are cholera toxin, the heat-labile *Escherichia coli* enterotoxins, and pertussis toxin. A third group modifies small GTPases; *Clostridium botulinum* ADP-ribosyltransferase C3 and other C3-like exoenzymes ADP-ribosylate small GTPases of the Rho family, and *Pseudomonas* exoenzyme S modifies Ras and Ras-like GTPases. A fourth group of ADP-ribosylating toxins also has to be mentioned. The target of these toxins is not a GTPase, but actin, an ATPase. Members of the fourth toxin family are *Clostridium botulinum* C2 toxin, *Clostridium perfringens* iota toxin, and various related toxins.

B. General Features of ADP-Ribosylating Toxins

The bacterial ADP-ribosylating toxins are polypeptides, and most are constructed according to the A–B model. The A subunit possesses ADP-ribosylating activity, whereas the B subunit is responsible for the binding of the toxin to the target cell. Thus, the intoxication of cells requires the holotoxin. However, the intoxication process is apparently more complicated. At least three steps appear to be involved in the toxin's action on eukaryotic cells: firstly, the binding of the toxin to a specific eukaryotic cell surface receptor; secondly, the transfer of the toxin into the target cell, a process which can be divided into the endocytotic uptake (diphtheria toxin and *Pseudomonas* exotoxin A) and the membrane translocation of the toxin (for review, see MADSHUS and STENMARK 1992); and finally, the enzyme reaction, namely the ADP-ribosylation of the GTPase. Recent reports on the atomic structure of *Pseudomonas* exotoxin A (ALLURED et al. 1986), diphtheria toxin (CHOE et al. 1992), and the heat-labile *E. coli* enterotoxins (SIXMA et al. 1991) indicate that in each case the three functions correspond to a three-domain structure of the toxin molecules. However, not all known toxins that ADP-ribosylate nucleotide-binding proteins are structured according to this model. The so-called C3-like exoenzymes that ADP-ribosylate small GTPases of the Rho family are puzzling. It appears that these bacterial enzymes only contain the catalytic subunit, since cells are poorly accessible for the exoenzymes (AKTORIES et al. 1990), although it is possible that the binding subunits have not yet been identified.

The general scheme of the mono-ADP-ribosylation reaction catalyzed by the bacterial toxins is as follows:

$$NAD^+ + \text{PROTEIN} \xrightarrow{\text{Toxin}} \text{ADP-ribosyl-PROTEIN} + \text{nicotinamide} + H^+$$

The enzyme component of the ADP-ribosylating protein toxins splits NAD^+ into ADP-ribose and nicotinamide and transfers the ADP-ribose moiety onto the GTPase. This probably occurs in a noncovalent ternary intermediate complex, as shown for diphtheria toxin (CHUNG and COLLIER 1977). In general, the equilibrium of the reaction greatly favors the reaction products under physiological conditions. However, in contrast to poly-ADP-ribosylation, in vitro the reactions are reversible in the presence of high concentrations of nicotinamide and at low pH, and result in native protein substrates (HONJO et al. 1968; CASSEL and PFEUFFER 1978; JUST et al. 1990; HABERMANN et al. 1991).

The toxins differ in respect to the acceptor amino acid modified: cholera toxin (MOSS and VAUGHAN 1977), the related *E. coli* enterotoxins (MOSS et al. 1979a), *Pseudomonas aeruginosa* exoenzyme S (COBURN et al. 1989b), and *C. botulinum* C2 toxin (VANDEKERCKHOVE et al. 1988) ADP-ribosylate arginine residues; pertussis toxin modifies cysteine (WEST et al. 1985); asparagine is the substrate for the C3-like enzymes (SEKINE et al. 1989); and diphtheria toxin and *Pseudomonas* exotoxin A ADP-ribosylate diphthamide (VAN NESS et al. 1980b), a posttranslationally modified histidine. However, the modification of the substrate proteins is clearly dependent on strictly defined structural requirements because most toxins selectively ADP-ribosylate a unique residue of the total amino acid sequence of the nucleotide-binding protein. In addition, all ADP-ribosylating toxins possess NAD glycohydrolase activity, cleaving NAD^+ into nicotinamide and ADP-ribose even in the absence of a protein substrate. However, this reaction is much slower than ADP-ribosylation and probably has no pathophysiological significance.

C. ADP-Ribosylation of Elongation Factor 2 by Diphtheria Toxin and *Pseudomonas aeruginosa* Exotoxin A

I. Introduction

EF-2 (M_r 95700) is a cytosolic GTPase that participates, together with elongation factor 1 (EF-1), in polypeptide chain elongation on ribosomes in eukaryotes (for review, see Moldave 1985). The basic reactions of the elongation cycle can be divided into three steps. Firstly, the ternary complex of the aminoacyl-tRNA with EF-1 and GTP binds to the A-site of the mRNA-programmed ribosome. This reaction is accompanied by hydrolysis of GTP (Ibuki and Moldave 1968). Secondly, the nascent peptide is transferred from the ribosomal P-site to the previously attached aminoacyl-t-RNA at the A-site. Finally, the elongated peptidyl-tRNA has to be transferred back from the A-site to the P-site. This translocation step of the peptidyl-tRNA requires the EF2–GTP complex. The reaction is coupled to the ribosome-regulated hydrolysis of GTP and results in the formation of an EF-2–GDP–ribosome complex (Lin et al. 1969; Henriksen et al. 1975; Nygard and Nilsson 1984; Nilsson and Nygard 1988). The posttranslocation ribosomes have a reduced affinity for EF-2–GDP and, subsequently, release the factor during a new cycle which starts with the interaction of the ribosome with the ternary EF-1–GTP–aminoacyl-tRNA complex.

EF-2 is homologous with other GTPases such as bacterial elongation factor Tu, elongation factor G, bacterial initiation factor 2α, and also heterotrimeric G-proteins and small GTPases of the Ras-like family (Kohno et al. 1986). The regions homologous with other GTPases are responsible for GTP binding and GTPase activity and encompass the first 160 amino acids. The C-terminal half contains the domains involved in the interaction of EF-2 with ribosomes and toxins. Diphtheria toxin and *Pseudomonas* exotoxin A have been shown to ADP-ribosylate EF-2 in this region (Honjo et al. 1968; Gill et al. 1969; Iglewski and Kabat 1991).

II. Diphtheria Toxin

Diphtheria toxin, which is produced by lysogenic *Corynebacteria diphtheriae*, the organism that causes diphtheria, is a *β-tox* phage-encoded protein of 58342 Da (mature protein) (Greenfield et al. 1983; Collier and Kandel

1971; for review, see Collier 1990). Trypsin treatment and mild disulfide bond reduction in vitro releases the enzymatically active fragment A (~21 kDa) that ADP-ribosylates EF-2, and the binding component fragment B (~37 kDa) (Collier and Kandel 1971; Drazin et al. 1971; Gill and Pappenheimer 1971). Recently, the atomic model of diphtheria toxin has been reported (Choe et al. 1992), showing it to consist of three domains, which are Y-shaped. One arm of the Y is the catalytic domain (C), which resembles the A fragment. The B fragment forms the other arm (R domain) and the base (T domain) of the Y. The R and T domains are suggested to be responsible for the cell surface recognition and the translocation of the toxin, respectively. Within the C domain a cleft is suggested to comprise the catalytic site of the toxin, with Glu-148 (Carroll and Collier 1984), His-21 (Papini et al. 1989), and Tyr-65 (Papini et al. 1991) (for detailed discussion of the structure, see Choe et al. 1992) being apparently involved in catalysis and NAD^+ binding.

III. *Pseudomonas aeruginosa* Exotoxin A

Pseudomonas aeruginosa exotoxin A is a single chain toxin (M_r 66580) (for further review, see Pastan and FitzGerald 1989; Wick and Iglewski 1990) that is produced as a 638 amino acid precursor with a 25 amino acid leader sequence at the N terminus (Gray et al. 1984). The single chain polypeptide is toxic to cells and animals but lacks in vitro ADP-ribosyltransferase activity. Either partial denaturing with urea and reduction of disulfide bonds or trypsin activation cause expression of the transferase activity (Leppla et al. 1978). Although the sites for receptor binding and enzymic activity of *Pseudomonas* exotoxin A and diphtheria toxin are located at opposite ends of the molecules – with the catalytic domain being at the C terminus of *Pseudomonas* exotoxin A and at the N terminus of diphtheria toxin – the crystalline structure resolved at 3 Å indicates a three-domain organization of the molecule (Allured et al. 1986) that shares the structural organization of diphtheria toxin (Choe et al. 1992). Domain I of *Pseudomonas* exotoxin A, which comprises the N-terminal half of the molecule, is involved in membrane binding; domain II, the central part of the molecule, is suggested to be responsible for the transfer of the toxin into the cell; and domain III, encompassing the C terminus, locates the enzyme activity (Pastan and FitzGerald 1989). Whereas the binding domain of *Pseudomonas* exotoxin A shows no homology with diphtheria toxin – both toxins appear to bind to different cell surface receptors – the enzyme component (domain III) is homologous with diphtheria toxin in amino acid sequence and steric structure. For example, an extended cleft is located in domain III that has been suggested to be the enzyme active site (Allured et al. 1986). Glu-553, which is located in the cleft of domain III, has been identified as the nicotinamide subsite of NAD binding (Carroll and Collier 1988) and is probably homologous to Glu-148 in diphtheria toxin (Carroll et al. 1985).

IV. Functional Consequences of the ADP-Ribosylation of Elongation Factor 2

Diphtheria toxin and *Pseudomonas* exotoxin A ADP-ribosylate EF-2 (HONJO et al. 1968; IGLEWSKI and KABAT 1991; GILL et al. 1969; for review, see PERENTESIS et al. 1992) in diphthamide (2-[3-carboxyamide-3-(trimethylammonio)propyl]histidine), a posttranslational modification of His-715 (VAN NESS et al. 1980a). At least six different enzymes are involved in the synthesis of diphthamide (BODLEY and VELDHAM 1990), the physiological function of which is still unclear, but which is apparently essential for the function of EF-2 (OMURA et al. 1989). This modified amino acid has only been found in the EF-2 protein in eukaryotes, and is not present in prokaryotes with the exception of archebacteria (KESSEL and KLINK 1980).

ADP-ribosylation reduces the affinity of EF-2 for ribosomes in the pretranslocation state by 2 orders of magnitude (NYGARD and NILSSON 1990) whereas the posttranslational binding of EF-2 in the presence of GDP is unaltered (BERMEK 1976; NYGARD and NILSSON 1985). The modification does not inhibit GTP-binding to EF-2, however, the ribosome-stimulated GTPase activity is inhibited (RAEBURN et al. 1986). Most important, it has been reported that the ADP-ribosylated elongation factor 2 is unable to promote translocation within the ribosome (DAVYDOVA and OVCHINNIKOV 1990). Therefore, the toxin-catalyzed ADP-ribosylation renders the elongation factor biologically inactive resulting in inhibition of polypeptide synthesis and subsequent cell death (HONJO et al. 1968; COLLIER 1968).

D. ADP-Ribosylation of G-Proteins

I. Introduction

In contrast to EF-2, G-proteins are heterotrimers consisting of a GTP-binding and hydrolyzing α subunit (39–52 kDa), one β subunit (35–36 kDa), and one γ subunit (7–10 kDa). G-proteins are the regulatory components of signal transduction pathways, and are reviewed in detail in other chapters of this volume.

II. Cholera Toxin

Cholera toxin, which is produced by *Vibrio cholerae*, is the principle cause of the watery diarrhea of cholera. The toxin stimulates adenylyl cyclase and increases cellular cAMP levels (and possibly additional factors), resulting in the alteration of electrolyte fluxes and fluid accumulation in the intestinal lumen (GILL 1977; FIELD 1980). Cholera toxin is composed of one A subunit and five B subunits (GILL 1976, 1977; FISHMAN 1990). Component A (M_r 27215) is proteolytically cleaved into two polypeptides, A_1 (~21 kDa) and A_2 (6 kDa), which are linked by a disulfide bridge (Moss et al. 1979b). A_1 is the enzyme component and possesses ADP-ribosyltransferase activity (Moss et al. 1979b; GILL and RAPPAPORT 1979), whereas the A_2 fragment may mediate the interaction of A_1 with the B subunits (GILL 1976, 1977; RIBI et al. 1988). Protomer B consists of five identical subunits with M_r 11677 that are linked by noncovalent bonds (GILL 1976, 1977). The B subunits, which are responsible for toxin binding, are apparently arranged in a pentameric ring-like structure containing the A_2 subunit. The A_1 subunit is believed to be located below the plane of the ring formed by the B protomers (GILL 1977; RIBI et al. 1988). Cholera toxin binds multivalently and with high affinity (1 n*M*) via the B subunits to the monosialoganglioside G_{M1} of the eukaryotic cell surface (VAN HEYNINGEN et al. 1971; FISHMAN et al. 1978), thereby allowing the hydrophobic domains of the A subunit to penetrate into the plasma membrane, where reduction of the disulfide bond between A_1 and A_2 induces the release of the enzymic A_1 subunit (RIBI et al. 1988; DWYER and BLOOMFIELD 1982; LUDWIG et al. 1986). The free A_1 subunit then ADP-ribosylates its substrates.

III. Heat-Labile *E. coli* Enterotoxins

Two major *E. coli* enterotoxin serotypes, including various subtypes, have been described that resemble cholera toxin in structure and function (DALLAS and FALKOW 1980; R.K. HOLMES et al. 1986; MOSS and RICHARDSON 1978). These toxins are suggested to be, at least in part, etiologically responsible for "traveller's diarrhea." *E. coli* enterotoxin type I is about 80% identical with cholera toxin, and similar in ganglioside binding and immunoreactivity (FUKUTA et al. 1988; YAMAMOTO et al. 1984, 1987). The homology of *E. coli* enterotoxin II on the nucleotide sequence level with cholera toxin is about 62% in A_1, 36% in A_2, and undetectable in the B subunit (PICKETT et al. 1987). Recently, the crystal structure of the enterotoxin toxin I from *E. coli* has been reported (SIXMA et al. 1991), confirming the high homology with cholera toxin. Furthermore, these studies revealed a significant structural homology of the enzymatic domain of the A subunits of the latter toxins with diphtheria toxin and *Pseudomonas exotoxin* A (SIXMA et al. 1991), although no sequence homology is detectable among the toxins. For example, Glu-112 of the A_1 fragment of the *E. coli* enterotoxin type I (and of cholera toxin), which has been shown by mutagenesis studies to be essential for activity (TSUJI et al. 1991), functionally corresponds to Glu-553 in *Pseudomonas* exotoxin A and Glu-148 in diphtheria toxin (SIXMA et al. 1991).

IV. Functional Consequences of the ADP-Ribosylation of G-Proteins by Cholera- and Heat-Labile *E. coli* Enterotoxins

GILL (1975) showed that the action of cholera toxin requires NAD^+. In 1978, several groups independently reported on the ADP-ribosylation of membrane proteins as the biochemical basis of adenylyl cyclase stimulation by cholera toxin (CASSEL and PFEUFFER 1978; JOHNSON et al. 1978; GILL and MEREN 1978), which were thereafter identified as the regulatory G_s-protein (Northup et al. 1980) (for early review, see Ross and GILMAN 1980). Similarly, *E. coli* heat-labile enterotoxins increase adenylyl cyclase activity by ADP-ribosylation (Moss and RICHARDSON 1978; CHANG et al. 1987). Quite early it was reported that several membrane and/or cytosolic factors increase cholera toxin-catalyzed ADP-ribosylation. The hypothesis of a second factor was substantiated with the purification of ARF (ADP-ribosylation factor) in 1984 (KAHN and GILMAN 1984a) and its identification as a GTPase (KAHN and GILMAN 1986). ARF proteins comprise a family of small GTPases with at least six mammalian members of about 180 amino acids (M_r ~20 kDa) and 60%–95% identity (SERVENTI et al. 1992; BOBAK et al. 1989) that have an extremely low GTPase activity (KAHN and GILMAN 1986) (for further details, see Chaps. 93 and 95). Originally, it was proposed that ARF interacts with G_s (KAHN and GILMAN 1986). However, because ARF stimulates all ADP-ribosyltransferase activities of cholera toxin as well as of *E. coli* toxins (for example, ARF increases the ADP-ribosylation of simple guanidino compounds and even enhances the NAD glycohydrolase activity of cholera toxin, NODA et al. 1990; TSAI et al. 1988), the following sequence is more likely: the GTP-bound form of ARF interacts with the thiol-cleaved A_1 subunit of cholera toxin, thereby greatly increasing the activity of the toxin. Thereafter, ARF-activated cholera toxin ADP-ribosylates G_s, which is then itself activated by binding of GTP (SERVENTI et al. 1992; BOBAK et al. 1990).

Cholera and *E. coli* enterotoxins ADP-ribosylate the GTP-binding α subunit of G_s (KAHN and GILMAN 1984b). Further substrates are the α subunits of G_t (transducin) (ABOOD et al. 1982; NAVON and FUNG 1984), which stimulate the retinal phosphodiesterase and G_{olf}, the olfactory G_s that is involved in the activation of olfactory adenylyl cyclase (JONES et al. 1990). The acceptor amino acid for the cholera toxin catalyzed ADP-ribosylation has been determined to be arginine-174 in G_t (VAN DOP et al. 1984). This amino acid corresponds to Arg-187, Arg-188, Arg-201, and Arg-202 of the four $G\alpha_s$-splicing variants (KAZIRO et al. 1991). CASSEL and SELINGER (1977) first proposed that the ADP-ribosylation of G_s activates $G\alpha_s$ by inhibition of its inherent GTPase activity. This view is in agreement with recent mutational studies showing that Arg-187 is crucial for the $k_{cat,GTP}$ of the G_s-protein. Exchange of Arg-187 with Ala, Glu, or Lys reduced the endogenous GTPase activity by about 100-fold. Interestingly, ADP-ribosylation by cholera toxin was greatly reduced, but not completely inhibited with these mutants, a finding possibly indicating that the toxin modifies neighboring arginines if the favored acceptor amino acid is not present (FREISSMUTH and GILMAN 1989). Furthermore, in certain human growth hormone-secreting pituitary tumours, mutations of $G_{s\alpha}$ have been observed that possess Cys or His instead of Arg-187 which, like ADP-ribosylation of Arg-187, show decreased GTPase activity and constitutive activation of adenylyl cyclase

(Landis et al. 1989). Further consequences of the ADP-ribosylation of G_s proteins by cholera toxins are the enhancement of GDP-release from G_s (Murayama and Ui 1984) and the promotion of the dissociation of its α and β–γ subunits by decreasing the affinity of the α subunit for β–γ (Kahn and Gilman 1984b). All these effects are in agreement with the activation of G_s and the increase in adenylyl cyclase activity by cholera toxin. On the other hand, it has been reported that the ADP-ribosylation of G_s by cholera toxin marks the protein for accelerated degradation (Chang and Bourne 1989). This effect appears not to be relevant for the activation of adenylyl cyclase.

Recently, it was reported that cholera toxin also ADP-ribosylates G_i and G_o proteins, an effect that was much greater with the receptor-activated G-proteins (Gierschik and Jakobs 1987; Iiri et al. 1989; Owens et al. 1985; Klinz and Costa 1989). Furthermore, considering that G_s is involved in the regulation of various effector systems other than adenylyl cyclase (Birnbaumer 1990) (see the respective chapters), further studies are required to elucidate the effects of cholera toxin in these systems.

V. Pertussis Toxin

Pertussis toxin, formerly also called islet-activating protein, histamine-sensitizing factor, lymphocytosis-promoting factor, or pertussigen, is one of the exotoxins produced by *Bordetella pertussis*, the causative organism of whooping cough. Pertussis toxin is a real A–B toxin of about 105 kDa with the enzymic component S1 (M_r 26220) and a binding protomer B that consists of the subunits S2 (M_r 21920), S3 (M_r 21860), S4 (M_r 12060), and S5 (M_r 10940) (Locht and Keith 1986). The binding component comprises the two dimers S2–S4 and S3–S4 and the subunit S5. The enzyme component (S1) of pertussis toxin exhibits significant homology with the A_1 subunit of cholera toxin and the heat-labile *E. coli* enterotoxin (Capiau et al. 1986; Nicosia et al. 1986). On the basis of mutagenesis studies and by photolabeling with NAD^+, it has been suggested that Glu-129 resembles the glutamic acid (Barbieri et al. 1989; Locht et al. 1989) that is crucial for the catalytic activities of diphtheria toxin and *Pseudomonas* toxin, again showing functional homology among the ADP-ribosylating toxins (for further references and discussion of this topic, see Rappuoli and Pizza 1991).

VI. ADP-Ribosylation of G_i, G_o, and G_t by Pertussis Toxin

Pertussis toxin was the key tool used to elucidate the bidirectional regulation of adenylyl cyclase, and allowed the G_i-protein to be identified and purified. To date, at least eight mammalian G-proteins ($G_{i1,2,3}$;$G_{o1,2}$;$G_{t1,2}$;G_{43}) that serve as substrates for ADP-ribosylation by pertussis toxin have been described (for review, see Gierschik 1992). The toxin catalyzes the ADP-ribosylation of a cysteine four residues from the carboxy-terminal end of the α subunit of the sensitive G-proteins (West et al. 1985). No additional factor is necessary for the modification by pertussis toxin. However, the ADP-ribosylation of G-proteins by pertussis toxin depends on the presence of $\beta-\gamma$ subunits. It appears that the intact heterotrimer is the favored substrate for ADP-ribosylation. ADP-ribosylation has no effect on basal guanine nucleotide binding and/or hydrolyzing properties of isolated G-proteins. However, multiple studies have shown that ADP-ribosylated G-proteins lose their ability to interact with the receptor (for review, see Ui 1990). Thus, ADP-ribosylation blocks the specific receptor-mediated pathway, which is why pertussis toxin is so useful in defining signal transduction pathways involving sensitive G-proteins.

E. ADP-Ribosylation of Small GTPases

I. Introduction

Recently, various small GTPases have been described that are regulated by GTPase cycles. The best studied example of these small GTPases is the Ras protein, which has been suggested as being involved in proliferation and/or differentiation of cells, and may have a role in cell transformation (see Chaps. 17, 18 and 20; recent reviews by Barbacid 1987, 1990). To date, more than 50 small GTPases have been identified that are especially homologous in four regions involved in GTP binding and/or GTPase activity (Bourne et al. 1990, 1991; Kaziro et al. 1991). Some of these regulatory proteins have been shown to be substrates for ADP-ribosylation by bacterial ADP-ribosyltransferases.

II. C3-Like ADP-Ribosyltransferases

Several bacterial exoenzymes that modify the GTP-binding Rho proteins have been described. Members of this family of transferases are *Clostridium botulinum* exoenzyme C3 (Aktories et al. 1987, 1988b; Rubin et al. 1988; for review, see Aktories et al. 1990, 1992a), *Clostridium limosum* exoenzyme (Just et al. 1992a), EDIN from *Staphylococcus aureus* (Inoue et al. 1991; Sugai et al. 1990), and the recently described exoenzyme produced by *Bacillus cereus* (Just et al. 1992b). The tranferase C3 is produced by certain strains of *Clostridium botulinum* type C and D that also produce botulinum neurotoxins C and D, as well as C2 toxin. In fact, C3 is encoded by phages that also encode for the structural gene of botulinum neurotoxins (Popoff et al. 1990, 1991). However, the botulinum neurotoxins are neither structurally, immunologically, nor functionally related to C3 (Aktories and Frevert 1987; Rösener et al. 1987). C3 is a basic (pI > 10), heat-stable (1 min, 95°C), and trypsin-resistant protein. At least two variants of C3 (A and B) have been described. The C3 exoenzyme A is produced as a polypeptide of 251 amino acids (27 823 Da) including a 40 amino acid signal peptide. The mature C3 exoenzyme is comprised of 211 amino acids with M_r of 23 546 (POPOFF et al. 1991). NARUMIYA and coworkers sequenced a C3 transferase (C3B) from a different *Clostridium botulinum* strain, revealing a protein of 244 amino acids (27 362 Da) with a leader sequence of 40 amino acids which was about 40% different in its amino acid sequence compared with C3 A (NEMOTO et al. 1991). Furthermore, JUST et al. purified an exoenzyme (~25 kDa) produced by *Clostridium limosum* which is closely related to C3 (JUST et al. 1992a). Although the precise sequence of the *Clostridium limosum* enzyme is still not known, partial amino acid sequence analysis of the *Clostridium limosum* exoenzyme revealed about 70% identity with C3 A. C3-like enzymes are not only restricted to clostridia. It has been shown that certain strains of *Staphylococcus aureus* produce an about 25 kDa Rho–ADP-ribosyltransferase that reportedly acts as an epidermal cell differentiation inhibitor and was therefore called EDIN (INOUE et al. 1991; SUGAI et al. 1992b). EDIN is produced as a 247 amino acid precursor protein having a 35 amino acid leader sequence. The mature protein (EDIN) of 212 amino acids shares about 35% homology (amino acid level) with C3 (INOUE et al. 1991). Moreover, JUST et al. recently described an about 28 kDa protein produced by *Bacillus cereus* that ADP-ribosylates Rho proteins. Polyclonal anti-C3 or anti-*C. limosum* exoenzyme antibody does not cross-react with the *B. cereus* exoenzyme, indicating that it is another more distantly related subtype of this transferase family (JUST et al. 1992b). Recently, the NAD-binding site of the C3-like exoenzymes has been determined by ultraviolet irradiation of *C. limosum* exoenzyme in the presence of [^{14}C-carbonyl]NAD showing a specific labelling of a glutamic acid residue corresponding to Glu-174 of C3 (M. JUNG, I. JUST, J. VAN DAMME, J. VANDEKERCKHOVE, K. AKTORIES, manuscript submitted).

All these transferases selectively ADP-ribosylate Rho proteins (BRAUN et al. 1989; AKTORIES et al. 1989; NARUMIYA et al. 1988; KIKUCHI et al. 1988; CHARDIN et al. 1989; JUST et al. 1992a; SUGAI et al. 1992). Rho proteins are present in all eukaryotic cells (MADAULE and AXEL 1985). At least three different Rho proteins (RhoA,B,C) of 193 (RhoA,C) and 196 (RhoB) amino acids with homology of about 85% with each other have been described in mammalian cells (CHARDIN et al. 1988; YERAMIAN et al. 1987). Rho proteins are about 35% homologous with Ras proteins. As is known for other GTPases the activity of the Rho proteins is determined by a regulatory GTPase cycle. At least three proteins have been described that are involved in the control of the GTPase cycle of Rho. The guanine nucleotide dissociation stimulator (GDS) increases GDP–GTP exchange (HIRAOKA et al. 1992; KAIBUCHI et al. 1991), while a guanine nucleotide dissociation inhibitor (GDI) (FUKUMOTO et al. 1990; UEDA et al. 1990) inhibits nucleotide exchange. Furthermore, a GTPase stimulatory protein (GAP) (GARRETT et al. 1989; MORII et al. 1991) stimulates the endogenous GTPase activity severalfold, thereby turning off the active state of the small GTPases (for details, see the respective chapters in this volume).

In contrast to cholera toxin and *Pseudomonas* exoenzyme S, no additional factors are necessary for the ADP-ribosylation of Rho proteins by C3-like enzymes. Also recombinant, non-polyisoprenylated Rho proteins are substrate for ADP-ribosylation. However, it appears that some cytosolic factors (OHTSUKA et al. 1989) and/or detergents greatly increase the ADP-ribosylation reaction, especially of non-polyisoprenylated Rho (JUST et al. 1992a). It has been shown that RhoA is ADP-ribosylated at asparagine-41 (SEKINE et al. 1989), and most likely Rho B and C are also modified at the same amino acid. The ADP-ribose–asparagine bond formed by C3 is particularly stable against neutral hydroxylamine and $HgCl_2$ (AKTORIES et al. 1988a), which cleave the ADP-ribose–arginine and ADP-ribose–cysteine bonds, respectively, formed by cholera toxin or pertussis toxin (MAYER et al. 1988).

III. Functional Consequences of the ADP-Ribosylation of Rho Proteins

So far the physiological role of the Rho proteins and the consequences of their modification by C3-catalyzed ADP-ribosylation are not completely understood. Studies are hampered by the fact that C3 is not easily accessible to the interior of cells, and high exoenzyme concentrations or cell permeabilization techniques are necessary. By using the osmotic shock method, it has been shown with 3T3 cells that C3 induces the rounding up of cells, formation of cell processes, and occurrence of binucleated cells (RUBIN et al. 1988). In PC12 cells, C3 causes the formation of neurite-like processes (RUBIN et al. 1988). In Vero cells, morphological changes are accompanied by the destruction of the microfilament network while the microtubule system remains intact (CHARDIN et al. 1989). In hepatoma cells, C3 additionally causes the redistribution of intermediate filaments, an effect possibly secondary to microfilament destruction (WIEGERS et al. 1991). Further insights into the role of Rho proteins and their modification by C3 was achieved by microinjection of Val-14 RhoA protein into Swiss 3T3 cells (PATERSON et al. 1990). This Rho mutant is characterized by inhibition of its endogenous GTPase activity, which makes the GTP-binding protein persistently active. Microinjection of Val-14 Rho into nonconfluent Swiss 3T3 cells causes rapid changes in cell morphology and the formation of finger-like processes containing large amounts of microfilaments. Microinjection into confluent and serum-starved Swiss 3T3 cells, which contain a rather scarce microfilament cytoskeleton, induces the formation of focal adhesions and of stress fibers within 10 to 15 min. Prior ADP-ribosylation of the Rho protein inhibits these effects (PATERSON et al. 1990; RIDLEY, HALL 1992). These data suggest that Rho proteins are involved in the regulation of the microfilament network and that ADP-ribosylation of Rho renders the GTPase biologically inactive. In agreement with this notion is the observation that C3 introduced into neutrophils by electro-permeabilization inhibits chemotaxis (Stasia et al. 1991). Recently, it has been shown that ADP-ribosylation of RhoA inhibits lymphocyte-mediated cytotoxicity (Lang et al. 1992). In bovine retina membranes, the ADP-ribosylation of Rho and/or Rho-like 21-kDa proteins by C3 in the presence of GTP[S] is reduced by light exposure (Wieland et al. 1990a,b). These data were interpreted as indicating that GTPases which are substrates for C3 also interact with the G-protein-regulated rhodopsin receptor in the retina. Whether the above described effects are somehow related to the regulatory function of Rho in cytoskeleton organization is not clear. Moreover, it has been reported that the Rho-ADP-ribosylating exoenzyme EDIN, which is produced by *S. aureus*, disassembles the Golgi apparatus in Vero cells and mimics the action of brefeldin A suggesting that Rho protein may be involved in membrane trafficking between golgi and the endoplasmatic reticulum (Sugai et al. 1992a).

The proposed inactivation of Rho proteins by ADP-ribosylation is apparently not caused by an effect on the endogenous GTPase cycle of Rho. The basal and GAP-stimulated GTPase single cycle GTPase activity of Rho is not affected by ADP-ribosylation (Paterson et al. 1990). In contrast, ADP-ribosylation increased the steady state GTPase activity of the recombinant RhoA and RhoB proteins in vitro by about 50%–100% (Mohr et al. 1992). This effect is rather small compared with the severalfold stimulation of GTPase activity by the Rho GAP protein (Garrett et al. 1989), and may be based on a small increase in GDP–GTP exchange of the ADP-ribosylated Rho protein. The apparent biological inactivation of Rho by ADP-ribosylation might be explained by inhibition of the interaction of ADP-ribosylated Rho with a putative effector. By analogy with Ras, Rho proteins are ADP-ribosylated in the so-called effector region (amino acids 32–42 in Ras) of small GTPases (Sekine et al. 1989; Pai et al. 1989). Provided that Ras and Rho are sterically homologous, steric hindrance of the interaction of Rho with its effector would be an explanation for the biological inactivation of Rho.

At least two other groups of small GTPases – CDC42Hs (also known as G_p, G-25K protein) (Shinjo et al. 1990) and Rac1,2 proteins (Didsbury et al. 1989; Polakis et al. 1989) – have asparagine, the ADP-ribose acceptor of Rho, at the same position as Rho proteins. Although Rac proteins have been postulated (Didsbury et al. 1989) to be substrates for ADP-ribosylation by C3, recent studies indicate that this family of small GTPases is rather poorly modified by C3-like enzymes. Only in the presence of 0.01% sodium dodecyl sulfate, C3 but not other C3-like enzymes modifies Rac1 protein maximally by about 10% (Just et al. 1992a). CDC42Hs protein is not a substrate for ADP-ribosylation under the same conditions. Although the possibility cannot be excluded that additional cellular factors, not known at present, may allow a significant ADP-ribosylation of Rac proteins, it appears that only RhoA,B, and C are the specific cellular substrates for ADP-ribosylation by C3-like enzymes.

IV. ADP-Ribosylation of Small GTPases by *Pseudomonas aeruginosa* Exoenzyme S

Many strains of *Pseudomonas aeruginosa* produce (besides exotoxin A) another ADP-ribosyltransferase called exoenzyme S, which shows no immunological cross-reactivity with the former transferase (IGLEWSKI et al. 1978; COBURN 1992). Exoenzyme S is secreted as a 53-kDa or 49-kDa protein. The 53-kDa protein is most likely the enzymatically inactive precursor of the active 49-kDa transferase. Exoenzyme S ADP-ribosylates several proteins in arginine (COBURN et al. 1989b). For example, exoenzyme S ADP-ribosylates the intermediate filament component vimentin and its fragments, a modification that inhibits the association of vimentin molecules to form filaments (COBURN et al. 1989a). Recently, it was demonstrated that exoenzyme S preferentially modifies GTPases. Substrates of the transferase are H-Ras, K-Ras, Rap1A, Ral, Rab3, and Rab4 (COBURN et al. 1989b). In contrast, Rho proteins, Rab1, ARF, and CDC42Hs (G_p, G-25K) are not modified by exoenzyme S. Similar to observations made for ADP-ribosylation by cholera toxin, the modification of GTPases by exoenzyme S depends on the presence of a cytosolic factor (COBURN et al. 1991). This factor activating exoenzyme S (FAS) has a M_r of about 28000 and belongs to a highly conserved, widely distributed eukaryotic protein family which members are known as 14-3-3 proteins (FU et al. 1993). So far studied, ADP-ribosylation of Ras proteins by exoenzyme S neither modifies the GTP-binding, GTP/GDP exchange or GTP hydrolysis by Ras proteins nor the interaction of Ras with Ras GTPase activating protein (RasGAP) or with the Ras guanine nucleotide dissociation stimulator (RasGDS) (COBURN 1992). Furthermore, it is not known whether the small GTPases mentioned above are pathophysiological substrates for ADP-ribosylation by *Pseudomonas* exoenzyme S in intact cells.

F. ADP-Ribosylation of Actin

I. Introduction

ADP-ribosylation catalyzed by bacterial protein toxins is not only observed with eukaryotic GTPases, but also with the ATPase actin (AKTORIES et al. 1986b; for review, see AKTORIES et al. 1990, 1992b; AKTORIES and WEGNER 1989; CONSIDINE and SIMPSON 1991). Actin is one of the most abundant proteins in eukaryotic cells. It is basically involved in the architecture of cells and, beside its functions in muscle cells, actin is important for various cellular motile functions such as cell migration, phagocytosis, endocytosis, and secretion in nonmuscle cells (BERSHADSKY and VASILIEV 1988). Actin consists of a single polypeptide chain with 375 residues that binds ATP or ADP at a high affinity. The physiological functions of actin largely depend on its ability to polymerize and to form filaments (see Chap. 5). These filaments are polar structures having two nonequivalent ends. At one end ("barbed end"), actin filaments have a higher affinity for monomeric actin than at the other end ("pointed end"). Therefore, actin filaments tend to polymerize faster at the barbed end than at the pointed end (Selve and Wegner 1986; Wegner 1976). Actin possesses ATPase activity, and polymerization of actin is accompanied by an increase in ATP hydrolysis (Brenner and Korn 1980). Thus, comparable with GTPases (especially with the GTP-binding tubulin, see Chap. 5), formation of protein complexes is regulated by nucleotide hydrolysis.

II. *Clostridium botulinum* C2 Toxin

Actin-ADP-ribosylating *Clostridium botulinum* C2 toxin is binary in structure and consists of a binding component and an enzymic component which are separate proteins (Ohishi et al. 1980). The binding component (C2II) of C2 toxin ($M_r \sim 100\,000$) has to be activated by trypsin treatment, which cleaves it to the active fragment of 74 000 (Ohishi 1987). The component binds to the cell surface, thereby inducing or demasking a binding site for the catalytic component C2I, which is then transferred into the cells, most likely by endocytosis (Simpson 1989). The enzyme component has a molecular weight of about 45 000 and ADP-ribosylates monomeric G-actin, but not polymerized F-actin (Aktories et al. 1986a,b). C2 toxin modifies actin at Arg-177 (Vandekerckhove et al. 1988). Although all vertebrate actin isoforms have arginine at this identical position (Vandekerckhove and Weber 1979), the toxin ADP-ribosylates cytoplasmic β–γ-actin and γ-smooth muscle actin, but not the α-actin isoforms from skeletal, cardiac, and smooth muscle (Mauss et al. 1990).

III. Other Actin-ADP-Ribosylating Toxins

In addition to toxin, various other actin-ADP-ribosylating toxins have been described (Aktories et al. 1992b; Considine and Simpson 1991) such as *Clostridium perfringens* iota toxin (Schering et al. 1988; Simpson et al. 1987), *Clostridium spiroforme* toxin (Simpson et al. 1989; Popoff et al. 1989), and *Clostridium difficile* ADP-ribosyltransferase (Popoff et al. 1988), which is clearly distinct from *Clostridium difficile* toxins A and B. All these toxins are binary in structure. Iota toxin is comprised of a ~47-kDa enzyme component and a ~71-kDa binding component. The enzyme component of *Clostridium spiroforme* is apparently heterogeneous, with M_r of 43 000–47 000. The ADP-ribosyltransferase of *Clostridium difficile* has a M_r of 43 000 (Popoff et al. 1988). So far, no binding component for this toxin has been found in *Clostridium difficile*; however, the binding components of *Clostridium spiroforme* and of *Clostridium perfringens* iota toxin are able to transport the *Clostridium difficile* enzyme into the cell. Apparently, all these toxins ADP-ribosylate actin in Arg-177 (VANDEKERCKHOVE et al. 1987; POPOFF et al. 1988). The toxins produced by *Clostridium spiroforme*, *Clostridium perfringens*, and *Clostridium difficile*, but not C2 toxin, are immunologically related (SIMPSON et al. 1989; POPOFF et al. 1988, 1989). Furthermore, the binding components of these toxins are interchangeable, whereas this does not extend to the binding component of *Clostridium botulinum* C2 toxin. Finally, in contrast to C2 toxin, iota toxin is able to modify all mammalian actin isoforms including α-skeletal, α-cardiac, and α-smooth muscle actin (SCHERING et al. 1988; MAUSS et al. 1990).

IV. Functional Consequences of the ADP-Ribosylation of Actin

C2-toxin-induced ADP-ribosylation causes gross alteration of actin functions. For example, the covalent modification of Arg-177 inhibits actin-ATPase activity (GEIPEL et al. 1989, 1990). Moreover, ADP-ribosylated actin loses its ability to polymerize (AKTORIES et al. 1986a,b; SCHERING et al. 1988). The ADP-ribose acceptor (Arg-177) is located at or near a subunit contact site of polymerized actin (HOLMES et al. 1990; KABSCH et al. 1990). Therefore, it is feasible that the bulky ADP-ribose group incorporated in G-actin inhibits actin polymerization by steric hindrance. This hypothesis explains why polymerized F-actin is a poor substrate for ADP-ribosylation (AKTORIES et al. 1986b), because the ADP-ribose acceptor amino acid is buried in the actin filament. It can be deduced from the recently reported atomic model of actin filaments (KABSCH et al. 1990) that the only position where F-actin can accept ADP-ribosylated actin is the barbed end of actin filaments. This view is in agreement with the observation that ADP-ribosylated actin behaves like a capping protein and binds to the fast growing ends ("barbed end") of actin filaments (WEIGT et al. 1989; WEGNER and AKTORIES 1988). The interaction of the ADP-ribosylated actin with the actin filament inhibits further polymerization of nonmodified actin at this site. In contrast, ADP-ribosylated actin does not interact with the pointed end of actin filaments (WEIGT et al. 1989). At the pointed end, the actin filaments preferentially depolymerize and the G-actin released is then substrate for ADP-ribosylation. This sequela may be important for the depolymerizing activity of the toxins in intact cells. On the other hand, in intact nonmuscle cells, the equilibrium between monomeric and polymerized actin is controlled by various actin-binding proteins such as gelsolin, profilin, vinculin, β-thymosin, and others (for review, see POLLARD and COOPER 1986; VANDEKERCKHOVE 1990; BERSHADSKY and VASILIEV 1988; HARTWIG and KWIATKOWSKI 1991). Therefore, it is of interest that the ADP-ribosylating toxins also modify actin in complexes with gelsolin, an actin-binding protein that functions as a severing, capping, and nucleating protein. ADP-ribosylation of the gelsolin–actin complex inhibits its nucleation activity, an action which may participate in the cytopathic effects of the toxins (WILLE et al. 1992).

G. Perspectives

During recent years various ADP-ribosylating toxins have been described which are important pathogenic factors. Surprisingly, all transferases identified so far modify nucleotide-binding proteins and all the toxins (with the exception of actin, which is an ATPase) ADP-ribosylate GTPases. The atomic structures of some of these toxins – *Pseudomonas* exotoxin A (Allured et al. 1986), diphtheria toxin (Choe et al. 1992), and heat-labile *E. coli* enterotoxin (Sixma et al. 1991) – have been analyzed and show a similar structural organization. These data are corroborated by NAD binding and mutational studies showing that the homology among the toxins has to be extended to cholera and pertussis toxins. All these data suggest a common ancestral origin of the bacterial ADP-ribosylating toxins.

Various studies reported recently indicate that the bacterial ADP-ribosylating toxins have eukaryotic counterparts which mirror the action of the bacterial transferases and share the same substrates. Endogenous ADP-ribosylation of EF-2 that modifies diphthamide has been reported (Lee and Iglewski 1984) and, moreover, it has been suggested that the endogenous activity is intrinsic to EF-2 itself (for review and discussion, see Iglewski and Fendrick 1990), although the data are not unequivocal. The endogenous ADP-ribosylation of G_s has been described in various systems, for example in adipocytes (Jacquemin et al. 1986) and platelets (Malencik and Anderson 1983). The ADP-ribosylation of G_i by an endogenous transferase was reported by Tanuma and coworkers (Tanuma et al. 1988). Recently, a C3-like enzyme from bovine brain has been characterized that modifies Rho proteins (Maehama et al. 1991). Furthermore, the ADP-ribosylation of actin by endogenous enzymes has been reported (Matsuyama and Tsuyama 1991; Terashima et al. 1992). The ADP-ribosylation of actin by the endogenous enzymes is, however, very different from the action of bacterial toxins, and it was not shown that actin is also the substrate for endogenous ADP-ribosylation in intact cells. In summary, while the importance of toxin-induced ADP-ribosylation as a pathogenic mechanism is well established and ADP-ribosylating toxins are valuable tools to elucidate the functions of their eukaryotic targets, the physiological significance of endogenous ADP-ribosylation and the evolutionary relations between the toxins and their possible eukaryotic counterparts are still unclear.

Acknowledgements. Work of the authors reported herein was supported by the Deutsche Forschungsgemeinschaft, Sonderforschungsbereich 246 and 249, and by the Fonds der Chemischen Industrie.

References

Abood ME, Hurley JB, Pappone M-C, Bourne HR, Stryer L (1982) Functional homology between signal-coupling proteins: cholera toxin inactivates the GTPase activity of transducin. J Biol Chem 257:10540–10543

Aktories K, Frevert J (1987) ADP-ribosylation of a 21–24 kDa eukaryotic protein(s) by C3, a novel botulinum ADP-ribosyltransferase, is regulated by guanine nucleotide. Biochem J 247:363–368
Aktories K, Wegner A (1989) ADP-ribosylation of actin by clostridial toxins. J Cell Biol 109:1385–1387
Aktories K, Ankenbauer T, Schering B, Jakobs KH (1986a) ADP-ribosylation of platelet actin by botulinum C2 toxin. Eur J Biochem 161:155–162
Aktories K, Bärmann M, Ohishi I, Tsuyama S, Jakobs KH, Habermann E (1986b) Botulinum C2 toxin ADP-ribosylates actin. Nature 322:390–392
Aktories K, Weller U, Chhatwal GS (1987) Clostridium botulinum type C produces a novel ADP-ribosyltransferase distinct from botulinum C2 toxin. FEBS Lett 212:109–113
Aktories K, Just I, Rosenthal W (1988a) Different types of ADP-ribose protein bonds formed by botulinum C2 toxin, botulinum ADP-ribosyltransferase C3 and pertussis toxin. Biochem Biophys Res Commun 156:361–367
Aktories K, Rösener S, Blaschke U, Chhatwal GS (1988b) Botulinum ADP-ribosyltransferase C3. Purification of the enzyme and characterization of the ADP-ribosylation reaction in platelet membranes. Eur J Biochem 172:445–450
Aktories K, Braun U, Rösener S, Just I, Hall A (1989) The rho gene product expressed in E. coli is a substrate of botulinum ADP-ribosyltransferase C3. Biochem Biophys Res Commun 158:209–213
Aktories K, Braun U, Habermann B, Rosener S (1990) Botulinum ADP-ribosyltransferase C3. In: Moss J, Vaughan M (eds) ADP-ribosylating toxins and G proteins. American Society for Microbiology, Washington, pp 97–115
Aktories K, Mohr C, Koch G (1992a) Clostridium botulinum C3 ADP-ribosyltransferase. Curr Top Microbiol Immunol 175:115–131
Aktories K, Wille M, Just I (1992b) Clostridial actin-ADP-ribosylating toxins. Curr Top Microbiol Immunol 175:97–113
Allured VS, Collier RJ, Carroll SF, McKay DB (1986) Structure of exotoxin A of Pseudomonas aeruginosa at 3,0-Angström resolution. Proc Natl Acad Sci USA 83:1320–1324
Barbacid M (1987) ras Genes. Annu Rev Biochem 56:779–827
Barbacid M (1990) ras Oncogenes: their role in neoplasia. Eur J Clin Invest 20: 225–235
Barbieri JT, Mende-Mueller M, Rappuoli R, Collier RJ (1989) Photolabeling of glu-129 of the S-1 subunit of pertussis toxin with NAD. Infect Immun 57:3549–3554
Bermek E (1976) Interactions of adenosine diphosphate-ribosylated elongation factor 2 with ribosomes. J Biol Chem 251:6544–6549
Bershadsky, AD, Vasiliev JM (1988) Cytoskeleton. Plenum, New York
Birnbaumer L (1990) G proteins in signal transduction. Annu Rev Pharmacol Toxicol 30:675–705
Bobak DA, Nightingale MS, Murtagh JJ, Price SR, Moss J, Vaughan M (1989) Molecular cloning, characterization, and expression of human ADP-ribosylation factor: two guanine nucleotide dependent activators of cholera toxin. Proc Natl Acad Sci USA 86:6101–6105
Bobak DA, Bliziotes MM, Noda M, Tsai S-C, Adamik R, Moss J (1990) Mechanism of activation of cholera toxin by ADP-ribosylation factor (ARF): Both low- and high-affinity interactions of ARF with guanine nucleotides promote toxin activation. Biochemistry 29:855–861
Bodley JW, Veldham SA (1990) Biosynthesis of Diphthamide: ADP-ribose acceptor for diphtheria toxin. In: Moss J, Vaughan M (eds) ADP-ribosylating toxins and G proteins: insights into signal transduction. American Society for Microbiology, Washington, pp 21–30
Bourne HR, Sanders DA, McCormick F (1990) The GTPase superfamily: a conserved switch for diverse cell functions. Nature 348:125–132
Bourne HR, Sanders DA, McCormick F (1991) The GTPase superfamily: conserved structure and molecular mechanism. Nature 349:117–127

Braun U, Habermann B, Just I, Aktories K, Vandekerckhove J (1989) Purification of the 22 kDa protein substrate of botulinum ADP-ribosyltransferase C3 from porcine brain cytosol and its characterization as a GTP-binding protein highly homologous to the rho gene product. FEBS Lett 243:70–76
Brenner SL, Korn ED (1980) The effects of Cytochalasins on actin Polymerization and actin ATPase provide insights into the mechanism of Polymerization. J Biol Chem 255:841–844
Camps M, Hou C, Sidiropoulos D, Stock JB, Jakobs KH, Gierschik P (1992) Stimulation of phospholipase C by guanine-nucleotide-binding protein β/v subunits. Eur J Biochem 206:821–831
Capiau C, Petre J, van Damme J, Puype M, Vandekerckhove J (1986) Protein-chemical analysis of pertussis toxin reveals homology between the subunits S as the haptoglobin-binding subunit. FEBS Lett 204:336–340
Carroll SF, Collier RJ (1984) NAD binding site of diphtheria toxin: identification of a residue within the nicotinamide subsite by photochemical modification with NAD. Proc Natl Acad Sci USA 81:3307–3311
Carroll SF, Collier RJ (1988) Amino acid sequence homology between the enzymic domains of diphtheria toxin and Pseudomonas aeruginosa exotoxin A. Mol Microbiol 2:293–296
Carroll SF, McCloskey JA, Crain PF, Oppenheimer NJ, Marschner TM, Collier RJ (1985) Photoaffinity labeling of diphtheria toxin fragment A with NAD: structure of the photoproduct at position 148. Proc Natl Acad Sci USA 82:7237–7241
Cassel D, Pfeuffer T (1978) Mechanism of cholera toxin action: covalent modification of the guanyl nucleotide-binding protein of the adenylate cyclase system. Proc Natl Acad USA 75:2669–2673
Cassel D, Selinger Z (1977) Mechanism of adenylate cyclase activation by cholera toxin: inhibition of GTP hydrolysis at the regulatory site. Proc Natl Acad Sci USA 74:3307–3311
Chang F-H, Bourne HR (1989) Cholera toxin induces cAMP-independent degradation of Gs. J Biol Chem 264:5352–5357
Chang PP, Moss J, Twiddy EM, Holmes RK (1987) Type II heat-labile enterotoxin of Escherichia coli activites adenylate cyclase in human fibroblasts by ADP-ribosylation. Infect Immun 55:1854–1858
Chardin P, Madaule P, Tavitian A (1988) Coding sequence of human rho cDNAs clone 6 and clone 9. Nucleic Acids Res 16:2717
Chardin P, Boquet P, Madaule P, Popoff MR, Rubin EJ, Gill DM (1989) The mammalian G protein rho C is ADP-ribosylated by Clostridium botulinum exoenzyme C3 and affects actin microfilament in Vero cells. EMBO J 8: 1087–1092
Choe S, Bennett MJ, Fujii G, Curmi PMG, Kantardjieff KA, Collier RJ, Eisenberg D (1992) The crystal structure of diphtheria toxin. Nature 357:216–222
Chuang DM, Weissbach H (1972) Studies on elongation factor II from calf brain. Arch Biochem Biophys 152:114–124
Chung DW, Collier RJ (1977) The mechanism of ADP-ribosylation of elongation factor 2 catalyzed by fragment A from diphtheria toxin. Biochim Biophys Acta 483:248–257
Coburn J (1992) Pseudomonas aeruginosa exoenzyme S. Curr Top Microbiol Immunol 175:133–143
Coburn J, Dillon ST, Iglewski BH, Gill DM (1989a) Exoenzyme S of Pseudomonas aeruginosa specifically ADP-ribosylates the intermediate filament protein vimentin. Infect Immun 57:996–998
Coburn J, Wyatt RT, Iglewski BH, Gill DM (1989b) Several GTP-binding proteins, including p21 c-H-ras, are preferred substrates of Pseudomonas aeruginosa exoenzyme S. J Biol Chem 264:9004–9008
Coburn J, Kane AV, Feig L, Gill DM (1991) Pseudomonas aeruginosa exoenzyme S requires a eukaryotic protein for ADP-ribosyltransferase activity. J Biol Chem 266:6438–6446

Collier RJ (1968) Effect of diphtheria toxin on protein synthesis: Inactivation of one of the transfer factors. J Mol Biol 25:83–98
Collier RJ (1990) Diphtheria Toxin: structure and function of a cytocidal protein. In: Moss J, Vaughan M (eds) ADP-ribosylating toxins and G proteins. American Society for Microbiology, Washington, pp 3–19
Collier RJ, Kandel J (1971) Structure and activity of diphtheria toxin: I. Thiol-dependent dissociation of a fraction of toxin into enzymically active and inactive fragments. J Biol Chem 246:1496–1503
Considine RV, Simpson LL (1991) Cellular and molecular actions of binary toxins possessing ADP-ribosyltransferase activity. Toxicon 29:913–936
Dallas WS, Falkow S (1980) Amino acid homology between cholera toxin and Escherichia coli heat-labile toxin. Nature 277:406–407
Davydova EK, Ovchinnikov LP (1990) ADP-ribosylated elongation factor 2 (ADP-ribosyl-EF-2) is unable to promote translocation within the ribosome. FEBS Lett 261:350–352
Didsbury J, Weber RF, Bokoch GM, Evans T, Snyderman R (1989) rac, a novel ras-related family of proteins that are botulinum toxin substrates. J Biol Chem 264:16378–16382
Drazin R, Kandel J, Collier RJ (1971) Structure and activity of diphtheria toxin: II. Attack by trypsin at a specific site within the intact toxin molecule. J Biol Chem 246:1504–1510
Dwyer JD, Bloomfield VA (1982) Subunit arrangement of cholera toxin in solution and bound to receptor-containing model membranes. Biochemistry 21:3227–3231
Field M (1980) Role of cyclic nucleotides in enterotoxic diarrhea. Adv Cyclic Nucleotide Res 12:267–277
Fishman PH (1990) Mechanism of action of cholera toxin. In: Moss J, Vaughan M (eds) ADP-ribosylating toxins and G proteins. American Society for Microbiology, Washington, pp 127–140
Fishman PH, Moss J, Osborne JC (1978) Interaction of choleragen with oligosaccharide of ganglioside GM: evidence for multiple oligosaccharide binding sites. Biochemistry 17:711–716
Freissmuth M, Gilman AG (1989) Mutations of Gsa designed to alter the reactivity of the protein with bacterial toxins. J Biol Chem 264:21907–21914
Fukumoto Y, Kaibuchi K, Hori Y, Fujioka H, Araki S, Ueda T, Kikuchi A, Takai Y (1990) Molecular cloning and characterization of a novel type of regulatory protein (GDI) for the rho proteins, ras p21-like small GTP-binding proteins. Oncogene 5:1321–1328
Fukuta S, Magnani JL, Twiddy EM, Holmes RK, Ginsburg V (1988) Comparison of the carbohydrate-binding specificities of cholera toxin and Escherichia coli heat-labile enterotoxins LTh-I, LT-IIa, and LT-IIb. Infect Immun 56:1748–1753
Garrett MD, Self AJ, van Oers C, Hall A (1989) Identification of distinct cytoplasmic targets for ras, R-ras and rho regulatory proteins. J Biol Chem 264:10–13
Geipel U, Just I, Schering B, Haas D, Aktories K (1989) ADP-ribosylation of actin causes increase in the rate of ATP exchange and inhibition of ATP hydrolysis. Eur J Biochem 179:229–232
Geipel U, Just I, Aktories K (1990) Inhibition of cytochalasin D-stimulated G-actin ATPase by ADP-ribosylation with Clostridium perfringens iota toxin. Biochem J 266:335–339
Gierschik, P (1992) ADP-ribosylation of signal-transducing guanine nucleotide-binding proteins by pertussis toxin. Curr Top Microbiol Immun 175:69–96
Gierschik P, Jakobs KH (1987) Receptor-mediated ADP-ribosylation of a phospholipase C-stimulating G protein. FEBS Lett 224:219–223
Gill DM (1975) Involvement of nicotinamide adenine dinucleotide in the action of cholera toxin in vitro. Proc Natl Acad Sci USA 72:2064–2068
Gill DM (1976) The arrangement of subunits of cholera toxin. Biochemistry 15: 1242–1248

Gill DM (1977) Mechanism of action of cholera toxin. Adv Cyclic Nucleotide Res 8:85–118
Gill DM, Meren R (1978) ADP-ribosylation of membrane proteins catalyzed by cholera toxin: basis of the activation of adenylate cyclase. Proc Natl Acad Sci USA 75:3050–3054
Gill DM, Pappenheimer AM (1971) Structure-activity relationships in diphtheria toxin. J Biol Chem 246:1492–1495
Gill DM, Rappaport SH (1979) Origin of the enzymatically active A fragment of cholera toxin. J Infect Dis 139:674–680
Gill DM, Pappenheimer AM, Brown R, Kurnick JT (1969) Studies on the mode of action of diphtheria toxin. VII. Toxin-stimulated hydrolysis of nicotinamide adenine dinucleotide in mammalian cell extracts. J Exp Med 129:1–21
Gray GL, Smith DH, Baldrige JS, Harkins RN, Vasil ML, Chen EY, Heyneker HL (1984) Cloning, nucleotide sequence, and expression in Escherichia coli of the exotoxin A structural gene of Pseudomonas aeruginosa. Proc Natl Acad Sci USA 81:2645–2649
Greenfield L, Bjorn MJ, Horn G, Fong D, Buck GA, Collier RJ, Kaplan DA (1983) Nucleotide sequence of the structural gene for diphtheria toxin carried by corynebacteriophage beta. Proc Natl Acad Sci USA 80:6853–6857
Habermann B, Mohr C, Just I, Aktories K (1991) ADP-ribosylation and de-ADP-ribosylation of the rho protein by Clostridium botulinum exoenzyme C3. Regulation by EDTA, guanine nucleotides and pH. Biochim Biophys Acta 1077: 253–258
Hartwig JH, Kwiatkowski DJ (1991) Actin-binding proteins. Curr Opin Cell Biol 3:87–97
Henriksen O, Robinson EA, Maxwell ES (1975) Interaction of guanosine nucleotides with elongation factor 2. J Biol Chem 250:720–724
Hescheler, Rosenthal W, Trautwein W, Schultz G (1987) The GTP-binding protein, Go regulates neuronal calcium channels. Nature 325:445–447
Hiraoka K, Kaibuchi K, Ando S, Musha T, Takaishi K, Mizuno T, Asada M, Ménard L, Tomhave E, Didsbury J, Snyderman R, Takai Y (1992) Both stimulatory and inhibitory GDP/GTP exchange proteins, smg GDS and rho GDI, are active on multiple small GTP-binding proteins. Biochem Biophys Res Commun 182:921–930
Holmes KC, Popp D, Gebhard W, Kabsch W (1990) Atomic model of the actin filament. Nature 347:44–49
Holmes RK, Twiddy EM, Pickett CL (1986) Purification and characterization of type II heat-labile enterotoxin of Escherichia coli. Infect Immun 53:424–433
Honjo T, Nishizuka Y, Hayaishi O (1968) Diphtheria toxin-dependent adenosine diphosphate ribosylation of aminoacryl transferase II and inhibition of protein synthesis. J Biol Chem 243:3553–3555
Ibuki F, Moldave K (1968) The effect of guanosine triphosphate, other nucleotides, and aminoacyl transfer ribonucleic acid on the activity of transferase I and its binding to ribosomes. J Biol Chem 243:44–55
Iglewski BH, Kabat D (1991) NAD-dependent inhibition of protein synthesis by Pseudomonas aeruginosa Toxin. Proc Natl Acad Sci
Iglewski WJ, Fendrick JL (1990) ADP-ribosylation of elongation factor 2 in animal cells. In: Moss J, Vaughan M (eds) ADP-ribosylating Toxins and G Proteins. American Society for Microbiology, Washington, pp 511–524
Iglewski BH, Sadoff J, Bjorn MJ, Maxwell ES (1978) Pseudomonas aeruginosa exoenzyme S: an adenosine diphosphate ribosyl transferase distinct from toxin A. Proc Natl Acad Sci USA 75:3211–3215
Iiri T, Tohkin M, Morishima N, Ohoka Y, Ui M, Katada T (1989) Chemotactic peptide receptor-supported ADP-ribosylation of a pertussis toxin substrate GTP-binding protein by cholera toxin in neutrophil-type HL-60 cells. J Biol Chem 264:21394–21400

Inoue S, Sugai M, Murooka Y, Paik S-Y, Hong Y-M, Ohgai H, Suginaka H (1991) Molecular cloning and sequencing of the epidermal cell differentiation inhibitor gene from Staphylococcus aureus. Biochem Biophys Res Commun 174: 459–464

Jacquemin C, Thibout H, Lambert B, Correze C (1986) Endogenous ADP-ribosylation of Gs subunit and autonomous regulation of adenylate cyclase. Nature 323:182–184

Johnson GL, Kaslow HR, Bourne HR (1978) Genetic evidence that cholera toxin substrates are regulatory components of adenylate cyclase. J Biol Chem 253: 7120–7123

Jones DT, Masters SB, Bourne HR, Reed RR (1990) Biochemical characterization of three stimulatory GTP-binding proteins: the large and small forms of G and the olfactory-specific G-protein G. J Biol Chem 265:2671–2676

Just I, Geipel U, Wegner A, Aktories K (1990) De-ADP-ribosylation of actin by Clostridium perfringens iota-toxin and Clostridium botulinum C2 toxin. Eur J Biochem 192:723–727

Just I, Mohr C, Schallehn G, Menard L, Didsbury JR, Vandekerckhove J, van Damme J, Aktories K (1992a) Purification and characterization of an ADP-ribosyltransferase produced by Clostridium limosum. J Biol Chem 267:10274–10280

Just I, Schallehn G, Aktories K (1992b) ADP-ribosylation of small GTP-binding proteins by Bacillus cereus. Biochem Biophys Res Commun 183:931–936

Kabsch W, Mannherz HG, Suck D, Pai EF, Holmes KC (1990) Atomic structure of the actin: DNase I complex. Nature 347:37–44

Kahn RA, Gilman AG (1984a) Purification of a protein cofactor required for ADP-ribosylation of the stimulatory regulatory component of adenylate cyclase by cholera toxin. J Biol Chem 259:6228–6234

Kahn RA, Gilman AG (1984b) ADP-ribosylation of Gs promotes the dissociation of its α and β subunits. J Biol Chem 259:6235–6240

Kahn RA, Gilman AG (1986) The protein cofactor necessary for ADP-ribosylation of G by cholera toxin is itself a GTP-binding protein. J Biol Chem 261: 7906–7911

Kaibuchi K, Mizuno T, Fujioka H, Yamamoto T, Kishi K, Fukumoto Y, Hori Y, Takai Y (1991) Molecular cloning of the cDNA for stimulatory GDP/GTP exchange protein for smg p21s (ras p21-like small GTP-binding proteins) and characterization of stimulatory GDP/GTP exchange protein. Mol Cell Biol 11:2873–2880

Kaziro Y, Itoh H, Kozasa T, Nakafuku M, Satoh T (1991) Structure and function of signal-transducing GTP-binding proteins. Annu Rev Biochem 60:349–400

Kessel M, Klink F (1980) Archebacterial elongation factor is ADP-ribosylated by diphtheria toxin. Nature 287:250–251

Kikuchi A, Yamamoto K, Fujita T, Takai Y (1988) ADP-ribosylation of the bovine brain rho protein by botulinum toxin type C1. J Biol Chem 263:16303–16308

Klinz F-J, Costa T (1989) Cholera toxin ADP-ribosylates the receptor-coupled form of pertussis toxin-sensitive G-proteins. Biochem Biophys Res Commun 165:554–560

Kohno K, Uchida T, Ohkubo H, Nakanishi S, Nakanishi T, Fukui T, Ohtsuka E, Ikehara M, Okada Y (1986) Amino acid sequence of mammalian elongation factor 2 deduced from the cDNA sequence: homology with GTP-binding proteins. Proc Natl Acad Sci USA 83:4978–4982

Landis CA, Masters SB, Spada A, Pace AM, Bourne HR, Vallar L (1989) GTPase inhibiting mutations activate the a chain of Gs and stimulate adenylyl cyclase in human pituitary tumours. Nature 340:692–696

Lang P, Guizani L, Vitté-Mony I, Stancou R, Dorseuil O, Gacon G, Bertoglio J (1992) ADP-ribosylation of the ras-related, GTP-binding protein RhoA inhibits lymphocyte-mediated cytotoxicity. J Biol Chem 267:11677–11680

Lee H, Iglewski WJ (1984) Cellular ADP-ribosyltransferase with the same mechanism of action as diphtheria toxin and Pseudomonas toxin A. Proc Natl Acad Sci USA 81:2703–2707

Leppla SH, Martin OC, Muehl LA (1978) The exotoxin of P. aeruginosa: a proenzyme having an unusual mode of activation. Biochem Biophys Res Commun 81:532–538

Lin SY, McKeehan WL, Culp W, Hardesty B (1969) Partial characterization of the enzymatic properties of the aminoacyl transfer ribonucleic acid binding enzyme. J Biol Chem 244:4340–4350

Litosch I, Wallis C, Fain JN (1985) 5-Hydroxytryptamine stimulates inositol phosphate production in a cell-free system from blowfly salivary glands. Evidence for a role of GTP in coupling receptor activation to phosphoinositide breakdown. J Biol Chem 260:5464–5471

Locht C, Keith JM (1986) Pertussis toxin gene: nucleotide sequence and genetic organization. Science 232:1258–1264

Locht C, Capian C, Feron L (1989) Identification of amino acid residues essential for the enzymatic activities of pertussis toxin. Proc Natl Acad Sci USA 86: 3075–3079

Ludwig DS, Ribi HO, Schoolnik GK, Kornberg RD (1986) Two-dimensional crystals of cholera toxin B-subunit-receptor complexes: projected structure at 17 A resolution. Proc Natl Acad Sci USA 83:8585–8588

Madaule P, Axel R (1985) A novel ras-related gene family. Cell 41:31–40

Madshus IH, Stenmark H (1992) Entry of ADP-ribosylating toxins into cells. Curr Top Microbiol Immunol 175:2–26

Maehama T, Takahashi K, Ohoka Y, Ohtsuka T, Ui M, Katada T (1991) Identification of a botulinum C3-like enzyme in bovine brain that catalyzes ADP-ribosylation of GTP-binding proteins. J Biol Chem 266:10062–10065

Malencik DA, Anderson SR (1983) Binding of hormones and neuropeptides by calmodulin. Biochemistry 22:1995–2001

Matsuyama S, Tsuyama S (1991) Mono-ADP-ribosylation in brain: purification and characterization of ADP-ribosyltransferases affecting actin from rat brain. J Neurochem 57:1380–1387

Mattera R, Yatani A, Kirsch GE, Graf R, Okabe K, Olate J, Codina J, Brown AM, Birnbaumer L (1989) Recombinant ai-3 subunit of G protein activates GK-gated K+ channels. J Biol Chem 264:465–471

Mauss S, Chaponnier C, Just I, Aktories K, Gabbiani G (1990) ADP-ribosylation of actin isoforms by Clostridium botulinum C2 toxin and Clostridium perfringens iota toxin. Eur J Biochem 194:237–241

Mayer T, Koch R, Fanick W, Hilz H (1988) ADP-ribosyl proteins formed by pertussis toxin are specifically cleaved by mercury ions. Biol Chem Hoppe-Seyler 369:579–583

Mohr C, Koch G, Just I, Aktories K (1992) ADP-ribosylation by Clostridium botulinum C3 exoenzyme increases steady state GTPase activities of recombinant rhoA and rhoB proteins. FEBS Lett 297:95–99

Moldave K (1985) Eukaryotic protein synthesis. Ann Rev Biochem 54:1109–11049

Morii N, Kawano K, Sekine A, Yamada T, Narumiya S (1991) Purification of GTPase-activating protein specific for the rho gene. J Biol Chem 266:7646–7650

Moss J, Richardson SH (1978) Activation of adenylate cyclase by heat-labile Escherichia coli enterotoxin. Evidence for ADP-ribosyltransferase activity similar to that of choleragen. J Clin Invest 62:281–285

Moss J, Vaughan M (1977) Mechanism of action of choleragen. Evidence for ADP-ribosyltransferase activity with arginine as an acceptor. J Biol Chem 252: 2455–2457

Moss J, Garrison SVC, Oppenheimer NJ, Richardson SH (1979a) NAD-dependent ADP-ribosylation of arginine and proteins by Escherichia coli heat-labile enterotoxin. J Biol Chem 254:11993–11996

Moss J, Stanley SJ, Lin MC (1979b) NAD glycohydrolase and ADP-ribosyltransferase activities are intrinsic to A peptide of choleragen. J Biol Chem 254:11993–11996
Murayama T, Ui M (1984) [3H]GDP release from rat and hamster adipocyte membranes independently linked to receptors involved in activation or inhibition of adenylate cyclase. J Biol Chem 259:761–769
Narumiya S, Sekine A, Fujiwara M (1988) Substrate for botulinum ADP-ribosyltransferase, Gb, has an amino acid sequence homologous to a putative rho gene product. J Biol Chem 263:17255–17257
Navon SE, Fung BK-K (1984) Characterization of transducin from bovine retinal rod outer segments. J Biol Chem 259:6686–6693
Nemoto Y, Namba T, Kozaki S, Narumiya S (1991) Clostridium botulinum C3 ADP-ribosyltransferase gene. J Biol Chem 266:19312–19319
Nicosia A, Perugini M, Franzini C, Casagli MC, Borri MG, Antoni G, Almoni M, Neri P, Ratti G, Rappuoli R (1986) Cloning and sequencing of the pertussis toxin genes: operon structure and gene duplication. Proc Natl Acad Sci USA 83:4631–4635
Nilsson L, Nygard O (1988) Structural and functional studies of the interaction of the eukaryotic elongation factor EF-2 with GTP and ribosomes. Eur J Biochem 171:293–299
Noda M, Tsai S, Adamik R, Moss J, Vaughan M (1990) Mechanism of cholera toxin activation by a guanine nucleotide-dependent 19 kDa protein. Biochim Biophys Acta 1034:195–199
Northup JK, Sternweis PC, Smigel MDC, Ross EM, Gilman AG (1980) Purification of the regulatory component of adenylate cyclase. Proc Natl Acad Sci USA 77:6516–6520
Nygard O, Nilsson L (1984) Nucleotide-mediated interactions of eukaryotic elongation factor EF-2 with ribosomes. Eur J Biochem 140:93–96
Nygard O, Nilsson L (1985) Reduced ribosomal binding of eukaryotic elongation factor 2 following ADP-ribosylation. Difference in binding selectivity between polyribosomes and reconstituted monoribosomes. Biochim Biophys Acta 824: 152–162
Ohishi I (1987) Activation of botulinum C2 toxin by trypsin. Infect Immun 55:1461–1465
Ohishi I, Iwasaki M, Sakaguchi G (1980) Purification and characterization of two components of botulinum C2 toxin. Infect Immun 30:668–673
Ohtsuka T, Nagata K, Iiri T, Nozawa Y, Ueno K, Ui M, Katada T (1989) Activator protein supporting the botulinum ADP-ribosyltransferase reaction. J Biol Chem 264:15000–15005
Omura F, Kohno K, Uchida T (1989) The histidine residue of codon 715 is essential for function of elongation factor 2. Eur J Biochem 180:1–8
Owens JR, Frame LT, Ui M, Cooper DMF (1985) Cholera toxin ADP-ribosylates the islet-activating protein substrate in adipocyte membranes and alters its function. J Biol Chem 260:15946–15952
Pai EF, Kabsch W, Krengel U, Holmes KC, John J, Wittinghofer A (1989) Structure of the guanine-nucleotide-binding domain of the Ha-ras oncogene product p21 in the triphosphate conformation. Nature 341:209–214
Papini E, Schiavo G, Sandona D, Rappuoli R, Montecucco C (1989) Histidine 21 is at the NAD^+ binding site of diphtheria toxin. J Biol Chem 264:12385–12388
Papini E, Santucci A, Schiavo G, Domenighini M, Neri P, Rappuoli R, Montecucco C (1991) Tyrosine 65 is photolabeled by 8-azidoadenine and 8-azidoadenosine at the NAD binding site of diphtheria toxin. J Biol Chem 266:2494–2498
Pastan I, FitzGerald D (1989) Pseudomonas exotoxin: chimeric toxins. J Biol Chem 264:15157–15160
Paterson HF, Self AJ, Garrett MD, Just I, Aktories K, Hall A (1990) Microinjection of recombinant p21 induces rapid changes in cell morphology. J Cell Biol 111:1001–1007

Perentesis JP, Miller SP, Bodley JW (1992) Protein toxin inhibitors of protein synthesis. Biofactors 3:173–184
Pickett CL, Weinstein DL, Holmes RK (1987) Genetics of type IIa heat-labile enterotoxin of Escherichia coli: operon fusions, nucleotide sequence and hybridization studies. J Bacteriol 169:5180–5187
Polakis PG, Weber RF, Nevins B, Didsbury JR, Evans T, Snyderman R (1989) Identification of the ral and racl gene products, low molecular mass GTP-binding proteins from human platelets. J Biol Chem 264:16383–16389
Pollard TD, Cooper JA (1986) Actin and actin-binding proteins. A critical evaluation of mechanisms and functions. Annu Rev Biochem 55:987–1035
Popoff MR, Rubin EJ, Gill DM, Boquet P (1988) Actin-specific ADP-ribosyltransferase produced by a Clostridium difficile strain. Infect Immun 56:2299–2306
Popoff MR, Milward FW, Bancillon B, Boquet P (1989) Purification of the Clostridium spiroforme binary toxin and activity of the toxin on HEp2 cells. Infect Immun 57:2462–2469
Popoff MR, Boquet P, Gill DM, Eklund MW (1990) DNA sequence of exoenzyme C3, an ADP-ribosyltransferase encoded by Clostridium butulinum C and D phages. Nucleic Acids Res 18:1291
Popoff MR, Hauser D, Boquet P, Eklund MW, Gill DM (1991) Characterization of the C3 gene of Clostridium botulinum types C and D and its expression in Escherichia coli. Infect Immun 59:3673–3679
Raeburn S, Goor R, Schneider JA, Maxwell ES (1968) Interaction of aminoacyl transferase II and guanosine triphosphate: inhibition by diphtheria toxin and nicotinamide adenine dinucleotide. Proc Natl Acad Sci USA 61:1428–1434
Rappuoli R, Pizza M (1991) Structure and evolutionary aspects of ADP-ribosylating toxins. In: Alouf JE, Freer JH (eds) Sourcebook of bacterial protein toxins. Academic, London, pp 1–21
Ribi HO, Ludwig DS, Mercer KL, Schoolnik GK, Kornberg RD (1988) Three-dimensional structure of cholera toxin penetrating a lipid membrane. Science 239:1272–1276
Ross EM, Gilman AG (1980) Biochemical properties of hormone-sensitive adenylate cyclase. Annu Rev Biochem 49:533–564
Rösener S, Chhatwal GS, Aktories K (1987) Botulinum ADP-ribosyltransferase C3 but not botulinum neurotoxins C1 and D ADP-ribosylates low molecular mass GTP-binding proteins. FEBS Lett 224:38–42
Rubin EJ, Gill DM, Boquet P, Popoff MR (1988) Functional modification of a 21-Kilodalton G protein when ADP-ribosylated by exoenzyme C3 of Clostridium botulinum. Mol Cell Biol 8:418–426
Schering B, Bärmann M, Chhatwal GS, Geipel U, Aktories K (1988) ADP-ribosylation of skeletal muscle and non- muscle actin by Clostridium perfringens iota toxin. Eur J Biochem 171:225–229
Sekine A, Fujiwara M, Narumiya S (1989) Asparagine residue in the rho gene product is the modification site for botulinum ADP-ribosyltransferase. J Biol Chem 264:8602–8605
Selve N, Wegner A (1986) Rate of treadmilling of actin filaments in vitro. J Mol Biol 187:627-631
Serventi IM, Moss J, Vaughan M (1992) Enhancement of cholera toxin-catalyzed ADP-ribosylation by guanine nucleotide-binding proteins. Curr Top Microbiol Immun 175:43–68
Shinjo K, Koland JG, Hart MJ, Narasimhan V, Johnson DI, Evans T, Cerione RA (1990) Molecular cloning of the gene for the human placental GTP-binding protein Gp (G25K): Identification of this GTP-binding protein as the human homolog of the yeast cell-division-cycle protein CDC42. Proc Natl Acad Sci USA 87:9853–9857
Simon M, Strathmann M, Gautam N (1991) Diversity of G proteins in signal transduction. Science 252:802–808

Simpson LL (1989) The binary toxin produced by Clostridium botulinum enters cells by receptor-mediated endocytosis to exert its pharmacologic effects. J Pharmacol Exp Ther 251:1223–1228
Simpson LL, Stiles BG, Zapeda HH, Wilkins TD (1987) Molecular basis for the pathological actions of Clostridium perfringens Iota toxin. Infect Immun 55: 118–122
Simpson LL, Stiles BG, Zepeda H, Wilkins TD (1989) Production by Clostridium spiroforme of an iotalike toxin that possesses mono(ADP-ribosyl)transferase activity: Identification of a novel class of ADP-ribosyltransferases. Infect Immun 57:255–261
Sixma TK, Pronk SE, Kalk KH, Wartna ES, van Zanten BAM, Witholt B, Hol WGJ (1991) Crystal structure of a cholera toxin-related heat-labile enterotoxin from E. coli. Nature 351:371–377
Stasia M-J, Jouan A, Bourmeyster N, Boquet P, Vignais PV (1991) ADP-ribosylation of a small size GTP-binding protein in bovine neutrophils by the C3 exoenzyme of Clostridium botulinum and effect on the cell motility. Biochem Biophys Res Commun 180:615–622
Sugai M, Enomoto T, Hashimoto K, Matsumoto K, Matsuo Y, Ohgai H, Hong Y-M, Inoue S, Yoshikawa K, Suginaka H (1990) A novel epidermal cell differentiation inhibitor (EDIN): purification and characterization from Staphylococcus aureus. Biochem Biophys Res Commun 173:92–98
Sugai M, Hashimoto K, Kikuchi A, Inoue S, Okumura H, Matsumota K, Goto Y, Ohgai H, Moriishi K, Syuto B, Yoshikawa K, Suginaka H, Takai Y (1992) Epidermal cell differentiation inhibitor ADP-ribosylates small GTP-binding proteins and induces hyperplasia of epidermis. J Biol Chem 267:2600–2604
Tang W-L, Gilman AG (1991) Type-specific regulation of adenylyl cyclase by G protein $\beta\gamma$ subunits. Science 254:1500–1503
Tanuma S, Kawashima K, Endo N (1988) Eukaryotic mono(ADP-ribosyl)transferase that ADP-ribosylates GTP-binding regulatory Gi protein. J Biol Chem 263: 5485–5489
Terashima M, Mishima K, Yamada K, Tsuchiya M, Wakutani T, Shimoyama M (1992) ADP-ribosylation of actins by arginine-specific ADP-ribosyltransferase purified from chicken heterophils. Eur J Biochem 204:305–311
Tsai SC, Noda M, Adamik R, Chang PP, Chen HC, Moss J, Vaughan M (1988) Stimulation of choleragen enzymatic activities by GTP and two soluble proteins purified from bovine brain. J Biol Chem 263:1768–1772
Tsuji T, Inoue T, Miyama A, Noda M (1991) Glutamic acid-112 of the A subunit of heat-labile enterotoxin from enterotoxigenic Escherichia coli is important for ADP-ribosyltransferase activity. FEBS Lett 291 2:319–321
Ueda T, Kikuchi A, Ohga N, Yamamoto J, Takai Y (1990) Purification and characterization from bovine brain cytosol of a novel regulatory protein inhibiting the dissociation of GDP from and the subsequent binding of GTP to rhoB p20, a ras p21-like GTP-binding protein. J Biol Chem 265:9373–9380
Ui M (1990) Pertussis toxin as a valuable probe for G-protein involvement in signal transduction. In: Moss J, Vaughan M (eds) ADP-ribosylating toxins and G proteins. American Society for Microbiology, Washington, pp 45–77
Van Dop C, Tsubokawa M, Bourne HR, Ramachandran J (1984) Amino acid sequence of retinal transducin at the site ADP-ribosylated by cholera toxin. J Biol Chem 259:696–698
van Heyningen WE, Carpenter CCJ, Pierce NF, Greenough WB (1971) Deactivation of cholera toxin by ganglioside. J Infect Dis 124:415–418
Van Ness BG, Howard JB, Bodley JW (1980a) ADP-ribosylation of elongation factor 2 by diphtheria toxin. J Biol Chem 255:10717–10720
Van Ness BG, Howard JB, Bodley JW (1980b) ADP-ribosylation of elongation factor 2 by diphtheria toxin. NMR spectra and proposed structures of ribosyl-diphthamide and its hydrolysis products. J Biol Chem 255:10710–10716

Vandekerckhove J (1990) Actin-binding proteins. Curr Opin Cell Biol 2:41–50
Vandekerckhove J, Weber K (1979) The complete amino acid sequence of actins from bovine aorta, bovine heart, bovine fast skeletal muscle and rabbit slow skeletal muscle. Differentiation 14:123–133
Vandekerckhove J, Schering B, Bärmann M, Aktories K (1987) Clostridium perfringens iota toxin ADP-ribosylates skeletal muscle actin in Arg-177. FEBS Lett 225:48–52
Vandekerckhove J, Schering B, Bärmann M, Aktories K (1988) Botulinum C2 toxin ADP-ribosylates cytoplasmic β/v-actin in arginine 177. J Biol Chem 263:696–700
Wegner A (1976) Head tail polymerization of actin. J Mol Biol 108:139–150
Wegner A, Aktories K (1988) ADP-ribosylated actin caps the barbed ends of actin filaments. J Biol Chem 263:13739–13742
Weigt C, Just I, Wegner A, Aktories K (1989) Nonmuscle actin ADP-ribosylated by botulinum C2 toxin caps actin filaments. FEBS Lett 246:181–184
West RE, Moss J, Vaughan M, Liu T, Liu T-Y (1985) Pertussis toxin-catalyzed ADP-ribosylation of transducin. J Biol Chem 260:14428–14430
Wick MJ, Iglewski BH (1990) Pseudomonas aeruginosa Exotoxin A. In: Moss J, Vaughan M (eds) ADP-ribosylating toxins and G proteins. American Society for Microbiology, Washington, pp 31–43
Wiegers W, Just I, Müller H, Hellwig A, Traub P, Aktories K (1991) Alteration of the cytoskeleton of mammalian cells cultured in vitro by Clostridium botulinum C2 toxin and C3 ADP-ribosyltransferase. Eur J Cell Biol 54:237–245
Wieland T, Ulibarri I, Aktories K, Gierschik P, Jakobs KH (1990a) Interaction of small G proteins with photoexcited rhodopsin. FEBS Lett 263:195–198
Wieland T, Ulibarri I, Gierschik P, Hall A, Aktories K, Jakobs KH (1990b) Interaction of recombinant rho A GTP-binding proteins with photoexcited rhodopsin. FEBS Lett 274:111–114
Wille M, Just I, Wegner A, Aktories K (1992) ADP-ribosylation of the gelsolin-actin complex by clostridial toxins. J Biol Chem 267:50–55
Yamamoto T, Nakazawa T, Miyata T, Kaji A, Yokota T (1984) Evolution and structure of two ADP-ribosylation enterotoxins: Escherichia coli heat-labile toxin and cholera toxin. FEBS Lett 169:241–246
Yamamoto T, Gojobori T, Yokota T (1987) Evolutionary origin of pathogenic determinants in enterotoxigenic Escherichia coli and Vibrio cholerae O1. J Bacteriol 169:1352–1357
Yatani A, Codina J, Brown AM, Birnbaumer L (1987) Direct activation of mammalian atrial muscarinic potassium channels by GTP regulatory protein Gk. Science 235:207–211
Yatani A, Imoto Y, Codina J, Hamilton SL, Brown AM, Birnbaumer L (1988) The stimulatory G protein of adenylyl cyclase, Gs, also stimulates dihydropyridine-sensitive Ca^{2+} channels. J Biol Chem 263:9887–9895
Yeramian P, Chardin P, Madaule P, Tavitian A (1987) Nucleotide sequence of human rho cDNA clone 12. Nucleic Acids Res 15:1869

Section II
Structures of the GTPase Switches

CHAPTER 9

Eukaryotic Translation Factors Which Bind and Hydrolyze GTP

J. CAVALLIUS and W.C. MERRICK

A. GTPase Factors

As is true for bacterial systems, GTP is the nucleotide which drives eukaryotic protein synthesis (MERRICK 1992). The primary use of ATP is in the activation of amino acids to provide aminoacyl-tRNAs, the chemically activated form of the amino acid. There are three discrete steps in eukaryotic protein synthesis: initiation, elongation, and termination. As indicated in Table 1, each of these steps is associated with specific factors which bind and hydrolyze GTP. The eukaryotic initiation factor 2 (eIF-2) is the central factor in initiation and binds GTP and the initiator tRNA, Met-tRNA$_i$. The ternary complex of eIF-2, GTP, and Met-tRNA$_i$ is bound to the 40S subunit and serves to locate the initiating AUG. Upon recognition of the AUG codon, the ternary complex is then susceptible to the action of eIF-5, which triggers the hydrolysis of GTP and leads to the release of eIF-2·GDP from the ribosome. A curious feature of eIF-5 is that it also possesses a ribosome-dependent GTPase activity which is required in model assays using eIF-2A in place of eIF-2 to form initiation complexes (ADAMS et al. 1975; MERRICK et al. 1975).

Because eIF-2 binds GDP quite tightly ($K_d \sim 10^{-8} M$), a recycling protein is required to facilitate nucleotide exchange, as was also noted for the bacterial elongation factor, EF-Tu. The recycling protein eIF-2B is composed of five subunits and facilitates and rapid exchange of free GTP for the bound GDP. The observation that the 30-kDa subunit of eIF-2B can be specifically cross-linked to GTP (DHOLAKIA et al. 1989) is consistent with an exchange mechanism which has a quaternary complex intermediate (eIF-2·GDP·eIF-2B·GTP) (DHOLAKIA and WAHBA 1989), although evidence has also been presented in support of an EF-Tu·Ts-like exchange mechanism (ROWLANDS et al. 1988).

The next step in translation requires elongation factor 1α (EF-1α), which binds GTP and aminoacyl-tRNAs in a manner similar to its bacterial counterpart EF-Tu (RIIS et al. 1990). Like EF-Tu, EF-1α is an abundant cellular protein comprising 5%–10% of the soluble protein in most systems. This abundance has made the protein relatively easy to purify and clone, and at the time of writing this article over 70 EF-Tu or EF-1α sequences had been reported. In fact, for all the translation factor sequences available, EF-1α represents the most highly conserved sequence of all translation factors.

Table 1. Eukaryotic translation factors which bind GTP

Factor	Subunit M_r (kDa)	GTPase stimulator
eIF-2	32, 38, 52	eIF-5, ribosomes
eIF-2B	26, 39, 58, 67, 82	?
eIF-5	125	ribosomes
EF-1α	52	aa-tRNA, ribosomes
EF-2	100	60S subunit
RF	55	ribosomes

See text for details. eIF, eukaryotic initiation factor; EF, elongation factor; RF, release factor.

Subsequent to the EF-1α-directed binding of aminoacyl-tRNAs to the A site on the ribosome, peptide bond formation occurs rapidly via the peptidyl transferase activity of the 60S subunit. Following this reaction, EF-2 catalyzes the GTP-dependent translocation of the peptidyl-tRNA fully to the P site to allow for the next cycle of EF-1α-directed binding of an aminoacyl-tRNA to the empty ribosomal A site. A unique feature of EF-2 in archebacteria, plants, and animals is the modification of a histidine residue (number 715 in mammals) to diphthamide, a posttranslational modification not known to occur to any other cellular protein. This renders EF-2 sensitive to mono-ADP ribosylation by diphtheria toxin (and presumedly cellular enzymes as well), although the diphthamide modification does not appear to be necessary for EF-2 enzymatic activity (OMURA et al. 1988).

Once the coding for the polypeptide is complete and one of three stop codons (UAA, UAG, or UGA) appears in the A site, the peptide chain termination factor RF (release factor) catalyzes the release of the completed chain from the ribosome in a GTP-dependent manner. The unique feature of the rabbit RF is that it is 86% identical (92% similar) to bovine tryptophanyl-tRNA synthetase (LEE et al. 1990; GARRET et al. 1991). Like the tryptophanyl-tRNA synthetase, RF contains sequence elements of class I aminoacyl-tRNA synthetases, although the conserved "HIGH" sequence is NVGH in rabbit RF and this difference may preclude RF serving as both a chain termination factor and aminoacyl-tRNA synthetase. At the same time, this raises the possibility that a tRNA-like molecule may be associated with termination, although there is no direct evidence to support this possibility.

What should be noted from the above description is that, although the translation factors all bind GTP, their use of GTP hydrolysis is different. For both eIF-2 and EF-1α, the hydrolysis of GTP is a signal that a correct positioning of the aminoacyl-tRNA has occurred. The conformational change that ensues "forces" the factor off the surface of the ribosome, thereby enhancing the kinetic rate at which the aminoacyl-tRNA is available for a subsequent reaction (for more detail, see Chap. 2). A protein which also "senses" the correct matching of the initiator tRNA with its AUG codon is eIF-5, which either triggers the GTP hydrolysis of eIF-2-bound GTP or triggers GTP hydrolysis directly when eIF-2A directs the binding of Met-tRNA$_i$ to 40S subunits in a nucleotide-independent manner (ADAMS et al. 1975). The result of the eIF-5-triggered GTP hydrolysis is the loss of all translation factors from the surface of the 40S subunit to allow 60S subunit joining.

The remaining translation factors exhibit different characteristics. eIF-2B facilitates nucleotide exchange, although it is not known if the GTP bound by eIF-2B is the one that is exchanged for eIF-2-bound GDP. This process may be similar to one of the cycles for the trimeric G-proteins (see chapters later in this book). EF-2 would seem to be the most unusual of the GTPase translation factors in that the movement of the peptidyl-tRNA and its matched codon would be seen more as "work" than as signal transduction. The RF is probably the best example of all the factors of signal transduction in protein synthesis, and is therefore perhaps the "classic G-protein." However, it is also the one protein whose sequence bears the least resemblance to other GTPases.

B. Consensus Sequences of GTPases Factors

Beyond the observation that the above initiation factors bind and hydrolyze GTP where sequences have become available, there is a very good match of the GTP consensus sequence elements (Table 2). While a portion of this reflects the general nature of the conserved GTP-binding domain as

Table 2. GTP Consensus sequences for translation factors

	GXXXXGK	DXXG	NKXD
PROKARYOTES			
EF-Tu	IGHVDHGKTTL	VDCPGH	LNKCDM
Sel B	AGHVDHGKTTL	IDVPGH	ATKADR
EF-2	SAHIDAGKTTT	IDTPGH	VNKMDR
IF-2	MGHVDHGKTSL	LDTPGH	VNKIDK
EUKARYOTES			
EF-1α	IGHVDSGKSTT	IDAPGH	VNKMDS
SUF-12	MGHVDAGKSTM	LDAPGH	VNKMDD
EF-2	IAHVDHGKSTL	IDSPGH	MNKMDR
eIF-2γ	IGHVAHGKSTV	VDCPGH	QNKVDL
CONSENSUS	GHVDHGK T	D PGH	NK D

The above sequences are from *E. coli* EF-Tu (ARAI et al. 1980), Sel B (FORCHHAMMER et al. 1989), EF-G (OVCHINNIKOV et al. 1982), IF-2 (SACERDOT et al. 1984), yeast EF-1α (NAGATA et al. 1984), SUF-12 (WILSON and CULBERTSON 1988), eIF-2γ (Dr. Ernie Hannig, personal communication), and hamster EF-2 (KOHNO et al. 1986).

evidenced by the crystal structures for EF-Tu and Ras (Kjeldgaard and Nyborg 1992; Pai et al. 1990), the high degree of conservation of the GHVDHGKST motif in eukaryotes (GHVDHGKTT in prokaryotes) appears to be diagnostic for translation factors. To date, only the Lep A protein of *Escherichia coli* has been identified as a protein that has this sequence but which is not a translation factor (March and Inouye 1985). A perhaps curious feature is reflected in one slight deviation from the consensus sequence. The proteins involved in the translocation step of elongation (EF-Tu, EF-2) have the conservative substitution of alanine (A) in place of the usual first glycine (G). The second consensus element is also reasonably well conserved with I, V, or L followed by DXPGH, where X is C, V, or T in prokaryotes and A, S, or C in eukaryotes. The final consensus element, NKXD, is least well conserved. Based upon the crystal structure (Kjeldgaard and Nyborg 1992) and site-directed mutagenesis (Hwang et al. 1989), these residues are associated with guanine specificity, and while they have been conserved, the immediate neighboring amino acids have not.

For those factors for which amino acid sequences have been determined, primarily from cDNA clones, the one factor not fitting the "GTP-binding protein" mold is RF. As noted earlier, this protein is 86% identical to mammalian tryptophanyl-tRNA synthetase (Lee et al. 1990; Garret et al. 1991). Like the synthetase, RF has an amino acid sequence consistent with class I synthetases and thereby the prediction of a Rossmann fold, a nucleotide-binding pocket best characterized for ATP-binding proteins. Second, like the synthetase, RF is functional as a dimer. The other GTPase translation factors are functional as monomers or oligomers with nonidentical subunits.

C. Evolution of EF-1α

While the conservation of the general GHVDHGK${}^{S}_{T}$T motif is impressive in being so highly conserved among translation factors of different functions, equally impressive is the conservation of sequence for the evolution of EF-Tu to EF-1α. As presented in Table 3, numerous sequences are available. These sequences present a continuum with few breaks, as can be seen either as the percentage of identity or, if very conservative substitution is allowed, as the percentage of similarity. The major breaks occur between prokaryotes and archebacteria and then between archebacteria and eukaryotes when just the percentage of homology is considered. Within eukaryotes there appears to be a smooth transition between plants, yeasts, and then higher eukaryotes.

A somewhat different picture is obtained when one looks at the protein sequences not by amino acid homology, but by the organization of the protein as indicated in Fig. 1. Here it can be seen that the loss of homology to bacterial EF-Tu is not only due to differences in amino acid sequences,

Table 3. Evolution of EF-Tu to EF-1α

	Amino Acid Sequence		# of Amino Acides	Accession #
	Simil.	Ident.		
Bacteria				
Coleochaete orbicularis (chloroplast)	52.3	27.5	414	M34286
Chlamydia trachomatis	51.9	29.9	393	M74221
Spirulina platensis	53.7	30.8	409	X15646
Cyanophora paradoxa	55.6	30.9	408	X52497
Synechococcus sp. (cyanobacterium)	56.9	31.0	408	S04430
Arabidopsis thaliana (chloroplast, tufA gene)	55.7	31.3	409	X52256
Mycoplasma hominis	56.3	32.0	396	M57675
Euglena gracilis (chloroplast)	54.1	32.0	408	X00044
Mycoplasma gallisepticum	53.5	32.2	393	X16462
Euglena gracilis (plastid genes)	55.1	32.5	408	X06254
Salmonella typhimurium (tufA)	55.6	32.6	393	X55116
Salmonella typhimurium (tufB)	55.6	32.6	393	X55117
Escherichia coli (tufA gene)	55.7	32.6	393	J01690
Escherichia coli (tufB gene)	55.6	32.6	393	J01717
Astasia longa (chloroplast)	54.8	32.7	408	X14385
Mycoplasma capricolum	51.3	32.8	393	X16463
Cryptomonas phi (chloroplast)	57.1	33.2	407	X52912
Chlamydomonas reinhardtii (tufA)	55.1	34.0	417	X52257
Saccharomyces cerevisiae (mitochondria, tufM)	55.8	34.4	436	K00428
Mycobacterium tuberculosis	60.0	35.3	395	X63539
Thermus thermophilus	54.8	35.5	405	X06657
Thermus aquaticus	55.0	35.6	405	P07157
Micrococcus luteus	57.5	36.8	405	D26956
Thermotoga maritima	57.0	37.3	399	M27479

Table 3. *Continued*

Bacteria	Amino Acid Sequence		# of Amino Acides	Accession #
Archeobacteria				
Halobacterium marismortui	68.5	50.7	421	X16677
Haloarcula sp.	68.6	50.8	421	S08060
Sulfolobus acidocaldarius	71.0	52.0	435	X52382
Methanococcus vannielii	71.0	53.7	428	X05698
Thermoplasma acidophilum	74.8	54.4	424	X53866
Thermococcus celer	72.4	55.1	428	X52383
Pyrococcus woesei	73.2	55.8	430	X59857
Eukaryota				
Plasmodium falciparum	83.8	69.6	443	X60488
Dictyostelium discoideum (Slime mold, EF1-I)	85.7	77.4	453	X55973
Dictyostelium discoideum (Slime mold, EF1-II)	85.9	77.9	453	X55972
Stylonychia lemnae	85.0	76.3	446	X57926
Phaseolus vulgaris (Bean, missing 1–49, Axelos et al. 1989)	85.5	76.8	incomp.	
Entamoeba histolytica	87.2	77.2	430	M92073
Arabidopsis thaliana (A1 gene)	86.2	77.8	449	X16430
Arabidopsis thaliana (A2 gene)	86.2	77.8	449	X16431
Arabidopsis thaliana (A3 gene)	86.2	77.8	449	X16431
Arabidopsis thaliana (A4 gene)	86.2	77.6	449	X16432
Daudus carota	86.9	78.0	449	X60302
Lycopersicon esculentum (Tomato, strain cv. VFNT cherry)	87.1	78.1	447	X53043
Lycopersicon esculentum (Tomato, strain caligrande)	87.1	78.2	448	X14449
Glycine max (Soybean)	86.4	78.3	447	X56856
Euglena gracilis	86.8	79.6	445	X16890

Saccharomyces cerevisiae (Yeast)(TEF1)	88.7	80.8	458	M15666
Saccharomyces cerevisiae (Yeast)(TEF2)	88.7	80.8	458	M15667
Candida albicans (Dimorphic Yeast)(TEF1)	88.9	81.9	458	M29934
Candida albicans (Dimorphic Yeast)(TEF2)	88.9	81.9	458	M29935
Mucor racemosus (TEF1 gene)	89.1	81.9	458	M16352
Mucor racemosus (TEF2 gene)	89.3	82.1	458	X17476
Mucor racemosus (TEF3 gene)	89.5	82.1	457	X17475
Absidia glauca (TEF-1 gene)	90.9	82.1	458	X54730
Mucor lusitanicus (2)	89.7	82.3	458	S06300
Mucor lusitanicus (3)	90.0	82.5	458	S08058
Artemia salina (Brine Shrimp)	90.5	82.5	462	X00546
Onchocerca volvulus	90.5	82.9	464	M64333
Drosophila melanogaster (Fruit Fly, F1 gene)	91.1	84.2	463	X06869
Drosophila melanogaster (Fruit Fly, F2 gene)	90.1	84.9	462	X06870
Apis mellifera (Honey bee, missing C-term. from aa #440)	91.8	85.9	incomp.	X52884
Xenopus laevis (Frog, X1EF1aO1, oocyte)	96.7	90.7	463	X52975
Xenopus laevis (Frog, X1EF-1aO, occyte)	96.7	91.1	461	X52976
Xenopus laevis (Frog, EF-1αS, somatic)	98.3	95.9	462	M25697
Rattus norvegicus (Rat)	99.8	99.8	462	X63561
Cricetulus longicaudatus (Hamster)	99.8	99.8	462	D00522
Mus musculus (Mouse)	99.8	99.8	462	M22432
Oryctologus cuniculus (Rabbit)	100.0	100.0	462	X62245
Homo sapiens (Human)	◆	◆	462	X03558

All sequences are compared to the human EF-1α sequence.

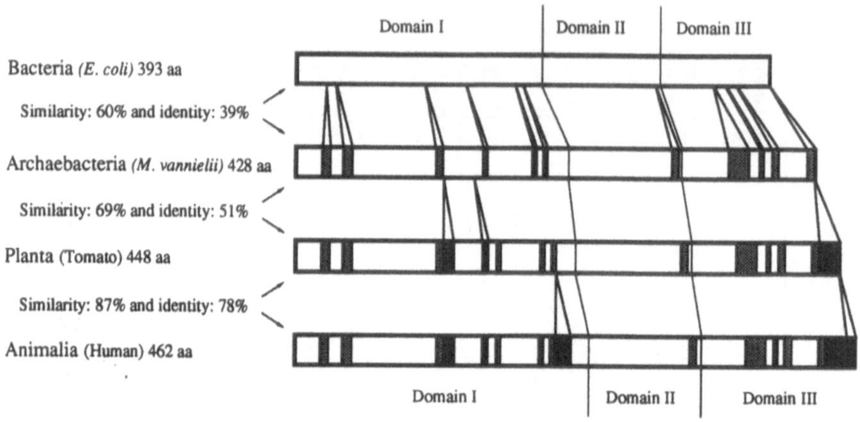

Fig. 1. Evolution of EF-1α. The schematic growth of EF-1α from *E. coli* EF-Tu. The domains of EF-Tu are: *I* residues 1–200; *II* residues 210–294; *III* residues 299–393. Domain *I* is the GTP-binding domain. The width of the boxes is representative of the number of amino acids inserted into the evolving EF-1α sequence. See also Table 3 for a comparison of sequence homology in the evolution of EF-1α

but also in the actual size of the emerging EF-1α protein. Thus, there is an increase in size of archebacterial EF-1α by the insertion of 35 amino acids at discrete locations into the EF-Tu sequence. All of these inserts are maintained in the subsequent evolution to plant EF-1α, which is now another 20 amino acids longer. Similarly, "higher" eukaryotes have an additional 14 amino acids, represented by two final inserts. The transitions among these four groups appear quite clear when the sequences are viewed by length as well as by homology in the distinction of bacteria from archebacteria from eukaryotes. However, the division of eukaryotes into a plant and an animal group is only seen by the length of the protein, as there is a reasonably smooth continuum in sequence homology going from plants to animals (Table 3).

D. The EF-Tu Family

Up to this point we have seen two different conservations of amino acid sequence. The first was the GTP consensus sequences which reflect an overall conservation of the GTP-binding domain of about 200 amino acids. The second was the conservation of sequence for the functionally identical proteins EF-Tu and EF-1α. There is yet a third conserved element, referred to as the EF-Tu family, which represents a group of proteins whose presumed or known function is to bind aminoacyl-tRNAs in a GTP-dependent manner. This group of proteins is presented in Table 4, which matches the sequences of the five proteins. Sel B is a bacterial protein which is required for the binding of selenocysteinyl-tRNA to ribosomes in response to a UGA termination codon and a specific sequence/structure in the mRNA 3′ to the UGA codon (FORCHHAMMER et al. 1989). The homology to EF-Tu is best in the GTP-binding domain (residues 1–200), reasonable in domain 2 (residues 200–300), but quite poor thereafter. The next 350 amino acids appear generally unrelated and presumedly it is this portion of the molecule which recognizes the RNA sequence/structure element which follows the UGA termination codon and distinguishes this codon for selenocysteine instead of an RF-2 termination event.

The next member listed in the EF-Tu family is the eukaryotic protein SUF-12, a protein which is characterized genetically and biochemically as being involved in the control of translational ambiguity in yeast, and which is essential for cell viability (WILSON and CULBERTSON 1988). While the first 250 amino acids appear unrelated to members of the EF-Tu family, the remainder of the molecule is quite similar to EF-1α. As with Sel B, the homology is best in what are the GTP-binding domain and domain 2 in EF-Tu. The subsequent match in sequence or length in domain 3 is poor, except for the carboxy-terminal 20 amino acids which match quite well (better than 50% similarity).

The final member of the EF-Tu family is the γ subunit of eIF-2. Like SUF-12, the amino terminal 90 amino acids do not appear to be related to any other sequences, although some poor matches can be made to portions of the amino terminal sequence of SUF-12. As might be anticipated, eIF-2γ shows most extensive homology to the other family members in the GTP-binding domain with one major exception from residues 142 to 185, which represents a unique insert into the GTP-binding domain. Past this first domain, eIF-2γ appears more similar to EF-Tu than EF-1α, both in sequence homology and the spacing of sequence elements. Thus, it would appear that eIF-2γ had a lineage separate from EF-1α in its evolution from EF-Tu.

It should be noted, however, that there is more to eIF-2 than just the EF-Tu-like γ subunit. Previous biochemical studies have implicated both the β and γ subunits in binding GTP and Met-tRNA$_i$ and most, but not all, affinity labeling studies showed modification of these two subunits as well (ANTHONY et al. 1990; BOMMER and KURZCHALIA 1989; DHOLAKIA et al. 1989; MERRICK et al. 1987; NYGÅRD et al. 1980; WESTERMANN et al. 1981). Second, mutations in the α or β subunit allow eIF-2 to misread UUG as an initiation codon, although this effect may more reflect binding to the 40S subunit than the positioning of the initiator tRNA on the surface of the eIF-2 molecule (CIGAN et al. 1989; DONAHUE et al. 1988). Third, preliminary studies indicate that the α or β subunit may influence the ability of the γ subunit to discriminate between the initiator tRNA and other aminoacyl-tRNAs (D.L. HUGHES and W.C. MERRICK, unpublished observation). Thus, while one would assume that eIF-2γ is the major component in binding the initiator tRNA in a GTP-dependent fashion, the other two subunits clearly influence this reaction and subsequent interactions with the ribosome.

Table 4. The EF-Tu family of GTP- and aminoacyl-tRNA binding proteins

```
eIF-2 gamma --------------------------------------------------------------------------------     0
EF-Tu       --------------------------------------------------------------------------------     0
Sel B       --------------------------------------------------------------------------------     0
SUF-12      MSDSNQGNNQQNYQQYSQNGNQQQGNNRYQGYQAYNAQAQPAGGYYQNYQGYSGYQQGGYQQYNPDAGYQQQYNPQGGYQ    80
EF-1 alpha  --------------------------------------------------------------------------------     0

eIF-2 gamma --------------------------------------------------------------------------------     0
EF-Tu       --------------------------------------------------------------------------------     0
Sel B       --------------------------------------------------------------------------------     0
SUF-12      QYNPQGGYQQQFNPQGGRGNYKNFNYNNNLQGYQAGFQPQSQGMSLNDFQKQQKQAAPKPKKTLKLVSSSGIKLANATKK   160
EF-1 alpha  --------------------------------------------------------------------------------     0

eIF-2 gamma MSDLQDQEPSIIINGNLEPVGEPDIVEETEVVAQETQETQDADKPKKKVAFTGLEEDGETEEEKRKREFEEGGGLPEQPL    80
EF-Tu       --------------------------------------------------------------------------------     0
Sel B       --------------------------------------------------------------------------------     0
SUF-12      VGTKPAESDKKEEEKSAETKEPTKEPTKVEEPVKKEEKPVQTEEKTEEKSELPKVEDLKISESTHNTNNANVTSADALIK   240
EF-1 alpha  --------------------------------------------------------------------------------     0

Tu struct.                      ββββββββββββ     αααααααααααααααααααα
eIF-2 gamma NPDFSKLNPLSAEIINRQATINIGTIGHVAHGKSTVVRAISGVQ----------------------------TVRFKDEL   132
EF-Tu       ---------SKEKFERTKPHVNVGTIGHVDHGKTTLTAAITTVLAKTYGGAARAFDQ---------------IDNAPEEK    56
Sel B       --------------------MIIATAGHVDHGKTTLLQAITGVN----------------------------ADRLPEEK    32
SUF-12      EQEEEVDDEVVNDMFGGKDHVSLIFMGHVDAGKSTMGGNLLYLTGSVDKRTIEKYEREAKDAGRQGWYLSWVMDTNKEER   320
EF-1 alpha  -------------MGKEKTHINIVVIGHVDSGKSTTTGHLIYKCGGIDKRTIEKFEKEAAEMGKGSFKYAWVLDKLKAER    67

Tu struct.     ββββββββββββββββ                                      ββββββββββββ  ααααααααααα
eIF-2 gamma ERNITIKLGYANAKIYKCQEPTCPEPDCYRSFKSDKEISPKCQRPGCPGRYKLVRHVSFVDCPGHDILMSTMLSGAAVMD   212
EF-Tu       ARGITINTSHVEYDTPT-------------------------------------RHYAHVDCPGHADYVKNMITGAAQMD    99
Sel B       KRGMTIDLGYAYWPQPDG------------------------------------RVPGFIDVPGHEKFLSNMLAGVGGID    76
SUF-12      NDGKTIEVGKAYFETEK-------------------------------------RRYTILDAPGHKMYVSEMIGGASQAD   363
EF-1 alpha  ERGITIDISLWKFETSK-------------------------------------YYVTIIDAPGHRDFIKNMITGTSQAD   110

Tu struct.  βββββββββ           αααααααααααααααα  βββββββββ           α→ ←αααααααααααααααααα    ββββ→
eIF-2 gamma AALLLIAGNESCPQ------PQTSEHLAAIEIMKLKHVIILQNKVDLRMEESALEHQKSILKFIR------GTIA-DGAP   279
EF-Tu       GAILVVAATDGPM-------PQTREHILLGRQVGVPYIIVFLNKCDMVDDEEL---LELVEMEVRELLSQYDFPGDDTPI   169
Sel B       HALLVVACDDGVM-------AQTREHLAILQLTGNPMLTVALTKADRVDE----ARVDEVERQVKEVLREYGFAEAKLFI   145
SUF-12      VGVLVISARKGEYETGFERGGQTREHALLAKTQGVNKMVVVVNKMDDPTVNWSKERYDQCVSNVSNFLRAIGYNIKTDVV   443
EF-1 alpha  CAVLIVAAGVGEFEAGISKNGQTREHALLAYTLGVKQLIVGVNKMDSTEPPYSQKRYEEIVKEVSTYIKKIGYNP-DTVA   189

Tu struct. ←ββββ   αααα   αααααααααααααααααααα→                    ←α   αααααα ββββββββββββββββ ββββββββββββ
eIF-2 gamma IVPISAQLKYNIDAVNEFIVKTIPVPPRDFMISPRLIVIRSFDVNKPGAEIEDL---------------KGGVAGGSILN   344
EF-Tu       VRGSALKALEGDAEWEAKILELAGFLDSY----------------IPEPERAIDKPFLLPIEDVFSISGRGTVVTGRVER   233
Sel B       TAATEGRGMDALREHLLQLPER---------------------------EHASQHSFRLAIDRAFTVKGAGLVVTGTALS   198
SUF-12      FMPVSGYSGANLKDHVDPKECPWYTGPTLLEYLDTMNHV----------DRHINAPFMLPIAAKMKD--LGTIVEGKIES   511
EF-1 alpha  FVPISGWNGDNMLEPSANMPWFKGWKVTRKDGNASGTTLLEALDCILPPTRPTDKPLRLPLQDVYKIGGIGTVPVGRVET   269
```

```
Tu struct.            ββββββββ    βββββββββ                   ββββββββββββ     ββββββββ      βββββββββββββββββββ
eIF-2 gamma  GVFKLGDEIEIRPGIVTKDDKGKIQCKPIFSNIVSLAFEQNDLKFAVPGGLIGVGTKVDPTLCRADRLVGQVVGAKGHLP   424
EF-Tu        GIIKVGEEVEIVGIKETQKSTCTGVEMFRKLLDEGRAGENVGVLLRGIKREEIERGQVLAKPGT--IKPHTKFESEVYIL   311
Sel B        GEVKVGDSLWLTGVNKPMRVRALHAQNQPTET--ANAGQRIALNIAGDAEKEQINRGDWLLADVPPEP-----------   264
SUF-12       GHIKKGQSTLLMPNKTAVEIQNIYNETENEVDM-AMCGEQVKLRIKGVEEEDISPGFVLTSPKN-PIKSVTKFVAQIAIV   589
EF-1 alpha   GVLKPGMVVTFAPVNVTTEVKSVEMHHEALSE--ALPGDNVGFNVKNVSVKDVRRGNVAGDSKNDPPMEAAGFTAQVIIL   347

Tu struct.                     βββββββββββ  ββββββββββββ          ββ→   ←ββββββββββββ     ββββ→
eIF-2 gamma  NIYTDIEINYFLLRRLLGVKTDGQKQAKVRKLEPNEVLMVNIGSTATGARVVAVKADMARLQLTSPACTEINEKIA----   500
EF-Tu        SKDEGGRHTPFFKGYRPQFYFRTTDVTGTIELPEGVEMVMPGD------NIKMVVTLIHPIAMDDGL------------   372
Sel B        ------------------------------------------------FTRVIVELQTHTPLTQWQPLHIHHAASHVTG   295
SUF-12       ELKSIIAAGFSCVMHVHTAIEEVHIVKLLHKLEKGTNRKSKKPPAFAKKGMKVIAVLETEAPVCVETYQDYPQLG-----   664
EF-1 alpha   NHPGQISAGYAPVLDCHTAHIACKFAELKEKIDRRSGKKLEDGPKFLKSGDAAIVDMVPGKPMCVESFSDYPPLG-----   422

eIF-2 gamma  --------------------------------------------------------------------------------   500
EF-Tu        --------------------------------------------------------------------------------   372
Sel B        RVSLLEDNLAELVFDTPLWLADNDRLVLRDISARNTLAGARVVMLNPPRRGKRKPEYLQWLASLARAQSDADALSVHLER   375
SUF-12       --------------------------------------------------------------------------------   664
EF-1 alpha   --------------------------------------------------------------------------------   422

eIF-2 gamma  --------------------------------------------------------------------------------   500
EF-Tu        --------------------------------------------------------------------------------   372
Sel B        GAVNLADFAWARQLNGEGMRELLQQPGYIQAGYSLLNAPVAARWQRKILDTLATYHEQHRDEPGPGRERLRRMALPMEDE   455
SUF-12       --------------------------------------------------------------------------------   664
EF-1 alpha   --------------------------------------------------------------------------------   422

Tu struct.                                                                       ←ββββββββ  ββββββ→
eIF-2 gamma  --------------------------------------------------------------------LSRRIEKHWRLI   512
EF-Tu        --------------------------------------------------------------------RFAIREGGRTVG   384
Sel B        ALVLLLIEKMRESGDIHSHHGWLHLPDHKAGFSEEQQAIWQKAEPLFGDEPWWVRDLAKETGTDEQAMRLTLRQAAQQGI   535
SUF-12       --------------------------------------------------------------------RFTLRDQGTTIA   676
EF-1 alpha   --------------------------------------------------------------------RFAVRDMRQTVA   434

Tu struct. ←ββββββββββββ
eIF-2 gamma  GWATIKKGTTLEPIA----------------------------------------------------------------   528
EF-Tu        AGVVAKVLS----------------------------------------------------------------------   393
Sel B        ITAIVKDRYYRNDRIVEFANMIRDLDQECGSTCAADFRDRLGVGRKLAIQILEYFDRIGFTRRRGNDHLLRDALLFPEK   615
SUF-12       IGKIVKIAE----------------------------------------------------------------------   686
EF-1 alpha   VGVIKAVDKKAAGAGKVTKSAQKAQKAK---------------------------------------------------   463
```

Presented in this table are the sequences of four proteins which have amino acid sequences similar to EF-Tu and which appear to function by binding GTP and aminoacyl-tRNA. The source of the amino acid sequences is given in the legend to Table 2, except the EF-1α sequence, which is from human (Accession # X03558). The repeated α's and β's represent the known structural elements (α helix, β sheet) of EF-Tu (Kjeldgaard and Nyborg 1992). The sequences were aligned by computer and subsequently by hand to optimize conserved amino acid sequences within the presumed structural elements. The alignment of carboxy-terminal sequences with residues 524–544 in sel B represents a poor match to the EF-Tu sequences and may not be valid. Thus, sel B may have either a 230 amino acid insert and a 70 amino acid carboxyteminal extension or just a 300 amino acid carboxy-terminal extension

E. Structures of the EF-Tu Family

At this point, several comments should be made as to the apparent structure of the EF-Tu family members, as it is not clear from sequence homology what parts of the EF-Tu structure have been conserved. By using the sequence matchups illustrated in Table 4 and the recently refined crystal structure of EF-Tu (KJELDGAARD and NYBORG 1992), it is possible to visualize regions of similarity and difference. The dark shaded regions in Fig. 2 indicate stretches of conserved sequences and, as indicated earlier, most of these are in domain 1, the GTP-binding domain. While most of the β strands, except for the last one, are conserved (residues 167–172), only three of the six helices are conserved (helices 1, 2, and 3). Helices 1 and 2 are directly adjacent to loops or turns containing the consensus sequence motifs GHVDHGKST and DXPGH discussed earlier. Although perhaps not evident from this "cartoon" of the EF-Tu structure, helix 2 is a major part of the interface between domain 1 and domain 3. The fact that this

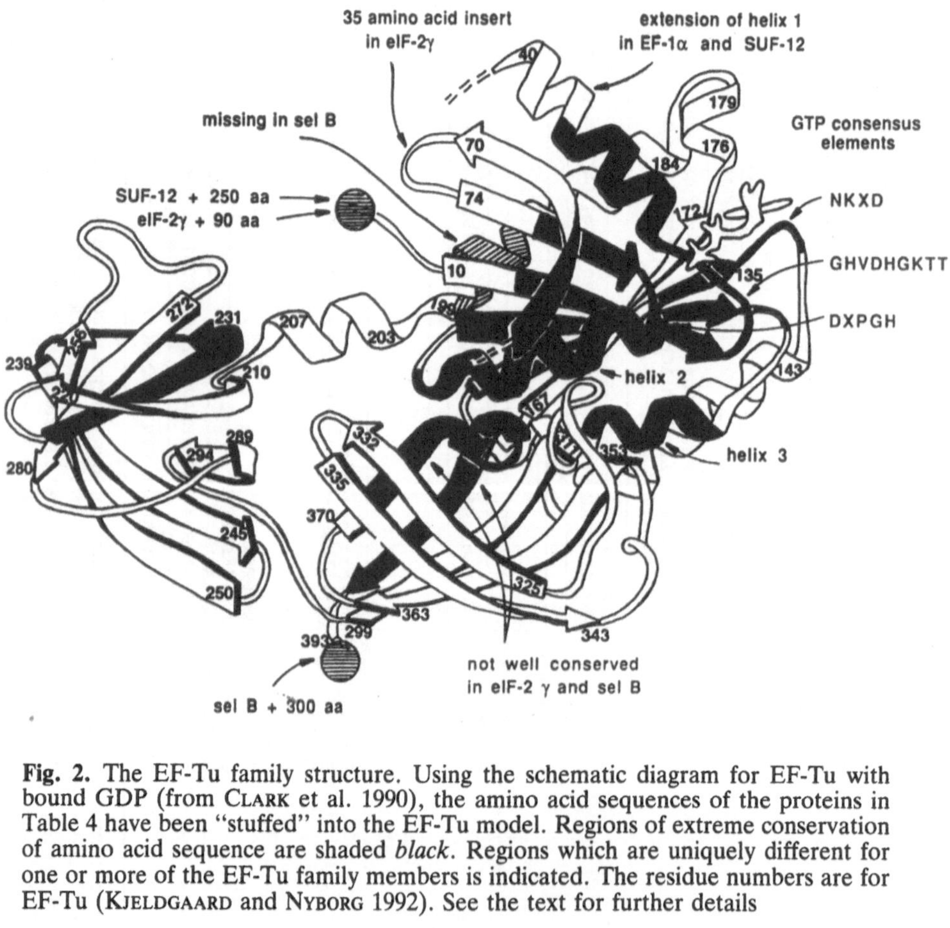

Fig. 2. The EF-Tu family structure. Using the schematic diagram for EF-Tu with bound GDP (from CLARK et al. 1990), the amino acid sequences of the proteins in Table 4 have been "stuffed" into the EF-Tu model. Regions of extreme conservation of amino acid sequence are shaded *black*. Regions which are uniquely different for one or more of the EF-Tu family members is indicated. The residue numbers are for EF-Tu (KJELDGAARD and NYBORG 1992). See the text for further details

interface is important is mirrored by the general conservation of amino acid sequence of the last two β strands in domain 3 (residues 370–378 and 381–392) which interact with helix 2. A third area of general conservation is the second β strand and following loop in domain 2 (residues 222–239).

At the same time it should be noted that for many of the family members, their "structures" can be generated by relatively small (2–6 amino acids) insertions into regions which lack structure (loops or turns), thus generally allowing the preservation of the orientation of α-helical and β-sheet elements. Perhaps the most interesting of these is eIF-2γ, with the apparent insertion of about 35 amino acids in the turn linking the β strands from residues 60–71 and 74–82 (see Fig. 2). This insert is very close to the region identified in Ras as the GAP (GTPase activating protein) binding site (BOURNE et al. 1991), which may reflect the requirement of two activators for eIF-2-dependent GTP hydrolysis, 40S subunits, and eIF-5. Secondly, while this stretch is generally hydrophilic, it does contain a rather high level of cysteines and prolines (5 and 6, respectively).

Another insert into the general structure is seen with EF-1α and SUF-12, which extend helix 1 by 6 or 7 amino acids. This conclusion is based upon the ability of these amino acids to form a helix (shown by computer analysis) and by maintenance of the hydrophilic and hydrophobic faces of the amphipathic helix. Presumedly to accommodate this extended helix, the amino acids linking this helix to the next β strand are increased from 20 to 28 (note: this connecting region is not present in the EF-Tu crystal structure).

For the molecules of the EF-Tu family, the inserts discussed above are small compared to the 250 (SUF-12) and 90 (eIF-2) amino acids added as amino terminal extensions and the 300 amino acid carboxy-terminal extension of sel B (see Table 4). Unfortunately, it is not possible to model the structure or function of these regions as independent entities or as modulators of the activity of the EF-Tu-like structural core.

In contrast to the large number of small insertions and the few large insertions, there is one unique shortening of a structural element. This occurs in helix 6 of sel B, which ends up 11 residues shorter. As this helix leads directly to the "bridge" that joins domain 1 with domain 2, it would appear that the orientation of domain 2 would be altered, either in its distance from domain 1 or in a change in the packing of the first β strand in domain 2. Other permutations would, of course, be possible. While this small deletion may seem trivial given the much larger insertions into some of the other members of the EF-Tu family, it is the only perceived change that would seem to have a rather dramatic consequence in the "core structure," which seems to have been preserved in other family members.

As noted earlier, there are several proteins associated with the translational apparatus which bind GTP for which amino acid sequence information is not yet available. This includes the 39000-da subunit of eIF-2B, eIF-5, and a mitochondrial ribosomal protein. It will be of interest to determine whether these proteins contain the translation factor GTP-binding sequence motifs, the characteristics of the EF-Tu family, homology to other proteins (i.e., such as RF, see above) or whether they will represent a new and unique sequence and structure. In addition, especially in eukaryotes, it is always possible that more proteins which bind GTP will be associated with the translational process (i.e., see Chap. 7).

Acknowledgements. The authors wish to thank Dr. Ernie Hannig for making the amino acid sequence of yeast eIF-2γ available prior to publication. The authors are grateful for many very useful conversations with Dr. JENS NYBORG and Dr. MORTEN KJELDGAARD concerning the three dimensional structure of EF-Tu and the substitution of the sequences of the other EF-Tu family members into the EF-Tu crystal structure. This manuscript was expertly prepared by Ms. TONI BODNAR. This research effort was supported in part by National Institutes of Health grant GM-26796, the Forskerakademiet (W. C. M.), and by the Northeast Ohio Chapter of the American Heart Association (J. C.).

References

Adams SL, Safer B, Anderson WF, Merrick WC (1975) Eukaryotic initiation complex formation: evidence for two distinct pathways. J Biol Chem 250:9083–9089

Anthony DD Jr, Kinzy TG, Merrick WC (1990) Affinity labelling of eukaryotic initiation factor 2 and elongation factor 1$\alpha\beta\gamma$ with GTP analogs. Arch Biochem Biophys 281:157–162

Arai K, Clark BFC, Duffy L, Jones MD, Kaziro Y, Laursen RA, L'Italien J, Miller DL, Nagarkatti S, Nakamura S, Nielsen KM, Petersen PE, Takahashi K, Wade M (1980) Primary structure of elongation factor Tu from Escherichia coli. Proc Natl Acad Sci USA 77:1326–1330

Axelos M, Bardet C, Liboz T, Thai ALV, Curie C, Lescure B (1989) The gene family encoding the Arabidopsis thaliana translation elongation factor EF-1α: Molecular cloning, characterization and expression. Mol Gen Genet 219: 106–112

Bommer U-A, Kurzchalia TV (1989) GTP interacts through its ribose and phosphate moieties with different subunits of the eukaryotic initiation factor eIF-2. FEBS Lett 244:323–327

Bourne HR, Sanders DA, McCormick F (1991) The GTPase superfamily: conserved structure and molecular mechanism. Nature 349:117–127

Cigan AM, Pabich EK, Feng L, Donahue TF (1989) Yeast translation initiation suppressor sui 2 encodes the α subunit of eukaryotic initiation factor 2 and shares sequence identity with the human α subunit. Proc Natl Acad Sci USA 86:2784–2788

Clark BFC, Kjeldgaard M, la Cour TFM, Thirup S, Nyborg J (1990) Structural determination of the functional sites of E. coli elongation factor Tu. Biochim Biophys Acta 1050:203–208

Dholakia JN, Wahba AJ (1989) Mechanism of the nucleotide exchange reaction in eukaryotic polypeptide chain initiation. J Biol Chem 264:546–550

Dholakia JN, Francis BR, Haley BE, Wahba AJ (1989) Photoaffinity labelling of the rabbit reticulocyte guanine nucleotide exchange factor and eukaryotic initiation factor 2 with 8-azidopurine nucleotides: identification of GTP- and ATP-binding domains. J Biol Chem 264:20638–20642

Donahue TF, Cigan AM, Pabich EK, Valavicius BC (1988) Mutations at a Zn (II) finger motif in the yeast eIF-2β gene alter ribosomal start-site selection during the scanning process. Cell 54:621–632

Forchhammer K, Leinfelder W, Böck A (1989) Identification of a novel translation factor necessary for the incorporation of selenocysteine into protein. Nature 342:453–456

Garret M, Pajot B, Trézequet V, Labouesse J, Merle M, Gandar J-C, Benedetto J-P, Sallafranque M-L, Alterio J, Gueguen M, Sarger C, Labouesse B, Bonnet J (1991) A mammalian tryptophanyl-tRNA synthetase shows little homology to prokaryotic synthetases, but near identity with mammalian peptide chain release factor. Biochemistry 30:7809–7817

Hwang Y-W, Sanchez A, Miller DL (1989) Mutagenesis of bacterial elongation factor Tu at lysine 136: a conserved amino acid in GTP regulatory proteins. J Biol Chem 264:8304–8309

Kjeldgaard M, Nyborg J (1992) The refined structure of elongation factor Tu from Eschericia coli. J Mol Biol 223:721–742

Kohno K, Uchida T, Ohkubo H, Nakanishi S, Nakanishi T, Fukui T, Ohtsuka E, Ikehara M, Okada Y (1986) Amino acid sequence of mammalian elongation factor 2 deduced from the cDNA sequence: homology with GTP-binding proteins. Proc Natl Acad Sci USA 83:4978–4982

Lee CC, Craigen WJ, Munzy DM, Harlow E, Caskey CT (1990) Cloning and expression of a mammalian peptide chain release factor with sequence similarity to tryptophanyl-tRNA synthetase. Proc Natl Acad Sci USA 87:3508–3512

March PE, Inouye M (1985) GTP-binding membrane protein of Escherichia coli with sequence homology to initiation factor 2 and elongation factors Tu and G. Proc Natl Acad Sci USA 82:7500–7504

Merrick WC (1992) Mechanism and regulation of eukaryotic protein synthesis. Microbiol Rev 56:291–315

Merrick WC, Kemper WM, Anderson WF (1975) Purification and characterization of homogeneous initiation factor M2A from rabbit reticulocytes. J Biol Chem 250:5556–5562

Merrick WC, Abramson RD, Anthony DD Jr, Dever TE, Caliendo AM (1987) Involvement of nucleotides in protein synthesis initiation. In: Ilan J (ed) Translational regulation of gene expression. Plenum, New York, p 265.

Merrick WC, Dever TE, Kinzy TG, Conroy SC, Cavallius J, Owens CL (1990) Characterization of protein synthesis factors from rabbit reticulocytes. Biochim Biophys Acta 1050:235–240

Nagata S, Nagashima K, Tsunetsugu-Yokota Y, Fujimura K, Miyazaki M, Karizo Y (1984) Polypeptide chain elongation factor 1α (EF-1α) from yeast: nucleotide sequence of one of the two genes for EF-1α from Saccharomyces cerevisiae. EMBO J 3:1825–1830

Nygård O, Westermann P, Hultin T (1980) Met-tRNA$_f^{Met}$ is located in close proximity to the β subunit of eIF-2 in the eukaryotic initiation complex eIF-2·Met-tRNA$_f$·GDPCP. FEBS Lett 113:125–128.

Omura F, Kohno K, Urchida T (1988) The histidine residue of codon 715 is essential for function of elongation factor 2. Eur J Biochem 180:1–7

Ovchinnikov YA, Alakhov YB, Bundulis YP, Bundule MA, Dovgas NV, Kozlov VP, Motuz LP, Vinokurov LM (1982) The primary structure of elongation factor G from Escherichia coli: a complete amino acid sequence. FEBS Lett 139:130–135

Pai EF, Krengel U, Petsko GA, Goody RS, Kabasch W, Wittinghofer A (1990) Refined crystal structure of the triphosphate conformation of H-ras p21 at 1.35 Å resolution: implications for the mechanism of GTP hydrolysis. EMBO J 9:2351–2359

Riis B, Rattan SIS, Clark BFC, Merrick WC (1990) Eukaryotic protein elongation factors. Trends Biochem Sci 15:420–424

Rowlands AG, Panniers R, Henshaw EC (1988) The catalytic mechanism of guanine nucleotide exchange factor action and competitive inhibition by phosphorylated eukaryotic initiation factor 2. J Biol Chem 263:5526–5533

Sacerdot C, Dessen P, Hershey JWB, Plumbridge JA, Grunberg-Manago M (1984) Sequence of the initiation factor IF-2 gene: unusual protein features and homologies with elongation factors. Proc Natl Acad Sci USA 81:7787–7791

Westermann P, Nygård O, Bielka H (1981) Cross-linking of Met-$tRNA_f$ to eIF-2β and to the ribosomal proteins S3a and S6 within the eukaryotic initiation complex, eIF-2·GMPPCP·Met-$tRNA_f$·small ribosomal subunit. Nucleic Acids Res 9:2387–2396

Wilson PG, Culbertson MR (1988) SUF 12 suppressor protein of yeast: a fusion protein related to the EF-1 family to elongation factors. J Mol Biol 199:559–573

CHAPTER 10

Heterotrimeric G-Proteins: α, β, and γ Subunits

H. ITOH and Y. KAZIRO

A. Introduction

Heterotrimeric G-proteins, which are involved in a variety of hormonal and sensory signal transductions in mammalian cells, consist of three subunits: α (39–52 kDa), β (35–36 kDa), and γ (7–10 kDa). They are widely distributed among different tissues and organisms and their structures are highly conserved. This chapter briefly reviews the structure of the cDNAs and the genes for various G-protein α, β, and γ subunits from mammalian cells and the occurrence of G-protein genes in lower eukaryotes.

B. Mammalian G-Proteins

Recent developments in molecular cloning have revealed the existence of multiple species of G-protein subunits. For α subunits, at least 16 genes have been found in mammalian cells. They code for at least 20 distinct species, since some of them given rise to more than one cDNA species by alternative splicing. The occurrence of four β subunits and seven γ subunits in mammalian cells has also been reported. The deduced amino acid sequences of the G-protein α subunits are compared in Fig. 1, and the organization of several G_α genes is illustrated in Fig. 2.

I. α Subunits

1. Isolation of cDNAs and Genomic DNAs

a) $G_{s\alpha}$

$G_{s\alpha}$ (stimulatory G_α) is involved in hormonal stimulation of adenylyl cyclase. Since there is a specific $G_{s\alpha}$ expressed only in olfactory cells (G_{olf}), we refer to the classical $G_{s\alpha}$ that is expressed in all kinds of tissues as $G_{s1\alpha}$, and to the one expressed exclusively in olfactory cells as $G_{s2\alpha}$. As shown in Fig. 2, human $G_{s1\alpha}$ gene contains 13 exons, of which exons 3 and 12 encode the unique sequences of $G_{s1\alpha}$ at N- and C-terminal regions, respectively (KOZASA et al. 1988). Four different $G_{s1\alpha}$ cDNAs (termed $G_{s1\alpha\text{-}1}$ to $G_{s1\alpha\text{-}4}$) are generated by alternative splicing of a single $G_{s\alpha}$ gene. $G_{s1\alpha\text{-}1}$ and $G_{s1\alpha\text{-}3}$ differ in that $G_{s1\alpha\text{-}3}$ lacks a stretch of 45 nucleotides for 15 amino acids

```
Consensus  MG...--------------------.LS.EEK.A...SK.IEK.L.ED.....R.VKLLLLGAGESGKSTIVKQM.IIH..GYS.E----      90
rGs        ..CLGN.SKT.E.............DQRN...AQREAN.K...Q.QK.KQVYRATHR.................R.L.VN.FNG.GGEE      75
rGolf      ..CLGNSSKTAE.............DQGVD..ERREAN.K...Q.QKERLAYKATHR.................R.L.VN.FNP.....      73
rGi1       ..C--....................T..A.D.A.VER..M.DRN.R..GEKAA.E...................K...EA...E.....      64
rGi2       ..C--....................TV.A.D.A.AER..M.D.N.R..GEKAA.E...................K...EA...E.....      64
rGi3       ..C--....................T..A.D.A.VER..M.DRN.R..GEKAAKE...................K...ED...E.....      64
rGo-1      ..C--....................T..A..RA.LER..A...N.K..GISAAKD...................K...ED...ED....      64
rGo-2      ..C--....................T..A..RA.LER..A...N.K..GISAAKD...................K...ED.F.G.....      64
bGt1       ..A--....................GA.A..----KH.REL..K.K..AEKDA.T...................K...QD...L.....      60
bGt2       ..S--....................GA.A.D.ELAKR..EL..K.Q..ADKEAKT...................K...QD...P.....      64
rGgust     ..S--....................GI.S.S.ESAKR..EL..K.Q..AERDA.T...................K...KN...KQ....      64
rGz        ..C--....................RQ.S...E.ARR.RR.DRH.RSESQRQR.EI......TSN.........K...SG.FNL.....      64
mGq        .TLESIMAC................C..E.A.E.RRINDE..RHVRR.KRDAR.EL......T.......FI...R...GS...D.....      70
mG11       .TLESMMAC................C..D.V.ESKRINAE...Q.RR.KRDAR.EL......T.......FI...R...GA...E.....      70
mG14       .AGCC....................C..A...ESQRI.AE..RHVRR.KKDAR.EL......T.......FI...R...GS...D.....      66
mG15       .ARSLTWGCCPW.............C.TE...T.ARIDQE.NRI.L.QKKQEREEL......P.......FI...R...GV...E.....      73
hG16       .ARSLTWRCCPW.............C.TED..A.ARVDQE.NRI.L.QKKQDRGEL......P.......FI...R...GA...E.....      73
mG12       .SGVVRTLSRCLLPAEAGARERRAGAARDA.RE.RRR.RD.DAL.ARERRAVR.L..I............FL...R...GREFDQKALLE      90
mG13       .ADFLP..SRSVLSV.....CFPGCV.TNG.AEQQRK..E.D.C.SREKTYVK.L..I............FL...R...GQDFDQRAREE      83

Consensus  ----------------.......VY.N...S..AI.RAM..L.I..........A.......................I..LW.D.GIQACF.R     180
rGs        DPQAARSNSDGEKATKVQDIKN.LKEAIET.VA..SN.VPPVELANPENQFRVDYILSVMNVPNFDFPPEFYEHAKA..E.E.VR..YE.     165
rGolf      ...........EKKQKILDIRK.VKDALVT.IS..STIIPPVPLANPENQFRSDYIKSIAPITDFEYSQEFFDHVKK..D.E.VK...E.     152
rGi1       ...........ECKQYKAV..S.TIQ.II..I...GR.K.DFGDAARADD.RQLFVLAGAAEEGFMTAE-LAGV.KR..K.S.V....N.     142
rGi2       ...........ECRQYRAV..S.TIQ.IM..VK..GN.Q.DFADPQRADD.RQLFALSCAAEEQGMLPEDLSGV.RR..A.H.V....G.     143
rGi3       ...........ECKQYKVV..S.TIQ.II..I...GR.K.DFGEAARADD.RQLFVLAGSAEEGVMTSE-LAGV.KR..R.G.V....S.     142
rGo-1      ...........DVKQYKPV..S.TIQ.LA..V...DT.GVEYGDKERKADSKMVCDVVSRMEDTEPFSAELLSAMMR..G.S...E..N.     143
rGo-2      ...........DVKQYKPV..S.TIQ.LA..V...DT.GVEYGDKERKADSKMVCDVVSRMEDTEPFSAELLSAMMR..G.S...E..N.     143
bGt1       ...........ECLEFIAII.G.TLQ.IL..V...TT.N.QYGDSARQDD.RKLMHMADTIEEGTMPKE-MSDI.QR..K.S......D.     138
bGt2       ...........ECLEYKAII.G.VLQ.IL..I...PT.G.DYAEVSCVDNGRQLNNLADSIEEGTMPPE-LVEV.RK..K.G.V....D.     142
rGgust     ...........ECMEFKAV..S.TLQ.IL..VK..TT.G.DYVNPRSREDQQLLLSMANTLEDGDMTPQ-LAEI.KR..G.P......E.     142
rGz        ...........ACKEYKPLIIY.AID.LTR.I..LAA.K.DFHNPDRAYD.VQLFALTGPAESKGEITPELLGVMRR..A.P.A....G.     143
mGq        ...........DKRGFTKL..Q.IFTAMQ.MI...DT.K.PYKYEHNKAH.QLVREVDVEKVSA--FENPYVDA.KS..N.P...E.YD.     147
mG11       ...........DKRGFTKL..Q.IFTAMQ.MV...ET.K.LYKYEQNKAN.LLIREVDVEKVTT--FEHQYVNA.KT..S.P.V.E.YD.     147
mG14       ...........DRKGFTKL..Q.IFTAMQ.MI...DT.R.QYMCEQNKEN.QIIREVEVDKVTA--LSRDQVAA.KQ..L.P...E.YD.     143
mG15       ...........DRRAFRLLI.Q.IFV.MQ.MID..DR.Q.PFSRPDSKQH.SLVMTQDPYKVST--FEKPYAVAMQY..R.A..R..YE.     150
hG16       ...........ERKGFRPL..Q.IFV.MR.MIE..ER.Q.PFSRPESKHH.SLVMSQDPYKVTT--FEKRYAAAMQW..R.A..R..YE.     150
mG12       FR.........DTIFDNILKGSRVLVDARDKLGIPWQHSENEKHGMFLMAFENKAGLPVE----PATFQLYVPALSA..R.S..REA.S.     167
mG13       FR.........PTIYSNVIKGMRVLVDAREKLHIPWGDNKNQLHGDKLMAFDTRAPMAAQGMVETRVFLQYLPA.RA..E.S...NAYD.     164

Consensus  .REYQL.DSA.YYL..LDRI....Y.PT.QDVLR.RV.TTGI.ET.F..K...FRMFDVGGQRSERKKWIHCFE.VTAIIF.VALS.YD.     270
rGs        SN....I.C.Q.F.DKI.V.KQAD.V.SD..L..C..L.S..F..K.QVDKVN.H.........D..R...Q..ND......V..S.S.NM     255
rGolf      SN....I.C.Q.F.ERI.SVSLVD.T..D..L..C..L.S..F..R.QVDKVN.H.........D..R...Q..ND.....YVA.C.S.NM     242
rGi1       S.....N...A...ND....AQPN.I..Q.....T..K....V..H.TF.DLH.K..................G......C....D..L     232
rGi2       S.....N...A...ND.E..AQSD.I..Q.....T..K....V..H.TF.DLH.K..................G......C....A..L     233
rGi3       S.....N...S...ND....SQTN.I..Q.....T..K....V..H.TF.ELY.K..................G......C....A..L     233
rGo-1      S.....N...K...DS....GAAD.Q..E..I..T..K....V..H.TF.NLH..L.................G......C....D..L     232
rGo-2      S.....N...K...DS....GAAD.Q..E..I..T..K....V..H.TF.NLH..L.................D......C....G..Q     233
bGt1       AS....N...G...SD.E.LVTPG.V..E.....S..K....I..Q.SF.DLN....................G..C...IA...A..M     228
bGt2       AA....N...S...NQ....TAPD.L.NE.....S..K....I..K.SV.DLN....................G..C...CA...A..M     232
rGgust     AS....N...A...ND...LTAPG.V.NE....HS..K....I..Q.SF.DLN....................G..C...CA...A..M     232
rGz        SS..H.E.N.A...ND.E..AAPD.I..VE.I..S.DM....V.NK.TF.ELT.K.V................G......C.E..G..L     233
```

```
mGq        R.....S..TK...ND...VADPS.L..Q.....V..P....I.YP.DLQSVI...V.........R.......N..S.M.L....E..Q      237
mG11       R..F..S...K...TDV...ATVG.L..Q.....V..P....I.YP.DLENII...V.........R.......N..S.M.L....E..Q      237
mG14       R.....S...K...TDIE..AMPSFV..Q.....V..P....I.YP.DLENII...V.........R.......S..S...L....E..Q      233
mG15       R..FH.L...V...SH.E..SEDS.I..A.....S.MP....N.YC.SV.KTKL.IV.........R.......N.I.L.YLAS..E..Q      240
hG16       R..FH.L...V...SH.E..TEEG.V..A.....S.MP....N.YC.SVQKTNL.IV.....K...........N.I.L.YLAS..E..Q      240
mG12       RS.F..GE.VK.F.DN....GQLN.F.SK..I.LA.KA.K..V.HD.VI.KIP.K.V.......Q.Q..FQ..DGI.S.L.M.SS.E..Q      257
mG13       R..F..GE.VK.F.DN..KLGVPD.I.SQ..I.LA.RP.K..H.YD.EI.NVP.K.V..........R.FE..DS..S.L.L.SS.EF.Q      254

Consensus  VL.ED...NRM.ESL.LF.SI.NN.WF..TSIILFLNKKDLLEEKI..--S.L...FPEY.G.......A...I...F...N.-------      360

rGs        .IR..NQT..LQ.A.N..K..W..R.LRTI.V......Q...A..VLAGK.KIEDY...FARYTTPEDATPEPGEDPRVTRAKYFIRDEF      345
rGolf      .IR..NNT..LR...D..E..W..R.LRTI........Q.M.A..VLAGK.KIEDY...ANYTVPEDATPDAGEDPKVTRAKFFIRDLF      332
rGi1       ..A..EEM...H..MK..D..C..K..TD...........F....KK...P.TICY...A.SNTYEE-.AAY.QCQ.EDL.K.......      312
rGi2       ..A..EEM...H..MK..D..C..K..TD...........F....TQ...P.TIC....T.ANKYDE-.ASY.QSK.EDL.K.......      313
rGi3       ..A..EEM...H..MK..D..C..K..TD...........F....KR...P.TICY...T.SNTYEE-.AAY.QCQ.EDL.R.......      312
rGo-1      ..H..ETT...H...M..D..C..KF.ID...........FG...KK...P.TIC....P.SNTYED-.AAY.QTQ.ESK.-.......      312
rGo-2      ..H..ETT...H.F.K..D..C..K..TD..........IF....KK...P.TIC....T.PSAFTE-.VAH.QGQYESK.-.......      312
bGt1       ..V..DEV...H...H..N..C.HRY.AT...V.......VFS...KK..AH.SIC..D.N.PNTYED-.GNY.KVQ.LEL.M.......      308
bGt2       ..V..DEV...H...H..N..C.HKF.AA...V...........F....KK..VH.SIC....D.NNSYED-.GNY.KSQ.LDL.M.......      312
rGgust     ..V..EEV...H...H..N..C.HKY.AT...V........FQ..VTK..VH.SIC....T.PNTFED-.GNY.KNQ.LDL.L.......      312
rGz        K.Y..NQTS..A...R..D..C..N..IN..L..........S...RR..IP.SVC....K.QNTYEE-.AVY.QRQ.EDL.R.......      313
mGq        ..V.SDNE...E..KA..RT.ITYP..QNS.V..............MY...H.VDY....D.PQRDAQA.REF.LKM.VDL.-.......      317
mG11       ..V.SDNE...E..KA..RT.ITYP..QNS.V...........D..LH...H.VDY...FD.PQRDAQA.REF.LKM.VDL.-.......      317
mG14       ..A.CDNE...E..KA..RT.ITYP..LNS.V..............MY...H.ISY....T.PKQDVKA.RDF.LKLYQDQ.-.......      313
mG15       C.E.NDQE...E...A..ST.LELP..KS..V......T.I..D..HT...H.ATY..SFQ.PRRDAEA.KSF.LDMYARVYASCAEP..      326
hG16       C.E.NNQE...K...A..GT.LELP..KS..V......T.I.....PT...H.ATY..SFQ.PKQDAEA.KRF.LDMYTRMYTGCVDG..      326
mG12       ..M..RRT..LV..MNI.ET.V..KL.FNV........M...V..VKS..VSIKKH..DFK.DPHRLEDVQRYLVQC.DRKR-.......      337
mG13       ..M..RQT..LT...NI.ET.V..RV.SNV........T......VQV..VSIKDY.L.FE.DPHCLRDVQKFLVEC.RGKRR.......      335

Consensus  --------....K.IY.HFTCATDT.NI.FVF.AV.D.I...NLK...L.                                            409

rGs        LRISTASGDGRHYC.P.....V..E..RR..NDCR.I.QRMH.RQYE.L                                             394
rGolf      LRISTATGDGKHYC.P.....V..E..RR..NDCR.I.QRMH..QYE.L                                             381
rGi1       .......RKDT.E..T........K.VQ...D..T.V.IKN...DCG.F                                             354
rGi2       .......RKDT.E..T........K.VQ...D..T.V.IKN...DCG.F                                             355
rGi3       .......RKDT.EV.T........K.VQ...D..T.V.IKN...ECG.Y                                             354
rGo-1      .......RSPN.E..C.M......N..QV..D..T.I.IAN..RGCG.Y                                             354
rGo-2      .......KSAH.EV.S.V......N..Q...D..T.V.IAK..RGCG.Y                                             354
bGt1       .......RRDV.E..S.M......Q.VK...D..T.I.IKE...DCG.F                                             350
bGt2       .......RKDV.E..S.M......Q.VK...D..T.I.IKE...DCG.F                                             354
rGgust     .......KKED.E..S.M......Q.VK...D..T.I.IKE...DCG.F                                             354
rGz        .......NKET.E..S........S..Q...D..T.V.IQN...YIG.C                                             355
mGq        .......PDSD.I..S........E..R...A..K.T.LQL...EYN.V                                             359
mG11       .......PDSD.I..S........E..R...A..K.T.LQL...EYN.V                                             359
mG14       .......PDKE.V..S........E..R...A..K.T.LQL..REFN.V                                             355
mG15       .QDGGRKGSRARRFFA........QSVRS..KD.R.SVLARY.DEIN.L                                             374
hG16       .PEGSKKGARSRRLFS.Y......Q..RK..KD.R.SVLARY.DEIN.L                                             374
mG12       .......RNRS.PLFH...T.I..E..R...H..K.T.LQE...DIM.Q                                             379
mG13       .......DQQQRPL.H...T.IN.E..RL..RD.K.T.LHD...QLM.Q                                             377
```

Fig. 1. Alignment of amino acid sequences of mammalian G protein α subunits. *Points* in all mammalian α sequences represent an amino acid residue that is identical to the residue in consensus sequence. *Hyphens* indicate the gaps introduced for optimal alignment. Sources for sequences are described in KAZIRO et al. (1991) and in

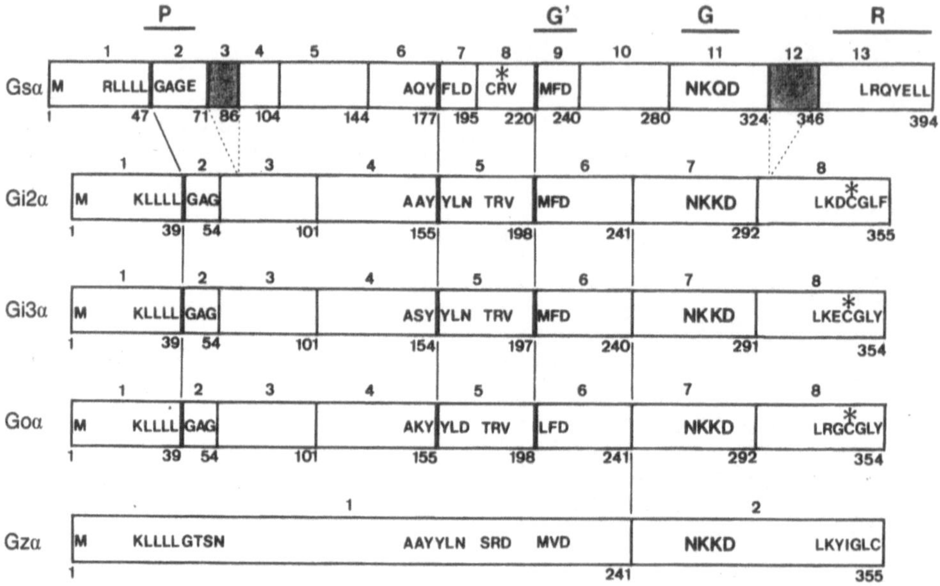

Fig. 2. Organization of human G-protein α subunit genes

coded by exon 3. $G_{s1\alpha\text{-}2}$ and $G_{s1\alpha\text{-}4}$ have three more nucleotides (CAG) than $G_{s1\alpha\text{-}1}$ and $G_{s1\alpha\text{-}3}$, respectively, 3′ to the above 45 nucleotides. $G_{s\alpha}$ has been purified as a mixture of 52-kDa and 45-kDa proteins on sodium dodecyl sulfate (SDS) polyacrylamide gels. $G_{s1\alpha\text{-}1}$ and $G_{s1\alpha\text{-}4}$, code for a 52-kDa and a 45-kDa protein, respectively, when they are expressed in COS-m6 cells (ROBISHOW et al. 1986) or in *Escherichia coli* (GRAZIANO et al. 1989). Functional differences between these two types of $G_{s1\alpha}$ proteins are unknown. $G_{s2\alpha}$ ($G_{olf\alpha}$) is expressed exclusively in olfactory sensory neurons and is therefore thought to be involved specifically in odorant signal transduction (JONES and REED 1989). $G_{s2\alpha}$ has 88% homology to $G_{s1\alpha}$ in amino acid sequence. $G_{s2\alpha}$ retains a sequence that corresponds to exon 12 of $G_{s1\alpha}$ but lacks a sequence corresponding to exon 3. $G_{s2\alpha}$ stimulates adenylyl cyclase when expressed in S49 cyc$^-$ kin$^-$ cells that lack $G_{s1\alpha}$ (JONES et al. 1990).

b) $G_{i\alpha}$

Studies of cDNA cloning of $G_{i\alpha}$ have revealed the existence of three kinds of distinct $G_{i\alpha}$ cDNAs, i.e., $G_{i1\alpha}$ (NUKADA et al. 1986), $G_{i2\alpha}$ (ITOH et al. 1986), and $G_{i3\alpha}$ (SUKI et al. 1987). Their amino acid sequences show more than 85% identity with each other (JONES and REED 1987). The three types of $G_{i\alpha}$ cDNAs are encoded by the distinct genes. So far the entire organization of human $G_{i2\alpha}$ and $G_{i3\alpha}$ genes and the partial structure of human $G_{i1\alpha}$

gene have been determined (ITOH et al. 1988a). The coding region of human $G_{i2\alpha}$ and $G_{i3\alpha}$ is encoded by eight exons, and their exon–intron organization is completely identical (Fig. 2). The same splice sites are also conserved in the partial sequence (exons 1–3) of human $G_{i1\alpha}$ gene. Two kinds of $G_{i\alpha}$ proteins, 40-kDa and 41-kDa, that have been purified from porcine brains correspond to $G_{i2\alpha}$ and $G_{i1\alpha}$, respectively, by sequence analysis (ITOH et al. 1988b), while $G_{i\alpha}$ purified from human erythrocyte (CODINA et al. 1988) and from bovine spleen (MORISHITA et al. 1989) as 41-kDa pertussis toxin substrate corresponds to $G_{i3\alpha}$. $G_{i2\alpha}$ and $G_{i3\alpha}$ are expressed in almost all kinds of tissues and cells, but the expression of $G_{i1\alpha}$ seems to be more restricted. $G_{i1\alpha}$ mRNA is predominantly expressed in brain (BRAY et al. 1987). Among several neuronal cell lines, $G_{i1\alpha}$ mRNA is detected in pheochromocytoma PC12 cells and neuroblastoma SK-N-SH cells, but not in glioma C6 cells and neuroblastoma-glioma hybridoma NG 108-15 cells (GARIBAY et al. 1991).

c) $G_{o\alpha}$

$G_{o\alpha}$ is (another) G-protein species which is abundant in brain. Recently, two types of $G_{o\alpha}$ cDNAs (designated here as $G_{o\alpha\text{-}1}$ and $G_{o\alpha\text{-}2}$, in other references as $G_{oA\alpha}$ and $G_{oB\alpha}$) were isolated from mouse (STRATHMANN et al. 1990), hamster (HSU et al. 1990), and rat (TSUKAMOTO et al. 1991). They differ only in the C-terminal amino acid sequences. The human $G_{o\alpha}$ gene spans more than 100 kb and contains 11 exons, including one at the 3′ noncoding region (TSUKAMOTO et al. 1991). It was found that exons 7 and 8 coding for amino acid residues 242–354 of $G_{o\alpha}$ protein were duplicated (referred to as exons 7A and 7B, and 8A and 8B, respectively). Exons 7A and 8A and exons 7B and 8B code for $G_{o\alpha\text{-}1}$ and $G_{o\alpha\text{-}2}$, respectively. Recently, KLEUSS et al. (1991) have demonstrated the functional difference of the two forms of $G_{o\alpha}$ in terms of the receptor coupling. By intranuclear injection of antisense oligonucleotides into rat pituitary GH3 cells, they found that $G_{o\alpha1}$ and $G_{o\alpha2}$ subtypes mediate the inhibition of the Ca^{2+} channel through muscarinic and somatostain receptors, respectively.

d) $G_{t\alpha}$ and $G_{gust\alpha}$

Two types of $G_{t\alpha}$ cDNA, $G_{t1\alpha}$ and $G_{t2\alpha}$, which possess 80% identity in their amino acid sequences have been isolated. $G_{t1\alpha}$ and $G_{t2\alpha}$ mRNAs are expressed in rods and cones, respectively (LEREA et al. 1986). The exon–intron organization of mouse $G_{t1\alpha}$ gene (RAPORT et al. 1989) is identical to that of human $G_{i2\alpha}$, $G_{i3\alpha}$, and $G_{o\alpha}$ genes. In $G_{i\alpha}$, $G_{o\alpha}$, and $G_{t\alpha}$, the cysteine residue fourth from the C terminus is ADP-ribosylated by pertussis toxin. More recently, a taste-cell-specific G-protein α subunit (α-gustducin, $G_{gust\alpha}$) has been cloned (MCLAUGHLIN et al. 1992). $G_{gust}\alpha$ is most closely related to the $G_{t\alpha}$s. Their amino acid sequences show 80% identity with each other.

```
Consensus   MSELEQLRQEAEQL.NQI.DARKAC.D.TL.QIT...D.VGRIQMRTRRT    50
beta1       ....D.........K...R......A.A..S...NNI.P...........    50
beta2       ..............R...R......G.S..T...AGL.P...........    50
beta3       .G.M..........KK..A......A.V..AELVSGLEV...V.......    50
beta4       ..............R...Q......N.A..V...SNM.S...........    50

Consensus   LRGHLAKIYAMHWGTDSRLLVSASQDGKLIIWDSYTTNKVHAIPLRSSWV   100
beta1       ..................................................   100
beta2       ..................................................   100
beta3       .............A...K............V...................   100
beta4       ..............Y........................M..........   100

Consensus   MTCAYAPSGN.VACGGLDNICSIYNLKTREGNVRVSREL.GHTGYLSCCR   150
beta1       ..........Y............................A..........   150
beta2       ..........F.............S..............P..........   150
beta3       ..........F........M.......S.....K.....SA.........   150
beta4       ..........Y....................D.......A..........   150

Consensus   FLDDNQI.TSSGDTTCALWDIETGQQTT.F.GH.GDVMSLSL.PD...FV   200
beta1       .......V....................T.T..T........A..TRL..   200
beta2       .......I...................VG.A..S........A..GRT..   200
beta3       .....N.V..................K.V.V..T..C...AVS..FNL.I   200
beta4       ....G..I....................T.T..S........S..LKT..   200

Consensus   SGACDAS.KLWDVR.GMCRQTFTGHESDINA..FFPNG.AF.TGSDDATC   250
beta1       .......A......E................IC.....N..A........   250
beta2       .......I......DS......I........VA.....Y..T........   250
beta3       .......A......E.T..............IC.....E.IC......S.   250
beta4       .......S....I.D.....S....I.....VS...S.Y..A........   250

Consensus   RLFDLRADQEL..YSHDNIICGITSVAFS.SGRLLLAGYDDFNCNVWDA.   300
beta1       ...........MT.............S..K...................L   300
beta2       ...........LM................R...............I...M   300
beta3       ...........ICF..ES...........L....F.............SM   300
beta4       ...........LL................K..............S....L   300

Consensus   K..R.GVLAGHDNRVSCLGVTDDGMAVATGSWDSFLKIWN             340
beta1       .AD.A...................................             340
beta2       .GD.A...................................             340
beta3       .SE.V.I.S............A..................             340
beta4       .GG.S...............................R...             340
```

Fig. 3. Alignment of amino acid sequences of mammalian G protein β subunits. *Points* in all mammalian β sequences represent an amino acid residue that is identical to the residue in consensus sequence. The WD-40 motifs are underlined

e) $G_{z\alpha}$

$G_{z\alpha}$ (FONG et al. 1988), also referred to as $G_{x\alpha}$ (MATSUOKA et al. 1988), is a new member of the G_α family and is expressed mainly in neuronal cells. In amino acid sequence, $G_{z\alpha}$ has the highest identity with $G_{i\alpha}$ (60%). The sequence of $G_{z\alpha}$ in the P region (see below) is different from other G-protein α subunits in that the consensus A-G-E is replaced by T-S-N (see Fig. 1), and the rate of GTP hydrolysis (about $0.05\,min^{-1}$) is about 200 times slower than that of other G_α subunits (CASEY et al. 1990). $G_{z\alpha}$ is ADP-ribosylated by neither pertussis toxin nor cholera toxin. The gene organization of $G_{z\alpha}$ is also quite different from other G_αs (Fig. 2). The human $G_{z\alpha}$ gene contained only two exons for its coding region. However, an additional exon codes for 170 base pairs far upstream (more than 20 kb) of the 5′-noncoding region (MATSUOKA et al. 1990).

f) $G_{q\alpha}$ and $G_{12\alpha}$

SIMON and his collaborators (STRATHMANN et al. 1989; STRATHMANN and SIMON 1990; SIMON et al. 1991) have isolated cDNAs for several additional G-protein α subunits by using the polymerase chain reaction (PCR). The new G_α cDNAs are classified into two groups by their sequence homology; the $G_{q\alpha}$ subgroup consisting of $G_{q\alpha}$, $G_{11\alpha}$, $G_{14\alpha}$, $G_{15\alpha}$, and $G_{16\alpha}$, and the $G_{12\alpha}$ subgroup consisting of $G_{12\alpha}$ and $G_{13\alpha}$ (STRATHMANN and SIMON 1991). The cDNAs for $G_{15\alpha}$ and $G_{16\alpha}$ were obtained from mouse (WILKIE et al. 1991) and human (AMATRUDA et al. 1991), respectively, and they probably code for the counterparts of the same G-protein α subunits.

Recently, it was found that $G_{q\alpha}$ and $G_{11\alpha}$ stimulated the activity of phospholipase C β1 subtype (TAYLOR et al. 1990, 1991). Thus, one of the putative G-protein α subunits involved in phospholipase C activation (G_p or GPLC) has been finally identified as $G_{q\alpha}$. It is likely that other members of the G_q subgroup might also be involved in the activation of other phospholipase Cβ subtypes.

2. Comparison of the Amino Acid Sequences

Amino acid sequences of G-protein α subunits deduced from the nucleotide sequences of several cDNAs and genomic DNAs are shown in Fig. 1. For $G_{s1\alpha}$, $G_{i1\alpha}$, $G_{i2\alpha}$, $G_{i3\alpha}$, $G_{o\alpha\text{-}1}$, $G_{o\alpha\text{-}2}$, and $G_{z\alpha}$, human sequences are shown. For $G_{s2\alpha}$, and $G_{t1\alpha}$, and $G_{t2\alpha}$, the sequences are derived from rat and bovine, respectively. G-protein α subunits contain several regions of highly conserved amino acid sequences. These regions are designated P, G′, G, and G″. They form a core structure of GTP binding, GTP hydrolysis, and conformational switch through the ligand change from GDP to GTP or vice versa (see other chapters for detailed discussions).

a) P Region

This region, with the consensus sequence motif G-X-X-X-X-G-K-S/T, is probably involved directly in the hydrolytic process. Mutation of Gly-12 at this region of Ha-ras p21 protein to Val or other amino acids results in deficient GTPase and increased transforming activities (BARBACID 1987). In the case of $G_{s1\alpha}$, the replacement of Gly-49 (which corresponds to Gly-12 of Ras) by Val also reduces the rate of k_{cat} of GTP hydrolysis (GRAZIANO and GILMAN 1989).

b) G′ Region

The consensus sequence motif of this region, D-X-X-G-Q, is also highly conserved in most GTPases. The sequence of EF-Tu at the corresponding region is D-C-P-G-H; this region of EF-Tu had been assigned as the site where the conformational change induced by the ligand change from GDP to GTP takes place (KAZIRO 1978). Supporting this assignment, comparison of the three-dimensional structures of Ras·GTP and Ras·GDP revealed the most remarkable structural change in this region (MILBURN et al. 1990; PAI et al. 1990; JURNAK et al. 1990). In the case of $G_{s1\alpha}$, the GTP-bound form is found to be more resistant than the GDP-bound form to tryptic cleavage at Arg-231 (located immediately downstream of G′ region). Also, H21a, a mutant of $G_{s1\alpha}$ in which Gly-226 (corresponding to Gly-60 of Ras protein) is replaced by Ala, is not protected by GTP from tryptic digestion, although the mutant can still bind and hydrolyze GTP (LEE et al. 1991). This indicates that the substitution in H21a may prevent a critical conformational change.

In Ras, replacement of Gln-61 by Leu shows the reduction of GTP hydrolysis and leads to malignant transformation (BARBACID 1987). Recent work has revealed the presence of oncogenic mutations of $G_{s1\alpha}$ in growth-hormone-secreting pituitary tumors (LANDIS et al. 1989; LYONS et al. 1990). One of the mutation sites is found at Gln-227 of $G_{s1\alpha}$, which is equivalent to Gln-61 of Ras. Site-directed mutagenesis studies (GRAZIANO and GILMAN 1989; MASTERS et al. 1989) have demonstrated that the Gln to Leu mutation at position 227 of $G_{s1\alpha}$ reduces the intrinsic GTPase activity and induces the constitutional activation of the $G_{s1\alpha}$.

c) G Region

This region interacts directly with the guanine ring of GTP. The consensus motif N-K-X-D is conserved in all GTPases. Earlier studies with elongation factor EF-Tu indicated that the modification of the N-K-C-D sequence at Cys-137 destroyed the interaction with GTP (KAZIRO 1978). Later, the four residues N-K-C-D were found by X-ray analysis to be situated close to the guanine ring (JURNAK 1985).

d) G″ Region

This is another region that is closely situated to the guanine ring of GTP in the three-dimensional structure of Ras protein. The conserved sequence motif of this region in Ras is E-T-S-A-K. In G-proteins, H-(F/M)-T-C-A-(T/V)-D-T may correspond to this region.

e) Cholera Toxin ADP-Ribosylation Site

Cholera toxin catalyzes the ADP-ribosylation of $G_{s1\alpha}$ and $G_{t\alpha}$, and the modification reduces GTPase activity. Arg-174 is the site of modification in $G_{t1\alpha}$ (VAN DOP et al. 1984). The corresponding amino acid residue in $G_{s1\alpha}$ is Arg-201. FREISSMUTH and GILMAN (1989) demonstrated that the replacement of this Arg by Ala, Glu, and Lys reduced k_{cat}GTP, and that these mutant proteins activated adenylyl cyclase constitutively in the presence of GTP.

Mutations of $G_{s1\alpha}$ in pituitary tumors inhibit GTP hydrolysis and cause constitutive elevation of adenylyl cyclase activity (LANDIS et al. 1989; LYONS et al. 1990). One mutation replaces an Arg-201 residue. All known G-protein α subunits contain an Arg residue at the position corresponding to Arg-201. In adrenal cortex tumors, mutations have been found in $G_{i2\alpha}$ that replace the conserved Arg-179 by Cys or His. The sequence around Arg-201 of $G_{s1\alpha}$ is very much homologous, not only to that of $G_{t\alpha}$, which is ADP-ribosylated by cholera toxin, but also to those of $G_{i\alpha}$ and $G_{o\alpha}$, which are not ADP-ribosylated by the toxin. This region may play an important role in the regulation of GTP hydrolysis.

3. Sequence Conservation

Availability of the complete amino acid sequences of G-protein α subunits from many mammalian species allows comparison of their primary structure. A strong conservation of the amino acid and the nucleotide sequences in each group of G-protein α subunit among different organisms was observed. For example, the amino acid sequence of $G_{s1\alpha}$ is strongly conserved between human and rat; only one out of 394 amino acid residues is different. The sequence of $G_{i1\alpha}$ is completely identical in bovine and human. For $G_{i2\alpha}$, $G_{i3\alpha}$, $G_{o\alpha}$, and $G_{z\alpha}$, more than 98% identity among amino acid sequences is maintained among different mammalian species. The strong conservation of the amino acid sequence of each G-protein α subunit among distant mammalian species may reflect evolutionary pressure to maintain the multiple and specific physiological functions of each G-protein gene product.

4. Evolutionary Tree

An evolutionary tree of G-protein α subunits can be drawn on the basis of the differences among the predicted amino acid sequences obtained from various mammalian sources (see Fig. 1 in Chap. 11). There are four subfamilies of G-protein α subunits, G_s, G_i, G_q, and G_{12}. It is remarkable that the homologies among the three $G_{i\alpha}$ and the G_q/G_{11} species are higher than those of the $G_{s\alpha}$ and $G_{t\alpha}$ subgroups.

II. $\beta\gamma$ Subunits

Initially, two distinct cDNAs for G-protein β subunits, $\beta1$ and $\beta2$, were cloned from a variety of tissues (SUGIMOTO et al. 1985; CODINA et al. 1986; FONG et al. 1987; GAO et al. 1987a). The $\beta1$ and $\beta2$ cDNAs encode 36-kDa and 35-kDa proteins, respectively (GAO et al. 1987b; AMATRUDA et al. 1988). Recently, a third cDNA, $\beta3$, has been cloned from retina (LEVINE et al. 1990). Using PCR, a new member of this family, $\beta4$, was also found (VON WEIZSACKER et al. 1992). These four polypeptides have more than 80% amino acid identity with each other. All of the β subunits conserve a repetitive segmental structure of about 40 amino acids, the WD-40 motif,

```
Consensus    M----....T.SIAQARK.VEQLK.EA.L.RIKVSKAAAD    40

gamma1       ..PVINIEDLTEKDKLKME.D...K.VT.E.ML...CCEE    39
gamma2       .....ASNN.A.......L.....M..NID..........    36
gamma3       .KGETPVNS.M..G....M.....I..S.C..........    40
gamma5       ......--SGSS.V.AMK.V.Q..RL..G.N.V...Q....   34
gamma7       ......--SA.NN......L....RI..GIE.......SSE   34


Consensus    LM.YCE.HA.EDPLLTGVP.SENPFREKK..C.IL         75

gamma1       FRD.V.ERSG....VK.I.EDK...K.L.GG.V.S         74
gamma2       ..A...A..K......P..A.........FF.A..         71
gamma3       ..T..DA..C....I.P..T.........FF.A..         75
gamma5       .KQF.LQN.QH.......SS.T....PQ.V-.SF.         68
gamma7       ..S...Q..RN....V...A.....KD..P-.I..         68
```

Fig. 4. Alignment of amino acid sequences of mammalian G protein γ subunits. *Points* in all mammalian γ sequences represent an amino acid residue that is identical to the residue in consensus sequence. *Hyphens* indicate the gaps introduced for optimal alignment

which is characterized by a tryptophan–asparatate pair (Fig. 3). Although the motif has been found in other proteins, its function remains unknown. Cross-linking experiments have shown that the amino terminal region of the β subunits may be involved in the interaction with γ subunit (BUBIS and KHORANA 1990). More recently, by intranuclear microinjection with antisense oligonucleotides, it has been shown that the $\beta 1$ and $\beta 3$ are selectively involved in the inhibition of voltage-dependent Ca^{2+} channels through muscarinic (M4) and somatostatin receptors, respectively (KLEUSS et al. 1992).

Five distinct full-length cDNAs coding for γ subunits have been isolated (Fig. 4) (HURLEY et al. 1984; YATSUNAMI et al. 1985; ROBISHOW et al. 1989; GAUTAM et al. 1989; FISHER and ARONSON 1992; CALI et al. 1992). Comparison of the partial amino acid sequences of the purified γ subunit proteins suggest the existence at least two more γ subunits. Northern analysis indicated that $\gamma 1$ was expressed only in retina, whereas $\gamma 2$ was expressed at different level in all tissues. The predicted amino acid sequences of bovine and rat $\gamma 5$ are completely identical (FISHER and ARONSON 1992). The N-terminal sequences among all members of γ subunits are divergent, while the four C-terminal amino acids of all γ subunits show the common sequence Cys-A-A-X, where A is an aliphatic acid and X is any amino acid. In the case of $\gamma 1$, the 15 carbon farnesylation on the Cys was identified (FUKADA et al. 1990). It is likely that some other γ subunits have the 20 carbon geranylgeranyl modification on the Cys (YAMANE et al. 1990; MUMBY et al. 1990). Analysis by site-directed mutagenesis suggests that the modification on the Cys is essential for the intracellular distribution of the $\beta\gamma$ complex (SIMONDS et al. 1991; MUNTZ et al. 1992).

C. G-proteins in Lower Eukaryotes

As described above, the structure of the G-protein gene family is strongly conserved among different mammalian species. Thus, G_α-homologous cDNAs and genes have been isolated from many sources, by using the cross-hybridization and PCR technique. Figure 5 compares the predicted amino acid sequences of G-proteins from lower eukaryotes including *Saccharomyces cerevisiae*, *Schizosaccharomyces pombe*, *Dictyostelium discoideum*, and *Arabidopsis thaliana*. Although not included in Fig. 5, G-protein cDNAs and genomic DNAs have also been isolated from *Drosophila melanogaster* and *Caenorhabditis elegans*. Some characteristic features of their structure and function in yeast, nematode, and plant are described below. Studies of G-proteins in *Drosophila melanogaster* and *Dictyostelium discoideum* will be described in Chaps. 63 and 64, respectively.

I. G-proteins from *Saccharomyces cerevisiae*

1. Two α Subunits, GPA1 and GPA2

Two genes, GPA1 and GPA2, were isolated from *S. cerevisiae* by cross hybridization with rat G_α cDNAs (NAKAFUKU et al. 1987, 1988). They code for proteins about 100 amino acid residues larger than mammalian G-protein α subunits. The additional stretches, 110 amino acids (residues 126–235) in Gpa1 and 83 amino acids (residues 37–119) in Gpa2, are inserted near the N-terminal portion at the different sites. The two extra sequences had no homology with each other or with any other proteins reported. Disregarding the unique stretches, the overall similarities of yeast G-proteins with mammalian G-proteins are remarkable. The four regions (P, G′, G, and G″ regions) discussed in the previous section are highly conserved from human to yeast. The calculated overall homology between GPA1 and GPA2 is 53% in nucleotide sequence and 60% in amino acid sequence. Northern blot analysis indicated that GPA1 is expressed only in haploid cells whereas GPA2 is expressed in both haploid and diploid cells. The expression of GPA1 is increased severalfold in response to mating factors. As will be discussed elsewhere (Chap. 62), GPA1 interacts with the products of STE2 and STE3 (α-receptor and a-receptor, respectively), and transmits mating factor signals downstream. The genes independently isolated by genetic analyses, *SCG1* and *CDC70*, were found to be identical with GPA1.

2. β and γ Subunits

The *STE4* and *STE18* genes of *S. cerevisiae* encode proteins homologous to mammalian G-protein β and γ subunits (WHITEWAY et al. 1989). The *STE4*-encoded protein is composed of 432 amino acids sharing 49% homology with two kinds of mammalian G_β. A unique repetitive structure of G_β is conserved in Ste4 protein. The deduced sequence of Ste18 exhibited partial similarities with mammalian G_γ. Ste18 protein of 110 amino acid residues shares 35% and 40% homologies with mammalian $\gamma 1$ and $\gamma 2$, respectively. Both *STE4* and *STE18* genes are expressed only in haploid

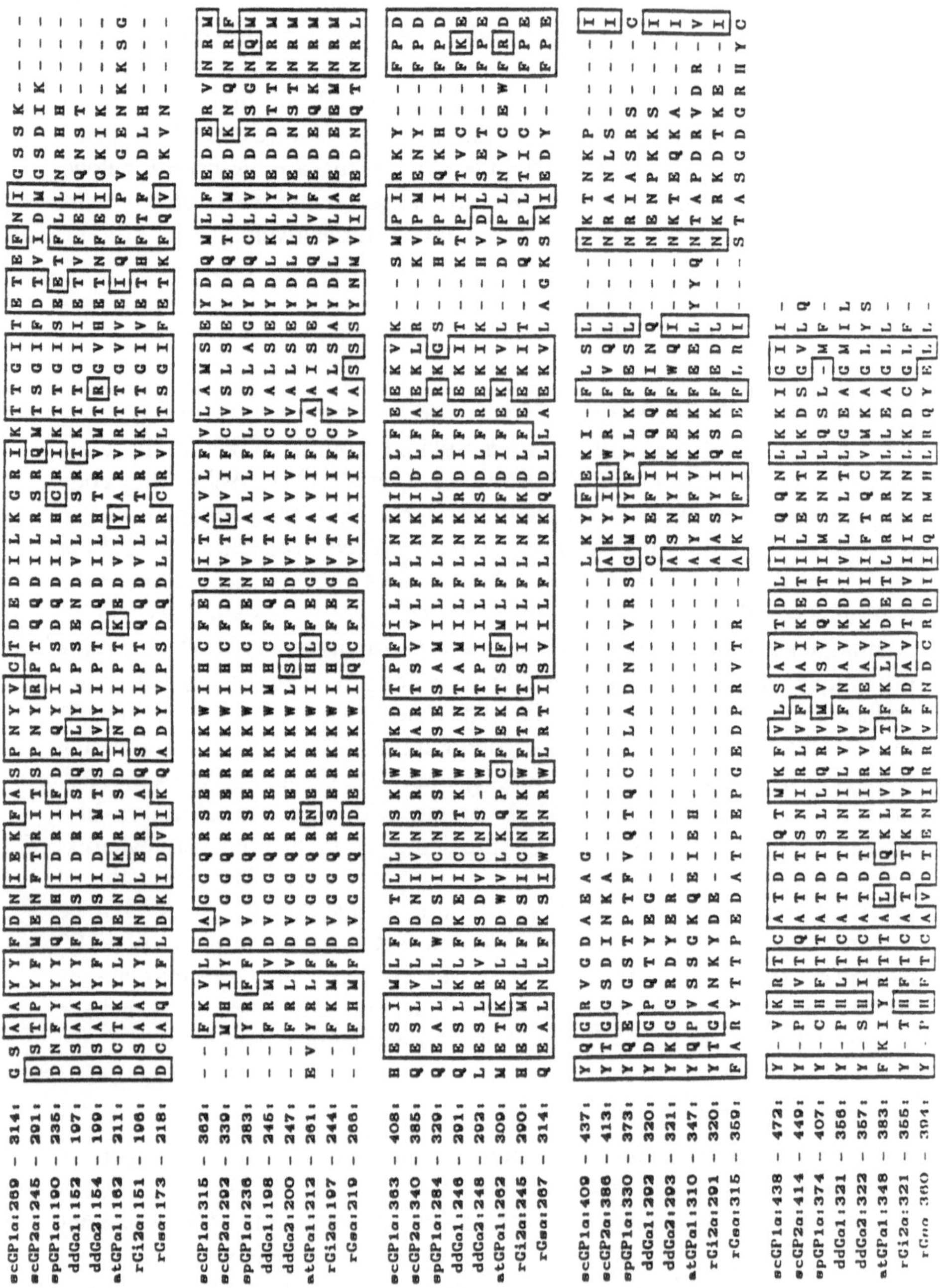

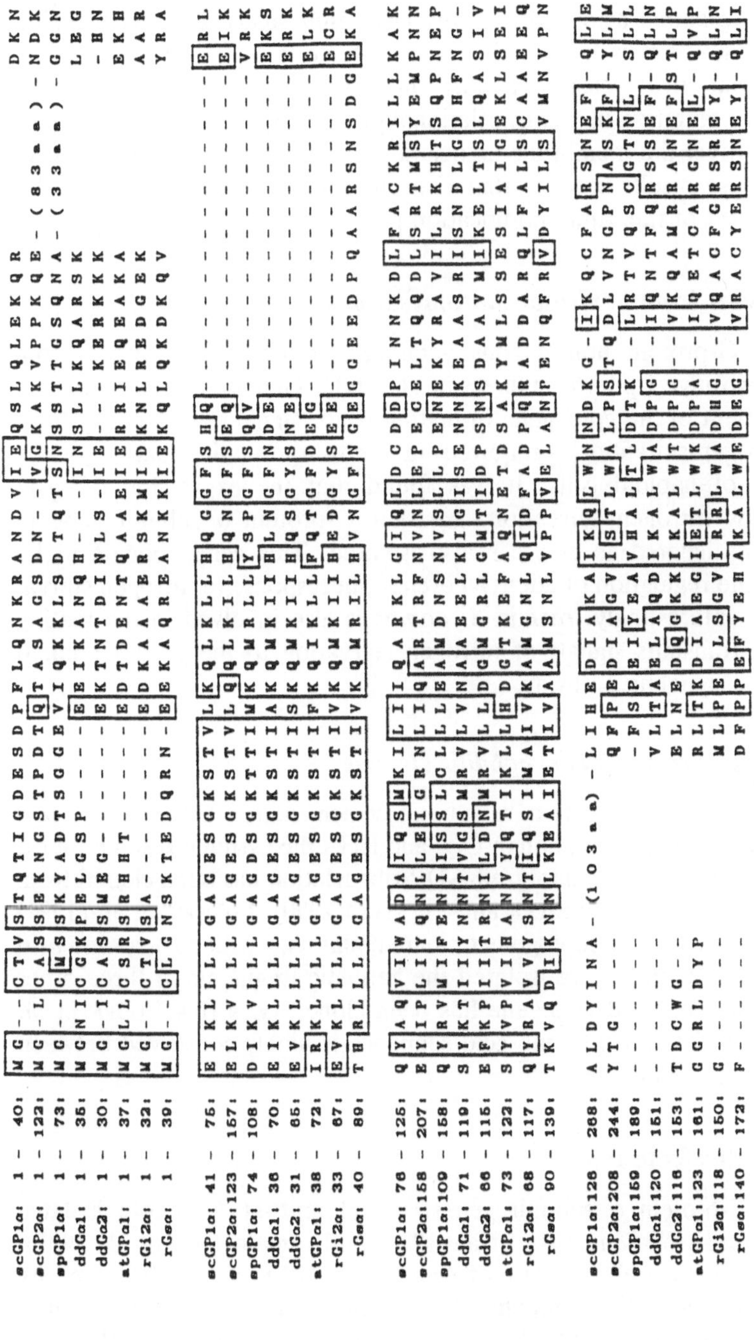

Fig. 5. Comparison of the deduced amino acid sequences of G-protein α subunits from different organisms. Identical and conservative amino acid residues are enclosed with solid lines. Aligned sequences are: $scGP_{1\alpha}$ and $scGP_{2\alpha}$, *Saccharomyces cerevisiae* $GP_{1\alpha}$ and $GP_{2\alpha}$; $spGP_{1\alpha}$, *Schizosaccharomyces pombe* $GP_{1\alpha}$; $ddG_{\alpha 1}$ and $ddG_{\alpha 2}$, *Dictyostelium discoideum* $G_{\alpha 1}$ and $G_{\alpha 2}$; $atGP_{\alpha 1}$, *Arabidopsis thaliana* $GP_{\alpha 1}$; $rG_{i2\alpha}$ and $rG_{s1\alpha}$, rat $G_{i2\alpha}$ and $G_{s1\alpha}$. Sources of the sequences are described in KAZIRO et al. 1991

cells, and the defective mutations of these genes results in the loss of mating abilities.

At the C terminus of Ste18 proteins, a Cys-A-A-X sequence (A represents aliphatic amino acid and X is the last amino acid) common to mammalian γ subunit is found. The conserved cysteine in the motif is the site of polyisoprenylation. A yeast *dpr1/ram1* mutation, which is originally isolated as a defective mutation in posttranslational processing of yeast Ras proteins, is found to affect the membrane association and biological activity of Ste18 proteins. These results indicate that G-protein γ subunits and Ras proteins may share a set of the same modification process.

II. G-proteins from *Schizosaccharomyces pombe*

Recently a G-protein homologous gene (*gpa1*) has been isolated from *Schizosaccharomyces pombe* (OBARA et al. 1991). Genetic analyses indicate the involvement of the *gpa1* gene in mating factor signal transduction. In *S. pombe*, nitrogen starvation is required, in addition to mating factor, for the conjugation of haploid cells. It was found that the expression of *gpa1* is induced by nitrogen starvation. Another G-protein α subunit gene in *S. pombe*, designated as *gpa2*, is also isolated. This gene appears to be involved in the regulation of adenylyl cyclase, in response to nitrogen starvation signals. Interestingly, *ras1* in *S. pombe* is also involved in the mating pathway. It transmits signals of nitrogen starvation to the *gpa1* mediated mating factor signaling pathway.

III. G-proteins from *Caenorhabditis elegans*

A gene coding for G-protein α subunit which shares 63% amino acid sequence identity with mammalian $G_{i2\alpha}$ and lacks the cysteine residue that is the site for pertussis-toxin-catalyzed ADP-ribosylation has been cloned from nematode *C. elegans* (FINO SILVA and PLASTERK 1990). LOCHRIE et al. (1991) have isolated the gene homologous to mammalian $G_{o\alpha}$ with 80% amino acid identity. By PCR, they have isolated the genomic fragments of three other G_α genes. The nematode G_β gene has been cloned (VAN DER VOORN et al. 1990). The predicted 340 amino acid sequence is highly conserved between mammalian and nematode.

IV. G-proteins from Plants

Two plant G-protein α subunit genes, GPA1 and TGA1, were isolated from *Arabidopsis thaliana* (MA et al. 1990) and tomato (MA et al. 1991), respectively. The overall structure of *Arabidopsis* GPA1 shares 36% identity and 73% homology with mammalian $G_{i\alpha}$. Comparison of the organization of the GPA1 gene, composed of 14 exons and 13 introns, with those of mammalian G_α genes, indicated that several exon–intron junctions are conserved. The third intron of GPA1 was at the same position as the first intron of the human $G_{s1\alpha}$ and $G_{i\alpha}$ genes. The positions of the fifth and sixth introns of GPA1 corresponded to those of the fourth intron of human $G_{s1\alpha}$ and the third intron of human $G_{i\alpha}$ genes, respectively. These observations again emphasize a strong structural conservation of G-proteins through evolution.

Notes and Acknowledgements. The citations are not comprehensive. The list of cDNAs and genomic DNAs coding for G-protein α subunits, which were cloned up to 1989, was described in Kaziro et al. (1991). We thank Dr. Katada for helpful discussions.

References

Amatruda TT, Gautam N, Fong HKW, Northup JK, Simon MI (1988) The 35- and 36-kDa beta subunits of GTP-binding regulatory proteins are products of separate genes. J Biol Chem 263(11):5008–5011

Amatruda TT, Steele DA, Slepak VZ, Simon MI (1991) Galpha16, a G protein alpha subunit specifically expressed in hematopoietic cells. Proc Natl Acad Sci USA 88:5587–5591

Barbacid M (1987) ras Genes. Annu Rev Biochem 56:779–827

Bray P, Carter A, Guo V, Puckett C, Kamholz J, Spiegel A, Nirenberg M (1987) Human cDNA clones for an α subunit of Gi signal-transducing protein. Proc Natl Acad Sci USA 84:5115–5119

Bubis J, Khorana HG (1990) Sites of interaction in the complex between β- and γ-subunits of transducin. J Biol Chem 265:12995–12999

Cali JJ, Balcueva EA, Rybalkin I, Robishaw JD (1992) Selective tissue distribution of G protein γ subunits, including a new form of the γ subunit identified by cDNA cloning. J Biol Chem 262:24023–24027

Casey PJ, Fong HKW, Simon MI, Gilman AG (1990) Gz, a guanine nucleotide-binding protein with unique biochemical properties. J Biol Chem 265:2383–2390

Codina J, Olate J, Abramowitz J, Mattera R, Cook RG, Birnbaumer L (1988) αi-3 cDNA encodes the α subunit of Gk, the stimulatory G protein of recepter-regulated K^+ Channels. J Biol Chem 263:6746–6750

Codina J, Stenge l D, Woo SLC, Birnbaumer L (1986) β-subunits of the human liver Gs/Gi signal transducing proteins and those of bovine retinal rod cell transducin are identical. FEBS Lett 207:187–192

Fino Silva I, Plasterk RH (1990) Characterization of a G-protein alpha-subunit gene from the nematoda Caenorhabditis elegans. J Mol Biol 215(4):483–487

Fisher KJ, Aronson NN (1992) Characterization of the cNDA and genomic sequence of a G protein gamma subunit (gamma5). Mol Cell Biol 12:1585–1591

Fong HKW, Amatruda TT, Birren BW, Simon MI (1987) Distinct forms of the β subunit of GTP-binding regulatory proteins identified by molecular cloning. Proc Natl Acad Sci USA 84:3792–3796

Fong HKW, Yoshimoto KK, Eversole-Cire P, Simon MI (1988) Identification of a GTP-binding protein α subunit that lacks an apparent ADP-ribosylation site for pertussis toxin. Proc Natl Acad Sci USA 85:3066–3070

Freissmuth M, Gilman AG (1989) Mutations of Gsα designed to alter the reactivity of the protein with bacterial toxins. Substitutions at Arg187 result in loss of GTPase activity. J Biol Chem 264(36):21907–21914

Fukada Y, Takao T, Ohguro H, Yoshizawa T, Akino T, Shimonishi Y (1990) Farnesylated gamma-subunit of photoreceptor G protein indispensable for GTP-binding. Nature 346:658–660

Gao B, Gilman AG, Robishaw JD (1987a) A second form of the β subunit of signal-transducing G proteins. Proc Natl Acad Sci USA 84(17):6122–6125

Gao B, Mumby S, Gilman AG (1987b) The G protein β2 complementary DNA encodes the beta35 subunit. J Biol Chem 262(36):17254–17257

Garibay JL, Kozasa T, Itoh H, Tsukamoto T, Matsuoka M, Kaziro Y (1991) Analysis by mRNA levels of the expression of six G protein alpha-subunit genes in mammalian cells and tissues. Biochim Biophys Acta 1094:193–199

Gautam N, Baetscher M, Aebersold R, Simon MI (1989) A G protein gamma subunit shares homology with ras proteins. Science 244:971–974

Graziano MP, Freissmuth M, Gilman AG (1989) Expression of Gsα in Escherichia coli. Purification and properties of two forms of the protein. J Biol Chem 264(1):409–418

Graziano MP, Gilman AG (1989) Synthesis in Escherichia coli of GTPase-deficient mutants of Gsα. J Biol Chem 264(26):15475–15482

Hsu WH, Rudolph U, Sanford J, Bertrand P, Olate J, Nelson C, Moss LG, Boyd A, Codina J, Birnbaumer L (1990) Molecular cloning of a novel splice variant of the alpha subunit of the mammalian Go protein. J Biol Chem 265:11220–11226

Hurley JB, Fong HKW, Teplow DB, Dreyer WJ, Simon MI (1984) Isolation and characterization of a cDNA clone for the γ subunit of bovine retinal transducin. Proc Natl Acad Sci USA 81:6948–6952

Itoh H, Katada T, Ui M, Kawasaki H, Suzuki K, Kaziro Y (1988b) Identification of three pertussis toxin substrates (41, 40 and 39 kDa proteins) in mammalian brain: comparison of predicted amino acid sequences from G-protein α-subunit genes and cDNAs with partial amino acid sequences from purified proteins. FEBS Lett 230:85–89

Itoh H, Kozasa T, Nagata S, Nakamura S, Katada T, Ui M, Iwai S, Ohtsuka E, Kawasaki H, Suzuki K, Kaziro Y (1986) Molecular cloning and sequence determination of cDNAs for α-subunits of the guanine nucleotide binding proteins G_s G_i, and G_o from rat brain. Proc Natl Acad Sci USA 83:3776–3780

Itoh H, Toyama R, Kozasa T, Tsukamoto T, Matsuoka M, Kaziro Y (1988a) Presence of three distinct molecular species of Gi protein α subunits: structure of rat cDNAs and human genomic DNAs. J Biol Chem 263:6656–6664

Jones DT, Masters SB, Bourne HR, Reed RR (1990) Biochemical characterization of three stimulatory GTP-binding proteins. The large and small forms of Gs and the olfactory-specific G-protein, Golf. J Biol Chem 265:2671–2676

Jones DT, Reed RR (1987) Molecular clonig of five GTP-binding protein cDNA species from rat olfactory neuroepithelium. J Biol Chem 262:14241–14249

Jones DT, Reed RR (1989) Golf: an olfactory neuron specific-G protein involved in odorant signal transduction. Science 244:790–795

Jurnak F, (1985) Structure of the GDP domain of EF-Tu and location of the amino acids homologous to ras oncogene proteins. Science 230:32–36

Jurnak F, Heffron S, Bergmann E (1990) Conformational changes involved in the activation of ras p21: implications for related proteins. Cell 60:525–528

Kaziro Y (1978) The role of guanosine 5′-triphosphate in polypeptide chain elongation. Biochim Biophys Acta 505:95–127

Kaziro Y, Itoh H, Kozasa T, Nakafuku M, Satoh T (1991) Structure and function of signal-transducing GTP-binding proteins. Annu Rev Biochem 60:349–400

Kleuss C, Hescheler J, Ewel C, Rosenthal W, Schultz G, Wittig B (1991) Assignment of G-protein subtypes to specific receptors inducing inhibition of calcium currents. Nature 353:43–48

Kleuss C, Scherubl H, Hescheler J, Schultz G, Wittig B (1992) Different β-subunits determine G-protein interaction with transmembrane receptors. Nature 358: 424–426

Kozasa T, Itoh H, Tsukamoto T, Kaziro Y (1988) Isolation and characterization of the human Gsα gene. Proc Natl Acad Sci USA 85:2081–2085

Landis CA, Masters SB, Spada A, Pace AM, Bourne HR, Valler L (1989) GTPase inhibiting mutations activate the α chain of Gs and stimulate adenylate cyclase in human pituitary tumours. Nature 340:692–696

Lee E, Taussig R, Gilman AG (1992) The G226A mutant of Gs alpha highlights the requirement for dissociation of G protein subunits. J Biol Chem 267:1212–1218
Lerea CL, Somers DE, Hurley JB, Klock IB, Bunt-Milan AH (1986) Identification of specific trasducin α subunits in retinal rod and cone photoreceptors. Science 324:77–80
Levine MA, Smallwood PM, Moen Pt Jr, Helman LJ, Ahn TG (1990) Molecular cloning of beta3 subunit, a third form of the G protein beta-subunit polypeptide. Proc Natl Acad Sci USA 87:2329–2333
Lochrie MA, Mendel JE, Sternberg PW, Simon MI (1991) Homologous and unique G protein alpha subunits in the nematode Caenorhabditis elegans. Cell Regul 2:135–154
Lyons J, Landis CA, Harsh G, Vallar L, Grünewald K, Feichtinger H, Duh QY, Clark OH, Kawasaki E, Bourne HR, McCormick F (1990) Two G protein oncogenes in human endocrine tumors. Science 249:655–659
Ma H, Yanofsky MF, Huang H (1991) Isolation and sequence analysis of TGA1 cDNAs encoding a tomato G protein alpha subunit. Gene 107:189–195
Ma H, Yanofsky MF, Meyerowitz EM (1990) Molecular cloning and characterization of GPA1, a G protein alpha subunit gene from Arabidopsis thaliana. Proc Natl Acad Sci Usa 87:3821–3825
Masters SB, Miller RT, Chi M-H, Chang F-H, Beiderman B, Lopez NG, Bourne HR (1989) Mutations in the CTP-binding site of Gsα alter stimulation of adenylyl cyclase. J Biol Chem 264:15467–15474
Matsuoka M, Itoh H, Kaziro Y (1990) Characterization of the human gene for Gxα, a pertussis toxin-insensitive regulatory GTP-binding protein. J Biol Chem 265: 13215–13220
Matsuoka M, Itoh H, Kozasa T, Kaziro Y (1988) Sequence analysis of cDNA and genomic DNA for a putative pertussis toxin-insensitive guanine nucleotide-binding regulatory protein α subunit. Proc Natl Acad Sci USA 85:5384–5388
McLaughlin SK, McKinnon PJ, Margolskee RF (1992) Gustducin is a taste-cell-specific G protein closely related to the transducins. Nature 357:563–569
Milburn M, Tong VL, deVos AM, Brunger A, Yamaizumi Z, Nishimura S, Kim S-H (1989) Molecular switch for signal transduction: structural differences between active and inactive forms of protooncogenic ras proteins. Science 247:939–945
Morishita R, Asano T, Kato K, Itoh H, Kaziro Y (1989) Purification and identification of two pertussis-toxin sensitive GTP-binding proteins of bovine spleen. Biochem Biophys Res Commun 161:1280–1285
Mumby SM, Casey PJ Gilman AG, Gutowski S, Sternweis PC (1990) G protein γ subunits contain a 20-carbon isoprenoid. Proc Natl Acad Sci USA 87(15):5873–5877
Muntz KH, Sternweis PC, Gilman AG, Mumby SM (1992) Influence of gamma subunit prenylation on association of guanine nucleotide-binding regulatory proteins with membranes. Cell Regul 3:49–61
Nakafuku M, Itoh H, Nakamura S, Kaziro Y (1987) Occurrence in Saccharomyces cerevisiae of a gene homologous to the cDNA coding for the α subunit of mammalian G proteins. Proc Natl Acad Sci USA 84:2140–2144
Nakafuku M, Obara T, Kaibuchi K, Miyajima I, Miyajima A, Itoh H, Nakamura S, Arai K, Matsumoto K, Kaziro Y (1988) Isolation of a second yeast Saccharomyces cerevisiae gene (GPA2) coding for guanine nucleotide-binding regulatory protein: studies on its structure and possible functions. Proc Natl Acad Sci USA 85:1374–1378
Nukada T, Tanabe T, Takahashi H, Noda M, Haga K, Haga T, Ichiyama A, Kangawa K, Hiranaga M, Matsuo H, Numa K (1986) Primary structure of the α subunit of bovine adenylate cyclase-inhibiting protein deduced from the cDNA sequence. FEBS Lett 197:305–310
Obara T, Nakafuku M, Yamamoto M, Kaziro Y (1991) Isolation and characterization of a gene encoding a G-protein alpha subunit from Schizosaccharomyces pombe: involvement in mating and sporulation pathways. Proc Natl Acad Sci USA 88:5877–5881

Pai EF, Krengel U, Petsko GA, Googy RS, Kabsch W, Wittinghofer A (1990) Refined crystal structure of the triphosphate conformation of H-ras p21 at 1.35 A resolution: implications for the mechanism of GTP hydrolysis. EMBO J 9:2351–2359
Raport CJ, Dere B, Hurley JB (1989) Characterization of the mouse rod transducin α subunit gene. J Biol Chem 264:7122–7128
Robishow JD, Kalman VK, Moomaw CR, Slaughter CA (1989) Existence of two γ subunits of the G proteins in brain. J Biol Chem 264:15758–15761
Robishow JD, Smigel MD, Gilman AG (1986) Molecular basis for two forms of the G protein that stimulates adenylate cyclase. J Biol Chem 261:9587–9590
Simon MI, Strathmann MP, Gautam N (1991) Diversity of G proteins in signal transduction. Science 252:802–808
Simonds WF, Butrynski JE, Gautam N, Unson CG, Spiegel AM (1991) G-protein betagamma dimers. Membrane targeting requires subunit coexpression and intact gamma C-A-A-X domain. J Biol Chem 266:5363–5366
Strathmann M, Simon MI (1990) G protein diversity: a distinct class of alpha subunits is present in vertebrates and invertebrates. Proc Natl Acad Sci USA 87:9113–9117
Strathmann M, Wilkie T, Simon MI (1989) Diversity of the G-protein family: sequences from five additional α subunits in the mouse. Proc Natl Acad Sci USA 86:7407–7409
Strathmann M, Wilkie TM, Simon MI (1990) Alternative splicing produces transcripts encoding two forms of the alpha subunit of GTP-binding protein Go. Proc Natl Acad Sci USA 87:6477–6481
Strathmann MP, Simon MI (1991) Galpha12 and Galpha13 subunits define a fourth class of G protein alpha subunits. Proc Natl Acad Sci USA 88:5582–5586
Sugimoto K, Nukada T, Tanabe T, Takahashi H, Noda M, Minamino N, Kangawa K, Matsuo H, Hirose T, Inayama S, Numa S (1985) Primary structure of the β-subunit of bovine transducin deduced from the cDNA sequence. FEBS Lett 191:235–240
Suki WN, Abramowitz J, Mattera R, Codina J, Birnbaumer L (1987) The human genome encodes at least three non-allellic G proteins with αi-type subunits. FEBS Lett 220(1):187–192
Taylor SJ, Chae HZ, Rhee SG, Exton JH (1991) Activation of the beta1 isozyme of phospholipase C by alpha subunits of the Gq class of G proteins. Nature 350:516–518
Taylor SJ, Smith JA, Exton JH (1990) Purification from bovine live membranes of a guanine nucleotide-dependent activator of phosphoinositide-specific phospholipase C. Immunologic identification as a novel G-protein alpha subunit. J Biol Chem 265:17150–17156
Tsukamoto T, Toyama R, Itoh H, Kozasa T, Matsuoka M, Kaziro Y (1991) Structure of the human gene and two rat cDNAs encoding the alpha chain of GTP-binding regulatory protein Go: two different mRNAs are generated by alternative splicing. Proc Natl Acad Aci USA 88:2974–2978
van der Voorn L, Gebbink M, Plasterk RH, Ploegh HL (1990) Characterization of a G-protein beta-subunit gene from the namatoda Caenorhabditis elegans. J Mol Biol 213(1):17–26
Van Dop C, Tsubokawa M, Bourne HR, Ramachandran J (1984) Amino acid sequences of retinal transducin at the site ADP-ribosylated by cholera toxin. J Biol Chem 259:696–698
von Weizsäcker, E, Strathmann MP, Simon MI (1992) Diversity among the beta subunits of heterotrimeric GTP-binding proteins: characterization of a novel beta-subunit cDNA. Biochem Biophys Res Commun 183:350–356
Whiteway M, Hougan L, Dignard D, Thomas DY, Beil L, Saari GC, Grant FJ, O'Hara P, Mackay VL (1989) The STE4 and STE18 genes of yeast encode potential β and γ subunits of the mating factor receptor-coupled G protein. Cell 56:467–477

Wilkie TM, Scherle PA, Strathmann MP, Slepak VZ, Simon MI (1991) Characterization of G-protein alpha subunits in the Gq class: expression in murine tissues and in stromal and hematopoietic cell lines. Proc Natl Acad Sci USA 88:10049–10053

Yamane HK, Farnsworth CC, Xie H, Howald W, Fung BK, Clarke S, Gelb MH, Glomset JA (1990) Brain G protein gamma subunits contain an all-trans-geranylgeranyl-cysteine methyl ester at their carboxyl termini. Proc Natl Acad Sci USA 87:5868–5872

Yatsunami K, Pandya BV, Oprian DD, Khorana HG (1985) cDNA-derived amino acid sequence of the γ subunit of GTPase rod outer segments. Proc Natl Acad Sci USA 82:1936–1940

CHAPTER 11

Molecular Diversity in Signal Transducing G-Proteins

L. BIRNBAUMER

Signal-transducing G-proteins are heterotrimers in which molecular diversity exists for each of the subunits. The genes known thus far (16 α, 4 β, and 5 γ) and splice variants for 2 α's allow for a calculated 400 combinations. Restriction in expression and lack of functional connotation of some of the splice variants decrease this number to a much smaller, but still considerable number of G-proteins that may be mediating receptor signals in a cell. Correlations between molecular diversity and function are being steadily worked out and an update follows.

A. The α Subunits

I. Molecular Diversity

The α subunits constitute not only a numerous, but also a complex family of proteins. This family belongs to the superfamily of GTP binding and hydrolyzing proteins. As a rule, G-proteins are named by the name of their α subunit, without regard to $\beta\gamma$ composition. The molecular diversity of G-protein α subunits has been extensively reviewed (e.g., BIRNBAUMER 1990 and references therein; SIMON et al. 1991 and references therein; STRATHMAN and SIMON 1991; AMATRUDA et al. 1991; WILKIE et al. 1991; McLAUGHLIN et al. 1992), and it is summarized in the left panel of Fig. 1, together with functional assignments that can presently be made.

There are 16 nonallelic mammalian α subunit genes known. Four relate to the senses: two for vision (transducins), one for olfaction (α_{olf}), and one for taste (gustducin). The remainder are α_s (four splice variants of uncertain functional difference), three α_i's, α_o (two splice variants), α_z, α_q, α_{11}, α_{12}, α_{13}, α_{14}, and α_{15} (mouse) or α_{16} (human). At 354–380 amino acids, they are similar, but on average only 47% homologous, the least homology being 36% (α_{olf} vs α_{12}), the most being 88% (α_q vs α_{11} or α_{11} vs α_{12}), and the most frequent homology value among them being 40–45%. Phylogenetic analysis (Fig. 1) subdivides α subunits into three major families: G_s, G_q, and G_i. α_s and α_{olf} are the sole members of the G_s family. The G_q family has two main branches, *branch I* containing α_q, α_{11}, α_{14}, and $\alpha_{15/16}$, and *branch II* containing α_{12} and α_{13} (homology between branches, 43%). The G_i family is

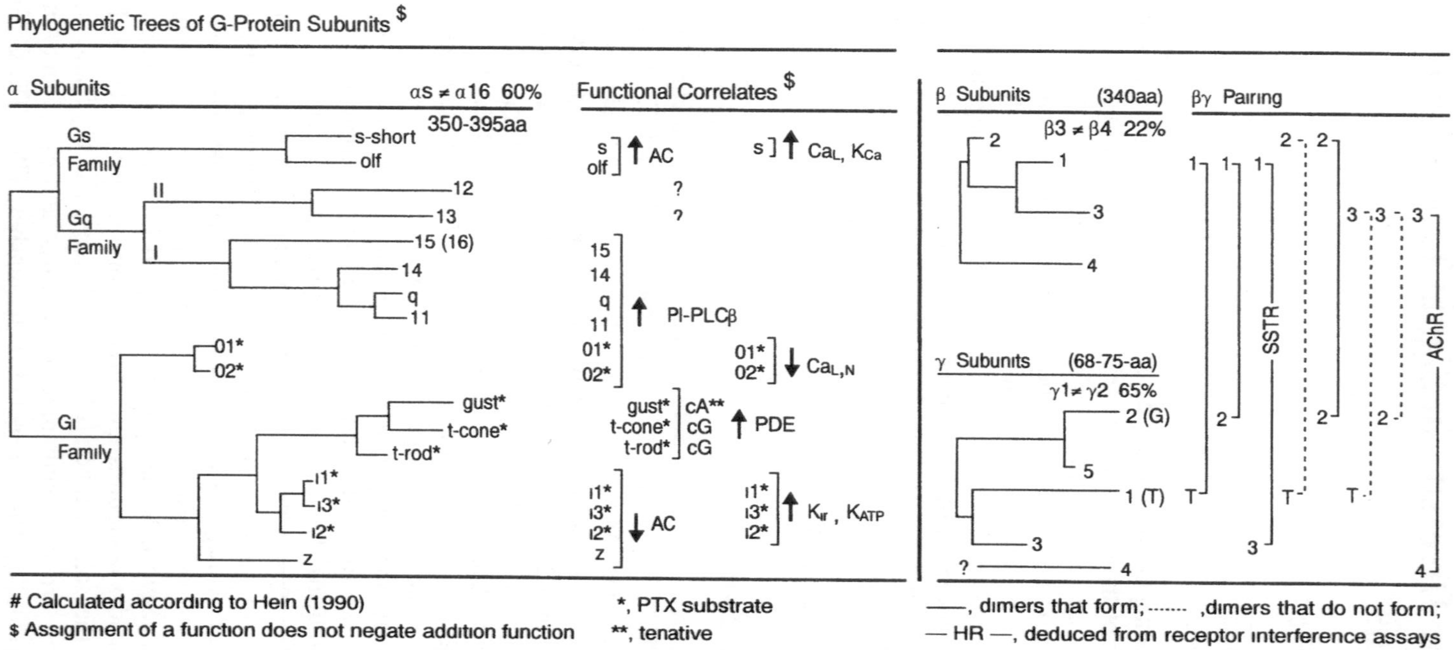

Fig. 1. Summary of molecular diversity in G-protein subunits, functional correlates for α subunits, and emerging rules of $\beta\gamma$ pairing

the most complex, containing the three α_i's, α_o, the three remaining sense transducers α_{t1} (rod), α_{t2} (cone), and α_{gust}, and α_z. Except for α_z, all of the members of this α subunit family are PTX substrates, and all of the known PTX substrates are in this family.

Of the α genes listed above, α_s, α_{i2}, and α_{11} appear to be expressed in all cells. Most cells also express α_q and either α_{i1} or α_{i3} (functional homologs of α_{11} and α_{i2}, respectively) and one or both the α_{12} and α_{13} genes. Thus, all cells express 6–8 "ubiquitous" α subunits.

Expression of α_o, α_{14}, α_{15}, and α_z is not ubiquitous, but it also is not exclusive to single cell types, as are α_t's, α_{olf}, and α_{gust}. Their products are found in groups of cells or tissues that often, but not always have a common embryonic origin. The α_o's (α_{o1} and α_{o2}) are preferentially expressed in cells derived from the neural crest and in endocrine cells (e.g., pituitary and pancreatic β cells), as well as in selected other cell types such as cardiac myocytes: α_{14} is expressed primarily in stromal and epithelial cells; and $\alpha_{15/16}$ is expressed in many, but not all cells with hematopoietic lineage; α_z is found primarily in neurons, but in small quantities also in red blood cells and other cell types. It follows that out of the repertoire of 16 α subunit genes, a standard cell expresses between 9 and 10 of these.

II. α Subunit Functions

G-protein research emerged from studies of function, was then taken over by the gathering of the structural features described above and is now returning to function. In broad terms there are correlates between the α gene families and their physiological roles.

The G_s family of α subunits stimulates adenylyl cyclases in all cells and hence mediates the effects of receptors that cause increases in intracellular cAMP (e.g., parathyroid hormone, PTH, and calcitonin, CT). In most cases the end effect is to stimulate the type of activity that is characteristic of the cell. In addition to stimulating adenylyl cyclase, α_s can stimulate calcium channels of the L-type and a subclass of Ca^{2+}-activated voltage-dependent K channels (K_{Ca} channels). These are functions that are expressed only in specific tissues and cell types, so that responses to receptors that activate G_s may vary among tissues. Thus, calcium channel stimulation by α_s occurs in tissues such as skeletal muscle and heart (Yatani et al. 1987, 1988a; Hamilton et al. 1991), where it potentiates the stimulatory effect of the cAMP-PKA system, but not in liver or endothelial cells that lack voltage-gated calcium channels; stimulation of the K_{Ca} channel occurs in a tissue such as coronary smooth muscle, where it potentiates the relaxing effect of the cAMP–protein kinase A (PKA) system, but not in atrial cells of the heart that lack K_{Ca} channels (Scornik et al. 1992).

The α_q/α_{11} subunits of the G_q family stimulate a phospholipase C of the β type (PLC-β) (Smrcka et al. 1991; Taylor et al. 1991; Wu et al. 1992; Lee et al. 1992) and mediate effects of receptors that cause cellular increases in inositol triphosphate (IP_3) and diacylglycerol. In addition, since the majority of cells express the IP_3-sensitive calcium release channel, activation of α_q/α_{11} leads to increases in intracellular Ca^{2+}. In general, the receptors that cause increases in protein kinase C (PKC) activity and intracellular Ca^{2+} are, like receptors that cause increases in cAMP, also of a "stimulatory" type. They cause an increase in the activity of the biochemical machineries for which the cells were made. These include stimulation of glucose production by hepatocytes, secretion of proteolytic enzymes, electrolytes and fluid from exocrine glands, secretion of hormones from endocrine cells, and stimulation of smooth muscle contraction. Like α_q and α_{11}, the structurally closely related α_{14} and $\alpha_{15/16}$ also stimulate a PLC of the β-type (see below). The function(s) of the more distant G_q relatives, α_{12} and α_{13}, are as yet unknown.

The G_i family of α subunits has as members α subunits that, like α_s and α_q, mediate endocrine, paracrine, and its autocrine regulation of cells, as well as the α subunits involved in vision and taste. All except α_z are PTX sensitive, and this toxin has served to gather indirect information about the possible functional roles of the members of the G_i family of α subunits. More direct information has come from studies with purified proteins in cell free systems. PTX, through ADP-ribosylation of a cysteine at position -4 from the C terminus of the G_i family of α subunits uncouples effects of receptors that cause inhibition of adenylyl cyclase, inhibition of Ca^{2+} currents, and stimulation of at least two classes of K channels. One or more of the α's of the G_i family directly or indirectly mediate these effects, which, from a physiological standpoint, counteract stimulatory effects triggered by the activation of α_s and α_q and their functional congeners. The subunits α_{i1}, α_{i2}, and α_{i3} all cause inhibition of adenylyl cyclase (Simonds et al. 1989; Wong et al. 1991) and activation of the "muscarinic"-type (inwardly rectifying), and of the ATP-sensitive K channels (also referred to as G-protein-gated K channels) (Yatani et al. 1988b; Kirsch et al. 1990). The α_o, in either one of its two splice variations, α_{o1} and α_{o2}, mediates receptor-induced inhibition of L- and N-type calcium channels (Kleuss et al. 1991); α_o also has the capacity of stimulating a PLC, as shown by the microinjection of the protein into *Xenopus* oocytes (Padrell et al. 1991). The calcium channel inhibition by α_o can either be a primary effect of α_o or an effect secondary of PLC stimulation (e.g., PKC-mediated phosphorylation). No specific role has as yet been reported for α_z, except that in transfection assays if inhibits cAMP accumulation.

Not all of the G-protein-dependent regulations of effector systems have a G-protein assigned to them. Stimulation of phospholipase A_2 (PLA_2), which causes cells to release arachidonic acid, a precursor of active autacoids such as prostaglandins, prostacyclin, thromboxanes, and leukotrienes, can be mediated by a PTX-sensitive, as well as a PTX-insensitive G-protein. In general, PLA_2 stimulation accompanies activation of PLC and could be secondary to increases in intracellular Ca^{2+} (see Lin et al. 1992), but these two effects have been separated by pharmacological studies (Burch et al. 1986). No cell-free assays for G-protein regulation of PLA_2 have been developed as yet. Indirect assays for PLA_2 activation involving measurement of ion channels indicate that PLA_2 can be activated either by specific α subunits, e.g., α_{13} in renal tubules (CANTIELLO et al. 1990), or $\beta\gamma$ dimers, e.g., in atrial membrane patches (KIM et al. 1989).

In cells of the hematopoietic lineage (platelets, neutrophils) and in various types of transformed cell lines, arachidonic acid production in response to agonists is PTX sensitive (BOKOCH and GILMAN 1984; NAKAMURA and UI 1985). Even though these are cell types that express α_{15}, this is a PTX-insensitive subunit and therefore not likely to be responsible for PTX-sensitive activation of PLA_2. Cells of this same lineage also exhibit PTX-sensitive activation of PLC, and here, too, the responsible G-protein cannot be named. The only PTX-sensitive G-protein known to stimulate PLC is G_o, but G_o is not expressed in these cells. The discovery of $\beta\gamma$-stimulated PLC-β_2 and PLC-β_3 makes it likely that most if not all of the PTX-sensitive stimulations of PLC are mediated by $\beta\gamma$'s derived from one or more of the PTX-sensitive Gi's.

A PTX-sensitive activation of N-type voltage-dependent Ca channels by receptors that activate PLC in a PTX-insensitive way in steroidogenic adrenal cells by angiotensin II and in pituitary cells by luteinizing hormone releasing hormone (LHRH) and thyrotropin releasing hormone (TRH) has been described (HESCHELER et al. 1988; ROSENTHAL et al. 1988; GOLLASCH et al. 1991), and the identity of this calcium channel-regulating G-protein(s), or the type of subunit that may mediate this effect, is(are) also unknown.

B. The $\beta\gamma$ Dimers

In addition to the many α subunits, four nonallelic β genes and at least five nonallelic γ genes are known (BIRNBAUMER 1990 and references therein; SIMON et al. 1991 and references therein; FISCHER and ARONSON 1992; VON WEIZSÄCKER et al. 1992). The β subunits, each of 340 amino acids, are between 78 and 86% homologous and constitute the most homologous set of G-protein subunits. The γ subunits, 68–75 amino acids long, are only 28–43% homologous and constitute the most heterogeneous set of the subunits that form G-proteins.

Combinatorial analysis gives the possibility of forming 20 distinct $\beta\gamma$-dimer proteins, or more if more β and or γ genes are discovered. However, the actual complexity to be found in any given cell is less staggering for two reasons. One is that not all these genes are expressed in any single cell; the other is that there appear to be rules that determine which β subunit can interact with which γ subunit. This last feature of G-protein subunit composition was discovered by SCHMIDT et al. (1992) and PRONIN and GAUTAM (1992), who studied the interaction of recombinant β_1, β_2, and β_3 with γ_1 (also γ_T) and γ_2. The results showed that β_1 interacts with both γ's, β_2 only with γ_2, and β_3 with neither. The receptor signaling interference assay of KLEUSS et al. (1992a,b) extended the listing of permissive dimer formations to include $\beta_1\gamma_3$ and $\beta_3\gamma_4$ (the full sequence of the latter has not yet been published). These results are summarized in the right panel of Fig. 1.

The type and complexity of signaling of the G-protein-coupled receptors depends on the G-protein complement expressed in the cells in which they are active. Although the tissue distribution of β and γ genes has been less well studied, it is clear that most cells express at least two β and two γ genes and more likely three of each. Together with the standard α subunit repertoire, this makes for 40–90 distinct G-proteins engaged in transducing receptor signals into modulated effector activities, if, as suggested by the studies of GRAF et al. (1992), α subunits do not discriminate among subsets of $\beta\gamma$ dimers.

It seems likely that regulation of effectors by $\beta\gamma$ dimers, like interaction with receptors, will exhibit at least some degree of specificity towards the dimer subunit makeup.

References

Amatruda TT III, Steele DA, Slepak VZ, Simon MI (1991) Gα16, a G protein α subunit specifically expressed in hematopoietic cells. Proc Natl Acad Sci USA 88:5587–5591

Birnbaumer L (1990) G proteins in signal transduction. Annu Rev Pharmacol Toxicol 30:675–705

Bokoch GM, Gilman AG (1984) Inhibition of receptor-mediated release of arachidonic acid by pertussis toxin. Cell 39:301–308

Burch RM, Luini A, Axelrod J (1986) Phospholipase A_2 and phospholipase C are activated by distinct GTP-binding proteins in response to alpha$_1$-adrenergic stimulation in FRTL-5 cells. Proc Natl Acad Sci USA 83:7201–7205

Cantiello HF, Patenaude CR, Codina J, Birnbaumer L, Ausiello DA (1990) Gα_{i3} regulates epithelial Na^+ channels by activation of phospholipase A_2 and lipoxygenase pathways. J Biol Chem 265:21624–21628

Fisher JK, Aronson NN (1992) Characterization of the cDNA and genomic sequence of a G protein γ subunit (γ_5). Mol Cell Biol 12:1585–1591

Gollasch M, Haller H, Schultz G, Hescheler J (1991) Thyrotropin-releasing hormone induces opposite effects on Ca^{2+} channel currents in pituitary cells by two pathways. Proc Natl Acad Sci USA 88:10262–10266

Graf R, Mattera R, Codina J, Evans T, Ho Y-K, Estes MK, Birnbaumer L (1992) Studies on the interaction of α subunits of G proteins with $\beta\gamma$ dimers. Eur J Biochem 210:609–619

Hamilton SL, Codina J, Hawkes MJ, Yatani A, Sawada T, Strickland FM, Froehner SC, Spiegel AM, Toro L, Stefani E, Birnbaumer L, Brown AM (1991) Evidence for direct interaction of $G_s\alpha$ with the Ca^{2+} channel of skeletal muscle. J Biol Chem 266:19528–19535

Hein J (1990) Unified approach to alignment and phylogenics. Methods-Enzymal 183:626–645

Hescheler J, Rosenthal W, Hinsch K-D, Wulfern M, Trautwein W, Schultz G (1988) Angiotensin II-induced stimulation of voltage-dependent Ca^{2+} currents in an adrenal cortical cell line. EMBO J 7:619–624

Kim D, Lewis DL, Graziadei L, Neer EJ, Bar-Sagi D, Clapham DE (1989) G protein $\beta\tau$-subunits activate the cardiac muscarinic K^+ channel via phospholipase A_2. Nature 337:557–560

Kirsch G, Codina J, Birnbaumer L, Brown AM (1990) Coupling of ATP-sensitive K^+ channels to purinergic receptors by G-proteins in rat ventricular myocytes. Am J Physiol 259:H820–H826

Kleuss C, Hescheler J, Ewel C, Rosenthal W, Schultz G, Wittig B (1991) Assignment of G-protein subtypes to specific receptors inducing inhibition of calcium currents. Nature 353:43–48
Kleuss C, Scherübl H, Hescheler J, Schultz G, Wittig B (1992a) Different β-subunits determine G protein interaction with transmembrane receptors. Nature 358:424–426
Kleuss et al. (1992b) SSTR and AChR interact with specific G protein γ subunits. Science (in press)
Lee CH, Park D, Wu D, Rhee SG, Simon MI (1992) Members of the G_q alpha subunit gene family activate phospholipase C-beta isozymes. J Biol Chem (in press)
Lin LL, Lin AY, Knopf JL (1992) Cytosolic phospholipase A_2 is coupled to hormonally regulated release of arachidonic acid. Proc Natl Acad Sci USA 89:6147–6151
Nakamura T, Ui M (1985) Simultaneous inhibitions of inositol phospholipid breakdown, arachidonic acid release, and histamine secretion in mast cells by islet-activating protein, pertussis toxin. A possible involvement of the toxin-specific substrate in the Ca^{2+}-mobilizing receptor-mediated biosignaling system. J Biol Chem 260:3584–3593
McLaughlin SK, McKinnon PJ, Margolskee RF (1992) Gustducin is a tast-cell-specific G protein closely related to the transducins. Nature 357:563–569
Padrell E, Carty DJ, Moriarty TM, Hildebrandt JD, Landau EM, Iyengar R (1991) Two forms of the bovine brain G_o that stimulate the inositol trisphospate-mediated Cl^- currents in Xenopus oocytes. J Biol Chem 256:9771–9777
Pronin AN, Gautham N (1992) Interaction between G protein β and γ subunit types is selective. Proc Natl Acad Sci USA 89:6220–6224
Rosenthal W, Hescheler J, Hinsch K-D, Spicher K, Trautwein W, Schultz G (1988) Cyclic AMP independent, dual regulation of voltage-dependent Ca^{2+} currents by LHRH and somastostatin in a pituitatry cell line. EMBO J 7:1627–1633
Schmidt CJ, Thomas TC, Levine MA, Neer EJ (1992) Specificity of G protein β and γ subunit interactions. J Biol Chem 267:13807–13810
Scornik F, Codina J, Birnbaumer L, Toro L (1992) G_s and β-adrenergic activation of K_{Ca} channel independent of cAMP-mediated phosphorylation (submitted)
Simon MI, Strathmann MP, Gautam N (1991) Diversity of G proteins in signal transduction. Science 252:802–808
Simonds WF, Goldsmith PK, Codina J, Unson CG, Spiegel AM (1989) G_{i2} mediates α_2-adrenergic inhibition of adenylyl cyclase in platelet membranes: In situ identification with G_α C-terminal antibodies. Proc Natl Acad Sci USA 86:7809–7813
Smrcka AV, Helper JR, Brown KO, Sternweis PC (1991) Regulation of polyphosphoinositide-specific phospholipase C activity by purified G_q. Science 251: 804–807
Strathmann MP, Simon MI (1991) Gα12 and Gα13 subunits define a fourth class of G protein α subunits. Proc Natl Acad Sci USA 88:5582–5586
Taylor SJ, Chae HZ, Rhee SG, Exton JH (1991) Activation of the β1 isozyme of phospholipase C by α subunits of the G_q class of G proteins. Nature 350:516–518
von Weizsäcker E, Strathmann MP, Simon MI (1992) Diversity among the beta subunits of heterotrimeric GTP-binding proteins: characterization of a novel beta-subunit cDNA. Biochem Biophys Res Commun 183:350–356
Wilkie TM, Scherle PA, Strathmann MP, Slepak VZ, Simon MI (1991) Characterization of G-protein α subunits in the G_q class: expression in murine tissues and in stromal and hematopoietic cell lines. Proc Natl Acad Sci USA 88:10049–10053
Wong UH, Federman A, Pace AM, Zachary I, Evans T, Pouysségur J, Bourne HR (1991) Mutant α subunits of G_{i2} inhibit cyclic AMP accumulation. Nature 351:63–65

Wu D, Lee CH, Rhee SG, Simon MI (1992) Activation of phospholipase C by the α subunit of the G_q and G_{11} protein in transfected Cos-7 cells. J Biol Chem 267:1811–1817

Yatani A, Codina J Imoto Y, Reeves JP, Birnbaumer L, Brown AM (1987) A G protein directly regulates mammalian cardiac calcium channels. Science 238:1288–1292

Yatani A, Imoto Y, Codina J, Hamilton SL, Brown AM, Birnbaumer L (1988a) The stimulatory G protein of adenylyl cyclase, G_s, directly stimulates dihydropyridine-sensitive skeletal muscle Ca^{2+} channels. Evidence for direct regulation independent of phosphorylation by cAMP-dependent protein kinase. J Biol Chem 263:9887–9895

Yatani A, Mattera R, Codina J, Graf R, Okabe K, Padrell E, Iyengar R, Brown AM, Birnbaumer L (1988b) The G protein-gated atrial K^+ channel is stimulated by three distinct $G_i\alpha$-subunits. Nature 336:680–682

CHAPTER 12

Structural Conservation of Ras-Related Proteins and Its Functional Implications

P. CHARDIN

A. Introduction: The Discovery of *ras* and *ras*-Related Genes

The H-*ras*, K-*ras*, and N-*ras* genes were discovered in the early 1980s (see Chap. 17). In 1983, soon after these first discoveries, an open reading frame located between the actin and tubulin genes of *Saccharomyces cerevisiae* was found to encode a protein sharing approximately 30% identity with mammalian ras proteins; it was named Ypt (see Chap. 27). One year later the *rho* genes were discovered, first in *Aplysia*, then in human. Rho proteins also shared approximately 30% identity with Ras or Ypt proteins (see Chap. 36). These discoveries of several proteins distantly related to Ras strongly suggested that Ras proteins belonged to a large family. Molecular genetic approaches were then undertaken to isolate new members of this family by homology probing. Two techniques have been highly successful in the discovery of new *ras*-related genes: the use of degenerate oligonucleotide mixes corresponding to conserved stretches of amino acids (usually the DTAGQE sequence around position 61) and low stringency hybridization with already isolated probes, sometimes from distant organisms. Another important strategy consisted in the biochemical isolation of small GTPases from various tissues on the basis of their GTP/GDP-binding ability and sequencing of short peptides to clone the corresponding cDNAs with oligonucleotide probes. Most of these proteins appeared closely related to the ones already discovered by molecular genetics. Several yeast mutants were also found to encode Ras-related proteins: SEC4, a secretion mutant (see Chap. 4), CDC42, a cell division cycle mutant (see Chap. 37), RSR1, a suppressor of CDC24, and SPI1, whose overexpression suppresses the "premature initiation of mitosis mutant", pim1. Some *ras*-related genes from *Dictyostelium*, expressed at a specific time in development, have also been isolated, such as *Ddras* or *sas1* and *sas2*. And a *Caenorhabditis elegans* mutant of vulval induction (let-60) was found to encode a Ras protein (HAN and STERNBERG 1990).

Taken together, all these approaches turned out to be so efficient that the primary structure of more than 60 Ras-related proteins is now available from cloned cDNA sequences. The purpose of this chapter is to compare the sequences of these various proteins, to establish their evolutionary relation-

a

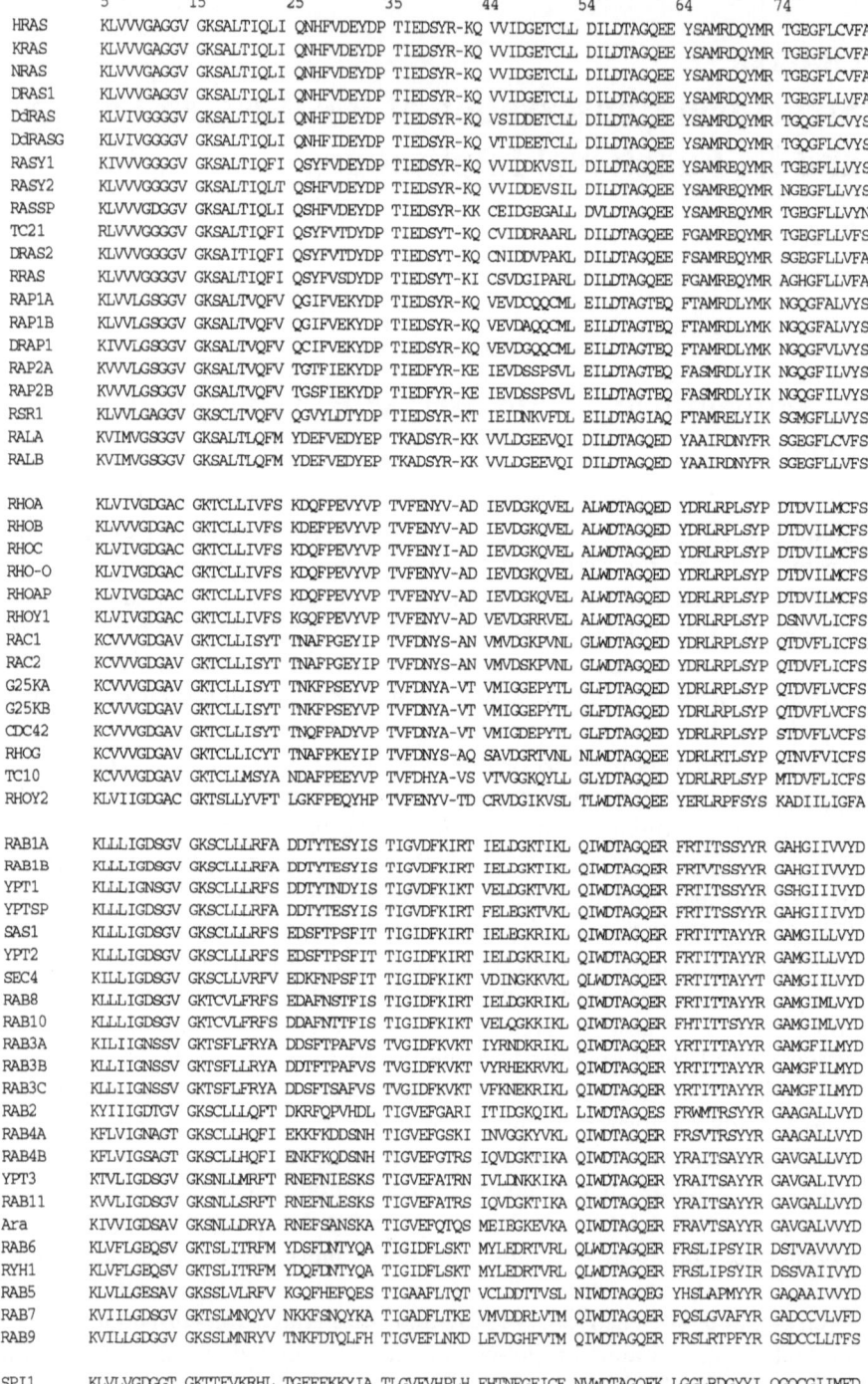

```
              5          15         25         35         44         54         64         74
HRAS      KLVVVGAGGV GKSALTIQLI QNHFVDEYDP TIEDSYR-KQ VVIDGETCLL DILDTAGQEE YSAMRDQYMR TGEGFLCVFA
KRAS      KLVVVGAGGV GKSALTIQLI QNHFVDEYDP TIEDSYR-KQ VVIDGETCLL DILDTAGQEE YSAMRDQYMR TGEGFLCVFA
NRAS      KLVVVGAGGV GKSALTIQLI QNHFVDEYDP TIEDSYR-KQ VVIDGETCLL DILDTAGQEE YSAMRDQYMR TGEGFLCVFA
DRAS1     KLVVVGAGGV GKSALTIQLI QNHFVDEYDP TIEDSYR-KQ VVIDGETCLL DILDTAGQEE YSAMRDQYMR TGEGFLLVFA
DdRAS     KLVIVGGGGV GKSALTIQLI QNHFIDEYDP TIEDSYR-KQ VSIDDETCLL DILDTAGQEE YSAMRDQYMR TGQGFLCVYS
DdRASG    KLVIVGGGGV GKSALTIQLI QNHFIDEYDP TIEDSYR-KQ VTIDEETCLL DILDTAGQEE YSAMRDQYMR TGQGFLCVYS
RASY1     KIVVVGGGGV GKSALTIQFI QSYFVDEYDP TIEDSYR-KQ VVIDDKVSIL DILDTAGQEE YSAMREQYMR TGEGFLLVYS
RASY2     KLVVVGGGGV GKSALTIQLT QSHFVDEYDP TIEDSYR-KQ VVIDDEVSIL DILDTAGQEE YSAMREQYMR NGEGFLLVYS
RASSP     KLVVVGDGGV GKSALTIQLI QSHFVDEYDP TIEDSYR-KK CEIDGEGALL DVLDTAGQEE YSAMREQYMR TGEGFLLVYN
TC21      RLVVVGGGGV GKSALTIQFI QSYFVTDYDP TIEDSYT-KQ CVIDDRAARL DILDTAGQEE FGAMREQYMR TGEGFLLVFS
DRAS2     KLVVVGGGGV GKSAITIQFI QSYFVTDYDP TIEDSYT-KQ CNIDDVPAKL DILDTAGQEE FSAMREQYMR SGEGFLLVFA
RRAS      KLVVVGGGGV GKSALTIQFI QSYFVSDYDP TIEDSYT-KI CSVDGIPARL DILDTAGQEE FGAMREQYMR AGHGFLLVFA
RAP1A     KLVVLGSGGV GKSALTVQFV QGIFVEKYDP TIEDSYR-KQ VEVDCQQCML EILDTAGTEQ FTAMRDLYMK NGQGFALVYS
RAP1B     KLVVLGSGGV GKSALTVQFV QGIFVEKYDP TIEDSYR-KQ VEVDAQQCML EILDTAGTEQ FTAMRDLYMK NGQGFALVYS
DRAP1     KIVVLGSGGV GKSALTVQFV QCIFVEKYDP TIEDSYR-KQ VEVDGQQCML EILDTAGTEQ FTAMRDLYMK NGQGFVLVYS
RAP2A     KVVVLGSGGV GKSALTVQFV TGTFIEKYDP TIEDFYR-KE IEVDSSPSVL EILDTAGTEQ FASMRDLYIK NGQGFILVYS
RAP2B     KVVVLGSGGV GKSALTVQFV TGSFIEKYDP TIEDFYR-KE IEVDSSPSVL EILDTAGTEQ FASMRDLYIK NGQGFILVYS
RSR1      KLVVLGAGGV GKSCLTVQFV QGVYLDTYDP TIEDSYR-KT IEIDNKVFDL EILDTAGIAQ FTAMRELYIK SGMGFLLVYS
RALA      KVIMVGSGGV GKSALTLQFM YDEFVEDYEP TKADSYR-KK VVLDGEEVQI DILDTAGQED YAAIRDNYFR SGEGFLCVFS
RALB      KVIMVGSGGV GKSALTLQFM YDEFVEDYEP TKADSYR-KK VVLDGEEVQI DILDTAGQED YAAIRDNYFR SGEGFLLVFS

RHOA      KLVIVGDGAC GKTCLLIVFS KDQFPEVYVP TVFENYV-AD IEVDGKQVEL ALWDTAGQED YDRLRPLSYP DTDVILMCFS
RHOB      KLVVVGDGAC GKTCLLIVFS KDEFPEVYVP TVFENYV-AD IEVDGKQVEL ALWDTAGQED YDRLRPLSYP DTDVILMCFS
RHOC      KLVIVGDGAC GKTCLLIVFS KDQFPEVYVP TVFENYI-AD IEVDGKQVEL ALWDTAGQED YDRLRPLSYP DTDVILMCFS
RHO-O     KLVIVGDGAC GKTCLLIVFS KDQFPEVYVP TVFENYV-AD IEVDGKQVEL ALWDTAGQED YDRLRPLSYP DTDVILMCFS
RHOAP     KLVIVGDGAC GKTCLLIVFS KDQFPEVYVP TVFENYV-AD IEVDGKQVEL ALWDTAGQED YDRLRPLSYP DTDVILMCFS
RHOY1     KLVIVGDGAC GKTCLLIVFS KGQFPEVYVP TVFENYV-AD VEVDGRRVEL ALWDTAGQED YDRLRPLSYP DSNVVLICFS
RAC1      KCVVVGDGAV GKTCLLISYT TNAFPGEYIP TVFDNYS-AN VMVDGKPVNL GLWDTAGQED YDRLRPLSYP QTDVFLICFS
RAC2      KCVVVGDGAV GKTCLLISYT TNAFPGEYIP TVFDNYS-AN VMVDSKPVNL GLWDTAGQED YDRLRPLSYP QTDVFLICFS
G25KA     KCVVVGDGAV GKTCLLISYT TNKFPSEYVP TVFDNYA-VT VMIGGEPYTL GLFDTAGQED YDRLRPLSYP QTDVFLVCFS
G25KB     KCVVVGDGAV GKTCLLISYT TNKFPSEYVP TVFDNYA-VT VMIGGEPYTL GLFDTAGQED YDRLRPLSYP QTDVFLVCFS
CDC42     KCVVVGDGAV GKTCLLISYT TNQFPADYVP TVFDNYA-VT VMIGDEPYTL GLFDTAGQED YDRLRPLSYP STDVFLVCFS
RHOG      KCVVVGDGAV GKTCLLICYT TNAFPKEYIP TVFDNYS-AQ SAVDGRTVNL NLWDTAGQEE YDRLRTLSYP QTNVFVICFS
TC10      KCVVVGDGAV GKTCLLMSYA NDAFPEEYVP TVFDHYA-VS VTVGGKQYLL GLYDTAGQED YDRLRPLSYP MTDVFLICFS
RHOY2     KLVIIGDGAC GKTSLLYVFT LGKFPEQYHP TVFENYV-TD CRVDGIKVSL TLWDTAGQEE YERLRPFSYS KADIILIGFA

RAB1A     KLLLIGDSGV GKSCLLLRFA DDTYTESYIS TIGVDFKIRT IELDGKTIKL QIWDTAGQER FRTITSSYYR GAHGIIVVYD
RAB1B     KLLLIGDSGV GKSCLLLRFA DDTYTESYIS TIGVDFKIRT IELDGKTIKL QIWDTAGQER FRTVTSSYYR GAHGIIVVYD
YPT1      KLLLIGNSGV GKSCLLLRFS DDTYTNDYIS TIGVDFKIKT VELDGKTVKL QIWDTAGQER FRTITSSYYR GSHGIIIVYD
YPTSP     KLLLIGDSGV GKSCLLLRFA DDTYTESYIS TIGVDFKIRT FELEGKTVKL QIWDTAGQER FRTITSSYYR GAHGIIIVYD
SAS1      KLLLIGDSGV GKSCLLLRFS EDSFTPSFIT TIGIDFKIRT IELEGKRIKL QIWDTAGQER FRTITTAYYR GAMGILLVYD
YPT2      KLLLIGDSGV GKSCLLLRFS EDSFTPSFIT TIGIDFKIRT IELDGKRIKL QIWDTAGQER FRTITTAYYR GAMGILLVYD
SEC4      KILLIGDSGV GKSCLLVRFV EDKFNPSFIT TIGIDFKIKT VDINGKKVKL QLWDTAGQER FRTITTAYYT GAMGIILVYD
RAB8      KLLLIGDSGV GKTCVLFRFS EDAFNSTFIS TIGIDFKIRT IELDGKRIKL QIWDTAGQER FRTITTAYYR GAMGIMLVYD
RAB10     KLLLIGDSGV GKTCVLFRFS DDAFNTTFIS TIGIDFKIKT VELQGKKIKL QIWDTAGQER FHTITTSYYR GAMGIMLVYD
RAB3A     KILIIGNSSV GKTSFLFRYA DDSFTPAFVS TVGIDFKVKT IYRNDKRIKL QIWDTAGQER YRTITTAYYR GAMGFILMYD
RAB3B     KLLIIGNSSV GKTSFLLRYA DDTFTPAFVS TVGIDFKVKT VYRHEKRVKL QIWDTAGQER YRTITTAYYR GAMGFILMYD
RAB3C     KLLIIGNSSV GKTSFLFRYA DDSFTSAFVS TVGIDFKVKT VFKNEKRIKL QIWDTAGQER YRTITTAYYR GAMGFILMYD
RAB2      KYIIIGDTGV GKSCLLLQFT DKRFQPVHDL TIGVEFGARI ITIDGKQIKL LIWDTAGQES FRWMTRSYYR GAAGALLVYD
RAB4A     KFLVIGNAGT GKSCLLHQFI EKKFKDDSNH TIGVEFGSKI INVGGKYVKL QIWDTAGQER FRSVTRSYYR GAAGALLVYD
RAB4B     KFLVIGSAGT GKSCLLHQFI ENKFKQDSNH TIGVEFGTRS IQVDGKTIKA QIWDTAGQER YRAITSAYYR GAVGALLVYD
YPT3      KTVLIGDSGV GKSNLLMRFT RNEFNIESKS TIGVEFATRN IVLDNKKIKA QIWDTAGQER YRAITSAYYR GAVGALIVYD
RAB11     KVVLIGDSGV GKSNLLSRFT RNEFNLESKS TIGVEFATRS IQVDGKTIKA QIWDTAGQER YRAITSAYYR GAVGALLVYD
Ara       KIVVIGDSAV GKSNLLDRYA RNEFSANSKA TIGVEFQTQS MEIEGKEVKA QIWDTAGQER FRAVTSAYYR GAVGALVVYD
RAB6      KLVFLGEQSV GKTSLITRFM YDSFDNTYQA TIGIDFLSKT MYLEDRTVRL QLWDTAGQER FRSLIPSYIR DSTVAVVVYD
RYH1      KLVFLGEQSV GKTSLITRFM YDQFDNTYQA TIGIDFLSKT MYLEDRTVRL QLWDTAGQER FRSLIPSYIR DSSVAIIVYD
RAB5      KLVLLGESAV GKSSLVLRFV KGQFHEFQES TIGAAFLTQT VCLDDTTVSL NIWDTAGQEG YHSLAPMYYR GAQAAIVVYD
RAB7      KVIILGDSGV GKTSLMNQYV NKKFSNQYKA TIGADFLTKE VMVDDRLVTM QIWDTAGQER FQSLGVAFYR GADCCVLVFD
RAB9      KVILLGDGGV GKSSLMNRYV TNKFDTQLFH TIGVEFLNKD LEVDGHFVTM QIWDTAGQER FRSLRTPFYR GSDCCLLTFS

SPI1      KLVLVGDGGT GKTTFVKRHL TGEFEKKYIA TLGVEVHPLH FHTNFGEICF NVWDTAGQEK LGGLRDGYYI QGQCGIIMFD
ran/TC4   KLVLVGDGGT GKTTFVKRHL TGEFEKKYVA TLGVEVHPLV FHTNRGPIKF NVWDTAGQEK FGGLRDGYYI QAQCAIIMFD
```

Fig. 1. Multiple alignments of Ras-related proteins sequences. **a** The sequences of the first highly conserved part are aligned, from residues 5 to 83, by blocks of ten amino acids. **b** The sequences of the second conserved region from residues 84 to 164, also in blocks of ten. Some insertions, indicated by *small black triangles*, have been omitted for clarity: one insertion in ralB between amino acids 104 and 105, an insertion of 12 or 13 amino acids, is found in all rho proteins between positions 122

b

```
 84         94         104        114        123        133        142        152        162
 INNTKSFEDI HQYREQIKRV KDSDDVPMVL VGNKCDLAA- RTVESRQAQD LARSYG-IPY IETSAKTRQG VEDAFYTLVR EIR
 INNTKSFEDI HHYREQIKRV KDSEDVPMVL VGNKCDLPS- RTVDTKQAQD LARSYG-IPF IETSAKTRQG VDDAFYTLVR EIR
 INNSKSFADI NLYREQIKRV KDSDDVPMVL VGNKCDLPT- RTVDTKQAHE LAKSYG-IPF IETSAKTRQG VEDAFYTLVR EIR
 VNSAKSFEDI GTYREQIKRV KDAEEVPMVL VGNKCDLAS- WNVNNEQARE VAKQYG-IPY IETSAKTRMG VDDAFYTLVR EIR
 ITSRSSYDEI ASFREQILRV KDKDRVPLIL VGNKADLDHE RQVSVNEGQE LAKD-S-LSF HESSAKSRIN VEEAFYSLVR EIR
 ITSRSSFDEI ASFREQILRV KDKDRVPMIV VGNKCDLESD RQVTTGEGQD LAKSFG-SPF LETSAKIRVN VEEAFYSLVR EIR
 VTSRNSFDEL LSYYQQIQRV KDSDYIPVVV VGNKLDLENE RQVSYEDGLR LAKQLN-APF LETSAKQAIN VDEAFYSLIR LVR
 ITSKSSLDEL MTYYQQILRV KDTDYVPIVV VGNKSDLENE KQVSYQDGLN MAKQMN-APF LETSAKQAIN VEEAFYTLAR LVR
 ITSRSSFDEI STFYQQILRV KDKDTFPVVL VANKCDLEAE RVVSRAEGEQ LAKSMH-CLY VETSAKLRLN VEEAFYSLVR TIR
 VTDRGSFEEI YKFQRQILRV KDRDEFPMIL IGNKADLDHQ RQVTQEEGQQ LARQLK-VTY MEASAKIRMN VDQAFHELVR VIR
 LNDHSSFDEI PKFQRQILRV KDRDEFPMLM VGNKCDLKHQ QQVSLEEAQN TSRNLM-IPY IECSAKLRVN VDQAFHELVR IVR
 INDRQSFNEV GKLFTQILRV KDRDDFPVVL VGNKADLESQ RQVPRSEASA FGASHH-VAY FEASAKLRLN VDEAFEQLVR AVR
 ITAQSTFNDL QDLREQILRV KDTEDVPMIL VGNKCDLEDE RVVGKEQGQN LARQWCNCAF LESSAKSKIN VNEIFYDLVR QIN
 ITAQSTFNDL QDLREQILRV KDTDDVPMIL VGNKCDLEDE RVVGKEQGQN LARQWNNCAF LESSAKSKIN VNEIFYDLVR QIN
 ITAQSTFNDL QDLREQILRV KDTDDVPMVL VGNKCDLEEE RVVGKELGKN LATQFN-CAF METSAKAKVN VNDIFYDLVR QIN
 LVNQQSFQDI KPMRDQIIRV KRYEKVPVIL VGNKVDLESE REVSSSEGRA LAEEWG-CPF METSAKSKTM VDELFAEIVR QMN
 LVNQQSSQDI KPMRDQIIRV KRYERVPMIL VGNKVDLEGE REVSYGEGKA LAEEWS-CPF METSAKNKAS VDELFAEIVR QMN
 VTDRQSLEEL MELREQVLRI KDSDRVPMVL IGNKADLINE RVISVEEGIE VSSKWGRVPF YETSALLRSN VDEVFVDLVR QII
 ITEMESFAAT ADFREQILRV KEDENVPFLL VGNKSDLEDK RQVSVEEAKN RADQWN-VNY VETSAKTRAN VDKVFFDLMR EIR
                       ▼
 ITEHESFTAT AEFREQILRV KEEDKIPLLV VGNKSDLEER RQVPVEEARS KAEEWG-VQY VETSAKTRAN VDKVFFDLMR EIR

                                            ▼
 IDSPDSLENI PEKWTPEVK- HFCPNVPIIL VGNKKDLRND EPVKPEEGRD MANRIGAFGY MECSAKTKDG VREVFEMATR AAL
 VDSPDSLENI PEKWVPEVK- HFCPNVPIIL VANKKDLRSD EPVRTDDGRA MAVRIQAYDY LECSAKTKEG VREVFETATR AAL
 IDSPDSLENI PEKWTPEVK- HFCPNVPIIL VGNKKDLRQD EPVRSEEGRD MANRISAFGY LECSAKTKEG VREVFEMATR AGL
 IDSPDSLENI PEKWTPEVK- HFCPNIPIIL VGNKKTAGDD EPVKPDDAKE MGSRIKAFGY LECSAKTKEG VREVFELASR AAL
 IDSPDSLENI PEKWTPEVR- HFCPNVPIIL VGNKKDLRND EPVRPEDGRA MAEKINAYSY LECSAKTKEG VRDVFETATR AAL
 IDLPDSLENV QEKWIAEVL- HFCQGVPIIL VGCKVDLRND QPVTSQEGQS VADQIGATGY YECSAKTGYG VREVFEAATR ASL
 LVSPASFENV RAKWYPEVR- HHCPNTPIIL VGTKLDLRDD TPITYPQGLA MAKEIGAVKY LECSALTQRG LKTVFDEAIR AVL
 LVSPASYENV RAKWFPEVR- HHCPSTPIIL VGTKLDLRDD APITYPQGLA LAKEIDSVKY LECSALTQRG LKTVFDEAIR AVL
 VVSPSSFENV KEKWVPEIT- HHCPKTPFLL VGTQIDLRDD KPITPETAEK LARDLKAVKY VECSALTQKG LKNVFDEAIL AAL
 VVSPSSFENV KEKWVPEIT- HHCPKTPFLL VGTQIDLRDD KPITPETAEK LARDLKAVKY VECSALTQRG LKNVFDEAIL AAL
 VISPPSFENV KEKWFPEVH- HHCPGVPCLV VGTQIDLRDD RPITSEQGSR LARELKAVKY VECSALTQRG LKNVFDEAIV AAL
 IASPPSYENV RHKWHPEVC- HHCPDVPILL VGTKKDLRAQ APITPQQGQA LAKQIHAVAY LECSALQQDG VKEVFAEAVR AVL
 VVNPASFQNV KEEWVPELK- EYAPNVPFLL IGTQIDLRDD KPICVEQGQK LAKEIGACCY VECSALTQKG LKTVFDEAII AIL
 VDNFESLINA RTKWADEAL- RYCPDAPIVL VGLKKDLRQE EMVPIEDAKQ VARAIGAKKY MECSALTGEG VDDVFEVATR TSL
                                            ▲

 VTDQESFNNV K-QWLQEIDR YASENVNKLL VGNKCDLTTK KVVDYTTAKE FADSLG-IPF LETSAKNATN VEQSFMTMAA EIK
 VTDQESYANV K-QWLQEIDR YASENVNKLL VGNKSDLTTK KVVDNTTAKE FADSLG-VPF LETSAKNATN VEQAFMTMAA EIK
 VTDQESFNGV K-MWLQEIDR YATSTVLKLL VGNKCDLKDK RVVEYDVAKE FADANK-MPF LETSALDSTN VEDAFLTMAR QIK
 VTDQDSFNNV K-QWLQEIDR YAVEGVNRLL VGNKSDMVDK KVVEYSVAKE FADSLN-IPF LETSAKDSTN VEQAFLTMSR QIK
 VTDEKSFGNI R-NWIRNIEQ HATDSVNKML IGNKCDMAEK KVVDSSRGKS LADEYG-IKF LETSAKNSIN VEEAFISLAK DIK
 VTDKKSFDNV R-TWFSNVEQ HASENVYKIL IGNKCDCEDQ RQVSFEQGQA LADELG-VKF LEASAKTNVN VDEAFFTLAR EIK
 VTDERTFTNI K-QWFKTVNE HANDEAQLLL VGNKSDM-ET RVVTADQGEA LAKQLG-IPF IESSAKNDDN VNEIFFTLAK LIQ
 ITNEKSFDNI R-NWIRNIEE HASADVEKMI LGNKCDVNDK RQVSKERGEK LALDYG-IKF METSAKANIN VENAFFTLAR DIK
 ITNGKSFENI S-KWLRNIDE HANEDVERML LGNKCDMDDK RVVPKGKGEQ IAREHG-IRF FETSAKVNIN IEKAFLTLAE DIL
 ITNEESFNAV Q-DWSTQIKT YSWDNAQVLL VGNKCDMEDE RVVSSERGRQ LADHLG-FEF FEASAKDNIN VKQTFERLVD VIC
 ITNEESFNAV Q-DWATQIKT YSWDNAQVIL VGNKCDMEEE RVVPTEKGQL LAEQLG-FDF FEASAKENIS VRQAFERLVD AIC
 ITNEESFNAV Q-DWSTQIKT YSWDNAQVIL VGNKCDMEDE RVISTERGQH LGEQLG-FEF FETSAKDNIN VKQTFERLVD IIC
 ITRRDTFNHL T-TWLEDARQ HSNSNMVIML IGIKSDLESR REVKKEEGEA FAREHG-LMF METSAKTASN VEEAFINTAK EIY
 ITSRETYNAL T-NWLTDARM LASQNIVIIL CGNKKDLDAD REVTFLEASR FAQENE-LMF LETSALTGED VEEAFVQCAR KIL
 IAKHLTYENV E-RWLKELRT LASPNIVVIL CGNKKDLDPE REVTFLEASR FAQENE-LMF LETSALTGEN VEEAFLKCAR TIL
 ITKQSSFDNV G-RWLKELRE HADSNIVIML VGNKTDLLHL RAVSTEEAQA FAE-NN-LSF IETSAMDASN VEEAFQTVLT EIF
 IAKHLTYENV E-RWLKELRD HADSNIVIML VGNKSDLRHL RAVPTDEARA FAEKNG-LSF IETSALDSTN VEAAFQTILT EIY
 ITRRTTFESV G-RWLDELKI HSDTTVARML VGNKCDLENI RAVSVEEGKA LAEEEG-LFF VETSALDSTN VKTAFEMVIL DIY
 ITNVNSFQQT T-KWIDDVRT ERGSDVIIML VGNKTDLADK RQVSIEEGER KAKELN-VMF IETSAKAGYN VKQLFRRVAA ALP
 ITNHNSFVNT E-KWIEDVRA ERGDDVIIVL VGNKTDLADK RQVTQEEGEK KAKELK-IMH METSAKAGHN VKLLFRKIAQ MLP
 ITNEESFARA K-NWVKELQR QASPNIVIAL SGNKADLANK RAVDFQEAQS YADDNS-LLF METSAKTSMN VNEIFMAIAK KLP
                          ▼
 VTAPNTFKTL D-SWRDEFLI QASPNFPFVV LGNKIDLENR QVATKRAQAW CYSKNN-IPY FETSAKEAIN VEQAFQTIAR NAL
 VDDSQSFQNL S-NWKKEFIY YADVSFPFVI LGNKIDISER QVSTEEAQAW CRDNGD-YPY FETSAKDATN VAAAFEEAVR RVL

 VTSRITYKNV P-HWWRDLVR VC-ENIPIVL CGNKVDVKER KVKAKAITFH RKKNLQ-YY- -DISAKSNYN FEKPFLWLAR KLV
 VTSRVTYKNV P-NWHRDLVR VC-ENIPIVL CGNKVDIKDS KVKAKSIVFH RKKNLQ-YY- -DISAKSNYN FEKPFLWLAR KLI
```

and 123 (see Sect. C.II.), and an insertion of four amino acids in rab7 between positions 107 and 108. **c** (p. 162) The C-terminal extensions, after the last highly conserved residue at position 164. These sequences were arbitrarily aligned to match the CaaX boxes or were aligned on the last cysteine when the consensus CaaX could not be found (rab and YPT proteins)

c

```
HRAS                               QHKLRKLNPPDESGPGCMSCKCVLS
KRAS (A)                           QYRLKKISKEEKTPGCVKIKKCIIM
KRAS (B)                            KHKEKMSKDGKKKKKKSKTKCVIM
NRAS                               QYRMKKLNSSDDGTQGCMGLPCVVM
DRAS1                              KDKDNKGRRGRKMNKPNCRFKCKML
DdRAS                                 KELKGDQSSGKAQKKKKQCLIL
DdRASG                              KDLKGDSKPEKGKKKRPLKACTLL
RASY1                  110 a.a. RSKQSAEPQKNSSANARKEYSGGCCIIC
RASY2               120 a.a. NAKQARKQQAAPGGNTSEASKSGSGGCCIIS
RASSP                                  RYNKSEEKGFQNKQAVCCVIC
TC21                                KFQEQECPPSPEPTRKEKDKKGCHCVIF
DRAS2                                KFQIAERPF--IEQDYKKKGKRKCCLM
RRAS                                 KYQEQELPPSPPSAPRKKGGGCPCVLL
RAP1A                                        RKTP-VEKKKPKKKSCLLL
RAP1B                                         RKTPVPGKARKKSSCQLL
DRAP1                                        KKSPEKKQKKPKKSLCVLL
RAP2A                                         YAAQPDKDDPCCSACNIQ
RAP2B                                         YAAQSNGDEGCCSACVIL
RSR1                             80a.a.SNRTGISATSQQKKKKKNASTCTIL
RALA                              ARKMEDSKEKNGKKKRKSLAKRIRERCCIL
RALB                                TKKMSENKDKNGKKSSKNKKSFERCCLL

RHOA                                          QARRGKKKSGCLVL
RHOB                                       QKRYGSQNGCINCCKVL
RHOC                                          QVRKNKRRRGCPIL
RHO-O                                         QAKKTKSKSPCLLL
RHOAP                                          QVKKKKKGGCVVL
RHOY1                              MGKSKTNGKAKKNTTEKKKKKCVLL
RAC1                                         CPPPVKKRKRKCLLL
RAC2                                         CPQPTRQQKRACSLL
G25KA                                         EPPEPKKSRRCVLL
G25KB                                         EPPETQPKRKCCIF
CDC42                                         EPPVIKKSKKCTIL
RHOG                                          NPTPIKRGRSCILL
TC10                                      TPKKHTVKKRIGSRCINCCLIT
RHOY2                                         LMKKEPGANCCIIL

RAB1A                    KRMGPGATAGGAEKSNVKIQSTPVKQSGGGCC
RAB1B                    KRMGPGAASGG-ERPNLKIDSTPVKSASGGCC
YPT1                 QSMSQQNLNETTQKKEDKGNVNLKGQSLTNTGGGCC
YPTSP                   ERMGNNTFASSNAKSSVKVGQGTNVSQSSSNCC
SAS1                      KRMIDTPNEQPQVVQPGTNLGANNNKKKACC
YPT2                        KQKIDAENEFSNQANNVDLGNDRTVKRCC
SEC4                   EKIDSNKLVGVGNGKEGNISINSGSGNSSKSNCC
RAB8                   AKMDKKLEGNSPQGSNQGVKITPDQQKRSSFFRCVLL
RAB10                       RKTPVKEPNSENVDISSGGGVTGWKSKCC
RAB3A                EKMSESLDTADPAVTGAKQGPQLSDQQVPPHQDCAC
RAB3B                DKMSDSLDT-DPSMLGSSKNTRLSDTPPLLQQNCSC
RAB3C                DKMSESLET-DPAITAAKQNTRLKETPPPPQPNCGC
RAB2      EIYEKIQEGVFDINNEANGIKIGPQHAATNATHAGNQGGQQAGGGCC
RAB4A          NKIESGELDPERMGSGIQYGDAALRQLRSPRRTQAPNAQECGC
RAB4B          NKIDSGELDPERMGSGIQYGDASLRQLRQPRSAQAVAPQPCGC
YPT3            RIVSNRSLEAGDDGVHPTAGQTLNIAPTMNDLNKKKSSSQCC
RAB11            RIVSQKQMSDRRENDMSPSNNVVPIHVPPTTENKPKVQCCQNI
Ara             NNVSRKQLNSDTYKDELTVNRVSLVKDDNSASKQSSGFSCCSST
RAB6                    GMESTQDRSREDMIDIKLEKPQEQPVSEGGCSC
RYH1                         GMENVETQSTQMIDVSIQPNENESSCNC
RAB5                       KNEPQNPGANSARGGGVDLTEPTQPTRNQCCSN
RAB7                    KQETEVELYNEFPEPIKLDKNERAKASAESCSC
RAB9                         ATEDRSDHLIQTDTVSLHRKPKPSSSCC

SPI1     GNPNLEFVASPALAPPEVQVDQQLLAQYQQEMNEAAAMPLPDEDDADL
ran/TC4  GDPNLEFVAMPALAPPEVVMDPALAAQYEHDLEVAQNPALPDEDD-DL
```

Fig. 1. *Continued*

ships, and to suggest functional similarities. Several sequences determined in chicken (WESTAWAY et al. 1986), *in Dictyostelium* (ROBBINS et al. 1990), in nematodes (HAN and STERNBERG 1990), in *Mucor racemosus* (CASALE et al. 1990), in *Neurospora crassa* (ALTSCHULER et al. 1990), in the electric ray (NGSEE et al. 1991), or in maize (PALME et al. 1992) were not included since they encoded proteins very similar to the ones already isolated from other organisms. A *ras*-related gene has also been found in *Aplysia* (SWANSON et al. 1986); partial sequences suggest that it might represent a new member of the *ras* branch; however, more sequence data would be needed to clarify this point and it was not included in the alignment. CHAVRIER et al. (1992) also obtained partial sequences for a large number of new proteins, most of them in the rab family. The cDNA for a protein closely related to rab3A, B, and C was recently isolated from mouse adipocytes and named rab3D (BALDINI et al. 1992), and several new rab cDNAs were recently isolated from a rat brain library (ELFERINK et al. 1992). The cDNAs they numbered 12–15 encode proteins more closely related to the rab8 and rab10 subbranches, and cDNA no. 16 encodes a protein that is probably the rat homolog of mouse rab3D.

Sequence comparisons enable us to classify ras-related proteins into four main branches and various subbranches, predicting functional similarities for each subclass of this family. Comparative analysis also defines conserved and variable regions that can be located on the three-dimensional model of the ras protein (see Chaps. 13, 14), suggesting where the major functional regions are located and how they may interact with other proteins.

B. Sequence Comparisons

Ras-related proteins might be subdivided into five main parts (reviewed in VALENCIA et al. 1991): (1) The N-terminal end before the first conserved residue (K5), variable in length and sequence, ranging from four amino acids in ras, rap1, and CDC42 to 30 amino acids in R-ras; (2) the phosphate-binding site from amino acid 5 to 83, which is the most highly conserved region; (3) the guanine-binding site from amino acid 84 to 164, also conserved but to a lower extent, with two main boxes of homology; (4) the C-terminal part, from amino acid 165 to the C terminus (except for the last four amino acids), which is highly variable, even among closely related proteins such as H-ras/K-ras/N-ras; and (5) the last four C-terminal amino acids that always include a cysteine and often fit the CaaX (C, cys; a, an aliphatic amino acid; and X, another amino acid) consensus for farnesylation or geranylgeranylation, cleavage and carboxymethylation, or the CxC or CC consensus for two geranylgeranylations (see Chap. 22; Fig. 1 and Table 1).

Table 1. Ras-related protein sequences; the genes and the sequence of the N-terminal extension of the protein before K5, the first strictly conserved residue

Gene	Reference			N-terminal extension
Hras	CAPON et al.	1983	HRAS	MTEY
Kras	MCGRATH et al.	1983	KRAS	MTEY
Nras	TAPAROWSKY et al.	1983	NRAS	MTEY
Dras1	NEUMAN-SILBERBERG	1984	DRAS1	MTEY
Ddras	REYMOND et al.	1984	DdRAS	MTEY
DdrasG	ROBBINS et al.	1989	DdRASG	MTEY
RASY1	DEFEO-JONES et al.	1983	RASY1	MQGNKSTMTEY
RASY2	POWERS et al.	1984	RASY2	MPLNKSNIREY
RASSP	FUKUI and KAZITO	1985	RASSP	MRSTYLREY
TC21	DRIVAS et al.	1990	TC21	MAAAAGGRLRQEKY
Dras2	BISHOP and CORCES	1988	DRAS2	MQMQTY
Rras	LOWE et al.	1987	RRAS	MSSGAASGTGRGRPRGGGPGPGDPPPSETH
rap1A	PIZON et al.	1988a	RAP1A	MREY
rap1B	PIZON et al.	1988b	RAP1B	MREY
Drap1	HARIHARAN et al.	1991	DRAP1	MREY
rap2A	PIZON et al.	1988a	RAP2A	MREY
rap2B	OHMSTEDE et al.	1990	RAP2B	MREY
RSR1	BENDER and PRINGLE	1989	RSR1	MRDY
ralA	CHARDIN and TAVITIAN	1986	RALA	MAANKPKGQNSLALH
ralB	CHARDIN and TAVITIAN	1989	RALB	MAANKSKGQSSLALH
rhoA	YERAMIAN et al.	1987	RHOA	MAAIRK
rhoB	CHARDIN et al.	1988	RHOB	MAAIRK
rhoC	CHARDIN et al.	1988	RHOC	MAAIRK
rho-O	NGSEE et al.	1991	RHO-O	MAAIRK
rhoApl	MADAULE and AXEL	1985	RHOAP	MAAIRK
RHOY1	MADAULE et al.	1987	RHOY1	MSQQVGNSIRR
rac1	DIDSBURY et al.	1989	RAC1	MQAI
rac2	DIDSBURY et al.	1989	RAC2	MQAI
G25KA	SHINJO et al.	1990	G25KA	MQTI

G25KB	Munemitsu et al.	1990	G25KB	MQTI
CDC42	Johnson and Pringle	1990	CDC42	MQTL
rhoG	Vincent et al.	1992	RHOG	MQSI
TC10	Drivas et al.	1990	TC10	MPGAGRSSMAHGPGALML
RHOY2	Madaule et al.	1987	RHOY2	MSEKAVRR
rab1A	Touchot et al.	1987	RAB1A	MSSMNPEYDYLF
rab1B	Vielh et al.	1989	RAB1B	MNPEYDYLF
YPT1	Gallwitz et al.	1983	YPT1	MNPEYDYLF
YPTSP	Fawell et al.	1989	YPTSP	MNPEYDYLF
SAS1	Saxe and Kimmel	1988	SAS1	MTSPATNKSAAYDYLI
YPT2	Haubruck et al.	1990	YPT2	MSTKSYDYLI
SEC4	Salminen and Novick	1987	SEC4	MSGLRTVSASSGNGKSYDSIM
rab8	Chavrier et al.	1990	RAB8	MAKTYDYLF
rab10	Chavrier et al.	1990	RAB10	MAKKTYDLLF
rab3A	Zahraoui et al.	1988	RAB3A	MASATDSRYGQKESSDQNFDYMF
rab3B	Zahraoui et al.	1989	RAB3B	MASVTDGKHGVKDASDQNFDYMF
rab3C	Matsui et al.	1988	RAB3C	MRHEAPMQMASAQDARYGQKDSSDQNFDYMF
rab2	Touchot et al.	1987	RAB2	MAYAYLF
rab4A	Zahraoui et al.	1988	RAB4A	MSETYDFLF
rab4B	Chavrier et al.	1990	RAB4B	MAETYDFLF
YPT3	Miyake and Yamamoto	1990	YPT3	MCQEDEYDYLF
rab11	Chavrier et al.	1990	RAB11	MGTRDDEYDYLF
Ara	Matsui et al.	1989	Ara	MSSDDEGREEYLF
rab6	Zahraoui et al.	1989	RAB6	MSTGGDFGNPLRKF
RYH1	Hengst et al.	1990	RYH1	MSENYSFSLRKF
rab5	Zahraoui et al.	1989	RAB5	MASRGATRPNGPNTGNKICQF
rab7	Bucci et al.	1988	RAB7	MLL
rab9	Chavrier et al.	1990	RAB9	MAGKSSLF
spi1	Matsumoto and Beach	1991	SPI1	MAQPQNVPTF
ran/TC4	Drivas et al.	1990	ran/TC4	MAAQGEPQVQF

D, *Drosophila*; Dd, *Dictyostelium*; Y, *Saccharomyces cerevisiae*; SP, *Schizosaccharomyces pombe*. Genes written in capitals are from *Saccharomyces cerevisiae*. Unless otherwise stated, all numberings refer to H-ras.

I. The N-Terminal Extension

In ras there are only four residues from the N-terminal end to the first highly conserved amino acid (K5); however, the three-dimensional structure shows that this short tail sticks out of the major globular domain, and the structure can accomodate much longer N-terminal parts. In some Ras-related proteins such as R-ras, rab3, rab5, and SEC4, 20–30 amino acid extensions are present at the N terminus, and they can probably form small additional domains that may be involved in the interaction with specific target proteins.

II. The Phosphate-Binding Part

The GXXXXGKS/T motif (where S/T means serine or threonine at this position) is found in all ras-related proteins, in most GTPases such as ARFs, Sar1p, α subunits of heterotrimeric G-proteins, elongation factors, SRP receptors, as well as in many other nucleotide-binding proteins. This motif forms a very rigid loop structure, with the side chain of lysine (K16) folding back to interact with the main chain carbonyl groups of G10 and A11. This seems to be an essential motif of many phosphate-binding sites. In ras, glycines are found in position 12 and 13, and replacement of G12 by any other amino acid except proline slows down the intrinsic GTP hydrolysis activity and leads to the acquisition of a transforming potential for ras proteins (see Chap. 18). However, some rab proteins, such as rab3, that have neither G12 nor G13 hydrolyze GTP with efficiencies comparable to p21, and Ras and Rac proteins that have an alanine instead of glycine 13 hydrolyze GTP faster than p21 ras, indicating that G12 and G13 are not absolute requirements for γ-phosphate hydrolysis. The reason why substitution of G12 by any other amino acid except proline impairs GTP hydrolysis in the ras–GAP complex and induces transforming activity is probably that ras cannot tolerate large side chains at this position without disturbing the whole structure of the phosphate-binding site. Amino acids 25–27 are at the beginning of loop L2, and the aromatic ring of F28 is perpendicular to the plane of the guanine ring and stabilizes its binding by hydrophobic interactions, a conformation often found in other nucleotide-binding proteins. Amino acids 30–34 are in the first "switch region," the most detached loop of the structure with all side chains pointing outwards, so it is difficult to explain why Y/F32 is strictly conserved; it seems to come close to the magnesium ion and β-phosphate in the GDP-bound conformation. The hydroxyl group of the T35 side chain is close to an oxygen of the γ-phosphate and is also involved in the coordination of the Mg^{2+} ion in the GTP-bound conformation. Surprisingly, the hydrophobic side chain of I 36 points outwards and is probably involved in the interaction with GAP or other target proteins. Amino acids 37–45 represent a β-sheet with many side chains sticking outwards. Amino acids 46–49 form a sharp turn with acidic side chains interacting with the basic side chains at positions 161 and 164. One can hypothesize that a protein interacting with this region could displace loop 46–49, and that the conformational change could propagate to the nucleotide binding site and "open" it to induce GDP release. Amino acids 57–62 are the second highly conserved region and have the DTAGQE motif. This region is located close to the β- and γ-phosphates of GTP. D57 is coordinated to magnesium in the GDP-bound form and to a water molecule that contacts magnesium in the GTP-bound form. G60 is hydrogen bounded to the γ-phosphate of GTP and participates in the conformational change following hydrolysis of this γ-phosphate. The side chain of Q61 is likely to play an important role in intrinsic GTP hydrolysis. The rap1 and rap2 proteins, which possess a T instead of this Q61, hydrolyze GTP slower than Ras, so it seems that the replacement of this glutamine by a threonine might significantly affect the GTPase rate. Region 61–68 comprises a loop, the beginning of an α helix, and several side chains pointing outwards. Mutations in this region impair GAP stimulation of GTP hydrolysis, but not GAP interaction in vitro, suggesting that this region is important for the hydrolytic process induced by GAP. It is also the second "switch region" changing from the GDP- to the GTP-bound conformation, and is thus a potential interaction site for target proteins specifically recognized by the GDP- or the GTP-bound forms.

III. The Guanine-Binding Part

There are two main stretches of amino acids highly conserved in this region, both involved in the binding of the guanine ring.

The first is the LVGNKXDL sequence, in which the N116 side chain interacts both with the hydroxyl group of the T144 side chain and with the N7 atom of the guanine ring. K117 interacts with an oxygen of the ribose, and the D119 side chain interacts with both the S145 side chain and the N1 atom and NH_2 group of the guanine ring. This motif is conserved in most GTPases and seems to be essential for the specificity of guanine recognition. The second is the ETSAK motif, which is highly conserved in all proteins of the ras family. E143 makes a hydrogen bond with Y141 and a salt bridge with R123. The T144 side chain OH interacts with N116 and it might be noted that the only other amino acids found in this position are serine (also with an OH group) or cysteine (with a SH group). S145 interacts with the side chain of D119. The long hydrophobic side chain of K147 is stacked on one site of the guanine ring and probably stabilizes its binding through hydrophobic interactions. Consistent with this interpretation, in some proteins it is replaced by leucine, which also has a long hydrophobic side chain. This ETSAK motif is not highly conserved in the other known GTPases and can be considered as a specific feature of ras-related proteins.

IV. The C-Terminal Extension

This part of the protein probably acts as a flexible spacer arm from the membrane-bound C terminus to the beginning of the globular nucleotide-binding domain in the cytoplasm. C-terminal extensions are variable in sequence, but are mainly homogeneous in length in each of the three branches, except in *Saccharomyces cerevisiae* proteins. The shortest C-terminal extensions are found in the rho branch (14–17 amino acids), and thus in these proteins the major globular domain is expected to be closer to the membrane. The proteins of the ras branch have extensions of 18–30 amino acids, while in the rab branch their length extends from 27 to 47 amino acids. It is clear that an extension of 47 amino acids can represent a small additional domain of the protein. In *S. cerevisiae* ras, the C-terminal domains are as large as the guanine nucleotide-binding domain. This C-terminal part, which comes close to the membrane and is the most specific of each protein, is believed to provide the specificity for the interaction with an integral membrane protein or with a transiently membrane bound protein that could be the "exchange factor."

V. The CaaX Motif

In ras proteins ending with a consensus CaaX, the cysteine is farnesylated, the last three residues are clipped, and the cysteine that becomes C terminus is carboxymethylated (see Chap. 22). These modifications help anchor the C terminus in the membrane. Among the many proteins ending with a CaaX consensus, those where X, the C-terminal residue, is Met, Ser, or Cys are farnesylated, while those where X is Leu (or Phe or Asn) are geranylgeranylated. Rab proteins ending with CC or CxC are also geranylgeranylated on both cysteines by a different enzymatic complex. However, the GGGCC sequence found in yeast Ypt is conserved in its mammalian homolog rab1, and the CNC sequence found in Ryh1 is CSC in its mammalian homolog rab6. The conservation of these sequences suggests some functional requirements which are not yet understood. The ran/TC4 and spi1 proteins do not end with a cysteine motif and are not membrane bound, but have a nuclear localization.

C. Evolutionary Relationships

I. Construction of a Homology Tree

The N-terminal and C-terminal parts, of variable length and sequence, cannot be aligned. Thus, only the core region, from amino acid 5 to 164, which constitutes the GTP/GDP-binding domain, has been taken into account for the alignments. The methods used have been described elsewhere

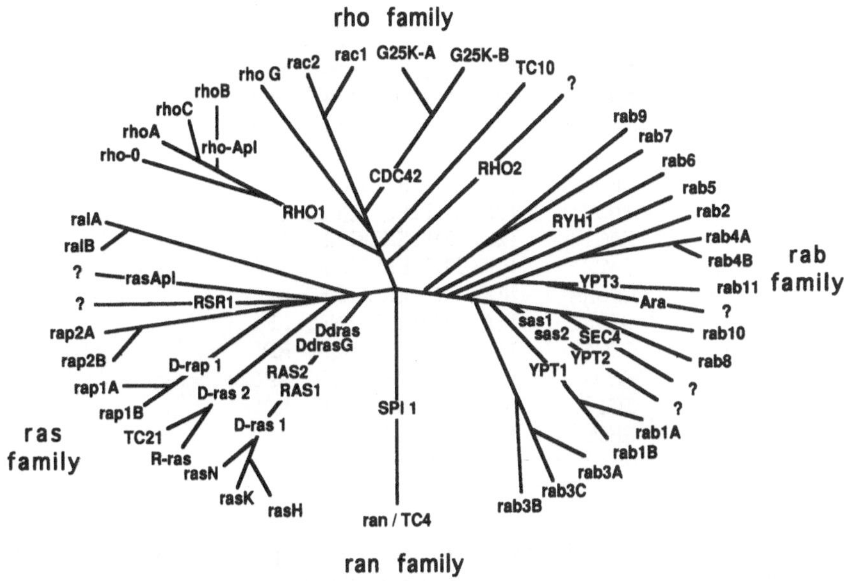

Fig. 2. Family tree of the Ras-related proteins. The names of the genes and their references are identical to those described in Table 1. Mammalian proteins are on the outer part, *Drosophila*, *Aplysia*, or plant proteins are underneath, while yeast proteins, indicated in *capital letters*, and *Dictyostelium* proteins are closest to the center

(Valencia et al. 1991). The consensus phylogenetic tree deduced from these results is represented in Fig. 2. The ras family is clearly divided in four main branches: ras, rho, rab, and ran. The ras and rho branches are the most homogeneous, with proteins always sharing more than 50% identity. The rab branch, which now includes the highest number of proteins, is also the most heterogeneous, some members of the rab branch sharing only 40% identity with other members of the same branch. The ran/TC4 and spil proteins have several unusual characteristics that place them as the first representives of a new branch; for instance, they do not end with a cysteine motif and have a nuclear localization. It would be interesting to know whether the Ran protein is unique or whether it represents the first member of a large branch such as the ras, rho, and rab branches.

II. Insertions and Deletions

Among all insertions and deletions, several are branch specific, providing additional arguments for the existence of four well-defined branches. A deletion between positions 122 and 123 appeared in the Ras subbranch before *Drosophila*, but it is not found in yeast Ras, *Dictyostelium* Ras, or in other subbranches of the Ras branch, so it appeared after the divergence of the phylums leading to yeast and insects/vertebrates. This also strongly

Table 2. rho insertions

RHOA	RND	EHTRRELAKMKQ	EPV
RHOB	RSD	EHVRTELARMKQ	EPV
RHOC	RQD	EHTRRELAKMKQ	EPV
RHO-0	AGD	EHTRRELAKMKQ	EPV
RHOAP	RND	ESTKRELMKMKQ	EPV
RHOY1	RND	PQTIEQLRQEGQ	QPV
RAC1	RDD	KDTIEKLKEKKL	TPT
RAC2	RDD	KDTIEKLKEKKL	API
G25KA	RDD	PSTIEKLAKNKQ	KPI
G25KB	RDD	PSTIEKLAKNKQ	KPI
CDC42	RDD	KVIIEKLQRQRL	RPI
RHOG	RAQ	PDTLRRLKEQSQ	API
TC10	RDD	PKTLARLNDMKE	KPI
RHOY2	RQE	AHFKENATDEM-	VPI

suggests that mammalian H-*ras*, K-*ras* and N-*ras* appeared by duplications of a single ancestral gene, followed by some divergence in mammals. In the rho branch, a single deletion is found at position 103, and one insertion between amino acids 138 and 139 (an insertion appears in the same position in rap1A, rap1B, and RSR1). Insertion of another 12 amino acids is found in all proteins of the rho branch between positions 122 and 123. The sequences of these insertions are given in Table 2. The size of this insertion is conserved, suggesting that it has appeared in a common ancestral gene. This insertion occurs in loop 116–126, where it is not expected to modify the overall structure of the protein, but certainly modifies the structure of this loop and of the following α helix. In the rab and ran branches one deletion is found in position 95, and one insertion between amino acids 41 and 42. Two additional deletions are found in the ran branch alone – one in position 106 and another of two amino acids in positions 141 and 142. Some insertions or deletions are only found in a single member of a subbranch: one deletion in Sec4 in position 120, one insertion in ralB between amino acids 104 and 105, and an insertion of four amino acids in rab7 between positions 107 and 108.

III. Estimation of the Number of Ras-Related Proteins in Mammals

Isolation of cDNAs for small GTPases by various approaches and from diverse cell types has increased exponentially from 1983 to 1989, but now leads more and more frequently to the rediscovery of already known proteins, suggesting that a major fraction of this family has now been discovered. At least 60 Ras-related proteins are present in mammals, including at least 30 Rab proteins, and from the present data we can predict that probably not much more than a hundred Ras-related proteins are encoded in the genome of man or other mammals.

D. Discussion

I. Internal Residues

Most conserved residues are involved in nucleotide binding or are internal residues. The main conserved regions involved in GDP/GTP-binding are shown in Fig. 1 and have been described in the previous paragraphs (see Sects. B.II. and B.III.). A few other amino acids such as K5, Y71, and F156 are strictly conserved. The aromatic rings of Y71 and F156 are on the inside of the protein and stabilize the structure through many hydrophobic interactions. In many other positions where side chains point inwards, the conservation is not strict, but only amino acids with hydrophobic side chains are found. The reason why K5 is conserved is not understood.

II. External Residues and Potential Targets for Interacting Proteins

Residues 25–43, 47–50, 61–67, [91–92 and 94–95], 85–88, 104–108, [121–124, 126–129, 131–133 and 135–137], 148–150, 161–164, and the whole C-terminal extension are the main regions where most of the side chains stick outwards. External regions provide the specificity for the interaction of each protein with its specific regulators and/or targets. Residues with side chains pointing outwards are usually the most variable, and it is noteworthy that the only exception to this correlation between "variable" and "external" regions are regions 36–48 and 61–67, where most of the side chains stick outwards in both the GDP and GTP structures of p21 ras, but where a rather high conservation is found among the diverse members of a given branch. The three-dimensional model of p21 ras cannot explain this conservation. However, residues 32–36 and 61–68 are the two parts of the protein core where the major changes from the GDP-bound to the GTP-bound conformations are observed. Since these two "switch regions" represent the site of GAP interaction and are quite well conserved, there is probably a family of GTPase activating proteins (GAPs), all of them having similarities in the sites of interaction with the various proteins of the ras family. However these structural similarities of GAP-like proteins might be difficult to recognize at the sequence level, since rasGAP and rapGAP have very little homology (see Chap. 23). The site for ADP-ribosylation of rho proteins by botulinum exo-enzyme C3 is at position 39 in ras numbering (N41 of rho). This modification has little effect on the biochemical properties of rho proteins, but results in a dramatic effect on rho function in vivo as demonstrated by the phenotypic changes in cytoskeletal organization (see Chaps. 8, 36, 43). This again argues for the major importance of this region in vivo, although its role is not yet understood from structural or biochemical studies on isolated proteins in vitro. The precise role of residues 61–67 is not clearly understood. The side chain of glutamine 61 (Q61) seems to be involved in GTP hydrolysis in isolated ras. The fact that heterotrimeric

G-proteins also have a Q in the position corresponding to ras Q61 and that the substitution of this glutamine by an arginine significantly reduces the rate of GTP hydrolysis is consistent with this notion. However, the hydrolysis of GTP in the ras–GAP complex is at least four orders of magnitude quicker than in isolated ras, suggesting a different mechanism. This hydrolytic mechanism in the presence of GAP is still unknown, and we do not know whether the side chain of Q61 is directly involved in GTP hydrolysis or whether it is mainly involved in the interaction with GAP. We can already predict that the proteins specifically interacting with the GDP-bound form (ras-GDP targets) will also interact with these two "switch regions," since they are the only ones where conformational differences between the GDP- or GTP-bound forms might be seen.

The N-terminal extension, the other external regions, and the C-terminal extension are potential targets for the exchange factors or other uncharacterized proteins. In these regions there are no major conformational changes from the GDP-bound state to the GTP-bound state; therefore, the proteins interacting with these regions are not expected to discriminate between the GDP- or GTP-bound forms. How could an exchange factor interacting with these regions promote GDP release? Loop L3 (amino acids 46–49) forms a sharp turn, with acidic side chains interacting with the basic side chains at positions 161 and 164 at the beginning of the C-terminal extension, which is supposed to provide a specific site for the interaction of the exchange factors. A protein interacting with this region could displace loop L3, and the conformational change could propagate to the nucleotide-binding site through β-sheets $\beta 2$ and $\beta 3$ to "open" it and induce GDP release. This hypothesis can now be tested with the various exchange factors recently characterized (see Chap. 24).

III. Relation to Other GTPase Families

At least seven distinct families of GTPases are present in eukaryotic cells: the tubulins (see Chap. 5), the α subunits of heterotrimeric G-proteins directly involved in signal transduction (see Chap. 3 and Sect. IV), the elongation factors of protein biosynthesis (see Chap. 9), the signal recognition particle (SRP) receptor family, the dynamin family (see Chap. 6), the ADP ribosylation factor (ARF) family (see Chap. 34), and the ras family. The consensus sequence of ras-related proteins has been aligned with the sequences of other GTPases representing these different families: ARF, Sar1p, α subunits of G-proteins, SRP receptor, and elongation factor Tu. The GX4GKS/T motif could easily be recognized in all of them, always followed by Y or F in the next 12 amino acids. The DXXG motif was also easily found in most of them, although in the SRP receptor two different regions matched this consensus. A threonine was always present approximately 20 amino acids upstream of this motif, corresponding to ras T35. The LVGNKXDL motif was also unambiguously found in all of them, but the following ETSAK motif was much more difficult to recognize, and its localization is only tentative for several proteins. This is not altogether surprising since only K147 of ras directly interacts with the guanine ring, and the other amino acid only interacts with side chains that form the guanine-binding pocket. It is interesting to note that the few strictly conserved amino acids of the ras family probably represent absolute requirements for GTP/GDP binding since most of them are also present in distantly related GTPases, with the only exception being this ETSAK sequence. Therefore the ras family cannot be defined on the absolute criteria of strictly conserved amino acids, but rather by the high homology (>30% identity) of any member to all other members of this family.

IV. Is There a Conserved Functional Mechanism for All Ras-Related Proteins?

The functional cycles of Ras-related proteins are discussed in several other chapters of this book. Structural studies define external regions that are potential targets for interacting proteins; since these regions are highly variable, a large variety of target proteins is also expected. Nevertheless, the general structures of all Ras-related proteins are highly analogous, suggesting a conserved functional mechanism. Rab proteins are involved in vesicular traffic and rho proteins are involved in the interaction of membrane compartments with actin or associated cytoskeletal components; preliminary results suggest that rap and ral proteins are also associated with vesicles and interact with cytoskeletal components. Furthermore, the ARF and Sar1p proteins, which are only distantly related to ras in sequence, but that might be analogous in structure, also seem to be involved in cellular compartmentation and traffic. Therefore, all Ras-related proteins are very homologous in structure and most of them seem to work as "switches" or "shuttles" involved in cytoskeletal dynamics, compartmentation, or vesicle traffic. Since ras and ras-related proteins have very similar structures, we suggest that in mammalian cells, H-ras, K-ras, and N-ras proteins also perform a similar "shuttle" function at the plasma membrane, controlling lateral segregation of multiprotein complexes involved in signal transduction (Chardin 1991).

Acknowledgement. P.C. is supported by INSERM.

References

Altschuler DL, Muro A, Schijman A, Bravo Almonacid F, Torres HN (1990) Neurospora crassa cDNA clones coding for a new member of the ras protein family. FEBS Lett 273:103–106

Baldini G, Hohl T, Lon HY, Lodish HF (1992) Cloning of a rab3 isotype predominantly expressed in adipocytes. Proc Natl Acad Sci USA 89:5049–5052

Bender A, Pringle JR (1989) Multicopy suppression of the cdc24 budding defect in yeast by CDC42 and three newly identified genes including the ras-related gene RSR1. Proc Natl Acad Sci USA 86:9976–9980

Bishop JG, Corces V (1988) Expression of an activated ras gene causes developmental abnormalities in transgenic Drosophila melanogaster. Genes Dev 2:567–577

Bucci C, Frunzio R, Chiariotti L, Brown A, Rechler M, Bruni C (1988) A new member of the ras gene superfamily identified in a rat liver cell line. Nucleic Acids Res 16:9979–9993

Capon DJ, Chen EY, Levinson AD, Seeburg PH, Goeddel DV (1983) Complete nucleotide sequences of the T24 human bladder carcinoma oncogene and its normal homologue. Nature 302:33–37

Casale WL, McConell DG, Wang S-Y, Lee Y-J, Linz JE (1990) Expression of a gene family in the dimorphic fungus Mucor racemosus which exhibits striking similarity to human ras genes. Mol Cell Biol 10:6654–6663

Chardin P (1991) Small GTP-binding proteins of the ras family, a conserved functional mechanism? Cancer Cells 3:117–126

Chardin P, Tavitian A (1986) The ral gene: a new ras-related gene isolated by the use of a synthetic probe. EMBO J 5(9):2203–2208

Chardin P, Tavitian A (1989) Coding sequences of human ralA and ralB cDNAs. Nucleic Acids Res 17:4380

Chardin P, Madaule P, Tavitian A (1988) Coding sequence of human rho cDNA clone 6 and clone 9. Nucleic Acids Res 16:2717

Chavrier P, Vingron M, Sander C, Simons K, Zerial M (1990) Molecular cloning of YPT/SEC4-related cDNAs from an epithelial cell line. Mol Cell Biol 10: 6578–6585

Chavrier P, Simons K, Zerial M (1992) The complexity of the rab and rho GTP-binding protein subfamilies revealed by a PCR cloning approach. Gene 112: 261–264

DeFeo-Jones D, Scolnick EM, Koller R, Dhar R (1983) ras-related gene sequences identified and isolated from Saccharomyces cerevisiae. Nature 306:707–709

Didsbury J, Weber RF, Bokoch GM, Evans T, Snyderman R (1989) rac, a novel ras-related family of proteins that are botulinum toxin substrates. J Biol Chem 264:16378–16382

Drivas GT, Shih A, Coutavas E, Rush MG, D'Eustachio P (1990) Characterization of four novel ras-like genes expressed in a human teratocarcinoma cell line. Mol Cell Biol 10:1793–1798

Elferink LA, Anzai K, Scheller R (1992) rab15, a novel low molecular weight GTP-binding protein specifically expressed in rat brain. J Biol Chem 267:5768–5775

Fawell E, Hook S, Armstrong J (1989) Nucleotide sequence of a gene encoding a YPT1-related protein from Schizosaccharomyces pombe. Nucleic Acids Res 17:4373

Fukui Y, Kaziro Y (1985) Molecular cloning and sequence analysis of a ras gene from Schizosaccharomyces pombe. EMBO J 4:687–691

Gallwitz D, Donath C, Sander C (1983) A yeast gene encoding a protein homologous to the human c-has/bas proto-oncogene product. Nature 306:704–707

Han M, Sternberg PW (1990) let-60, a gene that specifies cell fates during C. elegans vulval induction, encodes a ras protein. Cell 63:921–931

Hariharan IK, Carthew RW, Rubin GM (1991) The Drosophila roughened mutation: activation of a rap homolog disrupts eye development and interferes with cell determination. Cell 67:717–722

Haubruck H, Engelke U, Mertins P, Gallwitz D (1990) Structural and functional analysis of YPT2, an essential ras-related gene in the fission yeast Schizosaccharomyces pombe encoding a sec4 protein homologue. EMBO J 9:1959–1962

Hengst L, Lehmeier T, Gallwitz D (1990) The rhyl gene in the fission yeast Schizosaccharomyces pombe encoding a GTP-binding protein related to ras, rho and YPT: structure, expression and identification of its human homologue. EMBO J 9:1949–1955

Johnson DI, Pringle JR (1990) Molecular characterization of CDC42, a Saccharomyces cerevisiae gene involved in the development of cell polarity. J Cell Biol 111:143–152

Lowe D, Capon D, Delwart E, Sakaguchi A, Naylor S, Goeddel D (1987) Structure of the human and murine R-ras genes, novel genes closely related to ras protooncogenes. Cell 48:137–146

Madaule P, Axel R (1985) A novel ras-related gene family. Cell 41:31–40

Madaule P, Axel R, Myers A (1987) Characterization of two members of the rho gene family from the yeast Saccharomyces cerevisiae. Proc Natl Acad Sci USA 84:779–783

Matsui M, Sasamoto S, Kunieda T, Nomura N, Ishizaki R (1989) Cloning of ara, a putative Arabidopsis thaliana gene homologous to the ras-related gene family. Gene 76:313–319

Matsui Y, Kikuchi A, Kondo J, Hishida T, Teranishi Y, Takai Y (1988) Nucleotide and deduced amino acid sequence of a GTP-binding protein family with molecular weights of 25000 from bovine brain. J Biol Chem 263:11071–11074

Matsumoto T, Beach D (1991) Premature initiation of Mitosis in yeast lacking RCC1 or an interacting GTPase. Cell 66:347–360

McGrath JP, Capon DJ, Smith DH, Chen EY, Seeburg PH, Goeddel DV, Levinson AD (1983) Structure and organization of the human Ki-ras proto-oncogene and a related processed pseudogene. Nature 304:501–506

Miyake S, Yamamoto M (1990) Identification of ras-relared, YPT family genes in Schizosaccharomyces pombe. EMBO J 9:1417–1422

Munemitsu S, McInnis MA, Clark R, McCormick F, Ullrich A, Polakis P (1990) Molecular cloning and expression of a G25K cDNA, the human homolog of the yeast cell cycle gene CDC42. Mol Cell Biol 10:5977–5982

Neuman-Silberberg FS, Schejter E, Hoffmann FM, Shilo BZ (1984) The drosophila ras oncogenes: structure and nucleotide sequence. Cell 37:1027–1033

Ngsee JK, Elferink LA, Scheller RH (1991) A family of ras-like GTP binding proteins expressed in electromotor neurons. J Biol Chem 266:2675–2680

Ohmstede C-A, Farell FX, Reep BR, Clemetson KJ, Lapetina EG (1990) Rap2B: a ras related GTP-binding protein from platelets. Proc Natl Acad Sci USA 87:6527–6531

Palme K, Diefenthal T, Vingron M, Sander C, Schell J (1992) Molecular cloning and structural analysis of genes from Zea maïs. coding for members of the ras-related YPT gene family. Proc Natl Acad Sci USA 89:787–791

Pizon V, Chardin P, Lerosey I, Olofsson I, Tavitian A (1988a) Human rap1 and Rap2 cDNAs, homologous to the Drosophila melanogaster gene Dras 3, encode proteins closely related to ras in the effector region. Oncogene 3:210–204

Pizon V, Lerosey I, Chardin P, Tavitian A (1988b) Nucleotide sequence of a human cDNA encoding a ras-related protein (rap1B). Nucleic Acids Res 16:7719

Powers S, Kataoka T, Fasano O, Goldfarb M, Strathem J, Broach J, Wigler M (1984) Genes in S. cerevisiae encoding proteins with domains homologous to the mammalian ras proteins. Cell 36:607–612

Reymond CD, Gomer RH, Mehdy MC, Firtel RA (1984) Developmental regulation of a Dictyostelium gene encoding a protein homologous to mammalian ras proteins. Cell 39:141–148

Robbins SM, Williams JG, Jermyn KA, Spiegelman GB, Weeks G (1989) Growing and developing Dictyolstelium cells express different ras genes. Proc Natl Acad Sci USA 86:938–942

Robbins SM, Suttorp VV, Weeks G, Spiegelman GB (1990) A ras-related gene from the lower eucaryote Dictyolstelium that is highly conserved relative to the human rap genes. Nucleic Acids Res 18:5265–5269

Salminen A, Novick P (1987) A ras-like protein is required for a post golgi event in yeast secretion. Cell 49:527–538

Saxe SA, Kimmel AR (1988) Genes encoding novel GTP-binding proteins in Dictyostelium. Dev Genet 9:259–265

Shinjo K, Koland JG, Hart MJ, Narasimhan V, Johnson DI, Evans T, Cerione RA (1990) Molecular cloning of the gene for the human placental GTP-binding protein Gp (G25K): identification of this GTP-binding protein as the human homolog of the yeast cell division cycle protein CDC42. Proc Natl Acad Sci USA 87:9853–9857

Swanson M, Elste A, Greenberg S, Schwartz J, Aldrich T, Furth M (1986) Abundant expression of ras proteins in Aplysia neurons. J Cell Biol 103:485–492

Taparowsky E, Shimizu K, Goldfarb M, Wigler M (1983) Structure and activation of the human N-ras gene. Cell 34:581–586

Touchot N, Chardin P, Tavitian A (1987) Four additional members of the ras gene superfamily isolated by an oligonucleotide strategy: molecular cloning of YPT related cDNAs from rat brain. Proc Natl Acad Sci USA 84:8210–8214

Valencia A, Chardin P, Wittinghofer A, Sander C (1991) The ras protein family: evolutionary tree and role of conserved amino acids. Biochemistry (in press)

Vielh E, Touchot N, Zahraoui A, Tavitian A (1989) Nucleotide sequence of a rat cDNA: RAB1B, encoding a RAB1-YPT related protein. Nucleic Acids Res 17:1770

Vincent S, Jeanteur P, Fort P (1992) Growth-regulated expression of rhoG, a new member of the ras homolog gene family. Mol Cell Biol 12:3138–3148

Westaway D, Papkoff J, Moscovici C, Varmus HE (1986) Identification of a provirally activated c-Ha-ras oncogene in an avian nephroblastoma via a novel procedure: cDNA cloning of a chimaeric viral-host transcript. EMBO J 5: 301–309

Yeramian P, Chardin P, Madaule P, Tavitian A (1987) Nucleotide sequence of human rho cDNA clone 12. Nucleic Acids Res 4:1869

Zahraoui A, Touchot N, Chardin P, Tavitian A (1988) Complete coding sequences of the ras related rab3 and 4 cDNAs. Nucleic Acids Res 16:1204

Zahraoui A, Touchot N, Chardin P, Tavitian A (1989) The human rab genes encode a family of GTP binding proteins related to the Yeast YPT1 and SEC4 products involved in secretion. J Biol Chem 264:12394–12401

CHAPTER 13

Conformational Switch and Structural Basis for Oncogenic Mutations of *Ras* Proteins

S.-H. KIM, G.G. PRIVÉ, and M.V. MILBURN

A. Introduction

Ras proteins play an important role in the signaling of cell growth and differentiation. Mammalian *ras* proteins are made up of 188 or 189 amino acids with an approximate molecular mass of 21 kDa, and are hence named p21. Three *ras* genes have been identified in human cells, H-*ras*, K-*ras*, and N-*ras*, and they all show a high degree of sequence homology for the first 165 amino acids, after which the sequences diverge. The p21 proteins are attached to the cytoplasmic side of the membrane through a polyisoprenoid group which is covalently linked to Cys-186 (HANCOCK et al. 1989; BUSS et al. 1989; SCHAFER et al. 1989) and possess an inherent GTPase activity that is stimulated by GTPase activating protein (GAP; TRAHEY and MCCORMICK 1987; VOGEL et al. 1988; TRAHEY et al. 1988) and by the neurofibromatosis type 1 protein NF1 (MARTIN et al. 1990; XU et al. 1990).

Like other signaling proteins, such as G-proteins or transducin, a *Ras* protein in its GTP-bound state communicates to effector proteins and is deactivated by hydrolysis of GTP to GDP (reviewed in BARBACID 1987; BOS 1989; DER 1989; BOURNE et al. 1991). The molecular switching from the GDP- to the GTP-bound form is accompanied by extensive conformational changes in two parts of the protein, the switch I and switch II regions (Fig. 1a,b; MILBURN et al. 1990). Since signaling for cell growth and differentiation is one of the most essential steps in a cell's normal life cycle, dysfunction of *Ras* protein has catastrophic consequences.

Mutations at several positions in the protein appear to be especially critical: *ras* oncogenes isolated from human tumor cells often have single point mutations in the codons corresponding to amino acid positions 12, 13, 61, or 146 (BARBACID 1987; HIGASHI et al. 1990); both H and K rat sarcoma virus have a *ras* gene with mutations at codons 12 and 59; and in vitro mutagenesis at positions 12, 13, 59, 61, 63, 116, 117, or 119 endow p21 with cell-transforming capabilities (BARBACID 1987; DER 1989). As seen in Fig. 2c, these positions make up part of the guanine nucleotide binding pocket (MILBURN et al. 1990; DE VOS et al. 1988; PAI et al. 1989), and these can be mapped to either the phosphate-binding region (12, 13, 59, 61, 63) or guanine base-binding region (116, 117, 119, 146). To date, all of the biochemically characterized phosphate-binding region mutants show a decrease

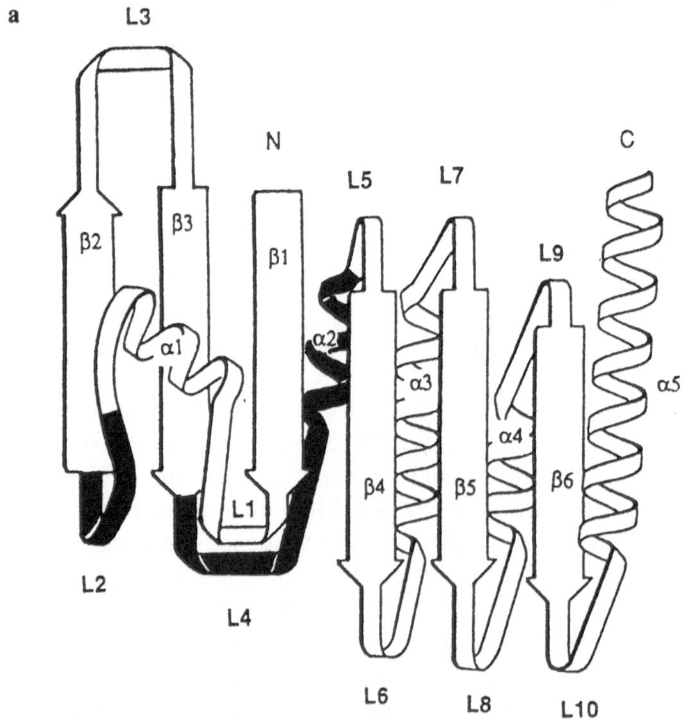

Fig. 1. a Topological structure of ras protein catalytic domain (residues 1–171). *β* strands, *α* helices, and loops are sequentially numbered from N terminus. Switch I in loop 2 and switch II consisting of loop 4 and *α*2 are shown in *black*. *β*-Phosphate-binding loop is marked L1. **b** Three-dimensional backbone structure, represented as a multistranded ribbon, of c-H-ras protein catalytic domain (residues 1–171) in complex with GDPCP, a nonhydrolyzable GTP analog, shown as *green space* filling model. Switch I and II regions are shown in *blue* and *gold* respectively

in the intrinsic GTP hydrolysis rate and are insensitive to GAP and NF1 stimulation, while all of the base-binding region mutations that have been tested show a decreased affinity for the guanine nucleotide and high GDP/GTP exchange rates. Either or both of these mechanisms, reduced hydrolysis rates and increased GDP/GTP exchange rates, are directly related to the switching mechanism and thought to be the biochemical reasons for cell transformation by oncogenic p21 proteins.

B. Conformational Switch

To understand the structural bases for the molecular switching, we have analyzed the conformational differences between crystal structures of GDP- and GTP-analog-bound forms of the normal *Ras* proteins, paying special attention to distinguish the changes caused by the nucleotide exchange from

b

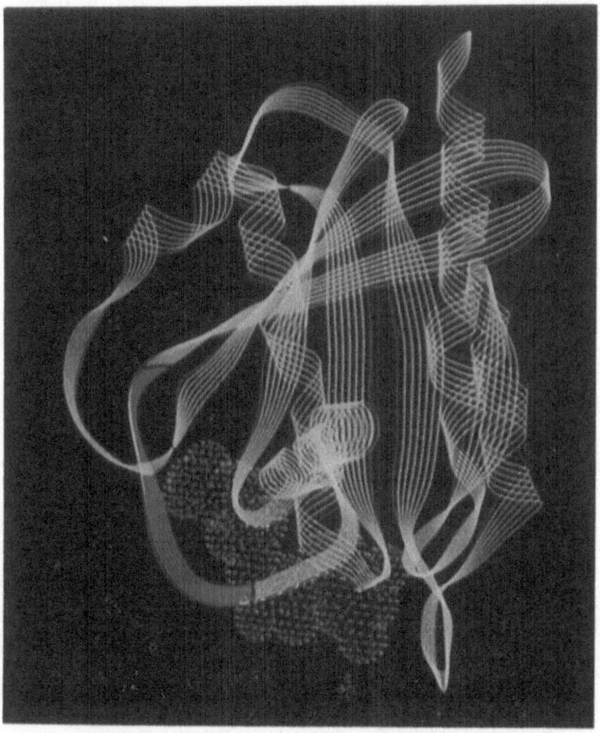

Fig. 1. *Continued*

those caused by crystal packing. To avoid the danger of misinterpretation, we compare crystal structures in several different crystal packing environments whenever possible. Table 1 lists the structures that are compared.

I. Confirmational Differences between GDP- and GTP-bound *Ras* Proteins: Switch I and II Regions

A superpositioning of Cα positions of the GDP and a nonhydrolyzable GTP analog-bound (GDP–CP: guanosine β,γ-methylene triphosphate) forms of the normal c-H-*Ras* protein indicated that the largest differences between the two structures are localized in two regions (MILBURN et al. 1990). The first region, named switch I, spans residues 30–37 and corresponds to most of the residues in loop 2, and the second region, switch II, consists of residues 60–76 of loop 4 and α helix 2 (Fig. 1a,b); these are two regions whose conformation "switches" when GTP replaces GDP in the protein. The details of the conformational differences are shown in Fig. 2. It should be noted that this change has been observed in two different crystal forms, and therefore, it is not an artifact caused by crystal packing: the conformation of the corresponding regions in the GTP-bound state are conserved (root-mean-square, r.m.s. differences of 0.8–0.9 Å) among the four independent molecules of the GDP–CP complex (Brünger et al. 1990), and also between the GDP–CP and GDP–NP (Pai et al. 1990) complexes.

The conformational difference in the switch I region is quite large, as indicated by relatively large r.m.s. differences in Cα positions (2.6–2.7 Å).

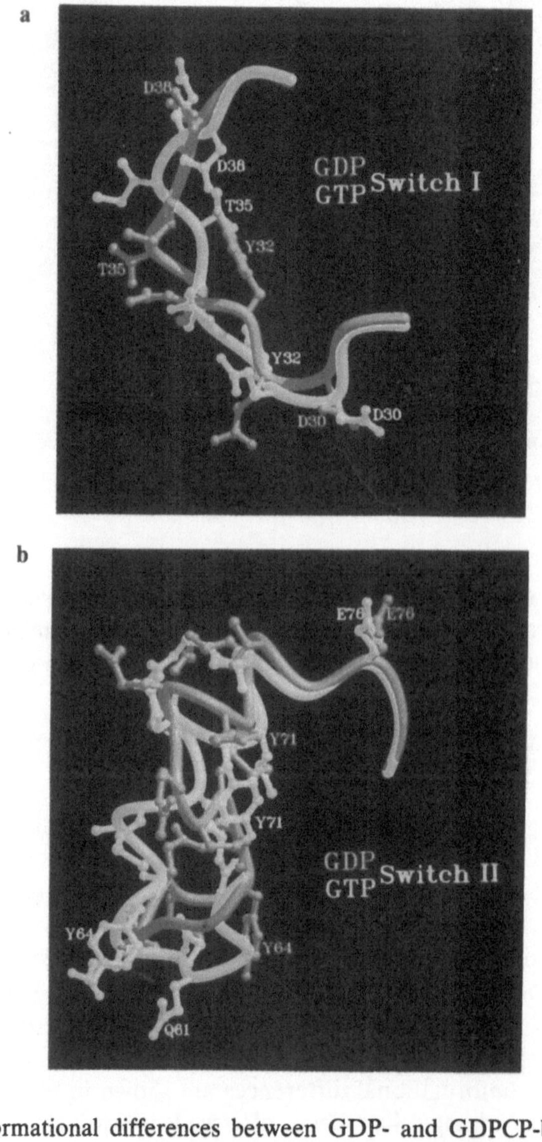

Fig. 2. Conformational differences between GDP- and GDPCP-bound states of **a** switch I and **b** switch II. **c** A stereo view of the superposition of GDP pocket and GTP pocket

c

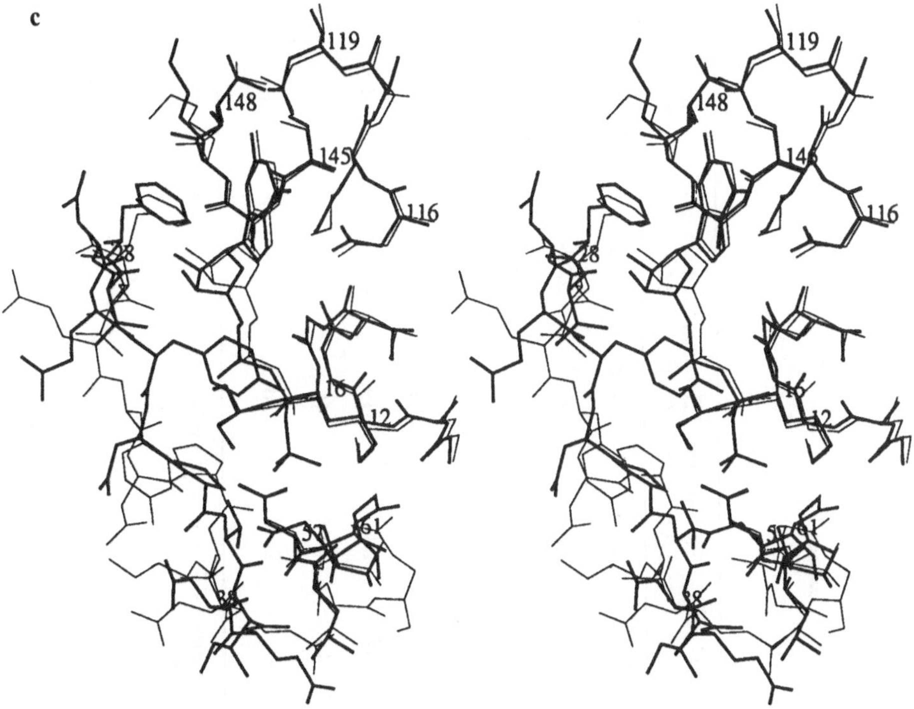

Fig. 2. *Continued*

This is clearly seen in the match between the GDP and GDP–CP complexes (Fig. 2a). Some of the interesting differences are: (a) the side chain of Thr-35 of loop 2 in the GTP-analog complex is coordinated to the Mg^{2+} ion and hydrogen bonded to the γ-phosphate, while the same residue points out toward the solvent in the GDP complexes; (b) the side chain orientations of residues 36 and 38 in the GDP–CP complex are substantially different from those in the GDP complexes. Such extensive conformational differences can certainly represent two different states of the molecular switch that can be distinguished by the effector protein(s). This notion is consistent with the earlier observation that mutations in this loop alter transforming activity of oncogenic mutants, thus suggesting that this region is a "putative effector recognition region" (SIGAL et al. 1986).

The other large difference between the GDP and GTP-analog complex structures appears in the switch II region (residues 60–76), corresponding to loop 4 and the following α helix, $\alpha 2$ (Fig. 2b). The difference in helix $\alpha 2$ is primarily manifested by the difference in orientation and the length of this helix. By far the largest difference is in the region of residues 60–68 (r.m.s. range of 3.6–7.4 Å). However, weak electron density is associated with a portion of this region in both the GDP complexes (residues 60–65 in the GDP complex of the intact protein) and the GDP–CP complexes (residues

Table 1. Refinement statistics for Ras mutants

Protein complex[a]	Space group	Unit cell	N[b]	Total refls	Resolution (Å)	R-factor (%)	rms bond (Å)	rms angle (°)
GDP · Gly-12	$P6_522$	a=b=83.2, c=105.1 $\alpha=\beta=90.0$, $\gamma=120.0$	1	8566	2.2	18.8	0.026	2.6
GDP · Val-12	$P6_522$	a=b=83.2, c=105.1 $\alpha=\beta=90.0$, $\gamma=120.0$	1	8690	2.2	19.2	0.024	2.5
GDP · Leu-61	$P6_522$	a=b=83.2, c=105.1 $\alpha=\beta=90.0$, $\gamma=120.0$	1	9447	2.3	20.7	0.027	2.9
GDP · Thr-59	$P6_522$	a=b=83.3, c=104.6 $\alpha=\beta=90.0$, $\gamma=120.0$	1	5640	2.7	18.4	0.027	2.9
GDP · Val-12/Thr-59	$P6_522$	a=b=82.9, c=101.8 $\alpha=\beta=90.0$, $\gamma=120.0$	1	8160	2.2	21.3	0.021	2.6
GDP · Intact	I4	a=b=97.8, c=41.8 $\alpha=\beta=\gamma=90.0$	1	10167	2.0	19.5	0.019	2.4
GDPCP · Gly-12	$P2_1$	a=42.0,b=79.9, c=130.5 $\alpha=\beta=90.0$, $\gamma=117.5$	4	41345	1.95	18.8	0.014	2.6
GDPCP · Leu-61	$P2_1$	a=42.0,b=79.9, c=129.2 $\alpha=\beta=90.0$, $\gamma=117.5$	4	31061	2.00	20.3	0.019	3.6

[a] All structures are of the truncated p21 "catalytic domain" fragment (residues 1-171), except GDP · Intact, which is from 1–188.
[b] Number of crystallographically independent molecules.

62–65 in four independent molecules), suggesting high flexibility of this portion.

Two specific interactions in switch II are noteworthy: the backbone NH groups of residues 60 and 61 are in a position to form hydrogen bonds to the γ-phosphate. This portion of the sequence is highly conserved among all *Ras* and *Ras*-related proteins. The functional importance of the switch II region is indicated by the observation that the second most commonly found oncogenic mutation site, residue 61, is located at the beginning of this region, and that a monoclonal antibody, Y13-259, which binds to the helix $\alpha2$ of this region (SIGAL et al. 1986), neutralizes the transforming activity of oncogenic mutants.

II. Conformational Domino Effect and Frozen Dynamic States

Although only one residue of loop 2, Thr-35, forms a hydrogen bond with the γ-phosphate in the GTP-analog complexes (see Fig. 2c), the entire switch I region has a conformation different from that in the GDP complexes. A similarly extended conformational difference is seen for the switch II region; only the backbone NH groups of residues 60 and 61 interact with the γ-phosphate, but the entire switch II region containing loop 4 and helix $\alpha2$ has a conformation or orientation different from that in the GDP complex. One can imagine two possible mechanisms to explain the observed conformational differences: "conformational domino effect" and "frozen dynamic states." The former suggests that the interaction of the γ-phosphate

a

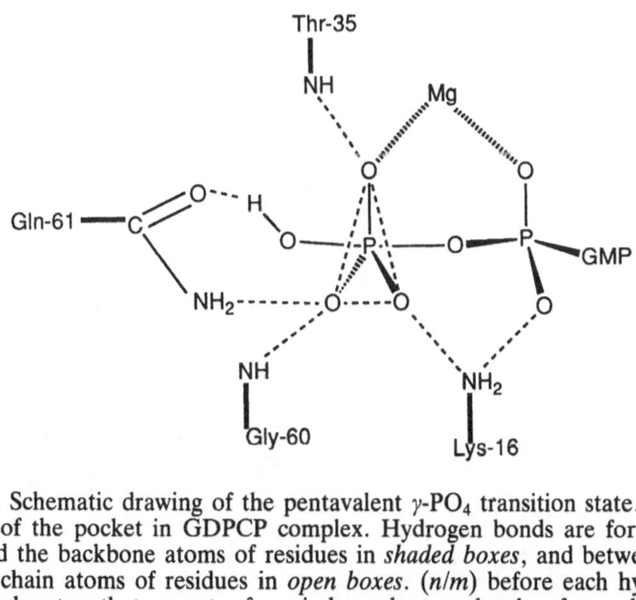

Fig. 3. **a** Schematic drawing of the pentavalent γ-PO_4 transition state. **b** Schematic drawing of the pocket in GDPCP complex. Hydrogen bonds are formed between GDP and the backbone atoms of residues in *shaded boxes*, and between GDP and the side chain atoms of residues in *open boxes*. (*n*/*m*) before each hydrogen bond distance denotes that *n* out of *m* independent molecules form that particular hydrogen bond (MILBURN 1991). **c** Schematic drawing of the pocket in GDP complex

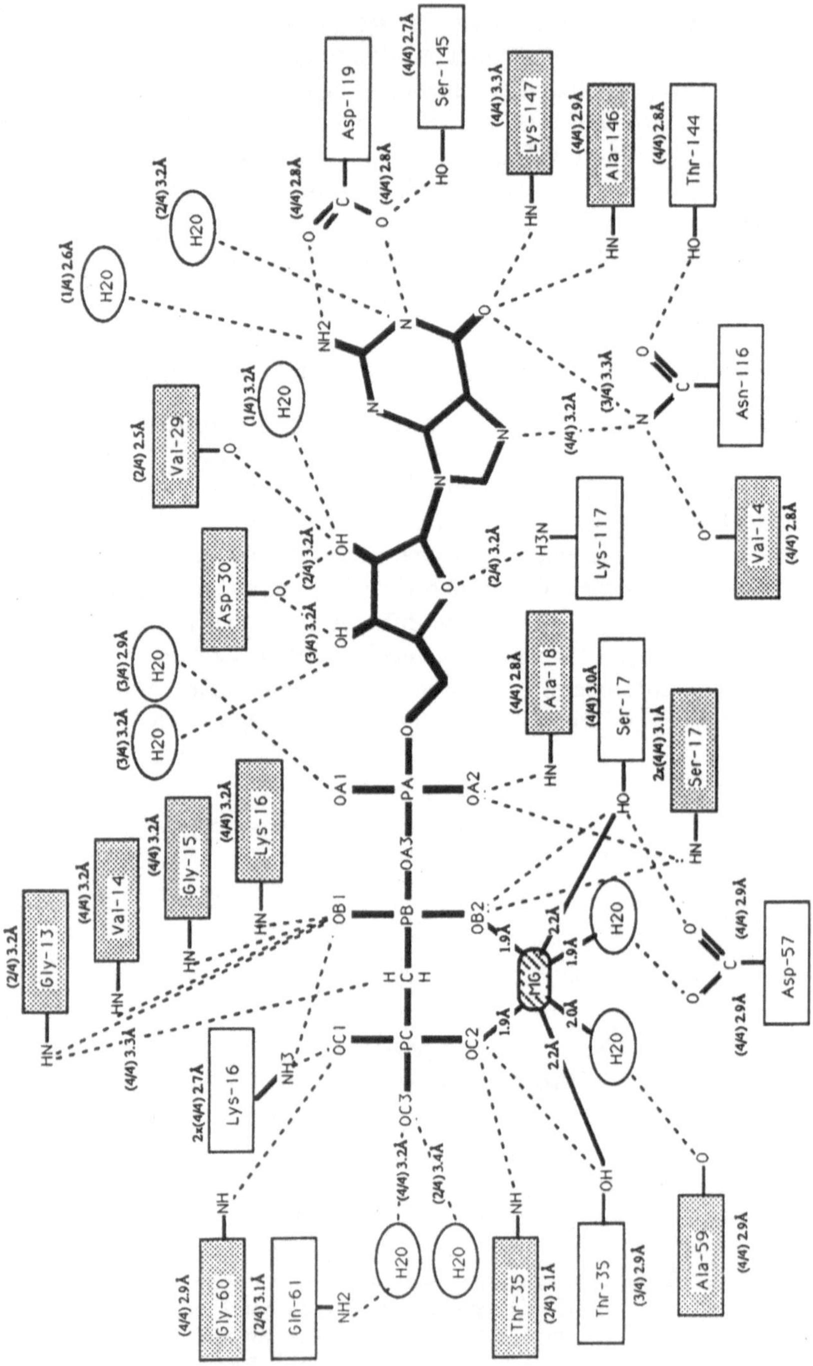
Asp-119
Ser-145
Lys-147
Ala-146
Thr-144
Asn-116
Val-29
Asp-30
Lys-117
Val-14
Ala-18
Ser-17
Ser-17
Gly-13
Val-14
Gly-15
Lys-16
Lys-16
Asp-57
Gly-60
Gln-61
Thr-35
Thr-35
Ala-59
H20
MG
OA1
PA
OA2
OA3
OB1
PB
OB2
OC1
PC
OC2
OC3

b

c

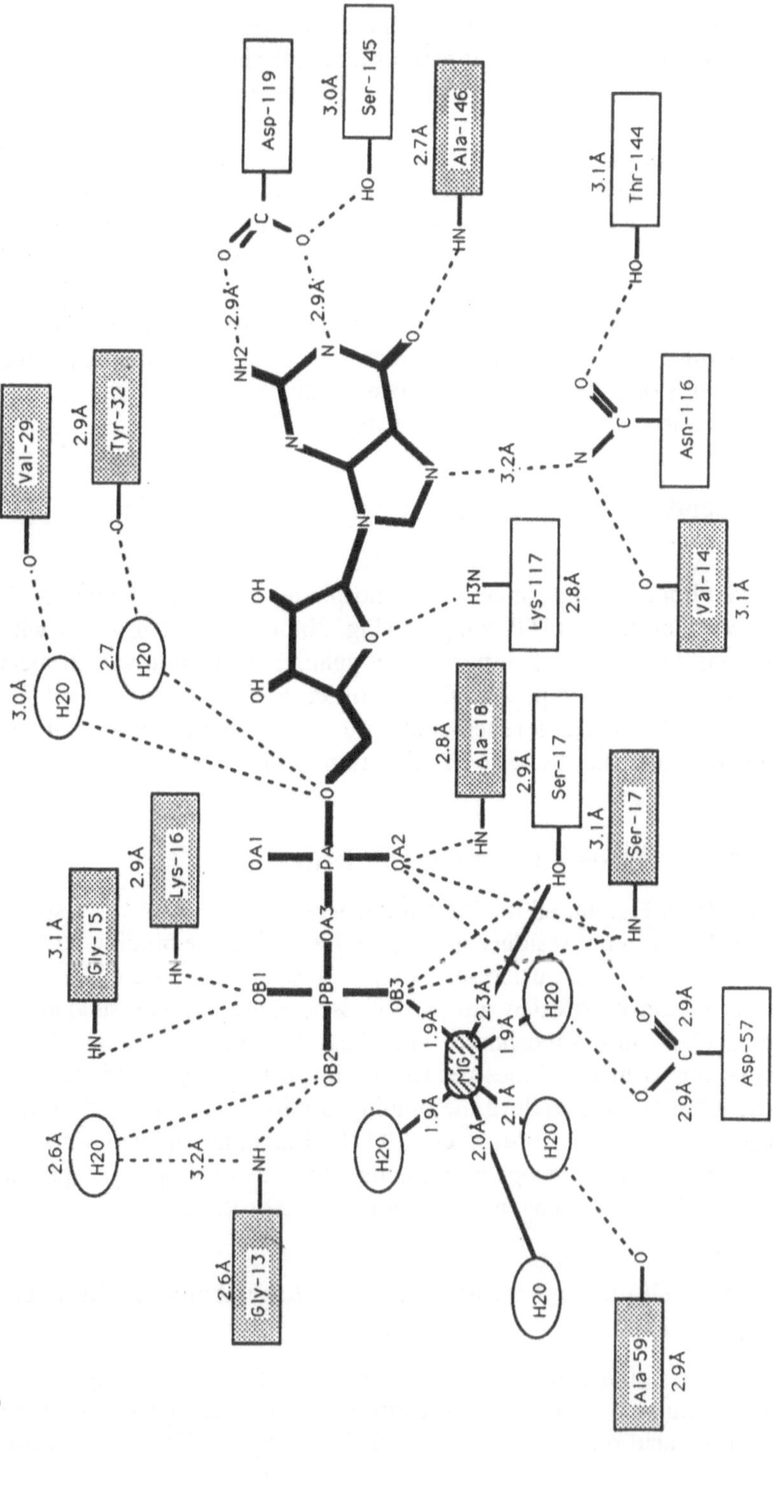

Fig. 3. *Continued*

of GTP with Thr-35 of switch I and the backbone of residues 60 and 61 of switch II triggers a cascade of changes, like dominoes, in the entire lengths of switch I and II, propagating the conformational changes over a span of 40 Å. It further suggests that the residues in the regions are conformationally "linked," so that change in one part can be "felt" by other parts of the regions. Recent evidence for such a domino effect was suggested by the work of VERROTTI et al. (1992), where the mutations at switch II residues affect GDP/GTP exchange by the exchange factor of yeast. The frozen dynamic state mechanism, on the other hand, proposes that, in the absence of GDP or GTP, both switch I and II regions are dynamically flexible and have multiple conformations. The regions "freeze" into one conformation on GDP binding and another conformation on GTP binding. It is likely that one mechanism is operative in some parts of switch I and II regions and the other in other parts.

III. Small Conformational Changes in the Phosphate-Binding Loop, L1

The hydrogen-bonding pattern of the β-phosphate in the GDP complex is different from that of the GTP complex (Fig. 3b,c). This is consistent with the observation that the relative position of guanine nucleotide with respect to loop 1 in the GDP complex is slightly different from that of the GTP complex (Fig. 4b). Such difference could alter electronic properties of the phosphates, thus altering the state of activation of GTP.

C. Structural Bases for Oncogenic Mutations

As is evident from Fig. 4a, all the known oncogenic mutations involve single amino acid substitution of residues in the guanine nucleotide binding pocket. The details of the guanine nucleotide binding pocket in stereo are shown in Fig. 4b,c. They can be divided into two classes: those residues associated with binding of β- and γ-phosphates and those binding the guanine base. Mutations of the former residues reduce the intrinsic GTPase activity, and those of the latter residues reduce the binding affinity of guanine nucleotides and increase the exchange rate of GDP/GTP. Examination of the crystal structures of normal and oncogenic mutants allows us to propose a specific structural basis for each oncogenic mutation of *Ras* protein.

I. Mutations at Gln-61 and the Stabilization of the Transition State of the γ-Phosphate of GTP

Residue 61 is one of two most commonly found activating point mutation sites of p21 in human tumors, and substitution of the normal Gln-61 with any other amino acid reduces its intrinsic GTPase activity (DER et al. 1986).

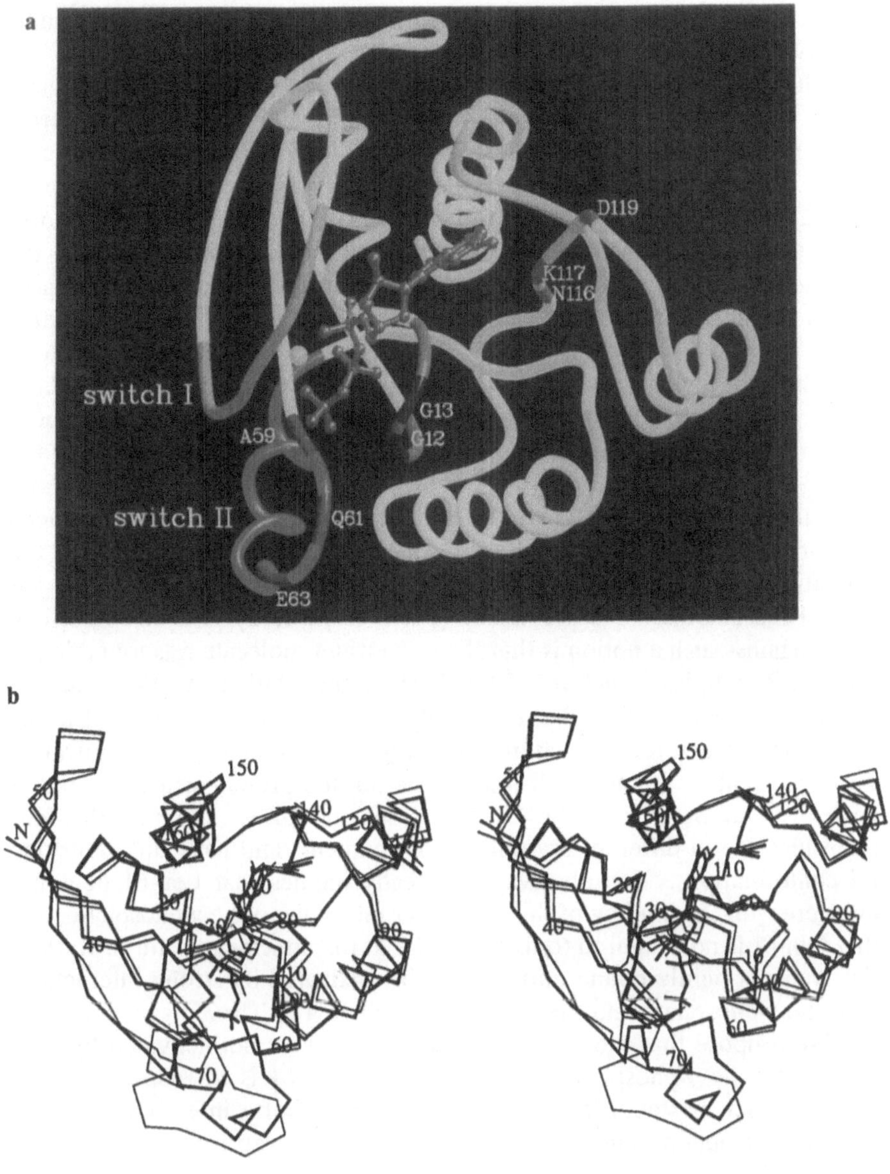

Fig. 4. **a** Backbone structure of ras protein looking into the guanine nucleotide pocket. Switch I and II regions are in *green* and *blue* respectively, the phosphate-binding loop in *purple*, guanine base binding loops in *pink*, GDPCP in *orange*, and magnesium ion in *yellow*. Oncogenic mutation sites are indicated by *darker tones* and *labeled*. **b** Stereo view of ras protein in GDP (*thin lines*) and GDPCP (*think lines*) bound states. The view is looking into the guanine nucleotide binding pocket. Every ten residues are numbered.

All of these except the proline and glutamate mutants have transforming activity.

The GTP hydrolysis reaction of p21 has been shown to proceed via a direct attack of a water molecule on the γ-phosphate of GTP (Feuerstein et al. 1989). Three factors could influence GTP hydrolysis by p21: (1) activation of the attacking water molecule by an increase of the negative partial charge of the water oxygen, (2) activation of the attacked γ-phosphorus by an increase of positive partial charge of the γ-phosphorous and the stabilization of the pentavalent transition state intermediate, and (3) activation of the leaving group, GDP, by decreasing the negative partial charge of the β,γ-bridging oxygen (see Fig. 3a). As can be seen in Fig. 3b, the γ-phosphorus is highly activated by an extensive network of hydrogen bondings to the γ-phosphate and its charge neutralization by Lys-16 and the Mg^{2+} ion. Likewise, the bridging oxygen is also activated by a hydrogen bond from the Gly-13 backbone (Fig. 3b,c). As for the activation of the attacking water molecule, Pai et al. (1990) proposed that Gln-61, in a hydrogen-bonded network with Glu-63, activates the water molecule. We argue that the activation of this water by Gln-61 is neither likely nor rate-limiting for several reasons discussed previously (Privé et al. 1992). The strongest evidence against such a notion is that the same water molecule was found in the GDP–CP complex structure of Q61L (Leu substitution at Q61) where a Leu-61 side chain is located at essentially the same position as Gln-61 in the normal p21 structure. Even more surprising is the finding that an equivalent water molecule is found in all four of the normal protein structures and at least two out of four of the Q61L mutant structures, although the four molecules are in different crystalline environments and have different loop L4 conformations. This observation directly implies that Leu-61 does not interfere with the binding of this key water molecule to the γ-phosphate, and that Gln-61 is not required to fix this water at the attacking position. Rather, this water is tightly bound and presumably activated by strong interactions with the γ-phosphate and the main-chain carbonyl of Thr-35.

We propose that the attacking water molecule is not activated by Gln-61, but by the γ-phosphate, and the role of Gln-61 is to reduce the activation barrier by stabilizing the pentavalent phosphorus intermediate of the hydrolysis reaction (Fig. 3a). The stabilization energy from the two proposed hydrogen bonds between the glutamine-61 and the transition state phosphate oxygens are sufficient to explain the rate enhancement of Gln-61 p21 relative to the position 61 mutants, which are unable to form this specific complex. We also suggest that the mutation of residue 63 may reduce the GTPase activity of the protein indirectly by interfering with the proper positioning of the catalytically important Gln-61 in stabilizing the reaction intermediate or by reducing the flexibility of the loop L4.

II. Mutations at Gly-12 and the Stabilization of the Transition State of the γ-Phosphate of GTP

Substitution of Gly-12 by any amino acid except proline was found to generate transforming mutants (SEEBURG et al. 1984), and of the mutants that have been tested, all have reduced intrinsic GTPase activity. The crystal structures at 2.2-Å resolution of the GDP-bound form of normal p21 and of a transforming mutant, G12V, have been determined (Table 1; TONG et al. 1991). The overall structures are identical except for residues 61–65 of the loop L4 region, which is flexible and has less well defined electron density than the rest of the protein.

We modeled the structure of the G12V mutant in the GTP-bound conformation by replacing residues 10–17 of the GDP–CP complex with the corresponding residues of the GDP complex of the G12V mutant. There are no significant conformational changes at or near the mutation site. However, the side chain of Val-12 comes into close contact with the γ-phosphate, and, thus, would interfere with the formation of the transition state. The role of Gln-61 as either a water-activating (PAI et al. 1990) or a transition state stabilizing (this article) residue, however, does not explain the GTPase activity of the G12P mutant. This mutant may use a different mechanism for hydrolyzing GTP.

III. Residues 12 and 13 Form a Type II β-Turn for Phosphate Binding

β-Turns are a common structural motif in proteins, and these are often found at the surface of proteins, where they play a role in reversing the polypeptide chain direction. In p21, the loop L1 region adopts a standard tetrapeptide type II β-turn conformation over the residues Ala-11, Gly-12, Gly-13, and Val-14 (Fig. 5b). β-Turns require specific backbone torsion angles for the amino acid residues at the second and third positions of the tetrapeptide. This turn is essentially identical in all the p21 structures, whether in the GDP or GTP form, and the average (ϕ,ψ) angles of residues 12 and 13 of all the molecules listed in Table 1 (including the Val-12 mutants) are (−63°, 140°) and (90°,−13°), respectively. These compare very well with the ideal values of (−60°,120°) and (80°,0°) for the classical $\beta_P\gamma$ type II turn, as described in a survey by WILMOT and THORNTON (1988). The type II β-turn directs all the backbone NH protons toward the β-phosphate to form hydrogen bonds (Fig. 3b,c). The survey also revealed strong amino acid preferences at the two central positions of type II turns. The second position (corresponding to residue 12) is six to seven times more likely to be a proline than any other amino acid (Fig. 5a), and this is easily explained by the preference of proline for the required (ϕ,φ) angles in this type of turn. Nonproline amino acids can adopt these angles, but the conformation is energetically strained. This observation may be important in explaining the

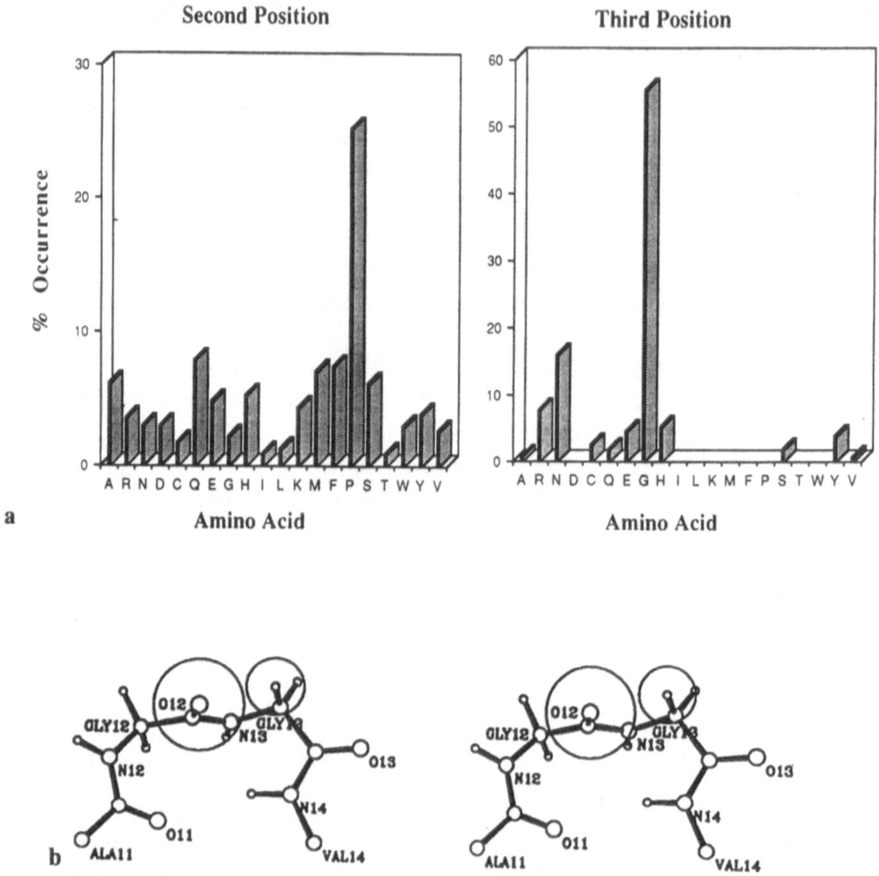

Fig. 5. **a** Occurence frequencies of amino acids at second and third position in type II β-turn of known protein structures. **b** Stereo view of type II β-turn as it appears in residues 11–14

near-normal GTP hydrolysis rates of the G12P mutant of p21. The reason for the G12P mutant being nontransforming, however, is probably not its normal intrinsic GTPase, but the role of proline in interfering with the effector interaction, as implied by the observation that G12P is GAP insensitive.

At the third position (corresponding to Gly-13 in p21), the amino acid restrictions for type II β-turns are even more severe, and glycine and (to a much lesser extent) asparagine are essentially the only two amino acids allowed (Fig. 5a). The structural basis for this restriction is that the carbonyl oxygen of the second residue will cause steric interference with any side chain larger than a hydrogen at the third position of the turn (Fig. 5b). This explains the conservation of glycine in the GXX**G**XGK (bold face G corresponding to Gly-13 in p21) sequence motif in the GTPases (BOURNE et al. 1991; VALENCIA et al. 1991), since nonglycine residues would distort the phosphate-binding loop conformation.

It is interesting to note that a turn with very similar conformation occurs in adenylate kinase, which also has glycine at this third position. In fact, the r.m.s. distance between the main-chain atoms of residues 10–16 in p21 and the corresponding atoms in adenylate kinase is only 0.5 Å (TONG et al. 1991). The conformation of the equivalent residues in bacterial EF-Tu (JURNAK 1985; LACOUR et al. 1985) is expected to be different, since this protein has an aspartate at the residue corresponding to Gly-13 in p21. Furthermore, EF-Tu has valine and histidine at the *Ras*-equivalent positions Gly-12 and Gln-61, respectively, suggesting that catalysis in this protein may involve a different mechanism.

IV. Mutation at Ala-59 and Switch II Conformation

Retrovirus-encoded *ras* oncogenes have two activating mutations: one at the codon for amino acid 12 (glycine to either serine or arginine), and the other for residue 59, where the alanine codon is changed to a threonine codon (DHAR et al. 1982; TSUCHIDA et al. 1982). Thr-59 mutants are unique in that they undergo an autophosphorylation reaction in which the GTP γ-phosphate is transferred to the threonine hydroxyl group. The study of A59T p21 has led to considerable confusion in the past, with various groups reporting different activities for mutants with this substitution (GIBBS et al. 1984; LACAL et al. 1986; HATTORI et al. 1987; JOHN et al. 1988; FEIG and COOPER 1988). A probable factor contributing to the conflicting results is that these workers studied mixtures of phosphorylated and nonphosphorylated protein. To resolve this situation, we purified the phosphorylated and nonphosphorylated forms of p21 (A59T) and p21 (G12V/A59T). All four forms of the protein showed reduced intrinsic GTPase activities and increased GTP exchange rates in vitro, while only the nonphosphorylated proteins were biologically active in a *Xenopus* oocyte geminal vesicle breakdown assay (CHUNG et al. 1992).

The crystal structures of the two nonphosphorylated p21 mutant proteins, A59T and the double mutant G12V/A59T in the GDP form, have been determined, and a model of G12V/A59T in the GTP form has been constructed. These structures and the model suggest (CHUNG et al. 1990; PRIVÉ et al. 1992) that the A59T mutant would cause the L4 region to shift and move the catalytically critical Gln-61 residue away from its normal position and prevent it from stabilizing the transition state, thus reducing the GTP hydrolysis rate. The model also suggests that the GTP form of the nonphosphorylated A59T mutant is transforming because it can still be recognized by its down-stream effector, but it is locked "on" because of its reduced GTPase activity. The phosphorylated mutant, on the other hand, would have a disrupted conformation around loops L2 and L4 due to the presence of the negatively charged phosphate at Thr-59. It is likely that this distorted protein would no longer be able to interact with the effector, and would thus be biologically inactive. The disrupting effect would be especially pronounced in the GTP complex of the phosphorylated protein, since this would bring a total of four phosphate groups into the phosphate-binding pocket.

D. Discussion

Elucidation of the three-dimensional structures, in various crystalline environments, of normal p21 protein and of several transforming mutants in the GDP-bound and the GTP-bound conformations has made it possible to analyze the detailed structural changes caused by GDP/GTP exchange and by oncogenic mutations, and provides a structural basis for understanding their switching mechanism and oncogenic effects. We found that the presence of a single phosphate (γ-PO_4) inside the protein causes dramatic conformational changes spanning the entire length of one side of the structure, that all the oncogenic mutation sites are in the guanine nucleotide binding pocket (Fig. 4a), and that, although the biological consequence of the oncogenic mutations are the same, the structural basis for the observed biochemical property of each mutation is different.

Our interpretation of the crystallographic results takes into account crystal-packing effects and crystallization condition effects by comparing crystallographically independent molecules in several different crystalline environments and solution conditions, because intermolecular contacts can sometimes influence local conformations (Faber and Matthews 1990). The value of comparing the crystal structures of the same molecule in several different crystalline environments is also evident in Fig. 3b, where the frequency of each hydrogen bond in the GTP-binding pocket is shown. One can safely assume that the hydrogen bonds observed in all molecules are important and not dependent on crystalline environment.

We suggest that the structures and interpretations presented here pertain only to purified p21 proteins, and that the in vivo behavior of p21 depends on interactions with many proteins, including nucleotide exchange factors, GAP, NF1, raf and other molecules including as yet unidentified effector molecules.

Acknowledgements. The crystal structures of *Ras* proteins described in this review have been determined in collaboration with the laboratories of Dr. Susumu Nishimura and Dr. Eiko Ohtsuka. This work has been supported in part by NIH (CA 45593), DOE (Director, Office of Energy Research, Office of Biological and Environmental Research, General Life Sciences Division under contract No. DE-AC03-76SF0098), and the Ministry of Health and Welfare for a Comprehensive Ten-Year Strategy for Cancer Control, Japan. We thank Xiangwei Weng for his help in producing some of the photographs used.

References

Barbacid M (1987) ras genes. Annu Rev Biochem 56:779–827

Bos JL (1989) ras oncogenes in human cancer: a review. Cancer Res 49:4682–4689

Bourne HR, Sanders DA, McCormick F (1991) The GTPase superfamily: conserved structure and molecular mechanism. Nature 349:117–127

Brünger AT, Milburn MV, Tong L, deVos AM, Jancarik J, Yamaizumi Z, Nishimura S, Ohtsuka E, Kim S-H (1990) Crystal structure of an active form of RAS protein, a complex of a GTP analogue and the HRAS p21 catalytic domain. Proc Natl Acad Sci USA 87:4849–4853

Buss JE, Solski PA, Schaeffer JP, McDonald MJ, Der CJ (1989) Activation of the cellular proto-oncogene product p21 Ras by addition of a myristylation signal. Science 243:1600–1603

Chung HH, Kim R, Kim S-H (1992) Biochemical and biological activity of phosphorylated and non-phosphorylated *ras* p21 mutants. Biochim Biophys Acta 1129:278–286

Der CJ (1989) ras and its relatives. Cancer Treat Res 47:73–119

Der CJ, Finkel T, Cooper GM (1986) Biological and biochemical properties of human ras-H genes mutated at codon 61. Cell 44:167–176

De Vos A, Tong L, Milburn MV, Matias PM, Jancarik J, Noguchi S, Nishimuras S, Miura K, Ohtsuka E, Kim S-H (1988) Three-dimensional structures of *ras* oncogene products, I: "catalytic domain" of normal human c-H-*ras* p21 protein. Science, 239:888

Dhar R, Ellis RW, Shih TY, Oroszlan S, Shapiro B, Maizel J, Lowy D, Scolnick E (1982) Nucleotide sequence of the p21 transforming protein of Harvey Murine Sarcoma Virus. Science 217:934–936

Faber HR, Matthews BW (1990) A mutant T4 lysozyme displays five different crystal conformations. Nature 348:263–268

Feig LA, Cooper GM (1988) Relationship among guanine nucleotide exchange, GTP hydrolysis, and transforming potential of mutated ras proteins. Mol Cell Biol 8:2472–2478

Feuerstein J, Goody RS, Webb MR (1989) The mechanism of guanosine nucleotide hydrolysis by p21 c-Ha-ras. J Mol Biol 264:6188–6190

Gibbs JB, Sigal IS, Poe M, Scolnick EM (1984) Intrinsic GTPase activity distinguishes normal and oncogenic ras p21 molecules. Proc Natl Acad Sci USA 81:5704–5708

Hancock JF, Magee AI, Childs JE, Marshall CJ (1989) All ras proteins are isoprenylated but only some are palmitoylated. Cell 57:1167–1177

Hattori S, Clanton DJ, Satoh T, Nakamura S, Kaziro Y, Kawakita M, Shih TY (1987) Neutralizing monoclonal antibody against ras oncogene product p21 which impairs guanine nucleotide exchange. Mol Cell Biol 7:1999–2002

Higashi T, Sasai H, Suzuki F, Miyoshi J, Ohuchi T, Takai S-I, Mori T, Kakunaga T (1990) Hamster cell line suitable for transfection assay of transforming genes. Proc Natl Acad Sci USA 87:2409–2413

John J, Frech M, Wittinghofer A (1988) Biochemical properties of Ha-ras encoded p21 mutants and mechanism of the autophosphorylation reaction. J Biol Chem 263:11792–11799

Jurnak F (1985) Structure of the GDP domain of EF-Tu and location of the amino acids homologous to ras oncogene proteins. Science 230:32–36

Lacal JC, Srivastava SK, Anderson PS, Aaronson SA (1986) Ras p21 proteins with high or low GTPase activity can efficiently transform NIH/3T3 cells. Cell 44:609–617

laCour TFM, Nyborg J, Thirup S, Clark BFC (1985) Structural details of the binding of guanosine diphosphate to elongation factor Tu from E. coli as studied by X-ray crystallography. EMBO J 4:2385–2388

Martin GA, Viskochil D, Bollag G, McCabe PC, Crosier WJ, Haubruck H, Conroy L, Clark R, O'Connell P, Cawthon RM, Innis MA, McCormick F (1990) The

GAP-related domain of the neurofibromatosis type 1 gene product interacts with ras p21. Cell 63:843–849
Milburn M (1991) High resolution crystal structures of the "ON" and "OFF" state of *ras* proteins. PhD thesis, University of California, Berkeley
Milburn MV, Tong L, deVos AM, Brünger A, Yamaizumi Z, Nishimura S, Kim S-H (1990) Molecular switch for signal transduction: structural differences between active and inactive forms of protooncogenic ras proteins. Science 247:939–945
Pai EF, Kabsch W, Krengel U, Holmes KC, John J, Wittinghofer A (1989) Structure of the guanosine-nucleotide-binding domain of the Ha-ras oncogene product p21 in the triphosphate conformation. Nature 341:209–214
Pai EF, Krengel U, Petsko GA, Goody RS, Kabsch W, Wittinghofer A (1990) Refined crystal structure of the triphosphate conformation of H-ras p21 at 1.35 Å resolution: implications for the mechanism of GTP hydrolysis. EMBO J 9:2351–2359
Privé GG, Milburn MV, Tong L, deVos AM, Yamaizumi Z, Nishimura S, Kim S-H (1992) X-ray crystal structures of transforming p21 ras mutants suggest a transition-state stabilization mechanism for GTP hydrolysis. Proc Natl Acad Sci USA 89:3649–3653
Schafer WR, Kim R, Sterne R, Thorner J, Kim S-H, Rine J (1989) Genetic and pharmacological suppression of oncogenic mutations in RAS genes of yeast and humans. Science 245:379–385
Seeburg PH, Colby WW, Capon DJ, Goeddel DV, Levinson AD (1984) Biological properties of human c-Ha-ras 1 genes mutated at codon 12. Nature 312:71–75
Sigal IS, Gibbs JB, D'Alonzo J, Scolnick EM (1986) Identification of effector residues and a neutralizing epitope of Ha-*ras*-encoded p21. Proc Natl Acad Sci USA 83:4725
Tong LA, deVos AM, Milburn MV, Kim S-H (1991) Crystal structures at 2.2 Å resolution of the catalytic domains of normal ras proteins and an oncogenic mutant complexed with GDP. J Mol Biol 217:503–516
Trahey M, McCormick F (1987) A cytoplasmic protein stimulates normal N-ras p21 GTPase, but does not affect oncogenic mutants. Science 238:542–545
Trahey M, Wong G, Halenbeck R, Rubinfeld B, Martin GA, Ladner M, Long CM, Crosier WJ, Watt K, Koths K, McCormick F (1988) Molecular cloning of two types of GAP complementary DNA from human placenta. Science 242:1697
Tsuchida N, Ryder T, Ohtsubo E (1982) Nucleotide sequence of the oncogene encoding the p21 transforming protein of Kirsten Murine Sarcoma Virus. Science 217:937–939
Valencia A, Chardin P, Wittinghofer A, Sander C (1991) The ras protein family: evolutionary tree and role of conserved amino acids. Biochemistry 30:4637–4648
Verrotti AC, Crechet JB, DiBlasi F, Seidita G, Mirisola MG, Kavounis C, Nastopolous V, Burder E, Vendittis E, Parmaggiani A (1992) Ras residues that are distant from GDP binding site play a critical role in dissociation factor – stimulated release of GDP. EMBO J 11:2855–2862
Vogel US, Dixon RA, Schaber MD, Diehl RE, Marshall MS, Scolnick EM, Sigal IS, Gibbs JB (1988) Cloning of bovine GAP and its interaction with oncogenic ras p21. Nature 335:90–93
Wilmot CM, Thornton JM (1988) Analysis and prediction of the different types of beta-turn in proteins. J Mol Biol 203:221–232
Xu GF, Lin B, Tanaka K, Dunn D, Wood D, Gestland R, White R, Weiss R, Tamanoi F (1990) The catalytic domain of the neurofibromatosis type 1 gene product stimulates *ras* GTPase and complements *ira* mutants of S. cervisiae. Cell 63:835

CHAPTER 14

Structural and Mechanistic Aspects of the GTPase Reaction of H-ras p21

A. WITTINGHOFER, E.F. PAI, and R.S. GOODY

A. Introduction

Guanine nucleotide binding proteins, such as the heterotrimeric G-proteins and the products of the *ras*-like genes, are involved in signal-transduction processes (BOURNE et al. 1990, 1991). The function of these proteins is to work as molecular switches in which the GTP-bound form represents the "on" state and the GDP-bound form, the "off" state of the molecule. The GTPase reaction of the molecule constitutes the actual switch mechanism, in which the protein goes from the active to the inactive conformation, the rate of the reaction determining the duration of the signal. The *ras* gene product p21 has been studied the most extensively by physicochemical methods and mutational analyses and is thus the system in which the molecular and atomic details of the conformational change and the GTPase reaction are best understood (BARBACID 1987; GRAND and OWEN 1991; WITTINGHOFER and PAI 1991). In this article, we summarize the results of crystallographic analyses of various p21–nucleotide complexes, discuss the kinetic mechanism of the GTPase reaction in detail, and relate these to the molecular mechanism of the intrinsic GTPase reaction. We further extend these considerations to the GTPase reaction in the presence of GTPase activating protein (GAP) and NF1, since it is this which represents the physiologically important switch reaction.

B. The Structure of the p21-Triphosphate State

The structure of the triphosphate, or "on" state, of p21 has been determined the a resolution of 1.35 Å using C-terminally truncated p21 and nonhydrolyzable analogs of GTP such as GppNHp and Gpp(CH_2)p (PAI et al. 1989, 1990; MILBURN et al. 1990; BRÜNGER et al. 1991; KRENGEL 1991). We have shown that removal of the 23 C-terminal amino acids does not have a significant effect on p21–nucleotide interaction or the interaction of p21–nucleotide complexes with either GAP or the nucleotide exchange factor SDC25 (JOHN et al. 1988; GIDEON et al. 1992; MISTOU et al. 1992). We have also determined the structure of the genuine p21–GTP complex, albeit at lower spatial resolution, using photoactivatable analogs of GTP and Laue diffraction methods (SCHLICHTING et al. 1989, 1990). These studies show that the GTP analogs, although their affinity relative to GTP is ten- to 100-fold lower, bind in a very similar way to the active site. The difference in affinity appears to be due to a hydrogen-bonding interaction of the oxygen bridging the β,γ-phosphate groups with the backbone NH of Gly-13. The corresponding interaction with the bridging NH group of GppNHp, although present (PAI et al. 1990), is presumably weaker than in GTP and cannot occur with $GppCH_2p$.

The three-dimensional analysis of the p21–GppNHp structure has given us the topology of the molecule, which we refer to as the G-domain fold (VALENCIA et al. 1991a). The comparison with the three-dimensional structure of elongation factor Tu (EF-Tu) has confirmed earlier models (JURNAK 1985; MCCORMICK et al. 1985) which showed that the G domain of EF-Tu and p21 have a common structural core (VALENCIA et al. 1991b). Thus, we can assume that the G domain of all GTPases with the characteristic sequence motifs GXXXXGKS/T, $TX_{ca\ 22}DXXG$, and NKXD have the same overall fold. The analysis has also shown that the conserved sequence elements, which are the footprint of GTPases, are involved in the binding of nucleotide or Mg^{2+}. Figure 1 gives a schematic view of the binding of the Mg–nucleotide complex to the active site of the molecule, highlighting the important hydrophobic contacts and hydrogen bonds.

The environment of the γ-phosphate is obviously of critical importance for the GTPase reaction and is formed by the three loops L1, L2, and L4 (see Fig. 2A). The structural analysis shows that loop L1 wraps around the β- and γ-phosphates, and this has been called the phosphate-binding loop or P-loop (SARASTRE et al. 1990). Loop L2 is part of the effector region which,

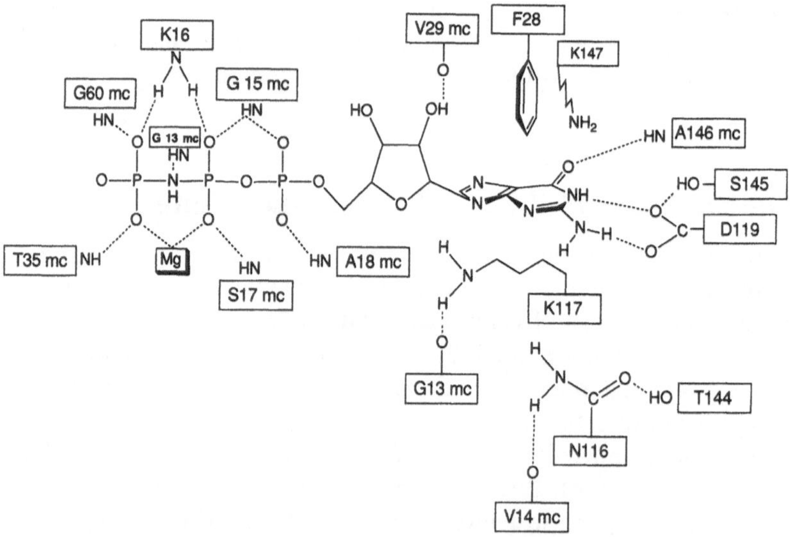

Fig. 1. The most important nucleotide–protein interactions. All hydrogen bonds (*dotted lines*) up to a length of 3.1 Å have been included

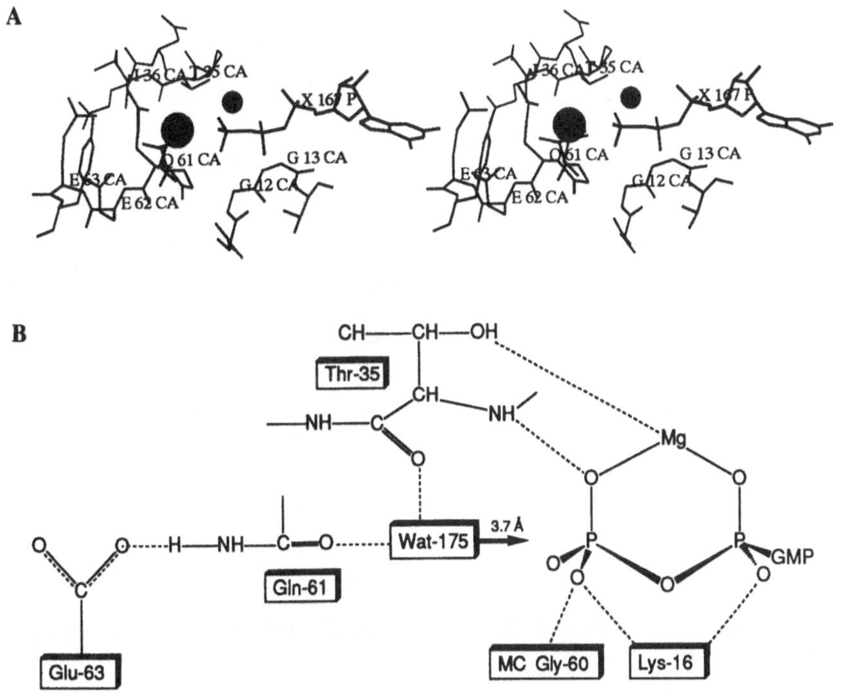

Fig. 2. **A** Stereo plot of the active site surrounding the γ-phosphate of GppNHp and Wat-175, which is in a position to attack the γ-P. Parts of loops L1 (amino acids 11–14), L4 (60–64), and of the effector region (34–38) are shown. The *small* and *big black circles* are Mg^{2+} and Wat-175, respectively. **B** The active conformation of the p21–triphosphate complex, which is postulated to be competent for GTP hydrolysis, together with the nucleophilic water molecule

based on mutational analysis, is involved in the interaction with the putative effector molecule GAP (McCormick 1989; Sigal et al. 1986; Adari et al. 1988; Calés et al. 1988). In the three-dimensional structure, L4 is located very close to L2 and the γ-phosphate. It is also critical for the interaction with GAP and for the biological activity of the protein (Srivastava et al. 1989; Gideon et al. 1992). Both L2 and L4 form hydrogen bonds to the γ-phosphate via Thr-35 and Gly-60, respectively. Both amino acids are completely conserved in guanosine nucleotide binding proteins. These interactions are lost as a result of the GTPase reaction and the ensuing release of P_i. The interacting amino acids and their respective loops have been postulated to trigger the conformational change as outlined in Fig. 3 (Wittinghofer and Pai 1991). Indeed, it has been shown by the structure determination of the GDP-bound form of p21 (Milburn et al. 1990; Tong et al. 1991) and by time-resolved X-ray crystallographic studies (Schlichting et al. 1990) that these two regions, and these regions only, change their conformation during the transition, although the details of this change are

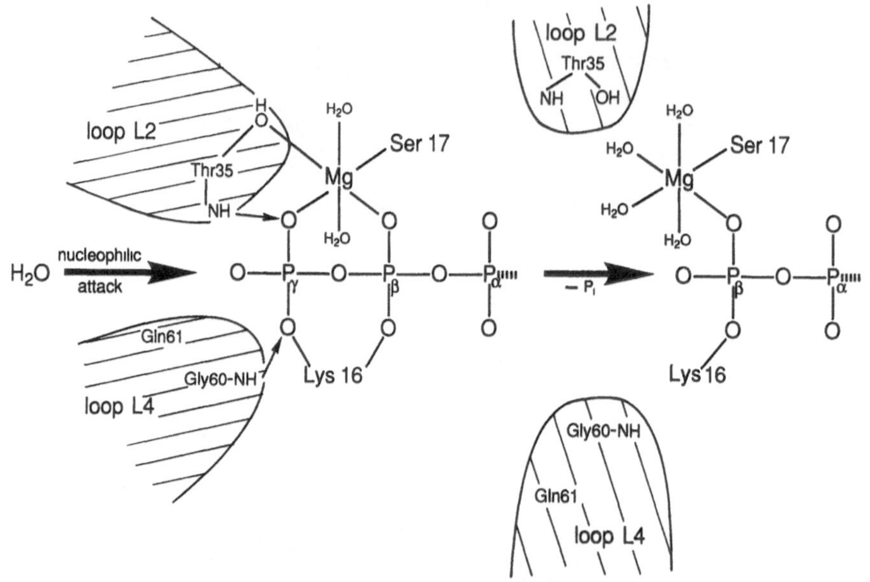

Fig. 3. Changes in interactions between nucleotide and protein occurring upon hydrolysis of GTP

somewhat controversial for the L2 region and cannot be defined for the L4/$\alpha 2$ region due to the high mobility of the latter.

C. The Structure and Biochemistry of p21 Mutants

Mutant p21s found in human tumors are believed to owe their oncogenicity to a slower GTPase reaction, so that they spend more time in the active GTP-bound state (BARBACID 1987). We have analyzed the three-dimensional structure of a number of these oncogenic mutants in their GTP-bound state to probe whether the three-dimensional structure could explain the differences in biochemical behavior. For the mutants of Gln-61, Q61H, and Q61L, we find the structures to be very similar to that of the wild-type protein (KRENGEL et al. 1990). On the basis of the proposed mechanism of GTP hydrolysis, which involves activation of the nucleophilic water by the side chain of Gln-61 (see PAI et al. 1990 and below), it would be expected that a mutation to Leu, because of the inert nature of the side chain, would lead to inhibition of the GTPase reaction. For the His mutant, it appears that a situation in which the side chain activates the attacking water molecule cannot be reached for steric reasons. We have also analyzed oncogenic mutants of p21 in which Gly-12 has been substituted by Arg, Val (KRENGEL et al. 1990), and Asp (FRANKEN et al. 1993). In each case we find structural differences confined to the active site. It appears that for each mutant, a different distortion of the conformation of the active site is found, which is presumably responsible for the decrease in its GTPase activity and oncogenicity.

As a positive control, we have determined the structure of the p21(G12P)–GppNHp complex (FRANKEN et al. 1993). It was shown by SEEBURG et al. (1984) that every mutation of Gly-12, except to proline, activates the transforming potential of p21. In the structure of the mutant protein we find, in contrast to other p21 mutants analyzed, that the nucleophilic water molecule is not only in the same position as in wild-type p21, but it is also apparently in contact with the side chain of Gln-61. This may provide an explanation for the biochemical finding that p21 (G12P) does not have a reduced, but on the contrary a somewhat higher GTPase rate than wild-type p21 (GIBBS et al. 1988).

D. The Kinetic Mechanism of the GTPase Reaction

Important clues to the manner in which GAPs activate the rate of GTP hydrolysis by Ras-like proteins can be obtained from studies on the kinetic mechanism of p21 both in the absence and presence of GAP (TRAHEY and MCCORMICK 1987). We review progress in this area at some length here.

In its isolated form p21 can hardly be considered an enzyme, since two extremely slow steps in the mechanism lead to very slow turnover of GTP. These steps are the GTP-cleavage reaction on the protein and the dissociation of one of the products of the reaction, GDP, from its binary complex with p21. The other hydroylsis product, inorganic phosphate, is released rapidly in comparison with these two processes. For wild-type p21, the rate constants for GTP cleavage (k_{cl}) and GDP release (k_{GDP}) are 4.6×10^{-4} and $1.3 \times 10^{-4} s^{-1}$ (10 mM Mg^{2+}, 37°C; JOHN et al. 1988), respectively. In transforming mutants of p21, one or both of these rate constants are modified. In general, the rate constant for GTP cleavage is reduced or the rate constant for GDP release (and thus the rate of GTP/GDP exchange) is either increased or decreased.

$$\text{p21} + \text{GTP} \underset{K_d}{====} \text{p21}\cdot\text{GTP} \overset{k_{cl}}{====} \text{p21}\cdot\text{GDP}\cdot\text{P}_i \quad (1)$$

$$\overset{k_{GDP}}{====} \text{p21}\cdot\text{GDP} ==== \text{p21} + \text{GDP}$$

Scheme 1 shows the minimal mechanism for the GTP hydrolysis reaction. These features of the kinetic mechanism were established using standard radioactive techniques and filter binding assays. A more detailed investigation has been made possible by fluorescence techniques. It was shown several years ago that GTP carrying a fluorescent label on the ribose moiety

(2′- and 3′- methylanthraniloyl esters) had good substrate properties with p21 (JOHN et al. 1989a). These analogs were used initially to characterize the kinetics of binding of GTP and GDP, and these experiments led to the conclusion that the binding reaction is at least a two-step process for both nucleotides, with a weak but rapid initial binding being followed by a quasi-irreversible isomerization step (JOHN et al. 1990). This leads to the following scheme for nucleotide binding by p21:

$$\text{p21} + \text{GXP} \overset{K_1}{====} \text{p21}\cdot\text{GXP} \overset{K_2}{====} \text{p21}^*.\text{GXP} \qquad (2)$$

This is a general scheme for the binding of nucleotides to p21. In the case of GDP, K_1 has a value of $5.7 \times 10^4 M^{-1}$, k_{+2} a value of $14.8\,s^{-1}$ and k_{-2} a value of $1.8 \times 10^{-5} s^{-1}$ at 25°C, with $K_2 = k_{+2}/k_{-2}$ (JOHN et al. 1990). Similar values are found for GTP. The nature and significance of the individual steps in the binding mechanism is not yet apparent. Interestingly, guanosine monophosphate (GMP) and guanosine both appear to bind in a one-step reaction, which seems to be analogous to the first step in GDP or GTP binding (JOHN et al. 1990). Thus, determination of the structure of the p21–GMP or p21–guanosine complexes should help to define the events occurring in the two steps. Such studies are in progress. Independent of this, the characteristics of the binding process suggest a model for the mechanism of action of exchange nucleotide factors, which could function by influencing the equilibrium between the first and second bound states of the p21–nucleotide complex.

Much of the kinetic work done on the interaction of p21 with nucleotides used 3′-methylanthraniloyl-2′-deoxy GDP or GTP. The fluorescent enhancement seen when p21 interacts with the analogs is identical, regardless of whether the parent nucleotide is dGTP or dGDP. In contrast, when the parent nucleotide is a ribonucleotide (GTP or GDP; the fluorescent nucleotide used is then a mixture of the 2′- and 3-isomers), it is seen that there is a slightly smaller enhancement with the GDP than with the GTP derivative. In keeping with this, after the initial large increase in fluorescence on interaction of p21 with mantGTP, there is a slow decrease (ca. 10%) occurring at the same rate as the GTPase reaction (NEAL et al. 1990). Since a similar fluorescence change was also seen using the nonhydrolyzable analog mantGppNHp, this change was interpreted to be due to a rate-limiting isomerization reaction which occurs before the actual cleavage step. This implies that the rate of the hydrolysis reaction is much faster than the effective rate of production of GDP. This model is attractive, since it suggests a simple mechanism for GAP activation of the GTPase. According to this hypothesis, GAP would accelerate the rate-limiting isomerization, after which GTP cleavage would occur by the same mechanism and with the same rate constant as in p21 alone. Moreover, this would suggest that in transforming mutants of p21 for which the rate of GTP hydrolysis cannot be accelerated by GAP, it might be the rate of the conformational change which is inhibited, and not neccessarily the rate of the GTP-cleavage step. It would also imply that the chemical mechanism of cleavage is the same for the catalyzed and the uncatalyzed reaction.

In principle, this model can be easily tested. If it is correct, GAP should accelerate the slow fluorescence change seen with both p21–mantGTP and p21–mantGppNHp. While it is clear that the required acceleration of the fluorescence transient occurs with mantGTP (Eccleston et al. 1991; Rensland et al. 1991), it is technically more difficult to perform the experiment with mantGppNHp. The main problem in this work is the small magnitude of the transient seen with mantGppNHp (less than 50% of that seen with mantGTP, meaning that the amplitude of the transient is less than 5%) and in our experiments the fact that the transient is already considerably faster than for mantGTP even in the absence of GAP. However, this rate is not accelerated by GAP, even under conditions where there is substantial acceleration of the fluorescence transient with p21–mantGTP. We have obtained additional evidence using mantGTP(γ-S) (Rensland et al. 1991). This analog is cleaved much more slowly than GTP (by at least a factor of 10), and the fluorescence transient is also slowed correspondingly. If the rate-limiting step were a precleavage conformational change, it would be expected to occur at about the same rate with GTP or GTP(γ-S), since experience with other systems has shown that this modification of GTP or ATP leads to a marked effect on the rate of the chemical cleavage step, but only to minor or negligible effects on steps involving interaction with the protein, including substrate-induced conformational changes (e.g., Bagshaw et al. 1974). Thus, the relatively large effect of the modification on the measured rate of cleavage and in parallel on the slow fluorescence change would be in keeping with the idea that the cleavage step, not a precleavage conformational change, is rate limiting. However, the effect of the thiophosphate modification on the rate of the cleavage reaction is smaller than has been seen with other enzymes (a factor of 400 for myosin ATPase, for example; Bagshaw et al. 1972), so that there is still some degree of uncertainty with this interpretation.

E. The Kinetic Mechanism of the GAP-Stimulated GTPase

The influence of GAP on the minimal mechanism shown in Scheme 1 can be described formally as an acceleration of k_{cl}, the rate constant for GTP cleavage on p21, without any of the other constants being affected. However, although the rate of GTP cleavage is accelerated by a factor of approximately 10^5 at saturation (i.e., the value of k_{cl} in the GAP–p21–GTP complex is 10^5 higher than in the p21–GTP complex; Gideon et al. 1992), even the p21/GAP system does not constitute what is normally regarded as an enzyme, since the rate of GDP dissociation is still very slow, so that a high rate of GTP turnover is not seen under steady-state conditions. The activation of the p21–GTPase reaction by GAPs can be described by the following simple scheme:

$$\text{p21}\cdot\text{GTP} + \text{GAP} \underset{K_m}{====} \text{GAP}\cdot\text{p21}\cdot\text{GTP} \overset{k_{cl}}{----} \text{GAP} + \text{p21}\cdot\text{GDP} + \text{P}_i \qquad (3)$$

The constant K_m defines the effective dissociation constant between p21–GTP and GAP under "steady-state" conditions (i.e., in the approximately linear phase in the kinetics of conversion of p21–GTP into p21–GDP in the presence of catalytic quantities of GAP). K_m could be equal to the dissociation constant for the p21–GTP–GAP interaction, k_{diss}/k_{ass}, or, in the general case, to $(k_{diss} + k_{cat})/k_{ass}$, where k_{ass} and k_{diss} are the rate constants for association and dissociation, respectively, of p21–GTP and GAP and k_{cat} refers to the rate-limiting step in the formation of p21–GDP, whether this is the cleavage reaction or not. For wild-type p21 and full-length GAP, values of $8.4\,\mu M$ for K_m and $19\,s^{-1}$ for k_{cat} (both at 25°C) have been measured, whereas the catalytic domain of GAP alone ("GAP_{334}") has given values of $19\,\mu M$ and $4.2\,s^{-1}$, respectively (GIDEON et al. 1992), at variance with earlier reports that the C-terminal domain of GAP is sufficient for full activity (MARSHALL et al. 1989). Although it has not been possible to measure the true dissociation constant for the GAP–p21–GTP interaction, the affinity of p21-GppNHp of $4.8\,\mu M$ suggests that it is close to the value measured for K_m. Recently the product of the neurofibromatosis type I gene, neurofibromin, has been identified as a second form of the p21ras specific GTPase stimulating protein with a 25% identity in the primary sequence along the catalytic domain (MARTIN et al. 1991; XU et al. 1991; BALLESTER et al. 1991). The interaction of p21ras–GTP with the catalytic fragment of neurofibromin has a k_{cat} of $4\,s^{-1}$ and K_m of $0.3\,\mu M$, so that the k_{cat}/K_m value is similar to that for full-length GAP. In contrast to GAP, the catalytic domain of neurofibromin alone is sufficient for full GTPase stimulating activity (WIESMÜLLER and WITTINGHOFER 1992; MCCORMICK unpublished).

The influence of GAP_{334} on the rate of the fluorescence transient with p21–mantGTP(γ-S) is easier to study than with mantGppNHp, due to the larger amplitude of the signal. It is found to be accelerated in a similar manner to the actual hydrolysis of mantGTP(γ-S), and the limiting rate (i.e., at saturating GAP concentration) is $0.006\,s^{-1}$ with a K_m of $3.5\,\mu M$. Comparing this with the rate constant for GTP (extrapolated from steady-state measurements), this is seen to be approximately 700 times slower. Thus, for the GAP-activated p21–GTPase, there is good agreement with the well-documented example of ATP and ATP(γ-S) cleavage by myosin ATPase (BAGSHAW et al. 1974), in contrast to the situation with p21 alone.

The structure of p21 with a mant nucleotide (mantdGppNHp) at the active site has recently been determined (GOODY et al. 1992). This shows that the nucleotides, in particular the base and the phosphate groups, are bound in a manner which is identical with that of GppNHp or GTP. The only obvious interaction with the protein is between the aromatic ring of the fluorescent group and Tyr-32, a residue in the effector-binding loop. The position of this residue changes dramatically when GTP is hydrolyzed to GDP (SCHLICHTING et al. 1990), and it therefore seems likely that the fluorescence change seen on the hydrolysis of GTP is associated with movement of this side chain. This observation prompted us to replace Tyr-32 by a tryptophan residue by site-directed mutagenesis, leading to a mutant protein which contains just one tryptophan, and this is in a position which participates in the hydrolysis-induced conformational change. Not unexpectedly, there is a large change in tryptophan fluorescence on GTP hydrolysis by this mutant, and since it has essentially wild-type properties with respect to activation by GAP, it has been used for a more detailed investigation of the kinetics of p21–GAP interaction (H. RENSLAND et al., unpublished). Using the stopped flow technique, it was possible to directly measure the rate of the fluorescence transient at saturating GAP_{334} concentration, which was found to be $2.1\,s^{-1}$ (cf. $4.5\,s^{-1}$ for GAP_{334} from steady-state measurements with wild-type p21; GIDEON et al. 1992). The dissociation constant of the GAP–p21(Y32W)–GTP complex was found to be $8\,\mu M$. It remains to be shown whether cleavage of GTP occurs at exactly the same rate as the fluorescence change, which would be the case if the fluorescence change were a direct signal of cleavage, or even if a precleavage isomerization were still exclusively rate limiting in the presence of GAP, but not if the conformational change model is correct and there is a change in the rate-limiting step (from the precleavage to the cleavage step) as the GAP concentration is increased. The results obtained so far can be explained by, but do not prove, the simple model in which GAP has a direct effect on the rate of the GTP-cleavage reaction.

F. GTPase Mechanism

The p21–GTP complex can be generated in situ in crystals of p21–cagedGTP after photolytic removal of the protecting group (SCHLICHTING et al. 1990). In the crystal, p21–GTP is hydrolyzed to p21–GDP with the same half-life as in solution, suggesting that the conformation of p21–GTP found in the crystal should reflect a conformation that is competent for cleavage. Since the structures of p21–GTP and p21–GppNHp are superimposable, it appears that the latter structure is also a reflection of an active conformation. It was also found that crystalline oncogenic p21(G12V)–GTP, generated from p21(G12V)–cagedGTP, has the same decreased GTPase activity as in solution (SCHLICHTING et al. 1989). These results encourage us to use the high resolution p21–GppNHp model and the information obtained from the time-resolved X-ray studies together with other biochemical and kinetic evidence to draw conclusions on the mechanism of the GTPase reaction.

FEUERSTEIN et al. (1989) showed that GTP hydrolysis proceeds by an associative mechanism involving in-line attack of the nucleophilic water on the γ-P. Such a water molecule situated on the opposite side of the scissile P–O–P bond was found in the structural analysis (PAI et al. 1990). It is 3.7 Å away from the γ-phosphorus and appears to be tightly bound, since it has a low temperature factor. This water molecule is held in place by hydrogen bonds to the main-chain carbonyl of residue Thr-35 and the side chain carbonyl of Gln-61 (Fig. 2A). Gln-61 is part of loop L4. With the exception of one of the four molecules in the assymetric unit of crystals of p21–GppCH$_2$P (BRÜNGER et al. 1990), residues 61–65 of L4 are highly flexible in all crystallized states of p21 examined, as evidenced from their high temperature factors, which are four times higher than the average B factors found for the rest of the molecule. Due to this flexibility, residues 61–65 are not well-defined in the three-dimensional structure and their side chains can adopt different positions. In one of at least two possible conformations, Gln-61 is located so as to interact with the nucleophilic water. The two hydrogen bonds can thus increase the nucleophilic potential of water and increase the probability of nucleophilic attack on the γ-P. This has led us to propose a mechanism for the phosphoryl transfer from GTP to water, which is shown schematically in Fig. 2B.

G. Arguments For and Against the Proposed Mechanism

The model proposed for the chemical mechanism of GTP hydrolysis has aroused considerable interest, but also criticism. The main argument against it has been that the glutamine side chain is a weak general base and that Gln-61 cannot be replaced by Glu without loss of activity, since it was reported earlier that p21(Q61E) is a less efficient GTPase (PRIVEE et al. 1992; DER et al. 1986a). On the basis of chemical considerations, it would be expected that a carboxylate side chain is a better activator for nucleophilic water than the carbamoyl side chain. We have reinvestigated the properties of p21(Q61E) and found it to be capable of hydrolyzing GTP at a rate that is at least as high as wild-type p21(FRECH et al., unpublished). The disagreement with the earlier results may have arisen from the fact that this mutant is thermally more unstable than wild-type protein, possibly due to the presence of the three consecutive glutamic acid residues E61, E62, and E63. The new results now lend further credibility to the chemical mechanism proposed for the GTPase reaction. It also argues against the proposed involvement (PRIVEE et al. 1992) of the side chain of Gln-61 in stabilizing the transition state of the GTPase reaction by interacting with an oxygen of the γ-phosphate, since substitution of the carboxy for the carbamoyl group would, because of the negative charge, certainly lead to a dramatic loss of transition-state stabilization and thus of GTPase activity.

We have also reinvestigated the properties of p21(Q61P), which had been reported to be nontransforming, thus suggesting that its GTPase might be normal (DER et al. 1986). We find that this mutant has a reduced GTPase and in fact appears to be strongly transforming, as analyzed by its ability to induce differentiation of PC12 cells (P. GIDEON, unpublished).

A disturbing finding is that, of all amino acid side chains involved in GTP binding and phosphoryl transfer, Gln-61 is not completely conserved in GTPases, the most notable exceptions being EF-Tu and its eukaryotic counterpart EF-1, which have His, and the rap-like small GTPases which have Thr/Ile in the corresponding positions (BOURNE et al. 1991; VALENCIA et al. 1991a). As far as EF-Tu/EF-1 are concerned, it seems likely that here the conserved His is the general base for proton abstraction. One argument in favor of this interpretation is the finding that the intrinsic GTPase of EF-Tu is the same as that of p21(Q61H). In addition, the latter mutant can even be partially stimulated by GAP, whereas other mutants, such as Q61L, cannot (BOLLAG and MCCORMICK 1991; GIDEON et al. 1992).

The case of the rap-like proteins (rap-1, rap-2, RSR-1) is more difficult to understand. However, it has been shown that mutating Thr-61 of p21rap1A to Gln increases the GTPase reaction of this mutant and makes it partially responsive to stimulation by GAP, whereas wild-type p21rap1A is not stimulated by GAP, but does bind tightly to it (FRECH et al. 1990). It has also been shown that the GTPase reaction stimulated by rap–GAP has sequence requirements different from those of ras–GAP: the GTPase reaction of p21rap1 is not affected by mutations in position 61, whereas that of p21ras is (MARUTA et al. 1991; HOLDEN et al. 1991; ZHANG et al. 1991). It appears that in the case of the rap proteins, a different chemical mechanism might be used for the GTPase reaction. Sequence analysis indicates that ras–GAP is unrelated to rap–GAP (RUBINFELD et al. 1991).

Apart from a few exceptions, Gln-61 is found in all GTPases, and it has been shown for some of these that its substitution by other amino acids renders the protein unable to hydrolyze GTP. Interestingly, G_α subunits of heterotrimeric G-proteins carrying the corresponding Q227R mutation have been found as oncogenes in pituitary tumors. In analogy to certain *ras* oncogenes, the decreased GTPase of this mutant was found to be responsible for its oncogenic potential (LANDIS et al. 1989).

In view of its complete conservation in small GTPases, it has been argued that Glu-62 might be a more likely candidate for a general base in the mechanism of hydrolysis (PRIVÉE et al. 1992). Speaking against this is the fact that Glu or even a carboxylate side chain is not conserved at this position in other GTPases. Furthermore, Glu-62 has been mutated to His in p21ras without causing drastic effects on the stimulated or nonstimulated GTPase reaction (GIDEON et al. 1992).

As already mentioned, it has also been speculated that the role of Gln-61 is not to activate the water molecule for nucleophilic attack, but rather to stabilize the transition state. These two roles are, in principle, not mutually exclusive. It is to be expected that a group which activates the attacking water molecule by interacting with one of its protons will stabilize the transition state (which is presumably structurally related to a pentacovalent phosphate group) by a hydrogen-bonding interaction between the hydrogen which originated from the water molecule and which is now on the "base" supplied by the enzyme and the oxygen of the hydroxyl group attached to the pentavalent phosphorus, which also originated from the attacking water. In the structure for the transition state proposed by PRIVÉE et al. (1992), there is a hydrogen-bonding interaction of the carbonyl group of the side chain of Gln-61 and the proton of the hydroxyl group of the pentavalent phosphate. No suggestion is made about the mechanism of proton abstraction, which must either be catalyzed by another group or must be spontaneous. Another criticism of the model proposed by PRIVÉE et al. (1992) is that it incorprates a hydrogen-bonding interaction between the backbone NH of Gly-60 and the pentavalent phosphorous hydroxyl group which originated from the attacking water. This is difficult to understand in mechanistic terms since the NH of Gly-60 already interacts with one of the oxygens of the γ-phosphate group in the ground state, i.e., in the p21–GTP state, so that it would have to switch from this (negatively charged) oxygen to the newly created (uncharged) hydroxyl group *after* attack of water.

It is conceivable that the GTPase reaction in the absence of GAP, since it is only approximately 1000 times faster than the spontaneous rate of GTP hydrolysis in water, does not require the assistance of a general base for nucleophilic attack. This would imply that other factors, such as the proper positioning of the catalytic water and increasing the susceptibility of the γ-phosphate to nucleophilic attack by the large number of electron-withdrawing interactions between the β- and γ-phosphates and the protein and Mg^{2+}, would be sufficient to speed up the reaction. For the GAP-stimulated reaction, however, which is at least 10^8 times faster than in water, it seems unlikely that such mechanisms are sufficient to explain the reaction-rate enhancement.

H. Role of GAP in the Chemical Mechanism

At present we can only speculate about the chemical mechanism by which the GAP molecules increase the GTP hydrolysis rate so dramatically. Two possible extreme classes of mechanism can be envisioned. In one, all the machinery for efficient catalysis is present in p21 itself. In such a scenario the catalytically active conformation of p21 would be energetically unfavorable or only reached at a very slow rate, and the role of GAP would be to stabilize or accelerate the rate of populating this conformational state. It would then be conceivable that the conformation of the active site identified in the three-dimensional structural analysis, which is shown schematically in Fig. 2 and which we refer to as the catalytic conformation, is such a conformation. It has been found that loop L4 in the crystal structure is highly flexible, and it is most likely that this part of the chain is also flexible in solution and could potentially be stabilized by GAP. Additionally, the role of GAP could be to prevent the catalytic water molecule, which in solution might be very mobile, from escaping before it has time for the nucleophilic reaction. This suggests a situation where the role of the enzyme (GAP) is to hold water in the active site, whereas in most kinases one of the roles of the enzyme is thought to be to exclude water from the active site such that the phosphoryl transfer occurs between the desired partners (JENCKS 1975).

It is not obvious that such an effect would lead to a large rate enhancement, since the very fact that electron density is found for the water molecule postulated to be that which attacks the γ-phosphate in the p21–GppNHp map suggests that the site is reasonably highly occupied even in the absence of GAP. Computer modeling of the active site of p21 has suggested that the lifetime of this water molecule at the active site is approximately 200 ps (FOLEY et al., 1992), consistent with a dissociation rate constant of approximately $10^9 s^{-1}$. If the association rate constant is assumed to be diffusion controlled (i.e., of the order of $10^9 M^{-1} s^{-1}$), this leads to a dissociation constant of more than $1 M$. This again leads to the conclusion that at prevailing water concentrations, the site would be well occupied. On the basis of these arguments, it seems unlikely that increasing the occupancy of the attacking water at the active site would be a major factor in the GAP-activation mechanism. However, an increase in the "rigidity" of binding of this loosely bound water could be of importance.

The other possible class of mechanism is one in which GAP supplies active site residues, suggesting that the chemical mechanism is different for the uncatalyzed and GAP-catalyzed reaction. Arguments against this hypothesis come from the observation that in cases where the intrinsic GTPase is impaired we also find the GAP-catalyzed GTPase reduced even further (MCCORMICK 1989; BOLLAG and MCCORMICK 1991; GIDEON et al. 1992). Other p21 mutants have been described where the intrinsic GTPase is as high or higher than wild-type p21, but is not increased by GAP as for p21(G12P) (GIBBS et al., 1988) and p21(Q61E) (FRECH et al., unpublished), but in these cases GAP may not be able to interact properly with p21 due to the mutations.

If GAP or neurofibromin did indeed supply catalytic residues, it would be expected that positively charged residues would participate in such a reaction. This is suggested by the structure of the adenylate-kinase–AP5A complex, which is thought to mimic the transition state of the reaction of a highly efficient phosphoryl transfer enzyme with a k_{cat} of 600–1000 s^{-1} (MÜLLER and SCHULZ 1992). In this structure, there are numerous lysine and arginine residues apparently stabilizing the transition state. In GAP and neurofibromin, one arginine and one lysine residue are completely conserved. Mutating the conserved arginine residue eliminates catalytic activity (SKINNER et al. 1991; WIESMÜLLER, unpublished), whereas the conserved lysine does not appear to be essential (WIESMÜLLER and WITTINGHOFER

1992). From the three-dimensional structure of p21 it appears that there is enough space for even a large molecule to come close to the active site. The crystal structure analysis has shown that a residue as bulky as tyrosine (from the neighboring molecule in the crystal) can come within hydrogen-bonding distance of the active site (PAI et al. 1990). Thus, based on structural considerations it is feasible that even a bulky residue from GAP, such as an arginine, could interact with the γ-phosphate of GTP while it is bound to p21.

A type of mechanism which does not seem very likely, but which does not appear to be excluded by available evidence, would involve phosphorylation of a residue on GAP by p21–GTP, followed by rapid (spontaneous?) dephosphorylation of GAP. The only potential site of phosphorylation in ras–GAP sequences appears to be a conserved glutamic acid residue. Evidence for or against this type of mechanism should be obtainable from experiments on the stereochemistry of the transfer of phosphate to water in the presence of p21 and GAP, of the type reported for p21 alone by FEUERSTEIN et al. (1989).

A distinction between the two general classes of mechanism would be helped by resolution of the conflicting interpretations of the kinetic mechanism of the GAP-activated p21–GTPase reaction.

I. Conclusion

Intense study of the structure and biochemistry of p21 has led to an understanding of the mechanism of GTP hydrolysis in general terms, but not yet in detail. More important, and possibly different, is the mechanism of GTP hydrolysis in the presence of GTPase activating molecules. At the present time, lack of structural information on such molecules and the incomplete understanding of the kinetic mechanism of the p21–GAP interaction precludes the formulation of detailed mechanistic hypotheses. It is of crucial importance to obtain the missing information in order to understand the mechanism of oncogenic transformation in this system at the level of atomic resolution.

References

Adari H, Lowy DR, Willumsen BM, Der CJ, McCormick F (1988) Guanosine triphosphatase activating protein (GAP) interacts with the p21 ras effector binding domain. Science 240:518–521

Bagshaw CR, Eccleston JF, Trentham DR, Yates D, Goody RS (1972) Transient kinetic studies of the Mg-dependent ATPase of myosin and its proteolytic subfragments. Cold Spring Harb Symp Quant Biol 37:127–1358

Bagshaw CR, Eccleston JF, Eckstein F, Goody RS, Gutfreund H, Trentham DR (1974) The Mg^{2+}-adenosine triphosphatase of myosin: two step processes of ATP association and ADP dissociation. Biochem J 141:351–364

Ballester R, Marchuk D, Boguski M, Saulino A, Letcher R, Wigler M, Collins F (1991) The NF1 locus encodes a protein functionally related to mammalian GAP and yeast IRA proteins. Cell 63:851–859

Barbacid M (1987) ras genes. Annu Rev Biochem 56:779–827
Bollag G, McCormick F (1991) Differential regulation of rasGAP and neurofibromatosis gene product activities. Nature 351:576–579
Bourne HR, Sanders DA, McCormick F (1990) The GTPase superfamily: a conserved switch for diverse cell functions. Nature 348:125–132
Bourne HR, Sanders DA, McCormick F (1991) The GTPase superfamily: conserved structure and molecular mechanism. Nature 349:117–127
Brünger AT, Milburn MV, Tong L, deVos AM, Jancarik J, Yamaizumi Z, Nishimura S, Ohtsuko E, Kim S-H (1990) Crystal structure of an active form of Ras protein, a complex of a GTP analog and the HRAS p21 catalytic domain. Proc Natl Acad Sci USA 87:4849–4853
Calés C, Hancock JF, Marshall CJ, Hall A (1988) The cytoplasmic protein GAP is implicated as the target for regulation by the ras gene product. Nature 332:548–551
Der CJ, Finkel T, Cooper GM (1986a) Biological and biochemical properties of human ras^H genes mutated at codon 61. Cell 44:167–176
Der CJ, Pan B-T, Cooper GM (1986b) ras^H mutants deficient in GTP binding. Mol Cell Biol 6:3291–3294
Eccleston JF, Moore KJM, Brownbridge GG, Webb MR, Lowe PN (1991) Fluorescence approaches to the study of the $p21^{ras}$ GTPase mechanism. Biochem Soc Trans 19:432–437
Feuerstein J, Goody RS, Webb MR (1989) The mechanism of guanosine nucleotide hydrolysis by p21 c-Ha-ras. J Biol Chem 264:6188–6190
Foley CK, Pedersen LG, Charifson PS, Darden TA, Wittinghofer A, Pai EE, Anderson MW (1992) Simulation of the solution structure of the H-ras p21-GTP complex. Biochemistry 31:4951–4959
Franken SM, Scheidig AJ, Krengel U, Rensland H, Lautwein A, Geyer M, Scheffzek K, Goody RS, Kalbitzer HR, Pai EF, Wittinghofer A (1993) Three-dimensional structures and properties of a transforming and a nontransforming Gly-12 mutant of $p21^{H\text{-}ras}$. Biochemistry (in press)
Frech M, John M, Pizon V, Chardin P, Tavitian A, Clark R, McCormick F, Wittinghofer A (1990) Inhibition of GTPase activating protein stimulation of ras-p21 GTPase by the Krev-1 gene product. Science 249:169–171
Gibbs JB, Schaber MD, Allard WJ, Sigal IS, Scolnick EM (1988) Purification of ras GTPase activating protein from bovine brain. Proc Natl Acad Sci USA 85:5026–5030
Gideon P, John J, Frech M, Lautwein A, Clark R, Scheffler JE, Wittinghofer A (1992) Mutational and kinetic analysis of the GTPase-activating protein (GAP)-p21 interaction: the C-terminal domain of GAP is not sufficient for full activity. Mol Cell Biol 12:2050–2056
Goody RS, Pai EF, Schlichting I, Rensland H, Scheidig A, Franken S, Wittinghofer A (1992) Studies on the structure and mechanism of H-ras p21. Philos Trans R Soc Lond [Biol] 336:3–11
Grand RJA, Owen D (1991) The biochemistry of ras p21. Biochem J 279:609–631
Holden JL, Nur-E-Kamal MSA, Fabri L, Nice E, Hammacher A, Maruta H (1991) Rsr1 and Rap1 GTPases are activated by the same GTPase activating protein and require threonine 65 for their activation. J Biol Chem 266:16992–16995
Jencks WP (1975) Binding enregy, specificity, and enzymic catalysis: the Circe effect. Adv Enzymol Relat Areas Mol Biol 43:219–403
John J, Frech M, Wittinghofer A (1988) Biochemical properties of Ha-ras encoded p21 mutants and mechanism of the autophosphorylation reaction. J Biol Chem 263:11792–11799
John J, Sohmen R, Feuerstein J, Linke R, Wittinghofer A, Goody RS (1990) Kinetics of interaction of nucleotides with nucleotide-free H-ras p21. Biochemistry 29:6059–6065
Jurnak F (1985) Structure of the GDP domain of EF-Tu and location of amino acids homologous to ras oncogene proteins. Science 230:32–36

Krengel U (1991) Struktur und Guanosintriphosphat-Hydrolysemechanimus des C-terminal verkürzten menschlichen Krebsproteins p21$^{H\text{-}ras}$. PhD thesis, University of Heidelberg
Krengel U, Schlichting I, Scherer A, Schumann R, Frech M, John J, Kabsch W, Pai EF, Wittinghofer A (1990) Three-dimensional structures of H-ras p21 mutants: molecular basis for their inability to function as signal switch molecules. Cell 62:539–548
Landis CA, Masters SB, Spada A, Pace AM, Bourne HR, Vallar L (1989) GTPase inhibiting mutations activate the chain of G_S and stimulate adenylyl cyclase in human pituitary tumours. Nature 340:692–696
Marshall MS, Hill WS, Ng. AS, Vogel US, Schaber MD, Scolnick EM, Dixon RAF, Sigal I, Gibbs JB (1989) A C-terminal domain of GAP is sufficient to stimulate ras p21 GTPase activity. EMBO J 8:1105–1110
Martin GA, Viskochil D, Bollg G, McCabe PC, Crosier WJ, Haubruck H, Conroy L, Clark R, O'Connel P, Cawthon RM, Innis MA, McCormick F (1991) The GAP-related domain of the neurofibromatosis type 1 gene product interacts with ras p21. Cell 63:843–849
Maruta H, Holden J, Sizeland A, D'Abaco G (1991) The residues of ras and rap proteins that determine their GAP specificities. J Biol Chem 266:11661–11668
McCormick F (1989) ras GTPase activating protein: signal transmitter and signal terminator. Cell 56:5–8
McCormick F, Clark BFC, la Cour TFM, Kjelgaard M, Norskov-Lauritsen L, Nyborg J (1985) A model for the tertiary structure of p21, the product of the ras oncogene. Science 230:78–82
Milburn MV, Tong L, DeVos AM, Brünger A, Yamaizumi Z, Nishimura S, Kim SH (1990) Molecular switch for signal transduction: Structural differences between active and inactive forms of protooncogenic ras proteins. Science 247:939–945
Mistou MY, Jacquet E, Poullet P, Rensland H, Gideon P, Schlichting I, Wittinghofer A, Parmeggiani A (1992) Mutations of H-ras p21 that define important regions for the molecular mechanism of the SDC25 C-domain, a guanine nucleotide dissociation stimulator. The EMBO J 11:2391–2397
Müller CW, Schulz GE (1992) Structure of the complex between adenylate kinase from Escherichia coli and the inhibitor Ap_5A refined at 1.9 Å resolution. J Mol Biol 224:159–177
Neal SE, Eccleston JF, Webb MR (1990) Hydrolysis of GTP by p21^{NRAS}, the NRAS protooncogene product, is accompanied by a conformational change in the wild type protein: use of a single fluorescent probe at the catalytic site. Proc Natl Acad Sci USA 87:3652–3565
Pai EF, Kabsch W, Krengel U, Holmes KC, John J, Wittinghofer A (1989) Structure of the guanine-nucleotide-binding domain of the Ha-ras oncogene product p21 in the triphosphate conformation. Nature 341:209–214
Pai EF, Krengel U, Petsko GA, Goody RS, Kabsch W, Wittinghofer A (1990) Refined crystal structure of the triphosphate conformation of H-ras p21 at 1.35 Å resolution: implications for the mechanism of GTP hydrolysis. EMBO J 9:2351–2359
Privée GG, Milburn MV, Tong L, DeVos AM, Yamaizumi Z, Nishimura S, Kim S-H (1992) X-ray crystal structures of transforming p21 ras mutants suggest a transition-state stabilization mechanism for GTP hydrolysis. Proc Natl Acad Sci USA 80:3649–3653
Rensland H, Lautwein A, Witinghofer A, Goody RS (1991) Is there a rate limiting step before GTP cleavage by H-ras p21? Biochemistry 30:11181–11185
Rubinfeld B, Munemitsu S, Clark R, Conroy L, Watt K, Crosier WJ, McCormick F, Polakis P (1991) Molecular cloning of a GTPase activating protein specific for the Krev-1 protein p21^{rap1}. Cell 65:1033–1042
Sarastre M, Sibbald PR, Wittinghofer A (1990) The P-loop, a common motif in ATP- and GTP-binding proteins. TIBS 15:430–434

Schlichting I, Rapp G, John J, Wittinghofer A, Pai EF, Goody RS (1989) Biochemical and crystallographic characterization of a complex of c-Ha-ras p21 and caged GTP with flash photolysis. Proc Natl Acad Sci USA 86:7687–7690

Schlichting I, Almo SC, Rapp G, Wilson K, Petratos K, Lentfer A, Wittinghofer A, Kabsch W, Pai EF, Petsko GA, Goody RS (1990) Time-resolved X-ray crystallographic study of the conformational change in Ha-ras p212 protein on GTP hydrolysis. Nature 345:309–315

Seeburg PH, Colby WW, Capon DJ, Goedel DV, Levinson AD (1984) Biological properties of human c-Ha-ras 1 genes mutated at codon 12. Nature 312:71–75

Sigal IS, Gibbs JB, D'Alonzo JS, Scolnick EM (1986) Identification of effector residues and a neutralizing epitope of Ha-ras encoded p21. Proc Natl Acad Sci USA 83:4725–4729

Skinner RH, Bradley S, Brown AL, Johnson NJE, Rhodes S, Stammers DK, Lowe P (1991) Use of the Glu-Glu-Phe C-terminal epitope for rapid purification of the catalytic doamin of normal and mutant ras GTPase-activating proteins. J Biol Chem 266:14163–14166

Srivastava SK, DiDonato A, Lacal JC (1989) H-ras mutants lacking the epitope for the neutralizing monoclonal antibody Y13-259 show decreased biological activity and are deficient in GTPase-activating protein interaction. Mol Cell Biol 9: 1779–1783

Tong L, DeVos AM, Milburn MV, Kim S-H (1991) Crystal structures at 2.2 Å resolution of the catalytic domains of normal and an oncogenic mutant complexed with GDP. J Mol Biol 217:503–516

Trahey M, McCormick F (1987) A cytoplasmic protein stimulates normal N-ras p21 GTPase, but does not affect oncogenic mutants. Science 238:542–545

Valencia A, Chardin P, Wittinghofer A, Sander C (1991a) The ras protein family: evolutionary tree and role of conserved amino acids. Biochemistry 30:4637–4648

Valencia A, Kjeldgaard M, Pai EF, Sander C (1991b) GTPase domains of ras p21 oncogene protein and elongation factor Tu: analysis of three-dimensional structures, sequence families, and functional sites. Proc Natl Acad Sci USA 88:5443–5447

Wiesmüller L, Wittinghofer A (1992) Expression of the GTPase activating domain of the neurofibromatosis Type 1 (NF1) gene in Escherischia coli and role of the conserved lysine residue. J Biol Chem 267:10207–10210

Wittinghofer A, Pai EF (1990) The structure of Ras protein: a model for a universal molecular switch. TIBS 16:382–387

Wolfman A, Macara I (1990) A cytosolic protein catalyzes the release of GDP frpm p21. Science 248:67–69

Xu G, Lin B, Tanaka K, Dunn D, Wood D, Getseland R, White R, Weiss R, Tamanoi F (1991) The catalytic domain of the neurofibromatosis type 1 gene product stimulates ras GTPase and complements ira mutants of S. cerevisiae, Cell 63:835–841

Zhang K, Papageorge AG, Martin P, Vass WC, Olah Z, Polakis P, McCormick F, Lowy DR (1991) Heterogeneous amino acids in ras and rap1A specifying sensitivity to GAP proteins. Science 254:1630–1634

CHAPTER 15

Analysis of Ras Structure and Dynamics by Nuclear Magnetic Resonance

S.L. CAMPBELL-BURK and T.E. VAN AKEN

A. Introduction

ras oncogenes are among the most prevalent found in human tumors (BARBACID 1987; BOS 1989). These genes encode highly related 21 kDa guanine nucleotide binding proteins which are located on the inner surface of the plasma membrane. They are believed to function as signal switch molecules which relay extracellular growth promoting signals to intracellular targets by cycling between the biologically active ras.GTP and the inactive ras.GDP forms. The on (ras.GTP) and off (ras.GDP) states of the ras signal switch are modulated by interactions with intracellular proteins. Factors which stimulate GDP dissociation are believed to promote GTP formation in vivo (DOWNWARD et al. 1990; WEST et al. 1990; HUANG et al. 1990). The inactive GDP form is generated via interaction with GTPase activating proteins, GAP and NF1, to promote hydrolysis of GTP to GDP (TRAHEY and MscCormick 1987; GIBBS et al. 1988; MARTIN et al. 1990; XU et al. 1990; Ballester et al. 1990). Ras proteins acquire oncogenic properties if activated by point mutations that affect binding, dissociation and/or hydrolysis of guanine nucleotide to favor the GTP form of the protein.

Despite the wealth of structural and biochemical information, little is known about the interaction between ras and modulators of ras function. Several high resolution crystal structures of truncated wild type and mutant ras proteins complexed to either GDP or stable GTP analogs (guanosine 5′-β,γ-imidotriphosphate (GMPPNP), guanosine 5′-β,γ-methylene triphosphate (GMPPCP)) have recently been solved (DEVOS et al. 1988; PAI et al. 1989, 1990; BRUNGER et al. 1990; MILLBURN et al. 1990; KRENGEL et al. 1990; TONG et al. 1991). However, efforts to crystallize ras with agents which modulate ras activity have been unsuccessful so far.

Nuclear magnetic resonance (NMR) spectroscopy is a useful tool for investigating protein structure and dynamics in solution. In addition to providing structural information that is complimentary to X-ray diffraction methods, NMR studies have the potential of providing insight into the mechanism of GTP hydrolysis and may shed some light on ras–modulator interactions. Another area where NMR should prove highly useful is in probing internal dynamics under conditions more physiologically relevant to the protein's native environment. Certain regions of the protein that are involved in ligand binding and catalysis has been shown to be highly mobile in the crystal. NMR dynamic studies may aid in our understanding of ras protein function since internal mobility is likely to play an important role in the ability of ras to convert between inactive and active forms.

B. NMR Studies of Proteins

I. NMR Structure Determination

1. NMR Methods: Larger Proteins

Prior to 1990, structure determination using NMR spectroscopy was limited to small proteins (<15 kDa). However significant progress, made recently using heteronuclear multidimensional approaches, has increased the potential limit of protein size to 20–40 kDa (Fesik and Zuiderweg 1990). Further development over the next few years may extend this even more. The strategy to determine protein structure by NMR is outlined in Fig. 1 and is summarized below. In order to obtain detailed structural information, most of the proton nuclei need to be identified or assigned according to both amino acid type and location in the polypeptide. Structural information is then obtained by quantitating distances between individual protons. The assignment task becomes increasingly difficult for proteins larger than 15 kDa since the number and line width of proton resonances increase nonlinearly with increasing molecular size in this range, leading to decreased resolution. Heteronuclear-edited multidimensional NMR techniques, developed recently, have proven highly successful for assignment of several proteins in the 15–32 kDa range (Ikura et al. 1990; Kay et al. 1990a, 1991; Montelione and Wagner 1990; Bax and Ikura 1991; Boucher et al. 1992a,b). These approaches require the use of ^{15}N- and/or ^{13}C-enriched proteins, generally produced by biosynthetic incorporation of ^{15}N and ^{13}C stable isotopes into a protein of interest using bacterial expression systems (Muchmore et al. 1989). The ^{15}N- and ^{13}C-enriched nuclei are used as a vehicle to relay magnetization between protons, to select subsets of protons attached to the heteronucleus, and to increase resolution by separating ^{1}H and ^{15}N and ^{13}C signals into two, three, and four dimensions (2D, 3D, and 4D).

2. NMR Resonance Assignments: Application to Ras

The assignment process consists of identifying the atoms in the ras protein responsible for each peak in the multidimensional spectrum. There are approximately 1600 ^{1}H, ^{15}N, and ^{13}C atoms in H-ras. GDP (1–166) that need to be assigned before solution structure determination, so the assignment task is not trivial for proteins of this size. Reported NMR assignments are restricted to selected amino acid *type* assignments based on specific

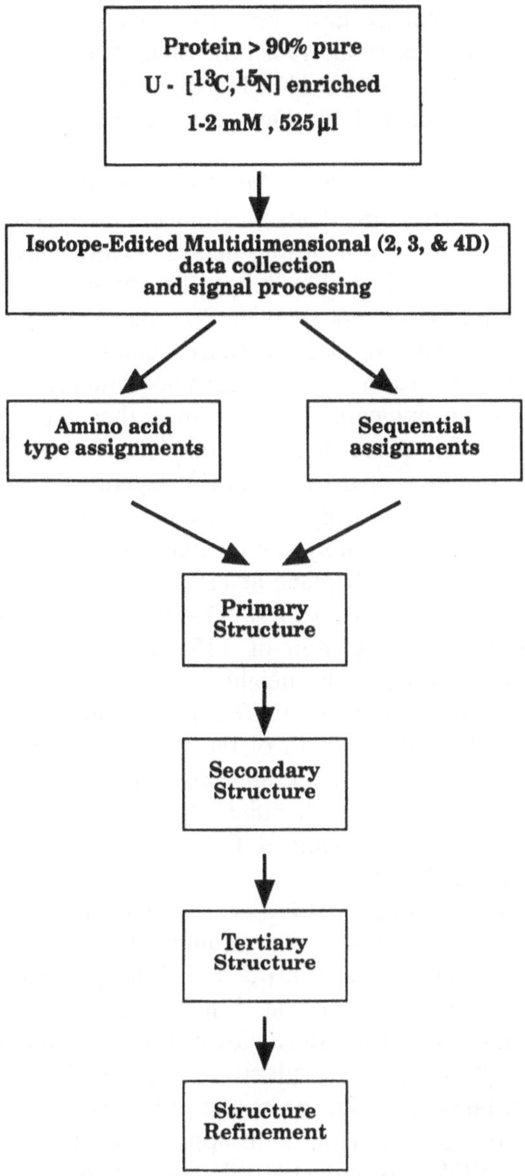

Fig. 1. Flow chart describing the steps involved in protein structure determination by NMR for proteins between 15 and 30 kDa

isotopic enrichment with ^{15}N, ^{13}C, and ^{2}H (CAMPBELL-BURK et al. 1989; CAMPBELL-BURK 1989; REDFIELD et al. 1990; YAMASAKI et al. 1989, 1992; SCHLICHTING et al. 1990). However, we have recently determined ^{1}H, ^{15}N and ^{13}C backbone assignments of [^{15}N,^{13}C]-enriched H-ras GDP (1–166) using multidimensional heteronuclear-edited NMR methods. Details of the backbone assignment strategy have recently been described (CAMPBELL-BURK et al. 1992; BOUCHER et al. 1992a,b). Side chain ^{1}H and ^{13}C nuclei were identified using ^{13}C-edited 3D HCCH experiments (BAX et al. 1990a,b; KAY et al. 1990b; CLORE et al. 1990; FESIK et al. 1990).

3. Secondary Structure Determination: Application to Ras

The next step toward structure determination entails quantitation of distances between individual protons using nuclear Overhauser enhancement spectrosopy (NOESY), a pivotal technique for studies involving protein structure determination (KUMAR et al. 1980; WAGNER et al. 1981). In NOESY spectra, proton nuclei that are close in space demonstrate an interaction, referred to as an NOE. Each NOE cross peak connects two previously assigned protons in the polypeptide chain when they are separated by a distance of approximately 4.5 Å or less in the 3D structure. The intensity of the NOE has an inverse sixth order ($1/r^6$) dependence on distance, so cross-peak intensity falls off sharply with increasing separation. NOEs thus identify stringent constraints on polypeptide conformation. Figure 2 shows a 2D slice corresponding to an ^{15}N plane at 115.5 ppm, taken from a 3D ^{15}N-edited NOESY spectrum of H-ras.GDP (1–166). NOE cross peaks are observed between the amide proton of T158 at 8.5 ppm and two other amide protons corresponding to the neighboring residues, L159 and Y157, indicating that the amide protons of L159 and Y157 are within 4.5 Å from the T158 amide proton. The strength of the NOE is proportional to $1/r^6$. That we see a fairly strong cross-peak intensity indicates that these protons are much less than 4.5 Å from each other, data consistent with the 2.9 Å distance (shown in Fig. 2) determined from the crystal structure of H-ras.GMPPNP (PAI et al. 1990).

Specific patterns of sequential NOEs can be correlated with secondary structure in proteins (WUTHRICH 1986). Sequential NOEs are cross peaks observed between neighboring atoms ($i,i+1$); the primary ones used to determine protein topology are illustrated in Fig. 3. The distance between the α-proton of residue i ($H_{\alpha i}$) and the amide proton of residue $i+1$ ($H_{N(i+1)}$) is short (~2.2 Å) in β-strands, consequently strong αN NOEs are observed. In α-helices, the distance between the amide proton of residue i (H_{Ni}) and its neighboring amide protons ($H_{N(i+1)}$) is approximately 2.5 Å, so medium to strong $H_{N(i+/-1)}$ NOEs are detected. Slowly exchanging amide protons can often be used to discriminte α-helix and β-strand structural elements from turns or loops because main chain amide protons that show a high degree of solvent protection (slow amide exchange rates) are normally involved in hydrogen bonds and buried in less mobile, hydrophobic regions of the protein. Tight turns or loops, on the other hand, are generally found to be exposed on the molecular surface (RICHARDSON 1981; SCHULZ and SCHIRMER 1979).

The crystal structure of H-ras p21.GDP (1–171) contains an α/β motif (TONG et al. 1991) with elements aligned as shown in Fig. 4. Also shown in

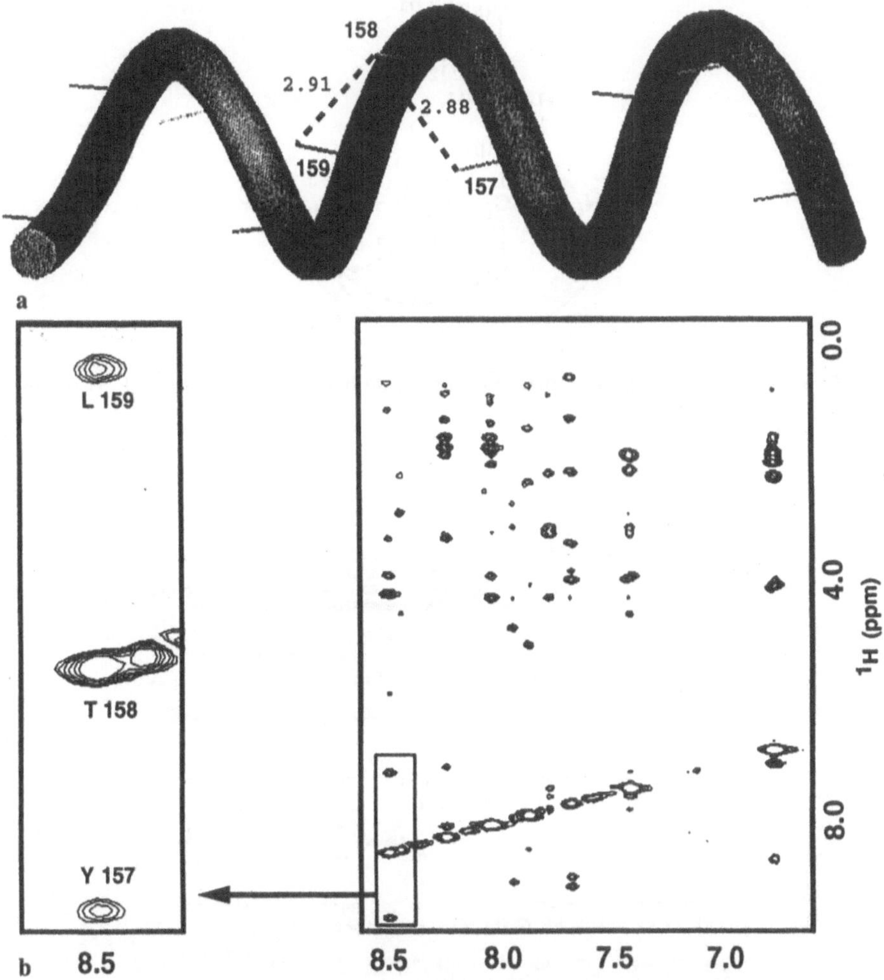

Fig. 2. a 2D slice of an ^{15}N plane at 115.5 ppm taken from a 3D ^{15}N-NOESY-HMQC data set acquired on H-ras.GDP (1–166). **b** Expanded region of the 2D slice showing proton-proton NOEs between the amide proton of T158 and the amide protons of L159 and Y157. **c** Portion of $\alpha 5$ taken from the X-ray coordinates of H-ras.GMPPNP showing distances between the amide protons of T158 and L159 and Y157

Fig. 4 is NMR derived short range sequential NN, αN connections and slowly exchanging amide protons as a function of amino acid sequence. The NOEs were quantitated on H-ras.GDP (1–166) for delineation of secondary structural features. βN and longer range αN and NN NOEs $(i, i + (2–4))$ are required to accurately determine the solution topolopy. The empirical pattern recognition approach, using αN and NN NOEs for secondary struc-

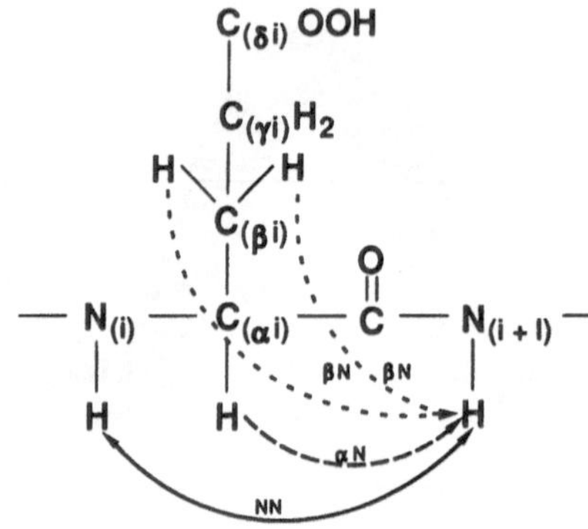

Fig. 3. Schematic drawing of a portion of a peptide backbone illustrating short range NN, αN and βN NOEs

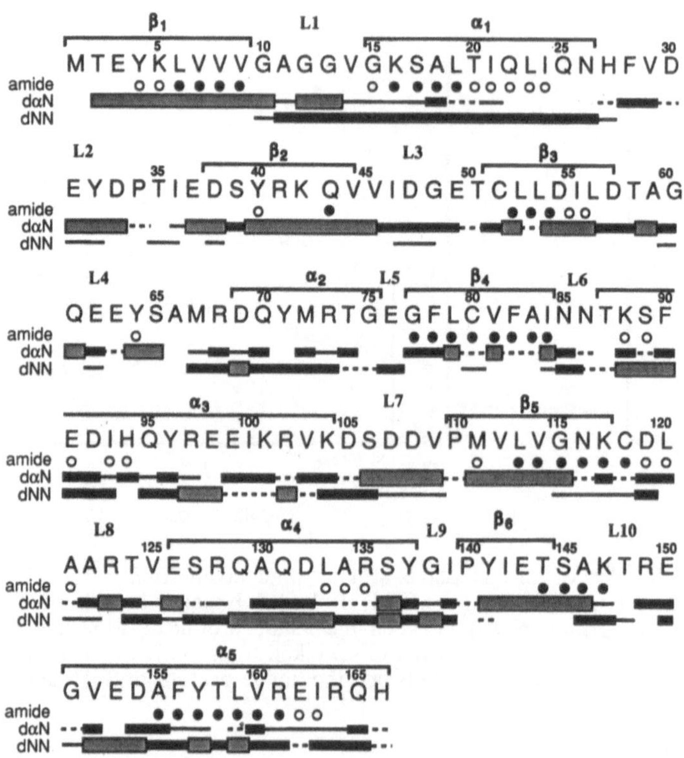

Fig. 4. Short range αN, NN NOEs and slowly exchanging amides as a function of the H-ras amino acid sequence. The intensity of the NOE (strong, medium, weak) is denoted by the *width* of the rectangle. *Hatched lines* indicate NOEs which were not quantitated due to degeneracy with i (+/−1) resonance. Slowly exchanging amide proton resonances (slowly exchanging amides) observed 2 and 48 h after sample exchange into a D_2O based buffer are denoted with *open* and *closed circles*, respectively. β-strand, α-helix, and loop regions determined from X-ray data are indicated

ture identification, has so far correlated well with the topology of the H-ras.GDP (1–171) X-ray structure (TONG et al. 1991). For example, strong αN NOEs, slowly exchanging amides, and the absence of NN NOEs are indicative of β-strand secondary topology for the first nine amino acids. The presence of both NN and αN NOEs and the absence of slowly exchanging amide protons from residue 10 to 14 are consistent with a loop or turn. A string of slowly exchanging amides and NN NOEs starting at residue 15 is consistent with the presence of an α-helix and so on. One interesting observation is the absence of any observable NN NOE at H94. In the crystal of H-ras.GDP (1–171), there is a break in helix $\alpha 3$ at residue 93 that may explain disruption in the string of NN connectivites in this region. A few main chain amide protons, located in loops 8 and 10, have slow solvent exchange rates that are generally uncharacteristic for residues located in loops. In the X-ray crystal structure of H-ras.GDP (1–171), several of these residues (N116, C118, D119, L120 and A121, S145, A146 and K147) comprise the guanine nucleotide binding pocket that is a deep narrow groove (TONG et al. 1991). The high degree of solvent protection associated with these residues can be explained by their location in the protein and interactions with the guanine base.

4. Tertiary Structure and Structure Refinement

A full solution structure determination requires quantitation of short, medium and long range NOEs. The NOEs are generally converted into semi-quantitative distance restraints for initial protein structure calculations using distance geometry algorithms (HAVEL and WUTHRICH 1984). When possible, additional restraints such as dihedral angles obtained from coupling constants are also used. Each restraint describes an allowed distance range or angle rather than a precise value. Molecular mechanics or molecular dynamic calculations are then used either as an alternative to distance geometry or as a supplementary technique for refinement of protein structures (CLORE and GRONENBORN 1989). These programs mimimize energy within the allowed parts of conformational space. For a review on other techniques for protein structure determination using NMR data, the interested reader can refer to BRAUN (1987).

II. Comparison of Solution and Crystal Structures

NMR spectroscopy and X-ray crystallography are the only two approaches available for complete protein structure determination. When used in combination, these methods can provide complementary information and serve to verify protein structures. One relevant example is an elegant NMR study on H-ras (1–171) conducted by YAMASAKI et al. (1989) that identified portions of an antiparallel β-sheet important in the conformational switch associated with GDP to GTP exchange. The study indicated that the topological arrangement of the two β-strands differed from that proposed in the original crystal structure (DEVOS et al. 1988). Concurrently, a higher resolution X-ray structure was determined and the correct topology elucidated (HOLBROOK and KIM 1989; TONG et al. 1991).

Differences observed between X-ray and NMR determined protein structures may reflect the fact that NMR studies are conducted in solution rather than in the crystalline state. To date, comparison of protein structures determined by NMR and X-ray diffraction methods show the most differences at molecular surfaces. In the 1.35 Å H-ras (1–166) crystal structure determined by PAI et al. (1990), the crystal unit cell is less solvated than normally observed in other protein crystals, which tends to increase the number of protein-protein contacts. Several crystal contacts are, in fact, found in the catalytic loop (loop 4) and effector domain.

1. Computer Simulation: Ras.GMPPNP Solution Structure

Important potential differences in ras protein structure and dynamics, between solution and the crystalline state, prompted computational dynamics studies of the GMPPNP-bound form of H-ras (1–166) in an aqueous environment (FOLEY et al. 1992). In the simulated solution structure, some intermolecular protein contacts due to crystal formation are lost in the solvation process, and this loss was thought to lead to localized unfolding of the molecule. In one such case, the antiparallel β-sheet (residues 38–57) is predicted to partially melt in agreement with earlier NMR studies (YAMASAKI et al. 1989). Another important difference is that the γ-phosphate of GTP posseses fewer contacts in the simulated solution structure than in the crystal structure. As with any model, its predictive powers are dependent on the experimental data and validity of the assumptions used to generate the model. NMR-based solution studies have the potential to test these predictions.

2. Protein Dynamics

Proteins are not rigid structures. X-ray crystal structures are generally represented as a single conformer. NMR experimental data, on the other hand, represent an incomplete set of intramolecular interatomic connectivities, so the result is commonly represented by a group of conformers. Each of the NMR structures represents one possible solution to the geometric problem of fitting the polypeptide chain to the ensemble of all experimental restraints. While variability in a group of NMR structures may indicate that too few experimental restraints were obtained to define a single structure, it may also point to conformational averaging. Fewer NOE restraints are generally observed in regions of high mobility. Careful analysis of NMR data, dynamic parameters, and the family of NMR-derived conformers can help differentiate beween the two possibilities.

The rate of GTP conversion to GDP is closely tied to ras protein function. Two distinct GTP hydrolysis mechanisms, based primarily on X-ray crystal structures of ras.GMPPCP and ras.GMPPNP, have been proposed recently (PAI et al. 1990; PRIVE et al. 1992). In the methanism proposed by PAI et al. (1990), Gln-61 activates a water molecule for nucleophilic attack through a charge relay network involving Glu-63. However, in the transition state mechanism proposed by PRIVE et al. (1992), Gln-61 and Glu-63 are not essential for positioning the attacking water molecule in GTP hydrolysis. The role proposed for Gln-61 is to reduce the activation barrier by stabilizing the pentavalent phosphate transition state intermediate of the hydrolysis reaction.

The region containing residues 61–65 has been shown to be highly mobile; consequently, this region is poorly defined in the crystal structures. In the H-ras.GMPPNP structure, it was assumed that the electron density could be accounted for by two or more configurations of each residue in loop 4. The conformation of the loop that places Q61 in a position which could activate the nucleophilic water was used for prediction of the GTP hydrolysis mechanism. On the other hand, in the GMPPCP bound p21 crystal structure, one of four molecules in an asymmetric unit is well ordered in the region of residues 61–65, and this molecule was analyzed to alternative model for the transition state. Perhaps, another description of this loop may be to consider it dynamic and define the conformational space sampled by residues in this domain.

C. Comparison of Full Length and Truncated ras Proteins

I. Protein Stability: Sample Preparation

Although our initial NMR studies (CAMPBELL-BURK et al. 1989; CAMPBELL-BURK 1989) utilized the full 189 residue form of N-ras, sample instability, most likely resulting from cysteine oxidation in the C-terminal portion of the protein was problematic. Light scattering data (not shown) indicates that the full length protein forms oligomers at concentrations required for NMR analysis whereas the truncated protein appears monomeric. Early attempts to crystallize the full length protein were unsuccessful but when the C-terminus was truncated by 18–23 amino acids, suitable crystals were obtained (DEVOS et al. 1988). Although the C-terminus is necessary for membrane attachment and overall biological activity, it is not required for nucleotide binding and GTP hydrolysis (WILLUMSEN et al. 1984; SEFTON et al. 1982; CHEN et al. 1985; BUSS and SHEFTON 1986; JOHN et al. 1989). Several glycine (^{15}N,$^{1}H_N$) resonances in the guanine nucleotide binding domain of the intact protein had previously been identified (CAMPBELL-BURK et al. 1989; CAMPBELL-BURK 1989; JOHN et al. 1989). Comparison of intact and truncated H-ras.GDP spectra showed similar glycine chemical shifts for residues near the active site. Our recent NMR assignments were therefore obtained on c-

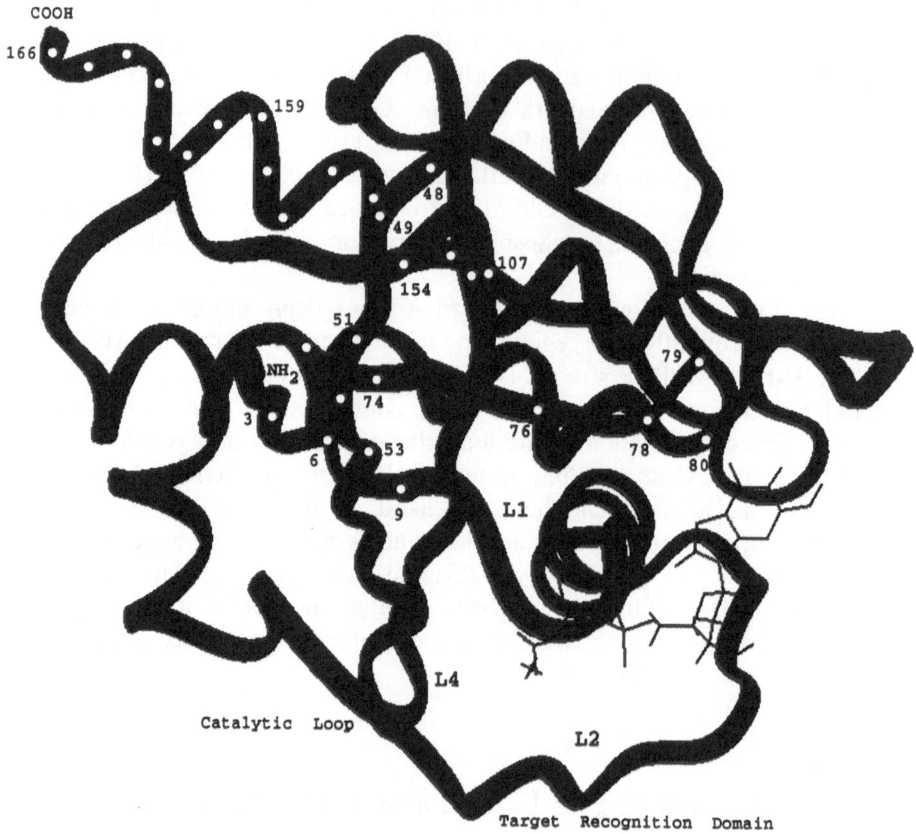

Fig. 5. Ribbon diagram of H-ras.GDP (1–166). NH signals whose chemical shifts (^{15}N, $^{1}N_N$) differ by more than 0.1 ppm between intact and truncated H-ras.GDP are illustrated with *white circles*. Chemical shift differences have been observed primarily in 6 regions: residues 3, 6 and 9 in β1, 46, 48–49 of loop 3, 51–53 of β3, 73, 74, 76, 78–80 of α2, loop 5 and β4, 107–108 of loop 7 and 154, 156–166 of α5. Switch 1 and switch 2 regions correspond to amino acids 30–40 and 58–76, respectively

H-ras (1–166). A more thorough comparison of intact versus truncated H-ras.GDP has recently been conducted with the results summarized below.

II. Chemical Shift Differences

A ribbon diagram of H-ras.GDP (1–166) is shown in Fig. 5. NH signals whose chemical shifts (^{15}N,$^{1}H_N$) differ by more than 0.1 ppm between intact in truncated p21.GDP (1–166) are denoted with white circles. The ^{15}N and $^{1}H_N$ chemical shifts of residues located in the guanine nucleotide binding consensus sequences (Halliday 1984; Valencia et al. 1991) and effector regions are not affected by removal of 23 amino acids in the C-terminal portion of the molecule. The data indicate that these important domains do not undergo global changes upon C-terminal truncation. However, certain residues located in regions peripheral to the guanine nucleotide binding and effector binding domains (i.e., $\beta 1$, $\beta 3$, $\alpha 2$, L5, $\beta 4$, L7 and $\alpha 5$) possess chemical shifts which differ between intact and truncated ras.GDP (1–166). Chemical shift changes in and of themselves are not well understood, but indicate differences in shielding resulting from differences in the electronic environment around the ^{1}H and ^{15}N nuclei, ^{1}H-^{1}H NOE measurements are required to determine whether the observed chemical shift changes are due to subtle rearrangements of side chains or a significant conformation change in backbone positions. Certain residues in the C-terminal portion of ras.GDP have been assigned to aid in structural and dynamic studies of the ras C-terminus. Some of this work is discussed in the next section.

III. Selective Isotope Enrichment Studies: Site Specific Probes

1. Identification of C-Terminal Peaks

Full length and truncated H-ras.GDP samples, specifically enriched with ^{15}N-glycine and ^{15}N-serine, have been used to assign glycine and serine resonances in the C-terminal portion of the molecule. The function of the ^{15}N-isotope is to remove NMR signals from all protons not attached to ^{15}N from the spectrum, resulting in considerable spectral simplification. The signals are further resolved by means of 2D methods in which the second dimension of dispersion is provided by the different ^{15}N chemical shifts. Through a combination of proton and ^{15}N pulsed excitation we can obtain a 2D NMR spectrum, shown in Fig. 6a, which contains peaks from only the 19 ^{15}NH glycine and serine amide groups in H-ras.GDP (1–166). Each of the 19 peaks comes from a single beckbone amide resonance of one of the 11 glycine and 8 labeled serines. The ^{1}H-^{15}N 2D spectral map of full length [^{15}N-gly,ser] H-ras.GDP (1–189), shown in Fig. 6b, contains five additional NH signals corresponding to S177, G178, G180, S183, S189. Both of the C-terminal glycine resonances possess similar chemical shifts under the conditions used to acquire the data, so their NH-sesonances cannot be distinguished on the spectrum. These C-terminal glycine and serine peaks were identified easily since glycine and serine $^{1}H_{N}$ resonances corresponding to residues 1–166, assigned previously (Campbell-Burk et al. 1989; Campbell-Burk 1989), possess similar peak positions in intact and truncated ^{1}H-^{15}N spectral maps. Selective labeling can classify resonances by amino acid type, but further methods are needed to assign resonances to specific amino acids in the sequence. Consequently, the precise sequence identities of the glycine and serine C-terminal resonances have not yet been defined.

2. Internal Dynamics

NMR spectroscopy is a useful tool for probing internal protein dynamics. Although quantitative determination of internal dynamics requires a series

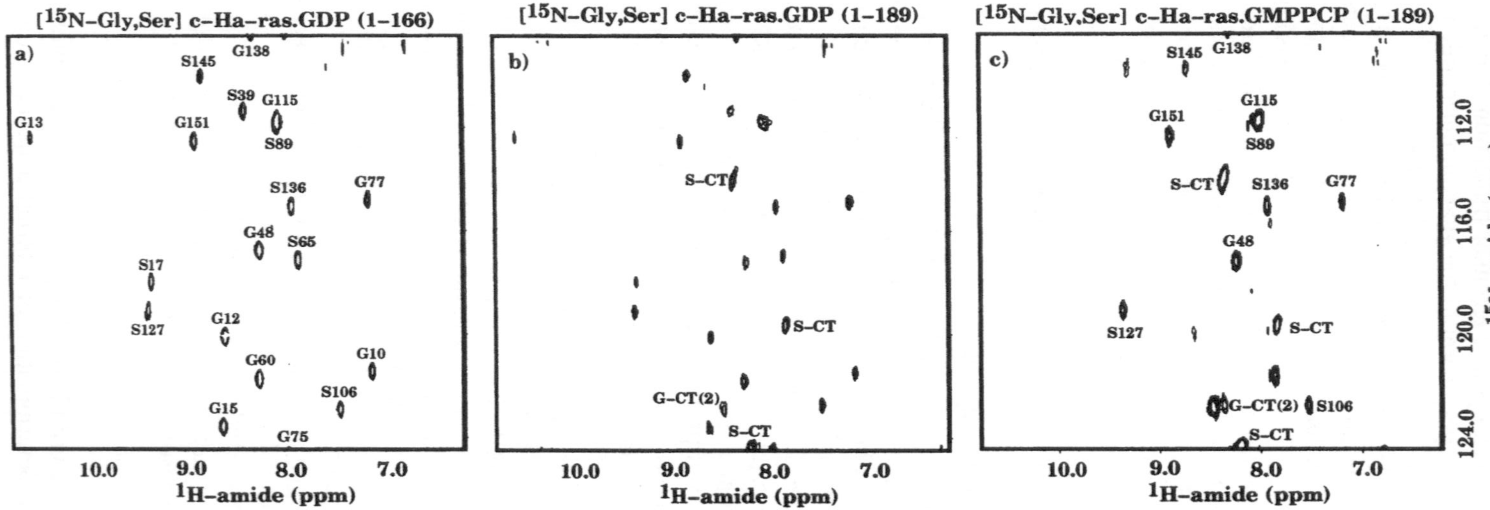

Fig. 6. **a** ^{1}H-^{15}H 2D spectrum of ^{15}N-[Gly,Ser] H-ras.GDP (1–166). The 19 glycine and serine peaks have been identified previously and are labeled on the 2D spectral map. **b** ^{1}H-^{15}N 2D correlation map of ^{15}N-[Gly,Ser] H-ras.GDP (1–189). The five glycine and serine NH resonances in the C-terminal portion of the molecule (177, 178, 180, 183, 189) are labeled. Glycine 178 and 180 NH-resonances possess similar chemical shifts and cannot be distinguished in the spectrum shown. **c** ^{1}H-^{15}H 2D correlation map of c-Ha-ras.GMPPCP (1–189). None of the NH resonance peak positions corresponding to glycine and serine residues in the C-terminal region differ from the spectrum shown in **b** of ^{15}N-[Gly,Ser] H-ras.GDP (1–189). Resonances whose chemical shifts do not differ between H-ras.GDP and H-ras.GMPPCP are labeled

of relaxation measurements (PENG and WAGNER 1992), regions containing considerable motion by conducting ^{1}H-$T_{1\rho}$ relaxation studies can be identified. Generally the longer the $T_{1\rho}$ relaxation time, the faster the rotational correlation time and the more mobile the NH group. $T_{1\rho}$ relaxation measurements, conducted on intact (1–189) and truncated (1–166)^{15}N-[gly,ser] H-ras.GDP indicate that the residues located in the C-terminal portion (S177, G178, G180, S183, S189) of the molecule possess, by far, the longest relaxation times suggesting that the C-terminal region is the most flexible part of the molecule. In fact, these measurements indicate that the C-terminal portion of the molecule is highly disordered, and most likely possesses little structure.

The thermal stability of H-ras (1–166) is lower than that of the intact protein (JOHN et al. 1989). In the X-ray crystal structure of H-ras.GDP (1–171) residues 151–171 form part of α5. This helix is shortened by five amino acids in the crystal structure of H-ras.GMPPNP (1–166). Our NMR data show perturbations in peaks corresponding to residues 154 and 156–166 in α5 upon removal of residues 167–189. It is possible that a structural change in this helix (α5) may cause less favorable contacts with other regions of the protein (i.e., β1, β3, α2, L5, β4, L7), also shown to be perturbed by C-terminal truncation, resulting in lower thermal stability.

3. Comparison of Intact ras.GDP and ras.GMPPCP

Comparison of the ^{1}H-^{15}N 2D spectral maps in Fig. 6(b,c) of full length H-ras.GDP (1–189) and H-ras.GMPPCP (1–189) shows 24 peaks in the GDP spectrum but only 19 peaks in the spectrum of ras.GMPPCP. It is possible that the five unobservable (glycine, serine) NH-resonances are broadened beyond detection by conformation averaging between two or more sites. The NH resonances of glycine and serine residues in the C-terminal region (177, 178, 180, 183, 189) of H-ras (1–189) appear unaffected by binding of GDP versus GMPPCP. Other residues shown to be insensitive to complexation of GDP versus GMPPCP are G77, S89, G115, S127, G138, S145, G151 whereas G48 and S106 show slight shifts. The remaining glycine and serine resonances of G10, G12, G13, G15, S17, S39, G60 and S65 differ between ras.GDP and ras.GMPPCP spectra. All of these residues are close to the γ-phosphate. Five out of eight of these resonances are not observable in the ras.GMPPCP spectrum. However, all of the glycine and serine NH resonances are detectable in ^{1}H-^{15}N spectra acquired on H-ras.GTP (data not shown). Our NMR data suggest that some of the residues located in loop 1, loop 2 and/or loop 4 of ras. GMPPCP undergo conformational averaging on a timescale different from that of ras.GTP.

D. Comparison of ras.GTP, ras.GTPγS, ras.GMPPCP and ras.GDP

I. Chemical Shift Differences

As described earlier, chemical shift changes in and of themselves are not well understood but can be highly informative when used to assess the relative differences between two states. For example, chemical shift differences between wild type and mutant ras proteins can indicate where perturbations, resulting from the mutation, occur in the molecule and whether the mutation causes a local or global change in the protein (Campbell-Burk 1989). Investigation of the exact nature of the perturbation requires more extensive analysis to determine whether the perturbation results from subtle changes around the nucleus or a significant structural alteration.

Figure 7 outlines (^1H_N,^{15}N) chemical shift differences between H-ras.GDP and H-ras.GTP, ras.GTPγS and ras.GMPPCP for residues 1–166.

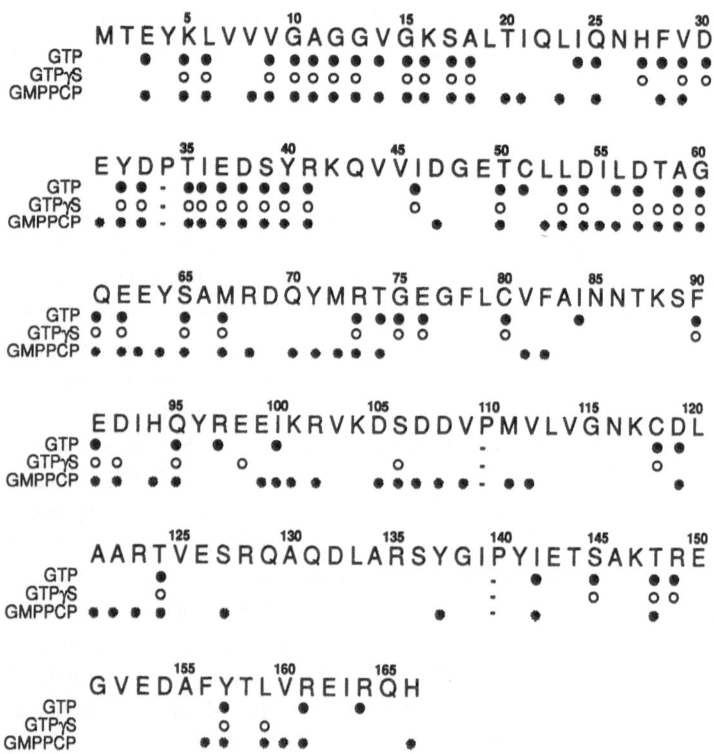

Fig. 7. Schematic showing NH chemical shift differences between the GTP, GTPγS, and GMPPCP forms of H-ras (1–166) as compared with the GDP form as a function of amino acid sequence. Prolines are not observed. Chemical shift differences greater than 0.1 ppm between ras.GDP and the various GTP analogs are denoted with *circles*

Analysis of relative chemical shift differences between GDP and various GTP and GTP analogs show certain trends. For example, amide (^1H_N and ^{15}N) chemical shifts corresponding to strings of residues located at positions 9–18, 32–41, and 57–62 differ between H-ras.GDP (1–166) and the all three of the various GTP (GTP, GTPγS and GMPPCP) forms. Two of these regions, 32–41 and 57–62, correspond to domains involved in the GDP to GTP conformational switch as determined by X-ray analysis (MILBURN et al. 1990; BRUNGER et al. 1990).

The chemical shift perturbations observed for residues 10–14 (loop 1) point to alterations in this domain. X-ray structural analysis of ras.GDP and ras.GTP (GTP analogs) revealed that several of the main chain amide protons of loop 1 form hydrogen bonds with the guanine nculeotide phosphates, However, loop 1 was not found to be involved in the conformational switch between the "active" ras.GTP and "inactive" ras.GDP state. In view of the X-ray diffraction data, the large ^1H_N chemical shift differences corresponding to residues in loop 1 may be due to changes in the electronic environment or geometry of the hydrogen bonds formed with the β-phosphate between GTP and GDP forms of ras. If so, differences in the electronic environment and dynamics around residues in this loop may prove to be important in understanding the mechanism of GTP hydrolysis. It is also possible that chemical shift changes observed in loop 1 may reflect structural changes which are not detected in the X-ray structure. Isolated chemical shift perturbations are also observed for residues located in β1, the beginning of α2, β3, as well as scattered shifts in other NH residues. Interpretation of these results will require a more detailed NMR structural analysis of H-ras.GTP.

Closer correspondence is observed between ^1H-^{15}N spectral maps of H-ras.GTP and H-ras.GTPγS than for H-ras.GMPPCP. Approximately 30% of the NH chemical shifts differ between ras.GDP as compared to ras.GTP and ras.GTPγS complexes, whereas >50% of the NH resonances differ between 2D spectral maps of ras.GDP and ras.GMPPCP. In ^1H-^{15}N spectrum of H-ras.GMPPCP, four regions (68–74, 99–112, 119–124 and 156–161) show NH-chemical shift differences relative to the ras.GDP spectrum that are not significantly perturbed in spectral maps of either H-ras.GTP or H-ras.GTPγS. It is conceivable that the mode of GMPPCP binding to H-ras differs from that of GTP and GTPγS. Comparison of ras.GDP and ras.GMPPCP crystal structures show a different conformation for switch 1 and switch 2 regions, which includes a change in attitude of helix-2 (66–74). However, a structural difference in α2 was not observed between crystal structures of H-ras.GDP and H-ras.GMPPNP. The relative affinity of GTP, GTPγS, GMPPNP and GMPPCP for H-ras.GDP (1–166) has been determined to be 1.1, 0.3, 0.07 and 0.01, respectively (JOHN et al. 1989). GMPPCP is not hydrolyzed by ras proteins and possesses a 100-fold lower affinity relative to GDP. Chemical shift differences observed between ras.GMPPCP and ras.GDP spectra for residues 99–112, 119–124, 156–161 and part of switch 2 (α2), but not in either ras.GTP and ras.GTPγS spectra, indicate that the perturbations are most likely due to differences in the binding mode of H-ras.GMPPCP relative to H-ras.GTP rather than differences between ras.GDP and ras.GTP complexes. In summary, our results on selectively and uniformly enriched samples suggest some differences in structure and dynamics between ras.GTP and ras.GMPPCP complexes indicating that the GMPPCP analog may possess a different mode of binding from GTP.

E. Kinetic Measurements

I. Kinetic and Fluorescence Studies

Guanine nucleotides modified by acetylation of the ribose moiety with the fluorophore N-methylanthranilic acid (mant) have been used for kinetic characterization of ras proteins (Neal et al. 1990; Eccleston et al. 1991). Kinetic studies with fluorescent analogs of GTP indicate the existence of a rate limiting isomerization reaction that precedes hydrolysis. The isomerization reaction possesses a rate constant of $2.2 \times 10^{-4}\,s^{-1}$ at 30°C. The ras-GTPase activating protein, GAP, was found to accelerate this isomerization step. However, fluorescence studies conducted by other labs are not in agreement. Studies conducted by Antonny et al. (1991), using a tryptophan H-ras mutant, also showed a fluorescence change that occcurs faster than the cleavage of GTPγS. However, the fluorescence change observed in their study was not accelerated by GAP. Fluorescence studies conducted by Rensland et al. (1991) with mant-guanine nucleotide derivatives do not support the existence of a conformational change which precedes the cleavage step.

NMR spectroscopy can be used to monitor the hydrolysis reaction on an atomic level by observing signals associated with ^{1}H, ^{31}P, ^{13}C, and ^{15}N nuclei as a function of time. NMR kinetic studies, some of which are discussed below, are currently being conducted to determine whether a conformation change occurs prior to GTP cleavage.

II. ^{31}P NMR: ras.GTP Hydrolysis

^{31}P spectra, showing the time course of GTP hydrolysis in full length H-ras (1–189) are shown in Fig. 8. ^{31}P resonances corresponding to the α- and β-phosphate of GDP, α-, β- and γ-phosphate of GTP and inorganic phosphate are designated. Each individual phosphorous resonance can be followed simultaneously. Spectra were acquired under conditions where only GTP hydrolysis to GDP occurs, starting with stoichiometric complexes of ras.GTP. Single turnover of GTP to GDP is essentially complete within 4 h at pH 6.5 and 21°C. The rate is consistent with previously reported GTP hydrolysis rates, when corrected for differences in temperature and pH (37°C, pH 7.5). No stable phosphate intermediates were detected during the time course of the experiment. The data are consistent with studies conducted by FEUERSTEIN et al. (1989), which demonstrated that GTP hydrolysis occurs with inversion of the γ-phosphorous, indicating that the mechanism most likely occurs via in-line transfer without a phosphorylated intermediate.

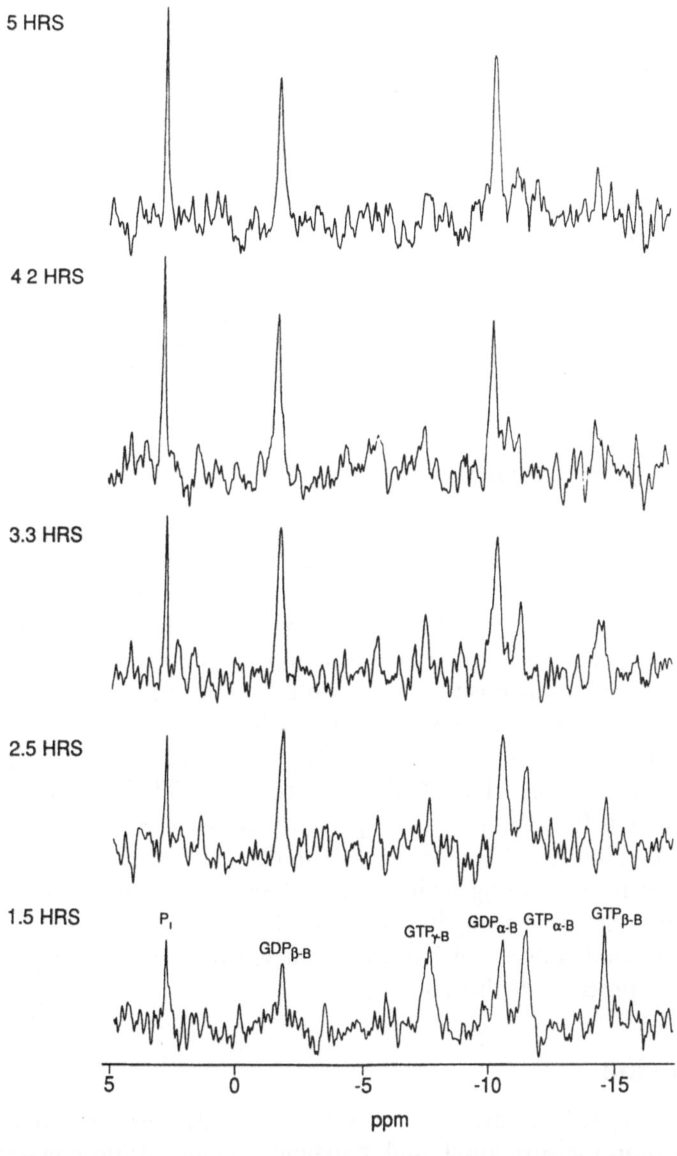

Fig. 8. ^{31}P spectra collected at various times after generation of 1-1 complex with GTP. Resonances corresponding to the α, β-phosphate of GDP, α-, β- and α phosphate of GTP and inorganic phosphate are designated on the spectrum

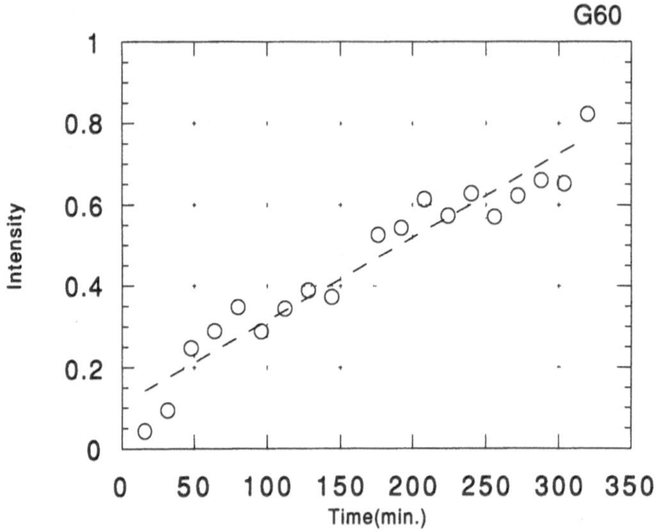

Fig. 9. Graph showing build-up in intensity of the H-ras.GDP (1–166) G60-NH resonance as a function of time during hydrolysis of GTP. Samples prepared as 1-1 stoichiometric complexes with GTP and production of GDP monitored

III. [^{1}H-^{15}N]-Edited NMR Spectroscopy: GTP Hydrolysis

Conversion of GTP to GDP was also monitored in H-ras (1–166) samples uniformly and specifically enriched with ^{15}N. ^{1}H-^{15}N correlation 2D spectra can be acquired faster than ^{31}P spectra (shown in Fig. 8) because the proton nucleus is more sensitive to NMR detection. In uniformly ^{15}N-enriched samples, the time course of every NH residue with the exception of proline can be followed simultaneously. For example, in Fig. 9 the intensity of the H-ras.GDP G60-NH resonance is plotted as a function of time. 2D spectra were acquired every 16 min under single turnover conditions starting with stoichiometric complexes of ras.GTP. The rate of GTP hydrolysis under these conditions is 0.002 min^{-1}. This approach can also be used to monitor protons attached to any ^{15}N or ^{13}C nucleus in the molecule. We are particularly interested in monitoring main chain and side chain atoms of residues believed to be important in hydrolysis of GTP to address the fluorescence results, summarized above, which question the existence of an isomerization reaction which preceeds hydrolysis.

F. Conclusion

In summary, very recent advances in NMR technology have rendered NMR a powerful tool for structural and dynamic studies of proteins in the 20 kDa molecular weight range. Our recent NMR assignments of H-ras.GDP (1–166) provide the basis for detailed investigation of ras proteins in solution. With the current NMR technology, analysis of H-ras.GTP (and slowly hydrolyzing GTP analogs) and protein variants by NMR may aid in understanding the structural and dynamic differences between the GTP and GDP forms of ras. NMR studies may also provide insight into the GTP hydrolysis mechanism.

Another area where NMR should prove useful is in probing interactions between ras proteins and agents which modulate ras signalling. Efforts to crystallize complexes of ras with other proteins (GAP and NF1) have proven unsuccessful so far. Probing the binding mode of agents which modulate or interfere with ras function should be valuable for the design of anti-ras agents.

Acknowledgements. We thank Rich DeLoskey, Jim Krywko, and Richard Yates for excellent technical assistance. The H-ras (1–166) bacterial expression vector and H-ras.GMPPNP X-ray coordinates were a generous gift from A. Wittinghofer. We think P. Domaille and E. Laue for developing heteronuclear multidimensional NMR methods which proved essential for H-ras.GDP (1–166) ^{1}H, ^{15}N and ^{13}C resonances assignments.

References

Antonny B, Chardin P, Roux M, Chabre M (1991) GTP hydrolysis mechanisms in ras p21 and in the ras-GAP complex studied by fluorescence measurements on tryptophan mutants. Biochemistry 30:8287–8295

Ballester R, Marchuk D, Boguski M, Saulino A, Letcher R, Wigler M, Collins F (1990) The NF1 locus encodes a protein functionally related to mammalian GAP and yeast IRA proteins. Cell 63:851–859

Barbacid M (1987) ras GENES. Annu Rev Biochem 56:779–827

Bax A, Clore GM, Driscoll PC, Gronenborn AM, Ikura M, Kay LE (1990a) Practical aspects of proton-carbon-proton 3 dimensional correlation spectroscopy of ^{13}C-labeled proteins. J Magn Reson 87:620–627

Bax A, Clore GM, Gronenborn AM (1990b) ^{1}H-^{1}H correlation via isotropic mixing of ^{13}C magnetization a new three-dimensional approach for assigning ^{1}H and ^{13}C enriched proteins. J Magn Reson 88:425–431

Bax A, Ikura M (1991) An efficient 3D NMR technique for correlating the proton and ^{15}N backbone amide resonances with the α-carbon of the preceding residue in uniformly $^{15}N/^{13}C$ enriched proteins. J Biomol NMR 1:99–104

Bos JL (1989) ras Oncogenes in human cancer: a review. Cancer Res 49:4682-4689

Boucher W, Laue ED, Campbell-Burk S, Domaille PJ (1992a) Four dimensional heteronuclear triple resonance NMR methods for the assignment of backbone nuclei in proteins. J Am Chem Soc 114:2262–2264

Boucher W, Laue ED, Campbell-Burk S, Domaille PJ (1992b) A 4D NMR strategy for the assignment of backbone nuclei in ^{13}C and ^{15}N labelled proteins: application to c-H-ras p21. J Biomol NMR in press

Braun W (1987) Distance geometry and related methods for protein structure and determination from NMR data. Q Rev Biophys 19:115–157

Brunger AT, Milburn MV, Tong L, DeVos AM, Jancarik J, Yamaizumi Z, Nishimura S, Ohtsuka E, Kim S-H (1990) Crystal structure of an active form of Ras Protein, a complex of a GTP analog and the HRas p21 catalytic domain. Proc Natl Acad Sci USA 87:4849–48530

Buss JE, Shefton BM (1986) Direct Identification of palmitic acid as the lipid attached to $p21^{ras}$. Mol Cell Biol 6:116–122

Campbell-Burk S (1989) Structural and dynamic differences between normal and transforming N-ras gene products: a^{31}P and isotope-edited NMR study. Biochemistry 28:9478–9484

Campbell-Burk SL, Papastavaros MA, McCormick F, Redfield AG (1989) Identification of resonances from an activation locus of human H-ras p21 protein using isotope edited NMR, Proc Natl Acad Sci USA 86:817–820

Campbell-Burk SL, Domaille PJ, Starovasnik MA, Boucher W, Laue ED (1992) Sequential assignment of the backbone nuclei (^{1}H, ^{15}N and ^{13}C) of c-H-ras (1–166) GDP using a novel 4D NMR strategy. J Biomol NMR in press

Chen ZQ, Ulsh LS, DuBois G, Shih TY (1985) Posttranslational processing of p21 ras proteins involves palmitylation of the c-terminal tetrapeptide containing cysteine-186. J Virol 56:607–612

Clore GM, Gronenborn AM (1989) Determination of three-dimensional structures of proteins and nucleic acids in solution by Nuclear Magnetic Resonance Spectroscopy. Critical Rev Biochem Biophys 24:479–564

Clore GM, Bax A, Driscoll PC, Wingfield PT, Gronenborn AM (1990) Assignment of the side-chain^{1}H and ^{13}C resonances of Interleukin-1β using double- and triple-resonance heteronuclear three-dimensional NMR spectroscopy. Biochemistry 29:8172–8184

DeVos AM, Tong L, Milburn MV, Marias PM, Jancarik J, Noguchi S, Nishimura S, Miura K, Ohtsuka E, Kim S-H (1988) Three-dimensional structure of an oncogene protein: catalytic domain of human c-H-ras p21. Science 239:888–893

Downward J, Riehl R, Wu L, Weinberg RA (1990) Identification of a nucleotide exchange-promoting activity for p21^{ras}. Proc Natl Acad Sci USA 87:5998–6002

Eccleston JF, Moore KJM, Brownbridge GG, Webb MR, Lowe PN (1991) Fluorescence approaches to the study of the p21ras GTPase mechanism. Biochem Soc Trans 19:432–437

Fesik SW, Zuiderweg ERP (1990) Heteronuclear three-dimensional NMR spectroscopy of isotopically labelled biological macromolecules. Q Rev Biophys 23:97–131

Fesik SW, Eaton HL, Olejniczak ET, Zuiderweg ERP, McIntosh LP, Dahlquist FW (1990) 2D and 3D NMR spectroscopy embloying ^{13}C-^{13}C magnetization transfer by isotropic mixing. Spin system identification in large proteins. J Am Chem Soc 112:886–888

Feuerstein J, Goody RS, Webb MR (1989) The mechanism of guanosine nucleotide hydrolysis by p21 c-H-ras. J Biol Chem 264:6188–6190

Foley CK, Pedersen LG, Charifoson PS, Darden TA, Wittinghofer A, Pai EF, Anderson MW (1992) Simulation of the solution structure of the H-ras p21-GTP complex. Biochemistry 31:4951–4959

Gibbs JB, Schaber MD, Allard WJ, Sigal IS, Scholnick EM (1988) Purification of ras GTPase activating protein from bovine brain. Proc Natl Acad Sci USA 85:5026–5030

Halliday KR (1984) Regional homology in GTP-binding proto-oncogene products and elongation factors. J Cyclic Nucleotide Protein Phosphorylation Res 99:435–448

Havel TF, Wuthrick K (1984) A distance geometry program for determining the structures of small proteins and other macromolecules from nuclear magnetic resonance measurements of intramolecular proton-proton proximities in solution. Bull Math Biol 46:673–698

Holbrook SR, Kim S-H (1989) Molecular model of the G protein α subunit based on the crystal structure of the HRAS protein. Proc Natl Acad Sci USA 86:7415

Huang YK, Kung HK, Kamata T (1990) P:urification of a factor capable of stimulating the guanine nucleotide exchange reaction of ras proteins and its effect on ras-related small molecular mass G proteins. Proc Natl Acad Sci USA 89:8008–8012

Ikura M, Kay LE, Bax A (1990) A novel approach for sequential assignments of ^{1}H, ^{13}C, and ^{15}N spectra of larger proteins: heteronuclear triple-resonance three-dimensional NMR spectroscopy: application to Calmodulin. Biochemistry 29: 4659–4667

John J, Schlichting I, Schiltz E, Rosch P, Wittinghofer A (1989) C-terminal truncation of $p21^H$ preserves crucial kinetic and structural properties. J Biol Chem 264: 13086–13092

Kaibuchi K, Mizuno T, Fujioka H, Yamamoto T, Kishi K, Fukumoto Y, Hori Y, Takai Y (1991) Molecular cloning of the cDNA for stimulatory GDP/GTP exchange protein for smg p21s (ras p21-like small GTP-binding proteins) and characterization of stimulatory GDP/GTP exchange proteins. Molc Cell Biol 11:2873–2880

Kay LE, Ikura M, Tschudin R, Bax A (1990a) Three-dimensional triple-resonance NMR spectroscopy of isotopically enriched proteins. J Magn Reson 89:496–514

Kay LE, Ikura M, Bax A (1990b) Proton-proton correlation via carbon-carbon coupling: A three-dimensional NMR approach for the assignment of aliphatic resonances in proteins labeled with carbon-13. J Am Chem Soc 112:888–889

Kay LE, Ikura M, Bax A (1991) The design and optimization of complex NMR experiments. Application to a triple-resonance pulse scheme correlates Hα, HN and ^{15}N chemical shifts in ^{15}N-^{13}C-labeled proteins. J Magn Reson 91:84–92

Krengel U, Schlichting I, Scherer, Schumann R, Frech M, John J, Kabsch W, Pai EF, Wittinghofer A (1990) Three-dimensional structures of H-ras p21 mutants: molecular basis for their inability to function as signal switch molecules. Cell 62:539–548

Kumar A, Ernst RR, Wuthrich K (1980) A Two-Dimensional Nuclear Overhauser Enhancement (2D NOE) experiment for the elucidation of complete proton-proton cross-relaxation networks in biological macromolecules. Biochem Biophys Rev Commun 95:1–6

Martin GA, Viskochil D, Bollag G, McCabe PC, Crosier WJ, Haubruk H, Conroy L, Clark R, O'Connell P, Cawthon RM, Innis MA, McCormick F (1990) The GAP-related domain of the Neurofibromatosis type 1 gene product interacts with ras p21. Cell 63:843–849

Milburn MV, Tong L, DeVos AM, Brunger A, Yamaizumi Z, Nishimura S, Kim S-H (1990) Molecular switch for signal transduction: structural differences between active and inactive forms of protooncogenic ras proteins. Science 247:939–945

Montelione G, Wagner G (1990) Conformation-independent sequential NMR connections in isotope-enriched polypeptides by ^{1}H-^{13}Cu2-^{15}N triple-resonance experiments. J Magn Reson 87:183–188

Muchmore DC, McIntosh LP, Russell CB, Anderson DE, Dahlquist FW (1989) Expression and nitrogen-15 labeling of proteins for proton and nitrogen-15 nuclear magnetic resonance. In: Oppenheimer NJ, James TL (eds) Methods in enzymology, vol. 177. Academic, San Diego, p 44

Neal SE, Eccleston JF, Webb MR (1990) Hydrolysis of GTP by p21 ^{Nras}, the Nras protooncogene product, is accompanied by a conformational change in the wild-type protein: Use of a single fluorescent probe at the catalytic site. Proc Natl Acad Sci USA 87:3562–3565

Pai EF, Kabash W, Krengel U, Holmes KC, John J, Wittinghofer A (1989) Structure of the Ha-ras oncogene product p21 in the triphosphate conformation. Nature 341:209–214

Pai EF, Krengel U, Petsko GA, Goody RS, Kabsch W, Wittinghofer A (1990) Refined crystal structure of the triphosphate conformation of H-ras p21 at 1.35A resolution: implications for the mechanism of GTP hydrolysis. EMBO J 9:2351–2359

Peng JW, Wagner G (1992) Mapping spectral density functions using heteronuclear NMR relaxation measurements. J Magn Reson 98:308–332

Prive GG, Milburn MV, Tong L, De Vos AM, Yamaizumi Z, Nishimura S, Kim S-H (1992) X-ray crystal structures of transforming p21 ras mutants suggest a transition-state stabilization mechanism for GTP hydrolysis. Proc Natl Acad Sci USA 89:3649–3653

Redfield AG, Papastavros MZ (1990) NMR study of the phosphoryl binding loop in purine nucleotide proteins: evidence for strong hydrogen bonding in human N-ras p21. Biochemistry 29:3509–3514

Rensland H, Lautwein A, Wittinghofer A, Goody RS (1991) Is there a rate-limiting step before GTP cleavage by H-ras p21? Biochemistry 30:11181–11185

Richardson J (1981) The anatomy and taxonomy of protein structure. Adv Protein Chem 34:167–339

Schlicting I, John J, Frech M, Chardin P, Wittinghofer A, Zimmerman H, Rosch P (1990) Proton NMR studies of transforming and nontransforming H-ras p21 mutants. Biochemistry 29:504–511

Schultz GE, Schrimer RH (1979) Principles of protein sturcture. Springer, Berlin Heildelberg New York

Sefton BM, Trowbridge IS, Cooper JA (1982) The transforming proteins of the Rous Sarcoma virus, Harvey Sarcoma virus and Abelson virus contain tightly bound lipid. Cell 31:465–474

Shou C, Farnsworth CL, Neel BG, Feig LA (1992) Molecular cloning of cDNAs encoding a guanine-nucleotide-releasing factor for Ras p21. Nature 358:351–354

Tong L, Devos AM, Milburn M, Kim S-H (1991) Crystal structure at 2.2A resolution of the catalytic domains of normal ras protein and an oncogenic mutant complexed with GDP. J Mol Biol 217:503–526

Trahey M, McCormick F (1987) A cytoplasmic protein stimulates normal N-ras p21 GTPase, but does not affect oncogenic mutants. Science 238:542–545

Valencia A, Kjeldgaard M. Pai EF, Sander C (1991) GTPase domains of ras p21 oncogene protein and elongation factor Tu: Analysis of three-dimensional structures, sequence families, and functional sites. Proc Natl Acad Sci USA 88: 5443–5447

Wagner G, Kumar A, Wuthrich K (1981) Systematic application of two-dimensional ^{1}H nuclear magnetic resonance techniques for studies of proteins. Eur J Biochem 114:375–384

West M, Kung HF, Kamata T (1990) A novel membrane factor stimulates guanine nucleotide exchange reaction of ras proteins. FEBS Lett 259:245–248

Willumsen BM, Christensen A, Hubbert NL, Papageorge AG, Lowy DR (1984) The p21 ras C-terminus is required for translocation and membrane association. Nature 310:583–586

Wolfman A, Macara IG (1990) A cytosolic protein catalyzes the release of GDP from $p21^{ras}$. Science 248:67–69

Wuthrich K (1986) NMR of proteins and nucleic acids. Wiley-Interscience, New York

Xu GF, O'Connell P, Viskochil D, Cawthon RM, Robertson M, Culver M, Dunn D, Stevens J, Gesteland R, White R, Weiss R (1990) The Neurofibromatosis type 1 gene encodes a protein related to GAP. Cell 62:599–609

Yamasaki K, Kawai G, Ito Y, Muto Y, Fujito J, Miyazawa T, Nishimura S, Yokoyama S (1989) Conformation change of effector-region residues in antiparallel β-sheet of human c-Ha-ras protein on GDP-GTPγS exchange: a two-dimensional NMR study. Biochem Biophys Res Commun 162:1054–1062

Yamasaki K, Muto Y, Ito Y, Walchli M, Miyazawa T, Nishimura S, Yokoyama S (1992) A ^{1}H-^{15}N NMR study of human c-Ha-ras protein: biosynthetic incorporation of ^{15}N-labeled amino acids. J Bio NMR 2:71–82

CHAPTER 16

Molecular Dynamics Studies of H-*ras* p21–GTP

C.K. FOLEY, L.G. PEDERSEN, T.A. DARDEN, P.S. CHARIFSON, A. WITTINGHOFER, and M.W. ANDERSON

A. Introduction

The simulation of the motion of macromolecules in realistic environments is made possible by the technique of molecular dynamics. Prior to application of the method is the development of a potential energy surface, or "force field", from which the forces on all atoms of the system can be derived. This force field is typically parameterized by resorting to spectroscopic methods and ab initio quantum mechanical computations on smaller molecules. Once the force field has been developed, Newton's equations of motion can be integrated to give the trajectory of the molecule. Most calculations on proteins to date have involved integrations of 20–200 ps, although we note a recent simulation of 550 ps on hen egg white lysozyme (MARK and VAN GUNSTEREN 1992). It is probably reasonable to state that the current level of simulation technique allows one to examine the early stages of unfolding events that might be anticipated as a molecule is solvated from a crystalline environment. Many aspects of the methodology are adequately discussed in recent reviews by VAN GUNSTEREN and MARK (1992) and KARPLUS and PETSKO (1990).

In this contribution we review our application of the molecular dynamics method to the H-*ras* p21 system based on the 1.35-Å resolution crystal coordinates of PAI et al. (1990). This paper is an extension of our earlier work (FOLEY et al. 1992) on this system. The p21–GTP system represents a substantial challenge for simulation in that the molecule and its ligands (GTP, Mg^{2+} ion) are highly charged and there are no disulfide bonds present to provide intrinsic stabilization. Other theoretical studies have been carried out by DYKES et al. (1992) based on the alpha carbon coordinates of the TONG et al. (1991) structure of p21–GTP and by PRIVE et al. (1992), who attempt to model the transition state of the hydrolysis of p21–GTP based on their crystal structure of p21–GTP (analog) and a similar crystal structure of the mutant protein in the GTP-bound state. The essential biological details of this molecule are more than adequately discussed in other papers of this volume; however, we do emphasize that our simulations pertain to the structure of p21–GTP which is responsible for the *intrinsic* GTP hydrolysis process.

B. Methods

All energy minimization and molecular dynamics computations were carried out with the AMBER 3A program (Weiner et al., 1984, 1986; Seibel 1989) using the 1.35-Å resolution x-ray crystallographic coordinates of H-*ras* p21 (residues 1-166) of Pai et al. (1990). The all-atom force field (Weiner et al. 1986) was employed for all calculations. A 9-Å nonbonded cutoff was employed for the energy minimization; however, a twin-range cutoff (9 Å, 22 Å) proved necessary for all dynamics calculations. A rectangular box of approximately 9500 water molecules of the TIP3P description (Jorgensen et al. 1983) was used to solvate the protein. A mass of 1 amu (atomic mass unit) was used for all hydrogen atoms and all bonds involving hydrogen were constrained (see below). A step length of 1 fs was used throughout with the nonbonded list updated every ten integration steps. At each list update, the force was corrected to include electrostatic and attractive van der Waals interactions in the range of 9.0–22.0 Å. A parallel computation utilizing only a single 9-Å cutoff, but with all other procedures the same, was performed to fully assess the importance of the twin-range cutoff. The frequent nonbonded list update and the twin cutoff were found to be necessary in this system to equilibrate the RMS deviation (simulation structure vs. crystal structure). Similar conclusions resulted from our simulations (TAD, PSC, LGP) performed on the calcium-bound bovine prothrombin fragment-1 molecular system, a highly charged ionic system (Hamaguchi et al. 1992). In the case of prothrombin, the twin cutoff was essential for the retention of an important ionic interaction in which the N terminus was folded back onto the protein. The simulations were performed at 300 K after very careful initial ramping of the temperature over the first 6 ps of the simulation. Care was taken to insure that the protein and water bath exhibited the same average temperatures throughout the simulations.

The crystal structure (Pai et al. 1990) was determined with a very slowly hydrolyzing GTP analog in which an isoelectronic N-H group replaced the etherial oxygen between the γ- and β-phosphate functional groups. The simulations were performed with the physiologically relevant oxygen in place. A single proton was added to the oxygen on the GTP γ-phosphate group that was not involved in any hydrogen-bonding interaction with the protein. The point charge parameters for the GTP were obtained by fitting the ab initio quantum mechanical electrostatic potential at the 3-21G* basis level (charges scaled to the 6-31G** level). The ab initio quantum chemical calculation was performed with the Gaussian 90 program (Frisch et al. 1990) and the subsequent electrostatic potential calculation was performed with the CHELP code (Chirlian and Francl 1987a,b). Computations were carried out on Cray Y-MPs at the National Cancer Center (Frederick, MD) and the North Carolina Supercomputing Center (Research Triangle Park, NC). Graphics analysis capability was provided by the MULTI program (Darden et al. 1991) on Silicon Graphics workstations.

A number of modifications to the distributed AMBER code were made to improve the efficiency of the computation; these are reviewed in FOLEY et al. (1992). Briefly, the original SHAKE algorithm was replaced a by more rapidly convergent code, a nonbond list update methodology was modified to employ a grid technique, preprocessing of the nonbond list interactions involving solvent to avoid null interactions was effected, and box recentering to the protein center at each step was added. These changes led to an approximately twofold increase in speed without compromising accuracy.

Based on a number of trial calculations, a protocol was establised for the simulations: (a) minimization of the protein hydrogens was performed with the heavy atoms fixed and with a distance-dependent dielectric constant (no solvent); (b) after placing the output from (a) in a large water box with protein atoms at least 13 Å from a box side, charge-neutralizing sodium ions were added sequentially to positions of highest negative electrostatic potential; the crystallographic waters and newly added sodium ions were then energy minimized in the constant field of the protein using the distance-dependent dielectric constant, and resolvation of the system followed; (c) the water was then equilibrated by first minimizing the water in the constant field of the protein and then performing 20 ps of dynamics on the water only; and (d) the output of (c) was then subjected to energy minimization of the water (only) followed by energy minimization of the entire system. The system was then subjected to the temperature ramping procedure (0–300 K) over 6 ps.

C. Results and Discussion

I. General Features of the Wild-Type Simulations

1. RMS

The simulation with the twin-range cutoff on the nonbonded interactions was carried out until the RMS deviation (backbone atoms for the simulation vs. crystal structure) became essentially constant (Fig. 1). The RMS for the average (110–130 ps) simulation structure was 1.13 Å, and when the simulation was extended to 200 ps, the corresponding RMS only rose to 1.32 Å. On the other hand, the simulation with only the single 9-Å cutoff has an RMS value of 2.46 Å (average structure from 110–130 ps) and is still rising with time at 130 ps (Fig. 1). These results graphically illustrate the importance of accounting for long-range ionic forces in this system. RMS (over all atoms) for the wild-type simulation was approximately 1.28 Å for the 110- to 130-ps average structure.

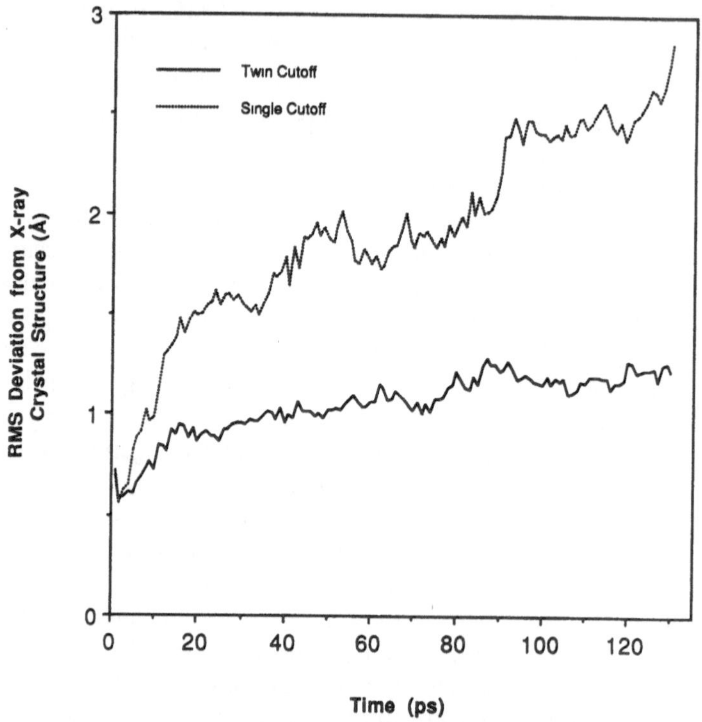

Fig. 1. Comparison of the RMS deviation (backbone atoms of simulation vs. X-ray) for the 9-Å nonbonded cutoff simulation (*dotted line*) and for the simulation employing the twin-range (9 Å, 22 Å) cutoff (*solid line*)

2. Protein–GTP Contacts

The protein forms a number of hydrogen bonds and ionic contacts with the GTP and with the magnesium ion in the crystal structure (PAI et al. 1990). The contacts that form the hexagonally coordinate environment for the magnesium ion (oxygens from the γ- and β-phosphate, side chain oxygens from Ser17 and Thr35, and the oxygens from two water molecules) are conserved in the two simulations (Table 1). On the other hand, the contacts of the backbone amide H of Thr35 and Gly60 with γ-phosphate oxygens, which are preserved in the twin cutoff simulation, are dissociated in the single cutoff simulation. Most of the major contacts with the β-phosphate and the guanine are also preserved in the twin cutoff simulation, although two of the GTP ribose–protein contacts (with Y29 and D30) are lost in the simulation. The γ-phosphate contact with A18 is also preserved. The side chain of Lys16 forms a hydrogen bond network with both terminal phosphate groups in the crystal structure, and this interesting interaction is stable in the twin-cutoff simulation but is lost in the single-cutoff simulation. This interaction is stable through 200 ps (we have seen it remain stable for at

Table 1. Significant protein–GTP contacts

Interactions	X-ray (Pai et al. 1990)	Wild-type (110–130 avg) twin-range cutoff	Wild-type (110–130 avg) single 9-Å cutoff
Mg^{2+}	γ- and β-phosphates 2 waters T35, S17	γ- and β-phosphates 2 waters T35, S17	γ- and β-phosphates 2 waters T35, S17
γ-Phosphate	Mg^{2+}, G13, K16 T35, G60	Mg^{2+}, K16 T35, [G60]	Mg^{2+}
β-Phosphate	Mg^{2+}, G13, G15 K16, S17	Mg^{2+}, G13, V14 G15, K16, S17	Mg^{2+}, [G13], V14 G15, K16, [S17]
α-Phosphate	A18	[G15], A18	[G15]
Ribose	Valine 29, D30, [K117]	[K117]	
Guanine	N116, D119 A146	N116, D119 [S145], A146 K147	N116, D119 A146, K147

The criterion for a hydrogen-bonded contact is [O,N]-[N,O] <3.5 Å and [O,N]-H <2.4 Å. Contacts shown in brackets indicate a near contact, usually the longer criterion met, but not the shorter. Ionic contact criterion: <2.5 Å. The water molecules bound to the Mg^{2+} are identical to those of the crystallographic structure.

least 400 ps in simulations on a mutant p21), and it is likely to be very important for defining the structure of the phosphate-binding domain of the protein. There can be little doubt that the 9-Å nonbond cutoff simulation is nonphysical.

3. Secondary Structure

The secondary structures of the various coordinate sets (crystal, wild-type 110–130 ps avg. twin- and single-cutoff structures) were computed using the algorithm of Kabsch and Sander (1983) and are given in Table 2. The crystal structure and the twin-cutoff simulation are very similar in the assignment of β-sheet and helix character, whereas the 9-Å cutoff simulation has somewhat less β-sheet character. The β-sheet in the N-terminal region appears to be weakened by going from crystal to solution, and the beginning of the antiparallel β-sheet as defined by residues 38–57 appears to be weakened by forming the solution (not shown). This observation is indicated by the large hydrogen-bond distance fluctuations experienced for the β-sheet hydrogen bond N_{38}–O_{57}.

II. Mechanism of Hydrolysis

What can simulations such as those described above say about GTP hydrolysis? The calculations use classical mechanics and therefore cannot accom-

Table 2. Secondary structure using the Kabsch-Sander algorithm (Kabsch and Sander 1983)

Seq. No.	1	2	3	4	5	6	7	8	9	0	1	2	3	4	5	6	7	8	9	0	1	2	3	4	5	6	7	8	9	0	1	2	3	4	5	6	7	8	9	0	1	2	3	4	5	6	7	8	9	0	–50
X-ray		E	E	E	E	E	E	E	E							H	H	H	H	H	H	H	H	H													E	E	E	E	E	E	E	E	E	E	T	T	E	E	
Wt.-twin				E	E	E	E	E	E			T	T			H	H	H	H	H	H	H	H	H	T												E	E	E	E	E	E	E	E	E	E	T	T	E	E	
Wt.-single			E	E			E	E	E			T	T			H	H	H	H	H	H	H	H	H	T														E	E	E	E	E	E	E	E	T	T	E	E	
	1	2	3	4	5	6	7	8	9	0	1	2	3	4	5	6	7	8	9	0	1	2	3	4	5	6	7	8	9	0	1	2	3	4	5	6	7	8	9	0	1	2	3	4	5	6	7	8	9	0	–100
X-ray	E	E	E	E	E	E	E	E										H	H	H	H	H	H	H			E	E	E	E	E	E	E	T	T		H	H	H	H	T	T	H	H	H	H	H	H	H	H	
Wt.-twin	E	E	E	E	E	E	E	E											H	H	H	H	H	H			E	E	E	E	E	E	E	T	T		H	H	H	H	H	H	H	H	H	H	H	H	H	H	
Wt.-single	E	E	E	E	E	E	E						T	T	T	H	H	H	H	H	H	H	H	H				E	E	E	E	E	T	T	T		T	H	H	H	H	H	H	H	H	H	H	H	H	H	
	1	2	3	4	5	6	7	8	9	0	1	2	3	4	5	6	7	8	9	0	1	2	3	4	5	6	7	8	9	0	1	2	3	4	5	6	7	8	9	0	1	2	3	4	5	6	7	8	9	0	–150
X-ray	H	H	H	T	T						E	E	E	E	E	E		T	T								H	H	H	H	H	H	H	H	H	H	H	T			E	E	E				T	T	T		
Wt.-twin	H	H	H	H							E	E	E	E	E	E		T	T								H	H	H	H	H	H	H	H	H	H	H	T			E	E	E			T	T	T	T		
Wt.-single	H	H	H	H								E	E	E	E	E		T	T								H	H	H	H	H	H	H	H	H	H	H	T			E	E	E			T	T	T	T		
	1	2	3	4	5	6	7	8	9	0	1	2	3	4	5	6																																			–166
X-ray		H	H	H	H	H	H	H	H	H	H	H	H	H	T																																				
Wt.-twin		H	H	H	H	H	H	H	H	H	H	H	H	H	H																																				
Wt.-single			H	H	H	H	H	H	H	H	H	H	H	H																																					

Wt.-twin, 110–130 ps average structure, twin-range (9 Å, 22 Å) cutoff on nonbonded interactions; Wt.-single, 110–130 ps average structure, single 9-Å cutoff on nonbonded interactions; E, β sheet; H, helix; T, hydrogen-bonded turn.

modate bond breakage or formation. Also, the most interesting biological observation concerning this system is that whereas mutations at positions 12 (e.g., G12V) and 61 (e.g., Q61L) can reduce the intrinsic hydrolysis rates by an order of magnitude, the relative reduction (mutant vs. wild-type) is much greater for the GAP-induced hydrolysis rates (BOLLAG and MCCORMICK 1991). These simulations can say little about the GAP-induced rates since there is no GAP present. On the other hand, it is possible to examine the time-dependent changes that occur in the catalytic region for possible conformational events that may provide for a mechanism for hydrolysis. It is presumed that hydrolysis takes place by an in-line attack of a water molecule, perhaps activated by protein contact, so as to form a trigonal bipyramidal transition state at the γ-phosphorus.

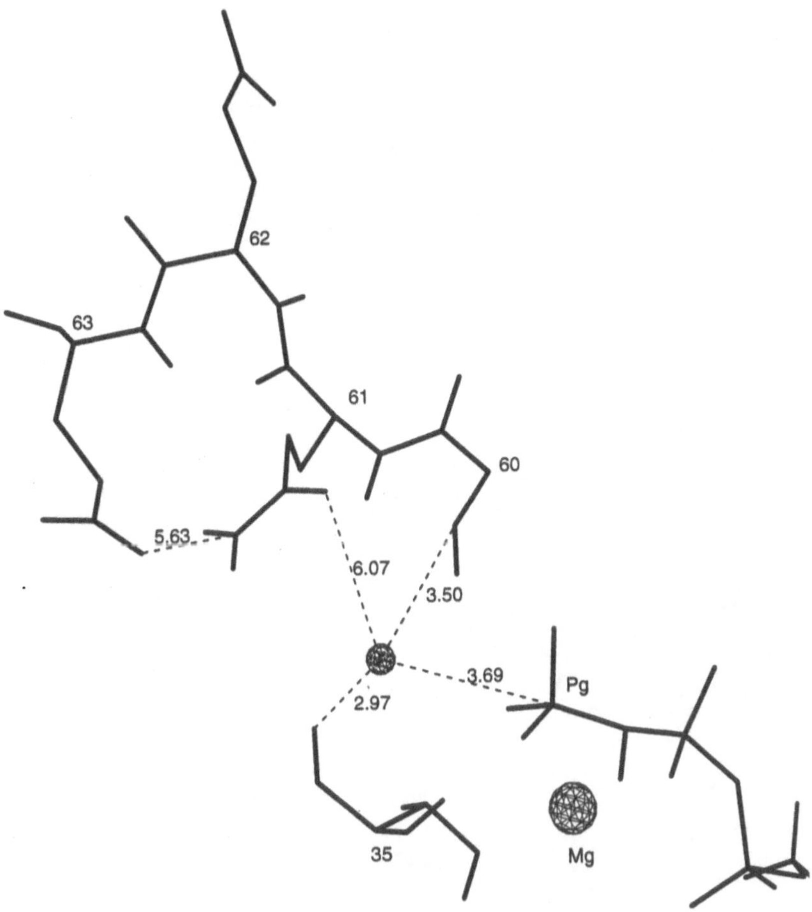

a

Fig. 2. The active site region for the hydrolysis of GTP by p21. **a** The X-ray crystal structure of p21; **b** the "alternate" X-ray crystal structure (10); **c** the 70-ps snapshot of the wild-type simulation

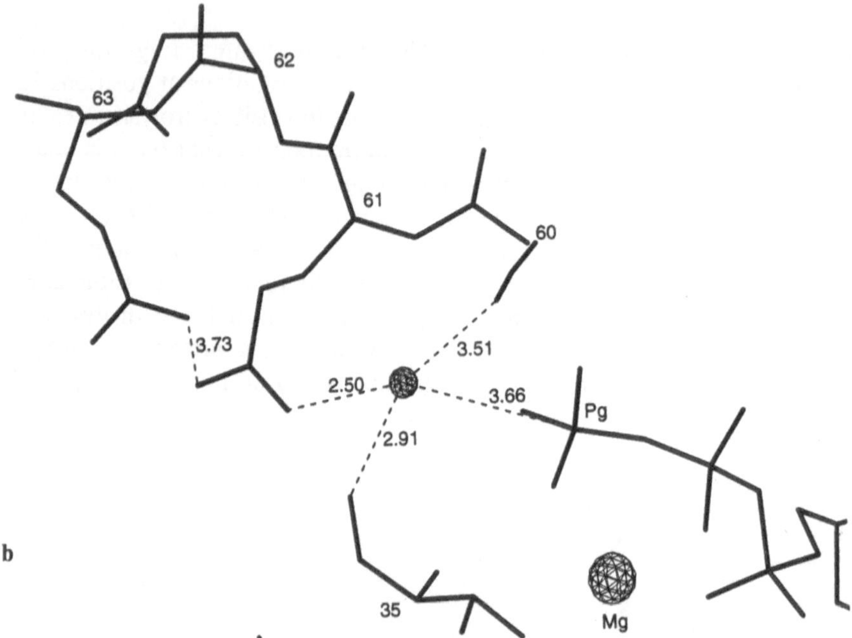

b

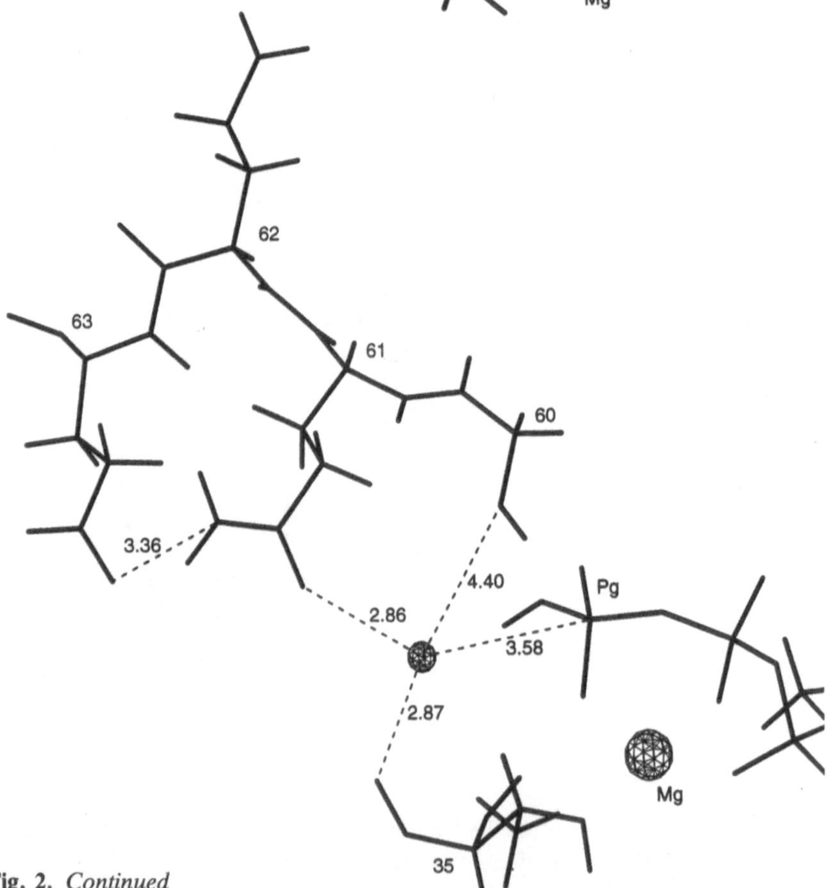

c

Fig. 2. *Continued*

PAI et al. (1990) suggest that the elements of the catalytic complex are a water molecule that is localized in the region of the three oxygens of the γ-phosphate, approximately bisecting the triangle that these oxygens form, Thr35, whose backbone oxygen interacts with this localized water, and the side chain oxygen of Q61, which also interacts with the water molecule. The side chain of Glu63 could also possibly stabilize the NH_2 group of the

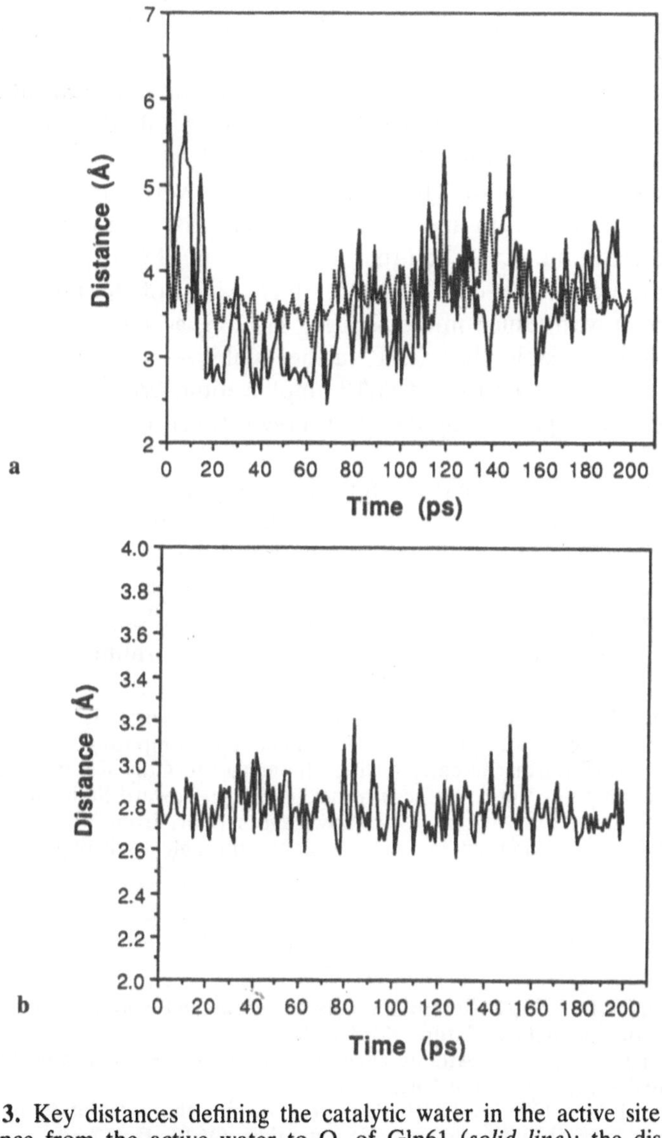

Fig. 3. Key distances defining the catalytic water in the active site of p21. **a** The distance from the active water to O_ε of Gln61 (*solid line*); the distance from the active water to the γ-phosphorus atom (*dotted line*). **b** The distance from the active water molecule to the backbone O of Thr35

Gln61 side chain through a charged hydrogen-bonding interaction. This proposal is based on the high mobility (large crystallographic B factors) of the loop defined by residues 61–64, which generates conformations of the side chain of Q61 either pointing away from the γ-phosphate or toward it. Figure 2 compares these possibilities to the simulation structure: Fig. 2a shows the catalytic region for the coordinate file that we used to begin our simulations (Pai et al. 1990), Fig. 2b shows a possible alternate structure suggested by Pai et al. (1990) for this region based on the crystal coordinates, and Fig. 2c shows the catalytic region 70 ps into the simulation. It is clear that even though the Glu61 side chain begins the simulation pointing away from the γ-phosphate, it moves into a position very similar to the proposed alternate position (Fig. 2b). Figure 3a shows the time dependence of the distance from the water molecule to the O_ε of Gln61 and to the γ-phosphorus. Figure 3b likewise shows the distance from the water molecule to the backbone atom of Thr35. These figures verify that a water molecule resides near the in-line attack position, that Thr35 stabilizes (activates?) this water molecule, and that the side chain of Q61 periodically interacts with the water. We note that the $-NH_2$ group of the Gln61 side chain does interact with Glu63 in a stabilizing fashion, as suggested by Pai et al. (1990). It is possible that GAP further stabilizes or modifies this interaction, leading to the enormous GAP amplification. We conclude that the catalytic complex suggested by Pai et al. (1990) is certainly dynamically feasible.

Molecular dynamics, in its current formulations, provides a method for estimating the structural changes of a macromolecule to be expected upon solvation from the crystal. It also provides a method for examining dynamics in key regions of a molecule, such as the catalytic domain of p21. Future methodology will hopefully improve the manner in which long-range interactions are treated as well as add reliable bond-breaking/forming capability.

Acknowledgements. We wish to thank the Frederick Cancer Research Center (NCI) for access to their Cray Y-MP, the North Carolina Supercomputing Center for use of their Cray Y-MP during its early testing phase, and to Cray Research for a generous grant of computer time. We wish to acknowledge Howard Smith for very able technical assistance. LGP wishes to acknowledge support from NIH for HL-27995 and NIEHS for making its facilities available through a visiting scientist appointment.

References

Bollag G, McCormick F (1991) Differential regulation of *ras*GAP and neurofibromatosis gene product activities. Nature 351:576–579

Chirlian LE, Francl MM (1987a) Atomic charges derived from electrostatic potentials: a detailed study. J Comput Chem 8:894

Chirlian LE, Francl MM (1987b) CHELP: a program for the calculation of net atomic charges from ab initio electrostatic potentials. QCPE Bull 7:39

Darden T, Johnson P, Smith H (1991) MULTI: a shared memory approach to cooperative molecular modeling. J Mol Graphics 9:18–23
Dykes DC, Brandt-Rauf P, Luster SM, Chung D, Friedman FK, Pincus MR (1992) Activated conformations of the *ras*-gene-encoded p21 protein. 1. An energy-refined structure for the normal p21 protein complexed with GDP. J Biomol Struct Dyn 9:1025–1044
Foley CK, Pedersen LG, Charifson PS, Darden TA, Wittinghofer A, Pai EF, Anderson MW (1992) Simulation of the solution structure of the H-*ras* p21–GTP complex. Biochemistry 31:4951–4959
Frisch MJ, Head-Gordon M, Trucks GW, Foresman JB, Schlegel HB, Raghavachari K, Robb MA, Binkley JS, Gonzalez C, Defrees DJ, Fox DJ, Whiteside RA, Seeger R, Melius CF, Baker J, Martin RL, Kahn LR, Sterwart JJP, Topiol S, Pople JA (1990) Gaussian 90. Gaussian, Pitttsburgh
Hamaguchi N, Charifson P, Darden T, Xiao L, Padmanabhan K, Tulinsky A, Hiskey R, Pedersen L (1992) Molecular dynamics simulation of bovine prothrombin fragment 1 in the presence of calcium ions. Biochemistry 31:8840–8848
Jorgensen WL, Chandrasekhar J, Madura JD, Impey RW, Klein ML (1983) Comparison of simple potential functions for simulating liquid water. J Chem Phys 79:926–935
Kabsch W, Sander C (1983) Dictionary of protein secondary structure: pattern recognition of hydrogen bonded and geometrical features. Biopolymers 22: 2577–2637
Karplus M, Petsko GA (1990) Molecular dynamics simulations in biology. Nature 347:631–639
Mark AE, van Gunsteren WF (1992) Simulation of the thermal denaturation of hen egg white lysozyme: trapping the molten globule state. Biochemistry 31: 7745–7748
Pai EF, Krengel U, Petsko GA, Goody RS, Kabsch W, Wittinghofer A (1990) Refined crystal structure of the triphosphate conformation of H-*ras* p21 at 1.35 Å resolution: implications for the mechanism of GTP hydrolysis. EMBO J 9:2351–2359
Prive GG, Milburn MV, Tong L, DeVos AM, Yamaizumi Z, Nishimura S, Kim S-H (1992) X-ray crystal structures of transforming p21 *ras* mutants suggest a transition-state stabilization mechanism for GTP hydrolysis. Proc Natl Acad Sci USA 89:3649–3653
Tong L, De Vos AM, Milburn MV, Kim S-H (1991) Crystal structures at 2.2 Å resolution of the catalytic domains of normal *ras* protein and an oncogenic mutant complexed with GDP. J Mol Biol 217:503–516
Seibel G (1989) AMBER 3.0 revision A. Dept Pharmaceut Chem, UCSF, San Francisco, CA 94143
van Gunsteren WF, Mark AE (1992) On the interpretation of biochemical data by molecular dynamics computer simulation. Eur J Biochem 204:947–961
Weiner SJ, Kollman PA, Case DA, Singh UC, Chio C, Alagona C, Profeta, Jr. S, Weiner P (1984) A new force field for molecular mechanical simulation of nucleic acids and proteins. J Am Chem Soc 106:765–784
Weiner SJ, Kollman PA, Nguyen DT, Case DA (1986) An all atom force field for simulations of proteins and nucleic acids. J Comput Chem 7:230–252

Section III
Small Ras-Related GTpases

A. Control of Growth and Differentiation by the Ras Family

CHAPTER 17

The Discovery of *ras* and Its Biological Importance

R.A. Weinberg

The roots of our knowledge about *ras* oncogenes are traceable directly back to the early days of tumor biology, indeed to the work that ultimately spawned the entire field of oncogene and human immunodeficiency virus (HIV) research and much of the current work on the transduction of mitogen-induced growth signaling. By uncovering a small group of genes that could transform normal cells into cancer cells, the tumor virologists opened the window that revealed the complex signal-processing circuitry used by cells to decide when they should remain quiescent, when they should grow or differentiate, and when they should die.

Peyton Rous's chicken sarcoma virus was the first of this class of viruses that passed before the microbiologist's eye, discovered by him in 1908 (Rous 1910). For reasons that remain unclear to this day, the field of tumor virology initiated by him remained in stasis for more than half a century. During this period came the discoveries of polyomavirus, SV40 virus, and Shope papillomas virus – now known to be DNA tumor viruses of the papovavirus class – mouse mammary tumor virus, and a group of mouse leukemia viruses (Gross 1970).

By the early 1970s, it had become clear that the Rous virus, the closely related avian leukosis viruses, and the mouse leukemia viruses differed greatly in structure and function from the much-studied DNA tumor viruses. Virions of the Rous virus and others of its class were found to carry their genetic information in the form of RNA molecules wrapped in a lipid bilayer coat. As we learned in 1970, these virus particles carried the enzyme reverse transcriptase for copying the virion-bound RNA into DNA form within the virus infected cells (Baltimore 1970; Temin and Mizutani 1970).

This discovery of reverse transcriptase was momentous for the field of tumor virology. Among the immediate fallouts was an enzymological test for the enzyme that yielded an extraordinarily sensitive assay for the presence of these RNA tumor viruses – retroviruses – in cell cultures. With this test in hand, a host of other known or suspected tumor viruses were identified as members of the retrovirus class (Weiss et al. 1984).

The early enthusiasm about reverse transcriptase obscured the fact that this discovery provided no information into the major question confronting this field of research: how do retroviruses cause cell transformation in vitro and, in turn, macroscopic tumors in vivo? In the case of Rous sarcoma virus

(RSV), this question seemed solved, at least in principle, with the discovery in 1969–1970 of RSV mutants that caused infected cells to show a temperature-dependent transformation phenotype, causing cells to grow as tumor cells at the permissive, low temperature but not at the nonpermissive, higher temperature (Toyoshima and Vogt 1969; Martin 1970). This discovery pointed to the existence of a discrete viral gene that could act within infected cells to elicit the changes associated with malignant cell transformation. A mutant form of this gene presumably encoded a thermolabile gene product whose inactivation at high temperature deprived the cell of a function needed for the continuous maintenance of the virus-initiated, transformed state. This virus-borne gene was termed the *src* oncogene.

The existence of the *src* oncogene as a discrete genetic and physical entity was further illustrated by the discovery of deletion mutants of RSV which could still replicate in susceptible cells but had lost all traces of transforming ability (Duesberg and Vogt 1973). By comparing these deletion mutants with their wild-type antecedents, one was able to map the *src* gene to a discrete region near the 3′ end of the viral genomic RNA (Wang et al. 1975).

Even this breakthrough left two substantial puzzles unaddressed. The first concerned the biochemical mechanisms of transformation: how does the viral gene product, still unknown at the time, elicit the transformed phenotype? The other dealt with the origin of the viral oncogene – where did it come from?

This second question was prompted by comparisons between the genomes of Rous virus and the closely related avian leukosis viruses. These viruses shared a virtually identical set of genes templating the process of virus replication. The additional presence of the distinctive *src* oncogene in the Rous genome set this virus apart, conferring on it the ability to transform cells in culture and induce malignancy with very short latency in infected animals. The widespread occurrence of leukosis viruses and the unique isolation of the related Rous virus suggested that RSV was somehow spawned from one or another leukosis virus. It was precisely this proposed relationship which brought into sharp focus the question of the origins of the viral oncogene.

Taking advantage of the physical differences between the genomes of the transformation-defective deletion mutant of RSV and the wild-type virus, Bishop and Varmus were able to generate a DNA probe that specifically recognized viral *src* sequences. With this *src*-specific probe in hand, they uncovered viral-oncogene-related sequences residing in the genomes of a variety of multicellular organisms found at all branches of the phylogenetic tree (Stehelin et al. 1976a,b). These related sequences constituted a gene whose great evolutionary conservation implied an essential role in normal organismic or cellular physiology. This gene was termed the *src* proto-oncogene. It was ostensibly involved in regulating normal cell proliferation, being subverted on rare occasions by retroviruses like the progenitor of RSV. This work established the paradigm that retroviruses can capture cellular proto-oncogenes, deregulate their activity, and exploit them to transform subsequently infected cells.

This model was extended with rapidity to a variety of other tumorigenic retroviruses known to be able to induce tumors in host animals with great rapidity. Each of these was found to carry an oncogene in its genome, and each of these viral oncogenes was found to have a normal cellular antecedent. Indeed, within months of the initial report, Arthur Frankel and Peter Fischinger reported that Moloney mouse sarcoma virus also carried host-cell-related sequences in the viral genome, ostensibly used by this virus for inducing cell transformation (FRANKEL and FISCHINGER 1976).

Initially, those working with the avian viruses presumed that the host-related oncogenic sequences found in other rapidly acting tumorigenic retroviruses would be versions of the previously discovered *src*-proto-oncogene, and accordingly spoke in terms of the Harvey sarcoma virus *src* gene. But it soon became apparent that many of these virus-associated genes, while acting like *src* in the induction of cell transformation, were not themselves related in any way to the RSV *src* oncogene. As a consequence, a nomenclature sprang up, following which each virus-associated oncogene was labeled with a three-letter italicized acronym reflecting the virus in which it was originally discovered. The oncogenes of Kirsten and Harvey rat sarcoma viruses, two independently isolated virus strains, were apparently unrelated, but the two genes were given related names – Ki-*ras* and Ha-*ras* respectively, each reflecting the cellular origin of the respective viral oncogenes. A flood of other newly named oncogenes followed in the late 1970s. Most well-known of these was the *myc* gene, named for its initial association with the genome of avian myelocytomatosis virus.

The origins of Harvey and Kirsten sarcoma viruses were in fact more complex than that of the Rous virus. Rous had sprung up spontaneously in a chicken delivered to Peyton Rous by an upstate New York poultry farmer. The Harvey and Kirsten viruses were the result of laboratory manipulation which involved injection of mouse leukemia virus into rats. The mouse leukemia virus genome, like that of avian leukosis virus, carries a full complement of replicative genes, but lacks any sequences that have specific oncogenic potential. After having been introduced into a rat, the progenitor of the Harvey virus underwent two successive types of recombination events with host cell chromosomal sequences (ELLIS et al. 1980, 1981). The first event led to the acquisition of endogenous retrovirus sequences – yielding fragments of endogenous rat proviruses that lack replicative potential when present in the rat genome and exert little effect on the functioning of the resulting chimeric retrovirus genome. A subsequent recombination event led to the acquisition of sequences, termed variously H-*ras* or Ha-*ras*.

The end product of these steps, as Jennifer Harvey first reported (1964), was a virus that now induced sarcomas and splenomegaly in infected mice and rats, unlike the parental virus, which caused only leukemias in these hosts. Similarly, 3 years later, KIRSTEN and MAYER (1967) reported a mouse lymphoma virus that yielded a sarcoma-inducing progeny stock after passage through rats.

The cellular progenitor of the acquired Ha-*ras* oncogene was found to encompass almost 6 kb of chromosomal DNA sequence; the derived viral oncogene was only one-tenth as large (CHANG et al. 1980). The difference, as was the case with the *src* gene, was attributable to the process of reverse transcription of spliced cellular mRNAs, which led to the disappearance of the introns which, in the case of the Ha-*ras* proto-oncogene, composed 90% of its sequence content.

There was, to be sure, one other significant distinction between RSV and these sarcoma viruses. Unlike Rous sarcoma virus, which in its most complete form is able to replicate and transform, these two sarcoma virus genomes were replication defective. The genetic events accompanying acquisition of cellular sequences caused a compensating deletion of leukemia virus sequences essential for replication. As a consequence, these sarcoma virus genomes could only be propagated in the presence of coinfecting, replication-competent murine leukemia viruses, which supplied complementing gene products essential for virus replication.

As it turned out, the use of a common term for the two rat sarcoma oncogenes was inadvertently prophetic, in that the two were found to derive from distinct, but clearly related, cellular genes. Use of antitumor antisera and genome sequencing showed that the Ha-*ras* and Ki-*ras* genes carried by the two viral genomes both encode 21-kDa proteins (SHIH et al. 1979). These proteins were also detectable in uninfected cells, although in far lower amounts than were ascertainable in virus-transformed cells. Indeed, this presence of such $p21^{ras}$ proteins in many normal cell types further strengthened the concept that they play a role in normal cell physiology.

These cellular sequences carried by the two rat sarcoma viruses had all the trappings of oncogenes, but the mechanism leading to their activation from antecedent proto-oncogenes was initially quite unclear. A most attractive explanation held that expression of the viral oncogenes was driven by the strong viral transcriptional promoter, in cont*ras*t to the relatively weak expression that characterized the counterpart proto-oncogenes residing in normal cells. Following this thinking, the overexpression of these acquired cellular gene products would suffice to induce malignant transformation of cells infected by either of these viruses. An opposing view held that the viral gene products were structurally and thus functionally distinct from the corresponding cellular gene products. This latter view was hardly encouraged by the early findings that both the viral oncogene-encoded and the prot-oncogene-encoded proteins have comparable molecular weights and ostensibly similar, if not identical, structures.

But before this question was resolved, another seminal experiment – carried out, like many others, in the laboratory of Edward Scolnick –

yielded an important surprise: the *ras* protein was found capable of binding guanine nucleotides (SHIH et al. 1980). By the time of this discovery in 1979, the RSV gene product of 60 kDa was known from work a year earlier to be a protein kinase (COLLETT and ERIKSON 1978), but the biochemical mechanisms of action of the few oncoproteins then known was totally mysterious.

The $p21^{ras}$ proteins were only one-third the mass of the *src* kinase, and thus, with great likelihood, members of a distinct family of proteins. Scolnick's group was struck by the fact that the Harvey and Kirsten oncoproteins were in fact phosphoproteins, and thus perhaps capable of autophosphorylation. Attempts at labeling these proteins in vitro by addition of radioactive ATP were unsuccessful, but GTP, used in parallel, ostensibly as a control, did contribute a phosphate group to the $p21^{ras}$ to which it bound (SCOLNICK et al. 1979).

In retrospect, we realize that this autophosphorylation by the Harvey and Kirsten proteins was an idiosyncrasy shown only by them and by none of the $p21^{ras}$ proteins described since then. We now know that these two viral proteins have sustained structural alterations that result in the presence of threonines at their 59th residues, and that these threonines capture the γ-phosphate of bound GTP covalently (SHIH et al. 1980). This fortuitous circumstance created the observed autophosphorylation, but any illusions that these proteins were active kinases were dispelled by the discovery that GDP could bind to $p21^{ras}$ with an affinity similar to that of the triphosphosphate. The analogy with, the heterotrimeric G-proteins soon became clear and striking.

Sequence analysis of the virus-borne oncogenes and the antecedent proto-oncogenes soon revealed subtle structural distinctions between the two versions of each of these genes. There was, to begin with the aforementioned base substitutions leading to the presence of threonines at residue 59. Further, the viral alleles carried point mutations affecting the 12th amino acid residue of the p21 proteins. The significance of all these changes was most obscure, and could be ascribed to the adventitious mutations occurring in the viral alleles during their transduction by the retrovirus genomes. Still, the striking parallelism between the mutations acquired independently by the Harvey and Kirsten oncogenes suggested some strong functional advantage which could select these particular mutations from among the thousands that were generated during the repeated rounds of viral replication undertaken by the progenitors of these two virus stocks. And there the puzzle lay until work from an ostensibly unrelated line of research converged on these viruses in a most unexpected way.

This other work, begun in my own laboratory, traced its roots back to work on Moloney (MoSV) and Harvey sarcoma virus (HaSV) genomes. In 1977, we found that through use of gene transfer (transfection), NIH3T3 mouse fibroblasts could be transformed by transfection of MoSV DNA, yielding foci of piled-up, refractile transformants in the otherwise flat monolayer culture (ANDERSSON et al. 1979). The work was soon extended to the viral DNAs of Harvey sarcoma virus as well (GOLDFARB and WEINBERG 1979).

This viral DNA could be obtained by several means. Initial experiments used the unintegrated viral DNAs synthesized in recently infected cells by the viral reverse transcriptase. This transfection-focus assay was soon found to be extraordinarily sensitive for detecting viral oncogenes present amid extraordinarily high excesses of unrelated DNA sequence. For this reason, we transfected the chromosomal DNA of virus-infected cells, each of which contained a small number (perhaps one) copy of the HaSV integrated into their chromosomal DNA (GOLDFARB and WEINBERG 1981). This DNA also yielded foci upon introduction into NIH3T3 cells. The conclusion was clear: we could detect a single viral oncogene residing amongst a 1 000 000-fold excess of cellular DNA sequences.

The power of the transfection-focus assay was soon brought to bear on a seemingly unconnected problem in cancer biology. This other problem concerned chemical carcinogenesis and its mechanisms. By 1978, when this work was begun, it had become clear from the work of Ames and many others that mutagenic chemicals were likely to be the causative agents for a variety of human and animal tumors (MCCANN and AMES 1976). Yet the mechanism of chemical carcinogenesis as it occurred in the body of a mammal seemed impossibly complex to us.

One possible simplification came from the work of Heidelberger who showed that C3H10T1/2 mouse fibroblasts growing in a culture dish could be transformed following exposure to the well-known carcinogen 3-methylcholanthrene (LANDOLPH and HEIDELBERGER 1979). These C3H10T1/2 mouse fibroblasts were biologically quite similar to the NIH3T3 cells and, following such chemical transformation, they showed phenotypes that were at least superficially similar to those of HaSV-transformed NIH3T3 cells. Yet these chemically transformed cells apparently had not acquired any viral genetic information.

Inspired by the discovery of proto-oncogenes and the notion that carcinogens were often mutagenic, we reasoned that these cells might carry a mutant oncogene, activated at the hands of the mutagenic carcinogen. Such an oncogene might then be detected by the transfection-focus assay which could pick up the presence of a single oncogene residing in a complex cellular genome. That was indeed the outcome (C. SHIH et al. 1979). The transfection-focus assay showed that these chemically transformed cells carried oncogenic information in their DNA. It was clear that this information could be passed from cell to cell via the simple artifice of the transfection procedure. Within several years, a number of other tumor cells, including human tumor cells, were shown to carry such transforming information in their DNA (COOPER et al. 1980; SHIH et al. 1981).

Taken together with the earlier work on retroviruses, this suggested that the cell carried two distinct sets of proto-oncogenes in its genome: the proto-oncogenes discovered through study of the viral oncogenes, and the genes activated during nonviral carcinogenesis by mutagens such as methylcholanthrene or spontaneously by random misreplication of DNA. Were these two repertoires of proto-oncogenes indeed related to one another?

A simple experiment, performed only in 1982, showed that the transfected cellular oncogenes and the viral oncogenes were indeed embarrassingly closely related to one another. In fact, an oncogene detected in the genome of a human bladder carcinoma line turned out to be none other than the HaSV *ras* oncogene that we had worked on 6 years earlier! This discovery, made simultaneously with others, greatly simplified the paradigm of oncogenes (Der et al. 1982; Parada et al. 1982; Santos et al. 1982). From then on, it became clear that proto-oncogenes could be activated by two distinctive routes: through somatic mutations induced by mutagenic carcinogens or spontaneous replication errors or, alternatively, by retroviruses that capture and transduce these proto-oncogenes.

This realization in turn provoked the question of the differences between the human bladder carcinoma oncogene and its proto-oncogene precursor. Like the *src* proto-oncogene, the Ha-*ras* gene was demonstrable in the genomes of all metazoa; ultimately a closely related gene was also found in the genome of *Saccharomyces cerevisiae* – common brewer's yeast. The carcinoma oncogene and its normal counterpart, both analyzed as molecular clones of genomic DNA, were indistinguishable in structure, and yet they were clearly very different in function. The proto-oncogene had no effect on cells into which it was introduced via transfection; the oncogene was potently transforming. Such a profound difference in function clearly pointed to some difference, albeit subtle, in genetic sequence, and required months of careful analysis of their respective sequences.

In the end, the difference in sequence turned out to be extraordinarily subtle – a single nucleotide difference affecting the 12th codon of the coding sequence, forcing replacement of the normally present glycine residue with a valine (Tabin et al. 1982; Reddy et al. 1982; Taparowsky et al. 1982). This difference took on even greater significance when it was realized that both the Harvey and Kirsten viral oncogenes also carried mutations in this critical 12th codon. This site in the *ras* reading frame now appeared to be a critical Achilles heel of the gene, where repeated, independently occurring mutational events succeeded in activating the latent oncogenic activity of these genes.

It was now clear that the virus-borne oncogenes present in the Harvey and Kirsten viral genomes differ from their normal antecedents in three respects. Each carries two critical lesions in its reading frame; one at codon 12 and a second at position 59. Beyond these alterations in protein structure, the expression of each is driven by the strong, constitutive retrovirus promoter, yielding relatively high levels of mutant gene products.

Only one further was required to close this first chapter in the history of *ras* oncogene investigation. Why did mutations in the 12th codon of the *ras* gene repeatedly lead to oncogenic activation? Here the clue and the solution came from Scolnick's earlier work, which showed that these $p21^{ras}$ proteins bound alternatively GDP and GTP in a fashion analogous to that of the biochemically well-studied G-protein associated with the adenylyl cyclase system. Scolnick's group soon showed that, while the oncogenic version of $p21^{ras}$ could bind both GDP and GTP like its wild-type counterpart, the ability to hydrolyze bound GTP was largely defective (Gibbs et al. 1984). The residue 12 substitutions substantially reduce the intrinsic GTPase activity, and this effect is potentiated by the independent alterations at position 59, where threonine has been introduced into the mutant viral proteins.

This discovery established a model which remains robust to this day, though still unadorned with many of the explanations that are required to describe the role of $p21^{ras}$ in a complex signaling pathway. Like the heterotrimeric G-proteins, $p21^{ras}$ normally acts as a signal transducer, acquiring afferent excitatory signals and passing them on to some downstream target. The acquisition of afferent signals by the oncogenic p21 proteins would appear to be normal. Equally normal is their ability to excite downstream targets, though these have yet to be definitively identified. The critical defect lies with the period of excitation. Once activated, normal p21 would seem to enjoy a period of activation which is rapidly terminated by its hydrolysis of bound GTP, forcing it back into a relaxed, nonsignaling state. Oncogenic p21, being largely or completely defective in this hydrolysis, remains in this excited state for extended periods of time, since it lacks the essential negative feedback mechanism. Thus, the cell is flooded with an unrelenting stream of excitatory signals, and an important step in the inception of cancer occurs.

The future still requires us to solve two important issues in this area which have proven most elusive: what normally excites the $p21^{ras}$ protein and what is its critical target, once it becomes excited? Both questions seem close to being solved. It has been a very long wait.

References

Andersson P, Goldfarb MP, Weinberg RA (1979) A defined subgenomic fragment of in vitro synthesized Moloney sarcoma virus DNA can induce cell transformation upon transfection. Cell 16:63–75

Baltimore D (1970) RNA-dependent DNA polymerase in virions of RNA tumor viruses. Nature 226:1209–1211

Chang EH, Maryak JM, Wei C-M, Shih TY, Shober R, Cheung HL, Ellis RW, Hager G, Scolnick EM, Lowy DR (1980) Functional organization of the Harvey murine sarcoma virus genome. J Virol 35:76–92

Collett MS, Erikson RL (1978) Protein kinase activity associated with the avian sarcoma virus *src* gene product. Proc Natl Acad Sci USA 75:2021–2024

Cooper GM, Okenquist S, Silverman L (1980) Transforming activity of DNA of chemically transformed cells and normal cells. Nature 248:418–421

Der C, Krontiris TG, Cooper GM (1982) Transforming genes of human bladder and lung carcinomas are homologous to ras genes of Harvey and Kirsten sarcoma viruses. Proc Natl Acad Sci USA 79:3637–3640

Duesberg PH, Vogt PK (1973) Gel electrophoresis of avian leukosis and sarcoma viral RNA in formamide: comparison with other viral and cellular RNA species. J Virol 12:594–599

Ellis RW, DeFeo D, Maryak JM, Young HA, Shih TY, Chang EH, Lowy DR, Scolnick EM (1980) Dual evolutionary origin for the rat genetic sequences of Harvey murine sarcoma virus. J Virol 36:408–420

Ellis RW, DeFeo D, Shih TY, Gonda MA, Young HA, Tsuchida N, Lowy DR, Scolnick EM (1981) The p21 *src* genes to Harvey and Kirsten sarcoma viruses originate from divergent members of a family of normal vertebrate genes. Nature 292:505–511

Frankel AE, Fischinger PJ (1976) Nucleotide sequences in mouse DNA and RNA specific for Moloney sarcoma virus. Proc Natl Acad Sci USA 73:3705

Gibbs JB, Sigal IS, Poe M, Scolnick EM (1984) Intrinsic GTPase activity distinguishes normal and oncogenic *ras* p21 molecules. Proc Natl Acad Sci USA 81:5704–5030

Goldfarb MP, Weinberg RA (1979) Physical map of biologically active Harvey sarcoma virus unintegrated linear DNA. J Virol 32:30–39

Goldfarb M, Weinberg RA (1981) Structure of the provirus within NIH 3T3 Cells transfected with Harvey sarcoma virus DNA. J Virol 38:125–135

Goldfarb M, Shimizu K, Perucho M, Wigler M (1982) Isolation and preliminary characterization of a human transforming gene from T24 bladder carcinoma cells. Nature 296:404–409

Kirsten WH, Mayer LA (1967) Morphological responses to a murine erythroblastosis virus. J Natl Cancer Inst 39:311–335

Gross L (1970) Oncogenic viruses. Pergamon, Oxford, p 959

Harvey JJ (1964) An unidentified virus which causes the rapid production of tumors in mice. Nature 204:1104–1105

Landolph JR, Heidelberger C (1979) Chemical carcinogens produce mutations to ouabain resistance in transformable C3H/10T1/2 Cl 8 mouse fibroblasts. Proc Natl Acad Sci USA 76:229–232

Martin GS (1970) Rous sarcoma virus: a function required for the maintenance of the transformed state. Nature 227:1021–1023

McCann J, Ames BM (1976) Detection of carcinogens as mutagens in the Salmonella/microsome tests: assay of 300 chemicals. Proc Natl Acad Sci USA 73:950–954

Parada LF, Tabin CJ, Shih C, Weinberg RA (1982) Human EJ bladder carcinoma oncogene is a homologue of Harvey sarcoma virus ras gene. Nature 297:474–478

Pulciani S, Santos E, Lauver AV, Long LK, Robbins KC, Barbacid M (1982) Oncogenes in human tumor cell lines. Molecular cloning of a transforming gene from human bladder carcinoma cells. Proc Natl Acad Sci USA 79:2845–2849

Reddy EP, Reynolds RK, Santos E, Barbacid M (1982) A point mutation is responsible for the acquisition of transforming properties by the T24 human bladder carcinoma oncogene. Nature 300:149–152

Rous P (1910) A transmissible avian neoplasm. (Sarcoma of the common fowl.) J Exp Med 12:696–705

Santos E, Tronick SR, Aaronson SA, Pulciani S, Barbacid M (1982) T24 human bladder carcinoma oncogene is an activated form of the normal human homologue of BALB- and Harvey-MSV transforming genes. Nature 298:343–347

Scolnick EM, Papageorge AG, Shih TY (1979) Guanine nucleotide activity as an assay for src protein of rat-derived murine sarcoma virus. Proc Natl Acad Sci USA 76:5355–5359

Shih C, Shilo BZ, Goldfarb M, Dannenberg A, Weinberg RA (1979) Passage of phenotypes of chemically transformed cells via transfection of DNA and chromatin. Proc Natl Acad Sci USA 76:5714–5718

Shih C, Padhy LC, Murray M, Weinberg RA (1981) Transforming genes of carcinomas and neuroblastomas introduced into mouse fibroblasts. Nature 290:261–264

Shih TY, Weeks MO, Young HA, Scolnick EM (1979) Identification of a sarcoma virus-coded phosphoprotein in nonproducer cells transformed by Kirsten or Harvey murine sarcoma virus. Virology 96:64–79

Shih TY, Papageorge AG, Stokes PE, Weeks MO, Scolnick EM (1980) Guanine nucleotide-binding and autophosphorylating activities associated with the $p21^{src}$ protein of Harvey murine sarcoma virus. Nature 287:686–691

Stehelin D, Guntaka RV, Varmus HE, Bishop JM (1976a) Purification of cDNA complementary to nucleotide sequences required for neoplastic transformation of fibroblasts by avian sarcoma viruses. J Mol Biol 101:349–365

Stehelin D, Varmus HE, Bishop JM, Vogt PK (1976b) DNA related to the transforming gene(s) of avian sarcoma viruses is present in normal avian DNA. Nature 260:170–173

Tabin CJ, Bradley SM, Bargmann CI, Weinberg RA, Papageorge AG, Scolnick EM, Dhar R, Lowy DR, Chang EH (1982) Mechanism, of activation of a human oncogene. Nature 300:143–149

Taparowsky E, Suard Y, Fasano O, Shimizu K, Goldfarb M, Wigler M (1982) Activation of the T24 bladder carcinoma transforming gene is linked to a single amino acid change. Nature 300:762–765

Temin HM, Mizutani S (1970) RNA-directed DNA polymerase in virions of Rous sarcoma virus. Nature 226:1211–1213

Toyoshima K, Vogt PK (1969) Temperature sensitive mutants of an avian sarcoma virus. Virology 39:930–931

Wang LH, Duesberg P, Beemon K, Vogt PK (1975) Mapping RNAse T1-resistant oligonucleotides of avian tumor virus RNAs: sarcoma-specific oligonucleotides are near the poly(A) end of Schmidt-Ruppin sarcoma virus. Proc Natl Acad Sci USA 73:447–451

Weiss R, Teich N, Varmus H, Coffin J (eds) (1984) Tumor viruses. Cold Spring Harbor Laboratory, Cold Spring Harbor

CHAPTER 18

Oncogenic Activation of Ras Proteins

G.J. CLARK and C.J. DER

A. Introduction

The Discovery that the viral oncogenes carried by the Harvey (v-H-*ras*) and Kirsten (v-K-*ras*) sarcoma viruses were transduced versions of normal cellular genes supported the idea that alterations in certain key cellular genes (cellular H-, K-, and N-*ras*) could contribute to human carcinogenesis (ELLIS et al. 1981; CHANG et al. 1982b). The identification of transforming versions of human *ras* genes in DNA isolated from human tumor cells in gene transfer assays confirmed this possibility (DER et al. 1982; PARADA et al. 1982; SANTOS et al. 1982; SHIMIZU et al. 1983). Molecular cloning and analyses of these tumor-derived *ras* sequences quickly established that conversion of the normal genes to the oncogenic versions was due simply to single amino acid substitutions, primarily at residues 12 (TABIN et al. 1982; REDDY et al. 1982; TAPAROWSKY et al. 1982), 13 (Bos et al. 1985), and 61 (YUASA et al. 1983). Concurrent with the development and application of sensitive molecular techniques for the detection of genetic point mutations in tumor DNAs was the finding that oncogenic *ras* sequences were frequently associated with a broad spectrum of human malignancies (reviewed in Bos 1988, 1989).

By analogy to the relatively well-studied heterotrimeric G-proteins (GILMAN 1984), GTPases which transduce extracellular signals via specific membrane receptors into cellular responses, it is now apparent that Ras proteins act as critical regulatory components in certain mitogenic signaling pathways. Like G-proteins, Ras proteins shuttle between an inactive GDP-bound state and an active GTP-bound state (reviewed in BOURNE et al. 1990a; HAUBRUCK and MCCORMICK 1991) and function as molecular switches to regulate as yet unidentified growth-regulatory pathways. Since oncogenic mutations freeze the proteins in the active GTP-complexed state, oncogenic ras proteins can easily be conceived to cause excessive, uncontrolled stimulation of these signaling pathways to trigger the aberrant growth properties of the malignant cell. In this chapter we will review the current status of the association of oncogenic ras proteins with carcinogenesis and provide an overview of the structural, biochemical, and biological properties that distinguish the oncogenic proteins from their normal counterparts. Finally, we will assess the clinical significance of these observations and discuss important aims for future research studies.

B. Oncogenic Versions of Cellular *ras* Genes Detected in Tumor Cells

While gene transfer assays were the original approach used to detect mutated *ras* genes in human tumor cells (reviewed in BARBACID 1987; LACAL and TRONICK 1988; DER 1989), nucleic acid based approaches were subsequently developed that provided more efficient assays for the detection of these mutated genes (BOS et al. 1984; WINTER et al. 1985). With the application of polymerase chain reaction (PCR) DNA amplification based methods, a range of approaches have been developed and applied to the screening of tumors for mutated *ras* sequences. These techniques vary in their ease of application, their usefulness in the screening of large numbers of tumors, their applicability to different tissue sources of DNA, and their sensitivity in detecting mutated *ras* sequences within heterogeneous cell populations.

I. Biological Detection of Activating *ras* Genes

The use of DNA-mediated gene transfer assays that allow biological selection for detection of transforming genes provided the first identification of activated *ras* genes in human tumor DNAs (DER et al. 1982; PARADA et al. 1982; SANTOS et al. 1982; SHIMIZU et al. 1983). These assays are based on the ability of transforming oncogenes to provoke morphologic transformation and a loss of density-dependent growth inhibition in various rodent fibroblast cell lines (SHIH et al. 1981; KRONTIRIS and COOPER 1981; PERUCHO et al. 1981). The NIH 3T3 mouse fibroblast cell line has been the most widely used recipient for such assays. Cells which acquire transfected activated *ras* sequences grow over the monolayer of surrounding normal cells over the course of a typical 2-week assay to form heaped, swirling piles of spindle-shaped cells. These foci of morphologically transformed cells are readily detectable on a background of density-arrested, morphologically flat cells.

The observation that the majority of tumor DNAs assayed in the NIH 3T3 assay were negative for focus-forming activity suggested that some activated cellular genes with oncogenic potential may not be detected via morphologic transformation. To overcome this limitation, NIH 3T3 cells transfected with tumor DNAs were injected subcutaneously into nude mice to select for transfected cells that had acquired tumorigenic potential (BLAIR et al. 1982; FASANO et al. 1984; BOS et al. 1985). This approach has identified codon-13-activated *ras* mutants (BOS et al. 1984), as well as other novel activated cellular oncogenes (FASANO et al. 1984) which exhibit limited or no morphologic transforming activity in the NIH 3T3 focus assay (BOS et al. 1985).

The principal advantage of the NIH 3T3 biological assay is that it utilizes biological activity for detection and hence is not restricted to a particular mutation or gene; it has been very useful for establishing the types of mutations that activate *ras*-transforming potential. However, disadvantages include the labor-intensive nature of the assay, the prolonged time required for results, the large amounts of high molecular weight tumor DNA required, the relative insensitivity of the assay, and the possible bias for detecting mutated *ras* sequences. Therefore, the limitations of the application of the NIH 3T3 assay to the screening of large numbers of tumors prompted the development of nucleic acid based approaches as described below.

II. Direct Detection of *ras* Mutations in Tumor DNA and RNA

The use of short synthetic oligonucleotide probes (12–17 bases) for selective hybridization to specifically recognize known mutations in codons 12, 13, and 61 of the three *ras* genes provided a rapid technique for screening large numbers of tumor DNAs (reviewed in Bos et al. 1984, 1986). The principal advantages of this approach are the ease of screening large sample numbers and the ability to determine the precise mutation responsible for activation. Drawbacks with this approach are the number of probes required for determining specific base changes at codons 12, 13, and 61 in three *ras* genes, the need to establish hybridization conditions for each probe, the relative insensitivity of the technique (5%–10% of the cells containing oncogenic *ras*), the inability to detect novel mutations, and the semi-quantitative nature of establishing the proportion of cells harboring the mutated gene.

A second approach was based on the ability of RNase A to recognize and cleave single-base mismatches in RNA–RNA duplexes (Winter et al. 1985). RNA probes complementary to the normal *ras* sequences can be annealled to tumor RNA to detect single-base substitutions in *ras* transcripts, and this technique can also provide an estimate of the ratio of normal and mutated sequences. An advantage of this approach is that detection of mutations is not restricted to codons 12, 13, or 61. Disadvantages of this approach include the failure to determine the precise base alteration, detection of base changes that do not affect the coding sequence, and false negatives due to the inability to detect some activating mutations.

III. Polymerase Chain Reaction Based Approaches to Screening Tumors

The advent of Taq polymerase chain reaction (PCR) DNA amplification has greatly improved the sensitivity of both the oligonucleotide hybridization (Verlaan-de Vries et al. 1986) and RNase mismatch cleavage (Almoguera et al. 1988) assays, and has also prompted the development of a variety of new methods for detection of point mutations. These methods have provided more sensitive assays that can be applied to situations in which only minute quantities of DNA are available (e.g., fine needle aspirates, stool specimens) (Shibata et al. 1990; Sidransky et al. 1992), or to specifically pinpoint tumor cells in heterogenous tissue samples. For example, allele-specific PCR amplification allows detection of rare mutant sequences within the tumor population as well as in surrounding normal tissue. This technique is very sensitive and can detect point mutations within tumor specimens when less than one out of 10^5 cells is affected (Kumar and Barbacid 1988; Ehlen and Dubeau 1989).

A number of the PCR-based techniques can be applied to archival pathology specimens. The analyses of formalin-fixed paraffin-embedded human tumor samples has greatly facilitated the retrospective studies of the prevalence of *ras* mutations in a variety of cancers as well as in premalignant and malignant tissues from the same patient. Other PCR-based techniques include restriction length polymorphism (Jiang et al. 1989), nonisotopic mismatched DNA amplification (Stork et al. 1991), direct DNA sequencing (Collins 1988; Gonzalez-Cadavid et al. 1989; Tada et al. 1990a), and single-strand conformation polymorphism analysis (Suzuki et al. 1990).

Finally, protein-based methods such as detection of altered mobilities of mutated ras proteins in SDS polyacrylamide gel electrophoresis (Der and Cooper 1983; Srivastava et al. 1985; Joyce et al. 1989) or use of mutation-specific anti-*ras* antibodies have been developed (Clark et al. 1985; Carney et al. 1986; La Vecchio et al. 1990; Wong et al. 1986; Bizub et al. 1989). However, these techniques are very restricted in their ability to detect most activating mutations, and the antibody-based techniques are complicated by their nonspecific recognition of unrelated proteins.

C. Frequent Occurrence of Mutated *ras* Genes in Human Tumors

The application of the above techniques has allowed the extensive examination of the occurrence of *ras* mutations in a wide range of human tumors. These results are summarized in Table 1. When examined, no activated sequences have been detected in the normal tissue surrounding the oncogenic *ras*-containing tumor tissue, indicating that somatic mutations are responsible for *ras* activation (Santos et al. 1984; Feig et al. 1984). Typically, cells which harbor the mutated *ras* gene also retain a normal copy, suggesting that oncogenic *ras* genes are dominant over their normal counterparts (Ricketts and Levinson 1988). As will be described below, the frequency of *ras* mutations, the specific *ras* gene activated, and type of mutations are nonrandom in different tumors. These specific patterns may reflect the action of specific environmental factors on specific tissues. Alternatively, specific mutations may reflect the different potency of different mutated forms in different tissues.

Table 1. *ras* activation in human tumors

Tumor type	Frequency (%)	Activated gene
Pancreatic adenocarcinoma	90	K
Cholangiocarcinoma	55	K,N
Colon adenoma, adenocarcinoma	50	K
Thyroid follicular adenoma, carcinoma	40	N,K,H
Seminoma		K,N
Embryonal rhabdomyosarcoma	35	K,N
Lung adenocarcinoma	30	K (N,H)
Myeloblastic syndromes		N
Multiple myeloma		N,K
Endometrial carcinoma		K,N
Acute myelogenous leukemia	25	N
Keranthoacanthoma		H
Lung carcinoma – large cell	20	K
Melanoma		N,K,H
Ovarian carcinoma	15	K
Thyroid papilary carcinoma		H,K,N
Acute lymphocytic leukemia	10	N
Bladder carcinoma		H
Medulloblastoma		N
Prostatic carcinoma		K,H
Squamous lung carcinoma, gastric carcinoma, Hodgkin's lymphoma, chronic lymphocytic leukemia, neuroblastoma, hepatocellular carcinoma, breast carcinoma, chronic myelogenous leukemia, cervical carcinoma, epidermoid carcinoma, renal cell carcinoma, chronic myelogenous leukemia.	5	K,N,H
Pituitary carcinomas, pheochromocytoma, medullary thyroid carcinoma, esophageal carcinoma, small cell lung carcinoma, mesothelioma, non-Hodgkin's lymphoma	<2	K,N,H

Frequency of ras mutations were compiled from Bos (1989) and from other selected studies (Ahuja et al. 1990; Albino et al. 1989; Burmer and Loeb 1989; Carter et al. 1990; Challen et al. 1992; Corominas et al. 1989; Enomoto et al. 1990, 1991a,b; Fukumoto et al. 1989; Gumerlock et al. 1991; Higaki et al. 1991; Imamura et al. 1992; Iolascon et al. 1991; Ireland 1989; Jiang et al. 1989; Karga et al. 1992; Konishi et al. 1992; LeMaistre et al. 1989; Lubbert et al. 1992; Metcalf et al. 1992; Miki et al. 1991; Mitsudomi et al. 1991; Moley et al. 1991; Nagata et al. 1990; Namba et al. 1990; Nanus et al. 1990; Neri et al. 1989; Oudejans et al. 1991; Rochlitz et al. 1989, 1992; Rodenhuis and Slebos 1990; Saglio et al. 1989; Shi et al. 1991; Shukla et al. 1989; Stork et al. 1991; Stratton et al. 1989; Suarez et al. 1990; Suzuki et al. 1990; Syvanen et al. 1992; Tada et al. 1990a,b, 1992; van Kamp et al. 1992; Wright et al. 1989, 1991).

Overall, approximately 30% of human tumors have been found to contain mutated *ras* sequences (reviewed in Bos 1988, 1989). However, the frequency is nonrandom (Table 1). For example, *ras* mutations have not been observed in pheochromocytomas and only infrequently observed (5%) in breast, cervical, hepatocellular, and renal cell carcinomas. In contrast, mutated *ras* sequences are frequently (30%–50%) detected in lung, thyroid, and colorectal carcinomas, and are present in virtually all pancreatic carcinomas. The frequency may also vary widely within different cell types of a particular tissue. For example, detailed studies of human lung cancers have identified frequent K-*ras* activations in adenocarcinomas (30%), less frequent activations in large cell (20%) or squamous cell (5%) carcinomas, and none have been detected in small cell carcinomas (VORBURGER et al. 1989; MAYER et al. 1991; RODENHUIS and SLEBOS 1992). Interestingly, the frequency of *ras* mutations in lung adenocarcinomas of smokers (30%) is significantly higher than in those of nonsmokers (5%), suggesting that *ras* mutations may be caused directly by exposure to tobacco smoke carcinogens (SLEBOS et al. 1991; RODENHUIS and SLEBOS 1992).

Although mutations in *ras* occur in a wide range of tumors, a degree of tissue specificity with respect to the type of *ras* gene may be observed (Table 1). K-*ras* is clearly the most commonly detected mutant gene, followed by N-*ras*, while mutated H-*ras* sequences are infrequently encountered. An overwhelming or exclusive detection of K-*ras* activations is observed in pancreatic, lung, and colon carcinomas. N-*ras* lesions are most commonly associated with acute myelogenous leukemia, while mutations of all three *ras* genes are observed with equivalent frequency in thyroid follicular adenomas and adenocarcinomas. Bearing in mind that the different functions, if any, of the three *ras* genes remain unclear, it is not possible as yet to explain the basis of this tissue-specific association. Thus, it is possible that the specificity merely represents a different susceptibility of each gene to mutation in different tissues. Alternatively, each ras protein may regulate distinct growth-regulatory pathways, and the specific perturbation of a particular ras protein may be required for malignant progression of a particular cell type.

Activating mutations of *ras* have been found in tumors most commonly at positions 12 and 61, and less frequently at position 13 (Fig. 1). Codon 12 seems to be the most common site but as this is the region of *ras* most commonly assayed, the occurrence of mutations at other sites may be underreported. For example, mutations observed with K-*ras* in pancreatic carcinomas have been found exclusively at codon 12 (ALMOGUERA et al. 1988; SMIT et al. 1988; MARIYAMA et al. 1989; SHIBATA et al. 1990; MOTOJIMA et al. 1991). In contrast, equivalent detection of 12, 13, and 61 mutations are observed with thyroid neoplasms (LEMOINE et al. 1988, 1989; NAMBA et al. 1990; SHI et al. 1991; SUAREZ et al. 1990; WRIGHT et al. 1991).

In addition to position 12, 13, and 61, rare substitutions have also been identified at positions 10, 11, 15, (IMAMURA et al. 1992) 18, 59 (MOLEY et al. 1991), 63 (HIGINBOTHAM et al. 1992), 117 (REYNOLDS et al. 1987), and 146 (ORITA et al. 1991; HIGASHI et al. 1990; SLOAN et al. 1990) in either human or (experimentally induced) rodent tumors. Furthermore, in vitro mutagenesis studies have also identified positions 28, 36, 116, 119, and 156

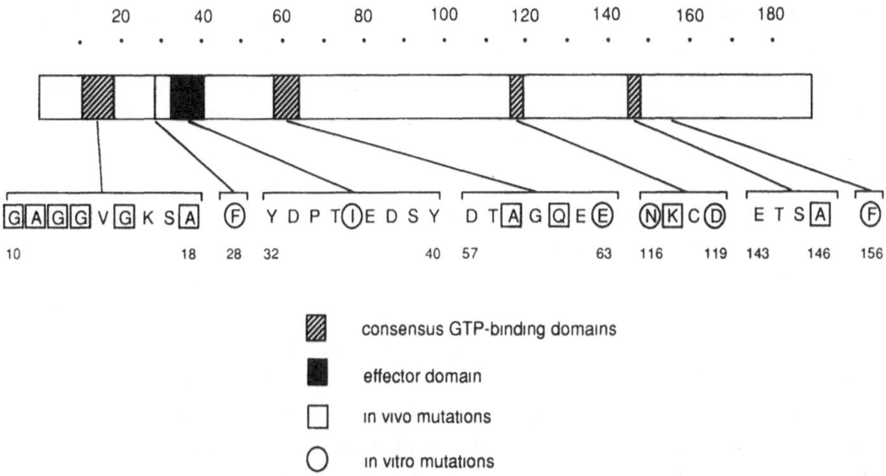

Fig. 1. Oncogenic *ras* mutations. Mutations that activate *ras*-transforming potential affect residues that regulate *ras* GDP–GTP exchange and GTP hydrolysis

as possible targets for activating mutations (Michel et al. 1986; Der et al. 1988; Reinstein et al. 1991). As shown in Figure 1, these residues typically correspond to *ras* sequences involved in the regulation of guanine nucleotide interactions. The very infrequent detection of these activating mutations in human tumors may be a reflection of the limited oncogenic activity of ras proteins that harbor these mutations (Bos et al. 1985; Corominas et al. 1991). Alternatively, these DNA sequences may not be susceptible to the mutagenic activity of environmental carcinogens. Finally, their underrepresentation may also reflect the fact that these mutations are typically not screened for in most tumor survey studies.

Whereas specific activating mutations are observed with certain tumors, it is not known whether the nature of the mutation has consequences for the clinical course of a particular tumor. For example, codon 12 mutations are the only activating substitutions associated with K-*ras* in pancreatic tumors, and it is not known whether mutations at residues 13 or 61 are insufficient to promote tumor progression in pancreatic tissue. As yet no unique characteristics have been apportioned to different tumors on the basis of which particular *ras* mutation they sport. However, it has been observed that colon carcinomas bearing codon-13 mutations arise later in life and are less aggressive than carcinomas bearing codon-12 mutations (M. Perucho, personal communication), and this seems to be borne out by a weaker focus-forming activity of K-*ras*-13 mutants relative to codon-12 mutants in NIH 3T3 cells (unpublished observation). Finally, a recent report showed that retroviruses that express different *ras* mutants initiated different patterns of tumors in infected chickens (GIVOL et al. 1992).

The nonrandom occurrence of *ras* mutations in different tumor types is consistent with a role for oncogenic *ras* in the development of tumors where activation is observed at a high frequency (e.g., pancreatic carcinomas). The incomplete association of *ras* mutations with a particular tumor (e.g., colorectal carcinomas) suggests that mutations in other cellular genes may confer an alternate, but analogous, selective growth advantage on the cell. Alternatively, aberrant *ras* function may occur in tumors that do not harbor mutated ras proteins as a consequence of defects in regulatory proteins that modulate *ras* function. These regulatory proteins include those which stimulate *ras*-GTPase activity to form the inactive, GDP-complexed protein (GAP and NF1) (HALL 1992a) and those which stimulate GDP–GTP exchange to form the active, GTP-complexed protein (designated GDSs, guanine nucleotide dissociation stimulators) (HAUBRUCK and MCCORMICK 1991). For example, a perturbation in NF1 neurofibromin (a negative regulator of *ras* function) activity in several different tumor cell lines correlates with the elevation of activated, GTP-complexed normal ras proteins (BASU et al. 1992; DECLUE et al. 1992; LI et al. 1992). Other ras regulatory proteins that either positively (e.g., GDSs) or negatively (e.g., GAPs) modulate *ras* activity represent additional targets for such mutations. While no GAP mutants have yet been identified in tumor cells, cell culture experiments with ras-GDS proteins suggest the possibility that constitutively activated GDS may potentiate the transforming activity of normal *ras* in tumors (SCHWEIGHOFFER et al. 1992). Finally, a number of tumor surveys have restricted their analyses to detection of only certain mutations, and in only certain *ras* genes. Thus it seems likely that some mutations have been missed and that, if anything, the present estimates of the frequency of *ras* mutations are conservative.

After the initial observation that mutated *ras* genes could transform NIH 3T3 cells, it was determined that elevated expression of the wild-type gene was also sufficient to cause an equivalent malignant transformation of NIH 3T3 cells (CHANG et al. 1982a; SANTOS et al. 1983; STACEY and KUNG 1984). Analysis of certain tumors (e.g., breast and gastric carcinomas) has shown that, while *ras* mutations are infrequently encountered, abnormally high levels of wild-type *ras* expression are observed in these tumors (SPANDIDOS 1987; FUJITA et al. 1987; MCKENZIE 1991). Enhanced expression may be facilitated by gene amplification (FUKUMOTO et al. 1989) or genetic alterations in regulatory sequences (COHEN and LEVINSON 1988). Thus, elevated *ras* expression, in the absence of structural mutations, may also contribute to tumor development. However, the inability of normal *ras* overexpression to transform Rat-1 fibroblasts suggests that elevated expression of the normal protein does not trigger the same biological consequences as oncogenic ras proteins (RICKETTS and LEVINSON 1988).

D. *Ras* Activation is Associated with Experimentally Induced Rodent Tumors

While the frequent detection of mutated *ras* genes in human tumors strongly implicates some role for oncogenic *ras* genes in tumor development, it is not yet certain whether this association is causative or consequential. Experimental chemical and radiation carcinogenesis studies in a variety of rodent systems have provided further support for a causative role of mutated *ras* in tumor development (reviewed in GUERRERO and PELLICER 1987; SUKUMAR 1989). Overall, where treatment with a specific carcinogenic agent induces reproducible formation of a specific tumor type, the consistent association of *ras* mutations with tumor development provides compelling support for a critical contribution of *ras* activation to tumorigenesis.

A summary of representative chemical carcinogenesis studies is presented in Table 2 and demonstrates that a variety of mutagenic agents are effective at provoking activating mutations in specific *ras* genes in specific tissues. Moreover, some of these activating events are specific to the base change induced. For example, a single nitroso-methylurea (NMU) treatment of female rats induced mammary carcinomas, 86% of which were found to be H-*ras* mutants, all of which had undergone a point mutation of G to A at the second G of codon 12 (SUKUMAR et al. 1983; ZARBL et al. 1985). In contrast, DMBA-induced mammary carcinomas harbored H-*ras* genes containing A to T mutations in the second base of codon 61 (ZARBL et al. 1985). These mutations are consistent with the mutagenic mechanism of these agents, suggesting that the mutagenic activity of each carcinogen is directly responsible for the *ras* mutation and that the activated *ras* gene is a critical trigger for tumor progression. However, since *ras* activation precedes development of the full malignant state, it is also clear that additional

Table 2. *ras* activation by chemical carcinogens

Tumor	Carcinogen	Gene	Mutation
Rabbit keratoacanthoma	12-dimethylbenzanthracene	H-*ras*	61 (A-T)
Rat kidney tumor	N-nitroso-N-methylurea	K-*ras*	12 (G-A)
Rat mammary carcinoma	N-nitroso-N-methylurea	H-*ras*	12 (G-A)
	12-dimethylbenzanthracene	H-*ras*	61 (A-N)
Rat lung carcinoma	tetranitromethane	K-*ras*	12 (G-A)
Mouse skin carcinoma	12-dimethylbenzanthracene	H-*ras*	61 (A-T)
Mouse hepatocellular carcinoma	N-hydroxy-2-acetylaminofluorene	H-*ras*	61 (C-A)
	vinyl carbamate	H-*ras*	61 (A-T/G)
	1′-hydroxy-2′,3′-dihydroestragole	H-*ras*	61 (A-T/G)
	12-dimethylbenzanthracene	K-*ras*	13 (G-C)
Mouse lung carcinoma	N-nitroso-N-methylurea	K-*ras*	12 (G-A)
	benzo[a]pyrene	K-*ras*	12 (G-T/A)
	ethyl carbamate	K-*ras*	61 (A-T/G)
	tetranitromethane	K-*ras*	12 (G-A)
Mouse thymic lymphoma	N-nitroso-N-methylurea	N-*ras*	61 (C-A)

genetic events must occur to complement the *ras* mutation for complete tumorigenic development in these rodent tumor models.

E. Biological Activities of Oncogenic ras Proteins

I. Malignant Transformation of Established Rodent Fibroblast Cell Lines

The detection of oncogenic *ras* genes in the NIH 3T3 assay demonstrated the striking biological difference between the normal and mutated *ras* genes (Table 3). NIH 3T3 cells expressing mutant *ras* proteins are morphologically transformed, proliferate in low serum or under anchorage-independent conditions, and form rapidly growing, progressive tumors when injected into the appropriate animals. Oncogenic, but not normal, *ras* expression also results in the malignant transformation of other rodent fibroblast cell lines such as Rat-1 and C3H10T1/2 cells and an immortalized human fibroblast cell line (Wilson et al. 1990; Fry et al. 1990). Altogether, the striking biological difference between normal and activated versions of *ras* genes in these established cell lines suggested the dominant nature of oncogenic *ras* genes.

Further characterization of the biological activity of normal versus oncogenic *ras* genes in a variety of cells has confirmed this difference, but

Table 3. Biological activity of oncogenic *ras*

Cells	Consequences on growth
NIH 3T3 mouse fibroblasts	malignant transformation
Rat-1 rat fibroblasts	malignant transformation
C3H10T1/2 mouse fibroblasts	malignant transformation
CHEF/18 hamster fibroblasts	malignant transformation
BALB/MK-2 mouse keratinocytes	EGF independent growth
BEAS-2B immortalized human bronchial epithelial cells	malignant transformation
MCF-10A immortalized human breast epithelial cells	malignant transformation
EBV-immortalized human B-cells	malignant transformation
MSU-1.1 immortalized human fibroblasts	malignant transformation
SV-HUC immortalized human uroepithelial cells	no malignant transformation
Normal human fibroblasts	none
Normal human B-lymphocytes	none
PC12 rat pheochromocytoma cells	growth arrest, neuronal differentiation
Primary rat Schwann cells	growth arrest
REF52 rat embryo fibroblasts	growth arrest
Primary human thyroid epithelial cells	growth arrest

has also demonstrated that oncogenic *ras* genes alone are not sufficient to convert a normal cell to the full malignant state. In contrast to results with established cell lines, the introduction of oncogenic *ras* into primary cells typically does not result in induction of the transformed phenotype. These results are consistent with the multistep nature of tumor progression, and suggest that *ras* activation is but one critical genetic event – together with the activation of other cellular oncogenes or inactivation of tumor suppressor genes – for tumor development (reviewed in RULEY 1990; HUNTER 1991). Finally, the inability of cells that express oncogenic *ras* proteins to form tumors in injected animals clearly demonstrates that *ras* alone is insufficient to promote the malignant growth of the tumor cell (CRAIG and SAGER 1985; FARR et al. 1988; GEISER et al. 1986).

II. *Ras* Requires Cooperation with Other Oncogenes for Transformation of Primary Cells

The use of primary rat embryo fibroblasts for *ras* transformation studies has provided strong evidence that other genetic events are required to cooperate with *ras* to trigger cellular transformation (Table 4; reviewed in RULEY 1990; HUNTER 1991). Primary fibroblasts with a limited ability to proliferate in tissue culture may readily be isolated by dissolving rat embryos in trypsin. These "normal" cells show considerable resistance to transformation by *ras*. Activated *ras* oncogenes can produce a transformed phenotype here only when they are cotransfected with another activated oncogene (reviewed in LAND et al. 1983). The fact that two or more oncogenes are required to

Table 4. Transformation by *ras* and cooperating oncogenes

Primary cell type	Cooperating gene
Rat embryo fibroblasts (REF)	c-*myc*
	L-*myc*
	N-*myc*
	SV40 large-T
	p53
	bovine leukemia virus tax
	bcl-2
	c-*jun*
Baby rat kidney epithelial cells (BRK)	c-*myc*
	adenovirus E1A
	polyoma virus large-T
	human papilloma virus 16 E7
Mouse B-lymphocytes	v-*myc*
Mouse keratinocytes	v-*fos*
Rat Schwann cells	SV40 large-T
	adenovirus E1A
	c-*myc*
Rabbit mammary epithelial cells	SV40 large-T

produce a fully transformed phenotype in these primary cells, although somewhat artificial, provides further evidence supporting the multistep nature of cancer.

Although much of the in vitro experimentation on the biological properties of activated *ras* mutants has been performed in fibroblast cell lines, *ras* mutations are only infrequently associated with sarcomas; they most frequently observed in tumors of epithelial origin. Thus, results from fibroblast cell systems may provide a somewhat inaccurate picture of oncogenic *ras* biological activity. The introduction of oncogenic *ras* into a variety of immortalized epithelial cell types has demonstrated the transforming potential of mutated *ras* sequences in these perhaps more relevant systems (HUBER and CORDINGLEY 1988; REDMOND et al. 1988; AMSTAD et al. 1988; CHRISTIAN et al. 1990; BASOLO et al. 1991). For example, oncogenic *ras* can facilitate the conversion of immortalized human bronchial, breast, or thyroid epithelial cells to the malignant state (AMSTAD et al. 1988; BASOLO et al. 1991; WYNFORD-THOMAS et al. 1990).

Since *ras* mutations are a frequent event in experimental and human skin tumors, primary rodent or human keratinocytes have been useful in vitro model systems for studying *ras* oncogene biology. These cells require exogenous epidermal growth factor (EGF) for growth in vitro, and may be induced to terminally differentiate by elevations in calcium levels. It has been found that introduction of oncogenic *ras* into keratinocytes (e.g., BALB–MK-2) produces cells that have lost the requirement for EGF and have become resistant to this calcium-induced terminal differentiation (WEISSMAN and AARONSON 1983; HARPER et al. 1986; HENRARD et al. 1990). Despite these biological effects, these *ras*-expressing cells do not exhibit tumorigenic potential, again emphasizing the requirement for other genetic events to complement *ras* for full conversion to the malignant state.

III. Induction of Differentiation and Growth Inhibition by Oncogenic *ras*

Whereas a growth-stimulatory response is typically associated with oncogenic ras proteins, an apparent role for *ras* in differentiation is observed in some cell types. Oncogenic *ras* mimics the action of nerve growth factor on PC12 rat pheochromocytoma cells and induces neuronal differentiation accompanied by cessation of growth (NODA et al. 1985; BAR-SAGI and FERAMISCO 1985; HAGAG et al. 1986). Similarly, introduction of viral H-*ras* into a human medullary thyroid carcinoma cell line results in decreased growth and induction of characteristics of normal thyroid C-cells (NAKAGAWA et al. 1987; NELKIN et al. 1990). Introduction of oncogenic H-*ras* into Epstein-Barr virus-immortalized B-cells was observed to promote plasmacytoid differentiation (SEREMETIS et al. 1989), while induction of oncogenic *ras* expression in the NIH 3T3 L1 preadipocyte line induces adipocyte differentiation (BENITO et al. 1991). Finally, oncogenic *ras* induces growth arrest of primary rat Schwann (RIDLEY et al. 1988) or human thyroid epithelial (WYNFORD-THOMAS et al. 1990) cells, while the coexpression of a second oncogene rescues this inhibitory response and now promotes transformation. Thus, oncogenic *ras* is likely to have distinct biological effects in different tissues, and hence, distinct roles in the pathogenesis of different malignancies.

IV. Transgenic Mouse Studies Establish *ras* Oncogenicity

Transgenic technology is one of the most powerful tools in modern molecular biology, and the increased incidence of tumors in mice carrying oncogenic *ras* genes has clearly demonstrated the ability of oncogenic *ras* to promote tumorigenesis de novo (reviewed in ADAMS and CORY 1991). A variety of studies have addressed the biological consequences of oncogenic *ras* expression in diverse cell types by using tissue specific or inducible promoter *ras* constructs. For example, mice carrying mouse mammary tumor virus (MMTV) promoter–ras transgenes develop mammary, salivary gland, and lung tumors (SINN et al. 1987; TREMBLAY et al. 1989; MANGUES et al. 1990), while those carrying an activated H-*ras* gene controlled by a pancreas-specific elastase promoter developed large, terminal pancreatic tumors in utero (QUAIFE et al. 1987), and mutant H-*ras* transgenes expressed in the skin elicited papillomas (BAILLEUL et al. 1990; LEDER et al. 1990).

Two themes of *ras* transformation established from cell culture experiments have also been supported by observations from transgenic studies. First, the expression of oncogenic *ras* alone is insufficient to induce tumor formation, with malignant progression requiring additional mutational events. The variable and long latency associated with the onset of neoplastic growth, and the greatly enhanced frequency of tumor formation when *ras* is accompanied by a second oncogene (e.g., *myc*), further document the fact that secondary genetic events are required for *ras* oncogenicity. Second, there is clearly a cell-type-specific consequence of oncogenic *ras* expression. Expression of oncogenic *ras* in preneoplastic tissue may reuslt in either no proliferative response (SINN et al. 1987; MANGUES et al. 1990), a limited hyperplasia (SINN et al. 1987), induction of differentiation (BAILLEUL et al. 1990), or blockage of cell proliferation (EFRAT et al. 1990). The different frequencies of *ras* mutations in different cell types presumably reflect this differential susceptibility of different cell types to oncogenic *ras*.

F. Structural and Biochemical Consequences of Oncogenic Mutations

In addition to the other members of the *ras* superfamily, the receptor coupled heterotrimeric G-proteins that are involved in signal transduction (e.g., G_s, G_i, G_o, and transducin) and the ribosomal protein synthesis elongation factor EF-Tu are among the best characterized proteins that utilize a GDP–GTP cycle to function as molecular switches (Bourne et al. 1990a,b). Ras proteins normally exist in the inactive, GDP-complexed form, and activation is promoted by guanine nucleotide dissociation stimulatory proteins (GDSs) which promote the exchange of the bound nucleotide for the more abundant GTP to form the active, GTP-complexed form. The active state is transient, and is terminated by the stimulation of the intrinsic GTPase activity by a GTPase activating protein (*ras* GAP) (Trahey et al. 1987) or NF1 (neurofibromin) (Xu et al. 1990; Martin et al. 1990; Ballester et al. 1990). As indicated in Figure 1, the known activating mutations occur in regions of the ras protein that are involved in the binding and hydrolysis of GTP. The net result of most activating mutations is to perturb the GDP–GTP cycle and to favor formation of a chronically active, GTP-complexed protein.

I. Activating Mutations at Residues 12, 13, or 61 Promote Active, GTP-Complexed ras Formation

The principal consequence of point mutations at the three hot spots for *ras* activation are to perturb the GDP–GTP cycle by reducing intrinsic and GAP–NF1-stimulated GTPase activities. The net consequence of these mutations is to favor formation of the active, GTP-bound ras proteins. The majority of amino acid substitutions at both position 12 (Seeburg et al. 1984) and position 61 (Der et al. 1986) are capable of activating transforming potential, while only some substitutions at position 13 result in transforming proteins. Residues 12 and 13 are located in a consensus GTP-binding motif (Gly-X-X-X-X-Gly-Ser–Thr; ras residues 10–16) which is present in all GTP-binding proteins, while residue 61 resides in a domain important for regulating intrinsic and GAP–NF1-stimulated GTPase activities (Bourne et al. 1990b; Valencia et al. 1991). Crystal structure analyses indicate that these mutations perturb surprisingly little of ras protein conformation, with alterations restricted primarily to two subregions of ras proteins corresponding to Loop 2 (switch region I; residues 32–40), where mutations cause loss of transforming activity, and Loop 4 (switch region II, residues 59–76), where mutations cause activation of transforming activity (Pai et al. 1989; Tong et al. 1989; Krengel et al. 1990; Milburn et al. 1990).

II. Other Activating Mutations Also Perturb the *ras* GDP–GTP Cycle

Ras residues 116–119 correspond to a consensus GTP-binding motif (Asn-Lys-X-Asp) that is conserved in essentially all GTP-binding proteins and interacts with the guanine ring of the bound GTP (reviewed in BOURNE et al. 1990b; VALENCIA et al. 1991). While only one mutation in this region has been observed in a rodent tumor (REYNOLDS et al. 1987), experimental mutagenesis studies have demonstrated that mutations at positions 116, 117, or 119 activate *ras*-transforming activity (SIGAL et al. 1986; WALTER et al. 1986; MICHEL et al. 1986; DER et al. 1988). However, in contrast to the more conventional 12, 13, or 61 mutations, mutations in this motif activate as a consequence of enhanced guanine nucleotide exchange, rather than perturbation in GTP hydrolytic activities. As intracellular GTP levels are in great excess (10- to 20-fold) relative to GDP, the net consequence of enhanced turnover of the bound nucleotide is to favor formation of active, GTP-bound *ras* (PATEL et al. 1992).

Substitutions at residue 28 (REINSTEIN et al. 1991), which interacts with the guanine ring, or residue 146 (FEIG and COOPER 1988) also result in enhanced rates of nucleotide exchange. A 28-Phe to Leu substitution results in a 100-fold increased nucleotide dissociation rate without perturbing GAP-stimulated GTPase activity, while 146-Ala to Val substitution results in a 1000-fold enhanced exchange rate. The activating substitution of 59-Ala with threonine alters both exchange rates and hydrolytic activity (FEIG and COOPER 1988), and also promotes autophosphorylation. Interestingly, the phosphorylated form of this mutant protein is not transforming, indicating that the primary consequence of this substitution is to perturb the GDP–GTP cycle, rather than to activate via phosphorylation (CHUNG et al. 1992). Finally, while some substitutions alone are not sufficient to activate *ras*-transforming activity (e.g., 14-Met, 59-Ile, 61-Pro), their influence on transforming potential has been demonstrated by complementation of these mutations with other weak activating mutations. Site-directed mutagenesis has been used to show that increased transforming activity may be obtained from a variety of such combinations that affect either GTP hydrolytic or nucleotide exchange rates (FEIG and COOPER 1988; DER et al. 1988).

Recent studies of *Drosophila* (HARIHATAN et al. 1991) and *Caenorhabditis elegans* (HAN and STERNBERG 1991) have identified mutant ras and ras-related proteins with single amino acid substitutions which result in dominant developmental phenotypes, and suggest that additional ras residues (e.g., 89-Ser, 156-Phe) may also be targets for activating mutations. For example, the 156-Phe residue is strictly conserved in all ras and ras-related proteins. A mutant *Drosophila* rap protein with a mutation at this conserved residue (157-Phe to Leu) results in a dominant gain-of-function that disrupts eye development and interferes with cell determination (HARIHATAN et al. 1991). We have recently observed that the equivalent mutation in human H-*ras* (156-Leu) is an activator of transforming activity (QUILLIAM and RABUN, unpublished observation). The mechanism of activation at 156-Phe, and whether it alters intrinsic properties or interactions with regulatory proteins, remains to be determined.

G. Clinical Implications of Oncogenic *ras* for Diagnosis and Treatment

One important aim of *ras* studies has been to establish whether the activation of *ras* genes correlates with important clinical features of a particular tumor that may provide useful diagnostic and prognostic markers for predicting the clinical course, and the response to therapy, of the disease. A second aim has been to determine whether pharmacologic intervention in oncogenic *ras* function will be an affective approach to cancer treatment. While only limited progress has been achieved to date for each of these aims, some information is emerging to suggest possible applications of oncogenic *ras* in both detection and treatment.

I. Diagnostic and Prognostic Applications of *ras* Mutations

While no clear consensus has been reached with regard to the contribution of *ras* to tumor progression, it is likely that oncogenic *ras* makes different contributions to the development of different malignancies. This possibility is supported by observations from transgenic studies where expression of oncogenic *ras* in different tissues results in a spectrum of different responses ranging from enhanced proliferation to cell death. The detection of *ras* mutations in premalignant tissues such as colorectal (Bos et al. 1987a; FORRESTER et al. 1987) or thyroid (LEMOINE et al. 1988, 1989; NAMBA et al. 1990; SHI et al. 1991; SUAREZ et al. 1990) adenomas, or in myelodysplastic syndrome (LIU et al. 1987; HIRAI et al. 1987; Bos et al. 1987b) suggests a role in initiation of tumor development. In contrast, the late association of *ras* activation with multiple myelomas (NERI et al. 1989) and chronic myelogenous leukemia (LEMAISTRE et al. 1989) suggests a role in tumor progression rather than initiation. While it remains possible that *ras* mutations occur as secondary events in the genetically unstable tumor cell, the considerable body of experimental evidence from in vitro and in vivo studies, together with the high frequency of oncogenic *ras* mutations observed in many malignancies, strongly argues that oncogenic *ras* contributes significantly to the development of human cancers.

Since many types of tumors consistently show a high degree of association with oncogenic *ras* mutations (Table 1), the screening for *ras* mutations may promise improved clinical treatment of these malignancies. In pilot studies on pancreatic cancers, the diagnosis of neoplasia proved roughly as accurate as standard histological methods, whilst having the potential to be more sensitive, more rapid, and less invasive (SHIBATA et al. 1990). PCR-based methods have also been used successfully to screen stool samples for the presence of *ras* mutations in the colon (SIDRANSKY et al. 1992).

Comparisons of *ras*-positive and *ras*-negative tumors of the same tissue origin have, for the most part, not identified differences in clinical aspects of the disease. Nevertheless, the presence of *ras* mutations has provided some indication of certain properties of some tumors. For example, it has been observed that leukemias which have oncogenic *ras* mutations may respond more favorably to chemotherapy than those which do not. Also, it appears that colon carcinomas with mutations in K-*ras* at codon 13 are less malignant than those with alterations at codon 12 (M. PERUCHO, personal communication). Several reports have associated *ras* activation with cellular resistance against radiation and a variety of drugs (e.g., cisplatin) (SKLAR 1988; MCKENNA et al. 1990), although the generality and the significance of these observations remain to be established (RODENHUIS and SLEBOS 1990). K-*ras* mutations in lung adenocarcinomas have been suggested to correlate with very poor prognosis and shortened disease-free survival (RODENHUIS and SLEBOS 1990, 1992). As more information is accumulated it may become possible to begin to select the most favorable treatment regime based on a molecular diagnosis of the tumor.

II. Protein Prenylation: Oncogenic ras Proteins as Targets of Therapy

The frequent association of mutated ras proteins with tumors has prompted the search for pharmacological agents which block the growth-perturbing effects of oncogenic *ras*. A number of pharmacologic approaches have been envisioned; prominent among these have been those which block the posttranslational modifications of ras proteins that are essential for their membrane association and oncogenicity (reviewed in GIBBS 1991). All ras proteins undergo a series of three closely linked carboxyl-terminal modifications that are triggered by a consensus cysteine-X-X-X sequence (X = any amino acid) (reviewed in COX and DER 1992; KHOSRAVI-FAR et al. 1992). Of these three modifications (farnesylation, proteolysis, and carboxyl methylation), it is the addition of the 15-carbon farnesyl isoprenoid (reviewed in MALTESE 1990; GLOMSET et al. 1991) which is the critical modification for *ras*-oncogenic potential (KATO et al. 1992). Since the farnesyl moiety is an essential intermediate of cholesterol biosynthesis, the existence of drugs (e.g., lovastatin) which interfere with the production of all isoprenoids, including farnesyl, and can inhibit the activity of *ras* by preventing this modification has stimulated considerable interest in the development of drugs for specifically blocking *ras* farnesylation (reviewed in GOLDSTEIN and BROWN 1990). The recent identification (REISS et al. 1990; SCHABER et al. 1990; MANNE et al. 1990) and molecular cloning (CHEN et al. 1991a,b; KOHL et al. 1991; HE et al. 1991) of the farnesyl transferase that catalyzes this modification has provided an excellent target for identifying and developing such drugs. However, since normal *ras* function is also dependent on farnesyl modification, it is presently not known whether such drugs will provide sufficient specificity to minimize toxicity to normal tissues.

A critical question that remains to be answered is whether pharmacologic intervention in oncogenic *ras* function will have any benefit for the cancer patient. For example, it has still not been established that persistence of these mutated proteins is required to maintain some aspect of the malignant phenotype or whether blocking oncogenic *ras* activity will retard the growth of oncogenic *ras*-containing tumors. Support for these ideas comes from limited studies with *ras* antisense, studies that suggest that the specific blockage of oncogenic *ras* function will perturb the growth of tumor cells in vitro (SAISON-BEHMOARAS et al. 1991). Thus, it is presently difficult to predict whether blockage of oncogenic *ras* function in pancreatic, lung, and colorectal carcinoma cells will have any clinical benefit for patients with these lethal diseases.

H. Future Questions

The frequent association of oncogenic *ras* genes with human malignancies is now well-established. While it remains possible that *ras* mutations in human tumors do not contribute to tumor formation and reflect merely tumor cell heterogeneity, or are a nonspecific consequence of the genetic instability of the cancer cell, the considerable body of experimental data provides strong support for some contribution to the malignant phenotype.

The consequences of these activating mutations to ras structure and biochemistry are well-characterized. Less clear are the specific growth-regulatory pathways, perturbed by oncogenic *ras*, that trigger the uncontrolled growth of the tumor cell. However, an emerging picture suggests that ras proteins are indeed critical regulatory components of many signal transduction pathways that regulate cell growth and differentiation. Therefore, critical questions that remain include the precise role of *ras* activation in multi-step carcinogenesis, and which critical cellular processes are perturbed by oncogenic *ras* that trigger the aberrant growth and behavior of the malignant cell.

The identification of mutations in the NF1 neurofibromin protein that result in deregulated GDP–GTP regulation of normal *ras* function suggests that the positive (e.g., GDS, CDC25 homologs) or negative (e.g., GAP, NF1) regulators of *ras* function may be candidates for oncogenes or anti-oncogenes that trigger carcinogenesis via deregulated *ras* function independently of *ras* mutations. Molecular cloning of the gene encoding a mammalian homologue of the yeast Ras CDC25 exchange protein (a *Ras* GDS) has been achieved (SHOU et al. 1992; WEI et al. 1992) and the possibility that deregulated expression of these positive mediators of *ras* activity will contribute to malignancy is an important question for future studies.

Finally, it is now apparent that the three *ras* genes represent but a subset of a large (>30 mammalian members) superfamily of *ras*-related genes that encode structurally (30%–55% amino acid identity) and biochemically related GTP-binding proteins (reviewed in DOWNWARD 1990;

Hall 1990; Valencia et al. 1991; Kahn et al. 1992). Present evidence suggests that these related proteins act as critical regulatory elements involved in a wide range of cellular functions such as signal transduction (ras and rap) (Kitayama et al. 1989), cytoskeletal organization (rho) (reviewed in Hall 1992b), intracellular transport (rab) (reviewed in Balch 1990), and cell cycle regulation (ran) (Matsumoto and Beach 1991; Bischoff and Ponstingl 1991). Whether aberrant function of these proteins also contributes to tumorigenesis will be an important question for future studies.

Acknowledgements. We thank Adrienne D. Cox, Roya Khosravi-Far, and Lawrence Quilliam for critical comments and suggestions, and Cecilia Rugliero for excellent preparation of tables and references. Our research was supported by grants from the National Institutes of Health to C.J.D. (CA42978, CA52072, and CA55008) and from the American Cancer Society (BE-29E) to M.S. C.J.D. is the recipient of an American Cancer Society Faculty Research Award.

References

Adams JM, Cory S (1991) Transgenic models of tumor development. Science 254:1161–1167

Ahuja HG, Foti A, Bar-Eli M, Cline MJ (1990) The pattern of mutational involvement of ras genes in human hematologic malignancies determined by DNA amplification and direct sequencing. Blood 75:1684–1690

Albino AP, Nanus DM, Mentle IR, Cordon-Cardo C, McNutt NS, Bressler J, Andreeff M (1989) Analysis of ras oncogenes in malignant melanoma and precursor lesions: correlations of point mutations with differentiation phenotype. Oncogene 4:1363–1374

Almoguera C, Shibata D, Forrester K, Martin J, Arnheim N, Perucho M (1988) Most human carcinomas of the exocrine pancreas contain mutant c-K-ras genes. Cell 53:549–554

Amstad P, Reddel RR, Pfeifer A, Malan-Shibley L, Mark III GE, Harris CC (1988) Neoplastic transformation of a human bronchial epithelial cell line by a recombinant retrovirus encoding viral Harvey ras. Mol Carcinogen 1:151–160

Bailleul B, Surani MA, White S, Barton SC, Brown K, Blessing M, Jorcano J, Balmain A (1990) Skin hyperkeratosis and papilloma formation in transgenic mice expressing a ras oncogene from a suprabasal keratin promoter. Cell 62:697–708

Balch WE (1990) Low molecular GTP-binding proteins (LMGPs) involved in vesicular transport: binary switches or biological transducers? Trends Biochem Sci 15:469–472

Ballester R, Marchuk D, Boguski M, Saulino A, Letcher R, Wigler M, Collins F (1990) The NF1 locus encodes a protein functionally related to mammalian GAP and yeast IRA proteins. Cell 63:851–859

Barbacid M (1987) ras Genes. Annu Rev Biochem 56:779–827

Bar-Sagi D, Feramisco JR (1985) Microinjection of the *ras* oncogene protein into PC12 cells induces morphological differentiation. Cell 42:841–848

Basolo F, Elliott J, Tait L, Chen XQ, Maloney T, Russo IH, Pauley R, Momiki S, Caamano J, Klein-Szanto AJP, Koszalka M, Russo J (1991) Transformation of human breast epithelial cells by c-Ha-ras oncogene. Mol Carcinogen 4:25–35

Basu TN, Gutmann DH, Fletcher JA, Glover TW, Collins FS, Downward J (1992) Aberrant regulation of ras proteins in malignant tumour cells from type 1 neurofibromatosis patients. Nature 356:713–715

Benito M, Porras A, Nebreda AR, Santos E (1991) Differentiation of 3T3-L1 fibroblasts to adipocytes induced by transfection of ras oncogenes. Science 253:565–568

Bischoff FR, Ponstingl H (1991) Mitotic regulator protein RCC1 is complexed with a nuclear ras-related polypeptide. Proc Natl Acad Acad Sci USA 88:10830–10834

Bizub D, Fischberg-Bender E, Heimer EP, Felix A, Skalka AM (1989) Detection of transforming ras proteins containing leucine at position 61 by a new mouse monoclonal antibody, ras(53–69)leu61. Cancer Res 49:6425–6431

Blair DG, Cooper CS, Oskarsson MK, Eader LA, Vande Woude GF (1982) New method for detecting cellular transforming genes. Science 218:1122–1125

Bos JL (1988) The ras gene family and human carcinogenesis. Mutat Res 195:255–271

Bos JL (1989) ras oncogenes in human cancer: a review. Cancer Res 49:4682–4689

Bos JL, Verlaan-de Vries M, Jansen AM, Veeneman GH, van Boom JH, van der Eb AJ (1984) Three different mutations in codon 61 of the human N-ras gene detected by synthetic oligonucleotide hybridization. Nucleic Acids Res 12:9155–9163

Bos JL, Toksoz D, Marshall CJ, Verlaan-de Vries M, Veeneman GH, van der Eb AJ, van Boom J, Janssen JWG, Steenvoorden ACM (1985) Amino-acid substitutions at codon 13 of the N-ras oncogene in human acute myeloid leukaemia. Nature 315:726–730

Bos JL, Verlaan-de Vries M, Marshall CJ, Veeneman GH, van Boom JH, van der Eb AJ (1986) A human gastric carcinoma contains a single mutated and an amplified normal allele of the Ki-ras oncogene. Nucleic Acids Res 14:1209–1217

Bos JL, Fearon ER, Hamilton SR, Verlaan-de Vries M, van Boom JH, van der Eb AJ, Vogelstein B (1987a) Prevalence of ras gene mutations in human colorectal cancers. Nature 327:293–297

Bos JL, Verlaan-de Vries M, van der Eb AJ, Janssen JWG, Delwel R, Lowenberg B, Colly LP (1987b) Mutations in N-ras predominate in acute myeloid leukemia. Blood 69:1237–1241

Bourne HR, Sanders DA, McCormick F (1990a) The GTPase superfamily: a conserved switch for diverse cell functions. Nature 348:125–132

Bourne HR, Sanders DA, McCormick F (1990b) The GTPase superfamily: conserved structure and molecular mechanism. Nature 349:117–126

Burmer GC, Loeb LA (1989) Mutations in the KRAS2 oncogene during progressive stages of human colon carcinoma. Proc Natl Acad Sci USA 86:2403–2407

Carney WP, Petit D, Hamer P, Der CJ, Finkel T, Cooper GM, Lefebvre M, Mobtaker H, Delellis R, Tischler AS, Dayal Y, Wolfe H, Rabin H (1986) Monoclonal antibody specific for an activated RAS protein. Proc Natl Sci USA 83:7485–7489

Carter BS, Epstein JI, Isaacs WB (1990) ras gene mutations in human prostate cancer. Cancer Res 50:6830–6832

Challen C, Guo K, Collier JD, Cavanagh D, Bassendine MF (1992) Infrequent point mutations in codons 12 and 61 of ras oncogenes in human hepatocellular carcinomas. J Hepatol 14:342–346

Chang EH, Furth ME, Scolnick EM, Lowy DR (1982a) Tumorigenic transformation of mammalian cells induced by a normal human gene homologous to the oncogene of Harvey murine sarcoma virus. Nature 297:479–483

Chang EH, Gonda MA, Ellis RW, Scolnick EM, Lowy DR (1982b) Human genome contains four genes homologous to transforming genes of Harvey and Kirsten murine sarcoma viruses. Proc Natl Acad Sci USA 79:4848–4852

Chen W-J, Andres DA, Goldstein JL, Brown MS (1991a) Cloning and expression of a cDNA encoding the α subunit of rat p21^{ras} protein farnesyltransferase. Proc Natl Acad Sci USA 88:11368–11372

Chen W-J, Andres DA, Goldstein JL, Russell DW, Brown MS (1991b) cDNA cloning and expression of the peptide-binding β subunit of rat p21ras farnesyltransferase, the counterpart of yeast DPR1/RAM1. Cell 66:327–334

Christian BJ, Kao C, Wu S-Q, Meisner LF, Reznikoff CA (1990) EJ/ras neoplastic transformation of simian virus 40-immortalized human uroepithelial cells: a rare event. Cancer Res 50:4779–4786

Chung H-H, Kim R, Kim S-H (1992) Biochemical and biological activity of phosphorylated and non-phosphorylated ras p21 mutants. Biochim Biophys Acta 1129:278–286

Clark R, Wong G, Arnheim N, Nitecki D, McCormick F (1985) Antibodies specific for amino acid 12 of the ras oncogene product inhibit GTP binding. Proc Natl Acad Sci USA 82:5280–5284

Cohen JB, Levinson AD (1988) A point mutation in the last intron responsible for increased expression and transforming activity of the c-Ha-ras oncogene. Nature 334:119–124

Collins SJ (1988) Direct sequencing of amplified genomic fragments documents N-ras point mutations in myeloid leukemia. Oncogene Res 3:117–123

Corominas M, Kamino H, Leon J, Pellicer A (1989) Oncogene activation in human benign tumors of the skin (keratoacanthomas): is HRAS involved in differentiation as well as proliferation? Proc Natl Acad Sci USA 86:6372–6376

Corominas M, Sloan SR, Leon J, Kamino H, Newcomb EW, Pellicer A (1991) ras Activation in human tumors and in animal model systems. Environ Health Perspect 93:19–25

Cox AD, Der CJ (1992) The ras/cholesterol connection: implications for ras oncogenicity. Crit Rev Oncogenesis (in press)

Craig RW, Sager R (1985) Suppression of tumorigenicity in hybrids of normal and oncogene-transformed CHEF cells. Proc Natl Acad Sci USA 82:2062–2066

DeClue JE, Papageorge AG, Fletcher JA, Diehl SR, Ratner N, Vass WC, Lowy DR (1992) Abnormal regulation of mammalian $p21^{ras}$ contributes to malignant tumor growth in von Recklinghausen (type 1) neurofibromatosis. Cell 69:265–273

Der CJ (1989) The ras family of oncogenes. In: Benz C, Liu E (eds) Oncogenes. Kluwer, Amsterdam, pp 74–119

Der CJ, Cooper GM (1983) Altered gene products are associated with activation of cellular rasK genes in human lung and colon carcinomas. Cell 32:201–208

Der CJ, Krontiris TG, Cooper GM (1982) Transforming genes of human bladder and lung carcinoma cell lines are homologous to the ras genes of Harvey and Kirsten sarcoma viruses. Proc Natl Acad Sci USA 79:3637–3640

Der CJ, Finkel T, Cooper GM (1986) Biological and biochemical properties of human rasH genes mutated at codon 61. Cell 44:167–176

Der CJ, Weissman B, MacDonald MJ (1988) Altered guanine nucleotide binding and H-ras transforming and differentiating activities. Oncogene 3:105–112

Downward J (1990) The ras superfamily of small GTP-binding proteins. Trends Biochem Sci 15:469–472

Efrat S, Fleischer N, Hanahan D (1990) Diabetes induced in male transgenic mice by expression of human H-ras oncoprotein in pancreatic *β* cells. Mol Cell Biol 10:1779–1783

Ehlen T, Dubeau L (1989) Detection of ras point mutations by polymerase chain reaction using mutation-specific, inosine-containing oligonucleotide primers. Biochem Biophys Res Commun 160:441–447

Ellis RW, DeFeo D, Shih TY, Gonda MA, Young HA, Tsuchida H, Lowy DR, Scolnick EM (1981) The p21 src genes of Harvey and Kirsten sarcoma viruses originate from divergent members of a family of normal vertebrate genes. Nature 292:506–511

Enomoto T, Inoue M, Perantoni AO, Terakawa N, Tanizawa O, Rice JM (1990) K-ras activation in neoplasms of the human female reproductive tract. Cancer Res 50:6139–6145

Enomoto T, Inoue M, Perantoni AO, Buzard GS, Miki H, Tanizawa O, Rice JM (1991a) K-ras activation in premalignant and malignant epithelial lesions of the human uterus. Cancer Res 51:5308–5314

Enomoto T, Weghorst CM, Inoue M, Tanizawa O, Rice JM (1991b) K-ras activation occurs frequently in mucinous adenocarcinomas and rarely in other common epithelial tumors of the human ovary. Am J Pathol 139:777–785
Farr CJ, Marshall CJ, Easty DJ, Wright NA, Powell SC, Paraskeva C (1988) A study of ras gene mutations in colonic adenomas from familial polyposis coli patients. Oncogene 3:673–678
Fasano O, Birnbaum D, Edlund K, Fogh J, Wigler M (1984) New human transforming genes detected by a tumorigenicity assay. Mol Cell Biol 4:1695–1705
Feig LA, Cooper GA (1988) Relationship among guanine nucleotide exchange, GTP hydrolysis, and transforming potential of mutated ras proteins. Mol Cell Biol 8:2472–2478
Feig LA, Bast RC Jr, Knapp RC, Cooper GM (1984) Somatic activation of rasK gene in a human ovarian carcinoma. Science 223:698–701
Forrester K, Almoguera C, Han K, Grizzle WE, Perucho M (1987) Detection of high incidence of K-ras oncogenes during human colon tumorigenesis. Nature 327:298–303
Fry DG, Milam LD, Dillberger JE, Maher VM, McCormick JJ (1990) Malignant transformation of an infinite life span human fibroblast cell strain by transfection with v-Ki-ras. Oncogene 5:1415–1418
Fukumoto M, Estensen RD, Sha L, Oakley GJ, Twiggs LB, Adcock LL, Carson LF, Roninson IB (1989) Association of Ki-ras with amplified DNA sequences, detected in human ovarian carcinomas by a modified in-gel renaturation assay. Cancer Res 49:1693–1697
Fujita K, Ohuchi N, Yao T, Okumura M, Fukushima Y, Kanakura Y, Kitamura Y, Fujita J (1987) Frequent overexpression, but not activation by point mutation, of ras genes in primary human gastric cancers. Gastroenterology 93:1339–1345
Geiser AG, Der CJ, Marshall CJ, Stanbridge EJ (1986) Suppression of tumorigenicity with continued expression of the c-Ha-ras oncogene in EJ bladder carcinoma-human fibroblast hybrid cells. Proc Natl Acad Sci USA 83:5209–5213
Gibbs JB (1991) ras C-terminal processing enzymes – new drug targets? Cell 65:1–4
Gilman AG (1984) G proteins and dual control of adenylate cyclase. Cell 36:577–579
Givol I, Greenhouse JI, Hughes SH, Ewert DL (1992) Retroviruses that express different ras mutants cause different types of tumors in chickens. Oncogene 7:141–146
Glomset J, Gelb M, Farnsworth C (1991) The prenylation of proteins. Curr Opin Lipidol 2:118–124
Goldstein JL, Brown MS (1990) Regulation of the mevalonate pathway. Nature 343:425–430
Gonzalez-Cadavid NF, Zhou D, Battifora H, Bar-Eli M, Cline MJ (1989) Direct sequencing analysis of exon 1 of the c-K-ras gene shows a low frequency of mutations in human pancreatic adenocarcinomas. Oncogene 4:1137–1140
Guerrero I, Pellicer A (1987) Mutational activation of oncogenes in animal model systems of carcinogenesis. Mutat Res 185:293–308
Gumerlock PH, Poonamallee UR, Meyers FJ, deVere White RW (1991) Activated ras alleles in human carcinoma of the prostate are rare. Cancer Res 51:1632–1637
Hagag N, Halegoua S, Viola M (1986) Inhibition of growth factor-induced differentiation of PC12 cells by microinjection of antibody to ras p21. Nature 319:680–682
Hall A (1990) The cellular functions of small GTP-binding proteins. Science 249:635–249
Hall A (1992a) Signal transduction through small GTPases-a tale of two GAPS. Cell 69:389–391

Hall A (1992b) ras-Related GTPases and the cytoskeleton. Mol Biol Cell 3:475–479
Han M, Sternberg PW (1991) Analysis of dominant-negative mutations of the Caenorhabditis elegans let-60 ras gene. Genes Dev 5:2188–2198
Harihatan IK, Carthew RW, Rubin GM (1991) The Drosophila roughened mutation: activation of a rap homolog disrupts eye development and interferes with cell determination. Cell 67:717–722
Harper JR, Roop DR, Yuspa SH (1986) Transfection of the EJ ras^{Ha} gene into keratinocytes derived from carcinogen-induced mouse papillomas causes malignant progression. Mol Cell Biol 6:3144–3149
Haubruck H, McCormick F (1991) Ras p21: effects and regulation. Biochim Biophys Acta 1072:215–229
He B, Chen P, Chen SY, Vancura KL, Michaelis S, Powers S (1991) RAM2, an essential gene of yeast, and RAM1 encode the two polypeptide components of the farnesyltransferase that prenylates a-factor and Ras proteins. Proc Natl Acad Sci USA 88:11373–11377
Henrard DR, Thornley AT, Brown ML, Rheinwald JG (1990) Specific effects of ras oncogene expression on the growth and histogenesis of human keratinocytes. Oncogene 5:475–481
Higaki J, Miya A, Miki T, Morishita R, Mikami H, Takai S, Ogihara T (1991) Contribution of the activation of the ras oncogene to the evolution of aldosterone- and renin-secreting tumors. J Hypertens 9:135–137
Higashi T, Sasai H, Suzuki F, Miyoshi J, Ohuchi T, Takai S-I, Mori T, Kakunaga T (1990) Hamster cell line suitable for transfection assay of transforming genes. Proc Natl Acad Sci USA 87:2409–2413
Higinbotham KG, Rice JM, Perantoni AO (1992) Activating point mutation in Ki-ras codon 63 in a chemically induced rat renal tumor. Mol Carcinogen 5:136–139
Hirai H, Kobayashi Y, Mano H, Hagiwara K, Maru Y, Omine M, Mizoguchi H, Nishida J, Takaku F (1987) A point mutation at codon 13 of the N-ras oncogene in myelodysplastic syndrome. Nature 327:430–432
Huber BE, Cordingley MG (1988) Expression and phenotypic alterations caused by an inducible transforming ras oncogene introduced into rat liver epithelial cells. Oncogene 3:245–256
Hunter T (1991) Cooperation between oncogenes. Cell 64:249–270
Imamura N, Kuramoto A, Ishihara H, Shimizu S (1992) H-ras point mutations in leukaemias and lymphomas. Lancet 339:1297–1298
Imamura T, Arima T, Kato H, Miyamoto S, Sasazuki T, Wake N (1992) Chromosomal delctions and K-ras gene mutations in human endometrial carcinomas. Int J Cancer 51:47–52
Iolascon A, Lania A, Badiali M, Pession A, Saglio G, Giangaspero F, Miraglia del Giudice E, Perrotta S, Cutillo S (1991) Analysis of N-ras gene mutations in medulloblastomas by polymerase chain reaction and oligonucleotide probes in formalin-fixed, paraffin-embedded tissues. Med Pediatr Oncol 19:240–245
Ireland CM (1989) Activated N-ras oncogenes in human neuroblastoma. Cancer Res 49:5530–5533
Jiang W, Kahn SM, Guillem JG, Lu S-H, Weinstein IB (1989) Rapid detection of ras oncogenes in human tumors: applications to colon, esophageal, and gastric cancer. Oncogene 4:923–928
Joyce AD, D'Emilia JC, Steele G Jr, Libertino JA, Silverman ML, Summerhayes IC (1989) Detection of altered H-ras proteins in human tumors using Western blot analysis. Lab Invest 61:212–218
Kahn RA, Der CJ, Bokoch GM (1992) The ras superfamily of GTP-binding proteins: guidelines on nomenclature. FASEB J 6:2512–2513
Karga HJ, Alexander JM, Hedley-Whyte ET, Klibanski A, Jameson JL (1992) ras Mutations in human pituitary tumors. J Clin Endocrinol Metab 74:914–919

Kato K, Cox AD, Hisaka MM, Graham SM, Buss JE, Der CJ (1992) Isoprenoid addition to ras protein is the critical modification for its membrane association and transforming activity. Proc Natl Acad Sci USA (in press)
Khosravi-Far R, Cox AD, Kato K, Der CJ (1992) Protein prenylation: key to ras function and cancer intervention? Cell Growth Diff (in press)
Kitayama H, Sugimoto Y, Matsuzaki T, Ikawa Y, Noda M (1989) A ras-related gene with transformation suppressor activity. Cell 56:77–84
Kohl NE, Diehl RE, Schaber MD, Rands E, Soderman DD, He B, Moores SL, Pompliano DL, Ferro-Novick S, Powers S, Thomas KA, Gibbs JA (1991) Structural homology among mammalian and Saccharomyces cerevisiae isoprenyl-protein transferases. J Biol Chem 266:18884–18888
Konishi N, Enomoto T, Buzard G, Ohshima M, Ward JM, Rice JM (1992) K-ras activation and ras p21 expression in latent prostatic carcinoma in Japanese men. Cancer 69:2293–2299
Krengel U, Schlichting L, Scherer A, Schumann R, Frech M, John J, Kabsch W, Pai EF, Wittinghofer A (1990) Three-dimentional structures of H-ras p21 mutants: molecular basis for their inability to function as signal switch molecules. Cell 62:539–548
Krontiris TG, Cooper GM (1981) Transforming activity of human tumor DNA. Proc Natl Acad Sci USA 78:1181–1184
Kumar R, Barbacid M (1988) Oncogene detection at the single cell level. Oncogene 3:647–651
La Vecchio JA, Hamer PJ, Ng SC, Trimpe KL, Carney WP (1990) Characterization of monoclonal antibodies specific to the activated ras p21 with aspartic acid at position 13. Oncogene Res 1173–1178
Lacal JC, Tronick SR (1988) The ras oncogene. In: Reddy EP, Skalka AM, Curran T (eds) The oncogene handbook. Elsevier, Amsterdam, pp 257–304
Land H, Parada LF, Weinberg RA (1983) Tumorigenic conversion of primary embryo fibroblasts requires at least two cooperating oncogenes. Nature 304:596–606
Leder A, Kuo A, Cardiff RD, Sinn E, Leder P (1990) v-Ha-ras transgene abrogates the initiation step in mouse skin tumorigenesis: effects of phorbol esters and retinoic acid. Proc Natl Acad Sci USA 87:9178–9182
LeMaistre A, Lee M-S, Talpaz M, Kantarjian HM, Freireich EJ, Deisseroth AB, Trujilo JM, Stass SA (1989) Ras oncogene mutations are rare late stage events in chronic myelogenous leukemia. Blood 73:889–891
Lemoine NR, Mayall ES, Williams ED, Thurston V, Wynford-Thomas D (1988) Agent-specific ras oncogene activation in rat thyroid tumours. Oncogene 3:541–544
Lemoine NR, Mayall ES, Wyllie FS, Williams ED, Goyns M, Stringer B, Wynford-Thomas D (1989) High frequency of ras oncogene activation in all stages of human thyroid tumorigenesis. Oncogene 4:159–164
Li Y, Bollag G, Clark R, Stevens J, Conroy L, Fults D, Ward K, Friedman E, Samowitz W, Robertson M, Bradley P, McCormick F, White R, Cawthon R (1992) Somatic mutations in the neurofibromatosis 1 gene in human tumors. Cell 69:275–281
Liu E, Hjelle B, Morgan R, Hecht F, Bishop JM (1987) Mutations of the Kirsten-ras proto-oncogene in human preleukemia. Nature 330:186–188
Lubbert M, Mirro J Jr, Kitchingman G, McCormick F, Mertelsmann R, Herrmann F, Koeffler HP (1992) Prevalence of N-ras mutations in children with myelodysplastic syndromes and acute myeloid leukemia. Oncogene 7:263–268
Maltese WA (1990) Posttranslational modification of proteins by isoprenoids in mammalian cells. FASEB J 4:3319–3328
Mangues R, Seidman I, Pellicer A, Gordon JW (1990) Tumorigenesis and male sterility in transgenic mice expressing a MMTV/N-ras oncogene. Oncogene 5:1491–1497

Manne V, Roberts D, Tobin A, O'Rourke E, De Virgilio M, Meyers C, Ahmed N, Kurz B, Resh M, Kung H-F, Barbacid M (1990) Identification and preliminary characterization of protein-cysteine farnesyltransferase. Proc Natl Acad Sci USA 87:7541–7545
Mariyama M, Kishi K, Nakamura K, Obata H, Nishimura S (1989) Frequency and types of point mutation at the 12th codon of the c-Ki-ras gene found in pancreatic cancers from Japanese patients. Jpn J Cancer Res 80:622–626
Martin GA, Viskochil D, Bollag G, McCabe PC, Crosier WJ, Haubruck H, Conroy L, Clark R, O'Connell P, Cawthon RM, Innis MA, McCormick F (1990) The GAP-related domain of the neurofibromatosis type 1 gene product interacts with ras p21. Cell 63:843–849
Matsumoto T, Beach D (1991) Premature initiation of mitosis in yeast lacking RCC1 or an interacting GTPase. Cell 66:347–360
Mayer BJ, Jackson PK, Baltimore D (1991) The noncatalytic src homology region 2 segment of abl tyrosine kinase binds to tyrosine-phosphorylated cellular proteins with high affinity. Proc Natl Acad Sci USA 88:627–631
McKenna WG, Weiss MC, Endich B, Ling CC, Bakanauskas VJ, Kelsten ML (1990) Synergistic effect of the v-myc oncogene with H-ras on radioresistance. Cancer Res 50:97–102
McKenzie SJ (1991) Diagnostic utility of oncogenes and their products in human cancer. Biochim Biophys Acta 1072:193–214
Metcalf RA, Welsh JA, Bennett WP, Seddon MB, Lehman TA, Pelin K, Linnainmaa K, Tammilehto L, Mattson K, Gerwin BI, Harris CC (1992) p53 and Kirsten-ras mutations in human mesothelioma cell lines. Cancer Res 52:2610–2615
Michel T, Winslow JW, Smith JA, Seidman JG, Neer EJ (1986) Molecular cloning and characterization of cDNA encoding the GTP-binding protein α_1 and identification of related protein, α_h. Proc Natl Acad Sci USA 83:7663–7667
Miki H, Ohmori M, Perantoni AO, Enomoto T (1991) K-ras activation in gastri epithelial tumors in Japanese. Cancer Lett 58:107–113
Milburn MV, Tong L, DeVos AM, Brunger A, Yamaizumi Z, Nishimura S, Kim S-H (1990) Molecular switch for signal transduction: structural differences between active and inactive forms of protooncogenic ras proteins. Science 247:939–945
Mitsudomi T, Viallet J, Mulshine JL, Linnoila RI, Minna JD, Gazdar AF (1991) Mutations of ras genes distinguish a subset of non-small-cell lung cancer cell lines from small-cell lung cancer cell lines. Oncogene 6:1353–1362
Moley JF, Brother MB, Wells SA, Spengler BA, Biedler JL, Brodeur GM (1991) Low frequency of ras gene mutations in neuroblastomas, pheochromocytomas, and medullary thyroid cancers: Cancer Res 51:1596–1599
Motojima K, Tsunoda T, Kanematsu T, Nagata Y, Urano T, Shiku H (1991) Distinguishing pancreatic carcinoma from other periampullary carcinomas by analysis of mutations in the Kirsten-ras oncogene. Ann Surg 214:657–662
Nagata Y, Abe M, Kobayashi K, Yoshida K, Ishibashi T, Naoe T, Nakayama E, Shiku H (1990) Glycine to aspartic acid mutations at codon 13 of the c-Ki-ras gene in human gastrointestinal cancers. Cancer Res 50:480–482
Nakagawa T, Mabry M, de Bustros A, Ihle JN, Nelkin BD, Baylin SB (1987) Introduction of v-Ha-ras oncogene induces differentiation of cultured human medullary thyroid carcinoma cells. Proc Natl Acad Sci USA 84:5923–5927
Namba H, Rubin SA, Fagin JA (1990) Point mutations of ras oncogenes are an early event in thyroid tumorigenesis. Mol Endocrinol 4:1474–1479
Nanus DM, Mentle IR, Motzer RJ, Bander NH, Albino AP (1990) Infrequent ras oncogene point mutations in renal cell carcinoma. J Urol 143:175–178
Nelkin BD, Borges M, Mabry M, Baylin SB (1990) Transcription factor levels in medullary thyroid carcinoma cells differentiated by Harvey ras oncogene: c-jun is increased. Biochem Biophys Res Commun 170:140–146

Neri A, Murphy JP, Cro L, Ferrero D, Tarella C, Baldini L, Dall-Favera R (1989) ras Oncogene mutation in multiple myeloma. J Exp Med 170:1715–1725
Noda M, Ko M, Ogura A, Liu D-G, Amano T, Takano T, Ikawa Y (1985) Sarcoma viruses carrying ras oncogenes induce differentiation-associated properties in a neuronal cell line. Nature 318:73–75
Orita S, Higashi T, Kawasaki Y, Harada A, Igarashi H, Monden T, Morimoto H, Shimano T, Mori T, Miyoshi J (1991) A novel point mutation at codon 146 of the K-ras gene in a human colorectal cancer identified by the polymerase chain reaction. Virus Genes 5:75–79
Oudejans JJ, Slebos RJC, Zoetmulder FAN, Mooi WJ, Rodenhuis S (1991) Differential activation of ras genes by point mutation in human colon cancer with metastases to either lung or liver. Int J Cancer 49:875–879
Pai EF, Kabsch W, Krengel U, Holmes KC, John J, Wittinghofer A (1989) Structure of the guanine-nucleotide-binding domain of the Ha-ras oncogene product p21 in the triphosphate conformation. Nature 341:209–214
Parada LF, Tabin CJ, Shih C, Weinberg RA (1982) Human EJ bladder carcinoma oncogene is homologue of harvey sarcoma virus ras gene. Nature 297:474–478
Patel G, MacDonald MJ, Khosravi-Far R, Hisaka MM, Der CJ (1992) Alternate mechanisms of ras activation are complementary and favor the formation of GTP-ras. Oncogene 7:283–288
Perucho M, Goldfarb M, Shimizu K, Lama C, Fogh J, Wigler M (1981) Human-tumor-derived cell lines contain common and different transforming genes. Cell 27:467–476
Quaife CJ, Pinkert CA, Ornitz DM, Palmiter RD, Brinster RL (1987) Pancreatic neoplasia induced by ras expression in acinar cells of transgenic mice. Cell 48:1023–1034
Reddy EP, Reynolds RK, Santos E, Barbacid M (1982) A point mutation is responsible for the acquisition of transforming properties by the T24 human bladder carcinoma oncogene. Nature 300:149–152
Redmond SMS, Reichmann E, Muller RG, Friis RR, Groner B, Hynes NE (1988) The transformation of primary and established mouse mammary epithelial cells by p21-ras is concentration dependent. Oncogene 2:259–265
Reinstein J, Schlichting I, Frech M, Goody RS, Wittinghofer A (1991) p21 with a phenylalanine 28-leucine mutation reacts normally with the GTPase activating protein GAP but nevertheless has transforming properties. J Biol Chem 266:17700–17706
Reiss Y, Goldstein JL, Seabra MC, Casey PJ, Brown MS (1990) Inhibition of purified $p21^{ras}$ farnesyl:protein transferase by Cys-AAX tetrapeptides. Cell 62:81–88
Reynolds SH, Stowers SJ, Patterson RM, Maronpot RR, Aaronson SA, Anderson MW (1987) Activated oncogenes in B6C3F1 mouse liver tumors: implications for risk assessment. Science 237:1309–1316
Ricketts MH, Levinson AD (1988) High-level expression of c-H-ras1 fails to fully transform Rat-1 cells. Mol Cell Biol 8:1460–1468
Ridley AJ, Paterson HF, Noble M, Land H (1988) ras-Mediated cell cycle arrest is altered by nuclear oncogenes to induce Schwann cell transformation. EMBO J 7:1635–1645
Rochlitz CF, Scott GK, Dodson JM, Liu E, Dollbaum C, Smith HS, Benz CC (1989) Incidence of activating ras oncogene mutations associated with primary and metastatic human breast cancer. Cancer Res 49:357–360
Rochlitz CF, Peter S, Willroth G, de Kant E, Lobeck H, Huhn D, Herrmann R (1992) Mutations in the ras protooncogenes are rare events in renal cell cancer. Eur J Cancer 28:333–336
Rodenhuis S, Slebos RJC (1990) The ras oncogenes in human lung cancer. Am Rev Respir Dis 142:S27–S30
Rodenhuis S, Slebos RJC (1992) Clinical significance of ras oncogene activation in human lung cancer. Cancer Res 52:2665s–2669s

Ruley HE (1990) Transforming collaborations between ras and nuclear oncogenes. Cancer Cells 2:258–268
Saglio G, Serra A, Novarino A, Falda M, Gavosto F (1989) N-ras mutations in myeloid leukemias. Tumori 7:337–340
Saison-Behmoaras T, Tocque B, Rey I, Chassignol M, Thuong NT, Helene C (1991) Short modified antisense oligonucleotides directed against Ha-ras point mutation induce selective cleavage of the mRNA and inhibit T24 cells proliferation. EMBO J 10:1111–1118
Santos E, Tronick SR, Aaronson SA, Pulciani S, Barbacid M (1982) T24 human bladder carcinoma oncogene is an activated form of the normal human homologue of BALB- and Harvey-MSV transforming genes. Nature 298:343–347
Santos E, Reddy EP, Pulciani S, Feldmann RJ, Barbacid M (1983) Spontaneous activation of a human proto-oncogene. Proc Natl Acad Sci USA 80:4679–4683
Santos E, Martin-Zanca D, Reddy EP, Pierotti MA, Della Porta G, Barbacid M (1984) Malignant activation of a K-ras oncogene in lung carcinoma but not in normal tissue of the same patient. Science 223:661–664
Schaber MD, O'Hara MB, Garsky VM, Mosser SD, Bergstrom JD, Moores SL, Marshall MS, Friedman PA, Dixon RAF, Gibbs JB (1990) Polyisoprenylation of ras in vitro by a farnesyl-protein transferase. J Biol Chem 265:14701–14704
Schweighoffer F, Barlat I, Chevallier-Multon MC, Tocque B (1992) Implications of GAP in ras-dependent transactivation of a polyoma enhancer sequences. Science 256:825–827
Seeburg PH, Colby WW, Hayflick JS, Capon DJ, Goeddel DV, Levinson AD (1984) Biological properties of human c-Ha-ras1 genes mutated at codon 12. Nature 312:71–75
Seremetis S, Inghirami G, Ferrero D, Newcomb EW, Knowles DM, Dotto G-P, Dalla-Favera R (1989) Transformation and plasmacytoid differentiation of EBV-infected human B lymphoblasts by ras oncogenes. Science 243:660–663
Shi Y, Zou M, Schmidt H, Juhasz F, Stensky V, Robb D, Farid NR (1991) High rates of ras codon 61 mutation in thyroid tumors in an iodide-deficient area. Cancer Res 51:2690–2693
Shibata D, Almoguera C, Forrester K, Dunitz J, Martin SE, Cosgrove MM, Perucho M, Arnheim N (1990) Detection of c-K-ras mutations in fine needle aspirates from human pancreatic adenocarcinomas. Cancer Res 50:1279–1283
Shih C, Padhy LC, Murray M, Weinberg RA (1981) Transforming genes of carcinomas and neuroblastomas introduced into mouse fibroblasts. Nature 290:261–264
Shimizu K, Goldfarb M, Suard Y, Perucho M, Li Y, Kamata T, Feramisco J, Stavnezer E, Fogh J, Wigler MH (1983) Three human transforming genes are related to the viral ras oncogenes. Proc Natl Acad Sci USA 80:2112–2116
Shou C, Farnsworth CL, Neel BG, Feig LA (1992) Molecular cloning of cDNAs encoding a guanine-nucleotide releasing factor for Ras-p21. Nature (in press)
Shukla VK, Hughes DC, Hughes LE, McCormick F, Padua RA (1989) ras Mutations in human melanotic lesions: K-ras activation is a frequent and early event in melanoma development. Oncogene Res 5:121–127
Sidransky D, Tokino T, Hamilton SR, Kinzler KW, Levin B, Frost P, Vogelstein B (1992) Identification of ras oncogene mutations in the stool of patients with curable colorectal tumors. Science 256:102–105
Sigal IS, Gibbs JB, D'Alonzo JS, Temeles GL, Wolanski BS, Socher SH, Scolnick EM (1986) Mutant ras-encoded proteins with altered nucleotide binding exert dominant biological effects. Proc Natl Acad Sci USA 83:952–956
Sinn E, Muller W, Pattengale P, Tepler I, Wallace R, Leder P (1987) Coexpression of MMTV/v-Ha-ras and MMTV/c-myc genes in transgenic mice: synergistic action of oncogenes in vivo. Cell 49:465–475
Sklar MD (1988) The ras oncogenes increase the intrinsic resistance of NIH3T3 cells to ionizing radiation. Science 239:645–647

Slebos RJC, Hruban RH, Dalesio O, Mooi WJ, Offerhaus GJA, Rodenhuis S (1991) Relationships between K-ras oncogene activation and smoking in adenocarcinoma of the human lung. J Natl Cancer Inst 83:1024–1027
Sloan SR, Newcomb EW, Pellicer A (1990) Neutron radiation can activate K-ras via a point mutation in codon 146 and induces a different spectrum of ras mutations than does gamma radiation. Mol Cell Biol 10:405–408
Smit VTHBM, Boot AJM, Smits AMM, Fleuren GJ, Cornelisse CJ, Bos JL (1988) KRAS codon 12 mutations occur very frequently in pancreatic adenocarcinomas. Nucleic Acids Res 16:7773–7782
Spandidos DA (1987) Oncogene activation in malignant transformation: a study of H-ras in human breast cancer. Anticancer Res 7:991–996
Srivastava SK, Yuasa Y, Reynolds SH, Aaronson SA (1985) Effects of two major activating lesions on the structure and conformation of human ras oncogene products. Proc Natl Acad Sci USA 82:38–42
Stacey DW, Kung HF (1984) Transformation of NIH/3T3 cells by microinjection of Ha-ras p21 protein. Nature 310:508–511
Stork P, Loda M, Bosari S, Wiley B, Poppenhusen K, Wolfe H (1991) Detection of K-ras mutations in pancreatic and hepatic neoplasms by non-isotopic mismatched polymerase reaction. Oncogene 6:857–862
Stratton MR, Fisher C, Gusterson BA, Cooper CS (1989) Detection of point mutations in N-ras and K-ras genes of human embryonal rhabdomyosarcomas using oligonucleotide probes and the polymerase chain reaction. Cancer Res 49:6324–6327
Suarez HG, de Villard JA, Severino M, Caillou B, Schlumberger M, Tubiana M, Parmentier C, Monier R (1990) Presence of mutations in all three ras genes in human thyroid tumors. Oncogene 5:565–570
Sukumar S (1989) ras Oncogenes in chemical carcinogeneis. In: Vogt PK (ed) Oncogenes and retroviruses, selected reviews. Springer, Berlin Heidelberg New York, pp 93–114
Sukumar S, Notario V, Martin-Zanca D, Barbacid M (1983) Induction of mammary carcinomas in rats by nitroso-methylurea involves malignant activation of H-*ras*-1 locus by single point mutations. Nature 306:658–662
Suzuki Y, Orita M, Shiraishi M, Hayashi K, Sekiya T (1990) Detection of ras gene mutations in human lung cancers by single-strand information polymorphism analysis of polymerase chain reaction products. Oncogene 5:1037–1043
Syvanen A-C, Soderlund H, Laaksonen E, Bengtstrom M, Turunen M, Palotie A (1992) N-ras gene mutations in acute myeloid leukemia: accurate detection by solid-phase minisequencing. Int J Cancer 50:713–718
Tabin CJ, Bradley SM, Bargmann CI, Weinberg RA, Papageorge AG, Scolnick EM, Dhar R, Lowy DR, Change EH (1982) Mechanism of activation of a human oncogene. Nature 300:143–149
Tada M, Omata M, Ohto M (1990a) Analysis of ras gene mutations in human hepatic malignant tumors by polymerase chain reaction and direct sequencing. Cancer Res 50:1121–1124
Tada M, Yokosuka O, Omata M, Ohto M, Isono K (1990b) Analysis of ras gene mutations in biliary and pancreatic tumors by polymerase chain reaction and direct sequencing. Cancer 66:930–935
Tada M, Omata M, Ohto M (1992) High incidence of ras gene mutation in intrahepatic cholangiocarcinoma. Cancer 69:1115–1118
Taparowsky E, Suard Y, Fasano O, Shimizu K, Goldfarb M, Wigler M (1982) Activation of T24 bladder carcinoma transforming gene is linked to a single amino acid change. Nature 300:762–765
Tong L, de Vos AM, Milburn MV, Jancarik J, Noguchi S, Nishimura S, Miura K, Ohtsuka E, Kim S-H (1989) Structural differences between a RAS oncogene protein and the normal protein. Nature 337:90–93
Trahey M, Milley RJ, Cole GE, Innis M, Paterson H, Marshall CJ, Hall A,

McCormick F (1987) Biochemical and biological properties of the human N-ras p21 protein. Mol Cell Biol 7:541–544

Tremblay PJ, Pothier F, Hoang T, Tremblay G, Brownstein S, Liszauer A, Jolicoeur P (1989) Transgenic mice carying the mouse mammary tumor virus ras fusion gene: distinct effect in various tissues. Mol Cell Biol 9:854–859

Valencia A, Chardin P, Wittinghofer A, Sander C (1991) The ras protein family: evolutionary tree and role of conserved amino acids. Biochemistry 80:4637–4648

van Kamp H, de Pijper C, Verlaan-de Vries M, Bos JL, Leeksma CHW, Kerkhofs H, Willemze R, Fibbe WE, Landegent JE (1992) Longitudinal analysis of point mutations of the N-ras proto-oncogene in patients with myelodysplasia using archived blood smears. Blood 79:1266–1270

Verlaan-de Vries M, Bogaard ME, van den Elst H, van Boom JH, van der Eb AJ, Bos JL (1986) A dot-blot screening procedure for mutated ras oncogenes using synthetic oligodeoxynucleotides. Gene 50:313–320

Vorburger K, Lehner CF, Kitten GT, Eppenberger HM, Nigg EA (1989) A second higher vertebrate B-type lamin cDNA sequence determination and in vitro processing of chicken lamin B_2. J Mol Biol 208:405–415

Walter M, Clark SG, Levinson AD (1986) The oncogenic activation of human $p21^{ras}$ by a novel mechanism. Science 233:649–652

Wei W, Mosteller RD, Sanyal P, Gonzales E, McKinney D, Dasgupta C, Li P, Liu B-X, Broek D (1992) Identification of a mammalian gene structurally and functionally related to the CDC25 gene of Sasccharomyces cerevisiae. Proc Natl Acad Sci USA 89

Weissman BE, Aaronson SA (1983) BALB and Kirsten murine sarcoma viruses alter growth and differentiation of EGF-dependent BALB/c mouse epidermal keratinocyte lines. Cell 32:599–606

Wilson DM, Yang D, Dillberger JE, Dietrich SE, Maher VM, McCormick JJ (1990) Malignant transformation of human fibroblasts by a transfected N-ras oncogene. Cancer Res 50:5587–5593

Winter E, Yamamoto F, Almoguèra C, Perucho M (1985) A method to detect and characterize point mutations in transcribed genes: amplification and overexpression of the mutant c-Ki-ras allele in human tumor cells. Proc Natl Acad Sci USA 82:7575–7579

Wong G, Arnheim N, Clark R, McCabe P, Innis M, Aldwin L, Nitecki D, McCormick F (1986) Detection of activated Mr 21,000 protein, the product of ras oncogenes, using antibodies with specificity for amino acid 12. Cancer Res 46:6029–6033

Wright PA, Lemoine NR, Mayall ES, Wyllie FS, Hughes D, Williams ED, Wynford-Thomas D (1989) Papillary and follicular thyroid carcinomas show a different pattern of ras oncogene mutation. Br J Cancer 60:576–577

Wright PA, Williams ED, Lemoine NR, Wynford-Thomas D (1991) Radiation-associated and "spontaneous" human thyroid carcinomas show a different pattern of ras oncogene mutation. Oncogene 6:471–473

Wynford-Thomas D, Bond JA, Wyllie FS, Burns JS, Williams ED, Jones T, Sheer D, Lemoine NR (1990) Conditional immortalization of human thyroid epithelial cells: a tool for analysis of oncogene action. Mol Cell Biol 10:5365–5377

Xu G, Lin B, Tanaka K, Dunn D, Wood D, Gesteland R, White R, Weiss R, Tamanoi F (1990) The catalytic domain of the neurofibromatosis type 1 gene product stimulates ras GTPase and complements ira mutants of S. cerevisiae. Cell 63:835–841

Yuasa Y, Srivastava SK, Dunn CY, Rhim JS, Reddy EP, Aaronson SA (1983) Acquisition of transforming properties by alternative point mutations within c-bas/has human proto-oncogene. Nature 303:775–779

Zarbl H, Sukumar S, Arthur AV, Martin-Zanca D, Barbacid M (1985) Direct mutagenesis of Ha-ras-1 oncogenes by N-nitroso-N-methylurea during initiation of mammary carcinogenesis in rats. Nature 315:382–385

CHAPTER 19

Dominant Inhibitory *Ras* Mutants: Tools for Elucidating *Ras* Function

L.A. FEIG

A. Introduction

ras proteins belong to a superfamily of guanine nucleotide binding proteins that cycle between the active GTP- and inactive GDP-bound states (BOURNE et al. 1991). They become activated by exchanging GTP for bound GDP, a process catalyzed by a guanine nucleotide exchange factor (GEF) (also called ras-GRF, guanine nucleotide releasing factor) (SHOU et al., in press). They become deactivated by hydrolyzing GTP to GDP, which is catalyzed by GTPase activating proteins, ras-GAP and NF-1 (BOLLAG and MCCORMICK 1991). Three *ras* genes, H-, N-, and K-*ras*, code for extremely similar 21-kDa proteins (p21s) that are localized to the inner surface of the plasma membrane. A role for p21 in the control of cell proliferation was first suggested by the observation that mutant *ras* genes are responsible for the oncogenicity of a set of rodent retroviruses. Subsequently, it was shown that these same mutations, which lock ras proteins in the active GTP-bound form, are present in endogenous *ras* genes from a wide variety of human tumor types (BARBACID 1987). Moreover, expression of the mutated genes in tissue culture cells or in transgenic animals promotes a neoplastic phenotype (QUAIFE et al. 1987).

ras proteins also appear to be involved in the control of differentiated function. For example, these proteins are present in neuronal cells that never divide (SWANSON et al. 1986). Moreover, in pheochromocytoma cells, expression of oncogenic ras induces neurite outgrowth (BAR-SAGI and FARAMISCO 1985), and inhibition of ras function in these cells blocks nerve growth factor (NGF) induction of this process (HAGAG et al. 1986; SZEBERENYI et al. 1990).

Since ras proteins play critical roles in these fundamental events, the biochemical processes that regulate p21 activity and those that are affected by an activated ras protein will undoubtedly be very complex. To understand them thoroughly will require the integration of information obtained from a variety of experimental techniques. One of these, which has proven to be quite powerful and is the topic of this review, is the inhibition of ras function in vivo by the expression of dominant inhibitory *ras* mutants. Two types of mutants have been identified. One, which contains a mutation in the protein's Mg^{2+}-binding site, appears to interfere with endogenous ras function by competing for a protein that activates ras. The other, which is rendered nonfunctional as a signal-transducing protein by removal of its membrane attachment site, appears to compete for a downstream target of activated ras. As such, these mutants are very specific reagents and may prove to be at least as powerful as gene inactivation, a technique used successfully in yeast to aid in unraveling complex biochemical pathways.

The first dominant inhibitory ras mutant was detected in the course of characterizing mutant ras proteins with defective interactions with guanine nucleotides (SIGAL et al. 1986). A mutation at amino acid 16 in the first GTP-binding domain of yeast Ras yielded a protein that blocked cell growth when expressed in *Saccharomyces cerevisiae*. Subsequently, a mutant with a Ser to Asn substitution in the neighboring position 17 of mammalian ras was detected by a functional screen for ras mutants defective in GTP binding following random mutagenesis (FEIG and COOPER 1988). This mutant blocked cell proliferation in NIH 3T3 cells. That this was due to inhibition of endogenous ras was supported by the fact that the effect could be reversed by coexpression of excess normal or constitutively activated H-*ras*-p21. Additional inhibitory mutants were then identified in yeast that mapped to this same region of Ras (POWERS et al. 1989). As discussed below, this class of mutants appears to compete with endogenous Ras for a guanine nucleotide exchange factor (GEF) that is responsible for activating Ras in cells.

Finally, a mutant ras protein, termed rasT, was designed to act as a competitor of ras-GAP, since ras-GAP may also function as a downstream target of ras (GIBBS et al. 1989). A similar mutant in yeast Ras was isolated by screening directly for dominant inhibitory mutants (MICHAELI et al. 1989). For all of these mutants, it is likely that all three ras proteins in mammalian cells are blocked, since ras-GAP and ras-GDS act on H-, K, and N-*ras*.

B. Mechanism of Inhibitory Action

Before discussing how these ras mutants have been used to identify biochemical pathways that are dependent upon p21, I will review what is known about their mechanism of action. This information will clearly help in interpreting results obtained using these novel reagents. As stated above, rasT was designed to be a competitive inhibitor of ras-GAP in mammalian cells. *S. cerevisiae* Ras1 was truncated at the carboxyl terminus to block its localization to the inner surface of the plasma membrane, and thus destroy its ability to function as a signal-transducing protein. An additional mutation (Gln to Leu) at position 61 was introduced to give the protein greater than 100-fold higher affinity for GAP than that of normal mammalian or yeast Ras (VOGEL et al. 1988). High levels of this mutant blocked cellular ras activity in *Xenopus* oocytes (GIBBS et al. 1989) and 3T3 cells (STACEY et al.

1991b). By competing with endogenous ras for GAP, RasT presumably interferes with downstream signaling (see Fig. 1A). This phenotype added support to the idea that ras-GAP functions as a downstream target of ras. This model may be oversimplistic, however. For example, more than one GTPase activating protein for ras has been identified (MARTIN et al. 1990; XU et al. 1990), and additional proteins may exist that are the true effectors of ras function. These proteins would be expected to interact with ras in a manner similar to GAP. It should be pointed out that because GAP is a resonably abundant protein in cells, RasT must be present at levels about a hundred times higher than endogenous ras in mammalian cells to achieve a significant inhibitory phenotype (STACEY et al. 1991a). As will be discussed

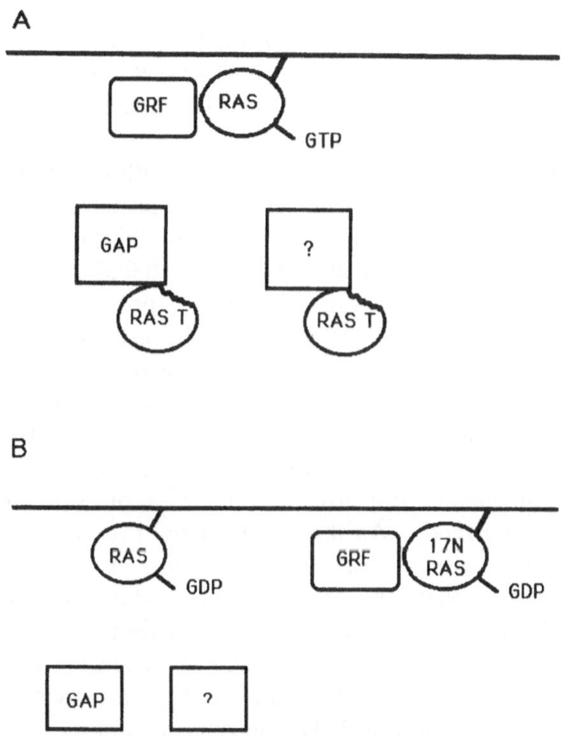

Fig. 1A,B. Proposed mechanism of RasT and 17N Ras inhibitory action. **A** RasT cannot localize to the plasma membrane because of a truncation of the membrane localization signal. For this reason it is not functional as a signal-transducing molecule. However, it has high affinity for GAPs and any other proteins that may interact with the "effector domain" of ras. Therefore, it is likely that it interferes with the normal ras pathway in cells by competing with endogenous ras for downstream targets. **B** 17N ras is known to be locked in an inactive conformation even when it is bound to GTP, because of a defect in complexing Mg^{2+} associated with bound nucleotide. It most likely interferes with the normal ras pathway in cells by competing with endogenous ras for a guanine nucleotide releasing factor (GRF; also called a guanine nucleotide exchange factor, GEF) that normally activates ras

later, RasT is actually more effective at inhibiting oncogenic ras than normal ras in cells.

Of the dominant inhibitory ras mutants that inhibit activation of ras, mutants at position 17 of mammalian ras have been the best characterized. The X-ray crystal structure of p21 has demonstrated that Ser 17 contributes to the binding of Mg^{2+} associated with bound guanine nucleotides (MILBURN et al. 1990; PAI et al. 1990). That improper complexing of Mg^{2+} is the fundamental defect in the inhibitory S17N protein was inferred from additional mutagenesis analysis at this site (FARNSWORTH and FEIG 1991). In particular, more subtle substitutions such as S17A, which removes the reactive hydroxyl group, or S17C, which merely substitutes a sulfur (known to be a poor binder of Mg^{2+}) for oxygen at this site also generated dominant inhibitory proteins. Consistent with this model, a S17T change retained normal ras activity.

A unique alteration in guanine nucleotide binding was found exclusively in these dominant inhibitory ras mutants (FARNSWORTH and FEIG 1991). Instead of displaying similar affinities for GTP and GDP, the inhibitory proteins all had preferential affinity for GDP. This was found to be a consequence of a drastic decrease in GTP affinity with only a mild decrease in the affinity for GDP. This suggested that these proteins may be inactive as signal-transducing proteins in cells, because they fail to bind GTP due to competition from GDP. Alternatively, the preferential affinity for GDP may reflect the fact that the mutations locked the proteins in an inactive conformation. According to this model, even if the proteins did bind GTP in vivo, they would be incapable of activating a downstream target. The latter hypothesis was proved to be correct by the demonstration that even if S17N p21 was preloaded with GTP, it failed to activate a model downstream target, yeast adenyl cyclase (FARNSWORTH and FEIG 1991).

These results document the critical role 17Ser plays in insuring that Mg^{2+} is complexed correctly. As the yeast adenyl cyclase assays showed, this appears to be necessary for correct coupling of active GTP-bound ras to downstream effector molecules. A molecular explanation for this may involve 35Thr, which resides in the region of ras thought to interact with effector molecules. This region has been shown to change its orientation when p21 replaces GDP with GTP (PAI et al. 1989; MILBURN et al. 1990). In particular, the hydroxyl group of 35Thr becomes complexed with Mg^{2+} in the active GTP-bound state. Apparently, in dominant inhibitory mutants incorrect complexing of Mg^{2+} prevents 35Thr from becoming tethered to Mg^{2+}, a process that may be neccessary for the effector domain of ras to adopt its correct configuration.

The likelyhood that the inhibitory phenotype of these mutants was due to an alteration in such a fundamental aspect of ras implies that comparable dominant inhibitory mutants can be made in any GTPase. This notion is supported by the fact that this region of ras (in particular 17Ser) is highly conserved in all GTPases. In fact, the comparable mutation has been made in the classic heterotrimeric GTPase $G_{\alpha s}$ and its properties suggest its expression in cells would inhibit endogenous $G_{\alpha s}$ function (HILDEBRANDT et al. 1991).

A variety of results support the idea that this class of inhibitory mutant interferes with upstream activation of endogenous ras by a guanine nucleotide exchange factor. Thus, endogenous ras would be prevented from reaching the active GTP-bound state which is required for its interaction with downstream targets (see Fig. 1B). First, the alternative that these mutants interfere with the ability of active ras to interact with downstream targets is not consistent with the finding that a second mutation known to block ras from binding downstream targets failed to relieve the inhibitory phenotype of S17N p21 (STACEY et al. 1991a). Second, this class of inhibitory mutants was much more effective in blocking normal ras than constitutively activated GTPase-deficient ras (STACEY et al. 1991a). This is significant because in yeast, normal Ras function is very dependent upon an upstream nucleotide releasing factor, while that of oncogenic Ras is not (BROEK et al. 1987). Third, the phenotype induced by a similar inhibitory mutant in yeast Ras could be overcome by overexpression of CDC25, a nucleotide exchange factor for Ras in *S. cerevisiae* (POWERS et al. 1989). Finally, guanine nucleotide exchange factor are known to be more effective on the inactive GDP-bound form of ras and thus might be expected to have a higher affinity for inhibitory mutants that are locked in the inactive state. Preliminary data with partially purified dissociation stimulator for ras confirmed this prediction. Because guanine nuceotide exchange factor appear to be present at very low levels in cells, 17N ras is effective in cells at levels comparable to endogenous ras (STACEY et al. 1991a). This level is about a hundred times lower than that required for RasT and about five times lower on a molar basis than that required for an anti-ras monoclonal antibody to inhibit ras function in vivo.

C. Defining Biochemical Pathways Dependent upon ras Function

A variety of studies have already been performed using dominant inhibitory ras expression to help define the role of ras proteins in signal transduction. A significant hurdle to overcome was that inhibition of ras function in cells usually blocks cell growth. Three approaches have been used to generate cells for biochemical studies. One very significant finding was that the pheochromocytoma cell line PC12 proliferates normally when ras function is blocked, although many signal transduction pathways, such as those activated by epidermal growth factor (EGF), NGF, and fibroblast growth factor (FGF) have been found to be altered (SZEBERENYI et al. 1990). Alternatively, cells expressing 17N ras on an inducible promoter have been generated (CAI et al. 1990; KREMER et al. 1991). Moreover, some experiments have been performed in transient expression systems (Cai et al. 1990; Medema et al. 1991). Finally, microninjection of the proteins into single cells has also been used (Farnsworth et al. 1991; Stacey et al. 1991a,b).

Since PC12 cells grow in the presence of 17N ras, most experiments have been performed with this cell line. First, it was shown that NGF and FGF failed to promote neurite outgrowth when 17N ras was expressed (Szeberenyi et al. 1990). This is consistent with earlier experiments where ras function was blocked by microinjection of monoclonal antibodies (Hagag et al. 1986). Then, by comparing the results of cell lines expressing different amounts of 17N p21, it was concluded that ras proteins are involved in at least two parallel signal transduction pathways. This was because low levels of the inhibitor blocked NGF or FGF induction of neurite outgrowth, but not of *fos* or *jun* gene expression. However, higher levels of 17N p21 expression blocked both. A similar result was obtained using transient expression of 17N H-*ras* and NIH 3T3 cells expressing 17N H-*ras* under the control of an inducible promoter (Cai et al. 1990). In these experiments, EGF stimulation of DNA synthesis, but not induction of *fos* expression, was blocked by low levels of 17N *ras* expression. However, high levels of the inhibitor blocked both.

Since 17N ras is likely to inhibit endogneous ras function by preventing its activation by a nucleotide releasing factor, the differential response to 17N levels observed above are probably due to differences in the percentage of ras reaching the active GTP-bound state upon stimulation by growth factors. For example, because neurite outgrowth induction by NGF in PC12 cells and DNA synthesis induction by EGF in NIH cells were found to be sensitive to low levels of 17N expression, these responses probably require high levels of ras-GTP. In contrast, because *fos* and *jun* induction by these same growth factors was not blocked by low 17N p21 levels, these responses required less ras-GTP. In this regard, it is intriguing that NF-1 has approximately 100-fold higher affinity for active ras than does ras-GAP. Since these two GTPases may function as downstream targets of ras, it is tempting to speculate that NF-1 functions in the gene induction pathway, since its higher affinity for ras might compensate for the partial supression of ras-GTP formation that occurs when low levels of 17N ras are present. Similarly, ras-GAPs involvement in the differentiation pathway is suggested, since the lower affinity of Ras-GAP would make this pathway very sensitive to modest supression of ras-GTP formation by growth factors.

This set of experiments also showed that 12-O-tetradecanoylphorbol-13-acetate (TPA) induction of DNA synthesis in 3T3 cells was dependent upon ras activation, but induction of *fos* and *jun* expression was not (Cai et al. 1990). Since TPA is thought to activate protein kinase C, this enzyme appears to function through both ras-dependent and ras-independent pathways.

Microinjection of ras antibodies gave the first indication of the biochemical pathway in which ras proteins function (Smith et al. 1986).

Subsequently, microinjection of both 17N p21 and rasT have confirmed the original findings that the growth-promoting effects of cell surface tyrosine kinases, such as Src, and PDGF receptor depend upon ras, whereas cytoplasmic ser/thr kinases such as Raf and Mos do not (Stacey et al. 1991b). In support of this notion that the former are upstream of ras and the latter downsteam in a signal transduction pathway, it has been found that activation of these tyrosine kinases leads to increased ras-GTP in cells, but activation of these ser/thr kinases does not (Gibbs et al. 1990; Satoh et al. 1991).

Studies on ras-blocked PC12 cells have expanded this concept. For example, NGF-induced Raf phosphorylation and autokinase activity is inhibited in these cells (Robbins et al. 1992). This, coupled with the observation that constitutively activated ras leads to hyperphosphorylation of this kinase, strongly argues that Raf mediates at least some of the actions of ras in these cells. A similar result was obtained when MAP kinases were investigated (Thomas et al. 1992; Wood et al. 1992; Robbins et al. 1992). This ser/thr and tyrosine kinase family, also called ERKs, are activated by a variety of signaling pathways, including those initiated by tyrosine kinase and G-protein receptors. An activator of MAP kinase has been detected in cells. Its activation by NGF is also dependent upon a functional ras protein (Robbins et al. 1992). A tentative signal transduction pathway is emerging where at least some of the actions of tyrosine kinase receptors are propogated by activation of ras-p21. This in turn leads to the activation of MAP-kinase activator and then MAP kinase. One of the substrates for MAP kinase is S6 kinase (also RSK) which phosphorylates the S6 subunit of ribosomes. Regardless of whether RAF is phosphorylated by MAP kinases or by some other mediator, this event is also a consequence of ras activation.

Protein kinase C can work by activating ras or by a ras-independent pathway. As with induction of DNA synthesis, activation of MAP kinases and Raf by TPA is dependent upon ras in these cells (Robbins et al. 1992; Thomas et al. 1992), yet induction of *fos* and *jun* is not (Szeberenyi et al. 1990). One explanation is that these two pathways are carried out by different protein kinase C isoforms.

It is likely that these two kinases are only a small subset of the potential targets of ras in cells that are responsible for the pleomorphic responses observed upon receptor activation of cells. For example, translation initiation factor eIF-4E phosphorylation in response to NGF is also blocked in these cells (Frederickson et al. 1992). This protein regulates the rate of translation, a process that appears to depend upon its phoshorylation. The importance of this protein in growth control is accented by the fact that its overexpression in cells can oncogenically transform tissue culture cells (Lazaris-Karatzas et al. 1990).

In agreement with the pattern that tyrosine kinase receptors act through ras, it has been observed that insulin receptor activation also leads to an enhancement of GTP-bound ras in cells (Burgering et al. 1991).

Furthermore, transient expression assays with inhibitory 17N ras and rasT have shown that induction of gene expression by insulin depends upon a functional ras pathway (MEDEMA et al. 1991).

Two major questions at hand are what steps are between active GTP-bound ras and activation of cytoplasmic kinases, and what steps are between activated tyrosine kinase receptors and activation of ras proteins?

Experiments with dominant inhibitory ras mutants have shwon that receptors that activate heterotrimeric G-proteins appear to act through ras-independent pathways. For example, in PC12 cells forskolin, a constitutive activator of adenyl cyclase, can activate MAP kinase (ROBBINS et al. 1992). However, this event is hardly influenced by ras inhibtion. Moreover, unlike cell transformation by activated src protein, transformation induced by a constitutively activated $G_{\alpha i}$-protein is not inhibited by coexpression of a dominant inhibitory ras mutant (GUPTA et al. 1992).

D. Some Surprises Revealed by Dominant Inhibitory ras Mutants

In addition to helping define the biochemical pathways in which ras proteins function, dominant inhibitory ras mutants have also revealed: (1) an important difference between the way in which ras and G-proteins function, (2) that an additional downstream Ras pathway exists in *S. cerevisiae*; and (3) a strategy for drug design that might lead to preferential inhibition of oncogenic ras over normal ras.

1. The phosphate binding loop of ras is conserved in all GTPases. This includes 17Ser, which is involved in Mg^{2+} binding. When a mutation comparable to S17N was introduced into the α subunit of the heterotrimeric G-protein G_s, the phenotype observed had both similarities to and differences from those oberved in ras (HILDEBRANDT et al. 1991). For example, like S17N p21, S54N $G_{\alpha s}$ displayed preferential affinity for GDP. This implied that the same biochemical defect in the protein, altered interactions with Mg^{2+}, was present. Moreover, altered interactions with its guanine nucleotide exchange factor, the β-adrenergic receptor, were also observed. However, unlike S17N p21, when S54N $G_{\alpha s}$ was bound to GTP it retained the capacity to stimulate its downstream target, adenyl cyclase. This implied that the role of Mg^{2+} in converting the protein to the active state is different in these two classes of GTPases. This may be due to the fact that the putative "effector domain" of ras (amino acids 25–40) is intimately involved in Mg^{2+} binding through threonine 35, as described earlier. In contrast, the comparable Thr (amino acid 204) in $G_{\alpha s}$ that is involved in Mg^{2+} binding is not thought to be in the putative effector domain that interacts with adenyl cyclase. This domain in G-proteins is

thought to be closer to the carboxy terminus (amino acids 236–356) (ITOH and GILMAN 1991; BERLOT and BOURNE 1992).

2. As stated previously, a dominant inhibitory mutant comparable to rasT was isolated in yeast Ras by screening directly for interference in the Ras pathway in *S. cerevisiae*. This cytoplasmic form of Ras1 had the capacity to block the phenotype induced by oncogenic Ras1 in these cells. It was noted, however, that overexpression of adenyl cyclase could not completely reverse the phenotype induced by this inhibitory Ras mutant (MICHAELI et al. 1989). This implied that oncogenic ras activates an additional pathway in this organism.

3. While microinjecting 17N p21 and rasT into 3T3 cells, it became obvious that although both could inhibit cell proliferation, clear differences in their mechanism of action existed. Based on their different biochemical properites, some of these differences were expected. For example, since 17N p21 is thought to interfere with activation of ras, it was not surprising that the inhibitory protein had to be located in the plasma membrane to function (FEIG and COOPER 1988). In contrast, rasT was designed to be exclusively in the cytoplasm. Moreover, 17N p21 was much more effective at inhibiting the function of normal ras than that of GTPase-deficient oncogenic ras. This is consistent with normal ras being very dependent upon activation rate, since it is deactivated very rapidly by ras-GAP in cells. Oncogenic ras is less dependent upon activation, since it remains in the GTP-bound state because of its insensitivity of ras-GAP.

The surprising result, however, was that rasT was a more effective inhibitor of oncogenic ras than normal ras (STACEY et al. 1991a). Such a phenotype is of course extremely exciting if one is trying to design strategies for inactivating oncogenic ras in tumor cells. One of the big challenges is to avoid inhibiting normal ras because of its critical role in the functions of all normal cells in the body.

Preferential inhibition of the oncogenic form of ras by rasT is actually consistant with what is known about the dual function of the putative "effector" domain of p21. This domain is responsible for both interactions with GAPs and downstream targets (which may in fact be GAPs). As such, when rasT suppresses the ability of ras and GAP to interact, one would predict that while the efficiency at which GTP-bound ras can couple to its downstream target is reduced, more of the ras proteins in cells would be in the active GTP-bound state in cells. This is because of the decrease in GAPs' ability to enhance the GTPase activity of ras. This could explain why rasT is a very weak inhibitor of normal ras function.

In contrast, oncogenic ras is already locked primarily in the active GTP-bound state because it is insensitive to the negative feedback of GAPs. By the above reasoning, rasT is a more potent inhibitor of this form of ras, because while it suppresses downstream signaling of oncogenic ras, the compensatory increase in GTP-ras found in normal ras would not occur in oncogenic ras. All of these predictions have actually been borne out by the analysis of ras proteins with mutations in the "effector domain" (Farnsworth et al. 1991).

E. Conclusions

The isolation of dominant inhibitory forms of proteins has helped illuminate the complex function of ras proteins in cells. This approach is analogous in some ways to gene knockout experiments perfected in yeast and now becoming more plausible in mammalian cells. However, inhibitory proteins can have the unique feature of blocking the function of proteins that interact with ras, rather than ras itself. Important differences can be noted between this technique and that of knocking out ras directly. One good example is the observation that downstream inhibition of p21 is more effective on oncogenic ras function than on normal ras function. Designing dominant inhibitory proteins may become even more important in future studies of larger signal-transducing proteins that contain multiple functional domains.

References

Bar-Sagi D, Faramisco JR (1985) Microinjection of the Ras oncogene protein into PC12 cells induces morphological differentiation. Cell 42:841–848

Barbacid M (1987) ras Genes. Annu Rev Biochem 56:779–827

Berlot CH, Bourne HR (1992) Identification of effector-activating residues of Gsa. Cell 68:911–922

Bollag G, McCormick F (1991) Regulators and effectors of ras proteins. Annu Rev Cell Biol 7:601–632

Bourne HR, Sanders DA et al. (1991) The GTPase superfamily: conserved structure and molecular mechanism. Nature 349:117–131

Broek D, Toda T et al. (1987) The S. cerevisiae CDC25 gene product regulates the Ral/adenylate cyclase pathway. Cell 48:789–799

Burgering BM, Medema RH et al. (1991) Insulin stimulation of gene expression mediated by p21 ras activation. EMBO J 10:1103–1109

Cai H, Szeberenyi J et al. (1990) Effect of a dominant inhibitory Ha-ras mutation on mitogenic signal transduction in NIH 3T3 cells. Mol Cell Biol 10:5314–5323

Farnsworth C, Feig LA (1991) Dominant inhibitory mutations in the Mg^{2+} binding site of ras blocks its activation by GTP. Mol Cell Biol 11:4822–4829

Farnsworth CL, Marshal MS et al. (1991) Preferential inihibition of the oncogenic form of RasH by mutations in the GAP binding/"effector" domain. Cell 64:625–633

Feig LA, Cooper GM (1988) Inhibition of NIH 3T3 cell proliferation by a mutant ras protein with preferential affinity for GDP. Mol Cell Biol 8/8:3235–3243

Frederickson RM, Mushynski WE et al. (1992) Phosphorylation of translation initiation factor eIF-4E is induced in a ras-dependent manner during nerve growth factor-mediated PC12 cell differentiation. Mol Cell Biol 12:1239–1247

Gibbs JB, Schaber M et al. (1989) Xenopus oocyte germinal vessicle breakdown induced by [val 12] ras is inhibited by a cytosol-localized ras mutant. Proc Natl Acad Sci USA 86:6630–6634

Gibbs JB, Marshall MS et al. (1990) Modulation of guanine nucleotides bound to Ras in NIH3T3 cells by oncogenes, growth factors, and the GTPase activating protein (GAP). J Biol Chem 265:20437–20442

Gupta SK, Gallego C et al. (1992) Mitogenic pathways regulated by G protein oncogenes. Mol Biol Cell 3:123–128
Hagag N, Halegoua S et al. (1986) Inhibition of growth factor-induced differentiation of PC12 cells by microinjection of antibody to ras p21. Nature 319/6055:680–682
Hildebrandt J, Day R et al. (1991) A mutation in the putative Mg^{2+} binding site of gsa prevents its activation by receptors. Mol Cell Biol 11:4830–4838
Itoh H, Gilman A (1991) Expression and analysis of gsa mutants with decreased ability to activate adenylyl cyclase. J Biol Chem 266:16226–16231
Kremer NE, D'Arcangelo G et al. (1991) Signal transduction by nerve growth factor and fibroblast growth factor in PC12 cells requires a sequence of src and ras actions. J Cell Biol 115:809–819
Lazaris-Karatzas A, Montine KS et al. (1990) Malignant transformation by a eukaryotic initiation factor subunit that binds to mRNA 5′ cap. Nature 345:544–547
Martin GA, Viskochil D et al. (1990) The GAP-related domain of the neurofibromatosis type-1 gene product interacts with ras p21. Cell 63:843–849
Medema RH, Wubbolts R et al. (1991) Two dominant inhibitory mutants of p21ras interfere with insulin-induced gene expression. Mol Cell Biol 11:5963–5967
Michaeli T, Field J et al. (1989) Mutants of H-ras that interfere with ras effector function in Saccharomyces cerevisiae. EMBO J 8:3039–3044
Milburn MV, Tong L et al. (1990) Molecular switch for signal transduction: structural differences between active and inactive forms of protooncogenic ras proteins. Science 247:939–945
Pai FE, Kabsch W et al. (1989) Structure of the guanine-nucleotide-binding domain of the Ha-ras oncogene product p21 in the triphosphate conformation. Nature 341:209–214
Pai EF, Krengel U et al. (1990) Refined crystal structure of the triphosphate conformation of H-ras p21 at 1.35 A resolution: implications for the mechanism of GTP hydrolysis. EMBO J 9:2351–2359
Powers S et al. (1989) Dominant yeast and mammalian Ras mutants that interfere with the CDC25-dependent activation of wild-type ras in Saccharomyces cerevisiae. Mol Cell Biol 9:390–395
Quaife CC, Pinkert CA et al. (1987) Pancreatic neoplasia induced by ras expression in acinar cells of transgenic mice. Cell 48:1023–1034
Robbins DJ, Cheng M et al. (1992) Evidence for a ras-dependent extracelluar signal-regulated protein kinase (ERK) cascade. Proc Natl Acad Sci USA (in press)
Satoh T, Nakafuka M et al. (1991) Involvement of rasp21 protein in signal-transduction pathways from interleukin 2, interleukin 3, and granulocyte/macrophage colony-stimulating factor, but not from interleukin 4. Proc Natl Acad Sci USA 88:3314–3318
Shou C, Farnsworth CL et al. Cloning of a cDNA encoding guanine-nucleotide releasing factor for ras p21. Nature (in press)
Sigal I, Gibbs JB et al. (1986) Mutant ras encoded proteins with altered nucleotide binding properties exert dominant biological effects. Proc Natl Acad Sci USA 83:952–956
Smith MR, DeGudicibus SJ et al. (1986) Requirement for c-ras proteins during viral oncogene transformation. Nature 320:540–543
Stacey DW, Feig LA et al. (1991a) Dominant inhibitory ras mutants selectively inhibit the activity of either cellular or oncogenic ras. Mol Cell Biol 11:4053–4064
Stacey DW, Roudebush M et al. (1991b) Dominant inhibitory ras mutants demonstrate the requirement for ras activity in the action of tyrosine kinase oncogenes. Oncogene 6:2297–2304
Swanson ME, Elste AM et al. (1986) Abundant expression of ras proteins in Aplysia neurons. J Cell Biol 103/2:485–492

Szeberenyi J, Cai H et al. (1990) Effect of a dominant inhibitory Ha-ras mutation on neuronal differentiation of PC12 cells. Mol Cell Biol 10:5324–5332

Thomas S, DeMarco M et al. (1992) Ras is essential for nerve growth factor- and phorbol ester-induced tyrosine phosphorylation of MAP kinases. Cell 68:1031–1040

Vogel U, Dixon RAF et al. (1988) Cloning of bovine GAP and its interactions with oncogenic ras p21. Nature 335:90–93

Wood KW, Sarnecki C et al. (1992) ras mediates nerve growth factor receptor modulation of three signal-transducing protein kinases: MAP kinase, RAf-1, and RSK. Cell 68:1041–1050

Xu G, Lin B et al. (1990) The catalytic domain of the neurofibromatosis type 1 gene product stimulates ras GTPase and complements ira mutants in S. cerevisiae. Cell 63:835–841

CHAPTER 20

The Involvement of Cellular ras in Proliferative Signaling

D.W. STACEY

A. Introduction

Proliferation is undoubtedly controlled by the flow of biochemical signals between various cellular molecules and classes of molecules, originating from an extracellular source and terminating at an as yet undefined intracellular target. In defining the relationship between various components of the proliferative signaling system, it will first be essential to determine the direction of information flow between them. It will then be essential to define how these molecules influence one another biochemically. Because the control of proliferation is so vital to the survival of a complex organism, it is likely that the signaling system will turn out to be subject to multiple levels of control. Neutralizing antibody and dominant inhibitory mutants of the oncogene *ras* have been used to make clear suggestions as to the hierarchy of information flow between at least two different classes of dominantly acting oncoproteins during proliferation. In addition, the biochemical means of this information flow has been investigated. These studies demonstrate one aspect of proliferative signaling and suggest a pattern of investigation which might be useful in studies of other components of proliferative signaling and related cellular phenomena.

B. The Relationship Between Tyrosine Kinase Oncogenes and Cellular ras

I. Neutralizing Anti-ras Antibody

Cellular ras proteins, which bind and hydrolyze GTP, are present within almost all cells. Mutations in *ras* genes have been found in a high proportion of spontaneous tumors of a variety of types. In order to understand the role of cellular ras in the control of proliferation, an attempt was made to inactivate ras activity within living cells and determine the consequences of this inactivation upon the proliferative capacity of the resulting cell. Accordingly, a variety of previously identified monoclonal antibodies were screened to find one able to neutralize the activity of endogenous ras proteins following microinjection into living cells (KUNG et al. 1986). The antibody identified, Y13–259, has since been shown to recognize an epitope which undergoes a critical alteration in three-dimensional configuration upon conversion from biologically inactive ras–GDP to the the active GTP-bound form (DE VOS et al. 1988; JURNAK et al. 1990; PAI et al. 1989). This fact might explain the high degree of efficiency with which the antibody inhibits the activity of ras proteins. The target epitope is conserved among different members of the ras family from a variety of species, but is not present in ras-related proteins (SANTOS and NEBREDA 1989), suggesting that this antibody is highly specific in its interaction with cellular ras proteins.

II. Inhibition in the Late G1 Phase of the Cell Cycle

When this anti-ras antibody was microinjected into a variety of cell types, including nontransformed and primary fibroblasts or epithelial cells, the cells were unable to initiate a new cycle of DNA synthesis. Cells in the S-phase at the time of antibody injection completed DNA synthesis and subsequently divided, but were unable to initiate a new S-phase until the antibody had been metabolized from within the injected cell (generally 40–80 h following antibody injection). The inhibition was observed late in the G1 phase of the cell cycle (MULCAHY et al. 1985). Interestingly, a number of cell lines derived from human tumors were not inhibited by the injected antibody, suggesting a fundamental difference between the proliferative characteristics of normal and tumor cells (STACEY et al. 1987).

III. *ras* and Other Oncogene Classes

The *ras* gene family is only one of several classes of dominantly acting oncogenes. The relationship between endogenous ras proteins and the activity of other types of oncogenes was tested by microinjecting anti-ras antibody into NIH3T3 cells transformed by other oncogenes. When such injections were made into cells transformed by membrane-associated tyrosine kinases, including oncogenes derived from growth factor receptor genes, the initially transformed cells reverted to the morphology characteristic of nontransformed cells and stopped proliferating. Cells transformed by the soluble serine kinase oncogenes *mos* or *raf*, on the other hand, were not affected by the injected antibody (SMITH et al. 1986). The dependence upon endogenous ras proteins had, therefore, distinguished the activity of two classes of oncogenes and suggested the fundamental characteristics of the proliferative signal transduction pathway within cells.

C. A Model for Proliferative Signal Transduction

On the basis of the antibody injection experiments described above, a model of proliferative signal tranduction was proposed in which the signal is initiated by the interaction between a growth factor molecule and its receptor. The occupied receptor then signals to cellular ras proteins, which then function to transfer the proliferative signal to a soluble serine kinase such as raf. No suggestion is made as to how the prolifertive signal received by cellular *raf* genes leads to induction of proliferation (STACEY et al. 1988). This model is weakened by the fact it was initially based only upon the relatively untested technique of antibody microinjection. Recently, however, a number of observations have been made which strengthen this model. First, two *ras* mutants identified in other laboratories have the ability to inhibit the activity of cellular ras proteins, and therefore provide an alternative means of inactivating cellular ras proteins (FEIG and COOPER 1988; GIBBS et al. 1989). In collaboration with the investigators who initially identified these mutants, the corresponding proteins were prepared in purified from and microinjected along with anti-ras antibody. In every detail described above, each of these dominant inhibitory mutants duplicates the effects observed in separate injections with anti-ras antibody (STACEY et al. 1991). With three separate means of inhibiting the activity of endogenous ras proteins, therefore, the results are indistinguishable and support the model described above. Interestingly, the injections with dominant inhibitory mutants also provide an insight into the interactions of ras with other cellular molecules required for its activity, as will be described later.

I. Other Studies Which Support the Model

In addition to microinjection experiments, several other types of investigations carried out in other laboratories support the general concepts of the model described above. First, the GTPase activity of ras is now known to be controlled by a GTPase activating protein (GAP) (TRAHEY and MCCORMICK 1987). When this protein is expressed at high levels within cells, it suppresses the activity of endogenous ras proteins. In such cells, tyrosine kinase oncogenes lose their transforming ability (ZHANG et al. 1990). Thus, a fourth means of inactivating endogenous ras proteins, one which does not involve microinjection, supports the observation that cellular ras is required for the action of tyrosine kinases. A direct biochemical connection between the action of growth factors and cellular ras activity was achieved when in a number of studies growth factors were shown to increase the proportion of cellular ras associated with GTP within 5 min of growth factor addition (SATOH et al. 1990). This result demonstrates that growth factors function to activate cellular ras, as predicted above.

Finally, support for the idea that ras functions to transduce signals initiated by tyrosine kinases was obtained in two separate genetic studies. These were designed to identify in either *Drosophila* or in *Caenorhabditis elegans* the genes required for the action of individual tyrosine kinases. In the nematode a signal for volval induction is received by the let-23 gene product, which is a growth factor receptor related tyrosine kinase (AROIAN et al. 1990). This observation is significant since it is known that the let-60 gene is required for, and functions downstream of, let-23. The let-60 gene product is a ras protein. The *Drosophila* studies have been even more dramatic. The tyrosine kinase gene, *sevenless*, is involved in eye development. A number of genes have been identified which affect the expression of *sevenless*. These genes include not only a *Drosophila ras* equivalant, but also other genes such as GAP or exchange factors shown in other systems to be required for ras activity (SIMON et al. 1991; ROGGE et al. 1991; GAUL et al. 1992; CAGAN et al. 1992). Taken together, these various types of evidence clearly establish that ras proteins function to transduce signals initiated by tyrosine kinases. This does not exclude the possibility that alternative means of transmitting such signals exist, or that ras might function in other ways, but the weight of evidence presently available leaves little doubt that a major function of ras is to receive and propagate signals initiated by tyrosine kinases. To this extent, therefore, one of the first objectives in elucidating the mechanism of proliferative signaling is satisfied: tyrosine kinases are *upstream* of ras.

D. Lipids and the Control of ras Activity

I. Dependence of Lipid Mitogens upon ras

The genetic types of experiments described above are powerful tools for determining the direction of information flow between oncogene classes, but provide little information concerning the biochemistry involved. It is known that one of the earliest and most dramatic biochemical effects of mitogenic treatments is the metabolism of certain types of lipids. Lipids identified as being produced following mitogenic treatments include diacylglycerol, phosphatidic acid, inositol phosphates, arachidonic acid (AA), and its metabolites. In addition, certain lipases are known to be controlled by GTPases. On the basis of these and other types of observations it had been predicted that ras might actually function to control the activity of a lipase (BAR-SAGI and FERAMISCO 1986).

Injections of anti-ras antibody were utilized to test the possibility that ras might control a lipase. We reasoned that if the proliferative requirement for ras involved its activation of a phospholipase, we could induce proliferation even in the absence of cellular ras activity by treating cells with molecules able to duplicate the effects of that phospholipase. NIH3T3 cells were injected with anti-ras antibody to inactivate cellular ras and then treated with phorbol esters and a calcium ionophore to duplicate the effects of diacylglycerol and inositol phosphates, the two second messengers produced by phospholipase C. Instead of overcoming the proliferative requirement for ras, however, the mitogenic effects of treatment with these two reagents were dramatically inhibited by the injected antibody. Two other phospholipases were therefore tested similarly to determine if ras might function to control their activity instead of phospholipase C. Anti-ras antibody was injected into cells which were then treated with phosphatidic acid or prostaglandin F2 alpha, the products of phospholipase D and phospholipase A2, respectively. Both of these lipids were able to induce proliferation of NIH3T3 cells in the absence of added serum, but as before this proliferation was completely blocked by the injected anti-ras antibody. While serum-induced proliferation was inhibited during these experiments by 75%–95% by the antibody, the proliferation induced by the lipid-related mitogens was inhibited by 98%–99.9%, even though in some cases the extent of proliferation induced by the lipids was as great as that induced by serum (Yu et al. 1988). The fact that the lipids were so dramatically dependent upon cellular ras activity for mitogenic stimulation suggested that ras does not control the action of any of the phospholipases tested. Instead, we postulated that the opposite might be true, that lipids might actually function to control cellular ras activity.

II. Biochemical Effects of Lipids upon ras

Experiments were designed to test the notion that lipids might function to control cellular ras activity. Biologically active ras is associated with GTP and can be inactivated by the endogenous GTPase activity of cellular ras proteins stimulated by GAP (Trahey and McCormick 1987) or neurofibromin (Xu et al.; Martin et al.; Ballister et al. 1991). Inactive ras associated with GDP can be activated by the endogenous nucleotide exchange activity most likely stimulated by a nucleotide exchange factor (Wolfman and Macara 1990). If lipids are normally involved in controlling ras activity, they must be able to control either the endogenous rates of GTPase or exchange, or the activity of proteins able to stimulate these functions. Consequently, attempts were made to determine if any lipid might alter the biochemical properties of ras proteins alone. This appeared not to be the case. Instead, certain lipids were found to control the activity of GAP. For example, in the presence of 100 μg/ml AA the ability of purified GAP to stimulate the GTPase activity of purified, bacterially synthesized ras was totally overcome. The rate of GTP hydrolysis in such reactions was indistinguishable from that of cellular ras in the absence of added GAP. At lower concentrations the AA induced only a partial inhibition of GAP activity. At no time was this fatty acid able to interfere with the endogenous rate of ras GTPase activity (Tsai et al. 1989a).

A number of lipids were tested for the ability to inhibit GAP activity. Lipids normally present at high concentrations within native membranes such as phosphatidylcholine, phosphatidylethanolamine, or phosphatidylserine were totally inactive. No saturated fatty acid was found to have the slightest inhibitory activity, while several unsaturated fatty acids did. Among these, the most inhibitory was AA (Tsai et al. 1989), although more recent studies suggest that the lipoxygenase metabolites of AA might have equal or greater inhibitory activity toward GAP (Yu et al. 1990).

E. Biochemical Analyses of the Interaction between ras and Lipids

At this point there is no conclusive biological evidence to support the hypothesis that lipids might normally function to control ras activity and thereby perform a critical role in the control of proliferation. Several predictions of this possibility, however, have been tested and will be described. First, if lipids normally function to control the activity of GAP, it might then be predicted that lipids would also function to control the activity of the many ras-related molecules which have their own GTPase stimulating proteins. Second, it was considered likely that lipids able to inhibit GAP activity could be identified soon after mitogenic stimulation of cells. Finally, it was important to determine if a physical interaction might exist between lipids and the GAP molecule. As described below, each of these predictions have been substantiated. In the process of this investigation, observations in this and other laboratories have added unexpected support to the notion that lipids might function naturally in the control of ras activity.

I. Lipids and ras-Related Proteins

Since the identification of ras, a number of small ras-related GTPases have been identified. These generally have about 50% sequence homology with ras and have a similar molecular mass (20–25 kDa) (SANTOS and NEBREDA 1989). GTPase activating proteins specific for several individual ras-related proteins have been identified along with molecules able to alter rates of nucleotide exchange. Ras-related proteins are apparently not involved in the control of proliferation, but are generally believed to be involved in intracellular membrane trafficking and cytoskeletal polymerization, among other functions (see HALL 1992; JEAN-PIERRE et al. 1991). None of these small GTP-binding and -hydrolyzing proteins associate with auxiliary proteins as trimeric G-proteins do. It is, therefore, unclear how their activity is controlled.

To test the possibility that these proteins might also be controlled by lipids, two separate ras-related proteins, rho and R-ras, were obtained in purified form from a bacterial expression system. The R-ras protein is one of the few to be stimulated by ras–GAP, while rho is stimulated only by a separate molecule, rho–GAP. Both these proteins were incubated along with the appropriate GAP molecule. As expected, the rate of GTPase activity was stimulated manyfold compared to the endogenous rate of GTPase associated with each protein. The incubation of these two proteins and their GAP molecules was then carried out in the presence of various lipids. As before, the lipids were able to inhibit the GAP activity associated with each protein, but did not inhibit the endogenous rate of GTPase activity (TSAI et al. 1989). Interestingly, the types of inhibitory lipids varied slightly from those seen with ras and ras–GAP. Even though ras–GAP stimulates the activity of both ras and R–ras, this stimulation is inhibited by phosphatidic acid composed of saturated fatty acids only for R-ras. Fully saturated phosphatidic acid also inhibited rho–GAP. The one constant in all the reactions was the observation that AA was inhibitory (TSAI et al. 1989b). While the details of inhibition might be of importance in the biological activity of different small GTPases, the fact that each is inhibited by lipids in its interaction with GTPase activating proteins is significant. This fact suggests that lipids might actually function to control this entire class of GTP-binding molecules through interaction with the corresponding GAP molecules.

II. Neurofibromin and Lipid Inhibition

The gene affected by the common human genetic disease, von Recklinghausen neurofibromatosis type 1, has recently been identified and characterized. Interestingly, this *NF-1* gene contains a region with approximately 50% sequence homology with the catalytic region of GAP. Both the catalytic region (XU et al.; MARTIN et al.; BALLISTER et al. 1991) and more recently the full-length neurofibromin molecule (GOLUBIC et al. 1992) have been shown to have GAP activity against purified ras molecules. Despite the homology between GAP and neurofibromin, the two molecules clearly have diverged dramatically. Outside the catalytic region there is little sequence homology between the two. In fact, neurofibromin is more closely related to the yeast GAP analogs IRA-1 and IRA-2 than to GAP itself. We reasoned that if the involvement of lipids in the control of GAP activity were of biological importance, then lipid inhibition might also be involved in the control of neurofibromin, despite the divergence of the two molecules. Initially only the catalytic domains of the two molecules were analyzed and compared. Despite the divergence of sequences even of their catalytic regions, they differed only subtly from one another in their susceptibility to lipid inhibition. Neurofibromin was approximately twofold more sensitive to inhibition by AA and was inhibited by phosphatidic acid (arachidonyl, stearoyl), while the catalytic fragment of GAP was not inhibited by phosphatidic acid (GOLUBIC et al. 1991). The fact that neurofibromin and GAP retain similar lipid inhibition characteristics despite their sequence divergence strongly suggests that this characteristic is an essential component in the biological function of each molecule.

With the observation that the GTPase activating proteins for rho and R-ras as well as neurofibromin all were susceptible to lipid inhibition, it became apparent that the yeast GAP analogs IRA-1 and IRA-2 might also be inhibited by lipids. The catalytic fragment of IRA was shown to stimulate the GTPase activity of yeast RAS as expected. When this stimulation was carried out in the presence of lipids, the ability of IRA-2 to stimulate GTPase activity was completely inhibited as above (GOLUBIC et al. 1991). In some ways this observation is surprising since, while lipid metabolism is well documented in animal cells at the time of mitogenic treatment, lipid metabolism is less well studied in yeast. The observation that IRA-2 is lipid sensitive not only supports the notion that lipids control ras activity in general, but also raises the possibility that lipid metabolism has a critically important control function even in yeast.

III. Production of GAP-Inhibitory Lipids by Mitogen Stimulation

The metabolism of a variety of lipids is dramatically altered at the time of mitogenic stimulation. In order to determine if any of the resulting lipids might be involved in the control of GAP activity, NIH3T3 cells were rendered quiescent by serum deprivation. Total cellular lipids were then collected at various times following mitogenic treatment. These were resolved into fractions on a silica gel TLC plate and individual fractions tested for the ability to inhibit GAP. Two fractions of GAP-inhibitory lipids were identified in this way. One migrated as expected for a fatty acid, while the second, less active lipid behaved like a phosphatidylinositide (Yu et al. 1990). Time–course studies indicated that the active lipids were produced within 3–5 min following serum addition, but that after 20 min they had been reduced to low levels. No activity was seen in parallel cultures of unstimulated cells. Significantly, these lipids were produced only in sparse cultures. If cells were near contact inhibition density, no GAP-inhibitory lipids were ever seen. This suggests that the production of these lipids is controlled positively by growth factors and negatively by cell density. These facts are consistent with the possibility that the lipid are critical in the control of proliferation.

The identity of the lipids produced within mitogen-stimulated cells is currently under investigation. Since the most active fraction had characteristics of a fatty acid, it was considered that metabolites of AA might be involved. Accordingly, several lipoxygenase or cyclooxygenase products of AA were tested for GAP-inhibitory activity. The lipoxygenase products were found to be highly inhibitory while the cyclooxygenase products were not (Yu et al. 1990). Current studies with inhibitors of each of these pathways are underway, as are attempts to purify these lipids.

IV. Physical Association Between GAP and Lipids

It is difficult to demonstrate conclusively a biological role of lipids in the control of cellular ras activity utilizing biochemical techniques. It was considered likely, however, that if lipids do function to control GAP activity, they might actually physically associate with GAP. To test this possibility, a novel type of lipid affinity column was developed. A fatty acid was covalently linked to a solid support. This linkage involved the carboxyl group and orientated the hydrophobic region of the molecule in the aqueous solvent. This column would not be expected to reproduce the lipid environment existing within a native membrane, and this column did not bind GAP. Such an immobilized lipid column, however, would be expected to retain a second lipid. The hydrophobic region of the second lipid would associate with the exposed nonpolar regions of the covalently linked lipid. Thus, the second lipid would become immobilized, with the polar groups oriented into the aqueous solvent phase as expected in a native membrane. When such a column was constructed with AA as the second lipid, the GAP within a crude cellular homogenate was quantitatively retained on the column. When the column was constructed with arachidic acid, which has the same number of carbon groups as AA but with no double bonds or the ability to inhibit GAP activity, no association with GAP was observed (Tsai et al. 1991).

The association between GAP and AA was also demonstrated in a more traditional mixed micelle system. In this assay micelles of the detergent Triton X100 were allowed to associate with a second lipid such as AA. These mixed micelles were then incubated with GAP and passed over a size exclusion column able to distinguished mixed micelles which had associated with a GAP molecule from those which had not. As above, those mixed micelles composed of the GAP-inhibitory AA were able to associate with GAP and migrate in size exclusion more rapidly than normal micelles. When the noninhibitory arachidic acid was placed in the mixed micelles, however, no association with GAP was observed (Tsai et al. 1991). Recent studies have identified a calcium-dependent lipid-binding domain in the GAP molecule (Clark et al. 1991).

The GAP bound to the AA–lipid affinity column above was not released with high salt or altered pH, but was quantitatively released when the column was treated with ethylene diamine tetraacetate (EDTA) to chelate divalent cations. When the GAP released was assayed for the ability to stimulate GTPase activity of ras, however, an unexpected observation was made. Along with an activity able to stimulate GTPase activity, an activity able to inhibit the GTPase activity of ras was also observed. This GTPase inhibitory activity was shown to be enhanced by certain lipids. Efforts are underway to purify this GTPase inhibitory protein which has a molecular mass of 60 kDa. It is interesting that this inhibitory activity is stimulated by mitogenically produced lipids, while the GTPase activating protein is inhibited by these lipids. In either case the lipids tend to increase the biological activity of ras.

V. Mutational Analysis of ras and the Lipid Inhibitory Phenotype

A final type of study undertaken in this laboratory relative to a possible role of lipids in the regulation of ras activity involved a survey of ras mutants for those with increased or decreased sensitivity to lipid inhibition. The study had to be confined to certain regions of the ras molecule identified as nonessential regions in which mutations fail to alter the activity of oncogenic *ras*. Such regions might he involved in the regulation of cellular ras, but this fact would not be apparent since oncogenic *ras* is not subject to normal regulation. A number of mutant *ras* genes were supplied in coded form by Hsiang-Fu Kung and analyzed in this laboratory by Fu-Sheng Wei. The mutant ras proteins were incubated with constant amounts of GAP and increasing amounts of an inhibitory lipid such as AA or phosphatidic acid (arachidonyl-stearoyl). The ability of the added GAP to stimulated GTPase activity was analyzed along with the ability of lipids to inhibit GAP activity. Several of the mutants were found to be unaltered in either activity (deletions 93–95, 166–183, or 166–179), but three were found to have increased sensitivity to lipid inhibition of GAP. In other words, while GAP worked normally on these three mutants, the ability of lipids to inhibit GAP activity was increased.

The three mutants with increased lipid sensitivity were shown to have overlapping mutations (deletions 97–103, 101–103, and 102–108). The results were particularly interesting with the 101–103 mutant since this small deletion altered no identifiable activity other than an enhancement in the ability of associated GAP to be inhibited by lipids (Fig. 1). It might be of significance that amino acids 101–104 are highly conserved not only in *ras* genes, but in a number of *ras*-related genes; yet this region has no previously identifiable function (Santos and Nebreda 1989). It is almost certain that a highly conserved region in a diverse gene family will have an important function. We propose that the function of this highly conserved region is involved in the regulation of these proteins by GAP and lipids. While such an observation requires more investigation, this possibility is consistent with the involvement of lipids in the normal control of these proteins.

VI. Other Studies of Lipids and GAP Activity

Studies in other laboratories support the notion that lipids control the activity of GAP. As noted above, several laboratories have shown the sensitivity of neurofibromin to lipid inhibition (Han et al. 1991; Bollag and McCormick 1991). In one of these studies, the ability of GAP to associate with ras protein has been related to its inhibition by lipids. It was shown that the region of GAP which associates with lipids is distinct from that which associates with ras. This lead to the proposal that when GAP comes to the membrane to interact with ras it might encounter a membrane-associated lipid which inhibits its GTPase stimulatory activity. Because its GTPase stimulatory activity is inhibited, GAP would remain associated with ras protein, while the ras protein would remain in the GTP-bound state, potentially forming a complex active in downstream signaling (Bollag and McCormick 1991). In this way, the GAP–cellular ras complex in the presence of inhibitory lipids would be similar to the complex which might form with oncogenic *ras*, whose GTPase activity cannot be stimulated by

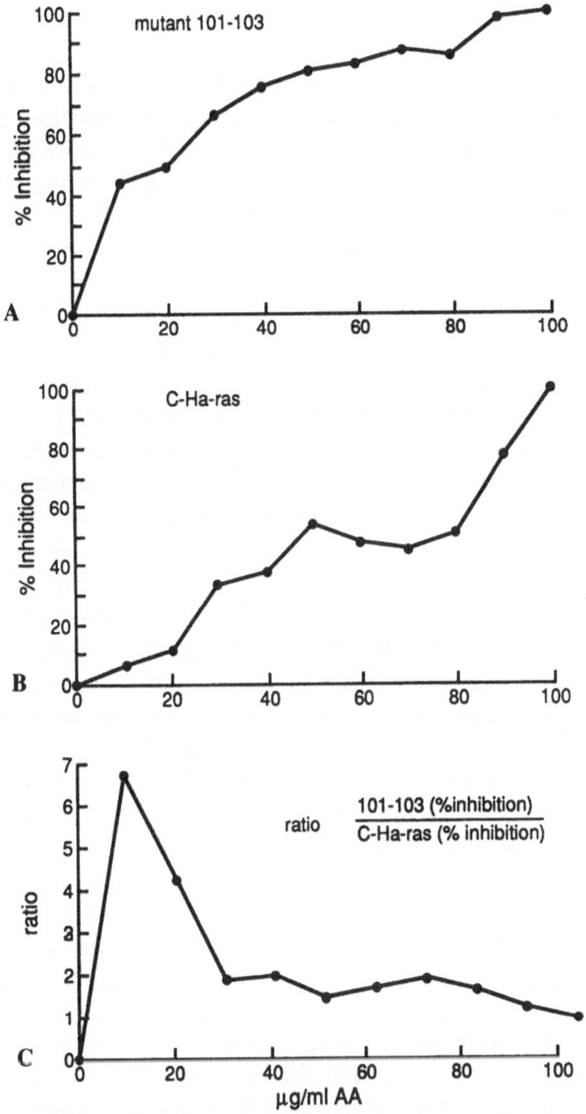

Fig. 1A–C. Mutations in the ras protein which affect the lipid sensitivity of GAP. **A, B** Increasing concentrations of arachidonic acid (AA) were utilized to inhibit the ability of GAP to stimulate the GTPase activity of two ras proteins. ras was allowed to associate with alpha-labeled GTP and then incubated with GAP and lipid. After the reaction, ras protein was immunoprecipitated with antibody and the proportion of GTP remaining associated with the protein was determined on a TLC plate. The percentage of inhibition reported above indicates the GTPase activity observed in the presence of lipid compared to reactions containing only ras and GAP (100% inhibition represents the endogenous rate of GTPase of ras in the absence of GAP). Mutant 101–103 (**A**) is stimulated by GAP activity (in the form of crude brain lysate) normally, but this stimulation is more sensitive to inhibition by low concentrations of AA than observed with native ras protein (**B**). The differences in lipid inhibition are particularly evident at low concentrations where little inhibition of GAP in the presence of native ras protein is observed, but where significant inhibition is observed in the presence of mutant protein. **C** Ratio of inhibition between mutant and native ras protein. (These unpublished data were obtained by Fu-Sheng Wei in collaboration with Hsiang-Fu Kung)

GAP. As further support for the potential role of lipid–GAP interactions, recent studies have identified a potential cation-dependent lipid association domain in the GAP molecule which is just outside the domain known to interact with ras (CLARK et al. 1991). The interpretation of this observation, however, is complicated by the fact the catalytic region of GAP, which does not include this lipid association domain, is nevertheless able to associate with lipids (SERTH et al. 1991).

In another study, an extensive survey of microbial broths for a molecule able to inhibit GAP activity was undertaken. Certain broths contained molecules able to inhibit GAP activity. The active molecules were isolated and found to be unsaturated fatty acids (SERTH et al. 1991). This study is important because no assumptions concerning the types of molecules which might be active were made prior to the initiation of the experiments; yet the study identified lipids as the active molecules. In the course of this study it was shown that, in many cases, fatty acids are able to inhibit GAP activity only at the critical micelle concentration. The differences in inhibition by fatty acids were found to relate to their ability to form micelles. This is an important observation and clearly adds insight into the possible mechanism of inhibition of GAP activity. Clearly, in a cell the lipids presented to GAP would be part of a membrane since GAP must move to the membrane to interact with ras. The suggestion that the lipids might function by *sequestering* GAP from the solution, however, is without foundation. To sequester GAP it would be necessary to physically separate it from ras protein. Micelles do not have the ability to surround and isolate a molecule in this way. Liposomes might be able to do this, but considerable effort would be required in their formation to ensure that the GAP and ras would be on opposite sides of the membranes formed. It would never happen in a solution in which all molecules were initially present together. Furthermore, as noted above, the inhibitory concentration of fatty acids upon catalytic fragments of GAP and neurofibromin is less than one-tenth the critical micelle concentration. It is likely that the physical presentation of lipids to GAP is critical. The formation of micelles, however, is neither essential to inhibition of GAP nor able to sequester it away from ras.

VII. Tyrosine Kinases and Lipid Metabolism

The first set of experiments described above demonstrated that ras proteins function to transduce the signals initiated by tyrosine kinases. Then it was suggested that GAP, and thereby ras, might be controlled by lipids. These two possibilities make the suggestion that tyrosine kinases might function by altering lipid metabolism in a way which eventually regulates GAP. It is interesting, therefore, to briefly review what is known about the action of tyrosine kinases to determine if there is a basis for this suggestion. Generalized studies of the substrates of tyrosine kinases have generally not yielded consistent results. No one substrate has been shown to be associated with the mitogenic activity of these molecules. On the other hand, the molecules which actually physically associate with active tyrosine kinases have been revealing. It has been reported that GAP, raf, phosphatidylinositol-3-kinase (PI-3 kinase), and phospholipase C gamma (PLC gamma) associate with tyrosine kinases, but only after their activation. The association between these kinases and raf and GAP is of a limited extent, and might not be important for mitogenic signaling. On the other hand, there are numerous studies which relate the ability to associate with PLC gamma, and more particularly PI-3 kinase, with the ability to function mitogenically (see CANTLEY et al. 1991; FANTL et al. 1992). Obviously these two enzymes control phospholipid metabolism, and their activity is closely related to the activity of the mitogenic tyrosine kinases. This observation is in complete agreement with the model presented (Fig. 2), suggesting that both tyrosine kinases and lipid metabolism are involved in regulating ras.

VIII. Model for the Control of Proliferation at the Level of ras Activity

Further study will be needed to determine if any of the metabolites of PI-3 kinase or those of PLC gamma are able to either directly control GAP activity or lead to the production of lipids which do. In this respect, it is interesting to consider again the production of lipids at the time of mitogenic stimulation. In confluent cells, the production of a GAP-inhibitory lipid was

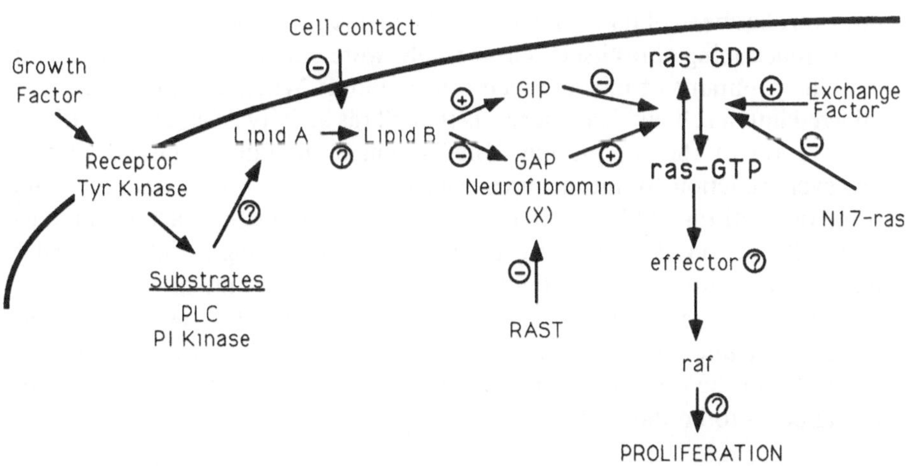

Fig. 2. Model for proliferative signal transduction. This model incorporates all the factors described in this communication concerning the interrelationship between different components of the proliferative signalling system. The involvement of factors such as the retinoblastoma protein, cyclins, nuclear oncogenes, etc. have not been discussed and are not mentioned in this model. Note also that the role of lipids in signaling between tyrosine kinases and ras is under investigation

not seen. It will be critical to determine if PI-3 kinase and PLC gamma are stimulated by tyrosine kinases in such confluent cells. If not, then the suggestion might be made that it is the activity of these enzymes which controls the production of GAP-inhibitory lipids, and that in confluent cells they are kept from functioning. If these two molecules are activated normally even in dense culture, however, then it would be assumed that their function is to initiate a cascade of lipid metabolism which eventually leads to the formation of a GAP-inhibitory lipid, whose production is subject to negative controls. The potential role of multiple levels of control in the production of mitogenic lipids leads to an attractive model for the control of proliferation. It is assumed that ras forms a critical switch in the control of proliferation. If so, a number of positive and negative proliferative signals might be expected to converge upon ras. If, as suggested, ras is controlled by lipid metabolism, positive signals would lead to the production of a given lipid or class of lipids able to inhibit GAP activity, while negative proliferative signals would inhibit the production of these lipids. Ras would therefore function to sense the end result of the positive and negative proliferative signals, as reflected in the production of certain critical lipids. Similarly, lipid metabolism would be expected to control the activity of other ras-related, small GTPases.

F. Cellular Factors Affecting ras Activity

The previous discussion has focused upon the role of ras in the control of proliferation as it relates to other classes of oncogenes. Some information, however, has been obtained relative to the metabolic steps which are involved in the functioning of ras itself. These results were obtained with the dominant inhibitory mutants of ras described briefly above. When anti-ras antibody is injected into a cell, evidence suggests that all ras activity is inhibited by direct inactivation of the ras protein. The dominant inhibitory mutants of ras, however, function by mechanisms which are distinct from each other and from the anti-ras antibody. It is assumed that dominantly acting mutants inhibit the action of the target gene product by disrupting an intermolecular interaction required for its activity. In this way, a given mutant might disrupt an essential interaction between ras and one molecule, while a second mutant might interrupt the interaction between ras and an entirely separate molecule. In the case of ras and the two dominant inhibitory mutants tested, this appears to be the case.

I. N17 ras Interferes with the Activation of Cellular ras

The asparagine substitution at position 17 of ras (N17) produces a molecule which has a greater than normal affinity for GDP (Feig and Cooper 1988). As described above, this molecule is able to inhibit cellular ras activity. When this mutant protein was microinjected into cells transformed by different types of *ras* genes, however, an interesting observation resulted. The protein was efficiently able to block the proliferation of NIH3T3 cells transformed by high levels of cellular ras, but did not inhibit the proliferation of cells transformed by oncogenic ras. This initial observation was carefully extended in a series of coinjection experiments. The minimally transforming concentrations of cellular and oncogenic ras were determined by microinjection. The oncogenic ras, as expected, transformed cells when injected at approximately 50-fold lower concentrations than was required for cellular ras. The minimally transforming concentrations of each protein were then doubled and coinjected with various concentrations of N17 ras. As little as 1 μg/ml N17 ras was able to block the ability of twice the minimally transforming concentration of cellular ras (150 μg/ml) to transform cells. On the other hand, 150 μg/ml N17 ras had only the slightest inhibitory activity upon oncogenic ras at twice its minimal transforming concentration (3 μg/ml). It was clear that N17 was almost 100-fold more inhibitory for cellular than oncogenic ras (Stacey et al. 1991a).

II. RAS^T is Preferentially Inhibitory for Oncogenic ras

In contrast to the results above, the second inhibitory mutant of ras exhibited a totally different inhibitory profile. This mutant was designed to bind GAP nonproductively (Gibbs et al. 1989). An activating mutation was introduced to incease GAP binding by 100-fold, while a mutation in the carboxyl terminus was introduced to block membrane association and thereby the ability to transmit a biological signal. This mutant (RAS^T) had to be injected at concentrations near 10 mg/ml to be inhibitory. As above, it was injected into cells transformed by either cellular or oncogenic ras. At high concentrations it was able to block the proliferation of each type of ras-transformed cell. Then, as above, this mutant was coinjected along with twice the minimally transforming concentrations of either cellular or oncogenic ras. The results were the opposite of those described above. The RAS^T mutant was able to block the transforming potential of oncogenic ras (3 μg/ml) when coinjected at 1 mg/ml. The transforming activity of cellular ras at twice the minimally transforming concentration, however, was only inhibited by concentrations of the inhibitor over 8 mg/ml (Fig. 3; Stacey et al. 1991a). The oncogenic ras was nearly tenfold more sensitive to inhibition by RAS^T than was the cellular ras.

The two inhibitory mutants were therefore selectively inhibitory to different types of ras protein. The N17, GDP-binding mutant, was nearly 100-fold more inhibitory for cellular ras, while RAS^T was nearly tenfold more inhibitory against oncogenic ras. In addition to this distinction, others were observed between the two mutants. The N17 ras required membrane association to be inhibitory, while the RAS^T did not. The effective concentrations of RAS^T were nearly 100-fold greater than N17 ras. The

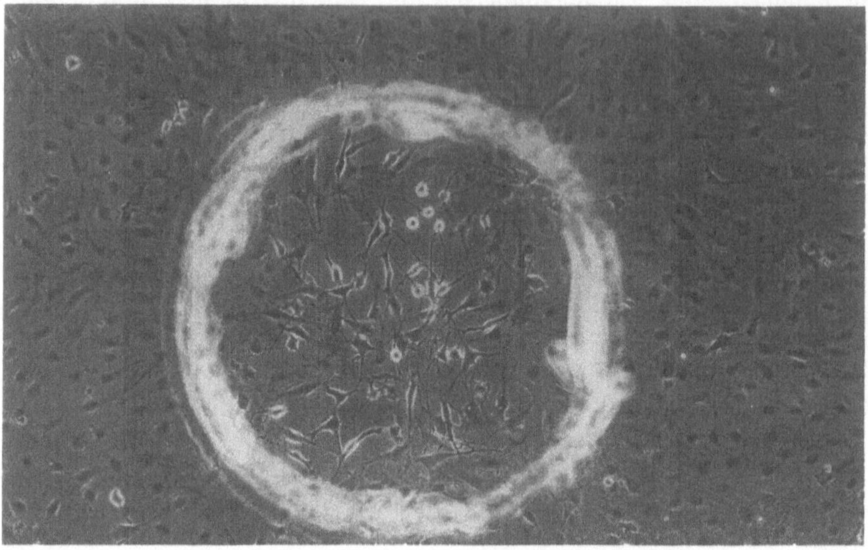

Fig. 3. Poor inhibition of cellular ras activity by RAS^T. Twice the minimal transforming concentration of cellular ras protein (150 μg/ml) was coinjected with the yeast inhibitory protein RAS^T at 3 mg/ml. This concentration of inhibitor is able to completely block the transforming capacity of oncogenic ras; but as observed above, is unable to block transformation of these NIH3T3 cells by cellular ras. Injections were performed 20 h earlier into all the cells within the area surrounded by a *circle* marked on the back of this coverslip. The morphology of transformation, a rounded and refractile appearance, is exhibited by the injected cells, but not by uninjected cells, outside the circular area of injection. This figure is a composite of many photographs so as to allow the visualization of the entire area of injection. (The phase-contrast photograph was prepared so as to minimize the appearance of boundaries between separate photographs)

effector-binding domain was required for RAS^T, but not for N17 (STACEY et al. 1991a).

III. Model for Inhibition of ras Activity by Dominant Inhibitory Mutants

To explain the differences in activity of the two inhibitory ras mutants, the following model was proposed. It is assumed that N17 interferes with a step required for the activation of cellular ras, since it is comparatively inactive against oncogenic ras, which is locked in the active conformation. Ras becomes active when the bound GDP is replaced by a GTP molecule. This is generally thought to require nucleotide exchange. If N17 inhibits activation of cellular ras, it is likely that it inhibits nucleotide exchange. On the other hand, the RAS^T mutant is considerably more inhibitory for oncogenic than cellular ras. This mutant was designed to bind GAP, which is known to downregulate cellular ras, but has no such effect upon oncogenic ras. The reduction of GAP activity within the cell, therefore, would tend to elevate the activity of cellular ras. Yet RAS^T must do more than eliminate an inhibitor of cellular ras activity since it inhibits the activity of both cellular and oncogenic ras. We therefore assume that in addition to binding and inactivating GAP, this mutant also binds and inactivates a molecule essential for the downstream signaling of all ras molecules (Stacey et al. 1991a). This factor would be required by both oncogenic and cellular ras and may well be GAP itself.

According to the above model, as the concentration of RAS^T is increased, two opposing effects are seen upon cellular ras. On the one hand, as the inhibitor binds and inactivates GAP, its ability to downregulate the activity of cellular ras as reduced. Since GAP does not interact with oncogenic ras, this effect selectively elevates the activity of cellular ras. On the other hand, however, the mutant binds the downstream effector of ras and thereby reduces the biological activity of both forms of ras. In the case of cellular ras, the inhibitor has both a stimulatory and an inhibitory effect, while with oncogenic ras its effect is simply inhibitory. In this way the mutant is comparatively more inhibitory for the oncogenic form of ras. This model does not indicate what target of RAS^T is involved in downstream signaling, but opens the possibility that it is GAP itself which is both involved in downregulation and downstream signaling.

IV. Biochemical Support for the Idea that RAS^T Binds an Effector

The data presented above suggest that RAS^T inactivates both GAP, which suppresses the activity of cellular ras, and a molecule which is essential for downstream signaling of all ras molecules. An attempt was therefore made to provide independent support for this model. Instead of directly attempting to inactivate GAP or a related molecule within the cell by injecting a dominantly acting mutant ras protein, mutations which are known to interfere with the interaction between GAP and ras were directly introduced into ras proteins. In this way, even though the activity of GAP within the cell would be unaffected, the ability of GAP to interact with the ras proteins introduced into these cells would be reduced due to the mutations. Thus, in two totally distinct ways the interaction between ras and GAP would be reduced and the results could be compared.

Mutations were introduced into the "effector-binding domain" of ras, which is required for the biological activity of ras, but not for any known biochemical function. Similar mutations were introduced into cellular and oncogenic ras. Mutations were selected to reduce, but not to eliminate the interaction between GAP and the mutant ras protein. Such mutant proteins were then microinjected to determine their transforming activity. Two such mutations had limited effect upon the transforming activity of cellular ras, but dramatically reduced the activity of oncogenic ras mutants. As expected on the basis of the data presented above, a mutation which interferes with GAP binding had resulted in a much greater interference with the transforming capacity of oncogenic than cellular ras. Similar results were seen when the corresponding *ras* genes were transfected and their ability to induce focus formation was determined (FARNSWORTH et al. 1991).

Attempts were then made to directly demonstrate the reduced interaction of the mutants described above with GAP and with a ras effector. No effector is known in higher cells, but this mutation reduced binding to adenylyl cyclase, the RAS effector of yeast. The mutant proteins were also shown to interact poorly with GAP, utilizing both GAP-binding assays and GTPase assays. The biological results of reduced GAP binding were also observed. In cellular ras-transformed cells an elevated proportion of ras was found in the GTP-bound state. These observations suggest why, in the microinjection experiments, the cellular ras with and without mutations had similar transforming activity. The reduced ability of GAP to stimulate GTP hydrolysis had apparently resulted in a higher proportion of ras bound to GTP. This increased ras–GTP was then able to produce a normal downstream signal despite reduced interaction with the effector. Consequently, with cellular ras the effector site mutation had resulted in a greatly increased proportion of the protein in the GTP-bound state, which was able to offset its reduced ability to induce downstream signaling. With the oncogenic ras containing effector mutations, however, only the downstream signaling function was apparently inhibited, resulting in reduced transforming efficiency (FARNSWORTH et al. 1991).

The GAP-binding domain mutants were therefore able to inhibit transforming capacity of the oncogenic ras much more efficiently than with the cellular ras protein. This observation supports the model presented above which suggests that GAP, or a related molecule, functions both to downregulate ras and also as its downstream effector molecule. Independent support for the possibility that GAP functions as a downstream effector of ras comes from mutation studies in this region mentioned above (ADARI et al. 1988; CALE'S et al. 1988). In addition, other biological studies suggest the function of GAP as a downstream effector for ras (SCHWEIGHOFFER et al. 1992; YANTI et al. 1990). In our studies it is clear that the RAS^T mutant binds both GAP and the downstream ras effector, and that mutations in the effector site reduce interaction with GAP and this effector similarly. Our data do not, however, directly demonstrate that GAP is the downstream ras effector.

G. Summary

In summary, there has been considerable progress in elucidating some of the interactions between cellular molecules which are critical for proliferative signal transduction. First of all, it is now clear that tyrosine kinase oncogenes transmit proliferative signals through cellular ras proteins. The evidence for this conclusion comes from many different types of experiments and different organisms. The evidence in support of this conclusion is constantly accumulating and there is at present little reason to question this result in general terms. The means by which tyrosine kinases signal through ras, however, is open to speculation. In our model, lipid metabolism is critical. Others suggest that binding between GAP and kinases is involved or that tyrosine phosphate and the SH-2 binding regions of certain proteins are involved. Further study will be required to determine which factors are important, but it is clear that several important predictions of the involvement of lipids in the control of ras activity have been verified.

In relationship to ras activity directly, studies with two dominant inhibitory molecules indicate that ras must interact with at least two different molecules to function biologically. One of these is required only for cellular ras and is apparently involved in its activation. The second is involved both in downregulating cellular ras and is essential for downstream signaling from ras.

References

Adari H, Lowy DR, Willumsen BM, Der GJ, McCormick F (1988) Guanosine triphosphatase activating protein (GAP) interacts with the p21 effector binding domain. Science 240:518–521

Aroian RV, Koga M, Mendel JE, Ohshima Y, Sternberg PW (1990) The let-23 gene necessary for C. elegans vulva induction encodes a tyrosine kinase of the EGF receptor subfamily. Nature 348:693–699

Ballester R, Michaeli T, Ferguson K, Xu HP, McCormick F, Wigler M (1989) Genetic analysis of Mammalina GAP expressed in yeast. Cell 59:681–686

Bar-Sagi C, Feramisco JR (1986) Induction of membrane ruffling and fluid-phase pinocytosis in quiescent fibroblasts by ras proteins. Science 233:1061–1068

Bollag G, McCormick F (1991) Differential regulation of rasGAP and neurofibromatosis gene product activities. Nature 351:576–579

Cagan RL, Kramer H, Hart AC, Zipursky SL (1992) The bride of sevenless and sevenless interaction: internalization of a transmembrane ligand. Cell 69:393–399

Cale's C, Hancock JF, Marshall CJ, Hall A (1988) The cytoplasmic protein GAP is implicated as the target for regulation by the ras gene product. Nature 332:548–551

Cantley LC, Auger KR, Carpenter C, Duckworth B, Graziani A, Kapeller R, Soltoff S (1991) Oncogenes and signal transduction. Cell 64:281–302

Clark JD, Lin LL, Kriz RW, Ramesha CS, Sultzman LA, Lin AY, Milona N, Knopf JL (1991) A novel arachidonic acid-selective cytosolic PLA_2 contains a Ca^{2+}-dependent translocation domain with homology to PKC and GAP. Cell 65:1043–1051

de Vos AM, Tong L, Milburn MV, Matias PM, Jancarik J, Noguchi S, Nishimura S, Miura K, Ohtsuka E, Kim SH (1988) Three dimensional structure of an oncogene protein catalytic domain of human c-H-vas p21. Science 239:888–893

Fanti WJ, Escobedo JA, Martin GA, Turck CW, del Rosario M, McCormick F, Williams LT (1992) Distant phosphotyrosines on a growth factor receptor bind to specific molecules that mediate different signaling pathways. Cell 69:413–423

Farnsworth CL, Marshall MS, Gibbs JB, Stacey DW (1991) Preferential inhibition of the oncogenic form of RasH by mutations in the GAP binding/"effector" domain. Cell 64:625–533

Feig LA, Cooper GM (1988) Inhibition of NIH3T3 cell proliferation by a mutant ras protein with preferential affinity for GDP. Mol Cell Biol 8:3235–3243
Gaul U, Mardon G, Rubin GM (1992) A putative ras GTPase activating protein acts as a negative regulator of signaling by the sevenless receptor tyrosine kinase. Cell 68:1007–1019
Gibbs JB, Schaber M, Schofield TL, Scolnick EM, Sigal IS (1989) Xenopus cocyte germinal vessicle break down induced by [Val 12] Ras is inhibited by a cytosol-localized Ras mutant. Proc Natl Acad Sci USA 86:6630–6634
Golubic M, Tanaka K, Dobrowolski S, Wood D, Tsai MH, Marshall M, Tamanoi F, Stacey DW (1991) The GTPase stimulatory activity of the neurofibromatosis Type 1 and yeast IRA proteins are inhibited by mitogenically responsive lipids. EMBO J 10:2897–2903
Golubic M, Roudebush M, Dobrowolski S, Wolfman A, Stacey DW (1992) Catalytic properties, tissue and intracellular distribution of the native neurofibromatosis type 1 protein. Oncogene 7:2151–2159
Gorvel JP, Chavrier P, Zerial M, Gruenberg J (1991) rab5 controls early endosome fusion in vitro. Cell 64:915–915
Hall A (1992) Signal transduction through small GTPases – a tale of two GAPS. Cell 69:389–391
Han JW, McCormick F, Macara IG (1991) Regulation of ras-GAP and the neurofibromatosis-1 gene product by eicosanoids. Science 252:576–579
Jurnak F, Heffron S, Bergmann E (1990) Conformational changes involved in the activation of ras p21: implications for related proteins. Cell 60:525–528
Kung H-F, Smith MR, Bekesi E, Manne V, Stacey DW (1986) Reversal of transformed phenotype by monoclonal antibodies against Ha-ras p21 proteins. Exp Cell Res 162:363–371
Mulcahy LS, Smith MR, Stacey DW (1985) Requirement for ras proto-oncogene function during serum-stimulated growth of NIH3T3 cells. Nature 313:241–243
Martin GA, Viskochil D, Bollag G, McCabe PC, Crosier WJ, Haubruck H, Conroy L, Clark R, O'Connell P, Cawthon RM, Innis MA, McCormick F (1990) The GAP-related domain of the neurofibromatosis type 1 gene product interacts with ras p12. Cell 63:843–849
Pai EF, Kabsh W, Krengel U, Holmes KC, John J, Wittinghofer A (1989) Structure of the quanine-nucleotide-binding domain of the Ha-ras oncogene product p21 in the triphosphate confirmation. Nature 341:209–214
Rogge RD, Karlovich CA, Banerjee U (1991) Genetic dissection of a neurodevelopmental pathway: son of sevenless functions downstream of the sevenless and EGF receptor tyrosine kinases. Cell 64:39–48
Santos E, Nebreda AR (1989) Structural and functional properties of ras proteins. FASEB J 3:2151–2163
Satoh T, Endo M, Nakafuku M, Nakafuku S, Kaziro Y (1990) Platelet derived growth factor stimulates formation of active T21 Ras GTP complex and Swiss mouse 3T3 cells. Proc Natl Acad Sci USA 87:5993–5997
Schweighoffer F, Bariat I, Chevallier-Multon MC, Tacque B (1992) Implication of GAP in Ras-dependent transactivation of a polyoma enhancer sequence. Science 256:825–827
Serth J, Lautwein A, Frech M, Wittinghofer A, Pingoud A (1991) The inhibition of the GTPase activating protein–Ha-ras interaction by acidic lipids is due to physical association of the c-terminal domain of the GTPase activating protein with micellar structures, EMBO J 10:1325–1330
Simon MA, Bowtell DDL, Dodson GS, Laverty TR, Tubin GM (1991) Ras1 and a putative quanine nucleotide exchange factor perform crucial steps in signaling by the sevenless protein tyrosine kinase. Cell 67:701–716
Smith MR, DeGudicibus SJ, Stacey DW (1986) Requirement for c-ras proteins during viral oncogene transformation. Nature 320:540–543
Stacey DW, DeGudicibus SJ, Smith MR (1987) Cellular *ras* proteins and tumor cell proliferation. Exp Cell Res 171:232–242

Stacey DW, Tsai MH, Yu CL, Smith JK (1988) Critical role of cellular ras proteins in proliferative signal transduction. Cold Spring Harbor Symp Quant Biol 53:871–881

Stacey DW, Feig LA, Gibbs JB (1991a) Dominant inhibitory Ras mutants selectively inhibit the activity of either cellular or oncogenic Ras. Mol Cell Biol 11: 4053–4063

Stacey DW, Roudebush M, Day R, Mosser SD, Gibbs JB, Feig LA (1991b) Dominant inhibitory Ras mutants demonstreate the requirements for Ras activity in the action of tyrosine kinase oncogenes. Oncogene 6:1197–2304

Trahey M, McCormick F (1987) A cytoplasmic protein stimulates normal N-ras p21 GTPase, but does not affect oncogenic mutants. Science 238:542–545

Tsai MH, Yu CL, Wei FS, Stacey DW (1989a) The effect of GTPase activating protein upon ras is inhibited by mitogenically responsive lipids. Science 243:522–526

Tsai MH, Hall A, Stacey DW (1989b) Inhibition by phospholipids of the interaction between R-ras and rho and their GTPase activating proteins. Mol Cell Biol 9:5260–5264

Tsai MH, Roudebush M, Gibbs JB, Stacey DW (1991) Ras GTPase-activating protein physically associates with mitogenically active phospholipids. Mol Cell Biol 11:2785–2793

Ullrich A, Schlessinger J (1990) Signal transduction by receptors with tyrosine kinase activity. Cell 61:203–212

Whitman M, Melton DA (1992) Involvement of $p21^{ras}$ in xenopus mesoderm induction. Nature 357:252–254

Wolfman A, Macara IG (1990) A cytosolic protein catalyzes the release of GDP from p21ras. Science 248:67–69

Xu G, Lin B, Tanaka K, Dunn D, Wood D, Gesteland R, White R, Weiss R, Tamanoi F (1990) The catalytic domain of the neurofibromatosis type 1 gene product stimulates Ras GTPase and complements *ira* mutants of S. cerevisiae., Cell 63:835–841

Yu CL, Tsai MH, Stacey DW (1988) Cellular ras activity and phospholipid metabolism. Cell 52:63–71

Yu CL, Tsai MH, Stacey DW (1990) A lipids is produced following serum stimulation of NIH3T3 cells which can inhibit GAP activity. Mol Cell Biol 10:6683–6689

Zhang K, DeClue JE, Vass WC, Papageorge AG, McCormick F, Lowy DR (1990) Suppression of c-ras transformation by GTPase-activating protein. Nature 346: 754–756

CHAPTER 21

Regulation of ras-Interacting Proteins in *Saccharomyces cerevisiae*

K. TANAKA, A. TOH-E, and K. MATSUMOTO

A. Introduction

The environment surrounding cells of multicellular organisms is rich in nutrients; cells can easily take up nutrients available in special spaces such as blood vessels. However, these cells require growth factor(s) to start the cell division cycle. In contrast, in unicellular microorganisms such as yeast, nutrients themselves regulate cell growth. Yeast cells continue to divide as long as there is a sufficient supply of extracellular nutrients. In poor nutrient conditions, cells stop growing and arrest at the G1 phase of the cell cycle. The physiology of the G1 phase in nutrient-deprived conditions is apparently

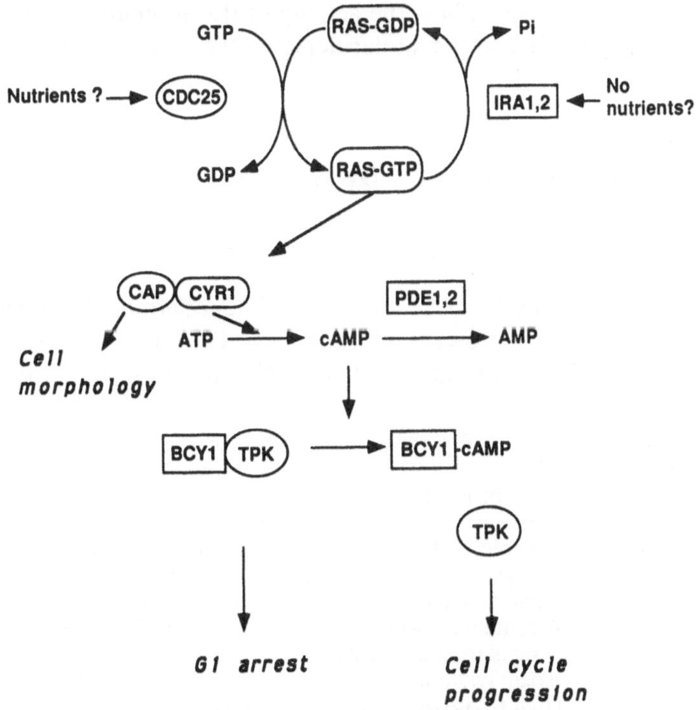

Fig. 1. The RAS-cAMP pathway. *Circled* genes exert positive functions, while *boxed* genes exert negative functions is the pathway. CAP/SRV2 has another function, the regulation of cell morphology

different from that of the G1 phase of growing cells. G1-arrested cells acquire resistance to environmental stress such as starvation or heat shock and become competent for the meiotic process.

The information on nutrient conditions seems to be conveyed through the signaling pathway involving Ras proteins which are counterparts of *ras* oncogenic proteins found in mammalian cells (MATSUMOTO et al. 1985; GIBBS and MARSHALL 1989; DESCHENES and BROACH 1989). This signal transduction pathway, the *RAS*-cAMP pathway, is schematically shown in Fig. 1 and the genes involved are listed in Table 1. The main feature of this signaling cascade is that Ras regulates adenylyl cyclase activity, and cAMP produced by adenylyl cyclase activates cAMP-dependent protein kinase (cAPK). In rich nutrient conditions, Ras proteins activate adenylyl cyclase to form cAMP. This second messenger in turn activates cAPK, which phosphorylates unknown protein(s) whose phosphorylation is crucial for cell cycle progression. Thus, mutants in which cAPK cannot be activated show a lethal phenotype. In contrast, in poor nutrient conditions, cAPK activity is kept low since Ras is inactivated. In this case, a low level of cAMP-dependent protein phosphorylation is essential for cells to stop at the G1 phase properly and to acquire stress resistance. Therefore, mutants with hyperactivated cAPK show sensitivity to environmental stresses.

In this review, we focus on how the activity of Ras proteins is regulated, and how Ras regulates its target molecule(s) in the yeast *Saccharomyces cerevisiae*.

B. Regulation of Ras Activity by Guanine Nucleotides

I. Biochemical Properties of Ras

There are two *ras* homologs, *RAS1* and *RAS2*, in *Saccharomyces cerevisiae* (DEFEO-JONES et al. 1983; POWERS et al. 1984). The first 80 amino acid residues of yeast Ras are more than 80% homologous with the

Table 1. Genes of the *RAS*-cAMP pathway

Genes	Gene product
RAS1, *RAS2*	Homologs of *ras* oncogene. GTP-binding regulatory proteins
IRA1, *IRA2*	GTPase activating proteins for Ras
CDC25	Guanine nucleotides exchange protein for Ras
CAP/SRV2	A protein associated with Cyr1. Also involved in the regulation of cell shape
CYR1	Adenylyl cyclase
PDE1	High K_m cAMP phosphodiesterase
PDE2	Low K_m cAMP phosphodiesterase
BCY1	Regulatory subunit of cAMP-dependent protein kinase
TPK1, *TPK2*, *TPK3*	Catalytic subunit of cAMP-dependent protein kinase

corresponding region of mammalian ras proteins. Like mammalian ras proteins, yeast Ras proteins bind guanine nucleotides, GTP or GDP, and slowly hydrolyze bound GTP to GDP and Pi (Tamanoi et al. 1984; Temeles et al. 1985). It is known that this intrinsic GTPase activity is impaired in representative oncogenic mutants of mammalian *ras* (Barbacid 1987). These mutations include G12V, A59T, and Q61L. Similar substitutions introduced into yeast Ras impaired the GTPase activity as well, and these mutations act as a dominant activating trait in vivo; these mutants are sensitive to heat shock and cannot arrest at the G1 phase properly upon nutrient starvation (Kataoka et al. 1984; Temeles et al. 1985). Accumulating evidence indicates that, like trimeric G-proteins such as transducin or G_s, the GTP-bound form of ras is an active conformation which transmits the signal downstream by activating a target molecule called an effector, while the GDP-bound form is an inactive conformation. In concogenic mutants, the conversion from the GTP-bound form to the GDP-bound form is inhibited, causing accumulation of an active, GTP-bound form. Thus, the activity of ras proteins as signal transducers is regulated by its ligands, GTP or GDP. Based on genetic techniques, two types of regulatory genes of Ras activity, negative and positive regulators, have been identified in *Saccharomyces cerevisiae*. The role of these proteins is to regulate the ratio of GTP to GDP bound to Ras.

II. The *CDC25* Gene

CDC25 is a positive regulator, and the effect of its mutation is lethal due to permanent arrest at the G1 phase. Dependency of cell growth on *CDC25* function is alleviated by a dominant activated mutation of *RAS2* such as *RAS2*(G19V), suggesting that *CDC25* is required for the formation of the GTP-bound form of Ras. Based on the fact that the exchange of bound GDP for GTP is an activation process for some trimeric G-proteins, it had been proposed that CDC25 promotes a similar reaction (Camonis et al. 1986; Broek et al. 1987; Robinson et al. 1987). Recently, it was shown that CDC25 protein overexpressed in yeast actually possesses this activity toward the Ras2 protein (Jones et al. 1991).

When glucose is added to yeast-derepressed cells, the cAMP level rises three- to five-fold within 30 s (Eraso and Gancedo 1985). It was demonstrated that this glucose-dependent cAMP formation requires intact *RAS* and *CDC25* genes (Mbonyi et al. 1988; Munder and Kuntzel 1989). It is likely that *CDC25* is activated upon glucose addition and in turn activates *RAS*, although the biochemical mechanism is yet to be revealed.

Three *CDC25*-like genes, *SCD25*, *LTE1*, and *BUD5*, have so far been identified in *Saccharomyces cerevisiae*. The *SCD25* gene was cloned as a truncated gene able to suppress the *cdc25* mutation when overexpressed (Boy-Marcotte et al. 1989). Subsequently, it was demonstrated that SCD25 produced by *Escherichia coli* promotes the GTP–GDP exchange reaction on the Ras2 protein (CRECHET 1990). However, in contrast to *CDC25*, *SCD25* is a nonessential gene (DAMAK et al. 1991). *BUD5* was identified in a collection of mutants which showed abnormal budding pattern (CHANT et al. 1991) or as a gene which suppressed a dominant negative mutation, *RAS2*(G22A) (POWERS et al. 1991). Genetic results suggest that *BUD5* functions upstream of a ras-like gene, *BUD1/RSR1*, as a positive regulator. The *LTE1* gene is required for growth at low temperature (WICKNER et al. 1987). The putative gene encoding a ras-like protein regulated by *LTE1* has not yet been identified. Because *CDC25*-like genes also occur as positive regulators of ras in other organisms such as *Drosophila melanogaster* (SIMON et al. 1991) and *Schizosaccharomyces pombe* (HUGHES et al. 1990), it is plausible that mammalian cells employ a similar protein to regulate *ras*.

III. *IRA1* and *IRA2* Genes

IRA1 and *IRA2* were characterized as genes involved in the negative regulation of *RAS* activity (TANAKA et al. 1989, 1990b). They encode proteins of 2938 and 3079 amino acids, respectively, and their primary structures are about 50% homologous. A mutation in either gene results in a phenotype similar to that of *RAS2*(G19V), and the *ira1 ira2* double mutant showed a stronger phenotype, suggesting that both genes function additively. The fact that the *cdc25 ira1* and *cdc25 ira2* double mutants are viable and heat-shock resistant suggests that *IRA1* and *IRA2* function antagonistically to *CDC25*. Moreover, it was demonstrated that GTP-bound Ras proteins accumulate in *ira* mutants (TANAKA et al. 1990a). Thus, Ira proteins seem to downregulate the level of the GTP-bound form of Ras.

Research into the homology with Ira1 in the sequence data bank revealed that there is a region homologous with the catalytic domain of GAP in Ira1. This homology also exists in Ira2. GTPase activating protein (GAP) was first identified in the cytosolic fraction of *Xenopus* oocytes as an activity which stimulates the intrinsic GTPase activity of normal N-ras protein about 1000-fold in vitro, but not that of N-ras carrying an oncogenic substitution (TRAHEY and MCCORMICK 1987). The hypothesis that Ira possesses GAP-like activity was consistent with the negative regulatory role of Ira on Ras. In fact, the GAP activity of Ira was demonstrated in vitro (TANAKA et al. 1991). GAP activity toward yeast Ras2 was found either in yeast extracts prepared from a strain overexpressing *IRA2* or the putative GAP catalytic domain of Ira2 purified from *E. coli* cells as a fusion protein with the glutathione-*S*-transferase. This activity stimulated the intrinsic GTPase activity of Ras2, but not that of Ras2(G19V).

GAP homologs have been identified in other organisms such as *D. melanogaster* and *S. pombe* (GAUL et al. 1992; IMAI et al. 1991). Genetic evidence indicates that they also function as downregulators of Ras activity. It was demonstrated that overexpression of GAP suppresses the transformed phenotype caused by overexpression of normal ras in mammalian cells, suggesting that GAP also functions as a negative regulator (ZHANG et al. 1990). However, there is a report which supports the claim that GAP is involved in the effector function (YATANI et al. 1990). Recently, it was reported that Ira1 is also required for effector function in yeast; adenylyl cyclase is not properly localized in the membrane fraction in the *ira1* mutant (MITTS et al. 1991). Further evidence must be accumulated before it may be concluded that Ira and GAP are involved in the effector function.

There is no homology between IRA and GAP other than the GAP catalytic domain. Recently, a protein showing higher homology with Ira than GAP was found in mammalian cells (XU et al. 1990a). This protein is neurofibromin, encoded by the gene whose mutation results in the clinically important disorder, neurofibromatosis type 1 (*NF1*). The primary structure of neurofibromin is much more like Ira than GAP (Fig. 2). In addition to the GAP catalytic domain, homology extends into the N-terminal and C-terminal directions. Moreover, the homology score in domains flanking the catalytic domain is as high as that in the GAP catalytic domain. Possibly these domains are involved in the regulation of the GAP catalytic activity. Subsequently, it was reported that NF1 actually has GAP activity towards ras (MARTIN et al. 1990; XU et al. 1990b; BALLESTER et al. 1990). Moreover, it was shown that the GTP-bound form of ras is accumulated in tumor cells isolated from NF1 patients. These results indicate that NF1 is structurally and functionally related to Ira (DECLUE et al. 1992; BASU et al. 1992).

In the nutrient-starved condition, it is likely that the Ras-bound GTP to GDP ratio is lowered by some mechanism. This may involve the downregulation of CDC25 activity or the activation of Ira. Thus, there might be regulatory proteins which function upstream of *IRA*. In this respect, the gene identified recently, *RPI1*, is interesting (KIM and POWERS

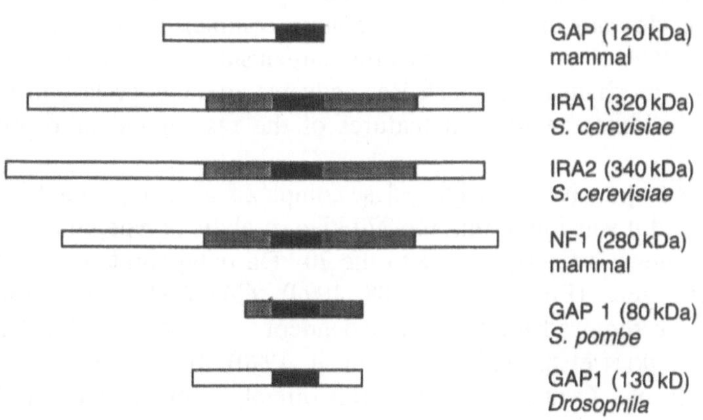

Fig. 2. Ras GTPase activating proteins. *Black boxes* indicate the catalytic domain. *Shaded boxes* are conserved between Ira and NF1 and partly conserved in S. pombe gap1. Ira1 and Ira2 are homologous throughout entire amino acid sequences with 50% identity

1991). *PRI1* was isolated as a multicopy suppressor of the heat-shock sensitive phenotype of a strain overexpressing *CYR1*, the structural gene for adenylyl cyclase. Since a disruption mutant of the *PRI1* gene shows the heat-shock sensitive phenotype, it is likely that *RPI1* downregulates the *RAS*-cAMP pathway. Overexpression of *RPI1* suppresses the heat-shock sensitive phenotype of either *ira1* or *ira2* mutation, although at least one of two intact *IRA* genes is required for this suppression. This result suggests that *RPI1* suppresses *ira* mutations through the activation of a residual wild-type *IRA* gene. Consistently, overexpression of *RPI1* fails to suppress the heat-shock sensitive phenotype of the *RAS2*(G19V) mutant. Further biochemical work will reveal how *RPI1* functions in the *RAS*-cAMP pathway.

C. Regulation of Adenylyl Cyclase by Ras

It has been established that the effector of Ras is the adenylyl cyclase complex in *Saccharomyces cerevisiae* (TODA et al. 1985; FIELD et al. 1988); only the GTP-bound form of Ras2 activates adenylyl cyclase (FIELD et al. 1988). Since mammalian ras activates yeast adenylyl cyclase both in vitro and in vivo (BROEK et al. 1985; KATAOKA et al. 1985a), and since modified yeast Ras transforms mammalian cells (DEFEO-JONES et al. 1985), it was postulated that the effector of ras in mammalian cells is also a protein related to adenylyl cyclase. However, it was reported that mammalian ras could not activate mammalian adenylyl cyclase (BECKNER et al. 1985). In addition, the *ras1* gene of fission yeast *S. pombe* was shown not to be involved in the regulation of adenylyl cyclase either (FUKUI et al. 1986). However, extensive homology between ras proteins from different organisms and their functional interchangeability suggest that the ras-interacting domain of the adenylyl cyclase complex is at least structurally related to that of effectors from other organisms. Thus, knowledge about the precise mechanism in which Ras activates adenylyl cyclase in yeast will help to elucidate the general features of the ras-dependent regulation of effectors.

A Ras-responsive adenylyl cyclase complex has been purified from yeast extracts, and it was found that the 270-kDa catalytic polypeptide encoded by the *CYR1* gene is tightly bound to the 70-kDa polypeptide encoded by the *CAP/SRV2* gene (FIELD et al. 1988, 1990). *CAP/SRV2* was subsequently shown to be required for the Ras-dependent activation of adenylyl cyclase (FEDOR-CHAIKEN et al. 1990; FIELD et al. 1990). It is unknown at present which molecule, Cyr1 or Cap/Srv2, Ras directly interacts with. If Cap/Srv2 directly interacts with Ras, it might be conserved in other organisms as a possible effector of ras.

Cap consists of at least two domains which have different functions; the N-terminal half is important for the Ras-dependent adenylyl cyclase activation, while the C-terminal half is important in maintaining normal cell morphology (FIELD et al. 1990; GERST et al. 1991). Mutation of the C-terminal region results in enlarged round cells with abnormal actin distribution and budding patterns. Recently, it was demonstrated that overexpression of the yeast homolog of profilin suppresses the mutation in the C-terminal region of *CAP/SRV2* (VOJTEK et al. 1991). Since profilin binds actin and polyphosphoinositide, this functional like between profilin and CAP suggests that the Ras pathway may have an alternative function: the regulation of cytoskeletal network in *Saccharomyces cerevisiae*.

D. Domains of Ras Interacting with Other Proteins

Mutational analysis and X-ray crystallographic analysis of mammalian ras revealed important domains of ras. Amino acid regions, 10–17, 57–60, and 116–119 are conserved in known G-proteins and essential for GDP–GTP exchange, GTP-induced conformational change, and GTP hydrolysis (BOURNE et al. 1991). The four C-terminal amino acids (CaaX, where C is Cys, a, a usually aliphatic amino acid and X, another amino acid) are essential for posttranslational processing involving carboxyl-terminal methylation and farnesylation of the cysteine residue (GIBBS 1991). This modification enables ras proteins to localize to the cytoplasmic membrane. A domain in which mutations do not alter intrinsic activity, such as nucleotides binding or GTPase activity, but affects transforming activity was identified in mammalian ras (SIGAL et al. 1986). This domain, the effector domain, encompasses amino acid residues 32–40 of H-*ras*. The effector domain of yeast Ras2, which is identical with that of mammalian ras, encompasses amino acids residues 39–47, due to seven extra amino acids at the N terminus. It was proposed that the effector domain interacts with and activates an effector molecule. Consistently, effector mutants of Ras2, Ras2(T42A) and Ras2(T42S) were shown to be unable to activate adenylyl cyclase in vivo and in vitro (MARSHALL et al. 1988).

Since GAP interacts directly with Ras proteins, it was interesting to examine interactions of GAP with effector mutants of Ras. GAP has been shown to be incapable of activating some of the effector mutants of ras (ADARI et al. 1988; CALES et al. 1988). Yeast Ira2 protein could not stimulate the GTPase activity of an effector mutant of Ras2, Ras2(D45N), either (TANAKA et al. 1991). An effector mutation which is deficient only in the GTPase stimulation by Ira would show a dominant activating trait. Such a mutation, *RAS2*(P41S), has been identified and characterized recently (TANAKA et al. 1992).

Biochemical and genetic results described above indicate that the effector domain of Ras2 is required for the interactions with the effector as well as with Ira2. Since both effector and GAP should distinguish the GTP-bound form of ras from the GDP-bound form, the effector domain may have different conformations in two different forms. Consistent with this idea is the observation that the effector domain is one of the two domains which change conformation upon GTP hydrolysis (MILBURN et al. 1990; SCHLICHTING et al. 1990). Further fine structural analysis will reveal how the interaction of ras proteins with effector(s) and regulator(s) overlaps on a single domain.

E. Conclusions

In this review we have discussed how Ras proteins play a key role as signal transducers for nutrient availability in *Saccharomyces cerevisiae*. It seems that the nutrient signal is translated into the Ras-bound GTP–GDP ratio through the regulation of Cdc25 or Ira activity. Cdc25 is a positive regulator which promotes the exchange reaction of Ras-bound GDP for GTP, while Ira1 and Ira2 proteins are negative regulators which stimulate the intrinsic GTPase activity of Ras proteins. It is not known how the signal from nutrients is conveyed to Cdc25 or/and Ira.

Only the GTP-bound form of Ras activates the adenylyl cyclase complex, which is an effector of Ras. A 70-kDa protein called Cap/Srv2 is tightly bound to the adenylyl cyclase catalytic polypeptide. It has been shown that Cap plays an essential role in the Ras-dependent activation of adenylyl cyclase. However, it remains to be revealed whether Ras binds to Cap, Cyr1, or both. Although *CYR1* is only involved in the cAMP-pathway, *CAP* has an additional function, the regulation of cell morphology. This raises the possibility that *RAS* transmits a signal to the regulatory pathway for cell morphology in *Saccharomyces cerevisiae*. It should be elucidated whether Cap/Srv2-like proteins are conserved and involved in the ras-mediated signal transduction pathway in other organisms.

Because mammalian ras activates yeast adenylyl cyclase, the ras-binding domain of mammalian effector(s) should be conserved at least structurally, with that of adenylyl cyclase. Clearly the identification of such a domain of adenylyl cyclase will provide important information about the mechanism of ras-dependent activation of effectors.

Acknowledgements. We thank R. Ruggieri for reading the manuscript.

References

Adari H, Lowy D-R, Willumsen B-M, Der C-J, McCormick F (1988) Guanosine triphosphatase activating protein (GAP) interacts with the p21 ras effector binding domain. Science 240:518–521

Ballester R, Marchuk D, Boguski M, Saulino A, Letcher R, Wigler M, Collins F (1990) The NF1 locus encodes a protein functionally related to mammalian GAP and yeast IRA proteins. Cell 63:851–859

Barbacid M (1987) ras Genes. Annu Rev Biochem 56:779–827

Basu T-N, Gutmann D-H, Fletcher J-A, Glover T-W, Collins F-S, Downward J (1992) Aberrant regulation of ras proteins in malignant tumor cells from type 1 neurofibromatosis patients. Nature 356:713–715
Beckner S-K, Hattori S, Shih T-Y (1985) The ras oncogene product p21 is not a regulatory component of adenylate cyclase. Nature 317:71–72
Boy-Marcotte E, Damak F, Camonis J, Garreau H, Jacquet M (1989) The C-terminal part of a gene partially homologous to CDC25 gene suppresses the cdc25-5 mutation in Saccharomyces cerevisiae. Gene 77:21–30
Bourne H-R, Sanders D-A, McCormick F (1991) The GTPase superfamily: conserved structure and molecular mechanism. Nature 349:117–127
Broek D, Samily N, Fasano O, Fujiyama A, Tamanoi F, Northup J, Wigler M (1985) Differential activation of yeast adenylate cyclase by wild-type and mutant ras proteins. Cell 41:763–770
Broek D, Toda T, Michaeli T, Levin L, Birchmeier C, Zoller M, Powers S, Wigler M (1987) The Saccharomyces cerevisiae CDC25 gene product regulates the RAS-adenylate cyclase pathway. Cell 48:789–800
Cales C, Hancock J-F, Marshall C, Hall A (1988) The cytoplasmic protein GAP is implicated as the targer for regulation by the ras gene product. Nature 332:548–551
Camonis J-H, Kalekine M, Gondre B, Garreau H, Boy-Marcotte E, Jacquet M (1986) Characterization, cloning and sequence analysis of the CDC25 gene which controls the cyclic AMP level of Saccharomyces cerevisiae. EMBO J 5:375–380
Chant J, Corrado K, Pringle J-R, Herskowitz I (1991) Yeast BUD5, encoding a putative GDP-GTP exchange factor, is necessary for bud site selection and interacts with bud formation gene BEM1. Cell 65:1213–1224
Crechet J-B, Poullet P, Mistou M-Y, Parmeggiani A, Camonis J, Boy-Marcotte E, Damak F, Jacquet M (1990) Enhancement of the GTP-GDP exchange of RAS proteins by the carboxyl-terminal domain of SCD25. Science 248:866–868
Damak F, Boy-Marcotte E, Le-Roscouet D, Guilbaud R, Jacquet M (1991) SDC25, a CDC25-like gene which contains a RAS-activating domain and is a dispensable gene of Saccharomyces cerevisiae. Mol Cell Biol 11:202–212
Declue J-E, Papageorge A-G, Fletcher J-A, Diehl S-R, Ratner N, Vass W-C, Lowy D-R (1992) Abnormal regulation of mammalian p21 ras contributes to malignant tumor growth in von Recklinghausen (type 1) neurofibromatosis. Cell 69:265–273
DeFeo-Jones D, Scolnick E, Koller R, Dhar R (1983) ras-Related gene sequences identified and isolated from Saccharomyces cerevisiae. Nature 306:707–709
DeFeo-Jones D, Tatchell K, Robinson L-C, Sigal I-S, Vass W-C, Lowy D-R, Scolnick E-M (1985) Mammalian and yeast ras gene products: biological function in their heterologous systems. Science 228:179–184
Deschenes R-J, Broach J-R (1989) The function of RAS genes in Saccharomyces cerevisiae. Adv Cancer Res 54:79–138
Eraso P, Gancedo J-M (1985) Use of glucose analogues to study the mechanism of glucose-mediated cAMP increase in yeast. FEBS Lett 191:51–54
Fedor-Chaiken M, Deschenes R-J, Broach J-R (1990) SRV2, a gene required for RAS activation of adenylyl cyclase in yeast. Cell 61:329–340
Field J, Nikawa J, Broek D, MacDonald B, Rodgeers L, Wilson I-A, Lerner R-A, Wigler M (1988) Purification of a RAS-responsive adenylyl cyclase complex from Saccharomyces cerevisiae by use of an epitope addition method. Mol Cell Biol 8:2159–2165
Field J, Vojtek A, Ballester R, Bolger G, Colicelli J, Ferguson K, Gerst J, Kataoka T, Michaeli T, Powers S, Riggs M, Rodgers L, Wieland I, Wheland B, Wigler M (1990) Cloning and characterization of CAP, the S. cerevisiae gene encoding the 70 kd adenylyl cyclase-associated protein. Cell 61:319–327
Fukui Y, Kozasa T, Kaziro Y, Takeda T, Yamamoto M (1986) Role of a ras homolog in the life cycle of Schizosaccharomyces pombe. Cell 44:329–336

Gaul U, Mardon G, Rubin G-M (1992) A putative Ras GTPase activating protein acts as a negative regulator of signaling by the Sevenless receptor tyrosine kinase. Cell 68:1007–1019
Gerst J, Ferguson K, Vojtek A, Wigler M, Field J (1991) CAP is a bifunctional component of the S. cerevisiae adenylyl cyclase complex. Mol Cell Biol 11:1248–1257
Gibbs J-B (1991) Ras C-terminal processing enzymes – new drug targets? Cell 65:1–4
Gibbs J-B, Marshall M-S (1989) The ras oncogene-an important regulatory element in lower eukaryotic organisms. Microbiol Rev 53:171–185
Hughes D-A, Fukui Y, Yamamoto M (1990) Homologous activators of ras in fission and budding yeast. Nature 344:355–357
Imai Y, Miyake S, Hughes D-A, Yamamoto M (1991) Identification of a GTPase-activating-protein homolog in Schizosaccharomyces pombe. Mol Cell Biol 11:3088–3094
Jones S, Vignais M-L, Broach J-R (1991) The CDC25 protein of Saccharomyces cerevisiae promotes exchange of guanine nucleotides bound to Ras. Mol Cell Biol 11:2641–2646
Kataoka T, Powers S, McGill C, Fasano O, Strathern J, Broach J, Wigler M (1984) Genetic analysis of yeast Saccharomyces cerevisiae RAS1 and RAS2 genes. Cell 37:437–445
Kataoka T, Powers S, Cameron S, Fasano O, Goldfarb M, Broach J, Wigler M (1985a) Functional homology of mammalian and yeast ras genes. Cell 40:19–26
Kataoka T, Broek D, Wigler M (1985b) DNA sequence and characterization of the Saccharomyces cerevisiae gene encoding adenylate cyclase. Cell 43:493–505
Kim J-H, Powers S (1991) Overexpression of RPI1, a novel inhibitor of the yeast Ras-cyclic AMP pathway, down-regulates normal but not mutationally activated Ras function. Mol Cell Biol 11:3894–3904
Martin G-A, Viskochil D, Bollag G, McCabe P-C, Crosier W-J, Haubruck H, Conroy L, Clark R, O'Connell P, Cawthon R-M, Innis M-A, McCormick F (1990) The GAP-related domain of the neurofibromatosis type 1 gene product interacts with ras p21. Cell 63:843–849
Marshall M-S, Gibbs J-B, Scolnick E-M, Sigal I-S (1988) An adenylate cyclase from Saccharomyces cerevisiae that is stimulated by ras proteins with effector mutations. Mol Cell Biol 8:52–61
Matsumoto K, Uno I, Kato K, Ishikawa T (1985) Genetic analysis of the role of cAMP in yeast. Yeast 1:15–24
Mbonyi K, Beullens M, Detremerie K, Geerts L, Thevelein J-M (1988) Requirement of one functional RAS gene and inability of an oncogenic RAS variant to mediate the glucose-induced cyclic AMP signal in the yeast Saccharomyces cerevisiae. Mol Cell Biol 8:3051–3057
Milburn M-V, Tong L, deVos A-M, Brunger A, Yamaizumi Z, Nishimura S, Kim S-H (1990) Molecular switch for signal transduction: structural differences between active and inactive forms of protooncogenic ras proteins. Science 247:939–945
Mitts M-R, Bradshaw-Rouse J, Heideman W (1991) Interactions between adenylate cyclase and the yeast GAP protein, IRA1. Mol Cell Biol 11:4591–4598
Munder T, Kuntzel H (1989) Glucose-induced cAMP signalling in Saccharomyces cerevisiae is mediated by the CDC25 protein. FEBS Lett 242:341–345
Powers S, Kataoka T, Fasano O, Goldfarb M, Strathern J, Broach J, Wigler M (1984) Genes in Saccharomces cerevisiae encoding proteins with domains homologous to the mammalian ras proteins. Cell 36:607–612
Powers S, Gonzales E, Christensen T, Cubert J, Broek D (1991) Functional cloning of BUD5, a CDC25-related gene from S. cerevisiae that can suppress a dominant-negative RAS2 mutant. Cell 65:1225–1231

Robinson L-C, Gibbs J-B, Marshall M-S, Sigal I-S, Tatchell K (1987) CDC25: a compoent of the RAS-adenylate cyclase pathway in Saccharomyces cerevisiae. Science 235:1218–1221

Schlichting I, Almo S-C, Rapp G, Wilson K, Petratos K, Lentfer A, Wittinghofer A, Kabsch W, Pai E-F, Petsko G-A, Goody R-S (1990) Time-resolved X-ray crystallographic study of the conformational change in Ha-Ras p21 protein on GTP hydrolysis. Nature 345:309–315

Sigal I-S, Gibbs J-B, D'Alonzo J-S, Scolnick E-M (1986) Identification of effector residues and a neutralizing epitope of Ha-ras-encoded p21. Proc Natl Acad Sci USA 83:4725–4729

Simon M-A, Bowtell D-D-L, Dodson G-S, Laverty T-R, Rubin G-M (1991) Ras1 and a putative guanine nucleotide exchange factor perform crucial steps in signalling by the Sevenless protein tyrosine kinase. Cell 67:701–716

Tamanoi F, Walsh M, Kataoka T, Wigler M (1984) A product of yeast RAS2 gene is a guanine nucleotide binding protein. Proc Natl Acad Sci USA 81:6924–6928

Tanaka K, Matsumoto K, Toh-e A (1989) IRA1, an inhibitory regulator of the RAS-cyclic AMP pathway in Saccharomyces cerevisiae. Mol Cell Biol 9:757–768

Tanaka K, Nakafuku M, Satoh T, Marshall M-S, Gibbs J-B, Matsumoto K, Kaziro Y, Toh-e A (1990a) Saccharomyces cerevisiae genes, IRA1 and IRA2 encode proteins that may be functionally equivalent to mammalian ras GTPase activating protein (GAP). Cell 60:803–807

Tanaka K, Nakafuku M, Tamanoi F, Kaziro Y, Matsumoto K, Toh-e A (1990b) IRA2, a second gene of Saccharomycess cerevisiae that encodes a protein with a domain homologous to mammalian ras GTPase-activating protein. Mol Cell Biol 10:4303–4313

Tanaka K, Lin B-K, Wood D-R, Tamanoi F (1991) IRA2, an upstream negative regulator of RAS in yeast, is a RAS GTPase activating protein (GAP). Proc Natl Acad Sci USA 88:468–472

Tanaka K, Wood D-R, Lin B-K, Khalil M, Tamanoi F, Cannon J-F (1992) A dominant activating mutation in the effector region of RAS abolishes IRA2 sensitivity. Mol Cell Biol 12:631–637

Temeles G-L, Gibbs J-B, D'Alonzo J-S, Sigal I-S, Scolnick E-M (1985) Yeast and mammalian ras proteins have conserved biochemical properties. Nature 313:700–703

Toda T, Uno I, Ishikawa T, Powers S, Kataoka T, Broek D, Cameron S, Broach J, Matsumoto K, Wigler M (1985) In yeast, RAS proteins are controlling elements of adenylate cyclase. Cell 40:27–36

Toda T, Cameron S, Saas P, Zoller M, Wigler M (1987) Three different genes in S. cerevisiae encode the catalytic subunits of the cAMP-dependent protein kinase. Cell 50:277–287

Trahey M, McCormick F (1987) A cytoplasmic protein stimulates normal N-ras p21 GTPase, but does not affect oncogenic mutants. Science 238:542–545

Vojtek A, Haarer B, Field J, Gerst J, Pollard T-D, Brown S, Wigler M (1991) Evidence for a functional link between profilin and CAP in the yeast S. cerevisiae. Cell 66:497–505

Wickner R-B, Koh T-J, Crowley J-C, O'neil J, Kaback D-B (1987) Molecular cloning of chromosome I DNA from Saccharomyces cerevisiae: isolation of the MAK16 gene and analysis of an adjacent gene essential for growth at low temperatures. Yeast 3:51–57

Xu G, O'Connell P, Viskochil D, Cawthon R, Robertson M, Culver M, Dunn D, Stevens J, Gesteland R, White R, Weiss R (1990a) The neurofibromatosis type 1 gene encodes a protein related to GAP. Cell 62:599–608

Xu G, Lin B, Tanaka K, Dunn D, Wood D, Gesteland R, White R, Weiss R, Tamanoi F (1990b) The catalytic domain of the neurofibromatosis type-1 gene product stimulates ras GTPase and complements ira mutants of S. cerevisiae. Cell 63:835–841

Yatani A, Okabe K, Polakis P, Halenbeck R, McCormick F, Brown A-M (1990) RAS p21 and GAP inhibit coupling of muscarinic receptors to atrial K^+ channels. Cell 61:769–776

Zhang K, DeClue J-E, Vass W-C, Papageorge A-G, McCormick F, Lowy D-R (1990) Suppression of c-ras transformation by GTPase activating protein. Nature 346:754–756

CHAPTER 22

Lipid Modifications of Proteins in the Ras Superfamily

J.B. GIBBS

A. Background

The *ras* oncogene protein, called Ras or p21, has served as a model system for the characterization of the small (ca. 20 kDa) GTPases. Among the early biochemical discoveries of Ras was the observation that this oncogene-encoded protein underwent posttranslational processing events which preceded the localization of Ras in the plasma membrane. The first specific modification identified was palmitoylation of a Cys residue somewhere near the C terminus of Ras. However, the full number and chemical nature of the modifications was not fully appreciated until recently. The steps of Ras processing (farnesylation, proteolysis, carboxyl methylation, and palmitoylation) have been reviewed extensively (for examples, see DER and Cox 1991; GIBBS 1991; SINENSKY and LITZ 1992), and a summary of these steps is shown in Table 1. These modifications occur in a C-terminal region that has acquired the acronym CaaX box (C, Cys; a, a usually aliphatic amino acid; X, another amino acid).

The prenylation of proteins was originally observed as a posttranslational modification by metabolites of mevalonic acid; recently, farnesyl diphosphate and geranylgeranyl diphosphate were identified as the isoprenoid precursors for protein prenylation (Table 2; see DER and Cox 1991; GIBBS 1991; SINENSKY and LITZ 1992). In most cases, the proteins modified by prenylation are involved in a signaling mechanism: visual transduction (transducin, cGMP phosphodiesterase, rhodopsin kinase), neutrophil

Table 1. Modifications observed for Ha-Ras

Step	Essential for transformation
Farnesylation of Cys_1	Yes
Proteolytic cleavage of Aaa-Aaa-Xaa	No
Methyl esterification of terminal COOH	No
Palmitoylation of Cys_2	No

The processing steps are indicated for a generic C-terminal CaaX region ($-Cys_2-(Xaa)_n-Cys_1$-Aaa-Aaa-Xaa). Aaa, a usually aliphatic amino acid; Xaa, another amino acid. The nonessential steps enhance the efficiency of Ras membrane localization and function but are not obligatory for cell-transforming activity.

Table 2. Metabolites of the isoprenoid biosynthetic pathway

Metabolite	Size	Utilization
Mevalonate	C5	Isoprene precursor
Dimethylallyl-PPi	C5	tRNA
Geranyl-PPi	C10	?
Farnesyl-PPi	C15	Proteins, cholesterol, heme a
Geranylgeranyl-PPi	C20	Proteins, dolicol, ubiquinones

Mevalonate, the product of a reaction catalyzed by HMG-CoA reductase, is the precursor for the isoprene unit that serves as the stepwise building block for the indicated isoprenoid diphosphates. Different enzymes serve as branchpoints for the various utilization steps.

activation (the GTPases of the Rac and Rap family), receptor coupled G-proteins (γ subunit) and vectorial vesicle fusion mechanisms involved in secretion (members of the various Rab families). However, structural proteins, such as the nuclear lamins, can also be prenylated. This review article will summarize more recent studies that have focused on the enzymology of each of the steps and on the importance of C-terminal modifications for the membrane association and biological function of members of the Ras superfamily.

B. Farnesylation

I. Farnesyl-Protein Transferase

The first modification that commits Ras to membrane localization is catalyzed by a farnesyl-protein transferase (FTase). This enzyme was discovered by several groups as a soluble activity which used farnesyl diphosphate as substrate to modify the CaaX Cys residue in Ras (see GIBBS 1991). A very simple enzyme assay (Scheme I) has led to a rapid purification and characterization of FTase.

$$[^3H]FPP + \underset{\displaystyle \text{SH}}{\text{Ras-CVLS}} \xrightarrow{\text{FTase, } Mg^{2+}/Zn^{2+}} \underset{\displaystyle \text{S-}[^3H]\text{farnesyl}}{\text{Ras-CVLS}} + PPi$$

Scheme I

In this assay, [^{3}H]farnesyl diphosphate is incubated in the presence of enzyme with an acceptor protein having a CaaX sequence. The product shown in Scheme I, [^{3}H]farnesyl-Ras, is isolated from the reactants by acid precipitation followed by collection on filter paper. Several methods for this and alternative assays have been described (MOORES et al. 1991; POMPLIANO et al. 1992; REISS et al. 1990).

Table 3. Molecular genetics of protein prenylation enzymes

Motif	Example	Enzyme	Mammalian	*S. cerevisiae*
CaaX	Ras-CVIM	FTase	$\alpha 1/\beta 1$	Ram2/Ram1
CaaL	Rap1-CVVL	GGTase-I	$\alpha 1/\beta 2$	Ram2/Cdc43
XXCC	Rab1-GGCC	GGTase-II	$\alpha 2?/\beta 3?/?$	?/Bet2
XCXC	Rab3-DCAC	GGTase-II?	$\alpha ?/\beta ?/\gamma ?$	?
CCXX	Rab5-CCSN	GGTase-?	?	?

Listed are examples of proteins that are substrates for farnesyl-protein transferase (FTase) or the indicated geranylgeranyl-protein transferases (GGTase). For FTase, X can be Ser, Met, Gln, Cys or Ala. The sequence requirements for X in the other motifs have not been defined. The proposed subunit composition of these enzymes is based upon the cloning of mammalian FTase, the identification of genes in *S. cerevisiae* which encode structural polypeptides of the enzymes, and the purification of the GGTase that modifies Rab3-DCAC. See text for further details.

The purification of FTase from rat brain using sodium dodecyl sulfate polycrylamide gel electrophoresis (SDS-PAGE) revealed that the enzyme consists of two subunits, termed α and β, which have molecular masses of 48 kDa and 46 kDa, respectively (Reiss et al. 1990). The cDNAs cloned for the *Saccharomyces cerevisiae*, rat, and bovine FTase reveal that the subunits are the products of different genes (Table 3; Chen et al. 1991a,b; He et al. 1991; Kohl et al. 1991). Expression and reconstitution of the subunits indicates that both are required for enzyme activity. The holoenzyme has a mass of about 100 kDa, as determined by gel filtration, suggesting that FTase is a heterodimer of structure α–β. Cross-linking studies have suggested that the β subunit binds protein-CaaX substrate (Reiss et al. 1991). The mechanism of farnesyl diphosphate binding and the function of the α subunit is not clear at this time. Two divalent cations, Mg^{2+} and Zn^{2+}, are required in the reaction (Reiss et al. 1990, 1992). Zn^{2+} appears to be required for substrate CaaX binding, perhaps interacting with the sulfhydryl moiety of the Cys. In contrast, Mg^{2+} is essential for catalysis, but does not seem to influence binding of either substrate.

The steady-state kinetic mechanism of FTase was determined to be sequential with the apparently random order of substrate addition (Pompliano et al. 1992). The critical conclusions from this study are: 1) there are distinct binding sites for substrates farnesyl diphosphate and Ras, 2) both substrates must be present in the active site for the reaction to proceed, and 3) the product of the reaction, S-farnesyl-Ras, is competitive with both farnesyl diphosphate and Ras, indicating that the substrate-binding sites are in close proximity. Reiss et al. (1991) have demonstrated that each substrate alone can bind to FTase, which is also consistent with independent substrate binding sites.

Specificity of the FTase reaction is a function of both recognition and catalytic utilization of substrates. In the case of isoprenoid diphosphates, FTase binds farnesyl diphosphate and geranylgeranyl diphosphate with similar affinities (MOORES et al. 1991; REISS et al. 1992; SCHABER et al. 1990). However, farnesyl diphosphate is a better substrate (MOORES et al. 1991; SEABRA et al. 1991). Other isoprenoids such as dimethylallyl diphosphate and geranyl diphosphate bind to FTase with poorer affinity (SCHABER et al. 1990). These two isoprenoids do not appear to be involved with protein modification (Table 2). The determinants for protein substrate recognition reside principally in the CaaX box itself (REISS et al. 1990; SCHABER et al. 1990). Tetrapeptides that mimic physiological CaaX sequences compete for Ras farnesylation in vitro by acting as substrates (GOLDSTEIN et al. 1991; MOORES et al. 1991; POMPLIANO et al. 1992). The affinity and efficiency of farnesylation observed with these tetrapeptides is similar to the authentic protein substrate. Although the particular amino acid at any of the four positions can affect binding affinity, the last amino acid is especially important for specific recognition by FTase. Interestingly, some substitutions render tetrapeptides as nonsubstrate or true inhibitors (GOLDSTEIN et al. 1991; POMPLIANO et al. 1992).

II. Function of Farnesylation

Farnesylation clearly confers hydrophobic properties on proteins for membrane localization (HANCOCK et al. 1989). However, it is curious that while farnesylation of Ras is essential for biological function, other lipidation mechanisms can serve as substitutes. The first described alternative was myristoylation at the N terminus of Ras (BUSS et al. 1989; LACAL et al. 1988). This result was unexpected, but one may rationalize the result from the Ras structure. Although the ultimate 20 residues are absent in the crystal structure of Ras, the C-terminal region is roughly near the N terminus (see Chaps. 13, 14). For the myristoylated form of Ras, one might envision that Ras is presented off the membrane so that it can still effectively interact with cellular proteins critical for Ras function. The second alternative modifications is C-terminal geranylgeranylation (COX et al. 1992; HANCOCK et al. 1991b). Substitution of Leu in the last amino acid of the Ras CaaX box results in geranylgeranylation of Ras. However, this mutant Ras protein retains full biological activity. These results indicate that although membrane localization is critical for Ras function, modification with a specific lipid moiety is not essential. For both the myristoylated and geranylgeranylated forms of Ras, it is assumed that because they can induce cellular transformation, they must be localized in the same subcellular compartment as farnesylated Ras. This possibility has yet to be verified by subcellular localization studies such as immunofluorescence. Since these mutant Ras proteins with different lipid modifications were transfected into cells that have endogenous normal Ras (which is farnesylated), one cannot rule out the possibility that the incorrectly modified Ras perturbs the regulatory pathways that modulate and activate the function of the endogenous farnesylated Ras.

C. Geranylgeranylation

Numerous proteins have been described to date that are modified by a prenyl group. Geranylgeranylation is far more abundant than farnesylation in terms of total mass as well as in the number of substrate proteins identified. Furthermore, the number of motifs that have been identified is also more extensive than farnesylation of substrates. Whereas CaaX is invariant (so far) for farnesylation, at least four motifs have been described for geranylgeranylation: CaaL, XXCC, XCXC, and CCXX. At least two different enzymes catalyze these reactions. Geranylgeranyl-protein transferase type I GGTase-I) modifies proteins that have the CaaL motif (CASEY et al. 1991; JOLY et al. 1991; MOORES et al. 1991; SEABRA et al. 1991; YOKOYAMA et al. 1991). GGTase-I clearly does not modify Rab1-GGCC or Rab5-CCSN, suggesting that proteins having the XXCC and CCXX motifs are not substrates (KINSELLA and MALTESE 1992; MOORES et al. 1991). A second enzyme, GGTase-II, was identified using as substrate the *S. cerevisiae* homolog of mammalian Rab1a, Ypt1 (which has the C-terminal sequence GGCC) (MOORES et al. 1991). This enzyme is chromatographically distinct from both FTase and GGTase-I. In addition, the analogous activities in yeast are dependent upon different gene products (Table 3; FINEGOLD et al. 1991; KOHL et al. 1991; MOORES et al. 1991).

GGTase-I has several similarities with FTase. Both Mg^{2+} and Zn^{2+} ions are required for activity (CASEY et al. 1991). In terms of substrate interaction, the enzyme is capable of binding and modifying a tetrapeptide having the sequence of a geranylgeranylation substrate (MOORES et al. 1991). A terminal Leu residue in the peptide is sufficient to lead to preferential interaction with GGTase-I and poorer binding to FTase. Geranylgeranylation specificity is also due to some 20-fold stronger binding of GGTase-I to geranylgeranyl diphosphate than to farnesyl diphosphate (MOORES et al. 1991). Genetic experiments in yeast (Table 3) and immunological studies with mammalian enzymes (FINEGOLD et al. 1991; KOHL et al. 1991; SEABRA et al. 1991) have shown that GGTase-I has a subunit in common with FTase(α). The β subunit of the yeast GGTase-I (Cdc43) is 30% identical with the yeast FTase β subunit (Ram1). Until the mammalian GGTase-I β subunit is cloned, we can only presume that it will share similar homology with the FTase β subunit. Nevertheless, it is striking that in spite of the structural similarities between FTase and GGTase-I, these enzymes are quite specific in the reactions that are catalyzed.

The GGTase-II appears to be a different enzyme completely (Table 3). Although it binds geranylgeranyl diphosphate stronger than farnesyl diphosphate, GGTase-II does not bind simple peptides of the XXCC prenylation site (MOORES et al. 1991). Using Ypt1 as substrate, both Cys residues appear to be geranylgeranylated. The putative β subunit of GGTase-II is different, as suggested by the observation that mutations in the yeast *BET2* gene, which encodes a protein 30% identical to Ram1 and

Cdc43, are specific for GGTase-II (KOHL et al. 1991). In addition, GGTase-II does not share the α subunit found in FTase and GGTase-I (KOHL et al. 1991).

A GGTase enzyme activity has also been reported using Rab3a-DCAC as substrate (HORIUCHI et al. 1991; SEABRA et al. 1992). As was observed for Ypt1 and GGTase-II, the recognition of the Rab3a substrate by this GGTase is dependent upon more structural information than the last few amino acids (SEABRA et al. 1992), and both Cys residues of the substrate appear to be geranylgeranylated (FARNSWORTH et al. 1991). The purification of the Rab GGTase indicates that this enzyme is more complex than FTase or GGTase-I (SEABRA et al. 1992). Whereas FTase and GGTase-I exist as a tight heterodimeric complex, Rab GGTase has at least three subunits. One of these subunits dissociates during purification and must be reconstituted with the other two subunits (60 kDa and 38 kDa) in order to yield active enzyme. The Rab GGTase also geranylgeranylates Rab1-GGCC at about 50% the rate observed with Rab3a-DCAC (SEABRA et al. 1992). This result raises the possibility that a single enzyme is responsible for modifying proteins having either the XCXC or XXCC motifs. The presence of Rab GGTase has not been described in yeast, and it is not known whether this activity is sensitive to *bet2* mutations (see Chap. 4).

The geranylgeranylation of proteins may be multipurpose. Some of the substrates are membrane bound, and geranylgeranylation serves as a hydrophobic anchor. Indeed, proteins modified by geranylgeranylation are more hydophobic and partition tighter into lipid and detergent phases than proteins modified by farnesylation (HANCOCK et al. 1991b). However, geranylgeranylation may also serve as part of a recognition sequence for protein–protein interaction. For example, the interaction between Ras-related proteins such as Rap1a–Krev-1 and Rab3a with guanine nucleotide dissociation inhibitors (GDIs) is dependent upon the geranylgeranyl modification (see Chap. 39).

D. Other Modifications

I. Proteolysis

Upon prenylation of proteins having a CaaX or CaaL box, proteolytic cleavage of the final three residues occurs (Table 1). Proteolytic cleavage is not required for proteins having the XXCC or XCXC motifs; it has not been reported whether geranylgeranylated CCXX is a substrate for proteolysis. Recently, an activity was described in the particulate fraction of mammalian and yeast cells which may be responsible for the CaaX proteolysis processing step (ASHBY et al. 1992; HANCOCK et al. 1991a; HRYCYNA and CLARKE 1992; MA and RANDO 1992). This activity appears to selectively recognize the prenylated CaaX peptide and remove the tripeptide aaX in an endoproteolytic reaction. While the proteolysis of Ras aaX sequence is important for the efficiency of Ras membrane localization and cell transforming activity, it is not essential. In a structure–function analysis of the Ras CaaX region, KATO et al. (1992) discovered a CaaX mutant, CVYM, that did not undergo proteolysis following farnesylation. This Ras mutant partitioned into a membrane fraction with about 50% efficiency and exhibited approximately 50% of the transforming activity of a Ras-CVIM wild-type control.

II. Methylation

Carboxyl methylation (Table 1) confers hydrophobic properties for membrane anchoring and modulates protein function. The carboxyl methyl transferase which catalyzes this reaction is associated with the particulate fraction of mammalian and yeast cells (see GIBBS 1991). However, it is clearly distinct from the protease, because *S. cerevisiae* cells which are defective in the methyl transferase (*ste14* mutants) have normal protease activity (HRYCYNA and CLARKE 1992). Fractionation studies of rat liver membranes suggest that little methyl transferase is associated with the plasma membrane; instead, the activity is located in the endoplasmic reticulum (STEPHENSON and CLARKE 1992). The methyl transferase specifically recognizes the prenylated Cys residue at the protein C terminus and catalyzes modification of both farnesyl-Cys and geranylgeranyl-Cys substrates (HRYCYNA et al. 1991; VOLKER et al. 1991a). A very minimal recognition sequence is required as evidenced by the simple compound *N*-acetyl-S-farnesyl cysteine (AFC) serving as a substrate for the methyl transferase (PEREZ-SALA et al. 1991; VOLKER et al. 1991b).

The function of carboxyl methylation for Ras is not yet understood. Yeast cells defective in this activity (*ste14*) are fully viable even though the Ras protein expressed in these cells is membrane bound, but is not methylated (HRYCYNA et al. 1991). In mammalian cells, AFC has been used to assess the importance of carboxyl methylation, since it blocks methylation of Ras and other cellular proteins in vitro and in vivo (PEREZ-SALA et al. 1991; VOLKER et al. 1991b). AFC does not affect the phenotype of *ras*-transformed cells, but AFC blocks methylation of Ras-related proteins and inhibits chemotactic responses in macrophages (VOLKER et al. 1991b). Thus, methylation may have different functions with different proteins.

III. Palmitoylation

Ras is the only mammalian protein of the small GTPase family which is palmitoylated. The palmitoyl modification of Ras enhances the efficiency of membrane localization and function, but it is not obligatory (Table 1). Farnesylation and cellular transformation can occur in the absence of palmitoylation. HANCOCK et al. (1989) demonstrated this point by first identifying distinct cysteine acceptors for farnesylation and palmitoylation and then subsequently observing cellular transformation with Ras proteins having mutations at the palmitoylation acceptor sites. This observation is consistent with the fact that the transforming Ki-4B Ras protein is not palmitoylated. The activity responsible for palmitolyation uses palmitoyl-CoA as substrate (GUTIERREZ and MAGEE 1991). Little is known about the protein substrate requirements other than that farnesylation of the CaaX Cys is a prerequisite for palmitoylation (HANCOCK et al. 1989). The activity is membrane associated; however, it is not associated with the plasma membrane (GUTIERREZ and MAGEE 1991). Thus, two different Ras-modifying enzymes, the palmitoyl-CoA transferase and the carboxyl methyl transferase, appear to be associated with internal membrane compartments. GUTIERREZ and MAGEE (1991) have suggested that Ras may be modified and localized to an intracellular vesicle prior to plasma membrane localization.

This point raises some intriguing possibilities regarding Rap1a–Krev-1 function. Under experimental conditions in which Rap1a–Krev-1 is overexpressed, it can suppress oncogenic Ras-transforming activity. It has been suggested that Rap1a–Krev-1 may compete for critical proteins to which oncogenic Ras binds (see Chap. 23). However, Rap1a–Krev-1 is localized to an intracellular membrane fraction near the nucleus, perhaps in the Golgi complex (BERANGER et al. 1991). If Ras is processed in a vesicle compartment prior to plasma membrane association, then Rap1a–Krev-1 and Ras may share a common localization at some point, perhaps interacting with a common set of proteins. This common localization may be part of the suppression mechanism of Rap1a–Krev-1.

E. Conclusions

The posttranslational modifications observed for members of the Ras superfamily are rapidly being recognized as critical aspects of the structure and function of these proteins. These modifications are important not only for membrane localization, but also for appropriate targeting and interactions with other proteins. A detailed understanding of the enzymes which catalyze these reactions may allow the development of inhibitors which can be used as pharmacological probes. The possible application of FTase inhibitors as potential anticancer agents in *ras*-mediated tumors has been noted by many (for examples, see DER and COX 1991; GIBBS 1991; SINENSKY and LUTZ 1992). These probes may ultimately have other clinical applications as the roles of small GTPases in disease processes are elucidated.

References

Ashby MN, King DS, Rine J (1992) Endoproteolytic processing of a farnesylated peptide in vitro. Proc Natl Acad Sci USA 89:4613–4617

Beranger F, Goud B, Tavitian A, DeGunzburg J (1991) Association of the ras-antagonistic rapl/krev-1 proteins with the golgi complex. Proc Natl Acad Sci USA 88:1606–1610

Buss JE, Solski PA, Schaeffer JP, MacDonald MJ, Der CJ (1989) Activation of the cellular proto-oncogene product p21 ras by addition of a myristylation signal. Science 243:1600–1603

Casey PJ, Thissen JA, Moomaw JF (1991) Enzymatic modification of proteins with a geranylgeranyl isoprenoid. Proc Natl Acad Sci USA 88:8631–8635

Chen W-J, Andres DA, Goldstein JL, Brown MS (1991a) Cloning and expression of a cDNA encoding the α subunit of rat $p21^{ras}$ protein farnesyltransferase. Proc Natl Acad Sci USA 88:11368–11372

Chen W-J, Andres DA, Goldstein JL, Russell DW, Brown MS (1991b) cDNA cloning and expression of the peptide-binding β subunit of rat $p21^{ras}$ farnesyltransferase, the counterpart of yeast DPR1/RAM1. Cell 66:327–334

Cox AD, Hisaka MM, Buss JE, Der CJ (1992) Specific isoprenoid modification is required for function of normal, but not oncogenic, ras protein. Mol Cell Biol 12:2606–2615

Der CJ, Cox AD (1991) Isoprenoid modification and plasma membrane association: critical factors for ras oncogenicity. Cancer Cells 3:331–340

Farnsworth CC, Kawata M, Yoshida Y, Takai Y, Gelb MH, Glomset JA (1991) C-terminus of the small GTP-binding protein smg p25A contains two geranylgeranylated cysteine residues and a methyl ester. Proc Natl Acad Sci USA 88:6196–6200

Finegold AA, Johnson DI, Farnsworth CC, Gelb MH, Judd SR, Glomset JA, Tamanoi F (1991) Protein geranylgeranyltransferase of Saccharomyces cerevisiae is specific for Cys-Xaa-Xaa-Leu motif proteins and requires the CDC43 gene product but not the DPR1 gene product. Proc Natl Acad Sci USA 88:4448–4452

Gibbs JB (1991) ras C-terminal processing enzymes – new drug targets? Cell 65:1–4

Goldstein JL, Brown MS, Stradley SJ, Reiss Y, Gierasch LM (1991) Nonfarnesylated tetrapeptide inhibitors of protein farnesyltransferase. J Biol Chem 266:15575–15578

Gutierrez L, Magee AI (1991) Characterization of an acyltransferase acting on $p21^{N\text{-}ras}$ protein in a cell-free system. Biochim Biophys Acta 1078:147–154

Hancock JF, Magee AI, Childs JE, Marshall CJ (1989) All *ras* proteins are polyisoprenylated but only some are palmitoylated. Cell 57:1167–1177

Hancock JF, Cadwallader K, Marshall CJ (1991a) Methylation and proteolysis are essential for efficient membrane binding of prenylated $p21^{K\text{-}ras(B)}$. EMBO J 10:641–646

Hancock JF, Cadwallader K, Paterson H, Marshall CJ (1991b) A CAAX or a CAAL motif and a second signal are sufficient for plasma membrane targeting of ras proteins. EMBO J 10:4033–4039

He B, Chen P, Chen S-Y, Vancura KL, Michaelis S, Powers S (1991) RAM2, an essential gene of yeast, and RAM1 encode the two polypeptide components of the farnesyltransferase that prenylates a-factor and ras proteins. Proc Natl Acad Sci USA 88:11373–11377

Horiuchi H, Kawata M, Katayama M, Yoshida Y, Musha T, Ando S, Takai Y (1991) A novel prenyltransferase for a small GTP-binding protein having a C-terminal Cys-Ala-Cys structure. J Biol Chem 266:16981–16984

Hrycyna CA, Clarke S (1992) Maturation of isoprenylated proteins in Saccharomyces cerevisiae. J Biol Chem 267:10457–10464

Hrycyna CA, Sapperstein SK, Clarke S, Michaelis S (1991) The Saccharomyces cerevisiae STE14 gene encodes a methyltransferase that mediates C-terminal methylation of a-factor and RAS proteins. EMBO J 10:1699–1709

Joly A, Popjak G, Edwards PA (1991) In vitro identification of a soluble protein: geranylgeranyl transferase from rat tissues. J Biol Chem 266:13495–13498

Kato K, Cox AD, Hisaka MM, Graham SM, Buss JE, Der CJ (1992) Isoprenoid addition to ras protein is the critical modification for its membrane association and transforming activity. Proc Natl Acad Sci USA 89:6403–6407

Kinsella BT, Maltese WA (1992) rab GTP-binding proteins with three different carboxyl-terminal cysteine motifs are modified in vivo by 20-carbon isoprenoids. J Biol Chem 267:3940–3945

Kohl NE, Diehl RE, Schaber MD, Rands E, Soderman DD, He B, Moores SL, Pompliano DL, Ferro-Novick S, Powers S, Thomas KA, Gibbs JB (1991) Structural homology among mammalian and Saccharomyces cerevisiae isoprenyl-protein transferases. J Biol Chem 266:18884–18888

Lacal PM, Pennington CY, Lacal JC (1988) Transforming activity of ras proteins translocated to the plasma membrane by a myristoylation sequence from the src gene product. Oncogene 2:533–537

Ma Y-T, Rando RR (1992) A microsomal endoprotease that specifically cleaves isoprenylated peptides. Proc Natl Acad Sci USA 89:6275–6279

Moores SL, Schaber MD, Mosser SD, Rands E, O'Hara MB, Garsky VM, Marshall MS, Pompliano DL, Gibbs JB (1991) Sequence dependence of protein isoprenylation. J Biol Chem 266:14603–14610

Perez-Sala D, Tan EW, Canada FJ, Rando RR (1991) Methylation and demethylation reactions of guanine nucleotide-binding proteins of retinal rod outer segments. Proc Natl Acad Sci USA 88:3043–3046

Pompliano DL, Rands E, Schaber MD, Mosser SD, Anthony NJ, Gibbs JB (1992) Steady-state kinetic mechanism of Ras farnesyl: protein transferase. Biochemistry 31:3800–3807

Reiss Y, Goldstein JL, Seabra MC, Casey PJ, Brown MS (1990) Inhibition of purified $p21^{ras}$ farnesyl: protein transferase by cys-AAX tetrapeptides. Cell 62:81–88

Reiss Y, Seabra MC, Armstrong SA, Slaughter CA, Goldstein JL, Brown MS (1991) Nonidentical subunits of $p21^{H\text{-}ras}$ farnesyltransferase. J Biol Chem 266:10672–10677

Reiss Y, Brown MS, Goldstein JL (1992) Divalent cation and prenyl pyrophosphate specificities of the protein farnesyltransferase from rat brain, a zinc metalloenzyme. J Biol Chem 267:6403–6408

Schaber MD, O'Hara MB, Garsky VM, Mosser SD, Bergstrom JD, Moores SL, Marshall MS, Friedman PA, Dixon RAF, Gibbs JB (1990) Polyisoprenylation of ras in vitro by a farnesyl-protein transferase. J Biol Chem 265:14701–14704

Seabra MC, Reiss Y, Casey PJ, Brown MS, Goldstein JL (1991) Protein farnesyltransferase and geranylgeranyltransferase share a common α subunit. Cell 65:429–434

Seabra MC, Goldstein JL, Sudhof TC, Brown MS (1992) Rab geranylgeranyl transferase. J Biol Chem 267:14497–14503

Sinensky M, Lutz RJ (1992) The prenylation of proteins. Bioessays 14:25–31

Stephenson RC, Clarke S (1992) Characterization of a rat liver protein carboxyl methyltransferase involved in the maturation of proteins with the -CXXX C-terminal sequence motif. J Biol Chem 267:13314–13319

Volker C, Lane P, Kwee C, Johnson M, Stock J (1991a) A single activity carboxyl methylates both farnesyl and geranylgeranyl cysteine residues. FEBS Lett 295:189–194

Volker C, Miller RA, McCleary WR, Rao A, Poenie M, Backer JM, Stock JB (1991b) Effects of farnesylcysteine analogs on protein carboxyl methylation and signal transduction. J Biol Chem 266:21515–21522

Yokoyama K, Goodwin GW, Ghomashchi F, Glomset JA, Gelb MH (1991) A protein geranylgeranyltransferase from bovine brain: Implications for protein prenylation specificity. Proc Natl Acad Sci USA 88:5302–5306

CHAPTER 23

GTPase Activating Proteins

F. McCormick

A. Introduction

Small GTPases such as ras p21 have low intrinsic GTPase activity and depend on GTPase activating proteins (GAPs) to convent their active GTP-bound forms to their inactive GDP-bound counterparts (Bollag and McCormick 1991b). GAPs therefore appear to be major negative regulators of these GTPases (Fig. 1). In addition, we and others have proposed the possibility that GAP-mediated down-regulation may be coupled to signal output so that GAPs comprise part of small GTPase effector systems (Adari et al. 1988; Calés et al. 1988; McCormick 1989; Hall 1990). This proposal is based mainly on the fact that GAPs interact only with the GTP-bound forms of ras p21 proteins (a criterion for a ras effector), and that GAPs bind ras p21 proteins at or near the so-called effector-binding site. The hypothesis has gained support from the recent observation that certain effectors of heterotrimeric G-proteins are also GAPs for these proteins (Berstein et al. 1992; Arshavsky et al. 1992), but remains controversial for small GTPases

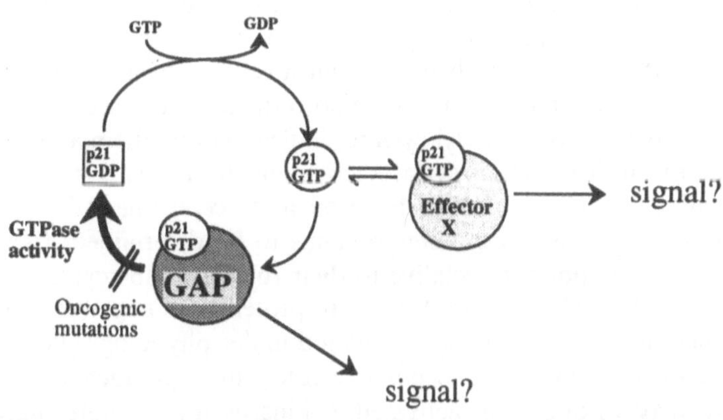

Fig. 1. Regulation of small GTPases by GTPase activating protein (*GAP*). Ras-like proeins cycle between inactive GDP states and active GTP states. Conversion of the GDP-bound to the GTP-bound form is mediated by GDP dissociation stimulators (GDSs), also referred to as guanine nucleotide releasing proteins (GNRPs) or GDP/GTP exchange factors. GTP-bound proteins interact with unknown effectors (*X*), some of which may be GAPs. Interaction with GAP promotes conversion to the inactive, GDP form. This step is blockd in oncogenic mutants

such as ras p21. In addition to the major issue of a possible role in effector functions for GAPs, a number of fundamental questions relating to their function have yet to be solved. For example, we have very little idea how, at the molecular level, GAPs stimulate GTPase activity or how GAP activities are regulated in cells. In this chapter, we will discuss known properties of GAPs for ras p21 and related proteins and speculate about their possible functions.

B. GTPase Activating Proteins for ras p21 Proteins

I. GTPase Activating Proteins in *Saccharomyces cerevisiae*

Saccharomyces cerevisiae contains at least two distinct genes for ras GAPs, referred to as *IRA1* and *IRA2* (for review, see Gibbs and Marshall 1989). Disruption of either gene results in elevated levels of RAS–GTP in growing yeast cells, indicating that the products of these genes normally share the task of maintaining Ras proteins in their inactive states (Tanaka et al. 1990a). The phenotype of cells defective in either *IRA* gene resembles that of cells containing an activated *RAS* gene, such as val-19 *RAS2*, in which mutant Ras proteins escape from *IRA* control and accumulate in their GTP states. This accumulation of RAS–GTP, whether from loss of *IRA* function or mutation in *RAS1* or *RAS2*, makes cells oblivious to sub-optimal or stressful growth conditions that normally provoke cell cycle arrest. As a result, stressed cells expressing excessive GTP-bound Ras proteins fail to accumulate normal levels of glycogen and trehalose, show reduced sporulation efficiency, and become sensitive to heat shock. These phenotypes can be attributed to overproduction of cAMP through constitutive activation of adenylyl cyclase, indicating that the major function of Ras proteins in these cells is cyclase activation. Ira proteins, then, are the major negative regulators of *RAS* function in *S. cerevisiae*. Whether they play additional roles in *S. cerevisiae* remains to be determined: such roles must be of minor importance relative to their roles as *RAS* regulators.

Genetic analysis clearly reveals that Ira proteins are negative regulators of *RAS* function, but how they are regulated under physiological conditions is not yet known. There are two possible roles: first, products of the *RAS* pathway (cAMP, products of activated A-kinases for example) may feed back to Ira proteins and thus regulate their activity. Such a feedback response would presumably increase IRA GAP activity, so that Ras proteins would be turned off. One prediction of this model is that cells expressing constitutively activated *RAS* mutants (such as val-19 *RAS1* or *RAS2*) would have higher GAP activity than normal cells. On the other hand, upstream signals might impinge on Ira proteins, inhibiting GAP activity and allowing Ras proteins to accumulate in their active GTP-states in a "feed-forward" signal. This model predicts a drop in GAP activity upon cellular activation.

Gurrently, arachidonic acid is the only known regulator of Ira proteins, this fatty acid inhibiting GAP activity of the Ira2 catalytic domain in vitro (GOLUBIC et al. 1991). The significance of this remains obscure since it is not clear whether arachidonic acid or its metabolites play a role in *S. cerevisiae* physiological regulation.

Both Ira1 and Ira2 are large proteins, encoding 3079 and 2938 amino acids, respectively (TANAKA et al. 1989, 1990b). The functions of amino acid sequences outside the GAP-related domain have not yet been fully investigated. The GAP-related domain itself is sufficient for Ras down-regulation and can even be replaced functionally by mammalian GAP or *NF1* (the gene responsible for the disease neurofibromatosis type I; BALLESTER et al. 1989; TANAKA et al. 1990a). Sequences outside the GAP-related domain could be involved in receiving signals from upstream or downstream elements or could have other functions not directly related to GTP hydrolysis. Identification of such functions is particularly important since these sequences are conserved in the human ira homolog neurofibromin, which plays a critical role in human pathologies, as described below. One possible additional function for Ira1 has been suggested by MITTS and coworkers, who provide evidence that the Ira1 protein is necessary for correct localization of adenylyl cyclase to the plasma membrane, so that the Ras regulator is intimately involved in effector function (MITTS et al. 1991). The exact significance of this intriguing report is not yet clear.

II. GTPase Activating Proteins in *Schizosaccharomyces pombe*

A GAP gene has been identified in the fission yeast *Schizosaccharomyces pombe* and is referred to as *sar1* (WANG et al. 1991) or *gap1* (IMAI et al. 1991). This GAP is a negative regulator of the *S. pombe ras* gene, *ras1*, which functions in the mating factor pathway in these cells. Loss of *sar1/gap1* gives a phenotype that is indistinguishable from that of activated *ras1*, showing clearly that *sar1/gap1* is not needed for *ras* effector function in these cells. The gene *sar1/gap1* encodes a protein of 766 amino acids, whose function can be replaced by human GAP or NF1. When expressed in *S. cerevisiae*, *sar1/gap1* can complement a defect in *ira1*. It therefore appears that with respect to RAS GTPase activation, all known GAPs are functionally interchangeable, albeit with varying efficiencies. These results are consistent with earlier observations that GAPs interact at the "effector-binding region" or Ras proteins that is perfectly conserved among yeast and mammalian species and, to some extent, defines small GTPase as a member of the ras family.

III. GTPase Activating Proteins in *Drosophila melanogaster*

Two types of GAPs have been identified in *Drosophila melanogaster*. One of these, *GAP1*, was identified genetically as a gene whose loss of function

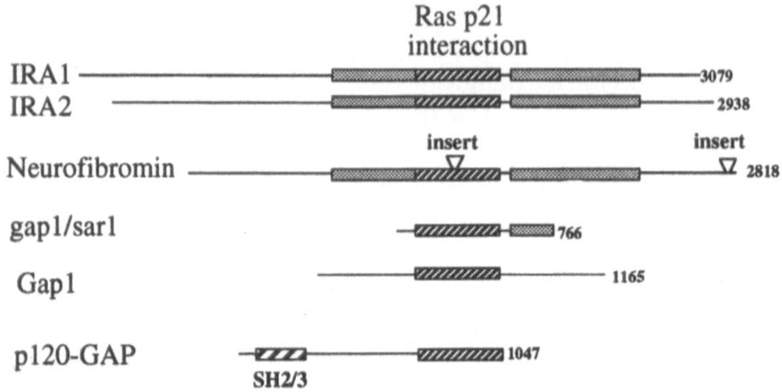

Fig. 2. Structures of GTPase activating proteins (GAPs) for ras p21 proteins

upregulates *ras* signaling in the sevenless pathway (Gaul et al. 1992). This pathway involves signaling from a tyrosine kinase receptor thorough *ras1*. The *GAP1* gene has homology in the GAP-related domain (Fig. 2), but otherwise does not resemble either p120-GAP or neurofibromin/IRA GAPs. The product of this gene appears to be a major negative regulator of *RAS1* in *D. melanogaster*, with no detectable effector function. As with GAPs from *S. cerevisiae*, loss of GAP function results in a phenotype that is indistinguishable from that of activated *ras* (i.e., a *ras* mutant with a GTPase-defective mutation such as valine 12). The second *D. melanogaster* GAP gene is the homolog of neurofibromin, cloned through sequence similarity to the human gene (Bernards, unpublished). The degree of similarity is striking: 70% amino acid sequence identity, but its function is not yet known. It will be of interest to determine whether its protein product is expressed in cells of the sevenless lineage and to determine whether it participates in this signaling pathway. Furthermore, we expect that *D. melanogaster* will express a gene homologous to mammalian p120-GAP, but this has yet to be identified.

IV. GTPase Activating Proteins in Mammalian Cells

Two types of GAP have been identified from mammalian cells (Fig. 2). The protein p120-GAP is the better characterized, having been the first GAP identified, purified, and cloned (Trahey and McCormick 1987; Trahey et al. 1988; Vogel et al. 1988). This protein has the interesting property of interacting with tyrosine phosphoproteins and with ras p21, suggesting a possible role in connecting signaling pathways involving tyrosine kinases with those involving ras. Such a connection has been known since the pioneering work of Stacey and coworkers showed that tyrosine kinase oncogenes and receptors depend on ras p21 for their activities (Mulcahy et al. 1985; Smith et al. 1986, 1989). This was demonstrated by injecting

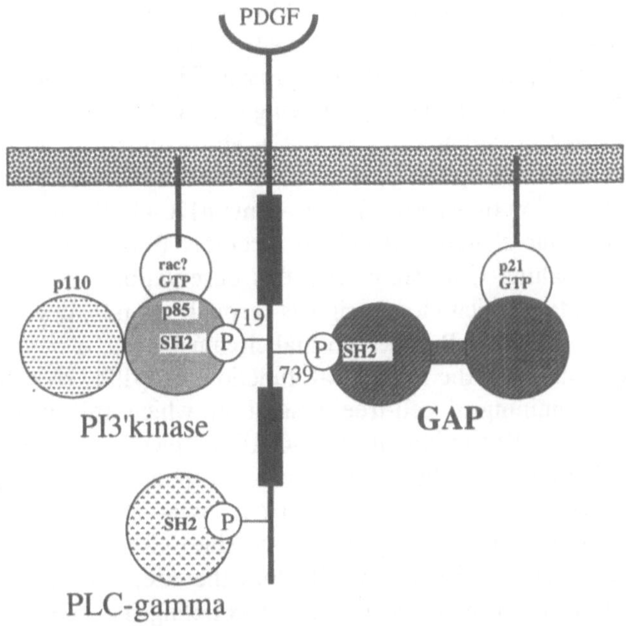

Fig. 3. Interaction of SH2-containing proteins with activated platelet-derived growth factor (*PDGF*) receptor. Activated receptors bind *PI3'kinase* and p120-GAP at distinct sites: the *numbers* used here are those of the mouse PDGF receptor (Fantl et al. 1992). The exact binding site for phospholipase C (*PLC*) is not known. GTPase activating protein (*GAP*) domains of p120-GAP and the p85 subunit of PI3′ kinase interact with ras p21 proteins and (probably) rac-like proteins respectively

anti-ras p21 antibodies into cells transformed by these kinases. In contrast, cells transformed by the serine kinases mos and raf are not affected by anti-ras p21 antibodies, suggesting a role downstream from ras. This epistasis has been confirmed and extended by Feig and coworkers using a dominant negative mutant of ras p21 that eliminates endogeous ras p21 function: this mutant prevents the action of tyrosine kinases, but not serine/threonine kinases (Stacey et al. 1992).

The suggestion that p120-GAP mediates signaling from tyrosine kinases to ras p21 gained support from the discovery that p120-GAP is itself a substrate for tyrosine phosphorylation (Molloy et al. 1989) and binds to certain activated tyrosine kinases, such as platelet-derived growth factor (PDGF) receptor (Kaplan et al. 1990; Kazlauskas et al. 1990), c-src (Brott et al. 1991; Park et al. 1992), Fyn, Lyn, and Yes (Cichowski et al. 1992), and, under some circumstances, insulin receptor (Pronk et al., submitted for publication). Currently, the possible role of p120-GAP in tyrosine kinase signaling is not clear. Signaling from PDGF receptor has been the most intensely studied system to date. Figure 3 shows the interaction between activated PDGF receptor, GAP, and other potential signaling molecules. SH2 domains of p120-GAP are necessary and sufficient components of p120-GAP for tight interaction. The receptor itself must be in an active state, phosphorylation of tyrosine 739 (in the mouse receptor sequence; 771 in the human receptor) being essential for p120-GAP binding (Fantl et al. 1992; Kashishian et al. 1992). However, the question remains as to whether binding of p120-GAP to the activated receptor plays any role in signaling. Mutation of tyrosine 739 prevents p120-GAP binding, but does not affect mitogenic signaling from the receptor. Binding is therefore not essential for signaling. Consistent with this observation, receptors such as the insulin receptor, which clearly signals through ras p21, does not appear to bind stably to p120-GAP under normal circumstances.

Some insights into the biological function of p120-GAP have been gained from examining a cell-free system in which this protein has a pronounced effect. Recombinant p120-GAP at picomolar concentrations inhibits the coupling of a heterotrimeric G-protein to its receptor in an excised patch of atrial cell membrane (Yatani et al. 1990). This effect appears to be dependent on ras p21 activity, since either antibodies against ras p21 or the ras antagonist rap1A p21 block the effects of p120-GAP. The region of p120-GAP responsible for this uncoupling has been localized to the SH2/3 region (Martin et al. 1992). This isolated domain uncouples the G-protein from its receptor, but, in contrast to full length p120-GAP, does so in a ras-independent manner. These observations led us to suggest a simple model for the role of ras p21 and p120-GAP in this sytem. According to this model, interaction of ras p21 with p120-GAP induces a conformational change that allows efficient interaction of the SH2/3 region with its target. The nature of the target is not yet known, but we speculate that a phosphorprotein component of the G-protein/receptor system is functionally inactivated by binding the p120-GAP SH2/3 region. The physiological significance of these results is unknown, but the model may apply to other p120-GAP–target interactions. For example, the model predicts that one role of ras p21 is to direct specific SH2/3 interactions in the cell, and that expression of the isolated SH2/3 domain may mimic some aspect of ras p21 function. This is indeed the case: transient expression of GAP SH2/3 domains activates transcription from the TPA-responsive element of the fos promoter, just as ras p21 activates this element (Medema et al. 1992). However, the isolated GAP domain does not replace ras p21 completely, since ras p21 function is still required for the effect. In support of this model, it has been reported recently that the C-terminal region of GAP can block ras-induced transactivation of the polyomavirus enhancer, whereas full-length GAP cannot. Furthermore, full-length GAP reversed the inhibitory effect of the C-terminal GAP domain, indicating a positive role for the full-length protein in signal transmission (Schweighoffer et al. 1992). Finally, biochemical examination of the properties of full-length GAP compared with the C-terminal domain reveal significant differences between these proteins, the C-terminal domain being 20 times less active than the full-length protein in stimulating ras p21 GTPase activity. This provides clear evidence for a physical interaction between GAP sequences outside the "GAP domain" at the C terminus and the domain itself, so that protein–protein interactions involving the SH regions of GAP, for example, are likely to be influenced by ras binding to the C terminus and vice versa (Gideon et al. 1992).

The discovery that p120-GAP contains SH2 domains raised the question of which cellular tyrosine phosphoproteins GAP interacts with. Clearly, PDGF receptor is one such protein, but, as we have discussed, the significance of this interaction remains unclear. Two other major tyrosine phosphoproteins are binded by p120-GAP in cells, designated p190 and p62 (Ellis et al. 1990). A substantial fraction of the p120-GAP in fibroblasts is bound to p190, whereas the p120-GAP–p62 complex accounts for a small proportion of the p120-GAP in cells (Moran et al. 1991). Both p190 and p62 have been purfied and cloned, and in both cases cloning revealed unexpected similarities to known proteins. The protein p190, as we will discuss below, has sequence similarity to the bcr protein and N-chimerin, both GAPs for rac p21 proteins (Settleman et al. 1992). The protein p62, on the other hand, showed significant similarity to a previously cloned putative hnRNP protein called GRP33 (Wong et al. 1992). This sequence similarity, as well as other indicative motifs (such as frequent RG pairs that are dimethylated at the arginine residues) suggest a role in some aspect of mRNA metabolism (such as splicing, transport, or stability).

In 1990, the gene responsible for the disease neurofibromatosis type I (von Recklinghausen's neurofibromatosis; NF1) was identified (Xu et al. 1990; Wallace et al. 1990). This gene had singificant homology to p120-GAP in the region known to be involved in ras p21 interaction (the C-terminal third of p120-GAP). Neurofibromin is similar in amino acid sequence to the yeast Ira1 and Ira2 proteins in regions that extend beyond the GAP-related domain. This suggests that these proteins all share functions additional to ras p21 GAP action. This function may be highly conserved during evolution: *Drosophila* neurofibromin, for example, is 70% identical to mammalian proteins.

These sequence similarities prompted an investigation of the biochemical properties of this domain (the GAP-related domain, GRD) of neurofibromin, and it was soon realized that the neurofibromin GRD does indeed have GAP activity against ras p21. The isolated domain expressed in insect cells or *Escherichia coli* stimulates ras p21 GTPase activity and, when expressed in *S. cerevisiae*, down-regulates RAS function (Xu et al. 1990; Ballester et al. 1990; Martin et al. 1990). The ras p21-binding properties of neurofibromin are summarized in Bollag and McCormick (1991a). Briefly, under in vitro assay conditions, neurofibromin exhibits a much higher affinity for ras proteins than does p120-GAP and is able to distinguish between N-ras p21 and H-ras p21. Furthermore, neurofibromin is much more sensitive to inhibition by lipids than p120-GAP. In addition, neurofibromin has the unexpected ability to bind tightly to tubulin, a property not shared by p120-GAP. While the biological significance of tubulin binding is not known, it is interesting that tubulin inhibits the GAP activity of neurofibromin, and its interaction with neurofibromin is inhibited by the antimitotic drug colchicine (Bollag et al., submitted for publication).

Neurofibromin may be regulated by differential splicing mechanisms. To date, two alternatively spliced variants of neurofibromin have been described (Marchuk et al. 1991; Suzuki et al. 1991). One of these results in an insertion of 21 amino acids in the GRD, and the other affects the carboxy terminus. Of these, the first is of more obvious interest. We have expressed both forms in insect cells and in *E. coli* and have not noted any dramatic differences in GAP activity between the two forms. However, a full biochemical investigation has not been completed. Neurofibromin type I (the mRNA species corresponding to the first published sequence, lacking the 21 amino acid insert) is expressed at higher levels than type II in brain, relative to brain-derived tumors (Suzuki et al. 1991), and the ratio of the two forms changes on differentiation (Nishi et al. 1991). These results are likely to provide valuable clues as to how neurofibromin interacts with ras p21 and regulates its function.

The disease neurofibromatosis type I appears to be caused by loss of one functional neurofibromin allele. The disease itself is characterized by abnormal growth of cells of neural crest origin, leading to "café-au-lait" spots on the skin (abnormal melanocytes), benign neurofibromas (abnormal Schwann cells, most likely), and other abnormalities. In affected cells, it appears that the normal neurofibromin allele is retained, suggesting that loss of half the normal complement of neurofibromin may be the basis of these phenotypic characteristics. Presumably, other mutational events are also necessary (second hits), but these have not been described and do not seem to be in the remaining neurofibromin gene. Reduced expression of neurofibromin could cause abnormal proliferation of neural cells by one of two mechanisms: first, less neurofibromin might lead to increases in ras p21 GTP levels, which, in turn, may drive uncontrolled growth. Second, neurofibromin might itself contribute directly to growth control, so that loss of function leads to loss of control, independent of effects on ras p21. In malignant tumors of neural crest origin that occur in NF1 patients at relatively high frequency, neurofibromin expression is lost altogether, possibly by a "second hit" in the remaining allele or by mutation of a gene that regulates neurofibromin expression. In cells derived from these tumors, ras p21 is indeed in the GTP state and contributes to the uncontrolled growth of these tumor cells. In these cells, ras p21 clearly acts through effectors that do not depend on neurofibromin. Interestingly, these cells express normal levels of p120-GAP, but, for unknown reasons, this form of GAP does not seem able to contribute to ras regulation in these cells.

Neurofibromin is ubiquitously expressed, and we anticipate a role for this protein in cells other than those of neural crest origin. In 3T3 fibroblasts, neurofibromin is the major form of GAP measured in cell extracts (BOLLAG and MCCORMICK 1991a), so that a role in ras regulation would seem likely. However, it has been shown that cells of this type contain sufficient p120-GAP to maintain ras p21 in its GDP state. A role for neurofibromin in epithelial cell growth has been implied from the recent observation that mutant neurofibromin alleles are detected as somatic mutants in certain human cancers (LI et al. 1992). Each of three mutations detected so far have been in the same codon of the neurofibromin gene (lysine-1423), and each results in loss of GAP activity from the altered protein. As usual, we can imagine two consequences of this mutation. First, ras p21–GTP might accumulate in these cells and contribute to abnormal growth. Second, the mutant protein might itself contribute to normal growth. The latter possibility is favored by the surprising restriction of "activating" mutations to a single codon on the very large gene: loss of function could be achieved by a large number of different mutations, deletions, rearrangements, and so on. There may well be a selection to retain the structure of the protein, while altering one aspect of its function, that is, ras regulation. The 1423 mutant proteins retain their ability to bind ras p21, even though no GTPase activation occurs, raising the possibility that a constitutive complex exists between wild-type ras p21–GTP and mutant neurofibromin. Such a complex might be functionally equivalent to the interaction between oncogenic *ras* mutants and wild-type neurofibromin, in which GTPase activation fails to occur because of the mutant ras protein rather than neurofibromin.

C. GTPase Activating Proteins for rap p21's

The "effector-binding region" of ras p21 has been defined as the region of ras p21 from amino acids 32–40: this is one of the regions of ras p21 with which p120-GAP interacts. This region of rap1 p21 is identical to that of ras p21, and p120-GAP does indeed bind to this region of rap1 p21 (FRECH et al. 1990). Unexpectedly, rap1 p21 bound to p120-GAP much more tightly than does ras p21, with a binding constant of 50 n*M*. Furthermore, rap1 p21 was completely unaffected in its GTPase activity by p120-GAP. This failure to stimulate rap1 GTPase is due, in part, to the fact that position 61 on rap1 p21 is a threonine residue; in all ras p21 proteins, position 61 is glutamine, and this glutamine is essential for intrinsic and GAP-mediated GTPase activity. Mutation of threonine-61 to glutamine allows p120-GAP to stimulate rap1 GTPase, albeit with an efficiency lower than ras p21 itself. Binding of p120-GAP to rap1 would account for its ability to revert cellular transformation, if indeed GAP is necessary for ras p21 function. Its tight binding to p120-GAP and failure to stimulate its GTPase make it attractive in this respect. If p120-GAP is not, in fact, necessary for ras p21 function, we expect that the true effector of ras function may well interact with rap1 p21 with these characteristics.

The discovery that p120-GAP fails to stimulate rap1 p21 GTPase raised the possibility that rap1 p21 has a GAP of its own, distinct from p120-GAP. An activity corresponding to rap1 GAP was indeed discovered, and the protein responsible for this activity was purified (Polakis et al. 1991) and cloned (Rubinfeld et al. 1991). Surprisingly rap1 GAP had no structural feature in common with ras p21 GAP, except that the domain responsible for GAP activity is similar in size to the ras p21 GAP domain. The protein rap1 GAP does contain motifs that suggest possible sites of regulation by protein kinases, and it is indeed modified by phosphorylation in vivo (with a substantial shift in electrophoretic migration) and by cAMP-dependent protein kinase and p34 cdc2 in vitro. Furthermore, the site of phosphorylation may be duplicated in an alternative splice of the gene. These modifications do not have detectable effects on rap1 GAP activity, but may well contribute to regulation within the cell. These possibilities are under investigation.

D. GTPase Activating Proteins for rho-Like Proteins

A GAP activity specific for rho proteins was first described by Hall and Garrett (Garrett et al. 1989). This activity, like that of ras p21 GAP, was shown to be sensitive to inhibition by fatty acids and lipids such as arachidonic acid (Tsai et al. 1989). A clone corresponding to rho-GAP is not available at the time of writing, but sufficient amino acid sequence has been obtained from rho GAP to suggest significant homology to the bcr protein, better known as part of the bcr/abl chimera that is encoded by the Philadelphia chromosome of chronic myelogenous leukemia (CML; Garrett et al. 1991). Based on this sequence similarity, the bcr protein was tested for GAP activity against a number of rho and rac proteins, and was indeed found to have GAP activity against rac1, but not against rho proteins themselves (Diekmann et al. 1991). The region of bcr responsible for ras GAP activity was localized to the C-terminal region of the protein, and consists of a domain similar in size to the equivalent domain of ras GAP molecules (Fig. 4). During genesis of the bcr/abl chimera, this domain is separated from the rest of the bcr protein, as shown in Fig. 4. Currently, we do not know whether loss of the rac GAP domain is necessary for the biological activity of bcr/abl protein. The possibility that the abl/bcr fusion that retains the rac GAP domain generated by the reciprocal chromosome translocation has biological activity itself has not been formally excluded, but it is known that the bcr/abl protein is sufficient to cause disease in transgenic mice, suggesting that the putative abl/bcr protein does not make a significant contribution to the development of CML.

N-chimerin is another protein with homology to rho GAP and bcr, and, indeed, this protein has GAP activity against racl, but not against rho proteins themselves (Diekmann et al. 1991). N-chimerin also has interesting

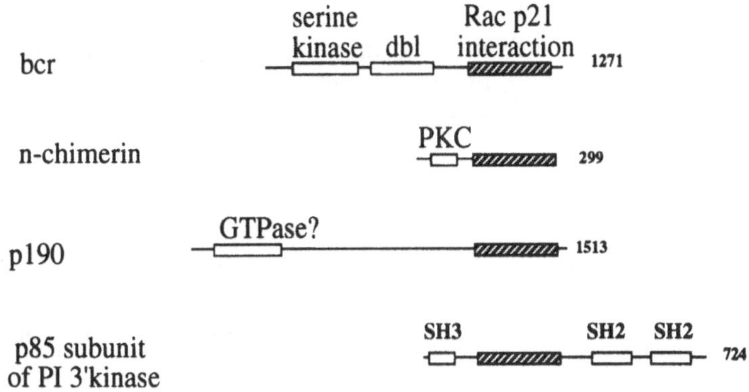

Fig. 4. Structures of GAPs for rho/rac proteins

sequence homology with the phorbolester-responsive regions of protein kinase C, suggesting that N-chimerin itself is regulated by diacylglycerol or related compounds (HALL et al. 1990).

Recently, the GAP-associated protein p190 was cloned and found to exhibit sequence similarity to the GAP domain of bcr and N-chimerin. Recombinant p190 has now been shown to possess GAP activity for rho, rac, and G25K proteins, a surprising range of specificities. Another member of the rho/rac GAP family has been identified by sequence comparison. This is the p85 subunit of PI 3′ kinase that contains SH2 and SH3 domains and associates with activated tyrosine kinase receptors, such as PDGF receptor. To date, no GAP activity has been ascribed to this protein.

E. GTPase Activating Proteins for Other Small GTPases

Small GTPase of the YPT/rab subgroup that are involved in protein trafficking have been shown to have GAPs, but molecular structures for these proteins are not yet available. However, there is great interest in determining their roles in cell physiology and, specifically, how they participate in the functions of rab proteins. Unlike ras p21 proteins, which are fully active when locked in their GTP states, rab proteins appear to need GTP hydrolysis to function, and therefore GAPs for these proteins are expected to be essential to their action (WALWORTH et al. 1992).

F. Concluding Remarks

A combination of biochemical, genetic, and molecular technologies have led to the identification of a class of proteins that regulate small GTPases by activating GTP hydrolysis. It seems likely that every small GTPase will be controlled by some form of GAP, with varying degrees of specificity. For example, p120-GAP stimulates GTPase of H-ras, K-ras, N-ras, and R-ras proteins and binds to rap1A p21. It has no effect on proteins of the rho/rac or rab sub-group. bcr and N-chimerin affect rac proteins but not rho, whereas the related p190 protein stimulates GTPase of all rac and rho proteins tested so far. To put these sorts of observation into a biological context, we need to resolve two major issues: first, are GAPs involved in signal transmission from small GTPases, and second, do they regulate small GTPases via feedback or "feed-forward" mechanisms? Do they, in other words, sense signals upstream or downstream from ras proteins? These issues may become clearer when pathways leading to and from small GTPases are better understood, so that the position of GAPs on these pathways may be clarified.

References

Adari H, Lowy DR, Willumsen BM, Der CJ, McCormick F (1988) Guanosine triphosphatase activating protein (GAP) interacts with the p21 ras effector binding domain. Science 240:518–521

Arshavsky VY, Bownds MD (1992) Regulation of deactiation of photoreceptor G protein by its target enzyme and cGMP. Nature 357:416–417

Ballester R, Marchuk D, Boguski M, Saulino A, Letcher R, Wigler M, Collins F (1990) The NF1 locus encodes a protein functionally related to mammalian GAP and yeast IRA proteins. Cell 63:851–859

Ballester R, Michaeli T, Ferguson K, Xu HP, McCormick F, Wigler M (1989) Genetic analysis of ammmalian GAP expressed in yeast. Cell 59:681–686

Bernstein G, Blank John DY, Exton JH, Rhee SG, Ross EM (1992) Phospholipase beta-1 is a GTPase Activating Protein for Gq/11, its physiological regulator. Cell 70:411–418

Bollag G, McCormick F (1991a) Differential regulation of rasGAP and neurofibromatosis gene product activities. Nature 351:576–579

Bollag G, McCormick F (1991b) Regulators and effectors of ras proteins. Annu Rev Cell Biol 7:601–632

Brott BK, Decker S, Shafer J, Gibbs JB, Jove R (1991) GTPase-activating protein interactions with the viral and cellular Src kinases. Proc Natl Acad Sci USA 88:755–759

Calés C, Hancock JF, Marshall CJ, Hall A (1988) The cytoplasmic protein GAP is implicated as the target for regulation by the ras gene product. Nature 332:548–551

Cichowski K, McCormick F, Brugge JS (1992) Three src-related protein tyrosine kinases, fyn, lyn, and yes coprecipitate with GAP in thrombin-activated platelets. J Biol Chem (in press)

Diekmann D, Brill S, Garett MD, Totty N, Hsuan J, Monfries C, Hall C, Lim L, Hall A (1991) Bcr encodes a GTPase-activating protein for $p21^{rac}$. Nature 351:400–402

Ellis C, Moran M, McCormick F, Pawson T (1990) Phosphorylation of GAP and GAP-associated proteins by transforming and mitogenic tyrosine kinases. Nature 343:377–381

Fantl WJ, Escobedo JA, Martin GA, Turck CW, del Rosario M, McCormick F, Williams LT (1992) Distinct phosphotyrosines on a growth factor receptor bind to specific molecules that mediate different signaling pathways. Cell 69:413–423

Frech M, John J, Pizon V, Chardin P, Tavitian A, Clark R, McCormick F, Wittinghofer A (1990) Inhibition of GTPase activating protein stimulation of Ras-p21 GTPase by the Krev-1 gene product. Science 249:169–171

Garrett MD, Self AJ, van Oers C, Hall A (1989) Identification of distinct cytoplasmic targets for ras/R-ras and rho regulatory proteins. J Biol Chem 264:10–13

Garrett MD, Major GN, Totty N, Hall A (1991) Purification and N-terminal sequence of the p21rho GTPase-activating protein, rho GAP. Biochem J 276 (pt3):833–836

Gaul U, Mardon G, Rubin GM (1992) A putative ras GTPase Activating Protein acts as a negative regulator of Signaling by the Sevenless receptor tyrosine kinase. Cell 68:1007–1019

Gibbs JB, Marshall MS (1989) The ras oncogene – an important regulatory element in lower eucaryotic organisms. Microbiol Rev 53:171–185

Gideon P, John J, Frech M, Lautwein A, Clark R, Scheffler JE, Wittinghofer A (1992) Mutational and kinetic analyses of the GTPase-activating protein (GAP)-p21 interaction: the C-terminal domain of GAP is not sufficient for full activity. Mol Cell Biol 12:2050–2056

Golubic M, Tanaka K, Dobrowolski S, Wood D, Tsai M-H, Marshall M, Tamanoi F, Stacey DW (1991) The GTPase stimulatory activities of the neurofibromatosis type 1 and the yeast IRA2 proteins are inhibited by arachidonic acid. EMBO J 10:2897–2903

Hall A (1990) ras and GAP – who's controlling whom? Cell 61:921–923

Hall C, Monfries C, Smith P, Lim HH, Ahmed RKS, Vanniasingham V, Leung T, Lim L (1990) Novel human brain cDNA encoding a 34000 Mr protein n-chimaerin, related to both the regulatory domain of protein kinase C and BCR, the product of the breakpoint cluster region gene. J Mol Biol 211:11–16

Imai Y, Miyake S, Hughes DA, Yamamoto M (1991) Identification of a GTPase-activating protein homolog in Schizosaccharomyces pombe. Mol Cell Biol 11:3088–3094

Kaplan DR, Morrison DK, Wong G, McCormick F, Williams LT (1990) PDGF beta-receptor stimulates tyrosine phosphorylation of GAP and association of GAP with a signaling complex. Cell 61:125–133

Kashishian A, Kazlauskas A, Cooper JA (1992) Phosphorylation sites in the PDGF receptor with different specificities for binding GAP and PI3 kinase in vivo. EMBO J 11:1373–1382

Kazlauskas A, Ellis C, Pawson T, Cooper JA (1990) Binding of GAP to activated PDGF receptors. Science 247:1578–1581

Li Y, Bollag G, Clark R, Stevens J, Conroy L, Fults D, Ward K, Friedman E, Samowitz W, Robertson M, Bradley P, McCormick F, White R, Cawthon R (1992) Somatic mutations in the neurofibromatosis 1 gene in human tumors. Cell 69:275–281

Marchuk DA, Saulino AM, Tavakkol R, Swaroop M, Wallace MR, Andersen LB, Mitchell AL, Gutmann DH, Boguski M, Collins FS (1991) cDNA cloning of the type 1 neurofibromatosis gene: complete sequence of the NF1 gene product. Genomics 11:931–940

Martin GA, Viskochil D, Bollag G, McCabe PC, Crosier WJ, Haubruck H, Conroy L, Clark R, O'Connell P, Cawthon RM, Innis MA, McCormick F (1990) The GAP-related domain of the neurofibromatosis type 1 gene product interacts with ras p21. Cell 63:843–849

Martin GA, Yatani A, Clark R, Conroy L, Polakis P, Brown AM, McCormick F (1992) GAP domains responsible for ras p21-dependent inhibition of muscarinic atrial K^+ channel currents. Science 255:192–194

McCormick F (1989) ras GTPase activating protein: signal transmitter and signal terminator. Cell 56:5–8

Medema R, de Laat W, Martin GA, McCormick F, Bos J (1992) GAP SH2-SH3 domains induce gene expression in a ras-dependent fashion. Mol Cell Biol (in press)

Mitts MR, Bradshaw-Rouse J, Heideman W (1991) Interactions between adenylate cyclase and the yeast GAP protein, IRA1. Science (in press)

Molloy CJ, Bottaro DP, Fleming TP, Marshall MS, Gibbs JB, Aaronson SA (1989) PDGF induction of tyrosine phosphorylation of GTPase activating protein. Nature 342:711–714

Moran MF, Polakis P, McCormick F, Pawson T, Ellis C (1991) Protein-tyrosine kinases regulate the phosphorylation, protein interactions, subcellular distribution, and activity of p21^{ras} GTPase-activating protein. Mol Cell Biol 11:1804–1812

Mulcahy LS, Smith MR, Stacey DW (1985) Requirements for ras proto-oncogene function during serum-stimulated growth of NIH3T3 cells. Nature 313:241–243

Nishi T, Lee PSY, Oka K, Levin VA, Tanase S, Morino Y, Saya H (1991) Differential expression of two types of the neurofibromatosis type 1 (NF1) gene transcripts related to neuronal differentiation. Oncogene 6:1555–1561

Park S, Marshall MS, Gibbs JB, Jove R (1992) Reconstitution of interactions between the Src tyrosine kinases and Ras GTPase-activating protein using a baculovirus expression system. J Biol Chem 267:11612–11618

Polakis PG, Rubinfeld B, Evans T, McCormick F (1991) Purification of a plasma membrane-associated GTPase-activating protein specific for rap1/K-rev-1 from HL60 cells. Proc Natl Acad Sci USA 88:239–243

Rubinfeld B, Munemitsu S, Clark R, Conroy L, Watt K, Crosier W, McCormick F, Polakis P (1991) Molecular cloning of a GTPase activating protein specific for the Krev-1 protein p21^{rap1}. Cell 65:1033–1044

Schweighoffer F, Bariat I, Chevallier-Multon M-C, Tocque B (1992) Implication of GAP in ras-dependent transactivation of a polyoma enhancer sequence. Science 256:825–827

Settleman J, Narasimhan V, Foster LC, Weinberg RA (1992) Molecular cloning of cDNAs encoding the GAP-associated protein p190: implications for a signaling pathway from Ras to nucleus. Cell 69:539–549

Smith MR, DeGudicibus SJ, Stacey DW (1986) Requirement for ras proteins during viral oncogene transformation. Nature 320:540–543

Smith MR, Ryu S-H, Suh P-G, Rhee SG, Kung H-F (1989) S-phase induction and transformation of quiescent NIH 3T3 cells by microinjection of phospholipase C. Proc Natl Acad Sci USA 86:3659–3663

Stacey DW, Feig LA, Gibbs JB (1991) Dominant inhibitor ras mutants selectively inhibit the activity of either cellular or oncogenic ras. Molecular and Cellular Biology 11:4053–4064

Suzuki Y, Suzuki H, Kayama T, Yoshimoto T, Shibahara S (1991) Brain tumors predominantly express the neurofibromatosis type 1 gene transcripts containing the 63 base insert in the region coding for GTPase activating protein related domain. Biochem Biophys Res Commun 181:955–961

Tanaka K, Matsumoto K, Toh-e A (1989) IRA1, an inhibitory regulator of the RAS-cyclic AMP pathway in Saccharomyces cerevisiae. Mol Cell Biol 9:757–768

Tanaka K, Nakafuku M, Satoh T, Marshall MS, Gibbs JB, Matsumoto K, Kaziro Y, Toh-e A (1990a) S. cerevisiae genes IRA1 and IRA2 encode proteins that may be functionally equivalent to mammalian ras GTPase activating protein. Cell 60:803–807

Tanaka K, Nakafuku M, Tamanoi F, Kaziro Y, Matsumoto K, Toh-e A (1990b) IRA2, a second gene of Saccharomyces cerevisiae that encodes a protein with a domain homologous to mammalian ras GTPase-activating protein. Mol Cell Biol 10:4303–4313

Trahey M, McCormick F (1987) A cytoplasmic protein stimulates normal N-ras p21 GTPase, but does not affect oncogenic mutants. Science 238:542–545

Trahey M, Wong G, Halenbeck R, Rubinfeld B, Martin GA, Ladner M, Long CM, Crosier WJ, Watt K, Koths K, McCormick F (1988) Molecular cloning of two types of GAP complementary DNA from human placenta. Science 242:1697–1700

Tsai M-H, Hall A, Stacey DW (1989) Inhibition by phospholipids of the interaction between Rras, rho, and their GTPase-activating proteins. Mol Cell Biol 9:5260–5264

Vogel US, Dixon RA, Schaber MD, Diehl RE, Marshall MS, Scolnick EM, Sigal IS, Gibbs JB (1988) Cloning of bovine GAP and its interaction with oncogenic ras p21. Nature 335:90–93

Wallace MR, Marchuk DA, Andersen LB, Letcher R, Odeh HM, Saulino AM, Fountain JW, Brereton A, Nicholson J, Mitchell AL, Brownstein BH, Collins FS (1990) Type 1 neurofibromatosis gene: identification of a large transcript disrupted in three NF1 patients. Science 249:181–186

Walworth NC, Brennwald P, Kabcenell AK, Garrett M, Novick P (1992) Hydrolysis of GTP by Sec4 protein plays an important role in vesicular transport and is stimulated by a GTPase-activating protein in Saccharomyces cerevisiae. Mol Cell Biol 12:2017–2028

Wang Y, Boguski M, Riggs M, Rodgers L, Wigler M (1991) sar1, a gene from Schizosaccharomyces pombe encoding a protein that regulates ras1. Cell Regul 2:453–465

Wong G, Muller O, Clark R, Conroy L, Moran MF, Polakis P, McCormick F (1992) Molecular cloning and nucleic acid binding properties of the GAP-associated tyrosine phosphoprotein p62. Cell 69:551–558

Xu GF, Lin B, Tanaka K, Dunn D, Wood D, Gesteland R, White R, Weiss R, Tamanoi F (1990) The catalytic domain of the neurofibromatosis type 1 gene product stimlates ras GTPase and complements ira mutants of S. cerevisiae. Cell 63:835–841

Yatani A, Okabe K, Polakis P, Halenbeck R, McCormick F, Brown AM (1990) ras p21 and GAP inhibit coupling of muscarinic receptors to atrial K^+ channels. Cell 61:769–776

CHAPTER 24

Guanine Nucleotide Dissociation Stimulators

I.G. MACARA and E.S. BURSTEIN

A. Introduction

Small Ras-like GTPases function as bipolar molecular switches to control a variety of cellular processes (BOURNE et al. 1991a,b). The switches cycle between GTP- and GDP-bound states. Evidence from oncogenic mutants of Ras identified the GTP-bound form as the active state, a conclusion that has been substantiated by the effects of dominant negative mutants and from observations that the Ras GTP to GDP ratio is increased following growth factor or cytokine stimulation of cells (e.g., STACEY et al. 1991; SATOH et al. 1991). Ras and other similar proteins possess slow intrinsic GTPase activities that can be catalytically augmented by GTPase activating proteins (GAPs) (TRAHEY and MCCORMICK 1987). The existence of a second catalytic factor which would convert Ras to the GTP-bound form was expected because of the very low k_{off} for GDP from the p21 Ras protein ($<0.01\,min^{-1}$) in the presence of physiologically relevant concentrations of Mg^{2+} (HALL and Self 1986; NEAL et al. 1988). This slow off-rate would lead to the irreversible accumulation of inactive Ras–GDP if no mechanism existed to catalyze conversion to the Ras–GTP state. Mutations which confer abnormally high nucleotide release rates can, however, abrogate the necessity for such a mechanism and activate the transforming potential of the Ras proteins (LACAL and AARONSON 1986; FEIG and COOPER 1988b).

B. Possible Mechanisms for Conversion to the GTP-Bound State

There are three mechanism by which protein-bound GDP might be converted to GTP (Fig. 1):

1) Direct exchange. An example is provided by the guanine nucleotide exchange factor (GEF) for eukaryotic initiation factor 2 (elF2). Both of these proteins are GTP-binding proteins and GEF appears to function by swapping its bound GTP for the GDP associated with elF2 (PANNIERS et al. 1988).

2) Phosphorylation of bound GDP. This mechanism remains controversial, except in the case of tubulin polymerization, in which GDP complexed

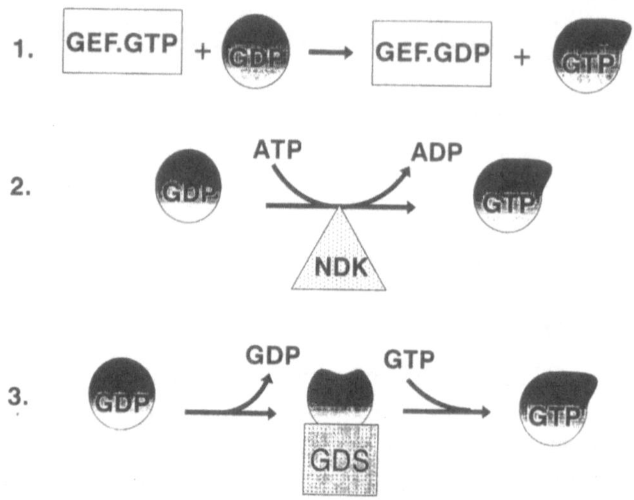

Fig. 1. Possible mechanisms for replacement of GDP by GTP on guanine nucleotide binding proteins. *1*. Direct exchange. The exchange factor is also a GTPase, and the bound GTP replaces the GDP on the target protein (*circle*). An example is the GEF for eukaryotic initation factor 2. *GEF*, guanine nucleotide exchange factor. *2*. Transphosphorylation. Nucleoside diphosphate kinase (*NDK triangle*) directly phosphorylates GDP which is bound to the target protein. Examples include tubulin and ARF (ADP ribosylation factor). *3*. Stimulation of guanine nucleotide dissociation. The factor stabilizes the empty state of the target protein, just as an enzyme stabilizes the transition state of its substrate. The k_{off} for GDP is catalyzed. GTP is more likely than GDP to rebind to the protein because of its higher concentration in cytosol (approximately 10:1). Examples are the seven-transmembrane domain receptor interactions with Gα subunits; and GDSs for several small GTPases

with tubulin is phosphorylated by nucleoside diphosphate kinase (NDK) (PENNINGROTH and KIRSCHNER 1977). NDK does copurify with a number of GTPases, and can be coupled functionally to these proteins in vitro, but it has been difficult to exclude local phosphorylation of free GDP by the NDK (KIKKAWA et al. 1991). Interest in NDK as a regulator of GTPases increased with the discovery that a gene called *nm32*, the expression of which correlates inversely with tumor metastasis, is identical to NDK (LEONE et al. 1991) and that the *awd* gene, which plays a role in *Drosophila* development, is also an NDK (BIGGS et al. 1990). Recently, RANDAZZO et al. (1991) have presented evidence that at least one small GTPase, ARF1, can be activated by NDK. But does NDK transphosphorylate GTPases in vivo? One problem with this hypothesis is the lack of specificity of NDK; the other is the difficulty in designing unambiguous tests that exclude local generation of free GTP within the intact cell.

3) GDP release. Catalysis of exchance by increasing the GDP k_{off} appears more general and is supported in some cases by compelling evidence.

Activation of heterotrimeric G-proteins is a consequence of accelerated release of GDP catalyzed by the seven-transmembrane domain receptors to which the G-proteins are coupled (GILMAN 1987). The high intracellular concentration of GTP then ensures that released nucleotide is replaced by GTP, rather than GDP. Factors that increase the k_{off} for guanine nucleotides and that operate on small GTPases have also been discovered. These factors are referred to as guanine nucleotide releasing factors or proteins (GRFs of GNRPs), or as guanine nucleotide dissociation stimulators (GDSs). It is with these factors that the remainder of this chapter will be concerned. One attractive hypothesis to explain acceleration of guanine nucleotide release is that a GDS stabilizes the apo-form of the GTPase, just as an enzyme stabilizes the transition state between substrate and product (Fig. 1). This hypothesis predicts that a GDS binds with higher affinity to the apo-protein than to guanine-nucleotide-associated protein. For several small GTPases the prediction has been validated directly, as will be discussed below. However, the molecular details of the interaction remain to be elucidated.

As described in Sect. II., Chap. 12, Vol. 108/I, the family of small GTPases includes more than 50 distinct members. Do specific GDS proteins exist for each? Or do multiple GTPases interact with a single type of GDS? Do different tissues express different species of GDS that interact with the same GTPase? How is GDS activity regulated? And do different GDS species interact with their target GTPases in the same way, through associations with cognate amino acid residues in the same region of the target protein? Is the mechanism for stimulation of k_{off} universal?

Most of these questions remain unanswered. However, several different GDS proteins have already been purified to homogeneity and/or cloned and sequenced. Within the next few years, much of the regulatory machinery for these small GTPases will be explained. This chapter will focus on a number of factors that have been described recently and which act on the Ras, Rab3A, and Ran proteins.

C. Nonspecific Guanine Nucleotide Dissociation Stimulators

Given the critical role small GTPases play in many essential cell processes, one might *a priori* expect a stringent specificity in the interaction of GDSs with their targets. However, two GDS proteins have been isolated that are surprisingly nonspecific.

The first, called Ras-guanine nucleotide exchange protein (rGEF), was isolated by WEST et al. (1990) from bovine brain membranes, but appears to be ubiquitous. It is active towards the oncogenic 12V Ras-mutant, and does not distinguish the GTP- and GDP-bound states of the protein. However, its effect was inhibited by the anti-Ras antibody Y13-259, which binds to loop 4 in the three-dimensional structure of Ras (see section II. 12–15 for the structure of Ras). The mass of rGEF was initially estimated by gel filtration to be about 100 kDa, but when purified to apparent homogeneity (HUANG et al. 1990) the factor eluted at 40 kDa on gel filtration, similar to its mass as estimated by sodium dodecyl sulfur polycrylamide gel electrophoresis (SDS-PAGE). The reason for this difference between the crude activity and the purified factor is unclear. Although rGEF had been purified using Ras as a substrate, the factor also catalyzes GDP–GTP exchange onto several other proteins, including Rap1A, R-Ras, Rho, and Rab1B (HUANG et al. 1990). Because these proteins show little sequence similarity outside the guanine nucleotide binding regions, rGEF presumably interacts with one of these conserved areas which is at least partially exposed on the surface of the folder polypeptide. The epitope for the Y13-259 antibody is specific to Ras and cannot, therefore, comprise a binding site for rGEF, but it may sterically hinder binding to nearby regions.

A second GDS has been purified and cloned by KAIBUCHI et al. (1990), which is active with a defined class of GTPases. This factor – discussed in detail in Sect. III., Chap. 39, Vol. 108/I – was first isolated as a GDS for the Rap1A (smg p21) protein. The smg p21 GDS cDNA encodes a 61-kDa protein, which stimulates release of GTPγS as well as GDP from the Rap1 proteins (YAMAMOTO et al. 1990). This GDS is active on c-Ki-Ras and RhoA proteins as well as on Rap1, but has no activity towards c-Ha-Ras, RhoB, or Rab3A proteins. The only common feature of the three substrates, absent in the unresponsive proteins, is a polybasic region in the hypervariable domain, immediately upstream of the C-terminal CaaX box (C, Cys; a, a usually aliphatic amino acid; X, another amino acid) (MIZUNO et al. 1991). In support of the idea that the GDS recognizes the C termini of its substrates, posttranslational processing of the CaaX box is essential for interaction with the GDS, and processed C-terminal peptides inhibit GDS function. Modified peptides in which the polybasic region is replaced by glutamate residues are tenfold less active (SHIRATAKI et al. 1991).

The Rap1–Ki-Ras–RhoA GDS forms equimolar complexes with the GDP-bound form of its GTPase substrates (MIZUNO et al. 1991), with an affinity for Rap1A of $2 \times 10^{-8} M$. This affinity is much higher than that of other known regulatory factors for small GTPases–guanine nucleotide complexes. The activity of the GDS in catalyzing GDP release is also smaller than that of other factors that have been examined to data and is not catalytic. It remains possible, therefore, that the role of the Rap1–Ki-Ras–RhoA GDS is related to the control of membrane association rather than to the exchange of GDP for GTP, and may be akin to the function of guanine nucleotide inhibitors (GDIs), which also form stable complexes complexes with the GDP-bound forms of small GTPases, and which can block their interaction with membranes (HORI et al. 1991; ARAKI et al. 1990).

D. Ras-Specific Guanine Nucleotide Dissociation Stimulators

I. Mammalian Guanine Nucleotide Dissociation Stimulators

A Ras-specific GDS was described in rat brain cytosolic extracts by WOLFMAN and MACARA (1990), and was called guanine nucleotide releasing factor (Ras-GRF). This activity is ubiquitously distributed, but is highest in brain (Fig. 2), a tissue which also expresses high levels of Ras. By gel filtration, the mass of GRF is shown to be 100–160 kDa. The dose response curve for GRF is sigmoidal, indicating that the interaction between Ras and GRF is highly cooperative. Unlike rGEF discussed above, the inhibitory antibody Y13-259 does not block GRF activity (WOLFMAN and MACARA, unpublished observations), suggesting that loop 4 is not involved in the interaction between GRF and Ras. GRF is active against recombinant Ha-Ras, which is not posttranslationally processed, and point mutations or insertion/deletions in the effector domain (loop 2) that interacts with GAP, or deletions of the C-terminal hypervariable domain, do not interfere with GRF action (STATHOPOLOUS, OKUHARA, LOWY, PAPAGEORGE and MACARA, unpublished observations). Interestingly, the dominant negative 17N mutant of Ras (Sect. III., Chap. 19, Vol. 108/I) competes for GRF more effectively than does the wild-type Ras (WOLFMAN, FEIG and MACARA, unpublished observations; Fig. 2B). This result is consistent with transfection experiments using 17NRas, which suggest that the 17N mutant sequesters an upstream activator, such as a GDS (FEIG and COOPER 1988a; STACEY et al. 1991), and it may indicate that the conformation of the 17N mutant resembles the apoform of Ras.

In other experiments, GRF was found to be equally effective against N-Ras and oncogenic Ha-Ras (V12) as against c-Ha-Ras, but Rap1B, R-Ras, Rab2, and Rab3A were unresponsive. Inspection of the regions of homology between these proteins (VALENCIA et al. 1991), together with information from the analysis of Ras mutants, suggests that the GRF might interact with loop 6 and/or loop 8 to effect a stimulation of guanine nucleotide release. A possible mechanism would involve disruption of the hydrogen bonds between the guanine base and D119 (loop 6) and/or A146 (loop 8). Mutation of these residues is known to increase dramatically the k_{fff} for guanine nucleotides from Ras (LACAL and AARONSON 1986; FEIG and COOPER 1988b); the conformation of the loops is not sensitive to the identity of the nucleotide (GDP or GTP) bound; and both loops are on the surface of the protein and accessible to interacting factors. Although D119 and A146 are highly conserved among the GTPase family members, the flanking sequences are variable and could provide the observed specificity for the interaction. The proximity of amino or carboxyl groups to GRF that could compete for hydrogen bond formation with D119 and A146 would stabilize

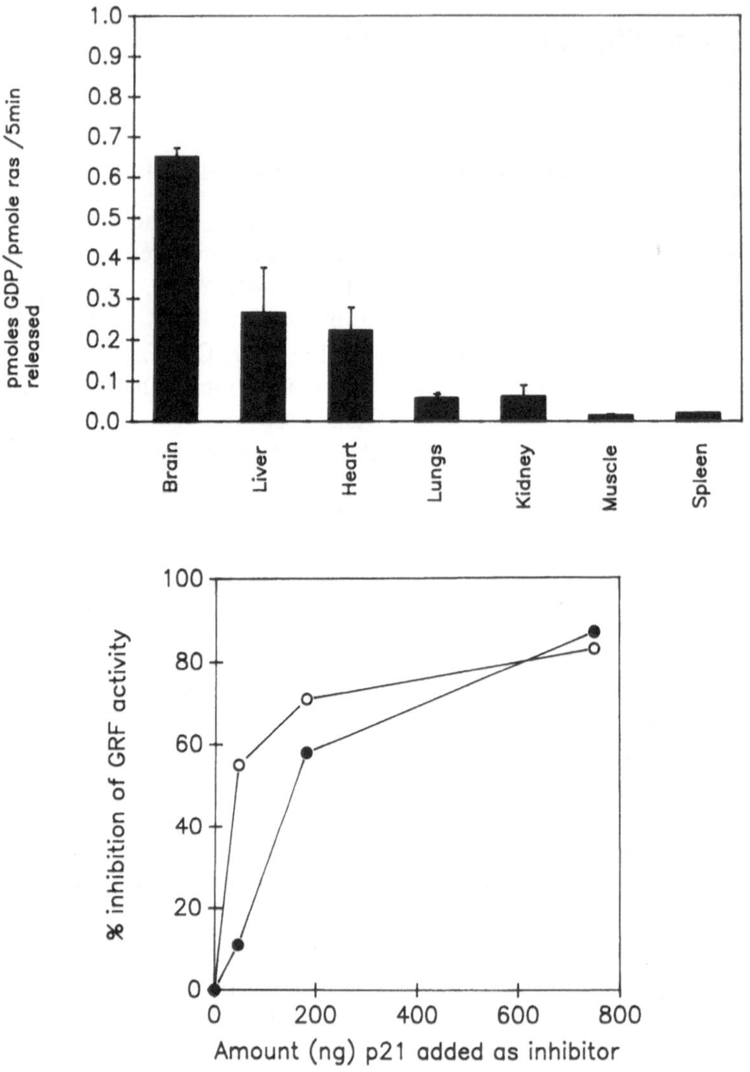

Fig. 2. **A** Distribution of Ras-GRF activity in different rat tissues. Cytosolic extracts were prepared and assayed for GRF activity as described in WOLFMAN and MACARA (1990), using equal protein concentrations for each tissue (15 mg/ml). **B** Competition assay for Ras-GRF. GDP release was assayed as described in WOLFMAN and MACARA (1990) using recombinant Ha-Ras prelabeled with [α-^{32}P]GDP, in the presence of increasing concentrations of either wild-type c-Ha-Ras or the 17N mutant of Ha-Ras (provided by L. Feig, Tufts University, Boston MA). The GRF was partially purified by ammonium sulfate precipitation from rat brain cytosol

the apo-form of Ras and provide increased affinity GRF and apo-Ras. Whether these predictions are borne out will, of course, require detailed analysis using purified GRF.

A second, mammalian ras-specific GDS has been reported by DOWNWARD et al. (1900a), and called ras exchange protein (REP). This activity was isolated from cytoplasmic extracts of human placenta, and gel filtration shows it to have a mass of about 60 kDa. Effector domain mutants of ras which were insensitive to stimulation by GAP retained partial sensitivity to REP. However, a deletion mutant of ras p21 lacking residues 165–183 was unresponsive to REP. This region corresponds to the hypervariable domain. In size and site of interaction, REP is therefore similar to the smg p21 GDS described by TAKAI's group, but REP function does not require posttranslational modification of the C-terminal Cys residue, and operates on both Ha-Ras and N-Ras. It is therefore most likely a different factor. Which determinants within the hypervariable domain REP interacts with remains puzzling, because there is little sequence similarity between these domains in N-Ras and Ha-Ras (BARBACID 1987). It would be interesting to test REP on other small GTPases, especially Ki-Ras which has a polybasic region in its carboxyl terminus.

One puzzling feature of all of the mammalian GDS factors identified to date – apart from the nonspecific rGEF – is that they are soluble, rather than membrane-bound. Yet evidence from studies of the 17N dominant-negative mutant of ras point to interactions with a membrane-bound component. Thus, double mutants of 17N that cannot be posttranslationally modified (186S) – and are therefore cytosolic – have a null phenotype (STACEY et al. 1991). One possible explanation is that the factors are present in large excess in the cytoplasm, and interact with a third, limiting component in the membrane that is required for stimulation of membrane-bound ras. Another possibility is that ras-GDS is normally membrane-bound, but is released into the soluble fraction during isolation. A membrane-associated GDS may exist in human placental tissues, sequestered in an inactive form, that can be released by trypsinization of placental membranes (DOWNWARD et al. 1990a) and the cytosolic, 60-kDa REP may therefore be a proteolytic fragment of a larger membrane-associated complex. Permeabilized T lymphocytes also contain a constitutively active GDS that may be membrane associated (DOWNWARD et al. 1990b).

Most of these issues are likely to be resolved with the cloning of mammalian ras-GDS factors. Several groups have obtained such clones, either by complementation of yeast mutants or by polymease chain reaction (PCR) amplification of sequences homologous to known GDS genes from yeast and *Drosophila*, and sequences are likely to be available in the near future [see Note Added in Proof].

II. Yeast Guanine Nucleotide Dissociation Stimulators: *CDC25*, *SCD25* and *ste6*

The yeast Ras pathway has been dissected with greater resolution than has yet been accomplished in mammalian systems. One example is the identification of the *S. cerevisiae* gene, *CDC25*, as an essential upstream regulator of Ras (BROEK et al. 1987; ROBINSON et al. 1987). The product of the *CDC25* gene is a 180-kDa protein that in overproducing strains is present mainly in the particulate fraction. Extracts from such strains exhibit increased GDP–GTP exchange activity towards Ras, and immunopurified Cdc25-β-galactosidase fusion protein also shows GDS activity (JONES et al. 1991).

A second gene, *SCD25*, the 3'-terminal end of which can suppress the requirement for *CDC25*, has been identified and shown to possess sequence similarity to *CDC25* (BOY-MARCOTTE et al. 1989). *SCD25* is not an essential gene in *S. cerevisiae*, and its function remains something of a mystery. However, the carboxy-terminal domain of *SCD25* can function as a GDS for Ras (CRECHET et al. 1990). It stimulates release of GDP from both yeast Ras2 and from Ha-Ras, and shows a marked preference for the GDP-bound forms of these proteins. However, it is not known whether this preference reflects a difference in affinity of the Scd25 between the GTP- and GDP-bound states (which would suggest that it recognizes a conformationally-sensitive domain of Ras such as loop 2 or loop 4), or can be explained simply because of the higher affinity of Ras for GTP. As for most of the other GDS factors that have been identified to date, no posttranslational processing of the Ras proteins is required for activity.

Little is yet known about the biochemistry of *ste6*, which is the fission yeast equivalent of *CDC25* (HUGHES et al. 1990). It most likely functions as a GDS because of significant sequence similarity to *CDC25* and *SCD25* in the C-terminal half of the gene, and because epistatic interactions place it upstream of Ras.

III. A Ras-Specific Guanine Nucleotide Dissociation Stimulator in *Drosophila*: SOS

Elegant genetic studies of compound eye development in *Drosophila* are providing new insight into signal transduction pathways that involve Ras. Indeed, it is likely that these studies, and similar work being undertaken on the development of the nematode, *C. elegans*, will reveal molecular details of the entire pathway from cell surface to the nucleus. The *Drosophila* compound eye is composed of repeating units, each of which contains eight photoreceptors and 12 accessory cells. A membrane-associated tyrosine kinase, Sevenless, controls the development of a single photoreceptor, R7. Genetic screens have identified a number of genes that interact with Sevenless, one of which is Ras, while another – Son of Sevenless (*SOS*) – is capable of encoding a 178-kDa protein with similarities to CDC25 and SCD25 (SIMON et al. 1991; BONFINI et al. 1992). Further studies have demonstrated that SOS operates upstream of Ras, and is highly likely therefore to represent a *Drosophila* Ras-GDS. It will be of interest to know whether SOS interacts directly with the Sevenless tyrosine kinase, or whether the interaction is mediated by adaptor proteins, such as those that contain SH2 and SH3 domains and can bind with high affinity to phosphorylated tyrosine residues (KOCH et al. 1991).

E. RAB3-Specific Guanine Nucleotide Dissociation Stimulator

Rab3A is a small GTPase with about 30% identity to Ras (TOUCHOT et al. 1987; MATSUI et al. 1988). It is expressed exclusively in exocrine secretory tissues such as neurones, exocrine pancreas, and adrenal medulla (MIZOGUCHI et al. 1989; DARCHEN et al. 1990), and associates with secretory granule membranes (MATTEOLI et al. 1991), although a significant fraction (about one third) is soluble (BURSTEIN and MACARA 1989). Rab3A may control the cycling of secretory vesicles to and from the plasma membrane, accompanied by a reversible dissociation of Rab3A from the vesicle membrane, as is discussed in Sect. III, Chap. 31, Vol. 108/I. The molecular mechanisms that control association and dissociation remain largely unknown. However, the C terminus of Rab3A is modified by the addition of two hydrophobic geranylgeranyl groups and a methyl group (FARNSWORTH et al. 1991). Transfer to the cytosol therefore requires that these groups be removed or be hidden from the aqueous phase. We have found (LINKO-STENTZ and MACARA, unpublished observation) that endogenous cytosolic Rab3 is both prenylated and methylated, and these groups must therefore be hidden. A candidate protein to accomplish this feat is the smg 25A guanine nucleotide inhibitor (GDI) (MATSUI et al. 1990). This factor associates specifically with the processed C-terminal domain of the GDP-bound form of Rab3, and can extract it from membranes into the aqueous phase (ARAKI et al. 1991). But this property raises a second problem: GDI is much more abundant than Rab3A, yet 70% of Rab3A is membrane associated in intact cells. However, GDI does not interact with the GTP-bound state of Rab3A. A cytosolic GDS could therefore function to convert the GDP–Rab3A–GDI complex to GTP–Rab3A. The GTP–Rab3A would then rebind to membranes. A Rab3A-specific GAP could catalyze hydrolysis of the Rab3A-bound GTP and complete the cycle. Such a hypothetical scheme is shown in Fig. 3.

A GDS specific for Rab3 called Rab3-GRF has recently been reported by BURSTEIN and MACARA (1992a), as well as a Rab3-GAP (BURSTEIN et al. 1991). Rab3-GRF was identified in rat brain cytosol. Based on sucrose density ultracentrifugation and gel filtration experiments, Rab3-GRF has a calculated molecular mass of 295 kDa, much larger than any other known GDS, but comparable to some of the GAP proteins, including Rab3-GAP. Rab3-GRF was in active towards the Ras and Rab2 proteins.

Uniquely, Rab3-GRF is specific for the GDP-bound form of Rab3A and does not catalyze release of bound GTP. However, when both the GRF and GAP are present, there is *apparent* GRF-activity towards GTP–Rab3A, because the GAP catalyzes hydrolysis of the bound GTP to GDP, which is then released by the GRF (BURSTEIN and MACARA, unpublished observations). Therefore it is possible, at least in vitro to demonstrate half of the cycle illustrated in Fig. 3. A GDS for the yeast Rab protein, Sec4p, has been cloned [see Note Added in Proof].

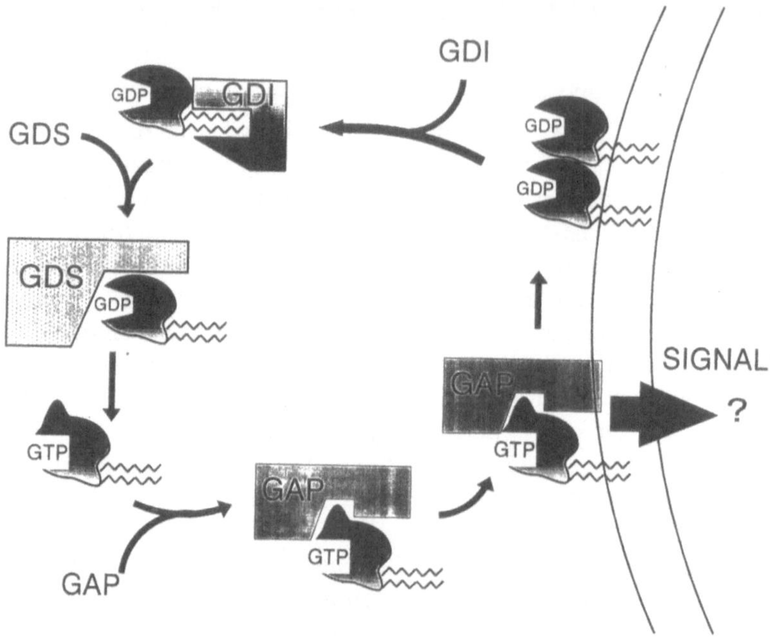

Fig. 3. Speculative model of the Rab3A cycle. GDP–Rab3A is extracted from synaptic granule membranes by cytosolic guanine nucleotide dissociation inhibitor (GDI) which associates with the prenylated C terminus of Rab3A. Cytosolic guanine nucleotide dissociation stimulator (*GDS*) then catalyzes exchange of GDP for GTP. GDI does not associate with the GTP-bound form of Rab3A and is therefore released. The exposed hydrophobic C terminus will rapidly associate with an adjacent membrane. Before or during membrane attachment, it interacts with GTPase activating proteins (*GAP*), which is distributed equally between membrane and particulate fractions. GAP catalyzes the hydrolysis of bound GTP to GDP, completing the cycle

F. Other Guanine Nucleotide Dissociation Stimulators

Two GTPases likely to occupy center stage over the next few years are Cdc42 and TC4, both of which interact with GDSs that play key roles in cell division. Cdc42 is a member of the Rac/Rho family of small GTPases, and in *S. cerevisiae* it controls budding orientation (JOHNSON and PRINGLE 1990). Its role in mammalian cells remains unknown, but a clue was provided by the discovery that the oncogene dbl can function as a Cdc42-specific GDS (HART et al. 1991). Dbl is a cytosolic protein without sequence similarity to other known oncogenes (EVA et al. 1988), but it may be the prototype of a family of GDS factors for Rac/Rho proteins.

TC4 identified as a cDNA with homology to Ras, but with an unusual C terminus lacking Cys residues (DRIVAS et al. 1990). Its function remained unknown until the discovery of a gene in fission yeast, *spi1*, that controls chromosome condensation and which is 80% similar in sequence to TC4

(MATSUMOTO and BEACH 1991). *Spil* interacts with another gene, *piml*, mutations in which lead to premature initiation of mitosis. *Piml* shows significant sequence similarity to RCC1, a mammalian 47-kDa nuclear protein that also regulates chromosome condensation (OHTSUBO et al. 1989). RCC1 has been isolated as a complex with an abundant protein of 25 kDa (BISCHOFF and PONSTINGL 1991b). When sequenced, peptides from this protein were virtually identical to those predicted by the TC4 sequence (BISCHOFF et al. 1990). The protein was called Ran (Ras-like nuclear protein).

Subsequent biochemical studies demonstrated that Ran is a GTPase, and that RCC1 catalyzes guanine nucleotide exchange on Ran (BISCHOFF and PONSTINGL 1991a). In the absence of RCC1, the rate of release of GDP from Ran is extremely slow ($k_{off} < 0.005\,min^{-1}$). RCC1 does not distinguish between GTP- and GDP-bound Ran, but is specific for Ran over Ras. RCC1 binds GDP–Ran complexes with quite high affinity (about 300 n*M*) compared to most other known GDSs but remarkably, in the absence of guanine nucleotides, RCC1 and apo-Ran form such a tight complex that it remains undissociated even in 3M guanidinium chloride. This observation supports the model described in Fig. 1, in which GDSs were proposed to work by stabilizing the empty "transition state" of GTPases.

Somehow, perhaps by decoration of the entire length of the chromosomal DNA at regular intervals, Ran and RCC1 are able to monitor the extent of DNA synthesis, and inhibit the synthesis of a mitotic inducer until it is complete, when RCC1 is presumably inactivated and Ran converts to the GDP-bound form. Studies with *spil* suggest that if insufficient Ran/*spil* protein is present in the cell, partial premature chromosome condensation occurs before DNA synthesis is complete. It will be of great interest to determine whether chromosomal instability in mammalian cells, for instance in malignant tumors, is a result of defects in Ran expression (MATSUMOTO and BEACH 1991).

G. Conclusions

Factors have been described which catalyze guanine nucleotide dissociation from several of the small GTPases. However, it is certain that scores of GDS factors remain to be discovered: there are at least 50, possibly 100, members of the Ras superfamily, and for each member there may be several species of GDS, as appears to be the case for Ras itself. Almost nothing is yet known about the regulation of any of the GDSs. No pattern has yet emerged as to their physical characteristics, but it is likely that common domains will be discovered as more GDSs are cloned. For instance, the Cdc25 catalytic domain shows significant homology not only to other putative Ras GDSs, but also to the smg p21 GDS described above (KAIBUCHI et al. 1990), to the smg p25A (Rab3A) GDI (MATSUI et al. 1990), to BUD5, which is involved in budding of *S. cerevisiae* (CHANT et al. 1991;

POWERS et al. 1991), and to JC310, a mammalian sequence of unknown function that interferes with Ras function (COLICELLI et al. 1991). Similarly it is likely that multiple *dbl*-related genes will be identified. One attractive possibility is that small GTPases such as Rho, Cdc42p and Ran function downstream of Ras, and that the targets or effects for Ras will interact in turn with these other GTPases. Cross talk between different GTPases is also likely to occur because of the promiscuity of some regulatory factors, although the in vivo significance of cross-regulation remains to be established. Overall, it is likely that the next few years will witness the cloning and characterization of a large number of GDS genes. The focus will then switch to the more complex question of how each GDS is regulated.

Acknowledgements. The research work of the authors described in this chapter was supported by NIH grants CA38888, CA56300, and CA43551. E. Burstein was supported by a pre-doctoral Toxicology Training grant, NIEHS. We thank Doug Johnson for reading and improving this review.

References

Araki S, Kikuchi A, Hata Y, Isomura M, Takai Y (1990) Regulation of reversible binding of *smg* p25A, a *ras* p21-like GTP-binding protein, to synaptic plasma membranes and vesicles by its specific regulatory protein, GDP dissociation inhibitor. J Biol Chem 265:13007–13015

Araki S, Kaibuchi K, Sasaki T, Hata Y, Takai Y (1991) Role of the C-terminal region of *smg* p25A in its interaction with membranes and the GDP/GTP exchange protein. Mol Cell Biol 11:1438–1447

Barbacid M (1987) Ras genes. Annu Rev Biochem 56:779–827

Biggs J, Hersperger E, Steeg PS, Liotta LA, Shearn A (1990) A Drosophila gene that is homologous to a mammalian gene associated with tumor metastasis codes for a nucleoside diphosphate kinase. Cell 63:933–940

Bischoff FR, Ponstingl H (1991a) Catalysis of guanine nucleotide exchange on Ran by the mitotic regulator RCC1. Nature 354:80–82

Bischoff FR, Ponstingl H (1991b) Mitotic regulator protein RCC1 is complexed with a nuclear *ras*-related polypeptide. Proc Natl Acad Sci USA 88:10830–10834

Bischoff FR, Maier G, Tilz G, Ponstingl H (1990) A 47-kDa human nuclear protein recognized by antikinetochore autoimmune sera is homologous with the protein encoded by RCC1, a gene implicated in onset of chromosome condensation. Proc Natl Acad Sci USA 87:8617–8621

Bonfini L, Karlovich CA, Dasgupta C, Banerjee U (1992) The Son of sevenless gene product: a putative activator of Ras. Science 255:603–605

Bourne HR, Sanders DA, McCormick F (1991a) The GTPase superfamily: a conserved switch for diverse cell functions. Nature 348:125–132

Bourne HR, Sanders DA, McCormick F (1991b) The GTPase superfamily: conserved structure and molecular mechanism. Nature 349:117–127

Boy-Marcotte E, Damak F, Camonis J, Garreau H, Jacquet M (1989) The C-terminal part of a gene partially homologous to CDC25 gene suppresses the cdc25-5 mutation in Saccharomyces cerevisiae. Gene 77:21–30

Broek D, Toda T, Michaeli T, Levin L, Birchmeier C, Zoller M, Powers S, Wigler M (1987) The S. cerevisiae CDC25 gene product regulates the RAS/adenylate cyclase pathway. Cell 48:789–799

Burstein ES, Macara IG (1989) The *ras*-like protein $p25^{rab3A}$ is partially cytosolic and is expressed only in nueral tissue. Mol Cell Biol 9:4807–4811

Burstein ES, Macara IG (1992a) Characterization of a guanine nucleotide-releasing factor and a GTPase-activating protein that are specific for the *ras*-related protein $p25^{rab3A}$. Proc Natl Acad Sci USA 89:1154–1158

Burstein ES, Macara IG (1992b) Interactions of the *ras*-like protein, $p25^{rab3A}$, with Mg^{2+} and Guanine Nucleotides. Biochem J 282:387–392

Burstein ES, Linko-Stentz K, Lu Z, Macara IG (1991) Regulation of the GTPase Activity of the *ras*-like Protein $p25^{rab3A}$: Evidence for a *rab3A*-specific GAP. J Biol Chem 266:2689–2692

Chant J, Corrado K, Pringle JR, Herskowitz I (1991) Yeast BUD5, encoding a putative GDP-GTP exchange factor, is necessary for bud site selection and interacts with bud formation gene BGM1. Cell 65:1213–1224

Colicelli J, Nicoletter C, Birchmeier C, Rodgers L, Riggs M, Wigler M (1991) Expression of three mammalian cDNAs that interfere with RAS function in Saccharomyces cerevisiae. Proc Natl Acad Sci USA 88:2913–2917

Crechet J-B, Poullet P, Mistou M-Y, Parmeggiani A, Camonis J, Boy-Marcotte E, Damak F, Jacquet M (1990) Enhancement of the GDP-GTP exchange of RAS proteins by the carboxyl-terminal domain of SCD25. Science 248:866–868

Darchen F, Zahraoui A, Hammel F, Monteils M-P, Tavitian A, Scherman D (1990) Association of the GTP-binding protein Rab3A with bovine adrenal chromaffin granules. Proc Natl Acad Sci USA 87:5692–5696

Downward J, Graves JD, Warne PH, Rayter S, Cantrell DA (1990a) Stimulation of $p21^{ras}$ upon T-cell activation. Nature 346:719–723

Downward J, Riehl R, Wu L, Weinberg RA (1990b) Identification of a nucleotide exchange-promoting activity for $p21^{ras}$. Proc Natl Acad Sci USA 87:5998–6002

Drivas GT, Shih A, Coutavas E, Rush MG, D'Eustachio P (1990) Characterization of four novel ras-like genes expressed in a human teratocarcinoma cell line. Mol Cell Biol 10:1793–1798

Eva A, Vecchio G, Rao CD, Tronick SR, Aaronson SA (1988) The predicted DBL oncogene product defines a distinct class of transforming proteins. Proc Natl Acad Sci USA 85:2061–2065

Farnsworth CC, Kawata M, Yoshida Y, Takai Y, Gelb MH, Glomset JA (1991) C terminus of the small GTP-binding protein smg p25A contains two geranylgeranylated cysteine residues and a methyl ester. Proc Natl Acad Sci USA 88:6196–6200

Feig LA, Cooper GM (1988a) Inhibition of NIH 3T3 cell proliferation by a mutant *ras* protein with preferential affinity for GDP. Mol Cell Biol 8:3235–3243

Feig LA, Cooper GM (1988b) Relationship among guanine nucleotide exchange, GTP hydrolysis, and transforming potential of mutated *ras* proteins. Mol Cell Biol 8:2472–2478

Gilman A (1987) G-proteins: transducers of receptor-generated signals. Annu Rev Biochem 56:615–649

Hall A, Self AJ (1986) The effect of Mg^{2+} on the guanine nucleotide exchange rate of $p21^{N\text{-}ras}$. J Biol Chem 261:10963–10965

Hart MJ, Eva A, Evans T, Aaronson SA, Cerione RA (1991) Catalysis of guanine nucleotide exchange on the CDC42Hs protein by the *dbl* oncogene product. Nature 354:311–314

Hori Y, Kikuchi A, Isomura M, Katayama M, Miura Y, Fujioka H, Kaibuchi K, Takai Y (1991) Post-translational modifications of the C-terminal region of the *rho* protein are important for its interaction with membranes and the stimulatory and inhibitory GDP/GTP exchange proteins. Oncogene 6:515–522

Huang Y, Kung H, Kamata T (1990) Purification of a factor capable of stimulating the guanine nucleotide exchange reaction of *ras* proteins and its effect of *ras*-related small molecular mass G proteins. Proc Natl Acad Sci USA 87:8008–8012

Hughes DA, Fukui Y, Yamamoto M (1990) Homologous activators of ras in fission and budding yeast. Nature 344:355–357

Johnson DI, Pringle JR (1990) Molecular characterization of CDC42, a Saccharomyces cerevisiae gene involved in the development of cell polarity. J Cell Biol 111:143–152

Jones S, Vignais M-L, Broach JR (1991) The CDC25 protein of Saccharomyces cervisiae promotes exchange of guanine nucleotides bound to Ras. Mol Cell Biol 11:2641–2646

Kaibuchi K, Mizuno T, Fujioka H, Yamamoto T, Kishi K, Fukumoto Y, Hori Takai Y (1990) Molecular cloning of the cDNA for stimulatory GDP/GTP exchange protein (GDS) for *smg* p21s, *ras* p21-like GTP-binding proteins, and characterization of GDS. J Biol Chem 265:16626–16634

Kikkawa S, Takahoshi K, Shimada N, Ui M, Kimura N, Katada T (1991) Conversion of GDP into GTP by nucleoside diphosphate kinase on the GTP-binding proteins [Erratum]. J Biol Chem 266:12795

Koch CA, Anderson D, Moran MF, Ellis C, Pawson T (1991) SH2 and SH3 domains: elements that control interactions of cytoplasmic signaling proteins. Science 252:668–674

Lacal JC, Aaronson SA (1986) Activation of *ras* p21 transforming properties associated with an increase in the release rate of bound guanine nucleotide. Mol Cell Biol 6:4212–4220

Leone A, Flatow U, King CR, Sandeen MA, Margulies IM, Liotta LA, Steeg PS (1991) Reduced tumor incidence, metastatic potential and cytokine responsiveness of nm23-transfected melanoma cells. Cell 65:25–35

Matteoli M, Takei K, Cameron R, Hurlbut P, Johnston PA, Sudhof TC, Jahn R, De Camilli P (1991) Association of Rab3A with synaptic vesicles at late stages of the secretory pathway. J Cell Biol 115:625–633

Matsui Y, Kikuchi A, Kondo J, Hishida T, Teranishi Y, Takai Y (1988) Nucleotide and deduced amino acid sequences of a GTP-binding protein family with molecular weights of 25 000 from bovine brain. J Biol Chem 263:11071–11074

Matsui Y, Kikuchi A, Araki S, Hata Y, Kondo J, Teranishi Y, Takai Y (1990) Molecular cloning and characterization of a novel type of regulatory protein (GDI) for smg p25A, a *ras* p21-like GTP-binding protein. Mol Cell Biol 10:4116–4122

Matsumoto T, Beach D (1991) Premature initiation of mitosis in yeast lacking RCC1 or an interacting GTPase. Cell 66:347–360

Mizoguchi A, Kim S, Ueda T, Takai Y (1989) Tissue distribution of smg p25A, a ras p21-like GTP-binding protein, studied by use of a specific monoclonal antibody. Biochem Biophys Res Commun 162:1438–1445

Mizuno T, Kaibuchi K, Yamamoto T, Kawamura M, Sakoda T, Fujioka H, Matsuura, Takai Y (1991) A stimulatory GDP/GTP exchange protein for smg p21 is active on the post-translationally processed form of c-Ki-*ras* p21 and *rhoA* p21. Proc Natl Acad Sci USA 88:6442–6446

Neal SE, Eccleston JF, Hall A, Webb MR (1988) Kinetic analysis of the hydrolysis of GTP by $p21^{N\text{-}ras}$: the basal GTPase mechanism. J Biol Chem 263:19718–19722

Ohtsubo M, Okazaki H, Nishimoto T (1989) The RCC1 protein, a regulator for the onset of chromosome condensation locates in the nucleus and binds to DNA. J Cell Biol 109:1389–1397

Panniers R, Rowlands AG, Henshaw EC (1988) The effect of Mg^{2+} and guanine nucleotide exchange factor on the binding of guanine nucleotides to eukaryotic to eukaryotic initiation factor 2. J Biol Chem 263:5518–5525

Penningroth SM, Kirschner MW (1977) Nucleotide binding and phosphorylation in microtubule assembly in vitro. J Mol Biol 115:643–673

Powers S, Gonzales E, Christensen T, Cubert J, Broek D (1991) Functional cloning of BUD5, a CDC25-related gene from S. cerevisiae that can suppress a dominant-negative RAS2 mutant. Cell 65:1225–1231

Randazzo PA, Northup JK, Kahn RA (1991) Activation of a small GTP-binding protein by nucleoside diphosphate kinase. Science 254:850–853

Robinson LC, Gibbs JB, Marshall MS, Sigal IS, Tatchell K (1987) CDC25: a component of the RAS-adenylate cyclase pathway in Saccharomyces cerevisiae. Science 235:1218–1221

Shirataki H, Kaibuchi K, Hiroyoshi M, Isomura M, Araki S, Sasaki T, Takai Y (1991) Inhibition of the action of the stimulatory GDP/GTP exchange protein for smg p21 by the geranylgeranylated synthetic peptides designed from its c-terminal region. J Biol Chem 266:20672–20677

Simon MA, Bowtell DDL, Dodson GS, Laverty TR, Rubin GM (1991) Ras1 and a putative guanine nucleotide exchange factor crucial steps in signaling by the sevenless protein tyrosine kinase. Cell 67:701–716

Stacey DW, Feig LA, Gibbs JB (1991) Dominant inhibitory Ras mutants selectively inhibit the activity of either cellular or oncogenic Ras. Mol Cell Biol 11:4053–4064

Touchot N, Chardin P, Tavitian A (1987) Four additional members of the ras gene superfamily isolated by an oligonucleotide strategy: molecular cloning of ypt-related cDNAs from a rat brain library. Proc Natl Acad Sci USA 84:8210–8214

Trahey M, McCormick F (1987) A cytoplasmic protein stimulates normal N-ras p21 GTPase, but does not affect oncogenic mutants. Science 238:542–545

Valencia A, Chardin P, Wittinghofer A, Sander C (1991) The *ras* protein family: evolutionary tree and role of conserved amino acids. Biochemistry 30:4637–4648

West M, Kung H, Kamata T (1990) A novel membrane factor stimulates guanine nucleotide exchange reaction of *ras* proteins. FFBS Lett 259:245–248

Wolfman A, Macara IG (1990) A cytosolic protein catalyzes the release of GDP from $p21^{ras}$. Science 248:67–69

Yamamoto T, Kaibuchi K, Mizuno T, Hiroyoshi M, Shirataki H, Takai Y (1990) Purification and characterization from bovine brain cytosol of proteins that regulate the GDP/GTP exchange reaction of *smg* p21s, *ras* p21-like GTP-binding roteins. J Biol Chem 265:16626–16634

Note Added in Proof

Since the submission of this review, rapid advances have occurred in our understanding of Ras signalling. Several Ras-specific GDS's have been cloned [SHOU et al. Nature 358:351–353 (1992); BOWTELL et al. Proc Natl Acad Sci USA 89:6511–6515 (1992); MARTEGANI et al. EMBOJ 11:2151–2157 (1992); CHARDIN et al. Science 260:1338–1343 (1993)]. The mechanisms of activation of Ras by several growth factors and by *src* have been elucidated [BUDAY and DOWNWARD Cell 73:611–620 (1993); SIMON et al. Cell 73:169–177 (1993); CHARDIN et al. Science 260:1338–1343 (1993); EGAN et al. Nature 363:45–51 (1993); ROZAKIS-ADCOCK et al. Nature 363: 83–85 (1993); LI et al. Nature 363:85–88 (1993); GALE et al. Nature 363:88–92 (1993); BALTENSPERGER et al. Science 260:1950–1952 (1993); SKOLNICK et al. Science 260:1953–1955 (1993)]. In hematopoietic cells it has been reported that the proto-oncogene VAV can also act as a Ras-specific GDS [GULBINS et al. Science 260: 822–825 (1993)].

Other exchange factors have also recently been cloned, including one for Ral [ALBRIGHT et al. EMBOJ 12:339–347 (1993)] and one for Sec4p [MOYA et al. Nature 361:460–463 (1993)].

CHAPTER 25

The Biology of Rap

G.M. Bokoch

A. Introduction

The Rap proteins are members of the Ras superfamily of GTPases. Two Rap families, designated Rap1 and Rap2, have been identified to date. Within each family are two members denoted as A and B, i.e., Rap1A and Rap1B (See Fig. 1). These small GTPases share 50% sequence homology with Ras and are expressed in a wide variety of tissues. As is the case with the majority of small GTPases, the actual role(s) of the Rap proteins in cellular function have not yet been elucidated. There is evidence, however, that Rap proteins exert biological activities in at least two cellular arenas – that of cellular growth and differentiation control, and in a phagocyte-specific enzyme system responsible for the generation of microbicidal oxygen radicals. (For a discussion of the possible roles of Rap in platelet function, see Chap. 38) The latter suggests that Rap can either play very specific biological roles in certain cells or, alternatively, that it serves a very general function which can involve different proteins and different effector systems depending on the cell involved. In this chapter we will examine current knowledge of the biology of the Rap proteins.

B. Cloning/Isolation of Rap(s)

The Rap proteins were cloned and/or purified by a number of laboratories within a very short period of time using a variety of strategies. Pizon et al.

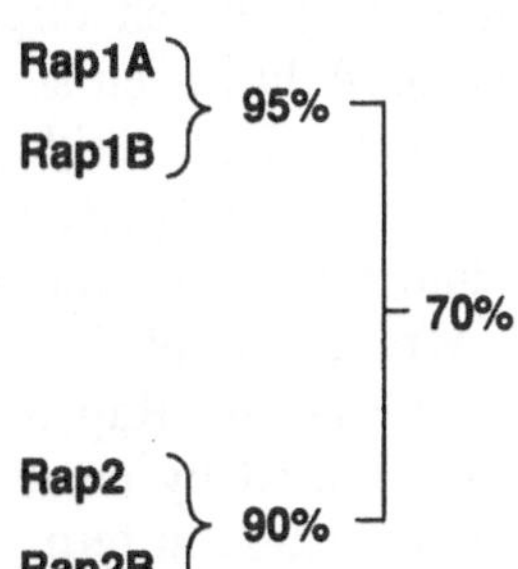

Fig. 1. The Rap proteins: four Rap subtypes have been identified. The percentage of sequence homology at the amino acid level between the indicated proteins is shown

(1988a) identified the *Rap1A* and *Rap2A* genes by screening a Raji human Burkitt lymphoma cell library using probes based upon the Dras3 protein previously identified in *Drosophila* (Schejter and Shilo 1985). This *Drosophila* Ras-related protein differed from the other members of the Ras superfamily known at that time in that it possessed a threonine for glutamine substitution at residue 61 of the highly conserved DTAGQE sequence found in positions 57–62 of the Ras proteins. Very soon afterward Pizon et al. (1988b) identified Rap1B using the same strategy. Rap1A (termed Krev-1) was isolated by Kitayama et al. (1989) from a human fibroblast cDNA expression library based on its ability to cause reversion of the transformed phenotype of v-Ki-Ras-transformed DT fibroblasts. Kawata et al. (1988) purified a 22-kDa GTPase (termed smg p21) from bovine brain, sequenced, and subsequently cloned the protein, which was identical to Rap1B. Takai's laboratory has subsequently purified Rap1B from human platelets (Ohmori et al. 1989) and bovine aortic smooth muscle (Kawata et al. 1989a). Rap1A has also been purified, sequenced, and cloned from human neutrophils (Bokoch et al. 1988; Quilliam et al. 1990). The identification of a fourth member of the Rap family, Rap2B, was achieved through molecular cloning (Ohmstede et al. 1990).

The Rap1 and Rap2 proteins are 70% homologous at the amino acid level (Fig. 1). Rap1A and 2B differ by only nine out of 184 amino acids (95% identity), with the only region of substantial nonsimilarity coming at positions 171 to 189 of the carboxyl terminus. Similarly, Rap2 and 2B differ by 18 out of 183 amino acids (90% identity), with the major area of divergence at residues 170–182 of the carboxyl terminus.

C. Posttranslational Modification of Rap Proteins

I. Isoprenylation

Like many of the other members of the Ras superfamily, the Rap proteins contain a COOH terminal CAAX consensus motif (C, Cys; A, an aliphatic amino acid; X, another amino acid), which directs posttranslational isoprenylation (Maltese 1990). Rap1A has been shown to be geranylgeranylated at Cys-181, with the loss of the LLL residues at the C terminus and subsequent carboxymethylation also taking place (Buss et al. 1991). Rap1B and Rap2, ending in CLQL and CVIL, appear to undergo similar modifications (Kawata et al. 1990; Winegar et al. 1991). Carboxymethylation of Rap1B (Huzoor-Akbar et al. 1991) and Rap1A (Quilliam and Bokoch 1992) has been found to be stimulated by guanosine 5′-(3-0-thio)triphosphate; the significance of this observation in terms of Rap modification and function in vivo is unknown.

Isoprenylated proteins terminating in leucine have been shown to be geranylgeranylated, while those ending in serine or methionine are farnesylated (Moores et al. 1991). It is of interest that Rap2B, which terminates in CVIL, is geranylgeranylated while the very closely related Rap2A, which terminates in CNIQ, is farnesylated (Farrell et al. 1992). These two structurally similar forms of Rap may thus partition into different membranes within cells, perhaps reflecting distinct biochemical activities.

II. Phosphorylation

After the purification of Rap1A and Rap1B, these two proteins were rapidly found to serve as substrates for phosphorylation by cAMP-dependent protein kinase in vitro (Bokoch and Quilliam 1990; Quilliam et al. 1991; Hoshijima et al. 1988; Kawata et al. 1989b; Lerosey et al. 1991). Both Rap proteins incorporate PO_4 to a level of 1 mol per mole, suggesting a single site of incorporation. Phosphorylation is not influenced by whether Rap is bound to GDP vs. GTP, and stoichiometric phosphorylation of Rap1A or 1B in vitro has no effect on the guanine nucleotide binding or hydrolysis properties of the two proteins, nor their responsiveness to Rap GAP (GTPase activating protein; Bokoch and Quilliam 1990; Hoshijima et al. 1988).

The site at which phosphate is incorporated has been shown to be Ser-180 in Rap1A (Quilliam et al. 1991). This is contained within the sequence KKKPKKK*S*C, which is similar to consensus cAMP-dependent protein kinase phosphorylation sites (Kemp and Pearson 1990). Rap1B is phosphorylated solely on Ser-179 within the sequence GKARKK*S*SC, again similar but not identical to "classical" cAMP-dependent kinase motifs (Hata et al. 1991). Both phosphorylation sites are adjacent to the Cys residue which is modified by the geranylgeranyl group. Possible influences of the phosphorylation on posttranslational processing of either Rap protein remain to be investigated. The Rap2 proteins lack potential consensus motifs for phosphorylation near the carboxyl terminal and do not serve as substrates for phosphorylation by cAMP-dependent or any known protein kinases.

Phosphorylation of Rap1A has been shown to occur in intact HL-60 cells which had been differentiated into neutrophil-like cells in response to dibutyryl cAMP, forskolin, PGE_1, or isoproternol (Quilliam et al. 1991). Phosphorylation of Rap1B occurs in human platelets in response to PGE_1 or the prostacyclin analog iloprost (Kawata et al. 1989b; Lapetina et al. 1989; Siess et al. 1990). A form of Rap1 has also been found to be phosphorylated in intact fibroblasts in response to 8-bromo cAMP (Lerosey et al. 1991). The possibility that Rap may mediate some of the cellular effects of cAMP will be considered later in this chapter (also see Chap. 38).

The Rap proteins clearly do not serve as substrates for protein kinase C, myosin light chain kinase, and insulin/epidermal growth factor (EGF) receptor tyrosine kinases in vitro (Bokoch and Quilliam 1990; Quilliam et al. 1991; Hoshijima et al. 1988; Kawata et al. 1989b). Recently, however,

Rap1B has been reported to be phosphorylated by cGMP-dependent protein kinase (Miura et al. 1991) and a neuronal calcium/calmodulin-dependent protein kinase (Sahyoun et al. 1991) in vitro (see Chap. 38). Both kinases appeared to phosphorylate at the same serine that is phosphorylated by cAMP-dependent kinase. Phosphorylation of Rap by such kinases has not been demonstrated in vivo.

D. Rap1 Regulatory Proteins

I. GTPase Activating Proteins

Several GAP activities that appear to be specific for Rap1 have been detected in the plasma membrane and cytosol of a number of cell types. Polakis et al. (1991) purified a membrane-associated 88-kDa GAP from HL-60 cells. This GAP was subsequently cloned and shown to be a unique protein which did not exhibit homology to any of the GAPs specific for Ras (Rubinfeld et al. 1991). This form of Rap GAP was not ubiquitously expressed and was most abundant in fetal tissues and certain tumor cell lines. Interestingly, the expression of the 88-kDa Rap GAP was decreased in HL-60 cells which had been differentiated with dimethylsulfoxide (DMSO). Characteristics of this Rap GAP are discussed in Chap. 23.

Two chromatographically resolvable peaks of Rap GAP activity have been observed in human platelets (Ueda et al. 1989) and bovine brain (Kikuchi et al. 1989) cytosol. One such cytosolic GAP was purified (Nice et al. 1992) as a 55-kDa protein. Limited amino acid sequence information indicates that this 55-kDa GAP is very closely related, if not identical, to the 88-kDa GAP purified by Polakis et al. It is thus not clear whether there are one or more forms of Rap GAP and what the significance of the differential subcellular localization of this protein might be. The changes in Rap GAP expression upon HL-60 cell differentiation and the identification of multiple, phosphorylated forms of this GAP (Polakis et al. 1992) suggest that Rap GAP may be regulated in a very specific manner in order to control Rap activity.

II. GDP/GTP Dissociation Stimulator

A guanine nucleotide exchange protein which is active on Rap1 has been identified and purified from bovine brain cytosol (Yamamoto et al. 1990). This 53-kDa protein has been cloned (Kaibuchi et al. 1991) and shown to be a unique protein with limited amino acid sequence homology to the CDC25 and SCD25 proteins which may regulate the GDP–GTP exchange reaction of the yeast Ras2 protein. Rap GDS appears to interact with Rap1A and 1B as a 1:1 stoichiometric complex (Kawamura et al. 1991a). This interaction involves, at least in part, the COOH terminal portion of Rap1, as indicated by: (a) the requirement for the posttranslationally processed form of Rap1 for guanine nucleotide dissociation stimulator (GDS) binding and activity (Hiroyoshi et al. 1991); (b) the demonstration that proteolytic removal of an approximately 1000-Da fragment of the COOH terminal prevents binding and activation of Rap1 by GDS (Hiroyoshi et al. 1991); (c) geranylgeranylated synthetic peptides representing the COOH terminus of Rap1 inhibit GDS action (Shirataki et al. 1991); and (d) phosphorylation of Rap1B at the COOH terminus enhances the interaction of Rap1B with GDS (Hata et al. 1991; Itoh et al. 1991). It is of interest that this "Rap1 GDS" is also able to catalyze GDP–GTP exchange for other posttranslationally processed small GTPases, including K-Ras, Rho A, and Rac1 (Mizuno et al. 1991). These proteins have in common a lysine/arginine-rich anionic region at their C terminus which may enable them all to interact with the GDS.

A number of anion phospholipids have been reported to antagonize the GDP–GTP exchange activity of Rap GDS, and can markedly reduce the ability of GDS to stimulate Rap1B GTP binding (Kawamura et al. 1991b). The effect of these lipids was reduced when Rap1B was phosphorylated by cAMP-dependent protein kinase (Itoh et al. 1991). Takai and associates (Itoh et al. 1991; Kawamura et al. 1991b) have suggested that Rap1B may bind to anionic lipids in the plasma membrane through the polycationic C-terminal domain, suppressing GDS action. Phosphorylation of Rap1B by cAMP-dependent protein kinase would decrease the ionic interaction of the lipids with Rap1B in this region, sensitizing Rap1B to the action of the GDS. GDS binding to Rap1B is associated with the release of Rap1B from the membrane as a GDS complex (Kawamura et al. 1991a), and this may occur in vivo as a consequence of Rap1B phosphorylation (Lapetina et al. 1989). Further details of the regulation of Rap1B subcellular localization by phosphorylation are presented in Chap. 38.

E. Biological Activities of Rap1 Protein

I. Antagonism of Ras by Rap1

The isolation of Rap1A by Noda and colleagues (Kitayama et al. 1989) as a cDNA which was able to suppress transformation of NIH/3T3 cells by v-Ki-Ras suggested that the Rap1A protein might directly antagonize Ras by competing for a common target or regulatory protein. Indeed, such a mechanism was postulated based upon the conservation in Rap1A and B of the putative "effector" domain region (amino acids 32–44) found to be crucial for GAP binding to Ras (Adari et al. 1988). It is possible, however, that Rap might antagonize Ras action by other means, such as directly activating a pathway which regulates cell growth and/or differentiation in a negative manner, or by indirect means, such as activation of other enzymes

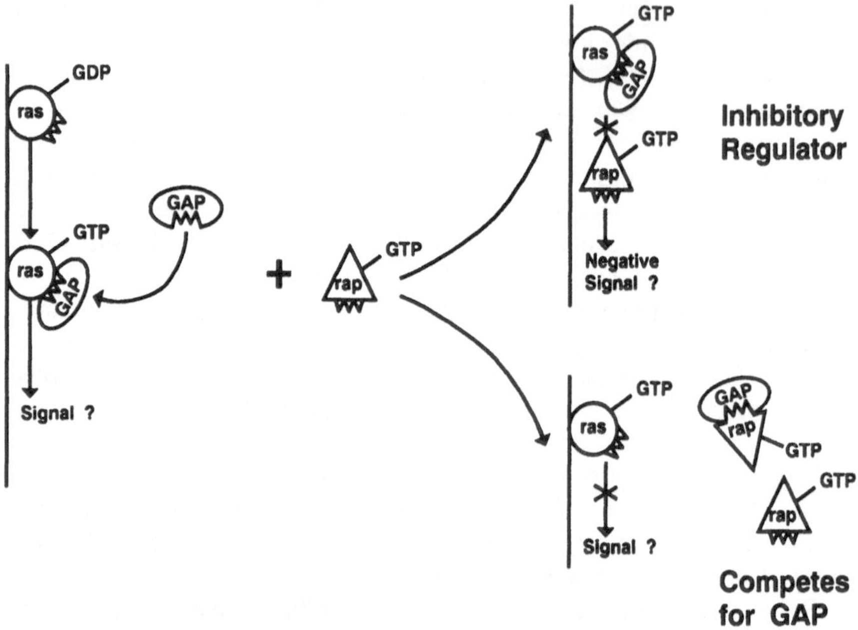

Fig. 2. Hypothetical mechanisms by which Rap might antagonize Ras downstream signaling. ʌʌʌ denotes the "effector" domain of Ras/Rap. GAP refers to Ras GAP

(kinases, Labadia et al. (1992); phosphatases, etc.) able to inhibit Ras growth signals (see Fig. 2).

In support of the "competition" hypothesis, Kitayama et al. (1990a) showed that an Asp 38 → Ala or Asn point mutation in the effector domain of Rap1A markedly inhibited the ability of Rap1A to cause phenotypic reversion of Ras-transformed cells. Mutations at position 12 (Gly → Val) and at position 59 (Ala → Thr), which are thought to activate Rap1 by maintaining it in the GTP-bound form (by analogy with Ras), substantially enhanced its reversion-inducing activity. The latter observation suggested that RapA was more "active" in the GTP-bound form.

Evidence that such a competitive model of Rap action was possible came from studies by Hata et al. (1990) and Frech et al. (1990). Both groups used purified components in vitro to demonstrate that Rap1A (Frech et al. 1990) or 1B (Hata et al. 1990) was able to compete with Ras for binding to Ras GAP. Binding to Ras GAP was more effective when Rap1 was bound with GTP rather than GDP, with Frech et al. reporting that the GDP form of Rap1A had an affinity of at least 100-fold less than the GTP form, while Hata et al. only reported a two- to threefold difference between the two forms of Rap1B. Ras GAP did not stimulate GTP hydrolysis by Rap1, as reported by Quilliam et al. 1990, suggesting that the Rap1 complex with GAP might be an unproductive one. Interestingly, the affinity of Rap1A GTP for Ras GAP was 50- to 100-fold greater than that of Ras GTP. Since the concentrations of Rap1 in cells such as platelets (OHMORI et al. 1989) and neutrophils (QUILLIAM et al. 1991) seems to be more than ten times that of Ras, Rap1A might be able to limit the amount of GAP available for interaction with Ras in vitro.

The ability of Rap1A to antagonize Ras action by an apparently competitive mechanism in an intact membrane system was demonstrated by YATANI et al. (1991). This study utilized the M_2-muscarinic receptor-regulated K^+ channel of atrial membranes, which is inhibited by the action of Ras and Ras GAP acting in concert (YATANI et al. 1990). Using a patch-clamp technique to directly manipulate the levels of Rap1A, Ras, and GAP in the system, it was possible to show that: (a) Rap1A antagonized the effect of Ras GAP on channel opening in a manner that was inversely proportional to the level of GAP added; (b) antagonism was dependent on an intact effector domain in Rap1A; (c) the inhibitory effect of Rap1A could be overcome by the addition of exogenous Ras; and (d) Rap1A did not antagonize a form of GAP (GAP32) whose ability to inhibit M_2-muscarinic receptor-regulated K^+ channels was independent of Ras. These results indicated that Rap1A was acting by a competitive mechanism, in this system at least, and that such inhibition could occur at picomolar concentrations of Rap1A, levels which are likely to be achieved in normal cells. The ability of Rap1A to antagonize Ras/GAP action in this system, and also that of Rap1B in *Xenopus* oocytes (CAMPA et al. 1991), indicates that this biological activity of Rap1 is not limited to antagonism of cell transformation by Ras.

The data discussed above suggest that the antagonism of Ras action by Rap1 occurs via a competitive mechanism involving the "effector domain" of Rap1. A number of studies have localized the suppressive activity of Rap1 to the NH_2 terminal portion of the molecule (BUSS et al. 1991; KITAYAMA et al. 1990b; ZHANG et al. 1990) and have identified residues 26, 27, 30, 31, and 45 as crucial in this regard (MARSHALL et al. 1991; NUR-E-KAMAL et al. 1992). It is of interest then to note that both of the Rap2 proteins, as well as R-Ras, also contain an "effector domain" identical to Ras, yet exhibit no ability to suppress the ability of Ras to transform cells (JIMENEZ et al. 1991; SCHWEIGHOFFER et al. 1990). Although R-Ras differs from Rap1 in the crucial amino acids adjacent to the "effector domain", Rap2 is identical to Rap1 in these positions. The consideration of these facts indicates that additional structural components need to be accounted for in order to understand antagonism of Ras by Rap1.

The physiological significance of the ability of Rap1 to inhibit Ras action is still uncertain. There is as yet little data to indicate that Rap1 plays such a regulatory role in normal cells. In fact, in some cell models of Ras action, Rap1 has been reported not to antagonize the actions of Ras (SCHWEIGHOFFER et al. 1990). In yeast, mammalian *rap* genes can *stimulate* some of the same effector pathways as does H-Ras (XU et al. 1990;

Ruggieri et al. 1992). It is tempting to speculate that, if Rap1 is indeed a physiological suppressor of Ras function, modulation of the guanine nucleotide state of Rap1 would play an important role in regulating the transduction of growth and differentiation signals via Ras. One could thus envision that changes in the activity of Rap1 GAPs, GDSs, etc., would also produce marked effects on Ras activity indirectly through their ability to regulate Rap1 GTP formation and thus binding to Ras GAP(s). Mutations in Rap1 or in Rap1-associated regulatory components might then play a significant role in the pathogenesis of mammalian tumors. A marked decrease in the levels of expression of Rap1 messenger RNA was reported in several types of tumors not normally associated with Ras mutations (i.e., salivary gland fibrosarcomas and adenocarcinomas, etc.) by Culine et al. (1989). Rap1B has been mapped to a chromosomal location near breakpoints associated with a number of malignant and benign neoplasms (Rousseau-Merck et al. 1990). Several investigations of Rap1 levels (Hong et al. 1990; Hsu and Gould 1991) and loss of heterozygosity (Young et al. 1992) in certain types of tumors have proven negative. However, Kyprianou and Taylor-Papadimitriou (1992) reported that Rap1 levels in azatyrosine-induced revertants of ras-transformed human mammary epithelial cells were significantly increased. Clearly further studies along such lines are warranted.

II. Interaction of Rap1A with the Phagocyte Reduced Nicotinamide Adenine Dinucleotide Phosphate Oxidase

The question of whether Rap1 serves as a physiological antagonist of Ras action remains unsettled. The abundance of Rap1 in a number of untransformed cells, such as platelets and neutrophils, the existence of multiple Rap-specific GAPs which can potentially serve unique effector pathways, and the possible regulation of Rap1 and Rap GAP by phosphorylation all suggest that Rap may normally regulate other functions in cells. There is evidence that Rap1A may play some role in the regulation of reduced nicotinamide adenine dinucleotide phosphate (NADPH) oxidase system of human neutrophils.

Human neutrophils and other phagocytic cells respond rapidly to contact with opsonized microorganisms by undergoing a "respiratory burst" in which molecular oxygen is reduced to form superoxide anion (O_2^-). O_2^- is subsequently converted to other toxic oxygen metabolites (Clark 1990). This respiratory burst is catalyzed by the NADPH oxidase, a multicomponent enzyme composed of at least four known proteins (see Fig. 3). A cytochrome b_{558}, consisting of 91-kDa ($gp91_{phox}$) and 22-kDa ($gp22_{phox}$) subunits, resides in the plasma membrane and specific granule membrane, where it serves as the terminal electron carrier of the oxidase. Two additional cofactors, $p47_{phox}$ and $p67_{phox}$, are cytosolic and may exist as preformed complexes which translocate to the plasma membrane upon

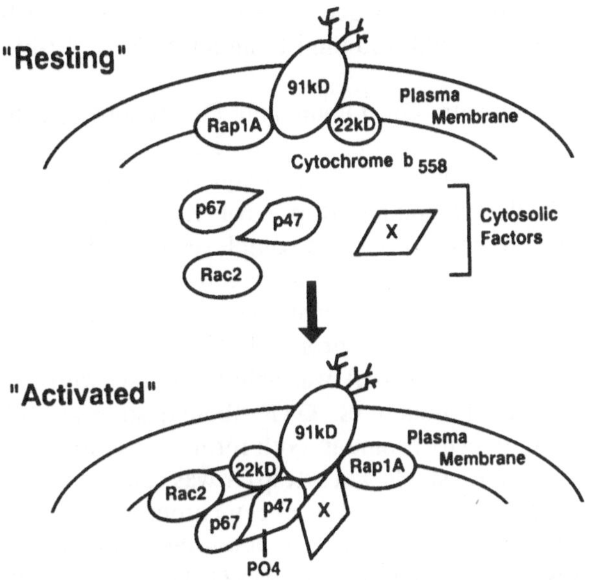

Fig. 3. The phagocyte respiratory burst oxidase: two GTPases, Rac2 and Rap1A, appear to be involved in this enzyme system. *91kD* and *22kD* represent the subunits of the cytochrome b_{558}; *p47*, *p67*, and *x*, cytosolic components of the reduced nicotinamide-adenine-dinucleotide phosphate (NADPH) oxidase. PO_4 indicates the known incorporation of phosphate groups into p47 during activation of the oxidase

phagocyte activation, becoming integral parts of the active oxidase (CLARK 1990; CURNUTTE 1992).

Guanine nucleotides, specifically GTP, have been shown to be absolutely required for NADPH oxidase activity in cell-free assay systems (PEVERI et al. 1992; UHLINGER et al. 1991), supporting the idea of a GTPase being involved in oxidase regulation (BOKOCH 1990). Our laboratory has also shown that treatment of HL-60 cells with drugs able to block isoprenoid metabolism totally prevents a respiratory burst response in these cells after they have been differentiated into a neutrophil-like form by DMSO (BOKOCH and PROSSNITZ 1992). This isoprenoid requirement is presumably due to the need for isoprenoids in the posttranslational processing of a small GTPase. Our laboratory has recently identified Rac2 as an important stimulatory component of the NADPH oxidase in human neutrophils (KNAUS et al. 1991). There are indications, though, that Rap1A must also play some modulatory role in this system.

The first indication of this came when QUINN et al. (1989) observed the coisolation of a 22-kDa protein distinct from the cytochrome b 22-kDa subunit during purification of the cytochrome b oxidase component. This protein had an amino-terminal amino acid sequence identical to that of Rap1. Evidence that this co-purification was not merely a fortuitous coincidence (since Rap1 could be separated from the cytochrome at a final sucrose gradient step) was provided by their observation that Rap1 present in neutrophil extracts would also bind to anticytochrome 91-kDa or 22-kDa antibody columns.

In collaboration with Dr. Quinn and Dr. Jesaitis, we have recently confirmed and extended this observation (Bokoch et al. 1991). Using a purified cytochrome preparation (free of contaminating Rap1) and purified baculovirus-Sf9 cell recombinant Rap1A or Rap1A purified from human neutrophils, we were able to observe complexes formed between these two proteins. Formation of complexes appeared to occur through the 1:1 association of Rap1A with the cytochrome. These results demonstrate that Rap1A binds directly to the cytochrome itself.

We examined whether Rap1A was able to interact with the cytochrome when it was in a GDP-bound vs. GTPγs-bound state and observed that Rap1A would bind to cytochrome in both forms. However, it appeared that the interaction was more stable when Rap1A was complexed with GTPγs, as our ability to observe the Rap1A-GDP–cytochrome complex was variable.

Agonists which elevate cAMP in neutrophils are able to markedly attenuate the respiratory burst in these cells (Rivkin et al. 1975; Sha'afi and Molski 1988; Mueller et al. 1988). We knew that Rap1A could be phosphorylated by cAMP-elevatory agents in intact HL-60 cells (Quilliam 1991) and that stoichiometric phosphorylation of Rap1A in vitro did not affect its ability to bind or hydrolyze guanine nucleotides or to respond to a cytoslic Rap GAP (Quilliam et al. 1990; Bokoch and Quilliam 1990). We wondered whether the interaction with the cytochrome could be modulated by phosphorylation of Rap1A. We found that the ability of phosphorylated Rap1A GTPγs to form complexes with cytochrome b was markedly reduced. This suggests that one potential mechanism by which elevations in neutrophil cAMP could inhibit the respiratory burst response is by disruption of Rap1A–cytochrome interactions.

It is of note that Rap1B phosphorylation has been shown to enhance its ability to bind and be stimulated by a GDS (Hata et al. 1991; Itoh et al. 1991). It is possible that the interaction of Rap1A and 1B with other regulatory macromolecules at the carboxyl terminal occurs and is modulated, either positively or negatively, by Rap1 phosphorylation. We further hypothesize that such regulation could also extend to K-Ras, which has a similar cationic region of its C terminus and which has been reported to undergo C-terminal phosphorylation as well (Ballester et al. 1987).

While it has been established that Rap1A can interact with the NADPH oxidase-associated cytochrome b, the significance of this interaction is not yet clear in terms of NADPH oxidase activity. We have observed no ability of Rap1A to modulate the activity of the cell-free NADPH oxidase system (unpublished observations). Eklund et al. (1991) have reported that antisera against a synthetic peptide corresponding to the effector region of Rap1A (amino acids 31–43) was able to completely inhibit oxidase activity in a cell-free system and that activity could be restored by addition of recombinant Rap1A. We have tried similar experiments with negative results (unpublished observations). The reason for these discrepancies is not yet clear. We (QUILLIAM et al. 1991) and others (MARIDONNEAU-PARINI and DE GUNZBURG 1992; QUINN et al. 1992a) find Rap1 solely in neutrophil membrane fractions when localized by specific immunoblotting, in contrast to EKLUND et al. (1991), who report Rap1 to be cytosolic and to translate from cytosol to membrane upon neutrophil activation by phorbol myristate acetate. The functional role of Rap1A in the NADPH oxidase system is thus not yet well understood. It is possible that Rap1A (unlike Rac2) may play a more subtle role in the system which may not be evident in studies using cell-free assays, which clearly rely on non-physiological activators in a structurally disrupted system. For example, based upon studies of the RSR1 Rap1A homolog in *Saccharomyces cerevesiae* which indicate it is crucial for positional information relevant to bud-site selection (RUGGIERI et al. 1992) and upon data which indicate an association of Rap1B with the platelet cytoskeleton upon thrombin activation (FISCHER et al. 1990), it is possible that Rap1A in the neutrophil might direct translocation of cytochrome b to select membrane sites upon cell activation or might mediate interactions between oxidase components and the cell cytoskeleton. Such interactions are thought to be important for oxidase activation and deactivation in vivo (CLARK 1990; HEYWORTH et al. 1991; NAUSEEF et al. 1991) and would be relatively undetectable in normal cell-free oxidase assays. Recently, QUINN et al. (1992b) reported that the ability of partially purified, Rap1-associated cytochrome b to reconstitute NADPH oxidase activity in CGD patient membranes lacking the cytochrome was decreased when the cytochrome was purified to essential homogeneity. This may reflect a requirement for Rap or another protein for proper reconstitutive capacity of the cytochrome. Other scenarios are also possible, as discussed in (QUILLIAM and BOKOCH 1992).

F. Conclusion

There are a number of inferences that can be drawn from the information presented above with regard to the biology of Rap. While it has not yet been shown to be of physiological significance that Rap1 can antagonize Ras action in vivo, such an action of Rap does occur in vitro and seems likely to be relevant to growth regulation and/or tumorigenesis. The fact that Rap1A and 1B serve as excellent substrates for cAMP-dependent protein kinase, as well as a number of other kinases, and that this phosphorylation can regulate the macromolecular interactions that Rap can undergo in vitro also implicates Rap in cellular regulation. Activation of these kinases via cAMP generation, etc., in intact cells is likely to exert significant effects on the GTP state of Rap (via GDS) and, therefore, on its suppression of Ras function.

It is known that in some cell types, elevations in cAMP levels can lead to tumorigenesis (DUMONT et al. 1989; VALLAR et al. 1987), but the mechanism of this effect has not been established. A possible role for Rap activation in this process deserves investigation. Stimulation of Rap1 GTP binding by increases in cAMP would also be expected to influence biological actions of Rap in other cellular systems as well. Rap may mediate some of the effects of cAMP in human neutrophils (Quilliam et al. 1991), where cAMP inhibits activity of the NADPH oxidase, and in human platelets (Lazarowski et al. 1990), where activation of phospholipase C is blocked by cAMP. Finally, there are indications that Rap activity may be regulated by hormones in vivo through actions on its GAP (Marti and Lapetina 1992).

The association of Rap1A with the NADPH oxidase cytochrome b component in human neutrophils is more difficult to assess. Unlike oxidase regulation by Rac 2, a GTPase which is expressed solely in cells of hemopoietic lineage (Didsbury et al. 1989), Rap1A is ubiquitously expressed and would seem unlikely to be solely a regulatory component for a phagocyte-specific enzyme. One hypothesis is that Rap1 plays a more general function in the cell, and that this function involves different protein components in different cells. Regulation of cytoskeletal interactions or protein–protein interactions in general come to mind as possibilities. This hypothesis predicts similar roles for Rap1 in other cell types, but the biological consequences of its action would differ completely.

References

Adari H, Lowy DR, Williamsen BM, Der CJ, McCormick F (1988) Guanosine triphosphatase activating protein (GAP) interacts with the p21 *ras* effector binding domain. Science 240:518–521

Ballester R, Furth ME, Rosen OM (1987) Phorbol ester- and protein kinase C-mediated phosphorylation of the cellular Kirsten ras gene product. J Biol Chem 262:2688–2695

Bokoch GM (1990) Signal transduction by GTP-binding proteins during leukocyte activation: phagocytic cells. In: Grinstein S, Rotstein OD (eds) Mechanisms of leukocyte activation: current topics in membranes and transport. Academic Press, San Diego, California, pp 65–101

Bokoch GM, Prossnitz V (1992) Isoprenoid metabolism is required for stimulation of the respiratory burst oxidase of HL-60 cells. J Clin Invest 89:402–408

Bokoch GM, Quilliam LA (1990) Guanine nucleotide binding properties of *rap1* purified from human neutrophils. Biochem J 267:407–411

Bokoch GM, Parkos CA, Mumby SM (1988) Purification and characterization of the 22,000-Dalton GTP-binding protein substrate for ADP-ribosylation by botulinum toxin, G_{22k}. J Biol Chem 263:16744–16749

Bokoch GM, Quilliam LA, Bohl BP, Jesaitis AJ, Quinn MT (1991) Inhibition of Rap1A binding to cytochrome b_{558} of the NADPH oxidase by phosphorylation of Rap1A. Science 254:1794–1796

Buss JE, Quilliam LA, Kato K, Casey PJ, Solski PA, Wong G, Clark R, McCormick F, Bokoch GM, Der CJ (1991) The COOH-terminal domain of the Rap1A (Krev-1) protein is isoprenylated and supports transformation by an H-ras: Rap1A chimeric protein. Mol Cell Biol 11:1523–1530

Campa MJ, Chang K-J, y Vedia LM, Reep BR, Lapetina EG (1991) Inhibition of ras-induced germinal vesicle breakdown in Xenopus oocytes by Rap1B. Biochem Biophys Res Commun 174:1–5

Clark RA (1990) The human neutrophil respiratory burst oxidase. J Infect Dis 161:1140–1147

Culine S, Olofsson B, Gosselin S, Honore N, Tavitian A (1989) Expression of the ras-related rap genes in human tumors. Int J Cancer 44:990–994

Curnutte JT (1992) Molecular basis of the autosomal recessive forms of chronic granulomatous disease. Immunodefic Rev 3:149–172

Didsbury J, Weber RF, Bokoch GM, Evans T, Snyderman R (1989) *Rac*, a novel *ras*-related family of proteins that are botulinum toxin substrates. J Biol Chem 264:16378–16382

Dumont JE, Jeuniaux JC, Roger PP (1989) The cyclic AMP-mediated stimulation of cell proliferation. Trends Biochem Sci 14:67–71

Eklund EA, Marshall M, Gibbs JB, Crean CD, Gabig TG (1991) Resolution of a low molecular weight G protein in neutrophil cytosol required for NADPH oxidase activation and reconstitution by recombinant Krev-1 protein. J Biol Chem 266:13964–13970

Farrell F, Torti M, Lapetina EG (1992) Rap proteins: Investigating their role in cell function. J Lab Clin Med 120:533–537

Fischer TH, Gatling MN, Lacal J-C, White GC II (1990) rap1B, a cAMP-dependent protein kinase substrate, associates with the platelet cytoskeleton. J Biol Chem 265:19405–19408

Frech M, John J, Pizon V, Chardin P, Tavitian A, Clark R, McCormick F, Wittinghofer A (1990) The protein product of the Krev-1 gene (rap1A) inhibits GAP activation of p21. Science 249:169–171

Hata Y, Kikuchi A, Sasaki T, Schaber MD, Gibbs JB, Takai Y (1990) Inhibition of the ras p21 GTPase-activating protein-stimulated GTPase activity of c-Ha-ras p21 by smg p21 having the same putative effector domain as ras p21's. J Biol Chem 265:7104–7101

Hata Y, Kaibuchi K, Kawamura S, Hiroyoshi M, Shirataka H, Takai Y (1991) Enhancement of the actions of smg p21 GDP/GTP exchange protein by the protein kinase A-catalyzed phosphorylation of smg p21. J Biol Chem 266:6571–6577

Heyworth PG, Curnutte JT, Nauseef WM, Volpp BD, Pearson DW, Rosen H, Clark RA (1991) Neutrophil nicotinamide adenine dinucleotide phosphate oxidase assembly: translocation of p47 [phox] and p67 [phox] requires interaction between p47 [phox] and cytochrome b_{558}. J Clin Invest 87:352–356

Hiroyoshi M, Kaibuchi K, Kawamura S, Hata Y, Takai Y (1991) Role of the C-terminal region of smg p21, a ras p21-like small GTP-binding protein, in membrane and smg p21 GDP/GTP exchange protein interactions. J Biol Chem 266:2962–2969

Hoshijima M, Kikuchi A, Kawata M, Ohmori T, Hashimoto E, Yamamura H, Takai Y (1988) Phosphorylation by cyclic AMP-dependent protein kinase of a human platelet Mr 22,000 GTP-binding protein having the same putative effector domain as the *ras* gene products. Biochem Biophys Res Commun 157:851–860

Hong HJ, Hsu L-C, Gould MN (1990) Molecular cloning of rat Krev-1 cDNA and analysis of the mRNA levels in normal and NMU-induced mammary carcinomas. Carcin 11:1245–1247

Hsu L-C, Gould MN (1991) Molecular cloning of Copenhagen rat Krev-1 and Rap1B cDNA's and study of their association with mammary tumor resistance in the Copenhagen rat. Carcin 12:533–536

Huzoor-Akbar, Winegar DA, Lapetina EG (1991) Carboxyl methylation of platelet rap1 proteins is stimulated by guanosine 5′-(3-0-Thio) triphosphate. J Biol Chem 266:4387–4391

Itoh T, Kaibuchi K, Sasaki T, Takai Y (1991) The smg p21 GDS-induced activation of smg p21 is initiated by cyclic AMP-dependent protein-kinase-catalyzed phosphorylation of smg p21. Biochem Biophys Res Commun 177:1319–1324

Jimenez B, Pizon V, Lerosey I, Beranger F, Tavitian A, de Gunzburg J (1991) Effects of the ras-related rap2 protein on cellular proliferation. Int J Cancer 49:471–479

Kaibuchi K, Mizuno T, Fujioka H, Yamamoto T, Kishi K, Fukumoto Y, Hori Y, Takai Y (1991) Molecular cloning of the cDNA for stimulatory GDP/GTP exchange protein for smg p21's (ras p21-like small GTP-binding proteins) and characterization of stimulatory GDP/GTP exchange protein. Mol Cell Biol 11:2873–2880

Kawamura S, Kaibuchi K, Hiroyoshi M, Hata Y, Takai Y (1991a) Stoichiometric interaction of smg p21 with its GDP/GTP exchange protein and its novel action to regulate the translocation of smg p21 between membrane and cytoplasm. Biochem Biophys Res Commun 173:1095–1102

Kawamura S, Kaibuchi K, Hiroyoshi M, Fujioka H, Mizuno T, Takai Y (1991b) Inhibition of the action of a stimulatory GDP/GTP exchange protein for smg p21 by acidic membrane phospholipids. Jpn J Cancer Res 82:758–761

Kawata M, Matsui Y, Kondo J, Hishida T, Teranishi Y, Takai Y (1988) A novel small molecular weight GTP-binding protein with the same putative effector domain as the *ras* proteins in bovine brain membranes. J Biol Chem 263:18965–18971

Kawata M, Kawahara Y, Araki S, Sumako M, Tsuda T, Fukuzaki H, Mizoguchi A, Takai Y (1989a) Identification of a major GTP-binding protein in bovine aortic smooth muscle membranes as *smg* p21. Biochem Biophys Res Commun 163:1418–1427

Kawata M, Kikuchi A, Hoshijima M, Yamamoto K, Hashimoto E, Yamamura H, Takai Y (1989b) Phosphorylation of smg p21, a ras p21-like GTP-binding protein, by cyclic AMP-dependent protein kinase in a cell-free system and in response to prostaglandin E_1 in intact human platelets. J Biol Chem 264:15688–15695

Kawata M, Farnsworth CC, Yoshida Y, Gelb MH, Glomset JA, Takai Y (1990) Posttranslationally processed structure of the human platelet protein smg p21B: evidence for geranylgeranylation and carboxyl methylation of the C-terminal cysteine. Proc Natl Acad Sci USA 87:8960–8964

Kemp BE, Pearson RB (1990) Protein kinase recognition sequence motifs. TIBS 15:342–346

Kikuchi A, Sasaki T, Araki S, Hata Y, Takai Y (1989) Purification and characterization from bovine brain cytosol of two GTPase-activating proteins specific for smg p21, a GTP-binding protein having the same effector domain as c-ras p21s. J Biol Chem 264:9133–9136

Kitayama H, Sugimoto Y, Matsuzaki T, Ikawa Y, Noda M (1989) A ras-related gene with transformation suppressor activity. Cell 56:77–84

Kitayama H, Matsuzaki T, Ikawa Y, Noda M (1990a) Genetic analysis of the Kristen-ras-revertant 1 gene: potentiation of its tumor suppressor activity by specific point mutations. Proc Natl Acad Sci USA 87:4284–4288

Kitayama H, Matsuzaki T, Ikawa Y, Noda M (1990b) A domain responsible for the transformation suppressor activity in Krev-1 protein. Jpn J Cancer Res 81:445–448

Knaus UG, Heyworth PG, Evans T, Curnutte JT, Bokoch GM (1991) Regulation of phagocyte oxygen radical production by the GTP-binding protein Rac2. Science 254:1512–1515

Kyprianou N, Taylor-Papadimitriou J (1992) Isolation of azatyrosine-induced revertants from ras-transformed human mammary epithelial cells. Oncogene 7:57–63

Labadia M, Bokoch GM, Huang C-K (1992) The low molecular weight G protein Rap1A modulates human neutrophil cytosolic protein kinase activity. FASEB J 6:6490 (abstract)

Lapetina EG, Lacal J-C, Reep BR, Molina y Vedia L (1989) A *ras*-related protein is phosphorylated and translocated by agonists that increase cAMP levels in human platelets. Proc Natl Acad Sci USA 86:3131–3134

Lazarowski ER, Winegar DA, Nola RD, Oberdisse E, Lapetina EG (1990) Effect of protein kinase A on inositide metabolism and rap1 G-protein in human erythroleukemia cells. J Biol Chem 265:13118–13123
Lerosey I, Pizon V, Tavitian A, de Gunzburg J (1991) The cAMP-dependent protein kinase phosphorylates the Rap1 protein *in vitro* as well as in intact fibroblasts, but not the closely related Rap2 protein. Biochem Biophys Res Commun 175:430–436
Maltese WA (1990) Posttranslational modification of proteins by isoprenoids in mammalian cells. FASEB J 4:3319–3328
Maridonneau-Parini I, de Gunzburg J (1992) Association of rap1 and rap2 proteins with the specific granules of human neutrophils. J Biol Chem 267:6396–6402
Marshall MS, Davis LJ, Keys RD, Mosser SD, Hill WS, Scolnick EA, Gibbs JB (1991) Identification of amino acid residues required for ras p21 target activation. Mol Cell Biol 11:3997–4004
Marti KB, Lapetina EG (1992) Epinephrine suppresses rap1B GAP-activated GTPase activity in human platelets. Proc Natl Acad Sci USA 89:2784–2788
Miura Y, Kaibuchi K, Itoh T, Corbin JD, Francis SH, Takai Y (1992) FEBS Lett 297:171–174
Mizuno T, Kaibuchi K, Yamamoto T, Kawamura M, Sakoda T, Fujioka H, Matsuura Y, Takai Y (1991) A stimulatory GDP/GTP exchange protein for smg p21 is active on the post-translationally processed form of c-Ki-ras p21 and rho A p21. Proc Natl Acad Sci USA 88:6442–6446
Moores SL, Schober MD, Mosser SD, Rands E, O'Hara MB, Garsky VM, Marshall MS, Pompliano DL, Gibbs JM (1991) Sequence dependence of protein isoprenylation. J Biol Chem 266:14603–14610
Mueller H, Motulsky HJ, Sklar LA (1988) The potency and kinetics of the β-adrenergic receptors on human neutrophils. Mol Pharmacol 34:347–353
Nauseef WM, Volpp BD, McCormick S, Leidal KG, Clark RA (1991) Assembly of the neutophil respiratory burst oxidase: protein kinase C-promoted cytoskeletal and membrane association of cytosolic oxidase components. J Biol Chem 266:5911–5917
Nice EC, Fabri L, Hammacher A, Holden J, Simpson RJ (1992) The purification of a Rap1 GTPase-activating protein from bovine brain cytosol. J Biol Chem 267:1546–1553
Nur-E-Kamel MSA, Sizeland A, D'Abaco G, Maruta H (1992) Asparagine 26, glutamic acid 31, valine 45, and tyrosine 64 of Ras proteins are required for their oncogenicity. J Biol Chem 267:1415–1418
Ohmori T, Kikuchi A, Yamamoto K, Kim S, Takai Y (1989) Small molecular weight GTP-binding proteins in human platelet membranes. J Biol Chem 264:1877–1881
Ohmstede C-A, Farrell FX, Reep BR, Clemetson KJ, Lapetina EG (1990) Rap2: a ras-related GTP-binding protein from platelets. Proc Natl Acad Sci USA 87:6527–6531
Peveri P, Heyworth PG, Curnutte JT (1992) Absolute requirement for GTP in activation of human neutrophil NADPH oxidase in a cell-free system: role of ATP in regenerating GTP. Proc Natl Acad Sci USA 89:2494–2498
Pizon V, Chardin P, Lerosey I, Olofsson B, Tavitian A (1988a) Human cDNA's rap 1 and rap 2 homologous to the Drosophila gene Dras 3 encode proteins closely related to ras in the "effector" region. Oncogene 3:201–204
Pizon V, Lerosey I, Chardin P, Tavitian A (1988b) Nucleotide sequence of a human cDNA encoding a ras-related protein (rap1B). Nucleic Acids Res 16:7719
Polakis PG, Rubinfeld B, Evans T, McCormick F (1991) Purification of a plasma membrane-associated GTPase-activating protein specific for rap1/Krev-1 from HL-60 cells. Proc Natl Acad Sci USA 88:239–243
Polakis PG, Rubinfeld B, McCormick F (1992) Phosphorylation of Rap1 GAP in vivo by cAMP-dependent kinase and the cell cycle p34 CDC kinase in vitro. J Biol Chem 267:10780–10785

Quilliam LA, Bokoch GM (1992) Structure and function of GTP-binding proteins in neutrophil signal transduction. In: Cochrane CG, Gimbrowe MA (eds) Signal transduction in inflammatory cells I. Academic Press, San Diego, California

Quilliam LA, Der CJ, Clark R, O'Rourke EC, Zhang K, McCormick FP, Bokoch GM (1990) Biochemical characterization of baculovirus-expressed Rap1A/Krev-1 protein and its regulation by GTPase activating proteins. Mol Cell Biol 10:2901–2908

Quilliam LA, Mueller H, Bohl BP, Prossnitz V, Sklar LA, Der CJ, Bokoch GM (1991) Rap1A is a substrate for cAMP-dependent protein kinase in human neutrophils. J Immunol 147:1628–1635

Quinn MT, Parkos CA, Walker L, Orkin SH, Dinauer MC, Jesaitis AJ (1989) Association of a Ras-related protein with cytochrome b of human neutrophils. Nature 342:198–200

Quinn MT, Mullen ML, Jesaitis AJ, Linner DG (1992a) Subcellular distribution of the Rap1A protein in human neutrophils: colocalization and cotranslocation with cytochrome b_{559}. Blood 79:1563–1573

Quinn MT, Curnutte JT, Parkos CA, Mullen ML, Scott PJ, Erickson RW, Jesaitis AJ (1992b) Reconstitution of defective respiratory burst activity with partially purified human neutrophil cytochrome b in two genetic forms of chronic granulomatous disease: possible role of Rap1A. Blood 79:2438–2445

Rivkin I, Rosenblatt J, Becker EL (1975) The role of cyclic AMP in the chemotactic responsiveness and spontaneous motility of rabbit peritoneal neutrophils. J Immunol 115:1126–1134

Rousseau-Merck MF, Pizon V, Tavitian A, Berger R (1990) Chromosome mapping of the human Ras-related Rap1A, Rap1B, and Rap2 genes to chromosomes 1p12 → p13, 12q14, and 13q34, respectively. Cytogenet Cell Genet 53:2–4

Rubinfeld B, Munemitsu S, Clark R, Conroy L, Watt K, Crossier WJ, McCormick F, Polakis P (1991) Molecular cloning of a GTPase activating protein specific for the Krev-1 protein p21 rap. Cell 65:1033–1042

Ruggieri R, Bender A, Matsui Y, Powers S, Takai Y, Pringle JR, Matsumoto K (1992) RSRI, a ras-like gene homologous to Krev-1 (smg 21A/Rap1A: role in the development of cell polarity and interactions with the ras pathway in *Saccharomyces cerevisiae*. Mol Cell Biol 12:758–766

Sahyoun N, McDonald OB, Farrell F, Lapetina EG (1991) Phosphorylation of a ras-related GTP-binding protein, Rap1B, by a neuronal Ca^{2+}/calmodulin-dependent protein kinase, CaM. Proc Natl Acad Sci USA 88:2643–2647

Schejter ED, Shilo BZ (1985) Characterization of functional domains of p21 ras by use of chimeric genes. EMBO J 4:407–412

Schweighoffer F, Rey I, Barlot I, Soubigou P, Mayaux JF, Tocque B (1990) Rap gene products mobilize a different metabolic pathway than p21 ras proteins. In: Nishizuka Y et al. (eds) The biology and medicine of signal transduction. Raven, New York, pp 329–334

Sha'afi RI, Molski TFP (1988) Activation of the neutrophil. Prog Allergy 42:1–64

Shirataki H, Kaibuchi K, Hiroyoshi M, Isomura M, Osaki S, Sasaki T, Takai Y (1991) Inhibition of the action of the stimulatory GDP/GTP exchange protein of smg p21 by the geranylgeranylated synthetic peptides designed from its C-terminal region. J Biol Chem 266:20672–20677

Siess W, Winegar DA, Lapetina EG (1990) Rap1B is phosphorylated by protein kinase A in intact human platelets. Biochem Biophys Res Commun 170:944–950

Ueda T, Kikuchi A, Ohga N, Yamamoto J, Takai Y (1989) GTPase activating proteins for the smg-21 GTP-binding protein having the same effector domain as the ras proteins in human platelets. Biochem Biophys Res Commun 159:1411–1419

Uhlinger DJ, Burnham DN, Lambeth JD (1991) Nucleoside triphosphate requirements for superoxide generation and phosphorylation in a cell-free system from human neutrophils. J Biol Chem 266:20990–20997

Vallar L, Spoda A, Giannattasio G (1987) Altered G_s and adenylate cyclase activity in human GH-secreting pituitary adenomas. Nature 330:566–568
Winegar DA, Molina y Vedia L, Lapetina EG (1991) Isoprenylation of rap2 proteins in human platelets and human erythroleukemia cells. J Biol Chem 266:4381–4386
Xu H-P, Wang Y, Riggs M, Rodgers L, Wigler M (1990) Biological activity of the mammalian rap genes in yeast. Cell Reg 1:763–679
Yamamoto T, Kaibuchi K, Mizuno T, Hiroyoshi M, Shirataki H, Takai Y (1990) Purification and characterization from bovine brain cytosol of proteins that regulate GDP/GTP exchange of smg p21's. J Biol Chem 265:16626–16634
Yatani A, Okabe K, Polakis P, Halenbeck R, McCormick F, Brown AM (1990) Ras p21 and GAP inhibit coupling of muscarinic receptors to atrial K^+ channels. Cell 61:769–776
Yatani A, Quilliam LA, Brown AM, Bokoch GM (1991) Rap1A antagonizes the ability of Ras and Ras-GAP to inhibit muscarinic K^+ channels. J Biol Chem 266:2222–22226
Young J, Searle J, Stitz R, Cowen A, Ward M, Chenevix-Trench G (1992) Loss of heterozygosity at the human Rap1A/Krev-1 locus is a rare event in colorectal tumors. Cancer Res 52:285–289
Zhang K, Noda M, Vass WC, Papageorge AG, Lowy DR (1990) Identification of small clusters of divergent amino acids that mediate the opposing effects of ras and Krev-1. Science 249:162–165

B. Vesicle Transfer/Vesicle Fusion

CHAPTER 26

GTPases and Interacting Elements in Vesicle Budding and Targeting in Yeast

C. BARLOWE and R. SCHEKMAN

A. Introduction

Protein transport between cellular organelles is mediated by a vesicle intermediate that forms from a donor compartment and fuses with a target compartment. A combined genetic and biochemical approach to examine protein transport in the yeast *Saccharomyces cerevisiae* has provided a wealth of information on the complex processes of protein transport. At least 30 genes have been identified in yeast that are required for the transit of a secreted protein to the cell surface (for a review, see PRYER et al. 1992). A number of these genes have been sequenced and at least four encode proteins with homology to GTPases of the *ras* family. Indeed, the first demonstration that a small GTPase functions in protein transport was reported by NOVICK and coworkers with the identification of Sec4p as a *ras*-like protein (SALMINEN and NOVICK 1987). More recently, other members of this family have been identified that participate in various stages of the yeast secretory pathway. Sar1p and Ypt1p function in transport from the endoplasmic reticulum (ER) to the Golgi apparatus (NAKANO and MURAMATSU 1989; D'ENFERT et al. 1991a; SEGEV et al. 1988; BACON et al. 1989). Yeast ADP ribosylation factor (Arf1p) is postulated to function in transport to or between Golgi compartments (STEARNS et al. 1990), while Sec4p is required for protein transport from the Golgi apparatus to the cell surface (SALMINEN and NOVICK 1987). Further genetic manipulation of yeast has facilitated the identification of genes that interact with or regulate these small GTPases.

This review is intended to provide an overview of known secretion-associated GTPases in yeast and possible interacting or regulatory elements. Because reviews of Sec4p, Ypt1p and ARF GTPases are provided in separate chapters of this volume, only a brief description of their identification and function in protein transport will be considered. The properties of Sar1p and its function in vesicle formation from the ER will be addressed in more detail.

B. Isolation and Characterization of Secretion Defective Yeast Strains

Several genetic screens and selections have been devised to isolate mutant yeast strains defective in protein transport to the cell surface. The first screen resulted in the isolation of 23 temperature sensitive (ts) secretion (sec) mutants based on density enrichment of mutagenized cells after shift to a nonpermissive temperature (NOVICK et al. 1980). Additional mutants, *bet1* and *bet2* (meaning "blocked early in transport"), were isolated by a suicide selection, in which cells blocked in secretion do not acquire lethal [^{3}H]mannose modifications to secretory glycoproteins (NEWMAN and FERRO-NOVICK 1987). NAKAJIMA and coworkers (1991) utilized the defect of aggregation in secretion-deficient cells to isolate genes that function in protein transport and have characterized *uso1*. Extragenic suppression analysis of existing secretory mutants has identified other interacting or regulatory proteins. For example, when a yeast genomic library was screened for suppressors of a *sec12* ts strain, two unique complementing genes were isolated. One encoded the authentic *SEC12*, while a second encoded a small GTPase named *SAR1* (secretion-associated and *ras*-superfamily related gene; NAKANO and MURAMATSU 1989). In this case, the increase in *SAR1* gene dosage compensates for the partial loss of Sec12p function. Similarly, *BOS1* and *BOS2* (*bet* one suppressor) were isolated as suppressors of *bet1* (NEWMAN et al. 1990; MANCINI et al. 1991). A different approach has been taken to investigate the cellular function of the ubiquitous small GTPase ARF. The conserved amino acid sequences among mammalian ARFs allowed SEWELL and KAHN (1988) to isolate the *ARF1* gene from yeast, and an *arf1* null strain was generated (STEARNS et al. 1990). This approach, termed "reverse genetics," implicated Arf1p in protein transport. Similarly, Ypt1p was initially identified as a *ras* homologue (GALLWITZ et al. 1983) and later found to function in secretion (SEGEV et al. 1988). The *Sly1* and *Sly41* genes were identified by virture of their ability to suppress loss of Ypt1p function in a *YPT1* deletion strain (DASCHER et al. 1991).

In secretory mutants, marker secretory proteins, such as carboxypeptidase Y (STEVENS et al. 1982), the mating pheromone α-factor (JULIUS et al. 1984), and invertase (ESMON et al. 1981), accumulate in incompletely processed forms reflecting the stage at which transport is blocked. Electron microscopic analyses of secretion mutants show accumulation of distinct membrane compartments (NOVICK and SCHEKMAN 1979). Thus, mutants in protein transport have been temporally ordered, based on which forms of secretory proteins and organelles accumulate at the restrictive condition. For example, in a *ypt1* strain, the core-glycosylated ER form of secretory proteins and enlarged ER compartments accumulate, indicating a block in transport from the ER (SEGEV et al. 1988; SCHMITT et al. 1988). In contrast, *sec4* strains accumulate Golgi-modified forms of secreted proteins, and

80-nm secretory vesicles appear near the cell surface at the emerging bud, indicating a post-Golgi apparatus block (Novick et al. 1980). *SEC* genes implicated in ER to Golgi transport have been subdivided into two classes after extensive analysis of mutant phenotype by electron microscopy. Class I genes (*SEC12*, *13*, *16*, *23*) are blocked prior to vesicle formation from the ER, while class II genes (*SEC17*, *18*, *22*) are required for fusion of ER-derived transport vesicles with the Golgi apparatus (Kaiser and Schekman 1990). Genetic analysis of secretory mutants has also identified subsets of genes that interact directly or function in concert to facilitate vesicle formation from the ER or fusion to the Golgi. For example, a double mutant strain containing both a *sec12* mutation and a *sec16* mutation has a more severe growth defect than either of the single mutant strains (Kaiser and Schekman 1990). This type of genetic interaction, known as synthetic lethality, has been useful in identifying a number of genes which interact with GTPases in the secretory pathway. For example, the *ypt1* and *bet2* mutations display synthetic lethality (Bacon et al. 1989). Recently *BET2* has been found to encode a subunit of an isoprenyltransferase required for Ypt1p isoprenlyation and proper membrane association (Rossi et al. 1991). A summary of the genetic analyses of *SAR1*, *YPT1*, *ARF1*, and *SEC4* are presented in Fig. 1, to provide a framework for understanding GTPase function in vesicular transport. However, a detailed understanding of GTPase function in the secretory pathway requires biochemical manipulation of these processes. To this end, vesicular transport from the ER to the Golgi apparatus has been reconstituted in vitro.

C. Biochemical Analysis of Protein Transport from the Endoplasmic Reticulum to the Golgi Apparatus

Protein transport from the ER to the Golgi apparatus has been reconstituted in vitro using gently lysed yeast cells (Baker et al. 1988; Ruohola et al. 1988) and in microsome preparations (Ruohola et al. 1988; Wuestehube and Schekman 1992). In cell-free systems, transport occurs between distinct donor and acceptor membrane compartments and requires cytosolic proteins, ATP, and incubation at physiological temperature. The secreted mating pheromone α-factor is used as a marker to measure transport from the ER to the Golgi apparatus. ^{35}S-labeled prepro-α-factor is translocated into the lumen of the ER where it receives N-linked core oligosaccharides. Upon addition of cytosol and ATP, core-glycosylated pro-α-factor is transferred to the Golgi apparatus where it acquires outerchain mannose residues in α1,6-linkages which can be visualized by autoradiography after sodium dodecyl sulfate polycrylamide gel electrophoresis (SDS-PAGE) or readily quantified by immune precipitation with specific anti-α1,6-mannose serum. Membranes and cytosol prepared from *sec23* cells are temperature sensitive for transport in vitro, thus confirming the faithful reproduction of

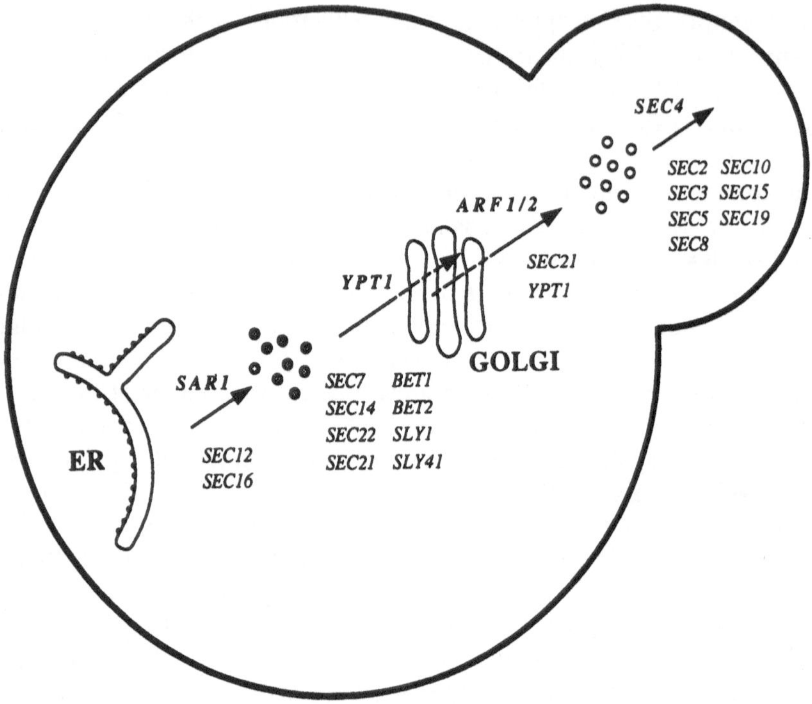

Fig 1. Small GTPases in the secretory pathway. The model shows proposed sites of *SAR1*, *YPT1*, *ARF1/2*, and *SEC4* function. GTPase genes are above *arrows* and interacting genes for each are listed below. Sar1p overproduction suppresses *sec12* and *sec16* (NAKANO and MURAMATSU 1989); *ypt1* displays synthetic lethality with *sec7*, *sec14*, *sec21*, *bet1*, and *bet2* (BACON et al. 1989). *SLY1*, *SLY41*, and *SEC22* genes are able to suppress the functional loss of *YPT1* (DASCHER et al. 1991). An *arf1* deletion strain is inviable in combination with *sec21* or *ypt1* (STEARNS et al. 1990), and *sec2*, *3*, *5*, *8*, *10*, *15*, and *19* display synthetic lethality in combination with a *sec4* strain (SALIMEN and NOVICK 1987)

transport in this cell-free system. Also, as reported for mammalian transport systems, ER to Golgi transport is inhibited by the guanine nucleotide analog GTPγS, confirming function of GTPases in intercompartmental transport in yeast (BAKER et al. 1988; RUOHOLA et al. 1988).

More recently, ER to Golgi transport in vitro has been divided into two distinct steps: vesicle formation and targeting/fusion to the Golgi. A vesicle intermediate derived from the ER freely diffuses and can be separated from the donor ER membranes by differential centrifugation. Manipulation of this in vitro assay indicates that Sec12p, Sar1p, and Sec23p function in vesicle formation from the ER (REXACH and SCHEKMAN 1991; D'ENFERT et al. 1991a). However, a block in Ypt1p function does not affect vesicle formation, but inhibits vesicle targeting/fusion to the Golgi (SEGEV 1991; REXACH and SCHEKMAN 1991). GTPγS is observed to inhibit both vesicle formation and targeting/fusion (REXACH and SCHEKMAN 1991), which is consistent with the proposed roles for Sar1p and Ypt1p involvement in vesicle formation and targeting/fusion, respectively.

The in vitro assay has been used to purify functional Sec proteins. Sec23p was purified using a complementation assay based on the ability of wild-type Sec23p to rescue the temperature-sensitive defect of *sec23* cells in vitro (HICKE and SCHEKMAN 1989). Functional Sec23p was purified from a complementing wild-type extract in a complex with a distinct 105-kDa protein (HICKE et al. 1992). The 105-kDa protein is required for vesicle formation from the ER, and the gene that encodes it is essential for growth (YOSHIHISA and SCHEKMAN, manuscript in preparation). Functional Sar1p has been purified to homogeneity and will be discussed next.

D. Sar1p Function in Vesicle Formation from the Endoplasmic Reticulum

SAR1 was initially identified as a multicopy suppressor of a *sec12* temperature-sensitive strain and, upon analysis of the corresponding gene, found to encode a 21-kDa protein with sequence homology to the *ras* family of GTPases (NAKANO and MURAMATSU 1989). NAKANO observed that Sar1p was a novel member with only 18%–23% identity to other family members, including Ypt1p and Sec4p. Sar1p shares highest identity with yeast Arf1p (35% over 168 amino acids). Also in common with the ARF proteins, Sar1p contains an aspartate residue at position 12 ($p21^{ras}$ numbering), in contrast to $p21^{ras}$, which has a glycine in this position, and the Ypt1p/Sec4p subfamily, which contain a serine or threonine residue at position 12. This residue is thought to be important for GTPase activity because many oncogenic *ras* gene products contain mutations at this site. Another unique feature of Sar1p is a histidine at position 61 ($p21^{ras}$ numbering), while other known small GTPases contain a glutamine residue at this position in the proposed nucleotide-binding motif DXAGQ. One model postulates that this glutamine is involved in the GTPase activity of $p21^{ras}$, since conversion of this conserved residue to different amino acids in $p21^{ras}$ results in reduced GTPase activity and cellular transformation (DER et al. 1986). Interestingly, a histidine residue is conserved in this position among the EF-Tu family of proteins. Thus, it has been suggested that Sar1p may function in a "proofreading" mechanism, similar to EF-Tu (NAKANO and MURAMATSU 1989). Membrane association of several small GTPases is mediated by a covalently bound lipid group. Ypt1p and Sec4p contain two C-terminal cysteines, one or both of which are thought to be isoprenylated for proper membrane association (WALWORTH et al. 1989; MOLENAAR et al. 1988). ARF proteins have an N-terminal glycine that is myristylated and proposed to mediate the membrane association of ARF (Sewell and Kahn 1988). Sar1p does not contain a C-terminal cysteine residue for potential isoprenlyation or an N-terminal glycine residue for myristylation. Sar1p expressed in yeast and *Escherichia coli* has the same SDS-PAGE electrophoretic mobility (Oka et al. 1991). Thus, Sar1p either does not contain a lipid modification or the modification could be of an uncharacterized type that is not detected by a mobility shift. The temperature sensitive phenotype of *sec12* cells is reproduced in the in vitro transport reaction (Rexach and Schekman 1991), and temperature sensitivity is suppressed by the addition of yeast Sarlp or recombinant Sar1p expressed in *E. coli*. Furthermore, a mutant form of Sar1p that is unable to bind GTP does not rescue the *sec12* temperature-sensitive phenotype in vitro (Oka et al. 1991). The ability of the *E. coli* expressed protein to function supports the idea that Sar1p is not post-translationally modified; however, modification may occur after the addition of the recombinant protein to the cell-free transport reaction.

Although posttranslational modification of Sar1p appears unlikely, a majority of Sar1p is tightly associated with membranes. Cell-fractionation studies show that Sar1p is peripherally associated with membranes (80%); however, a smaller soluble pool of Sar1p is detectable (d'Enfert et al. 1991b). Sar1p partially dissociates from membranes in the presence of 2 *M* urea, but is not extracted by elevated salt or pH (Nishikama and Nakano 1991). Overproduction of Sar1p increases the amount of the solube form, perhaps indicating saturation of a membrane-binding site (d'Enfert et al. 1991b; Nishikawa and Nakano 1991). Conversely, overproduction of Sec12p specifically titrates Sar1p from the soluble pool to the membrane, while not altering the distribution of other peripherally associated ER proteins such as Sec23p or Sec13p (d'Enfert et al. 1991b). These observations, coupled with the fact that elevated levels of Sar1p suppress the transport defect of *sec12* cells, suggests a direct interaction between Sar1p and Sec12p.

Sec12p is a 70-kDa integral membrane glycoprotein composed of a 40-kDa N-terminal cytoplasmic domain and a 30-kDa domain that projects into the lumen of the ER. Approximately 20 kDa of the lumenal domain can be accounted for by carbohydrate (Nakano et al. 1988; d'Enfert et al. 1991b). Based on the observation that overproduction of Sec12p depletes the cytosol of Sar1p, cell-free transport assays have been devised that require the addition of Sar1p to a depleted cytosol. The transport defect in a Sar1p-depleted cytosol occurs at the vesicle formation step, indicating a role for Sar1p function in vesicle formation from the ER (d'Enfert et al. 1991a; Barlowe et al. 1993).

This Sar1p-dependent vesicle budding assay has been used to purify functional Sar1p to homogeneity. Purified Sar1p binds GTP in solution, as determined by a filter-binding assay, and has GTPase activity (100 μmol mol^{-1} min^{-1}). GTP binding and hydrolysis require the presence of

Mg^{2+} and nonionic detergent or phospholipids (BARLOWE et al. 1993). The Mg^{2+} requirement is common among GTPases; however, the detergent requirement is somewhat unusual, as mammalian ARFs are the only other small GTPases exhibiting a similar requirement. ARF proteins require both detergent and phospholipids for maximal guanine nucleotide binding and do not have detectable GTPase activity (WEISS et al. 1989). Sarlp, like most other small GTPases, has low GTPase activity. As described elsewhere (BOURNE et al. 1991), it is likely that binding and hydrolysis of guanine nucleotides by small GTPases, including Sarlp, are tightly regulated in vivo by interacting proteins. Yeast Sec proteins are good candidates for such regulatory roles, and recently it has been observed that purified Sec23p stimulates the GTPase activity of Sarlp to increase more than tenfold (YOSHIHISA et al. 1993).

Vesicle formation from the ER is inhibited by GTPγS, raising the possibility that Sarlp is the site of this inhibition (REXACH and SCHEKMAN 1991). NAKANO and coworkers reported that GTPγS-bound Sarlp is less active in promoting transport from the ER to the Golgi apparatus (OKA et al. 1991). Recently, we have found that GTPγS-bound Sarlp inhibits vesicle formation from the ER, thus implicating GTP hydrolysis by Sarlp in this event (BARLOWE et al. 1993). Stimulation of Sarlp GTPase by Sec23p suggests that these proteins work in concert to facilitate vesicle budding; it is also consistent with the observed role of Sec23p in vesicle formation (REXACH and SCHEKMAN 1991). Details of Sec12p, Sarlp, and Sec23p functional interaction in this process remain obscure. Since Sec12p is a membrane-spanning protein, it could transmit a signal from the lumen to the cytosolic surface of the ER, perhaps employing Sarlp to initiate vesicle formation (NAKANO and MURAMATSU 1989). The transmembrane segment of Sec12p is essential for function; however, the lumenal domain is dispensable (D'ENFERT et al. 1991b). Thus, the transmembrane portion of Sec12p could serve as a membrane anchor for Sec12p that promotes vesicle formation by providing a site for the binding and assembly of a protein complex. In both cases, Sec12p-specific membrane attachment of Sarlp could result in the regulated formation of ER-derived transport vesicles at specific sites on the ER membrane. By analogy to the EF-Tu cycle, Sarlp-GTP could interact with Sec12p to promote assembly of the correct complex in association with Sec23p. Hydrolysis of GTP would occur concomitant with vesicle release and disassociation of Sarlp-GDP from the ER membrane (Fig. 2). The fact that Sec23p stimulates Sarlp GTP hydrolysis and is also required for vesicle formation is intriguing. Sec23p purifies in a complex with a distinct 105-kDa protein that is also required for vesicle budding from the ER. The Sec23p monomer contains Sarlp-specific GTPase stimulatory activity, perhaps modulated by the 105-kDa protein (YOSHIHISA et al. 1993). Further examination of this interaction in the in vitro reaction should elucidate the function of these proteins in vesicle formation.

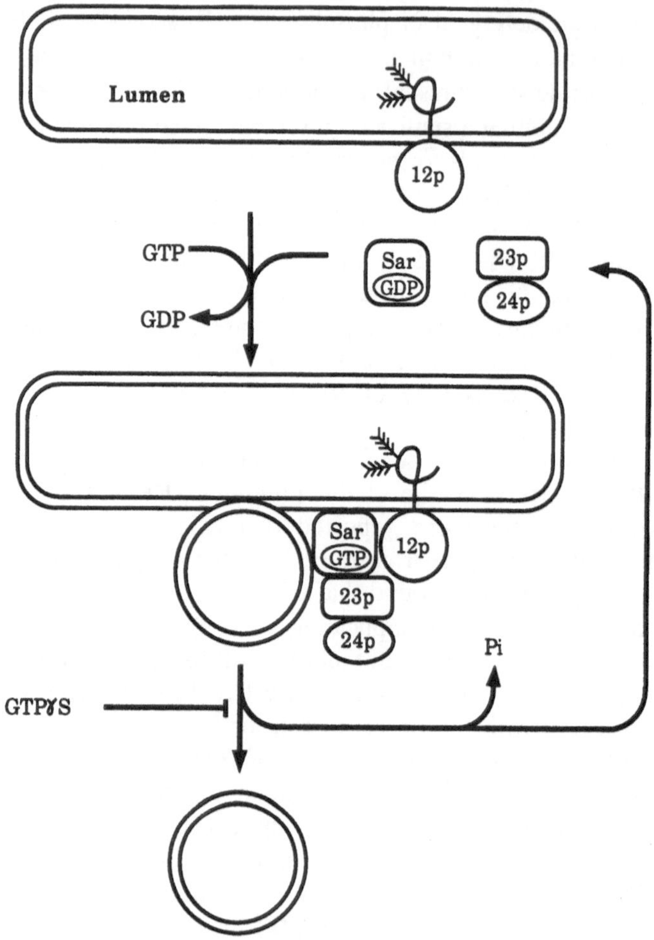

Fig. 2. A proposed model for Sec12p (12p), Sar1p (Sar), Sec23p (23p), and p105 (24p) function in vesicle budding from the endoplasmic reticulum (*ER*). GTPγS inhibits vesicle formation

E. Concluding Remarks

The complexity of interorganellar transport revealed by genetic analysis is likely to underlie the required regulation and specificity in formation of a vesicle intermediate from a donor compartment and transport/fusion of the vesicle to the correct target compartment. At least four small GTPases are required for protein transport from the ER to the cell surface in yeast. Deletion of any one is lethal, indicating nonoverlapping specificity of these GTPases. In the case of ER to Golgi transport, there appears to be at least one GTPase required for vesicle formation (Sar1p) and at least one for vesicle targeting/fusion (Ypt1). This multiple requirement for a single round of transport could be a paradigm for vesicle-mediated transport to insure specificity at different steps of the reaction. Sec4p and Ypt1p appear to function similarly in promoting vesicle attachment or fusion to an acceptor compartment. Interestingly, Sec4p and Ypt1p also share significant sequence identity, perhaps diverging from a common ancestor to mediate mechanistically conserved reactions. Indeed, a chimeric Sec4p/Ypt1p protein that contains a large internal portion of Ypt1p (between the GX_4GKS/T and NKXD motifs) substituted into the analogous region of Sec4p is able to fully complement a *ypt1* strain (Brennwald and Novick 1993; Dunn et al. 1993). *SEC1*, which is required for vesicle transfer from the Golgi complex to the cell surface, and *SLY1*, required for ER to Golgi transport, also encode homologous proteins and display genetic interaction with *SEC4* and *YPT1*, respectively (Aalto et al. 1992). Again, it appears that a conserved machinery is found in vesicle fusion steps at the Golgi complex and the cell surface. Similarity between Arf1p and Sar1p may underlie common function in vesicle formation; however, this remains more speculative. The GTP–GDP allosterism of small GTPases seems well suited for the cyclic processes and unidirectionality required at each step of protein transport. An understanding of small GTPase functions in ER to Golgi transport through a combined genetic and biochemical approach in yeast should provide a general mechanism for this conserved process in eukaryotes.

References

Aalto MK, Keranen S, Ronne H (1992) A family of proteins involved in intracellular transport. Cell 68:181–182

Bacon RA, Salminen A, Ruohola H, Novick P, Ferro-Novick S (1989) The GTP-binding protein Ypt1 is required for transport in vitro: the Golgi apparatus is defective in ypt1 mutants. J Cell Biol 109:1015–1022

Baker D, Hicke L, Rexach M, Schleyer M, Schekman R (1988) Reconstitution of SEC gene product-dependent intercompartmental protein transport. Cell 54:335–344

Barlowe C, d'Enfert C, Schekman R (1993) Purification and chacterization of Sar1p, a small GTP-binding protein required for transport vesicle formation from the endoplasmic reticulum. J Biol Chem 268:873–879

Bourne HR, Sanders DA, McCormick F (1991) The GTPase superfamily: conserved structure and molecular mechanism. Nature 349:117–127

Brennwald P, Novick P (1993) Interactions of three domains distinguishing the ras-related GTP-binding proteins Ypt1 and Sec4. Nature 362:560–563

Dascher C, Ossig R, Gallwitz D, Schmitt HD (1991) Identification and structure of four yeast genes (SLY) that are able to suppress the functional loss of YPT1, a member of the RAS superfamily. Mol Cell Biol 11:872–885

d'Enfert C, Wuestehube LJ, Lila T, Schekman R (1991a) Sec12p-dependent membrane binding of the small GTP-binding protein Sar1p promotes formation of transport vesicles from the ER. J Cell Biol 114:663–670

d'Enfert C, Barlowe C, Nishikawa S, Nakano A, Schekman R (1991b) Structural and functional dissection of a membrane glycoprotein required for vesicle budding from the endoplasmic reticulum. Mol Cell Biol 11:5727–5734

Der CJ, Finkel T, Cooper GM (1986) Biological and biochemical properties of human ras^H genes mutated at codon 61. Cell 44:167–176

Dunn B, Stearns T, Botstein D (1993) Specificity domains distinguish the ras-related GTPases Ypt1 and Sec4. Nature 362:563–565

Esmon B, Novick P, Schekman R (1981) Compartmentalized assembly of oligosaccharides on exported glycoproteins in yeast. Cell 25:451–460

Gallwitz D, Donath C, Sander C (1983) A yeast gene encoding a protein homologous to the human c-has/bas proto-oncogene product. Nature 306:704–707

Hicke L, Schekman R (1989) Yeast Sec23p acts in the cytoplasm to promote protein transport from the endoplasmic reticulum to the Golgi complex in vivo and in vitro. EMBO J 8:1677–1684

Hicke L, Yoshihisa T, Schekman R (1992) Sec23p and a novel 105 kD protein function as a multimeric complex to promote vesicle budding at protein transport from the ER. Mol Biol Cell (in press)

Julius D, Schekman R, Thorner J (1984) Glycosylation and processing of prepro-α-factor through the yeast secretory pathway. Cell 36:309–318

Kaiser C, Schekman R (1990) Distinct sets of SEC genes govern transport vesicle formation and fusion early in the secretory pathway. Cell 61:723–733

Mancini P, Zarate V, Ferro-Novick S (1991) The identification of the new genes that suppress the yeast secretory mutants bet1 and sec22. J Cell Biol 115:62a

Molenaar CMT, Prange R, Gallwitz D (1988) A carboxyl-teminal cysteine residue is required for palmitic acid binding and biological activity of the ras-related yeast YPT1 protein. EMBO J 7:971–976

Nakajima H, Hirata A, Ogawa Y, Yonehara T, Yoda K, Yamasaki M (1991) A cytoskeleton-related gene, USO1, is required for intracellular protein transport in Saccharomyces cerevisiae. J Cell Biol 113:245–260

Nakano A, Muramatsu M (1989) A novel GTP-binding protein, Sar1p, is involved in transport form the endoplasmic reticulum to the Golgi apparatus. J Cell Biol 109:2677–2691

Nakano A, Brada D, Schekman R (1988) Membrane glycoprotein, Sec12p, required for transport from the endoplasmic reticulum to the Golgi apparatus in yeast. J Cell Biol 107:851–863

Newman AP, Ferro-Novick S (1987) Characterization of new mutants in the early part of the yeast secretory pathway isolated by a [3H] mannose suicide selection. J Cell Biol 105:1587–1594

Newman AP, Shim J, Ferro-Novick S (1990) BET1, BOS1 and SEC22 are members of a group of interacting yeast genes required for transport from the endoplasmic reticulum to the Golgi complex. Mol Cell Biol 10:3405–3414

Nishikawa S, Nakano A (1991) The GTP-binding Sar1 protein is localized to the early compartment of the yeast secretory pathway. Biochim Biophys Acta 1093:135–143

Novick P, Schekman R (1979) Secretion and cell-surface growth are blocked in a temperature-sensitive mutant of Saccharomyces cerevisiae. Proc Natl Acad Sci USA 76:1858–1862

Novick P, Field C, Schekman R (1980) Identification of 23 complementation groups required for post-translational events in the yeast secretory pathway. Cell 21:205–215

Oka T, Nishikawa S, Nakano A (1991) Reconstitution of GTP-binding Sar1 protein function in ER to Golgi transport. J Cell Biol 114:671–679

Pryer NK, Wuestehube LJ, Schekman R (1992) Vesicle-mediated protein sorting. In: Richardson CC (ed) Annual review of biochemistry. Annual Reviews, Palo Alto, p 471

Rexach MF, Schekman RW (1991) Distinct biochemical requirements for the budding, targeting, and fusion of ER-derived transport vesicles. J Cell Biol 114:219–229

Rossi G, Jiang Y, Newman AP, Ferro-Novick S (1991) Dependence of Ypt1 and Sec4 membrane attachment on BET2. Nature 351:158–161

Ruohola H, Kastan Kabcenell A, Ferro-Novick S (1988) Reconstitution of protein transport from the endoplasmic reticulum to the Golgi complex in yeast: the acceptor Golgi compartment is defective in the sec23 mutant. J Cell Biol 107:1465–1476

Salminen A, Novick PJ (1987) A ras-like protein is required for a post-Golgi event in yeast secretion. Cell 49:527–538

Schmitt D, Puzicha M, Gallwitz D (1988) Study of a temperature-sensitive mutant of the ras-related YPT1 gene product in yeast suggests a role in the regulation of intracellular calcium. Cell 53:635–647

Segev N (1991) Mediation of the attachment or fusion step in vesicular transport by the GTP-binding Ypt1 protein. Science 252:1553–1556

Segev N, Mulholland J, Botstein D (1988) The yeast GTP-binding Ypt1 protein and a mammalian counterpart are associated with the secretion machinary. Cell 52:915–924

Sewell JL, Kahn RA (1988) Sequences of the bovine brain and yeast ADP-ribosylation factor and comparison to other GTP-binding proteins. Proc Natl Acad Sci USA 85:4620–4624

Stearns T, Willingham MC, Botstein D, Kahn RA (1990) ADP-ribosylation factor is functionally and physically associated with the Golgi complex. Proc Natl Acad Sci USA 87:1238–1242

Stevens T, Esmon B, Schekman R (1982) Early stages in the yeast secretory pathway are required for transport of carboxypeptidase Y to the vacuole. Cell 30:439–448

Walworth NC, Goud B, Kabcenell AD, Novick PJ (1989) Mutational analysis of SEC4 suggests a cyclical mechanism for the regulation of vesicular traffic. EMBO J 8:1685–1693

Weiss O, Holden J, Rulka C, Kahn RA (1989) Nucleotide binding and cofactor activities of purified bovine brain and bacterially expressed ADP-ribosylation factor. J Biol Chem 264:21066–21072

Wuestehube LJ, Schekman R (1992) Reconstitution of transport from the endoplasmic reticulum to the Golgi complex using an ER-enriched membrane fraction from yeast. Methods Enzymol 219:124–136

Yoshihisa T, Barlowe C, Schekman R (1993) Requirement for a GTPase-Activating Protein in Vesicle Budding from the Endoplasmic Reticulum. Science 259: 1466–1468

CHAPTER 27

Ypt Proteins in Yeast and Their Role in Intracellular Transport

M. STROM and D. GALLWITZ

A. Introduction

Members of the superfamily of guanine nucleotide binding proteins (GTPases) mediate diverse cellular processes that are important for both prokaryotic and eukaryotic organisms. The monomeric Ras oncoproteins constitute one GTPase family, which like heterotrimeric GTPases are thought to be involved in signal transduction. The identification of the yeast *Saccharomyces cerevisiae YPT1* gene in the actin–β-tubulin gene cluster as belonging to the Ras superfamily initiated new studies on the cellular functions of small GTPases (GALLWITZ et al. 1983). Since the initial isolation of Ypt1p, the number of related proteins in *S. cerevisiae* and the corresponding Rab proteins in mammalian systems has grown so dramatically that now they constitute their own family, called Ypt in yeast and Rab in mammals. There are now six Ypt proteins in *S. cerevisiae* and more than 20 Rab proteins identified in mammals. Their small size and high degree of sequence conservation facilitates the isolation of new family members using the polymerase chain reaction and low stringency DNA hybridization techniques (TOUCHOT et al. 1987; HAUBRUCK et al. 1987, 1990; HENGST et al. 1990; MIYAKE and YAMAMOTO 1990; CHAVRIER et al. 1992; ELFERINK et al. 1992).

This independent classification of a Ypt/Rab protein family is not only justified based on primary sequence similarity among these proteins, but also since their cellular function is significantly different from Ras or Rho proteins. Members of the Ypt/Rab protein family function in directing protein transport in the secretory and endocytic pathways in *S. cerevisiae* and mammals. The importance of the Ypt proteins in *S. cerevisiae* for cellular function is shown by the fact that deletion of these genes is often fatal for the cells or leads to growth inhibition at increased temperatures (SCHMITT et al. 1986; SALMINEN and NOVICK 1987). Deletion or mutations in the genes encoding the various Ypt proteins in *S. cerevisiae* results in accumulation of endoplasmic reticulum (ER) and ER-derived vesicles (Ypt1p) (SEGEV et al. 1988; SCHMITT et al. 1988; BECKER et al. 1991), accumulation of Golgi-derived vesicles (Sec4p) (SALMINEN and NOVICK 1987), and fragmentation of the vacuole (Ypt6p and Ypt7p) (unpublished results in this laboratory). The use of antibodies raised against the

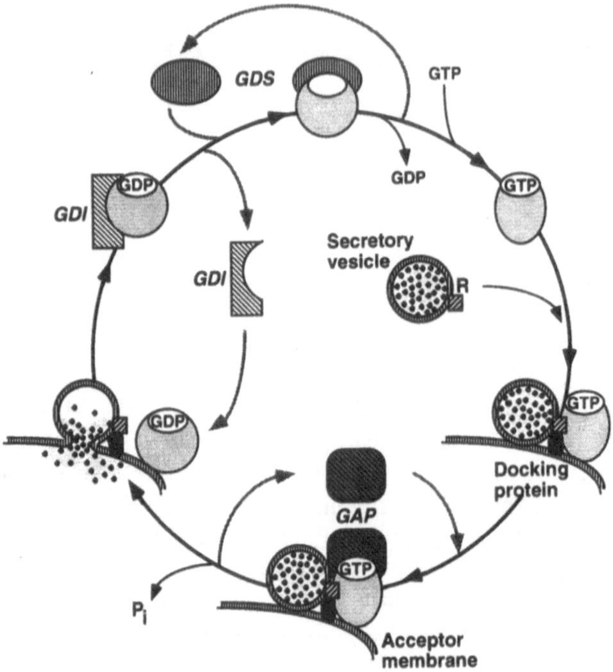

Fig. 1. Schematic representation of the putative functional cycle of Ypt proteins. It is assumed that the GTP-bound form of a given Ypt protein binds to specific carrier vesicles via a receptor protein (*R*). This complex recognizes a docking protein at the surface of the proper acceptor membrane. GTP hydrolysis mediated by a specific GTPase activating protein (*GAP*) is required either for vesicle fusion or for recycling of the Ypt protein. The removal of the GDP-bound protein from the acceptor membrane involves the association with a GDP dissociation inhibitor (*GDI*). Reactivation of the Ypt protein is mediated by a GDP dissociation stimulator (*GDS*) that catalyzes GDP/GTP exchange

mammalian Rab proteins also shows their association with various secretory and endocytic compartments including the ER, the Golgi apparatus, early and late endosomes, and various vesicular structures (Chavrier et al. 1990a; Goud et al. 1990; Plutner et al. 1991). The combined evidence from yeast and mammalian systems leaves little doubt that these proteins are involved in intracellular protein transport.

Figure 1 is a schematic representation depicting how Ypt/Rab proteins might function to mediate unidirectional flow of proteins through the cell, based on the original proposal by Bourne (1988). Although each interacting protein has not been identified for individual Ypt/Rab family members, this scheme considers the proteins that functionally interact with Ras, Rho, or Ypt proteins. In general it is thought that GTP-bound Ypt proteins mediate attachment of secretory vesicles at the acceptor membrane and that the hydrolysis of GTP occurs after the secretory vesicle has docked at the acceptor membrane. The subsequent GTP hydrolysis event either promotes vesicle fusion and consequent release of the vesicular contents into the acceptor organelle or is required for the protein to be competent to enter a new vesicle transport cycle. GAP, short for GTPase activating protein, accelerates the slow intrinsic activity of the GTPase and therefore may play a critical role in the vesicular fusion event. GDI, or guanine nucleotide dissociation inhibitor, dissociates the GDP-bound protein from the acceptor membrane and inhibits the release of GDP, thereby keeping the protein in its inactive form (SASAKI et al. 1991), but the effect of GDI on the protein transport mechanism is still unknown. GDS, or guanine nucleotide dissociation stimulator (YAMAMOTO et al. 1990), facilitates the exchange of GDP for GTP which is necessary for the "reactivation" of the Ypt protein so that it can reenter the vesicle targeting cycle. Nothing is known about the hypothetical receptor proteins on the secretory vesicles or the docking proteins on the acceptor membranes that are assumed to convey target specificity for the various Ypt/Rab proteins. In this review we will discuss the Ypt proteins in *S. cerevisiae*, their structure, their role in protein transport, and the recent identification of interacting proteins.

B. Ypt Proteins in *Saccharomyces cerevisiae*

The easy handling and the amenability of yeast to classical and molecular genetics are obvious advantages that this unicellular eukaryote offers for studying the cellular function of a given protein. Due to an efficient system of homologous recombination, chromosomal genes can be easily disrupted or replaced by in vitro mutagenized variants. Phenotypic alterations and metabolic defects can also be examined using conditionally lethal mutants or cells depleted of the protein under study. Protein depletion is achieved by placing the desired gene under transcriptional control of a regulable promoter, for example *GAL10*, that is easily turned on and off at any point in cell growth. These powerful methods are routinely used to supplement traditional genetic approaches and have been employed to investigate the role of the various Ypt proteins in the budding yeast *S. cerevisiae*.

I. Ypt1 Protein

The *YPT1* gene encodes a Ras-like protein of 206 amino acids (GALLWITZ et al. 1983). Disruption of the gene is lethal (SCHMITT et al. 1986; SEGEV and BOTSTEIN 1987). Cells depleted of Ypt1p as well as conditionally lethal *ypt1* mutants display various phenotypic alterations, including defects in the organization of cytoskeletal elements, abnormal calcium regulation, and a block in protein secretion (SEGEV and BOTSTEIN 1987; SEGEV et al. 1988; SCHMITT et al. 1986, 1988). Such mutants are characterized by an accumulation of ER membranes, 50-nm vesicles presumably derived from the ER, and of ER core-glycosylated proteins destined to reach the plasma membrane or the vacuole (SEGEV et al. 1988; SCHMITT et al. 1988; BACON et al. 1989; BECKER et al. 1991). The additional finding that the membrane-associated Ypt1p seems to be localized primarily in the Golgi apparatus led SEGEV et al. (1988) to suggest that the primary function of this protein is to regulate ER-to-Golgi protein transport. This was corroborated by in vitro protein transport studies which showed that anti-Ypt1p antibodies effectively block the protein flow from the ER to the Golgi apparatus and lead to the accumulation of ER-derived vesicles to which Ypt1p appears to be bound (BAKER et al. 1990; REXACH and SCHEKMAN 1991; SEGEV 1991). The Ypt1 protein is most likely required for targeting these vesicles to the *cis*-Golgi compartment.

A *GAL10–YPT1* fusion has been used to search for yeast genes capable of suppressing the loss of Ypt1p function. Four genes, *SLY1–20*, *SLY2*, *SLY12*, and *SLY41*, were identified whose protein products allow yeast cells to grow in the absence of Ypt1p (DASCHER et al. 1991). These genes are *YPT1*-specific suppressors, although they can also rescue two other secretion-defective *S. cerevisiae* mutants (*sec21* and *sec22*), albeit with varying efficiency (OSSIG et al. 1991). *SLY1–20* is a single-copy suppressor that encodes a mutant protein with a single amino acid substitution from the wild-type Sly1p. *SLY2*, *SLY12*, and *SLY41* are genes that act as multicopy suppressors. Most interestingly, recent sequencing studies have shown that *SLY2* is allelic with *SEC22* (NEWMAN et al. 1992a) and that *SLY12* is identical to *BET1* (FERRO-NOVICK, personal communication). Both *SEC22* and *BET1* are involved in ER-to-Golgi protein transport and had been isolated previously by completely different strategies (NOVICK et al. 1980; NEWMAN and FERRO-NOVICK 1987). Although the precise mode of action of the *YPT1* suppressors has not yet been elucidated, the strategy used for their isolation resulted in identifying gene products acting in the same pathway as the GTPase itself.

II. Sec4 Protein

Conditionally lethal *sec4* mutants were shown to be blocked at a late stage in the secretion pathway and to massively accumulate Golgi-derived vesicles well before Sec4p was identified as a Ypt1-related protein (NOVICK et al. 1980, 1981). The *SEC4* gene encodes a GTPase of 215 amino acids whose primary structure is 47.5% identical with that of Ypt1p (SALMINEN and NOVICK 1987).

Currently, Ypt1p and Sec4p are the best studied proteins of the Ypt family. Sec4p cycles between, and is associated with, post-Golgi vesicles and the plasma membrane. In temperature-sensitive *sec4* mutants, the protein is primarily bound to the accumulated vesicles (GOUD et al. 1988). As seen with Ypt1p, an asparagine-to-isoleucine substitution in the conserved G-4 region (NKXD – discussed in Sect. C.I) of Sec4p results in a block of secretion in a dominant fashion (WALWORTH et al. 1989).

A detailed discussion of the function of Sec4p is presented in Chap. 4. It is clear now that Ypt1p and Sec4p both act in unidirectional vesicle transport using the same basic mechanism. The fact that mammalian Rab1p can complement *YPT1*-defective mutants and ypt2p of *Schizosaccharomyces pombe* (*ypt2* is the *S. pombe* homolog to *SEC4*) is able to rescue conditional *sec4* mutants suggests that most Ypt/Rab proteins fulfill similar or identical functions in all eukaryotic cells (Haubruck et al. 1989, 1990).

III. Ypt3, Ypt6 and Ypt7 Proteins

Using low stringency hybridization and the polymerase chain reaction (PCR) technology, we have recently isolated several genes from the budding yeast *S. cerevisiae* that encode other members of the Ypt protein family. These are listed in Table 1 together with their mammalian and *S. pombe* homologs. That a particular Rab protein is the homolog to a *S. cerevisiae* Ypt protein is first indicated by the high degree of sequence conservation and an identical effector region (Table 2). Additionally, it has been shown for *YPT1* and *YPT6* that mutants defective in these genes can be complemented by their structural homologs from the fission yeast *S. pombe* (Miyake and Yamamoto 1990; Hengst et al. 1990) and from mammals (Haubruck et al. 1989; Hengst et al. 1990).

YPT3A and *YPT3B* encode proteins that are more than 80% identical and share the highest similarity with the *S. pombe* Ypt3p (Miyake and Yamamoto 1990) and the mammalian Rab11p (Chavrier et al. 1990b). Whereas the deletion of either *YPT3A* or *YPT3B* has no apparent phenotype, disruption of both genes is lethal. The deletion of *YPT6* or *YPT7* does not interfere with cell viability, but *ypt6* null mutants are

Table 1. *S. cerevisiae YPT* genes and their mammalian and *S. pombe* homologs

S. cerevisiae		Mammals	*S. pombe*	
Gene	Gene disruption (growth phenotype)	Gene	Gene	Gene disruption (growth phenotype)
YPT1	Lethality	*rab1A,B**	*ypt1**	Lethality
SEC4	Lethality	?	*ypt2**	Lethality
YPT3A *YPT3B*	Lethality (when both genes are disrupted)	*rab11*	*ypt3*	Lethality
?	?	*rab5*	*ypt5*	?
YPT6	Temperature sensitivity	*rab6**	*ryh1**	Temperature sensitivity
YPT7	Like wild-type	*rab7*	?	?

The table lists the known *YPT* genes of the budding yeast *S. cerevisiae* and the growth phenotypes following gene disruption.
* Homologous genes from either mammals or the fission yeast *S. pombe* that have been shown to complement the functional loss of the respective *S. cerevisiae* genes.

Table 2. Sequence comparison of Ypt/Rab family proteins

Organism	Protein	Effector region				C Terminus	
S. cerevisiae	Ypt1	35	ND	YISTIGVDF	KI	202	GGGCC
S. pombe	ypt1	35	ES	YISTIGVDF	KI	199	SSNCC
H. sapiens	rab1	38	ES	YISTIGVDF	KI	201	GGGCC
S. cerevisiae	Sec4	47	PS	FITTIGIDF	KI	211	KSNCC
S. pombe	ypt2	36	PS	FITTIGIDF	KI	196	VKRCC
S. cerevisiae	Ypt3A	40	MD	SKSTIGVEF	AT	219	GNNCC
S. cerevisiae	Ypt3B	40	IE	SKSTIGVEF	AT	218	SSNCC
S. pombe	ypt3	37	IE	SKSTIGVEF	AT	210	SSQCC
C. familiaris	rab11	38	LE	SKSTIGVEF	AT	212	KVQCCQNI
S. cerevisiae	Ypt6	37	DH	YQATIGIDF	LS	212	SACQC
S. pombe	ryh1	38	NT	YQATIGIDF	LS	197	SSCNC
H. sapiens	rab6	40	NT	YQATIGIDF	LS	204	GGCSC
S. cerevisiae	Ypt7	35	QQ	YKATIGADF	LT	204	NSCSC
C. familiaris	rab7	35	NQ	YKATIGADF	LT	203	ESCSC

S. cerevisiae, *Saccharomyces cerevisiae*; *S. pombe*, *Schizosaccharomyces pombe*; *H. sapiens*, *Homo sapiens*; *C. familiaris*, *Canida familiaris*.
The table shows the effector region and C-terminal sequences of the known *S. cerevisiae* Ypt proteins and their mammalian and *S. pombe* counterparts. Identical effector regions and C-terminal cysteine residues are highlighted. Numbers to the left indicate the position of the first residue of the sequence segments shown in the respective polypeptide.

temperature sensitive, as are *ryh1* deletion strains of *S. pombe* (HENGST et al. 1990).

S. cerevisiae mutants defective in *YPT3A*, *YPT3B*, *YPT6*, or *YPT7* are impaired, to a different extent, in vacuolar protein processing and sorting and in α-factor degradation, and they are characterized by an aberrant vacuolar morphology (WICHMANN et al. 1992). Protein secretion is not affected in either of these mutants. It seems likely therefore that all of these GTPases have a role in protein transport in the endocytic pathway which is ill defined in yeast. The transport steps in which these Ypt proteins might act is indicated in Fig. 2, taking into account the available data.

C. Ypt Protein Structure

I. Nucleotide Binding

Monomeric GTPases including the Ypt proteins have five highly conserved short stretches of amino acids that have been named G-1 through G-5 (BOURNE et al. 1991; VALENCIA et al. 1991). These five regions are important for nucleotide binding, as mutations in amino acids within these regions lead to deleterious effects on the protein's ability to bind guanine nucleotides. The conserved sequence segments G-1 through G-5 were also shown by

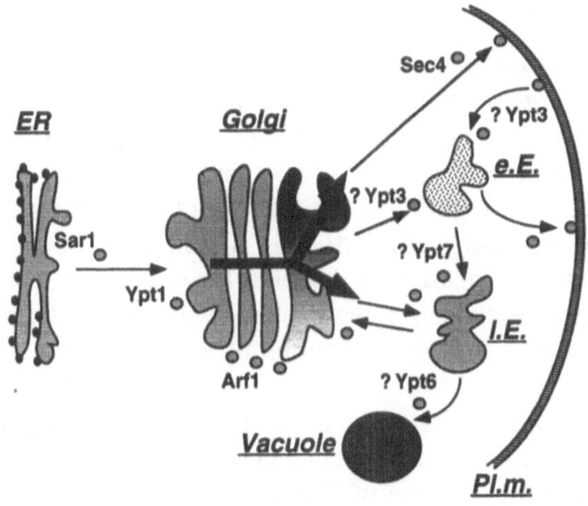

Fig. 2. Involvement of *S. cerevisiae* Ypt proteins in protein transport. It is assumed that protein transport through different compartments of the secretory and endocytic pathway takes place via vesicular intermediates and that Ypt proteins are responsible for vesicular targeting to specific acceptor membranes. According to preliminary data, Ypt3p, Ypt6p, and Ypt7p are likely to regulate various post-Golgi transport steps. The role of Ypt1p and Sec4p in ER-to-Golgi and Golgi-to-plasma membrane, respectively, is supported by data mentioned in the text. *ER*, endoplasmic reticulum; *Pl.m.*, plasma membrane; *e.E.* and *l.E.*, early and late endosome, respectively

X-ray crystallographic analyses of the human H-Ras protein to interact with the bound nucleotide (Pai et al. 1989). Although there have been no crystallographic studies on any Ypt or Rab protein to date, the basic structure of these proteins is likely to be very similar to that of Ras proteins. The high degree of structural conservation among the Ras and Ypt/Rab protein families is demonstrated by mutations in the G-4 region (NKXD). The identical amino acid substitution (isoleucine for asparagine) in G-4 of the Ypt1p, Sec4p, Rab5p, and mammalian Ras proteins leads to a severe impairment in the protein's nucleotide binding capability and results in dominant negative effects (Walter et al. 1986; Wagner et al. 1987; Schmitt et al. 1986, 1988; Walworth et al. 1989; Gorvel et al. 1991).

In addition, substitution of lysine with methionine in G-1 (GXXXXGK$\frac{S}{T}$) renders Ypt1p nonfunctional, and substitution of alanine with threonine in G-3 (DTAG) leads to autophosphorylation of Ypt1p (Wagner et al. 1987) in exactly the same way as the equivalent substitutions in oncogenic viral Ras proteins (Shih et al. 1982). The nucleotide-bound H-Ras p21 crystal structure shows contacts between the lysine in G-1 and the β- and γ-phosphates and also the juxtaposition of the alanyl residue in G-3 with the γ-phosphate of the bound nucleotide (Pai et al. 1989). Therefore, Ypt1p G-1 and G-3 mutation results support the belief that the three-dimensional structure of Ypt is similar to Ras.

II. Effector Region

One of the highly conserved regions, G-2, is also known as the effector region. Table 2 compares the sequences of this region and the C termini of the known Ypt proteins of the budding yeast *S. cerevisiae* and their counterparts in the fission yeast *S. pombe* and in mammals. The sequence identity of the effector region indicates the functional similarity of proteins from the various organisms studied. This is supported by the observation that, in all cases tested, yeast cells depleted for one of the Ypt proteins are only complemented by those foreign proteins with identical effector regions (MIYAKE and YAMAMOTO 1990; HAUBRUCK et al. 1989, 1990; HENGST et al. 1990).

In the crystallographic studies on H-Ras, this region was shown to undergo a drastic conformational change upon GTP hydrolysis (SCHLICHTING et al. 1990). The effector region of Ypt proteins is thought to be the site of interaction with the GAP. This is supported by mutations in the effector region of Ypt1p that either decrease or abolish the protein's sensitivity to partially purified Ypt-GAP from yeast and porcine liver (BECKER et al. 1991; TAN et al. 1991). These mutations were either lethal or inhibited cell growth at increased temperatures (BECKER et al. 1991). Furthermore, the addition of synthetic peptides of the Rab3p effector region has been shown to inhibit protein transport from the ER to the Golgi in a mammalian cell-free system (PLUTNER et al. 1990).

The intrinsic GTPase activity of Ypt proteins is very slow. Values of $0.006\,\text{min}^{-1}$ and $0.0012\,\text{min}^{-1}$ have been determined at 30°C for Ypt1p and Sec4p, respectively (WAGNER et al. 1987; KABCENELL et al. 1990). GTPase activating proteins therefore are likely to play a critical role for the cycling of the Ypt proteins between an active and inactive state.

III. C Terminus

All members of the Ras superfamily of proteins have one or two cysteine residues at or near their carboxyl terminal end. Most of the mammalian Rab proteins and all known yeast Ypt proteins terminate with either two consecutive cysteines or the sequence CXC (Table 2). As was first shown for yeast Ypt1p, at least one of the C-terminal cysteine residues is required for membrane association. Deletion of both cysteines or their substitution with serines result in cytoplasmically soluble, biologically inactive proteins (MOLENAAR et al. 1988). Similar observations have been made with Sec4p (WALWORTH et al. 1989).

It now seems certain that in all members of the Ypt/Rab protein family either one or both C-terminal cysteines are isoprenylated (FARNSWORTH et al. 1991; KHOSRAVI-FAR et al. 1991; KOHL et al. 1991). Apparently specific modification by palmitic acid in the thioester linkage of one of the two cysteines in *S. cerevisiae* Ypt1p (MOLENAAR et al. 1988) and *S. pombe* Ypt3p

(Newman et al. 1992b) has also been observed. The significance of this type of lipidation is not entirely clear, but it should be noted that it has been found only when two consecutive cysteines are present. Geranylgeranylation of Ypt1p and Sec4p is likely to be performed by the same transferase. A mutant allele of the *S. cerevisiae BET2* gene, originally isolated from a secretion-defective mutant (Newman and Ferro-Novick 1987), has been shown to encode a protein related to the β subunit of yeast and mammalian farnesyl transferases (Rossi et al. 1991).

Although the hydrophobic C-terminal adducts mediate membrane association of the proteins, they are neither sufficient for membrane anchoring, nor specific enough to explain the selectivity of membrane attachment of the multiple Ypt/Rab family members. The specific binding of different Ypt proteins for different cellular membranes seems to be directed in part by sequences upstream of the cysteine residues, since exchanging the C-terminal 30–40 amino acids between different Rab proteins changes their cellular localization (Chavrier et al. 1991).

D. GTPase Activating Proteins for YPT Family Members

Since protein secretion is a complex process involving several intracellular compartments and vesicles, it is often speculated that the different Ypt proteins act by directing the formation of protein complexes that mediate vesicle fusion/budding at the appropriate target membrane, thereby mediating the flow of proteins through the cell. Identification of proteins that interact directly with GTPases is another approach to isolating proteins involved directly in the Ypt protein pathway. As described in the introduction (Sect. A) and depicted in Fig. 1, various proteins have been identified that affect the hydrolysis and/or binding of guanine nucleotides by various Ras or Ras-related proteins. In our laboratory, we have focused on the identification and isolation of GAPs. Studies discussed in Sect C.II showed that the effector region is a site of interaction with GAP. Complementation studies with yeast mutants defective in different *YPT* genes have shown that the effector region is an important determinant for the functional specificity of various Ypt proteins. It seems therefore likely that specific GAPs exist for different members of the Ypt/Rab protein family.

Preliminary evidence in our laboratory indicates that this is indeed the case. By ion exchange chromatography of cytosolic yeast proteins, GAP activities for Ypt1p, Ypt3Ap, and Ypt6 can clearly be distinguished. A GAP of about 40 kDa has been partially purified from the *S. cerevisiae* cytosol using Ypt1p as a substrate (Tan et al. 1991). This protein accelerates the intrinsic GTPase activity of Ypt1p and mammalian Rab1p, but an effector mutant of Ypt1 protein and H-Ras p21 do not serve as substrates (Tan et al. 1991; Becker et al. 1991). We have recently cloned the gene encoding the

GAP for Ypt6p (STROM et al. 1993). The primary structure of the Gyp6 protein (GAP of Ypt protein 6) is unrelated to known GAPs specific for either Ras, Rho, or Rac proteins. It enhances the GTP hydrolyzing activity of Ypt6p and to a limited extent that of Ypt7p. It is, however, not reactive with Ypt1, Ypt2 (from *S. pombe*), or Ypt3A proteins.

Evidence for the existence of multiple GAPs with apparent specificity for different Rab proteins has also been obtained from our and other laboratories. From porcine liver and intestine, GAPs for Rab1 (TAN et al. 1991) and Rab2 (W. LAUFER and D. GALLWITZ, unpublished results) have been partially purified, and a GAP for Rab3 has been identified from rat brain (BURSTEIN et al. 1991).

E. Summary

Figure 2 represents a summary of where in the secretory and endocytic pathways the known Ypt proteins most likely function or, based on experimental evidence, are postulated to function. Through genetic studies with conditional mutants and null alleles and biochemical analyses of proteins that are secreted or modified in the secretion pathway, critical steps for the functional involvement of the various Ypt proteins in yeast can be pinpointed. In addition, in vitro protein transport systems have been successfully used to define the role of Ypt1p in the targeting of ER-derived vesicles to the *cis* Golgi compartment (REXACH and SCHEKMAN 1991; SEGEV 1991) and of Sar1p, a GTPase belonging to the Arf family (NAKANO and MURAMATSU 1989), in the budding of transport vesicles from the ER (D'ENFERT et al. 1991; OKA et al. 1991).

Within the *trans* Golgi compartment, vacuolar proteins are sorted from proteins destined to reach the plasma membrane and most likely pass through endosomal compartments. In studies on the transport of the a-factor peptide from the plasma membrane to the vacuole, the existence of such endosomal compartments in yeast has recently been documented (SINGER and RIEZMAN 1990). Preliminary studies from our laboratory suggest that Ypt3p, Ypt6p, and Ypt7p promote several transport steps in the endocytic pathway.

Acknowledgements. We are indebted to many colleagues in our laboratory for helpful discussions and especially to Dr. Warren Kibbe for critical reading of this manuscript. The work from the laboratory of D. Gallwitz was supported from the Max Planck Society, the Deutsche Forschungsgemeinschaft, and Fonds des Chemischen Industrie. M.S. is supported by a Human Frontier Science Program postdoctoral fellowship.

References

Bacon RA, Salminen A, Ruohola H, Novick P, Ferro-Novick S (1989) The GTP-binding protein Ypt1 is required for transport *in vitro*: the Golgi apparatus is defective in Ypt1 mutants. J Cell Biol 109:1015–1022

Baker D, Wuestehube L, Schekman R, Botstein D, Segev N (1990) GTP-binding Ypt1 protein and Ca^{2+} function independently in a cell-free protein transport reaction. Proc Natl Acad Sci USA 87:355–359

Becker J, Tan TJ, Trepte H-H, Gallwitz D (1991) Mutational analysis of the putative effector domain of the GTP-binding Ypt1 protein in yeast suggests specific regulation by a novel GAP activity. EMBO J 10:785–792

Bourne H (1988) Do GTPases direct membrane traffic in secretion? Cell 53:669–671

Bourne HR, Sanders DA, McCormick F (1991) The GTPase superfamily: conserved structure and molecular mechanism. Nature 349:117–127

Burstein ES, Linko-Stentz K, Lu Z, Macara IG (1991) Regulation of the GTPase Activity of the ras-like Protein p25 rab3A. J Biol Chem 266:2689–2692

Chavrier P, Parton RG, Hauri HP, Simons K, Zerial M (1990a) Localization of low molecular weight GTP binding proteins to exocytic and endocytic compartments. Cell 62:317–329

Chavrier P, Vingron M, Sander C, Simons K, Zerial M (1990b) Molecular cloning of *YPT1/SEC-4*-related cDNAs from an epithelial cell line. Mol Cell Biol 10:6578–6585

Chavrier P, Gorvel J-P, Stelzer E, Simons K, Gruenberg J, Zerial M (1991) Hypervariable C-terminal domain of Rab proteins acts as a targeting signal. Nature 353:769–772

Chavrier P, Simons K, Zerial M (1992) The complexity of the Rab and Rho GTP binding protein subfamilies revealed by a PCR cloning approach. Gene 112:261–264

d'Enfert C, Wuestehube LJ, Lila T, Schekman R (1991) Sec12p-dependent membrane binding of the small GTP-binding protein Sar1p promotes formation of transport vesicles from the ER. J Cell Biol 114:663–670

Dascher C, Ossig R, Gallwitz D, Schmitt HD (1991) Identification and structure of four yeast genes (*SLY*) that are able to suppress the functional loss of *YPT1*, a member of the Ras superfamily. Mol Cell Biol 11:872–885

Elferink LA, Anzai K, Scheller RH (1992) Rab15, a novel low molecular weight GTP-binding protein specifically expressed in rat brain. J Biol Chem 267:5768–5775

Farnsworth CL, Marshall MS, Gibbs JB, Stacey DW, Feig LA (1991) Preferential inhibition of the oncogenic form of Ras^{H} by mutations in the GAP binding/effector domain. Cell 64:625–633

Gallwitz D, Donath C, Sander C (1983) A yeast gene encoding a protein homologous to the human c-Has/Bas proto-oncogene product. Nature 306:704–707

Gorvel J-P, Chavrier P, Zerial M, Gruenberg J (1991) Rab5 controls early endosome fusion in vitro. Cell 64:915–925

Goud B, Salminen A, Walworth NC, Novick PJ (1988) A GTP-binding protein required for secretion rapidly associates with secretory vesicles and the plasma membrane in yeast. Cell 53:753–768

Goud B, Zahraoui A, Tavitian A, Saraste J (1990) Small GTP-binding protein associated with Golgi cisternae. Nature 345:553–556

Haubruck H, Disela C, Wagner P, Gallwitz D (1987) The Ras-related Ypt protein is an ubiquitous eukaryotic protein: isolation and sequence analysis of mouse cDNA clones highly homologous to the yeast *YPT1* gene. EMBO J 6:4049–4053

Haubruck H, Prange R, Vorgias C, Gallwitz D (1989) The Ras-related mouse Ypt1 protein can functionally replace the *YPT1* gene product in yeast. EMBO J 8:1427–1432

Haubruck H, Engelke U, Mertins P, Gallwitz D (1990) Structural and functional analysis of *YPT2*, an essential Ras-related gene in the fission yeast *Schizosaccharomyces pombe* encoding a Sec4 protein homologue. EMBO J 9:1957–1962

Hengst L, Lehmeier T, Gallwitz D (1990) The *RYH1* gene in the fission yeast *Schizosaccharomyces pombe* encoding a GTP-binding protein related to Ras,

Rho and Ypt: structure, expression and identification of its human homologue. EMBO J 9:1949–1955
Kabcenell AK, Goud B, Northrup JK, Novick PJ (1990) Binding and hydrolysis of guanine nucleotides by Sec4p, a yeast protein involved in the regulation of vesicular traffic. J Biol Chem 265:9366–9372
Khosravi-Far R, Lutz RJ, Cox AD, Conroy L, Bourne JR, Sinensky M, Balch WE, Buss JE, Der CJ (1991) Isoprenoid modification of rab proteins terminating in CC or CXC motifs. Proc Natl Acad Sci USA 88:6264–6268
Kohl NE, Diehl RE, Schaber MD et al. (1991) Structural homology among mammalian and *Saccharomyces cerevisiae* isoprenyl-protein transferases. J Biol Chem 266:18884–18888
Miyake S, Yamamoto M (1990) Identification of Ras-related, *YPT* family genes in *Schizosaccharomyces pombe*. EMBO J 9:1417–1422
Molenaar CMT, Prange R, Gallwitz D (1988) A carboxyl-terminal cysteine residue is required for palmitic acid binding and biological activity of the Ras-related yeast Ypt1 protein. EMBO J 7:971–976
Nakano A, Muramatsu M (1989) A novel GTP-binding protein, Sar1p, is involved in transport from the Endoplasmic Reticulum to the Golgi Apparatus. J Cell Biol 109:2677–2691
Newman AP, Ferro-Novick S (1987) Characterization of new mutants in the early part of the yeast secretory pathway isolated by a [^{3}H]mannose suicide selection. J Cell Biol 105:1587–1594
Newman AP, Mancini P, Rossi G, Lian JP, Ferro-Novick S (1992a) *SEC22* and *SLY2* are identical. Mol Cell Biol (in press)
Newman CMH, Giannakouros T, Hancock JF, Fawell EH, Armstrong J, Magee AI (1992b) Post-translational processing of *Schizosaccharomzces pombe* Ypt proteins. J Biol Chem 267:11329–11336
Novick P, Field C, Schekman R (1980) Identification of 23 complementation groups required for post-translational events in the yeast secretory pathway. Cell 21:205–215
Novick P, Ferro S, Schekman R (1981) Order of events in the yeast secretory pathway. Cell 25:461–469
Oka T, Nishikawa S, Nakano A (1991) Reconstitution of GTP-binding Sar1 protein function in ER to Golgi transport. J Cell Biol 114:671–679
Ossig R, Dascher C, Trepte H-H, Schmitt HD, Gallwitz D (1991) The yeast *SLY* gene products, suppressors of defects in the essential GTP-binding Ypt1 protein, may act in Endoplasmic Reticulum-to-Golgi transport. Mol Cell Biol 11:2980–2993
Pai EF, Kabsch W, Krengel U, Holmes KC, John J, Wittinghofer A (1989) Structure of the guanine-nucleotide-binding domain of the Ha-ras oncogene product p21 in the triphosphate conformation. Nature 341:209–214
Plutner H, Schwaninger R, Pind S, Balch WE (1990) Synthetic peptides of the Rab effector domain inhibit vesicular transport through the secretory pathway. EMBO J 9:2375–2383
Plutner H, Cox AD, Pind S, Khosravi-Far R, Bourne JR, Schwaninger R, Der CJ, Balch WE (1991) Rab1b regulates vesicular transport between the Endoplasmic Reticulum and successive Golgi compartments. J Cell Biol 115:31–43
Rexach MF, Schekman RW (1991) Distinct biochemical requirements for the budding, targeting, and fusion of ER-derived transport vesicles. J Cell Biol 114:219–229
Rossi G. Jiang Y, Newman AP, Ferro-Novick S (1991) Dependence of Ypt1 and Sec4 membrane attachment on *BET2*. Nature 351:158–161
Salminen A, Novick P (1987) A Ras-like protein is required for a post-Golgi event in yeast secretion. Cell 49:527–538
Sasaki T, Kaibuchi K, Kabcenell AK, Novick PJ, Takai Y (1991) A mammalian inhibitory GDP/GTP exchange protein GDI Dissociation Inhibitor) for Smg p25A is active on the yeast Sec4 protein. Mol Cell Biol 11:2909–2912

Schlichting I, Almo SC, Gert R et al. (1990) Time-resolved X-ray crystallographic study of the conformational change in Ha-Ras p21 protein on GTP dydrolysis. Nature 345:309–315

Schmitt HD, Wagner P, Pfaff E, Gallwitz D (1986) The Ras-related *YPT1* gene product in yeast: A GTP-binding protein that might be involved in microtubule organization. Cell 47:401–412

Schmitt HD, Puzicha M, Gallwitz D (1988) Study of a temperature-sensitive mutant of the Ras-related *YPT1* gene product in yeast suggests a role in the regulation of intracellular calcium. Cell 53:635–647

Segev N (1991) Mediation of the attachment of fusion step in vesicular transport by the GTP-binding of Ypt1 protein. Science 252:1553–1556

Segev N, Botstein D (1987) The Ras-like *YPT1* gene is itself essential for growth, sporulation and starvation response. Mol Cell Biol 7:2367–2377

Segev N, Mulholland J, Botstein D (1988) The yeast GTP-binding Ypt1 protein and a mammalian counterpart are associated with the secretion machinery. Cell 52:915–924

Shih TY, Stokes PE, Smythers GW, Dhar R, Oroszlan S (1982) Characterization of the phosphorylation sites and the surrounding amino acid sequences of the p21 transforming proteins coded for by the Harvey and Kirsten strains of murine sarcoma viruses. J Biol Chem 257:11767–11773

Singer B, Riezman H (1990) Detection of an intermediate compartment involved in transport of *α*-factor from the plasma membrane to the vacuole in yeast. J Cell Biol 110:1911–1922

Strom S, Vollmer P, Tan TJ, Gallwitz D (1993) A GTPase-activating protein that interacts specifically with a member of the Ypt/Rab family. Nature 361:736–739

Tan TJ, Vollmer P, Gallwitz D (1991) Identification and partial purification of GTPase-activating proteins from yeast and mammalian cells that preferentially act on Ypt1/Rab1 proteins. FEBS Lett 291:322–326

Touchot N, Chardin P, Tavitian A (1987) Four additional members of the ras gene superfamily isolated by an oligonucleotide strategy: molecular cloning of *YPT*-related cDNAs from a rat brain library. Proc Natl Acad Sci USA 84:8210–8214

Valencia A, Chardin P, Wittinghofer, Sander C (1991) The Ras protein family: evolutionary tree and role of conserved amino acids. Biochemistry 30:4637–4648

Wagner P, Molenaar CMT, Rauh AJG, Brökel R, Schmitt HD, Gallwitz D (1987) Biochemical properties of the Ras-related Ypt protein in yeast: a mutational analysis. EMBO J 6:2373–2379

Walter M, Clark SG, Levinson AD (1986) The oncogenic activation of human $p21^{ras}$ by a novel mechanism. Science 233:649–652

Walworth NC, Goud B, Kabcenell AK, Novick PJ (1989) Mutational analysis of *SEC4* suggests a cyclical mechanism for the regulation of vesicular traffic. EMBO J 8:1685–1693

Wichmann H, Hengst L, Gallwitz D (1992) Endocytosis in Yeast: Evidence for the Involvement of a Small GTP-Binding Protein (Ypt7p). Cell 71:1131–1142

Yamamoto T, Kaibuchi K, Mizuno T, Hiroyoshi M, Shirataki H, Takai Y (1990) Purification and characterization from bovine brain cytosol of proteins that regulate the GTP exchange reaction of Smg p21s, Ras p21-like GTP proteins. J Biol Chem 265:16626–16634

CHAPTER 28

Compartmentalization of rab Proteins in Mammalian Cells

V.M. OLKKONEN, P. DUPREE, L.A. HUBER, A. LÜTCKE, M. ZERIAL, and K. SIMONS

A. Subcellular Compartmentalization and Membrane Traffic

Classically, membrane-bound compartments have been viewed as comprising physically distinct entities each having a specific protein composition and characteristic functions. By this criterion, compartment boundaries should be easily delineated by ultrastructural localization of resident proteins. This simple view, however, is complicated by several factors. For example, even the localization of two proteins to different regions might not unequivocally demonstrate that they are present in physically discontinuous compartments. Such a distribution of proteins could reflect the existence of subdomains in an otherwise continuous membrane array. An example is the lateral heterogeneity found between the rough and the smooth endoplasmic reticulum (ER). Conversely, markers may be segregated into physically distinct membranes which nevertheless remain functionally continuous by repeated membrane fissions and fusions or transient tubular interconnections. The existence of such interactions among "like" elements would play an important role in preserving continuity among equivalent, but physically separated organelles. Furthermore, the definition of a compartment becomes more complicated if one considers organelles involved in membrane traffic. Given that membrane material is continuously transported and recycled between two organelles, precise biochemical definition of the compartments may be difficult. Some components are, of course, known to be resident, due perhaps to interactions with compartment-specific structural frameworks, and these can be used for compartment characterization. However, precise definition of the subcellular compartments requires detailed knowledge of the machinery regulating the entry and the exit of material at compartment boundaries, as well as the retention of molecules within the compartment (for further discussion see MELLMAN and SIMONS 1992). One family of proteins that may play an important role in defining compartment boundaries are the rab proteins, small GTPases that will be discussed in this chapter.

I. Membrane Trafficking

Membrane transport between subcellular compartments is thought to take place via vesicular carriers which bud off from one compartment and fuse with another (PALADE 1975) or via tubular transport intermediates (see HOPKINS et al. 1990; KREIS 1992). Despite the extensive intracellular flux of molecules, the cell maintains the compartmentalization required for vital biochemical processes to be executed. Furthermore, the cell is capable of large-scale organelle rearrangements, such as mitotic fragmentation and subsequent reassembly of the nuclear membrane, the ER, and the Golgi apparatus (for a review see WARREN 1985). All these processes require strict control of membrane fission and fusion events, precise targeting of vesicular or tubular carrier structures, and efficient recycling of components of the transport machineries.

1. Indications for a Role of Sec4/Ypt1/rab GTPases

A number of common components generally involved in intracellular membrane fusion events are known (for review see ROTHMAN and ORCI 1992). However, the answers to the key questions of the molecular principles of transport pathway organization reside in the isolation and characterization of specific components regulating events at compartment boundaries. At the moment, rab proteins belonging to the ras superfamily of small GTPases are among the best candidates for such regulatory factors. The idea of rab proteins being involved in regulation of membrane traffic was originally based on the findings that mutations in the homologous Sec4 (see Chap. 4) and Ypt1 proteins in *Saccharomyces cerevisiae* led to transport blocks in the secretory pathway. The Sec4 protein was found on the cytoplasmic surface of the plasma membrane and on secretory vesicles, and SEC4 mutants showed accumulation of post-Golgi transport vesicles (SALMINEN and NOVICK 1987; GOUD et al. 1988). The Ypt1 protein was shown to function in an early transport step, between the ER and the Golgi apparatus (SEGEV et al. 1988; SCHMITT et al. 1988). Based on these findings and the known capacity of GTPases to function as molecular switches in a variety of biological processes, BOURNE (1988) proposed that these small GTPases may monitor the specificity of interaction between targeting molecules on the transport vesicle and the acceptor membrane. The vectorial nature of the transport process would be ensured by cycles of GTP binding and hydrolysis by the small GTPase.

Several in vitro reconstituted mammalian cell transport processes were known to be inhibited by a nonhydrolyzable GTP analog, GPTγS (MELANCON et al. 1987; GODA and PFEFFER 1988; BECKERS and BALCH 1989; MAYORGA et al. 1989; ORCI et al. 1989). This led to the idea that GTPases, perhaps rab proteins by analogy to the yeast Sec4p and Ypt1p, could play a central role in regulating membrane traffic and organelle function in mammalian cells. This hypothesis was supported by the first studies on the intracellular distribution of the mammalian rab proteins (CHAVRIER et al. 1990b; GOUD et al. 1990), which indicated that these proteins are specifically localized to membrane compartments along the endo- and exocytic transport pathways.

In mammalian cells, the rab proteins form a large family of small GPTases. At the moment, cDNAs for more than 30 rab proteins have been cloned, and the proteins encoded are being characterized. The first rab sequences, rab1–rab4, were cloned by the groups of Tavitian (TOUCHOT et al. 1987) and Gallwitz (HAUBRUCK et al. 1987). Later, more rab proteins were identified by these and other groups (rab3b and c, MATSUI et al. 1988; rab5 and rab6, ZAHRAOUI et al. 1989; rab7, BUCCI et al. 1988; rab4b, rab8–rab11, CHAVRIER et al. 1990a; rab1b, VIELH et al. 1989; rab5b, WILSON and WILSON 1992). Scheller's group isolated cDNAs for rab12–rab16 (ELFERINK et al. 1992). Furthermore, the polymerase chain reaction (PCR) based approach of CHAVRIER et al. (1992) improved the efficiency of cloning sequences expressed at low levels, allowing the identification of rab5c and rab17–rab24. In our laboratories, a comprehensive study on the rab family of small GTPases is underway. This review will focus on the compartmentalization of rab proteins and their function in compartment dynamics.

B. Localization of rab Proteins on Subcellular Compartments

I. The rab Proteins Associated with the Biosynthetic Route

1. Endoplasmic Reticulum and Golgi Apparatus

A mammalian homolog of the yeast Ypt1 protein, rab1b, which is involved in ER to Golgi transport (see Sect. A.I.1.), was localized to the smooth ER and the Golgi, as shown by immunofluorescence and by subcellular fractionation (PLUTNER et al. 1991). The protein rab2 was shown to reside on one *cis* cisterna of the Golgi apparatus and in the intermediate region between the ER and the Golgi (CHAVRIER et al. 1990b). This region is under intensive study and its division into distinct compartments is unclear. It comprises structures on the smooth ER previously called transitional elements and a tubular network on the *cis* side of the Golgi complex now called the *cis* Golgi network. In addition, tubulo-vesicular structures involved in transport between the ER and the Golgi complex are included (for a discussion see HAURI and SCHWEIZER 1992; KLAUSNER et al. 1992; KREIS 1992; MELLMAN and SIMONS 1992; SARASTE and KUISMANEN 1992). rab1b and rab2 colocalized in this intermediate region (PLUTNER et al. 1991). Functional studies on rab1 and rab2 are discussed in Sect. C.II.1.

In contrast to the rab1 and rab2 proteins, rab6 was shown to be concentrated on the Golgi medial and *trans* cisternae in several cell types (GOUD et al. 1990). Thus, this protein may control transport between the Golgi cisternae and/or the cisternae and the *trans* Golgi network (TGN).

The TGN is an organelle responsible for the proper sorting of material transported from the Golgi apparatus to a variety of destinations: vesicles of the constitutive secretory route, of regulated exocytic processes, and ones destined directly to endosomes (for reviews see GRIFFITHS and SIMONS 1986; MELLMAN and SIMONS 1992). The sequences of rab8 and rab10 (CHAVRIER et al. 1990a) showed higher homology to the yeast Sec4 protein (see Sect. A.I.1.) than any other mammalian rab proteins. These proteins were thus regarded as potential post-Golgi pathway regulators. Immunofluorescence on rab8 indicates that the protein is localized to the Golgi region and to the plasma membrane (HUBER et al. 1993a). In filter-grown Madin-Darby-canine kidney (MDCK) cells, the protein is mainly associated with the Golgi complex and the basolateral plasma membrane. Furthermore, analysis of immunoisolated apical and basolateral post-Golgi vesicles (see WANDINGER-NESS et al. 1990) by ^{32}P [GTP] ligand blots of two-dimensional gels suggests that rab8 is enriched in the basolateral transport vesicles. These morphological and biochemical data suggest that the protein is a potential regulator of traffic from the TGN to the basolateral plasma membrane. Interestingly, the rab8 protein is exclusively seen on the somatodendritic plasma membrane domain of rat hippocampal neurons (HUBER et al. 1993b). Several proteins found on the basolateral plasma membrane of epithelial cells are routed to the dendrites of the hippocampal neurons, while apical proteins seem to be routed to the axon (for review see SIMONS et al. 1993). The localization of rab8 to the basolateral as well as to the dendritic routes might reflect similar mechanisms for polarized sorting in the two cell types. So far, a rab protein specific for the apical exocytic route has not been identified (see also Sect. D.III.).

2. The rab3a Protein on Regulated Exocytic Vesicles (see Chapter 31)

Membrane-bound rab3A/*smg* p25A was shown to be associated with synapses in the brain; more specifically, it was found on synaptic vesicles, secretory organelles that store and release neurotransmitters in a regulated manner, and on the synaptic plasma membranes (FISCHER VON MOLLARD et al. 1990; MIZOGUCHI et al. 1990). The protein was also found in the endocrine adrenal medulla chromaffin cells, on microvesicles having similarities to synaptic vesicles in neurons (FISCHER VON MOLLARD et al. 1990), and on chromaffin granules responsible for catecholamine secretion (DARCHEN et al. 1990). Furthermore, the protein was detected in the endocrine cells of rat pancreatic islets, the acinar cells of rat exocrine pancreas, and in the exocrine cells of rat submaxillary gland (MIZOGUCHI et al. 1989). The subcellular localizations of *smg* p25B and C (rab3b and c) have not been established.

A fourth isotype of rab3 has been cloned (BALDINI et al. 1992). This protein, rab3d, is predominantly expressed in adipocytes. The subcellular localization of this protein has not been established, but the authors suggest that, by analogy to rab3a in synaptic vesicles, it may be involved in the insulin-induced exocytosis of vesicles containing an adipocyte-specific glucose transporter.

II. The rab Proteins on Endocytic Compartments

1. The rab5 and rab4 Proteins on Early Endosomes

CHAVRIER et al. (1990b) raised antibodies against synthetic peptides from the C-terminal region of the deduced amino acid sequence of rab5. They localized the protein in baby hamster kidney (BHK) and MDCK cells on the plasma membrane, on a fine tubular network in the cell periphery, and on small vesicular structures of different sizes throughout the cytoplasm. The intracellular structures were confirmed by immunoelectron microscopy to represent early endosomes. Occasionally, labeling was also seen on coated pits. No label was seen on late endocytic compartments. When the protein was overexpressed in BHK cells from the cDNA clone using the T7 RNA polymerase recombinant vaccinia virus system (FUERST et al. 1986), increased labeling on the plasma membrane and the early endosomes was seen.

Using cell fractionation techniques and specific antibodies, VAN DER SLUIJS et al. (1991) showed that most of the rab4 protein in chinese hamster ovary (CHO) cells is associated with early endosomes and endocytic vesicles containing internalized ^{125}I-labeled transferrin (Tfn). The localization was confirmed by immunofluorescence after expression of the human rab4 cDNA in HeLa cells. The data on the functions of rab5 and rab4 are discussed in Sect. C.II.3.

2. The rab Proteins on Late Endocytic Compartments

Antibodies against a synthetic peptide from the C-terminal region of rab7 were shown to stain the endogenous protein on large vesicular structures in the perinuclear region of cells and colocalize with bovine serum albumin (BSA) gold internalized under conditions in which late endosomes are labeled (CHAVRIER et al. 1990b). The protein also colocalized with the cation-independent mannose-6-phosphate receptor (CI-MPR). It is noteworthy that the antibodies only stained CI-MPR-positive late structures, the CI-MPR-negative lysosomes being devoid of label. So far, no rab protein localized on lysosomes has been found.

Rab9 was colocalized with the CI-MPR in the perinuclear region of BHK cells (LOMBARDI et al. 1993). The staining pattern did not change upon brefeldin A treatment, confirming that the structures represented late endosomes and/or the TGN (see CHEGE and PFEFFER 1991). The experimental data on rab9 function are discussed in Sect. C.II.1.

III. The Molecular Basis of rab Compartmentalization

1. The C-Terminal Modifications

GTPases of the rab subfamily are synthesized as soluble cytoplasmic proteins which are posttranslationally modified by lipid moieties (for a review see MAGEE and NEWMAN 1992). All rab proteins, as well as most of the other ras-related GTPases, display at their very carboxy terminus a motif containing one or two cysteine residues. In the rab subfamily proteins, the following C-terminal sequences are found: -CaaX, -CC, -CXC, -CCXX, -CXXX, and -CCXXX (C, cysteine, a, aliphatic amino acid; X, any amino acid). Cysteine residues in these motifs are posttranslationally modified by thioether-linked polyisoprenyl groups (see JOHNSTON et al. 1991; KHOSRAVI-FAR et al. 1991; KINSELLA and MALTESE 1991; SEABRA et al. 1992; PETER et al. 1992). Rab3a (with a CXC motif) has been shown to aquire geranylgeranyl groups on both cysteines as well as a carboxymethyl group esterified on the terminal cysteine (FARNSWORTH et al. 1991; HORIUCHI et al. 1991). Truncated rab5 and rab4 proteins lacking the C-terminal cysteine motif do not seem to associate with cellular membranes (CHAVRIER et al. 1990b; VAN DER SLUIJS et al. 1992b) and are functionally inert (GORVEL et al. 1991; BUCCI et al. 1992). However, contradictory results have been reported for rab1b (TISDALE et al. 1992; Sect. C.II.1.). The biochemistry of the posttranslational modifications of the ras superfamily proteins is discussed in more detail elsewhere in this volume.

2. Role of the C-Terminal Variable Region

Even though the C-terminal cysteine motifs are a prerequisite for membrane association and vary between different rab proteins, these motifs do not determine the specific subcellular destination of the proteins. CHAVRIER et al. (1991) constructed chimeric forms of rab2, rab5, and rab7, with C-terminal stretches of varying length replaced by the corresponding amino acids from another rab protein (Fig. 1). After transient expression in BHK cells, the localization of the chimeric proteins was determined by subcellular fractionation and by confocal immunofluorescence microscopy. Rab5 with the C-terminal CCSN motif replaced by CSC (normally found, e.g., in rab7) was still targeted to early endosomes. Substituting the eight or 13 C-terminal amino acids of rab5 with those of rab7 did not affect the localization either. However, changing the 34 C-terminal amino acids directed the rab5/rab7 hybrid protein to late endosomes. Likewise, replacing 35 amino acids at the C terminus of rab2 with those of rab5 or rab7 redirected the protein to early

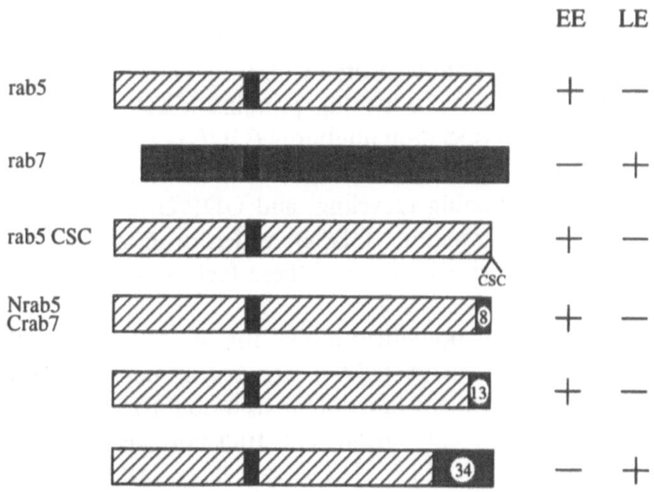

Fig. 1. The hypervariable C-terminal domain of rab proteins acts as a targeting signal. Hybrid proteins consisting of the N-terminal sequences of rab5 and C-terminal stretches from rab7 were expressed in baby hamster kidney (BHK) cells and the subcellular localization of the proteins was determined (CHAVRIER et al. 1991). The rab5 and rab7 sequences are presented as *bars* aligned on the highly conserved WDTAGQE motif (see VALENCIA et al. 1991) indicated by a *black box*. The first chimeric protein has the CCSN motif of rab5 replaced by the rab7 CSC motif; the others have an increasing number of C-terminal amino acids of rab5 (indicated in the *bars*) replaced by those of rab7. The subcellular localization of each protein is indicated on the right; *EE*, early endosomes; *LE*, late endosomes. (Modified from CHAVRIER et al. 1991)

or late endosomes, respectively. These data strongly indicate that much of the signal for the specific membrane targeting of rab proteins resides in the region close to their C termini. This region is will suited for the targeting function since it is the part of rab proteins that displays the greatest sequence variability. However, no defined signals responsible for subcellular localization of rab GTPases in this or any other part of the proteins have so far been identified.

C. The Function of rab Proteins in Membrane Trafficking

I. The Present Model for rab Function

By analogy to ras and several other types of GTPases (see BOURNE et al. 1991), rab proteins are thought to cycle between the GDP- and GTP-bound forms. The GTP-bound forms are assumed to promote downstream effector functions such as vesicle targeting and/or fusion. After a specific recognition event, hydrolysis of the GTP to GDP has to occur (see WALWORTH et al.

1992). Subsequently, the inactive GDP-bound rab should be released, recharged wtih GTP, and start a new cycle. Several types of molecules interacting with ras superfamily proteins and regulating this GTPase cycle have been isolated: GTPase activating proteins (GAPs), which stimulate GTP hydrolysis; GDP dissociation inhibitors (GDIs), which inhibit exchange of bound GDP for GTP and remove the GTPase from membranes, making it soluble and thus facilitating recycling; and GDP/GTP exchange proteins (or GDP dissociation stimulators, GDSs), which promote the exchange of GDP for GTP (for further discussion of these factors, see Chap. 39; EVANS et al. 1991).

Figure 2 presents a hypothetical model for the functional cycle of rab proteins modified from PFEFFER (1992). The recruitment of a rab protein onto a membrane compartment functioning as a donor in a transport step is shown in the upper part, and fusion of a transport vesicle with a target membrane in the lower part of Fig. 2. The specificity in the recruitment step could, in principle, be created by a rab-specific exchange protein. In the case of the *smg* p21B protein, geranylgeranylated synthetic peptides from the C terminus of the protein inhibit the action of the appropriate exchange protein on the GTPase (SHIRATAKI et al. 1991), indicating that the domain of the GTPase interacting with the exchange protein is at least in close proximity of the C-terminal region conferring compartment specificity (see Sect. B.III.2.). However, if exchange proteins served as specific rab receptors, this would predict that there would be perhaps 30 such proteins on the different membrane compartments of a mammalian cell. A multisubunit system with a common catalytic exchange protein combined with rab-specific receptor subunits would seem more reasonable. Recently, a putative receptor (M_r 85–86 kDa) for rab3a was isolated by a cross-linking approach from bovine brain crude membranes (SHIRATAKI et al. 1992). It is puzzling that the cross-linking was done using bacterially produced rab3a, which is not C-terminally modified and thus should not associate with membranes. The protein isolated does not display GAP, GDI, or exchange activities, and could thus represent the type of rab receptor indicated in Fig. 2. The molecule did not bind rab11, but a number of other rab proteins must, of course, be tested before one can characterize this protein as a specific rab3a receptor. After interaction with the receptor and the exchange protein, the activated rab might be released into the membrane, where it could remain bound by its hydrophobic isoprenyl tail inserted in the lipid bilayer.

The intrinsic GTPase activity of rab proteins is very low, so a GAP activity has to be present on the target membrane. By analogy to the ras proteins, the interaction with GAPs is thought to occur mainly at the so-called effector loop of rab proteins (see VALENCIA et al. 1991). Rab proteins have, however, very similar effector loop sequences, suggesting that the number of different rab GAPs low (for review see PFEFFER 1992). Accordingly, it is conceivable that a directed fusion event may require

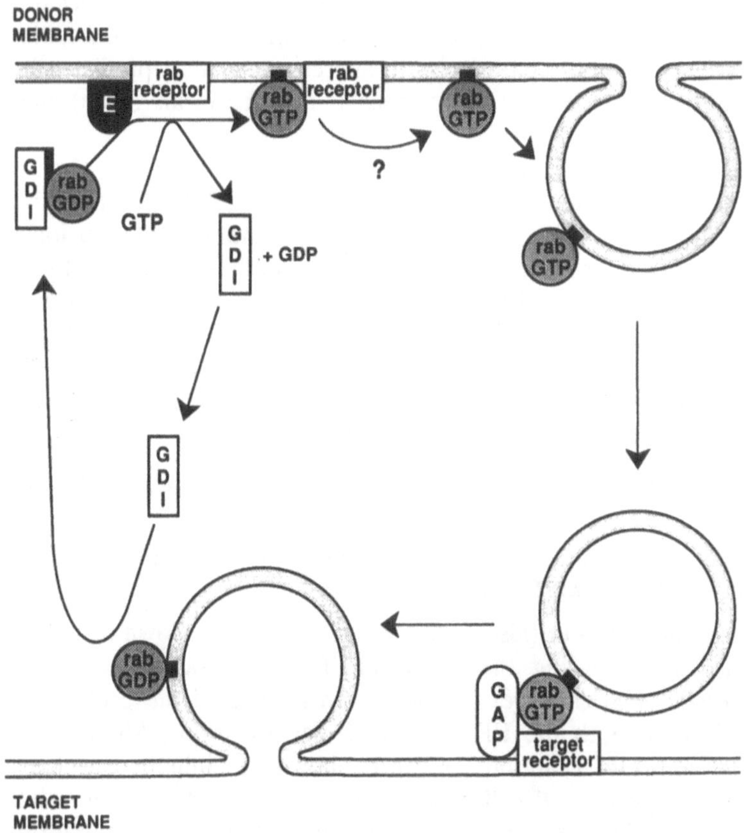

Fig. 2. A model for the functional cycle of rab proteins in membrane traffic. A cytosolic rab protein bound by a GDP dissociation inhibitor (*GDI*) is recognized by a putative receptor on the donor membrane. A GDP/GTP exchange protein (*E*) catalyzes the exchange of the GDP for a GTP. The exchange protein may itself be the specific rab receptor or be coupled to it. The rab-receptor complex may be stable or the GTP-bound rab may be released into the membrane, where it can remain bound by its isoprenyl tail. The GTP-bound rab is included in a transport vesicle or other intermediate. At the target membrane, a ternary complex between the rab, a specific receptor, and a *GAP* (GTPase activating protein) is formed. The GAP enhances the GTP hydrolysis activity of the rab, which ensures the directionality of the transport process. The GDP-bound rab is recycled to the donor membrane by a GDI. Note: the actual recognition between the transport vesicle and the target compartment may take place via non-rab molecular components. (Modified from PFEFFER 1992)

formation of a ternary complex between a rab protein, a GAP, and a specific rab receptor on the target membrane. However, there is no evidence for a role of the rab proteins in the specificity of the transport vesicle docking process. It is also possible that the rab proteins give specificity to the events activating fusion after a vesicle has docked onto a target membrane using a non-rab recognition mechanism. After the fusion, the GDP rab associated with the target membrane would be recycled to the donor membrane, probably by a GDI (see ARAKI et al. 1990). Distribution of rab proteins in two separate membrane compartments predicted by this model has been observed, e.g., for rab5 (CHAVRIER et al. 1990b) and rab8 (HUBER et al. 1993a).

For a rab protein involved in a fusion process between "like" compartments, the model would, of course, be slightly different. In that situation, both the GAP and the exchange protein would obviously be present on both interacting membranes, and recycling of the rab by a GDI would not be necessary. The process would not have directionality in the sense that the above model for directed transport events does, and regulation of the overall activity of the similar fusion machineries on the interacting membranes would be a key feature of the model.

II. Experimental Evidence for rab Function in Membrane Trafficking

1. The rab1, rab2, and rab9 Proteins are Involved in Transport Steps on the Biosynthetic Route

The mouse rab1 protein (also called mouse ypt1) was shown to be capable of functionally replacing the yeast YPT1 gene product when strongly overexpressed (HAUBRUCK et al. 1989). An isotype of the protein, rab1b, was localized to both the ER and the Golgi apparatus. Monoclonal and polyclonal anti-rab1b antibodies were shown to inhibit ER to Golgi transport in a perforated cell assay (PLUTNER et al. 1991). The inhibition seemed to take place at an early stage in the process, i.e., during formation of transport intermediates at the ER. However, the antibodies were also shown to inhibit trafficking into the medial Golgi compartment. In a recent study, TISDALE et al. (1992) investigated ER to Golgi transport in HeLa cells overexpressing mutant forms of rab1a, rab1b, or rab2 by monitoring the glycosylation of vesicular stomatitis virus G-protein and by immunofluorescence. Substitutions in the GTP-binding domains of rab1a and rab1b were potent *trans* dominant inhibitors of transport. Mutations affecting guanine nucleotide exchange or GTP hydrolysis of rab2 caused similar inhibition. Mutations in the putative effector domains (see Sect. C.I.) of the proteins showed only partial inhibition. Interestingly, a *trans* dominant rab1b mutant also showed inhibition in a truncated form lacking the C terminus. This suggests that a protein devoid of the isoprenyl modification can effectively interfere with the transport machinery. Furthermore, the same mutant protein carrying C-terminal sequences of rab5 had no inhibitory effect, suggesting that it was efficiently segregated from factors involved in ER to Golgi traffic, perhaps by localization to early endosomes.

During mitosis, the rab1 (or rab1a) protein undergoes phosphorylation and its distribution between membrane-bound and cytosolic forms changes

(BAILLY et al. 1991). This phenomenon may play a role in the mitotic block of membrane transport (see WARREN 1985).

The rab9 protein has been colocalized with the cation-independent CI-MPR on the surface of the TGN and/or late endosomes (LOMBARDI et al. 1993). A cell-free transport assay reconstituting the recycling of CI-MPR from late endosomes to TGN was specifically stimulated by cytosol containing overexpressed rab9 or purified *Escherichia coli* expressed rab9, indicating that the protein is involved in regulation of transport between these compartments. Maximal stimulation with the bacterially produced protein was only observed if the reactions were supplemented with geranylgeranyl diphosphate, suggesting that the nonmodified protein was C-terminally modified in the transport reaction mixture.

2. The rab3a Protein and Regulated Secretion

The rab3a protein seems to associated with the synaptic vesicles at a late stage in the cell periphery, probably as part of the regulatory machinery assembled onto the vesicles prior to exocytosis (MATTEOLI et al. 1991). In experiments in which exocytosis of neurotransmitter from isolated synaptosomes was artifically induced, FISCHER VON MOLLARD et al. (1991) showed that rab3a disappeared from the synaptic vesicle fraction upon secretion. The vesicles devoid of rab3a were suggested to represent vesicles retrieved from the plasma membrane. During secretion, the protein becomes associated with the synaptosomal plasma membrane, from which it may be retrieved and recycled to synaptic vesicles by a GDI protein (ARAKI et al. 1990; see also Sect. C.I.). Synthetic peptides of the rab3a effector domain have been shown to stimulate regulated secretion events (PADFIELD et al. 1992; SENYSHYN et al. 1992; OBERHAUSER et al. 1992). This is surprising since, according to the model by BOURNE (1988), blocking of the effector molecule on the target membrane should inhibit the secretory function. On the other hand, blocking the effector molecule should keep the rab bound to secretory vesicles in the GTP-bound form. In an earlier study, PLUTNER et al. (1990) showed that cell-free ER to Golgi and intra-Golgi transport reactions were inhibited by the same peptides. These results are difficult to interpret, but they could indicate that, in regulated secretion, the sequence of events on the target membrane may deviate from that predicated by the model in Fig. 2. The rab3a functional cycle is discussed more specifically elsewhere in this volume.

3. Functional Studies on rab5 and rab4

The first actual evidence for rab involvement in membrane fusion events in mammalian cells was provided by the in vitro endosome fusion experiments of GORVEL et al. (1991). The experimental system measures fusion of purified early endosomal fractions incubated in differently modified cytosols by assaying complex formation between avidin internalized in one fraction and biotinylated horseradish peroxidase (HRP) in the other. Including anti-rab5 antiserum in the cytosol strongly inhibited the fusion reaction. Furthermore, cytosols with overexpressed rab5 or the mutant protein rab5Ile^{133} (in the yeast Sec4 and Ypt1 proteins the corresponding mutation is lethal; SCHMITT et al. 1986; WALWORTH et al. 1989), which does not bind GTP on ligand blots, showed opposite effects in the assay. The overexpressed wild-type protein stimulated the fusion while the mutant protein inhibited fusion activity. Addition of cytosol containing overexpressed rab5 to endosomes pretreated with the anti-rab5 antibodies restored fusion activity. These results indicate that rab5 is required for lateral fusion between early endosomes in vitro. It is worth noting that the membrane-bound form of overexpressed rab5 fractionated in the early endosome fraction. It seems that the mechanisms responsible for receiving rab5 are able to handle considerable overexpression of the protein. Similar results have been reported for rab4 (VAN DER SLUIJS et al. 1992b).

The function of rab5 has recently also been studied in vivo. BUCCI et al. (1992) introduced wild-type rab5 or rab5Ile^{133} into BHK cells using vaccinia-virus-based expression systems. In cells expressing the mutant protein, the rate of receptor-mediated endocytosis of Tfn and of fluid-phase internalization was decreased by about 50%, whereas the rate of recycling was not signifiantly affected. Also, the morphology of early endocytic compartments was drastically changed: electron microscopy (EM) and confocal microscopy revealed that the internalized HRP was restricted to peripheral tubular structures and vesicles. Surprisingly, overexpression of the wild-type protein led to accelerated uptake of the markers and to the appearance of abnormally large early endosomes. These studies demonstrate that rab5 is a component of the regulatory machinery in the early part of the endocytic pathway. Rab5 seems not only to regulate the transport from coated vesicles to early endosomes, but also to control early endosome morphology, perhaps by an interplay between fusion–fission events and incoming traffic from the cell surface.

This work is complemented by the study of VAN DER SLUIJS et al. (1992a) on the function of rab4. Using stably transfected CHO cells overexpressing the protein, the authors measured its effects on endocytosis and recycling. The results suggest for rab4 a role clearly different from that of rab5. Initial rates of endocytosis were not significantly affected, but overexpression of the wild-type protein substantially reduced intracellular accumulation of HRP (fluid-phase marker), and the steady-state distribution of Tfn receptor (Tfn-R) in the cells was changed, the majority of the receptor being found on cell surface. Overexpression of the wild-type protein also slowed down the release of internalized ^{125}I-Tfn from the cells and prevented iron discharge from the protein. This could be explained by a failure of the Tfn to enter acidic endosomes, or could be due to its exit from the endosomes before the iron has dissociated from the Tfn. In EM, internalized HRP-TFn was seen in small tubular clusters, which were morphologically different from typical early endosomes, perhaps representing a nonacidic recycling structure.

In the light of these observations, it seems that rab5 and rab4 may form a balancing pair of proteins, rab5 regulating the uptake of material and fusion of incoming vesicles with early endosomes and rab4 being a regulator of the recycling route.

The rab4 protein contains a $p34^{cdc2}$ kinase phosphorylation site and is hyperphosphorylated during mitosis, leading to redistribution of the protein from the membrane-bound state to a cytosolic one (BAILLY et al. 1991). Accordingly, the authors suggested that rab4 might play a role in the phosphorylation-dependent inhibition of endocytic processes during mitosis (see TUOMIKOSKI et al. 1989; WARREN 1989; WOODMAN et al. 1992). Recently, VAN DER SLUIJS et al. (1992b) have shown that rab4 is phosphorylated in vitro by purified $p34^{cdc2}$ on Ser^{196}. Protein mutated at this residue did not appear in the cytosol in mitosis, showing that the phosphorylation truly regulates the redistribution.

4. Conclusion from the Functional Data

The functional studies performed on several mammalian rab proteins show that each of the proteins investigated regulates a transport step (or steps) taking place in the subcellular region where the protein has been localized. The studies provide biochemical evidence for the effects of overexpression of wild-type or mutant proteins, or of inactivation of the proteins by specific antibodies, on the kinetics of transport reactions. The in vivo studies of BUCCI et al. (1992) and VAN DER SLUIJS et al. (1992a) also demonstrate that overexpression of rab5 or rab4 results in accumulation of specific structures associated with early endocytic events. These observations mostly support the suggested model for the rab functional cycle (Sect. C.I.). However, since only a small number of rab proteins have been studied in more detail, one has to abstain from generalizations at this stage. It may well be that the rab proteins can be divided into subgroups with different functions in membrane traffic (see PADFIELD et al. 1992; SENYSHYN et al. 1992; OBERHAUSER et al. 1992).

D. The Novel rab Proteins

I. Why Clone More rab Sequences?

Assuming that one or more rab GTPases are involved in each specific intracellular transport step, characterizing more rab proteins will provide a more comprehensive picture of the overall organization of the transport pathways. The localization of rab proteins will define the potential trafficking routes in the cell, in other words, the arrows that connect compartments by intracellular membrane transport.

Characterized rab proteins can be used as specific subcellular markers in morphological studies and also for determining the identity or purity of given subcellular fractions. In fact, we have developed an intracellular "mapping" system based on ^{32}P-GTP ligand blots of two-dimensional gels (HUBER et al. 1993). Here, the position of each rab in the blots is identified using proteins overproduced from cloned cDNAs. The reference map thus created and the data on the localization of the proteins can be used to define subcellular fractions.

CHAVRIER et al. (1992) cloned from mouse kidney cDNA fragments of 11 novel rab sequences by a PCR approach based on the use of degenerate oligonucleotides corresponding to the amino acid stretches highly conserved in rab and rho GTPases (see VALENCIA et al. 1991). CHAVRIER et al. (1992) named the novel PCR fragment sequences PCR-rab12 to PCR-rab20. Since the numbering from rab12 to 16 was used for other new rab proteins in the simultaneously published study of ELFERINK et al. (1992), we have renamed five of the PCR-fragment-derived sequences: PCR-rab12=rab21, PCR-rab13=rab12 (it is the same sequence as the rab12 of ELFERINK et al. 1992), PCR-rab14=rab22, PCR-rab15=rab23, and PCR-rab16=rab24. We have now cloned the full-length cDNAs of these proteins and are in the process of characterizing their subcellular localization (OLKKONEN et al. 1993; LÜTCKE et al. 1993a, 1993b).

II. Subcellular Localization

1. Novel Proteins on the Biosynthetic Pathway

We have analyzed the subcellular localization of the novel proteins by expressing epitope-tagged proteins carrying at their N termini the c-*myc* epitope recognized by the monoclonal antibody 9E10 (EVAN et al. 1985; MUNRO and PELHAM 1987) in several cell lines. For the expression of the proteins we have used the T7 RNA polymerase recombinant vaccinia virus (FUERST et al. 1986) and Semliki Forest virus (LILJESTRÖM and GAROFF 1991) vector systems.

We have identified two novel proteins on the biosyntetic pathway (OLKKONEN et al. 1993). Rab24 is visualized by immunofluorescence as a network in the perinuclear region of the cell. The staining colocalizes partially with ER markers and quite well with rab2, indicating a localization in the intermediate region between the ER and the Golgi. There is also some colocalisation with late endosomal – markers. Rab12 is found on the Golgi apparatus, colocalizing with TGN38 (LUZIO et al. 1990) and with β-Cop (DUDEN et al. 1991).

2. Novel Proteins on Early Endocytic Compartments

Our results on the novel cDNAs suggest that among the proteins encoded there are several rabs associated with early endosomes. These include two novel isotypes of rab5, which we call rab5b (see also WILSON and WILSON

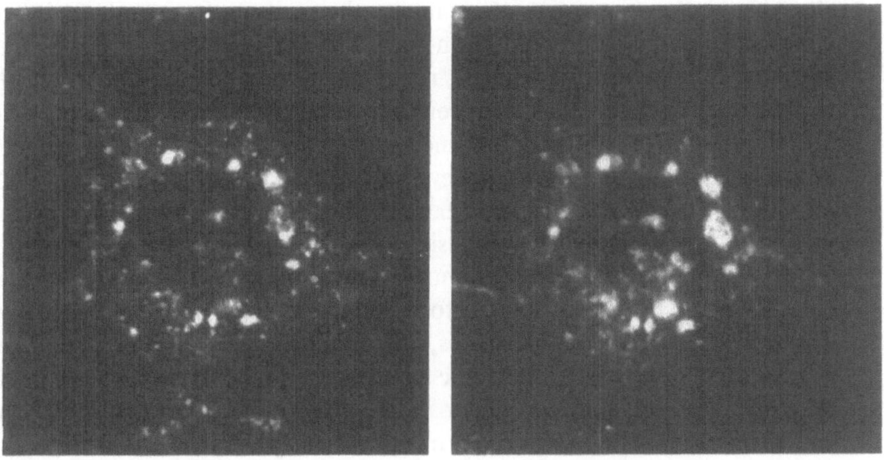

A B

Fig. 3A,B. Localization of c-*myc* epitope-tagged rab22 in BHK cells. The cells were infected with T7 RNA polymerase recombinant vaccinia virus (FUERST et al. 1986) and then cotransfected with plasmids containing the rab22 cDNA (OLKKONEN et al. 1993) or human transferrin receptor cDNA (ZERIAL et al. 1986) under the T7 promoter. After a 4-h transfection period, fluorescein-conjugated human transferrin (FITC-Tfn) was bound on the plasma membrane receptors on ice, followed by internalization at 37°C for 5 min. **A** The FITC-Tfn; **B** the tagged rab22 visualized using the 9E10 monoclonal antibody (EVAN et al. 1985)

1992) and rab5c (CHAVRIER et al. 1992). rab5c seems to colocalize to a great extent with rab5, and when overexpressed in BHK cells, it has functional effects similar to those of rab5 (C. BUCCI et al., unpublished). All three isotypes are, as judged from northern blots, ubiquitously expressed in the mouse tissues that we have studied. An interesting question is whether these proteins have different functions in the early endosomes or whether they are functionally redundant.

One of the most interesting new proteins is rab22 (OLKKONEN et al. 1993). When expressed in BHK cells, the epitope-tagged protein localizes mainly on perinuclear large vesicle-like structures (Fig. 3). The structures, which are not seen in nontransfected cells, contain fluorescein-conjugated Tfn after a 5-min internalization period, thus representing an early endocytic compartment. The staining also includes the plasma membrane and small vesicles in the cell periphery. The structures stained, however, do not match those labeled by antibodies against endogenous rab5, suggesting that they represent a subcompartment of early endosomes, the function of which is still unclear.

How can this multitude of rab proteins associated at least partially with early endosomes be explained? The functional studies on rab5 and rab4 (see Sect. C.II.3.) give us some idea of the fine-tuning of early endosome function. Early endosomes perform a multitude of sorting activities. They receive material from the plasma membrane and recycle receptor molecules to the cell surface. In this context one has to remember that more than one route has been suggested to connect the cell surface and the early endosomes: not only the clathrin-coated vesicle pathway, but also an alternative non-clathrin-mediated route (see VAN DEURS et al. 1989). Furthermore, early endosomes receive material from the TGN (LUDWIG et al. 1991), and, of course, sort endocytosed molecules destined to later endocytic compartments (for reviews see GRUENBERG and HOWELL 1989; RODMAN et al. 1990; WATTS and MARSH 1992). Early endosomes might also function in sorting molecules to vesicles involved in the regulated recycling of glucose transporters, water channels, and proton ATPases (see BLOK et al. 1988; BROWN 1989). Taking into account the complexity of the sorting tasks this compartment performs, the number of distinct, regulated transport steps in early endocytic compartments is bound to be high. Therefore, the high number of endosomal rab proteins is not surprising, and these proteins will be useful tools to dissect early endosome structure and function.

The subcellular localizations of the rab proteins discussed in this review are summarized in Fig. 4. Considering the distribution of rab proteins to the different sites within the cell, it is striking that the tubular networks in the ER-Golgi intermediate region, the TGN, and the early endosomes have the highest numbers of different rab proteins. Does this imply that these compartments have key roles in traffic regulation in the cell? This is in keeping with the known complexity of sorting functions of the early endosomes (see above) and the TGN (GRIFFITHS and SIMONS 1986; MELLMAN and SIMONS 1992). Perhaps the *cis* Golgi network has more sorting functions than presently surmised?

III. Epithelial-Specific rab Proteins?

Epithelial cells have a structurally and functionally polarized phenotype. Their plasma membrane is divided into two domains displaying distinct protein and lipid compositions: the apical domain faces the outside environment, and the basolateral domain forms the interface towards other cells of the epithelial sheet, the basement membrane, and the bloodstream (for reviews see SIMONS and FULLER 1985; RODRIGUEZ-BOULAN and NELSON 1989; SIMONS and WANDINGER-NESS 1990; MATLIN 1992; NELSON 1992). Epithelial cells target membrane proteins and lipids to the distinct plasma membrane domains, secrete proteins in a polarized manner, and transcytose material from one plasma membrane domain to the other. This obviously requires many more traffic routes than those found in fibroblast-type cells. Accordingly, one can envision that there are transport steps that are specific for epithelial cells and that these transport events require specialized machinery, including perhaps rab proteins. These considerations imply that it should be possible to detect rab proteins that are restricted to tissues with a high proportion of epithelial cells, such as intestine and kidney. This search has led to the identification of rab17 as the first candidate for

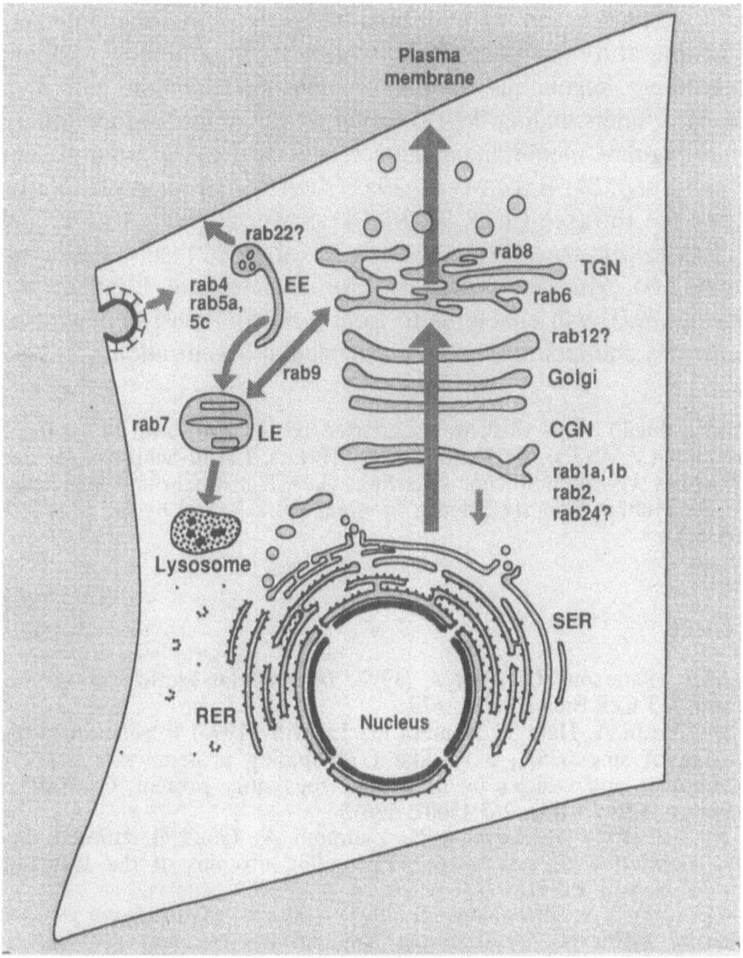

Fig. 4. A summary of the subcellular localization of ubiquitously expressed rab proteins (rab3 only present in cells performing regulated secretion is not included). *RER*, rough endoplasmic reticulum; *SER*, smooth endoplasmic reticulum; *CGN*, *cis* Golgi network; *TGN*, *trans* Golgi network; *EE*, early endosomes; *LE*, late endosomes. (Modified from Bucci et al. 1991)

an epithelial-specific member of the subfamily (Lütcke et al. 1993a). Identification of such rab proteins specific for polarized cells should provide insights into the mechanisms responsible for generating the epithelial phenotype.

E. Conclusion

Specific rab proteins have by now been found on most of the subcellular compartments along the endocytic and biosynthetic routes, and knowledge about their function and the molecular components that they interact with is accumulating at a rapid pace. Rab proteins are undoubtedly components of the machinery regulating cellular compartmentalization and membrane trafficking. Understanding their function, in the context of the other factors known to regulate membrane dynamics, such as the ADP-ribosylation factor (ARF see Chap. 34; BALCH et al. 1992; LENHARD et al. 1992; TAYLOR et al. 1992) and the rho (see Chap. 36; MCCAFFREY et al. 1991; RIDLEY and HALL 1992; RIDLEY et al. 1992; ADAMSON et al. 1992) subfamilies of small GTPases, and trimeric G-proteins (for reviews see BARR et al. 1992; BURGOYNE 1992), will enable us to form a comprehensive view of how the cell generates and maintains its compartmental organization.

Acknowledgements. This work was supported by the European Molecular Biology Organization (V.M.O.), the Royal Society (P.D.), Erwin Schroedinger Fonds zur Förderung der wissenschaftlichen Forschung (L.A.H.), Boehringer Ingelheim Fonds (A.L.), the Human Frontier Science Program (M.Z.), and by the SFB 352 of the DFG (K.S.).

References

Adamson P, Paterson HF, Hall A (1992) Intracellular localization of the $p21^{rho}$ proteins. J Cell Biol 119:617–627

Araki S, Kikuchi A, Hata Y, Isomura M, Takai Y (1990) Regulation of reversible binding of smg p25A, a ras-like GTP-binding protein, to synaptic plasma membranes and vesicles by its specific regulatory protein, GDP dissociation inhibitor. J Biol Chem 265:13007–13015

Bailly E, McCaffrey M, Touchot N, Zahraoui A, Goud B, Bornens M (1991) Phosphorylation of two small GTP-binding proteins of the Rab family by $p34^{cdc2}$. Nature 350:715–718

Balch WE, Kahn RA, Schwaninger R (1992) ADP-ribosylation factor is required for vesicular trafficking between the endoplasmic reticulum and the cis-Golgi compartment. J Biol Chem 267:13053–13061

Baldini G, Hohl T, Lin HY, Lodish HF (1992) Cloning of a rab3 isotype predominately expressed in adipocytes. Proc Natl Acad Sci USA 89:5049–5052

Barr FA, Leyte A, Huttner WB (1992) Trimeric G proteins and vesicle formation. Trends Cell Biol 2:91–94

Beckers CJM, Balch WE (1989) Calcium and GTP: essential components in vesicular trafficking between endoplasmic reticulum and Golgi apparatus. J Cell Biol 108:1245–1256

Blok J, Gibbs EM, Lienhard GE, Slot JW, Geuze, HJ (1988) Insulin-induced translocation of glucose transporters from post-Golgi compartments to the plasma membrane of 3T3-L1 adipocytes. J Cell Biol 106:69–76

Bourne HR (1988) Do GTPases direct membrane traffic in secretion? Cell 53:669–671

Bourne HR, Sanders DA, McCormick, F (1991) The GTPase superfamily: a conserved switch for diverse cell functions. Nature 348:125–132

Brown D (1989) Membrane recycling and epithelial cell function. Am J Physiol 256:F1–12

Bucci C, Frunzio R, Chiarotti R, Brown L, Rechler MM, Bruni CB (1988) A new member of the ras gene superfamily identified in a rat liver cell line. Nucleic Acids Res 16:9979–9993

Bucci C, Parton R, Mather I, Stunnenberg H, Simons K, Zerial M (1991) Involvement of low molecular weight GTP-binding proteins in membrane traffic in mammalian cells. In: Verna R, Nishizuka Y (eds) Biotechnology of cell regulation. Raven, New York, p 103

Bucci C, Parton RG, Mather IM, Stunnenberg H, Simons K, Hoflack B, Zerial M (1992) The small GTPase rab5 functions as a regulatory factor in the early endocytic pathway. Cell 70:715–728

Burgoyne RD (1992) Trimeric G proteins in Golgi transport. Trends Biochem Sci 17:87–88

Chavrier P, Vingron M, Sander C, Simons K, Zerial M (1990a) Molecular cloning of YPT1/SEC4-related cDNAs from an epithelial cell line. Mol Cell Biol 10:6578–6585

Chavrier P, Parton RG, Hauri HP, Simons K, Zerial M (1990b) Localization of low molecular weight GTP-binding proteins to exocytic and endocytic compartments. Cell 62:317–329

Chavrier P, Gorvel J-P, Stelzer E, Simons K, Gruenberg J, Zerial M (1991) Hypervariable C-terminal domain of rab proteins acts as a targeting signal. Nature 353:769–772

Chavrier P, Simons K, Zerial M (1992) The complexity of the rab and rho GTP-binding protein subfamilies revealed by a PCR cloning approach. Gene 112:261–264

Chege NW, Pfeffer SR (1991) Compartmentalization of the Golgi complex: Brefeldin-A distinguishes trans-Golgi cisternae from the trans-Golgi network. J Cell Biol 111:893–899

Darchen F, Zahraoui A, Hammel F, Monteils M-P, Tavitian A, Scherman D (1990) Association of the GTP-binding protein rab3A with bovine adrenal chromaffin granules. Proc Natl Acad Sci USA 87:5692–5696

Duden R, Griffiths G, Frank R, Argos P, Kreis TE (1991) β-COP, a 110 kd protein associated with non-clathrin-coated vesicles and the Golgi complex, shows homology to β-adaptin. Cell 64:649–665

Elferink LA, Anzai K, Scheller RH (1992) rab15, a novel low molecular weight GTP-binding protein specifically expressed in rat brain. J Biol Chem 267:1–8

Evan GI, Lewis GK, Ramsay G, Bishop JM (1985) Isolation of monoclonal antibodies specific for human c-myc proto-oncogene product. Mol Cell Biol 5:3610–3616

Evans T, Hart MJ, Cerione RA (1991) The Ras superfamilies: regulatory proteins and post-translational modifications. Curr Opin Cell Biol 3:185–191

Farnsworth CC, Kawata M, Yoshida Y, Takai Y, Gelb MH, Glomset JA (1991) C terminus of the small GTP-binding protein smg p25A contains two geranylgeranylated cysteine residues and a methyl ester. Proc Natl Acad Sci USA 88:6196–6200

Fischer von Mollard G, Mignery GA, Baumert M, Perin MS, Hansson TJ, Burger PM, Jahn R, Südhof T (1990) rab3 is a small GTP-binding protein exclusively localized to synaptic vesicles. Proc Natl Acad Sci USA 87:1988–1992

Fischer von Mollard G, Südhof T, Jahn R (1991) A small GTP-binding protein dissociates from synaptic vesicles during exocytosis. Nature 349:79–81

Fuerst TR, Niles EG, Studier FW, Moss B (1986) Eukaryotic transient-expression system based on recombinant vaccinia virus that synthesizes bacteriophage T7 RNA polymerase. Proc Natl Acad Sci USA 83:8122–8126

Goda Y, Pfeffer SR (1988) Selective recycling of the mannose 6-phosphate/IGF-II receptor to the trans Golgi network in vitro. Cell 55:309–320

Gorvel J-P, Chavrier P, Zerial M, Gruenberg J (1991) rab5 controls early endosome fusion in vitro. Cell 64:915–925

Goud B, Salminen A, Walworth NC, Novick PJ (1988) A GTP-binding protein required for secretion rapidly associates with secretory vesicles and the plasma membrane in yeast. Cell 53:753–768

Goud B, Zahraoui A, Tavitian A, Saraste J (1990) Small GTP-binding protein associated with Golgi cisternae. Nature 345:553–556

Griffiths G, Simons K (1986) The trans-Golgi network: sorting at the exit site of the Golgi complex. Science 234:438–443

Gruenberg J, Howell KE (1989) Membrane traffic in endocytosis: insights from cell-free assays. Annu Rev Cell Biol 5:453–481

Haubruck H, Disela C, Wagner P, Gallwitz D (1987) The ras-related ypt protein is an ubiquitous eukaryotic protein: isolation and sequence analysis of mouse cDNA clones highly homologous to the yeast YPT1 gene. EMBO J 6:4049–4053

Haubruck H, Prange R, Vorgias C, Gallwitz D (1989) The ras-related mouse ypt1 protein can functionally replace the YPT1 gene product in yeast. EMBO J 8:1427–1432

Hauri H-P, Schweizer A (1992) The endoplasmic reticulum-Golgi intermediate compartment. Curr Opin Cell Biol 4:600–608

Hopkins CR, Gibson A, Shipman M, Miller K (1990) Movement of internalized ligand-receptor complexes along a continuous endosomal reticulum. Nature 346:335–339

Horiuchi H, Kawata M, Katayama M, Yoshida Y, Musha T, Ando S, Takai Y (1991) A novel prenyltransferase for a small GTP-binding protein having a C-terminal Cys-Ala-Cys structure. J Biol Chem 266:16981–16984

Huber LA, Pimplikar S, Virta H, Parton RG, Zerial M, Simons K (1993a) Rab8 a small GTPase involved in vesicular traffic between the TGN and the basolateral plasma membrane. J Cell Biol 123: (in press)

Huber LA, de Hoop M, Dupree P, Zerial M, Simons K, Dotti C (1993b) Dendritic anterograde transport of newly synthesized membrane proteins is regulated by rab8p. J Cell Biol 123: (in press)

Huber LA, Virta H, Dupree P, Olkkonen VM, Lütcke A, Sztul E, Peter ME, Bennett MK, Zerial M, Simons K (1993) Two-dimensional mapping of small GTP-binding proteins: a novel approach for identification and intracellular localization in preparation

Johnston PA, Archer III BT, Robinson K, Mignery GA, Jahn R, Südhof TC (1991) rab3A attachment to the synaptic vesicle membrane mediated by a conserved polyisoprenylated carboxy-terminal sequence. Neuron 7:101–109

Khosravi-Far R, Lutz R, Cox AD, Clark R, Kato K, Bourne JR, Casey PJ, Sinensky M, Balch WE, Buss JE, Der CJ (1991) Isoprenylation of rab proteins terminating in CC or CXC. Proc Natl Acad Sci USA 88:6264–6268

Kinsella B, Malteese WA (1991) rab GTP-binding proteins implicated in vesicular transport are isoprenylated in vitro at cysteines within a novel carboxy-terminal motif. J Biol Chem 266:8540–8544

Klausner RD, Donaldson JG, Lippincott-Schwartz J (1992) Brefeldin A: insights into the control of membrane traffic and organelle structure. J Cell Biol 116:1071–1080

Kreis TE (1992) Regulation of vesicular and tubular membrane traffic on the Golgi complex by coat proteins. Curr Opin Cell Biol 4:609–615

Lenhard JM, Kahn RA, Stahl PD (1992) Evidence for ADP-ribolylation factor (ARF) as a regulator of in vitro endosome-endosome fusion. J Biol Chem 267:13047–13052

Liljeström P, Garoff H (1991) A new generation of animal cell expression vectors based on the Semliki Forest virus replicon. Bio/Technology 9:1356–1361

Lombardi D, Soldati T, Riederer MA, Goda Y, Zerial M, Pfeffer SR (1993) Rab9 functions in transport between late endosomes and the trans Golgi network. EMBO J 12:677–682

Ludwig T, Griffiths G, Hoflack B (1991) Distribution of newly synthesized lysosomal enzymes in the endocytic pathway of normal rat kidney cells. J Cell Biol 115:1561–1572

Lütcke A, Jansson S, Parton RG, Chavrier P, Valencia A, Huber LA, Lehtonen E, Zerial M (1993a) rab17, a novel small GTPase, is specific for epithelial cells and is induced during cell polarization. J Cell Biol 121:553–564

Lütcke A, Valencia A, Olkkonen VM, Dupree P, Huber LA, Griffiths G, Simons K, Zerial M (1993b) Molecular cloning and characterization of three novel members of the rab subfamily of small GTPases in preparation
Luzio JP, Brake B, Banting G, Howell KE, Braghetta P, Stanley KK (1990) Identification, sequencing and expression of an integral membrane protein of the trans-Golgi network (TGN38). Biochem J 270:97–102
Magee T, Newman C (1992) The role of lipid anchors for small G proteins in membrane trafficking. Trends Cell Biol 2:318–323
Matlin KS (1992) W(h)ither default? Sorting and polarization in epithelial cells. Curr Opin Cell Biol 4:623–628
Matsui Y, Kikuchi A, Kondo J, Hishida T, Teranishi Y, Takai Y (1988) Nucleotide and deduced amino acid sequences of a GTP-binding protein family with molecular weights of 25,000 from bovine brain. J Biol Chem 263:11071–11074
Matteoli M, Takei K, Cameron R, Hurlbut P. Johnston PA, Südhof TC, Jahn R, De Camilli P (1991) Association of rab3a with synaptic vesicles at late stages of the secretory pathway. J Cell Biol 115:625–633
Mayorga LS, Diaz R, Stahl PD (1989) Regulatory role for GTP-binding proteins in endocytosis. Science 244:1475–1477
McCaffrey M, Johnson JS, Goud B, Myers AM, Rossier J, Popoff MR, Madaule P, Boquet P (1991) The small GTP-binding protein rho1p is localized on the Golgi apparatus and post-Golgi vesicles in Saccharomyces cerevisiae. J Cell Biol 115:309–319
Melancon P, Glick BS, Malhotra V, Weidman PJ, Serafini T, Gleason ML, Orci L, Rothman JE (1987) Involvement of GTP-binding "G" proteins in transport through the Golgi stack. Cell 51:1053–1062
Mellman I, Simons K (1992) The Golgi complex: in vitro veritas? Cell 68:829–840
Mizoguchi A, Kim S, Ueda T, Takai Y (1989) Tissue distribution of smg p25A, a ras p21-like GTP-binding protein, studied by use of a specific monoclonal antibody. Biochem Biophys Res Commun 162:1438–1445
Mizoguchi A, Kim S, Ueda T, Kikuchi A, Yorifuji H, Hirokawa N, Takai Y (1990) Localization and subcellular distribution of smg p25A, as ras p23-like GTP-binding protein, in rat brain. J Biol Chem 265:11872–11879
Munro S, Pelham RB (1987) A C-terminal signal prevents secretion of luminal ER proteins. Cell 48:899–907
Nelson WJ (1992) Regulation of cell surface polarity from bacteria to mammals. Science 258:948–955
Oberhauser AF, Monck J, Balch WE, Fernandez JM (1992) Exocytic fusion is activated by rab3a peptides. Nature 360:270–273
Olkkonen VM, Dupree P, Killisch I, Lütcke A, Zerial M, Simons K (1993) Molecular cloning and subcellular localization of three GTP-binding proteins of the rab subfamily (submitted)
Orci L, Malhotra V, Amherdt M, Serafini T, Rothman JE (1989) Dissection of a single round of vesicular transport: sequential intermediates for intercisternal movement in the Golgi stack. Cell 56:357–368
Padfield PJ, Balch WE, Jamieson JD (1992) A synthetic peptide of the rab3a effector domain stimulates amylase release from permeabilized pancreatic acini. Proc Natl Acad Sci USA 89:1656–1660
Palade G (1975) Intracellular aspects of the process of protein synthesis. Science 189:347–358
Peter M, Chavrier P, Nigg EA, Zerial M (1992) Isoprenylation of rab proteins on structurally distinct cysteine motifs. J Cell Sci 102:857–865
Pfeffer SR (1992) GTP-binding proteins in intracellular transport. Trends Cell Biol 2:41–46
Plutner H, Schwaninger R, Pind S, Balch WE (1990) Synthetic peptides of the rab effector domain inhibit vesicular transport through the secretory pathway. EMBO J 9:2375–2383

Plutner H, Cox AD, Pind S, Khosravi-Far R, Bourne JR, Schwaninger R, Der CJ, Balch B (1991) rab1b regulates vesicular transport between the endoplasmic reticulum and successive Golgi compartments. J Cell Biol 115:31–43

Ridley AJ, Hall A (1992) The small GTP-binding protein rho regulates the assembly of focal adhesions and actin stress fibers in response to growth factors. Cell 70:389–399

Ridley AJ, Paterson HF, Johnston CL, Diekman D, Hall A (1992) The small GTP-binding protein rac regulates growth factor-induced membrane ruffling. Cell 70:401–410

Rodman JS, Mercer RW, Stahl PD (1990) Endocytosis and transcytosis. Curr Opin Cell Biol 2:664–672

Rodriguez-Boulan E, Nelson WJ (1989) Morphogenesis of the polarized epithelial cell phenotype. Science 245:718–725

Rothman JE, Orci L (1992) Molecular dissection of the secretory pathway. Nature 355:409–415

Salminen A, Novick P (1987) A ras-like protein is required for a post-Golgi event in yeast secretion. Cell 49:527–538

Saraste J, Kuismanen E (1992) Pathways of protein sorting and membrane traffic between the rough endoplasmic reticulum and the Golgi complex. Semin Cell Biol 3 (in press)

Schmitt HD, Wagner P, Pfaff E, Gallwitz D (1986) The ras-related YPT1 gene product in yeast: a GTP-binding protein that might be involved in microtubule organization. Cell 47:401–412

Schmitt HD, Puzicha M, Gallwitz D (1988) Study of a temperature-sensitive mutant of the ras-related YPT1 gene product in yeast suggests a role in the regulation of intracellular calcium. Cell 53:635–647

Seabra MC, Goldstein JL, Südhof TC, Brown MS (1992) Rab geranylgeranyl transferase. J Biol Chem 267:14497–14503

Segev N, Mulholland J, Botsein D (1988) The yeast GTP-binding YPT1 protein and a mammalian counterpart are associated with the secretion machinery. Cell 52:915–924

Senyshyn J, Balch WE, Holz RW (1992) Synthetic peptides of the effector-binding domain of rab enhance secretion from digitonin-permeabilized chromaffin cells. FEBS Lett 309:41–46

Shirataki H, Kaibuchi K, Hiroyoshi M, Isomura M, Araki S, Sasaki T, Takai Y (1991) Inhibition of the action of the stimulatory GDP/GTP exchange protein smg p21 by the geranylgeranylated synthetic peptides designed from its C-terminal region. J Biol Chem 266:20672–20677

Shirataki H, Kaibuchi K, Yamaguchi T, Wada K, Horiuchi H, Takai Y (1992) A possible target protein for smg-25A/rab3A small GTP-binding protein. J Biol Chem 267:10946–10949

Simons K, Fuller SD (1985) Cell surface polarity in epithelia. Annu Rev Cell Biol 1:243–288

Simons K, Wandinger-Ness A (1990) Polarized sorting in epithelia. Cell 62:207–210

Simons K, Dupree P, Fiedler K, Huber L, Kobayashi T, Kurzchalia T, Olkkonen V, Pimplikar S, Parton R, Dotti C (1993) The biogenesis of cell surface polarity in epithelial cells and neurons. Cold Spring Harbor Symp Quant Biol Vol. LVII: 611–619

Taylor TC, Kahn RA, Melancon P (1992) Two distinct members of the ADP-ribosylation factor family of GTP-binding proteins regulate cell-free intra-Golgi transport. Cell 70:69–79

Tisdale EJ, Bourne JR, Khosravi-Far R, Der CJ, Balch WE (1992) GTP-binding mutants of rab1 and rab2 are potent inhibitors of vesicular transport from the endoplasmic reticulum to the Golgi complex. J Cell Biol 119:749–761

Touchot N, Chardin P, Tavitian A (1987) Four additional members of the ras gene superfamily isolated by an oligonucleotide strategy: molecular cloning of YPT-related cDNAs from a rat brain library. Proc Natl Acad Sci USA 84:8210–8214

Tuomikoski T, Felix M-A, Dorée M, Gruenberg J (1989) Inhibition of endocytic vesicle fusion in vitro by the cell-cycle control protein kinase cdc2. Nature 342:942–945

Valencia A, Chardin P, Wittinghofer A, Sander C (1991) The ras protein family: evolutionary tree and role of conserved amino acids. Biochemistry 30:4637–4648

van der Sluijs P, Hull M, Zahraoui A, Tavitian A, Goud B, Mellman I (1991) The small GTP-binding protein rab4 is associated with early endosomes. Proc Natl Acad Sci USA 88:6313–6317

van der Sluijs P, Hull M, Webster P, Mâle P, Goud B, Mellman I, (1992a) The small GTP-binding protein rab4 controls an early sorting event on the endocytic pathway. Cell 70:729–740

van der Sluijs P, Hull M, Huber LA, Male P, Goud B, Mellman I (1992b) Reversible phosphorylation-dephosphorylation determines the localization of rab4 during the cell cycle. EMBO J 11:4379–4389

van Deurs B, Petersen OW, Olsnes S, Sandvig K (1989) The ways of endocytosis. Intern Rev Cytol 117:131–176

Vielh E, Touchot N, Zahraoui A, Tavitian A (1989) Nucleotide sequence of a rat cDNA: rab1B, encoding a rab1l-Ypt1 related protein. Nucleic Acids Res 17:1770

Walworth NC, Goud B, Kabcenell AK, Novick PJ (1989) Mutational analysis of SEC4 suggests a cyclical mechanism for the regulation of vesicular traffic. EMBO J 8:1685–1693

Walworth NC, Brennwald P, Kabcenell AK, Garrett M, Novick P (1992) Hydrolysis of GTP by sec4 protein plays an important role in vesicular transport and is stimulated by a GTPase-activating protein in Saccharomyces cerevisiae. Mol Cell Biol 12:2017–2028

Wandinger-Ness A, Bennett MK, Antony C, Simons K (1990) Distinct transport vesicles mediate the delivery of plasma membrane proteins to the apical and basolateral domains of MDCK cells. J Cell Biol 111:987–1000

Warren G (1985) Membrane traffic and organelle division. Trends Biochem Sci 10:439–443

Warren G (1989) Mitosis and membranes. Nature 342:857–859

Watts C, Marsh M (1992) Endocytosis: what goes in and how? J Cell Sci 103:1–8

Wilson DB, Wilson MP (1992) Identification and subcellular localization of human rab5b, a new member of the ras-related superfamily of GTPases. J Clin Invest 89:996–1005

Woodman PG, Mundy DI, Cohen P, Warren G (1992) Cell-free fusion of endocytic vesicles is regulated by phosphorylation. J Cell Biol 116:331–338

Zahraoui A, Touchot N, Chardin P, Tavitian A (1989) The human rab genes encode a family of GTP-binding proteins related to yeast YPT1 and SEC4 products involved in secretion. J Biol Chem 264:12394–12401

Zerial M, Melancon P, Schneider C, Garoff H (1986) The transmembrane segment of the human transferrin receptor functions as a signal peptide. EMBO J 5:1543–1550

CHAPTER 29

GTPases in Transport Between Late Endosomes and the *Trans* Golgi Network

S.R. PFEFFER

A. Small GTPases in Membrane Traffic

One of the most exciting recent discoveries in the area of protein trafficking is the finding that almost every organelle involved in exocytic and endocytic processes bears on its surface at least one small, organelle-specific, ras-like GTPase (PFEFFER 1992). The potential role of GTPases in regulating vesicular transport was first discovered when NOVICK and coworkers determined the primary structure of the yeast *SEC4* gene product (SALMINEN and NOVICK 1987). Temperature-sensitive mutations in *sec4* lead to the accumulation of secretory vesicles containing invertase at the nonpermissive temperature. The sequence of Sec4p indicated that the protein is ras-like GTPase, and subsequent analyses revealed that this protein is present on the surface of secretory vesicles (GOUD et al. 1988) and can bind and hydroylze GTP (KABCENELL et al. 1990). Another yeast protein, Ypt1p, is also a ras-related GTPase that is 48% identical to Sec4p (GALLWITZ et al. 1983). Mutations in this protein inhibit vesicular transport between the endoplasmic reticulum (ER) and the Golgi complex, and lead to the proliferation of the ER (SEGEV et al. 1988; SCHMITT et al. 1988). The accumulation of transport vesicles in yeast strains harboring mutant Sec4p strongly suggests that Sec4p functions in the targeting and/or fusion of secretory vesicles with the plasma membrane. Functional analyses of ypt1p have demonstrated an analogous role for this protein in ER-to-Golgi transport (see below). Together, these observations underscore the potential importance of GTPases in regulating vesicular transport events.

B. In Vitro Assays to Analyze the Role of GTP in Membrane Traffic

I. Introduction

In recent years, a number of cell-free systems have been devised that reconstitute the vesicular transport of proteins between membrane-bound organelles of the secretory and endocytic transport pathways. Thus, it is now possible to study the transport of proteins from the ER to the Golgi complex (BECKERS et al. 1987; BAKER et al. 1988; RUOHOLA et al. 1988), between Golgi cisternae (BALCH et al. 1984; ROTHMAN 1987), as well as the budding of transport vesicles from the *trans* Golgi network (TGN; BENNETT et al. 1989; DECURTIS and SIMONS 1990; TOOZE and HUTTNER 1990), and their subsequent fusion with the plasma membrane (WOODMAN and EDWARDSON 1986; SALAMERO et al. 1990). In addition, a number of events in the endocytic pathway have also been reconstituted (see GRUENBERG and HOWELL 1989 for review). The availability of these systems has permitted a biochemical analysis of the molecular mechanisms that underlie vesicular transport processes (BALCH 1989; GODA and PFEFFER 1989; ROTHMAN and ORCI 1990; WATTENBERG 1990).

II. Transport of Mannose 6-Phosphate Receptors From Late Endosomes to the *trans* Golgi Network In Vitro

We have recently described a cell-free system that reconstitutes the transport of mannose 6-phosphate receptors (MPRs) from late endosomes to the TGN (Fig. 1; GODA and PFEFFER 1988). MPRs carry newly synthesized, soluble lysosomal hydrolases from the TGN to late endosomes and are then transported back to the TGN to complete a cycle of biosynthetic, lysosomal enzyme transport (KORNFELD and MELLMAN 1989). Our endosome-to-TGN transport assay relies upon the unique localization of sialyltransferase to the *trans* Golgi and TGN and utilizes a mutant cell line in

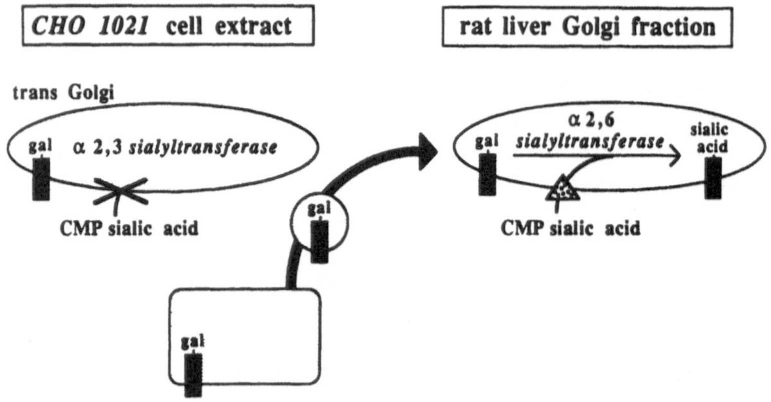

Fig. 1. Complementation scheme to detect transport of man6P receptors from endosomes back to the *trans* Golgi network. An extract prepared from ^{35}S-labeled Chinese hamster ovary (*CHO^{1021}*) cells is incubated with wild-type Golgi membranes, cytosol, ATP, and an ATP-regenerating system at 37°C. CHO^{1021} cell MPRs (*rectangles*) possess galactose-terminating oligosaccharides and will acquire *sialic acid* if transported to the wild-type *trans* Golgi network. At the end of the reactions, receptors are isolated and sialic acid acquisition is monitored by slug lectin chromatography

which glycoproteins are not sialylated (CHO clone 1021; BRILES et al. 1977). Radiolabeled MPRs, present in late endosomes in a mutant cell extract acquire sialic acid residues when they are transported to the TGN of wild-type Golgi complexes present in reaction mixtures. Sialic acid acquisition by MPRs in this system reflects a vesicular transport process, since it is dependent on time, temperature, ATP and cytosol and also requires GTP hydrolysis (GODA and PFEFFER 1988). Furthermore, MPRs and sialyltransferase remain in sealed membrane compartments throughout the reaction, and nonspecific membrane fusion is ruled out by several criteria (GODA and PFEFFER 1988).

III. GTPγS Inhibits Endosome-to-TGN Transport In Vitro

ROTHMAN and coworkers, stimulated by the discovery of the Sec4p primary structure, were the first to investigate the role of GTP hydrolysis in vesicular transport in vitro (MELANÇON et al. 1987). Data presented by these workers suggested that the ability of the Golgi membranes to serve as transport vesicle recipients was blocked in the presence of the nonhydrolyzable GTP analog, GTPγS. Since the target of transport vesicles in the endosome-to-TGN transport assay is also a Golgi cisterna, we tested the effect of GTPγS on this vesicular transport process. GTPγS inhibited the transport of MPRs to the TGN by at least 70%; half-maximal inhibition was seen at $10\,\mu M$ GTPγS (GODA and PFEFFER 1988). MELANÇON et al. (1987) found that inhibition of intra-Golgi transport was only complete when reactions were carried out at high ratios of cytosolic proteins to Golgi membranes. This is also the case for endosome-to-TGN transport (GODA and PFEFFER 1991). GTPγS inhibited the ability of cytosol to stimulate transport, analogous to other vesicular transport steps that require GTP hydrolysis (MELANÇON et al. 1987; BECKERS and BALCH 1989; MAYORGA et al. 1989; WESSLING-RESNICK and BRAELL 1990). These data indicate that one or more cytosolic factors, which may or may not themselves be the target of GTPγS, are used to generate an intermediate that requires GTP hydrolysis.

The ability of GTPγS to inhibit the transport of MPRs from endosomes to the TGN in vitro is likely to reflect a requirement for GTP hydrolysis in transport, since excess GTP or GDP, but not excess adenosine triphosphate (ATP) or cytidine triphosphate (CTP), abrogated the inhibitory effect of GTPγS (GODA and PFEFFER 1991). Excess GTP blocked the ability of GTPγS to inhibit transport only if it was added within the first few minutes of incubation to transport reactions containing GTPγS. Similar kinetics were observed in ER-to-Golgi and intra-Golgi transport (BECKERS and BALCH 1989; MELANÇON et al. 1987). Since GTPγS might bind tightly (and rapidly) to a later-acting component, the inability of GTP to block GTPγS inhibition after a few minutes of incubation cannot be assumed to reflect an early requirement for GTP hydrolysis. Thus, this experiment did not distinguish the point in transport at which GTP hydrolysis was required.

During intra-Golgi transport, GTP hydrolysis is required at a stage prior to transport vesicle fusion (MELANÇON et al. 1987). We carried out kinetic experiments to map the point along the vesicular transport reaction pathway at which the endosome-to-TGN reaction becomes resistant to inhibition by GTPγS. We have previously shown that the in vitro transport of MPRs from endosomes to the TGN displays an initial lag of ~18 min, and then proceeds linearly for the next 150 min (GODA and PFEFFER 1988). GTPγS inhibited transport throughout the course of incubation; resistance to GTPγS addition was achieved with a halftime of ~45 min (GODA and PFEFFER 1991).

IV. A GTPγS-Sensitive Transport Component Requires Late Endosomes for Its Activity

It has been postulated that GTPases are required for accurate targeting of transport vesicles to the acceptor membrane. Accordingly, results from ER-to-Golgi, intra-Golgi, and endocytic vesicle fusion assays indicate that GTPases are required at a late stage in transport, most likely preceding vesicle fusion (MELANÇON et al. 1987; ORCI et al. 1989; WESSLING-RESNICK and BRAELL 1990; BECKERS and BALCH 1989; BAKER et al. 1990; BECKERS et al. 1990). In addition, the ability of GTPγS to inhibit vesicle transport has been shown to require membrane components (MELANÇON et al. 1987; BECKERS and BALCH 1989; MAYORGA et al. 1989; WESSLING-RESNICK and BRAELL 1990).

We carried out preincubation experiments with individual reaction components to determine the site (or sites) at which GTPγS can inhibit endosome-to-TGN transport. Semi-intact CHO1021 cell extracts and/or purified Golgi membranes were incubated separately in the presence of GTPγS, ATP, and cytosol at 37°C. Transport assays were then carried out in the presence of excess GTP to prevent any subsequent transport inhibition by GTPγS. Preincubation of the ^{35}S-labeled semi-intact cell extract, which provides the donor late endosome compartment, was sufficient to significantly inhibit the transport reaction. In contrast, pretreatment of the acceptor Golgi membranes did not reduce their ability to support transport. Since the donor semi-intact cells are more complex than purified Golgi membranes, it was important to rule out the possibility than nonendosomal components in the semi-intact cell extract were responsible for the GTPγS inhibition observed. In control experiments, inclusion of non-radiolabeled, semi-intact cells during the GTPγS pre-incubation of Golgi complexes did not alter their insensitivity to the nonhydrolyzable GTP analog.

Kinetic experiments have suggested that GTPγS binds rapidly and tightly to transport components, by a process that is facilitated by cytosolic factors. The above experiments showed that the ability of GTPγS to inhibit transport is also dependent upon the presence of the donor membranes containing radiolabeled MPRs. In other words, GTPγS only inhibits when incubated with the membranes from which assayable transport vesicles are formed. Although we have no information regarding the precise point in transport at which bound GTP is hydrolyzed, our data strongly suggest that a guanine nucleotide is stably recruited onto nascent, budding transport vesicles at the beginning of transport. GTP hydrolysis could then accompany the budding process (TOOZE and HUTTNER 1990) and/or fusion of transport vesicles with their target membrane.

C. Role of rab Proteins in Endosome to *trans* Golgi Network Transport

A search for the mammalian counterparts of the Sec4 and Ypt1 proteins has led to the identification of a large number of ras-related GTPases that may include more than 20 different gene products (see TAKAI et al. 1992; PFEFFER 1992, for review). The Sec4p and Ypt1p-related proteins have been grouped into a family termed "rab" proteins ("ras-like proteins from rat brain"). The rab proteins are 21–25 kDa in mass and are approximately 30% identical to the proto-oncogene product, ras. Most highly conserved are amino acid residues that comprise the GTP-binding domain.

We undertook a search for rab proteins that might function in endosome-to-TGN transport. We have found that rab9, a recently identified rab protein, resides on the surfaces of late endosomes and possibly also, the TGN (LOMBARDI et al. 1993). In addition, rab9 protein stimulates the recycling of MPRs from late endosomes to the TGN in our cell-free system (LOMBARDI et al. 1993).

The protein rab9 was first identified in a screen for Ypt1p and Sec4p-related cDNA clones (CHAVRIER et al. 1990a). The partial available sequence was found to be 54% identical to that of rab7, a constituent of late endosomes (CHAVRIER et al. 1990b). The initial cDNA library was rescreened to obtain clones encompassing the missing amino terminal portion; the sequence of the full length protein is 57% identical to rab7, and 39% to Ypt1p.

Anti-rab9 antibody immunofluorescence experiments revealed a striking perinuclear staining which matched the distribution of MPRs in these cells. Colocalization of rab9 with MPRs strongly suggests that most rab9 is also present on the surface of that organelle, since the vast majority of MPRs reside in late endosomes (GRIFFITHS et al. 1988).

The localization of rab9 to late endosomes and possibly the TGN suggested that this protein may play a role in membrane traffic between these two compartments. To test this possibility, we employed our cell-free endosome-to-TGN transport system. First, anti-rab9 antibodies inhibited endosome-to-TGN transport by approximately 50% in a concentration-dependent manner, under conditions in which anti-rab7 antibodies had no effect. The inability of anti-rab7 antibodies to block transport suggests that the anti-rab9 antibody inhibition was specific, because both rab7 and rab9 are likely to be present on the surface of late endosomes. Although anti-rab9 antibodies inhibited the overall reaction by approximately 50%, cytosol-stimulated transport was almost completely inhibited under these conditions.

To further analyze the possible role of rab9 in endosome-to-TGN transport, we expressed the protein in baby hamster kidney (BHK) cells using the vaccinia virus/T7 RNA polymerase expression system (Fuerst et al. 1986). Cytosols containing overexpressed rab9 protein stimulated the overall transport rection by approximately 2.5-fold. In contrast, no stimulation was observed with cytosols containing overexpressed rab7 or rab4b. Similarly, cytosols from mock-transfected cells had control levels of activity. To confirm that rab9 protein was responsible for the increase in cytosol activity, we expressed rab7 and rab9 in *Escherichia coli* and purified the proteins to greater than 95% homogeneity. Addition of ≤3 ng of purified rab9 protein was sufficient to yield the level of stimulation observed with BHK cytosol containing overexpressed rab9. Maximal stimulation was only observed if the reactions were supplemented with geranylgeranyl diphosphate. Addition of geranylgeranyl diphosphate alone had no effect. Control experiments demonstrated that the *E. coli*-produced rab9 protein became prenylated in the in vitro transport reaction mix, a carboxy-terminal modification that is essential for rab protein function (Molenaar et al. 1988; Takai et al. 1992). It is important to note that neither a C-terminally truncated rab9 protein (rab9ΔC), which lacks approximately 30 amino acid residues at its carboxy terminus, nor rab7 protein stimulated endosome-to-TGN transport in vitro.

Immunoblot analyses indicated that in vitro reactions contain approximately 1–2 ng endogenous rab9 protein, about 90% of which is membrane associated (not shown). The ability of an additional nanogram of exogenous rab9 to stimulate transport demonstrates that it is a limiting constituent in transport reactions, and as such is poised to function as a key regulator of vesicular transport. Other components become limiting for transport in reactions containing more than 1 ng exogenous rab9 protein. The finding that a small increase in the level of rab9 protein yields a large increase in transport implies that some of the endogenous rab9 protein may be present in an inactive form. Nevertheless, physiological levels of purified rab9 are sufficient to stimulate transport.

In summary, rab9 is localized to the surface of late endosomes and possibly the *trans* Golgi network, where it acts to facilitate the process of vesicular transport from late endosomes to the TGN. Of the two rab proteins now localized to late endosomes, only rab9 stimulates this transport event. This finding strongly suggests that multiple rab proteins on the surface of a given organelle have different physiological roles. In analogy with other rab proteins, rab9 may be recruited onto late endosome membranes and become incorporated into nascent transport vesicles. After these vesicles deliver their cargo to the TGN, rab9 must be recycled to the membrane from which it originated. The ability of rab9 to stimulate transport in an in vitro assay will permit us to investigate the mechanism by which this protein acts to facilitate the recycling of membrane proteins from late endosomes to the *trans* Golgi network of eukaryotic cells.

D. A Model for rab Protein Function

I. Recruitment of rab Proteins onto Nascent Transport Vesicles

1. Newly Synthesized rab Proteins are Cytosolic

Fiture 2 presents a model for the recruitment of rab proteins into the vesicular transport machinery. The rab proteins are initially synthesized as soluble proteins and then become geranylgeranylated. Prenylated rab proteins may occur in the cytoplasm complexed with a protein termed GDI (or "GDP dissociation inhibitor"). Both rab3A and SEC4p form a soluble complex with this protein, and dissociation of GDP from rab3A and SEC4p can be inhibited by a mammalian GDI (Sasaki et al. 1991). In addition to regulating rab protein function, GDI binding may also serve to ensure the solubility of the cytoplasmic pool of prenylated rab proteins.

2. Membrane Association

Overexpression of rab proteins does not lead to their mislocalization, but rather to their cytoplasmic accumulation (cf. Chavrier et al. 1990b). This result would be consistent with the existence of a saturable rab receptor. However, it is important to note that this result might also be explained by

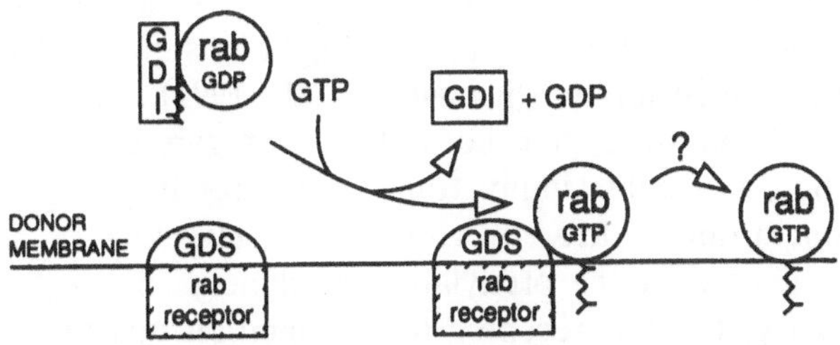

Fig. 2. Model for the recruitment of rab proteins onto *donor membranes*. The rab proteins may exist in the cytosol complexed with a *GDI* (GDP dissociation inhibitor). A putative *rab receptor* binds the rab protein; a *GDS* (GDP dissociation stimulator) triggers GDP release and GTP binding. The GDS may represent the rab receptor or be coupled to it. The rab protein, when *GTP*-bound, may have a lower affinity for the rab receptor, yet be sufficiently hydrophobic to stably insert into the adjacent membrane. Alternatively, the rab protein–receptor complex may be stable. The rab protein and rab receptor may or may not be incorporated into the newly forming transport vesicle (see text)

saturation of the prenylation machinery which must act upon rab proteins to permit membrane association. Thus, localization specificity might be achieved by virtue of some organelle-specific, non-saturable process (see below).

Specific membrane association would require release of GDI and interaction with the rab protein-recognition machinery. As shown in Fig. 2, this could be catalyzed by a membrane-associated, nucleotide exchange factor or "GDS" protein (GDP dissociation stimulator; TAKAI et al. 1992). GDS proteins have been characterized for other classes of ras-like GTPases (YAMAMOTO et al. 1990). If a given rab protein has a specific, cognate GDS protein, the GDS protein could itself serve as the rab receptor.

It would not be unreasonable for a rab-specific GDS protein to utilize the carboxy-terminal domain for substrate identification. Indeed, it is the C-terminal region of another class of ras-like GTPases that is recognized by a cognate GDS protein (HIROYOSHI et al. 1991). But if GDS proteins serve as rab receptors, then each rab must have a distinct GDS present on the appropriate organelle. Accordingly, there might need to be more than 20 different GDS proteins for the presently identified collection of rab proteins. Additional biochemistry and genetics will be required to resolve the nature and complexity of rab receptors and GDS proteins.

The organelle-specific, rab protein-recognition machinery may recognize a prenyl moiety in addition to the carboxy-terminal 34 amino acid residues, since only prenylated rab proteins display membrane association. A similar type of recognition has been described for the src receptor, which recognizes the amino terminal portion of the src protein only when it is myristylated (RESH 1989).

An attractive scenario would be one in which the rab receptor (or GDS protein itself) bound preferentially to a rab–GDP–GDI complex, thereby recruiting the rab protein to the appropriate organelle. After GDS-catalyzed nucleotide exchange, rab-GTP could be released from the receptor; the inherent hydrophobicity of the rab protein might be sufficient to result in its stable membrane association (SERAFINI et al. 1991). According to this scenario, prenylation need not be utilized for rab protein recognition purposes; rather, it would simply trap a rab protein at the appropriate membrane compartment. Moreover, organelle targeting of rab proteins would only be limited by the prenylation machinery and organelle surface area. In summary, the rab receptor (or recognition machinery) need not accompany the rab protein to the target membane, and rab protein localization need not reflect a stable, stoichiometric, receptor interaction.

II. Action of rab Proteins After Transport Vesicle Formation

The role of rab proteins in vesicle targeting and/or fusion is strongly supported by the accumulation of transport vesicles in yeast cells harboring *sec4* mutations (SALMINEN and NOVICK 1987) and the ability of anti-ypt1p antibodies to block ER-to-Golgi transport after transport vesicle formation

(Rexach and Schekman 1991; Segev 1991). The time at which anti-rab1B antibodies block ER-to-Golgi transport is not inconsistent with rab1B functioning in vesicle targeting (Plutner et al. 1991). In addition, anti-rab5 antibodies block in vitro endosome fusion, and overexpressed rab5 protein stimulates this process (Gorvel et al. 1991). Finally, a role for mammalian rab proteins in membrane fusion events in vivo is underscored by the observation that overexpression of rab5 increases the size of early endosomes, and overexpression of a rab5 point mutant which cannot bind GTP leads to fragmentation of the early endosome compartment (Marino Zerial, personal communication).

As shown in Fig. 3, a GAP ("GTPase activating protein") present on the target membrane may initiate a series of reactions that lead to accurate vesicle fusion. The involvement of a GAP is implied by the extremely low rate of intrinsic GTPase activity of purified rab proteins (Wagner et al. 1987; Kabcenell et al. 1990), the detection of YPT1p and rab3A GAP activities in cytosolic and membrane fractions (Becker et al. 1991; Burstein et al. 1991), and the existence of GAPs for other ras-related proteins (Takai et al. 1992).

Although target membrane-specific GAPs may suffice to ensure accurate vesicle targeting, the limited data available suggest that an organelle-specific "docking" receptor is also required. This tentative conclusion is based upon the localization of rab proteins to multiple compartments. For example, rab1B is likely to function in ER-to-Golgi and intra-Golgi transport events (Plutner et al. 1991).

As described above, rab GAP activites have been detected in both cytosolic and membrane fractions. A general rab-GAP activity would not need to be restricted to any specific membrane compartment. Perhaps GAPs are recruited onto rab-bearing transport vesicles, and rab proteins are recognized by the target organelle in the context of a rab–GAP complex. Formation of a ternary complex of rab, GAP, and a target membrane docking receptor might trigger GAP activity and lead to the release of rab and GAP from the transport vesicle (Fig. 3). Membrane fusion would then ensue.

After membrane fusion, rab proteins in their GDP-bound forms are likely to be released from the membrane target and eventually recycle back to their respective membranes of origin. Release from target membranes after exocytosis has been demonstrated directly for rab3A (Fischer-von-Mollard et al. 1991). The release process may be stimulated by cytosolic GDI, which has been shown to be capable of releasing rab3A-GDP, but not rab3A-GTP, from synaptic plasma membranes and vesicles (Araki et al. 1990).

The models presented in Figs. 2 and 3 can explain the presence of rab proteins on both donor and target organelles. For example, SEC4p resides primarily on the surface of secretory granules, but is also associated with the plasma membrane. Similarly, rab1B is present on both the ER and the Golgi complex (Plutner et al. 1991) and rab5 is concentrated on early

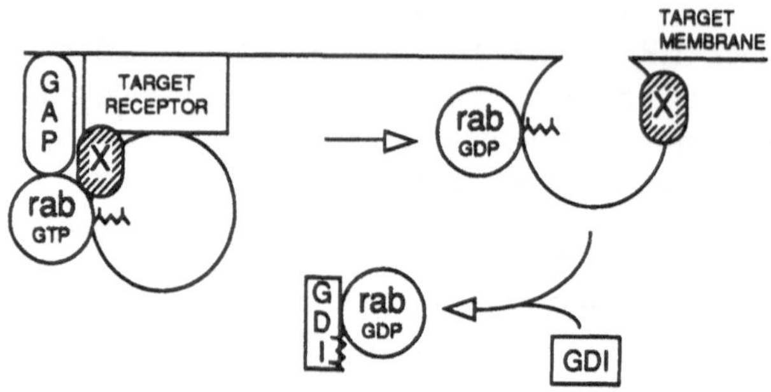

Fig. 3. Model for rab–GAP interaction as a signal for accurate vesicle targeting. Vesicles bearing rab proteins on their surfaces may recognize a target membrane-specific *GAP* (GTPase activating protein) which leads to hydrolysis of the rab-bound GTP. This recognition event may trigger a series of reactions that result in membrane fusion. Alternatively, a more general GAP may be utilized together with a target organelle "docking" receptor to signal accurate vesicle targeting. Preferentially, rab–GDP interacts with *GDI* (GDP dissociation inhibitor), which may catalyze its release from the membrane

endosomes, but can also be detected on the inner surface of the plasma membrane (GORVEL et al. 1991). These results suggest that rab proteins which function in the "late" events of vesicle targeting and/or fusion may be incorporated into transport vesicles during the vesicle formation process.

E. Future Perspectives

Although rab9 participates in endosome-to-TGN transport, it is very likely that additional GTPases will also be required. For example, in ER-to-Golgi transport, GTP hydrolysis is coupled to vesicle formation (D'ENFERT et al. 1990). The ADP-ribosylation factors represent another family of small GTPases that are utilized in some way during intra-Golgi vesicle transport (STEARNS et al. 1990; SERAFINI et al. 1991). Finally, recent experiments also suggest a role for trimeric G-proteins in numerous vesicular transport events.

To date, we have shown that GTPγS inhibits endosome-to-TGN transport. We must now determine each of the targets of this nonhydrolyzable analog. Hopefully, by learning how GTP hydrolysis is coupled to vesicle transport steps, we will better understand the molecular basis of this event in intracellular protein transport.

References

Araki S, Kikuchi A, Hata Y, Isomura M, Takai Y (1990) Regulation of reversible binding of smgp25A, a ras p21-like GTP-binding protein, to synaptic plasma membranes and vesicles by its specific regulatory protein, GDP dissociation inhibitor. J Biol Chem 265:13007–13015

Baker D, Hicke L, Rexach M, Schleyer M, Schekman R (1988) Reconstitution of SEC gene-product-dependent intercompartmental protein transport. Cell 54:335–344

Baker D, Wuestehube L, Schekman R, Botstein D, Segev N (1990) GTP-binding Ypt1 protein and Ca^{2+} function independently in a cell-free protein transport reaction. Proc Natl Acad Sci USA 87:355–359

Balch WE (1989) Biochemistry of interorganelle transport: a new frontier in enzymology emerges from versatile in vitro model systems. J Biol Chem 264:16965–16968

Balch WE, Dunphy WG, Braell WA, Rothman JE (1984) Reconstitution of the transport of protein between successive compartments of the Golgi measured by coupled incorporation of N-acetylglucosamine. Cell 39:405–416

Becker J, Tan TJ, Trepte HH, Gallwitz D (1991) Mutational analysis of the putative effector domain of the GTP-binding Ypt1 protein in yeast suggests specific regulation by a novel GAP activity. EMBO J 10:785–792

Beckers CJM, Balch WE (1989) Calcium and GTP: essential components in vesicular trafficking between the endoplasmic reticulum and the Golgi apparatus. J Cell Biol 108:1245–1256

Beckers CJM, Keller DS, Balch WE (1987) Semi-intact cells permeable to macromolecules: use in reconstitution of protein transport from the ER to the Golgi complex. Cell 50:523–534

Beckers CJM, Plutner H, Davidson HW, Balch WE (1990) Sequential intermediates in the transport of protein between the endoplasmic reticulum and the Golgi. J Biol Chem 265:18298–18310

Bennett MK, Wandinger-Ness A, Simons K (1989) Release of putative exocytic transport vesicles from perforated MDCK cells. EMBO J 7:4075–4085

Briles EB, Li E, Kornfeld S (1977) Isolation of wheat germ agglutinin-resistant clones of CHO cells deficient in membrane sialic acid and galactose. J Biol Chem 252:1107–1116

Burstein ES, Linko-Stentz ZL, Macara IG (1991) Regulation of the GTPase activity of the ras-like protein p25 rab3A. J Biol Chem 266:2689–2692

Chavrier P, Vingron M, Sander C, Simons K, Zerial M (1990a) Molecular cloning of YPT1/SEC4-related cDNAs from an epithelial cell line. Mol Cell Biol 10:6578–6585

Chavrier P, Parton RG, Hauri H-P, Simons K, Zerial M (1990b) Localization of low molecular weight GTP binding proteins to exocytic and endocytic compartments. Cell 62:317–325

Chavrier P, Gorvel J-P, Stelzer E, Simons K, Gruenberg J, Zerial M (1991) Hypervariable C-terminal domain of rab proteins acts as a targeting signal. Nature 353:769–771

d'Enfert C, Wuestehube LJ, Lila T, Schekman R (1991) Sec12p-dependent membrane binding of the small GTP binding protein Sar1p promotes formation of transport vesicles from the ER. J Cell Biol 114:663–670

deCurtis I, Simons K (1989) Isolation of exocytic carrier vesicles from BHK cells. Cell 58:719–727

Fischer-von-Mollard G, Südhof T, Jahn R (1991) A small GTP-binding protein dissociates from synaptic vesicles during exocytosis. Nature 349:79–81

Fuerst TR, Niles EG, Studier FW, Moss B (1986) Eukaryotic transient-expression system based on recombinant vaccinia virus that synthesizes bacteriophage T7 RNA polymerase. Proc Natl Acad Sci USA 83:8122–8125

Gallwitz D, Donath C, Sander C (1983) A yeast gene encoding a protein homologous to the human c-ras/h-ras proto-oncogene product. Nature 306:704–707

Goda Y, Pfeffer SR (1988) Selective recycling of the mannose 6-phosphate/IGF-II receptor to the TGN in vitro. Cell 55:309–320

Goda Y, Pfeffer SR (1989) Cell-free systems to study vesicular transport along the secretory and endocytic pathways. FASEB J 3:2488–2495

Goda Y, Pfeffer SR (1991) Identification of a novel, N-ethylmaleimide-sensitive cytosolic factor required for vesicular transport from endosomes to the TGN in vitro. J Cell Biol 112:823–831
Gorvel J-P, Chavrier P, Zerial M, Gruenberg J (1991) rab5 controls early endosome fusion in vitro. Cell 64:915–925
Goud B, Salminen A, Walworth NC, Novick PJ (1988) A GTP-binding protein required for secretion rapidly associates with secretory vesicles and the plasma membrane in yeast. Cell 53:753–768
Griffiths G, Hoflack B, Simons K, Mellman I, Kornfeld S (1988) The mannose 6-phosphate receptor and the biogenesis of lysosomes. Cell 52:329–341
Gruenberg J, Howell KE (1989) Membrane traffic in endocytosis: insights from cell-free assays. Annu Rev Cell Biol 5:453–481
Hiroyoshi M, Kaibuchi K, Kawamura S, Hata Y, Takai Y (1991) Role of the C-terminal region of smgp21, a ras p21-like small GTP binding protein, in membrane and smg p21 GDP/GTP exchange protein interactions. J Biol Chem 266:2962–2969
Kabcenell AK, Goud B, Northup JK, Novick PJ (1990) Binding and hydrolysis of guanine nucleotides by Sec4p, a yeast protein involved in regulation of vesicular traffic. J Biol Chem 265:9366–9372
Kornfeld S, Mellman I (1989) The biogenesis of lysosomes. Annu Rev Cell Biol 5:483–525
Lombardi D, Soldati T, Riederer MA, Goda Y, Zerial M, Pfeffer SR (1993) Rab 9 functions in endosome-to-TGN transport. EMBO J 12:677–682
Mayorga LS, Diaz R, Colombo MI, Stahl PD (1989) GTPγS stimulation of endosome fusion suggests a role for a GTP binding protein in the priming of vesicles before fusion. Cell Regul 1:113–124
Melançon P, Glick BS, Malhotra V, Weidman PJ, Serafini T, Gleason ML, Orci L, Rothman JE (1987) Involvement of GTP-binding "G" proteins in transport through the Golgi stack. Cell 51:1053–1062
Molenaar CMT, Prange R, Gallwitz D (1988) A carboxy-terminal cysteine residue is required for palmiotic acid binding and biological activity of the ras-related YPT-1 protein. EMBO J 7:971–976
Orci L, Malhotra V, Amherdt M, Serafini T, Rothman JE (1989) Dissection of a single round of vesicular transport: sequential intermediates for intercisternal movement in the Golgi stack. Cell 56:357–368
Pfeffer, SR (1992) GTP binding proteins in intracellular transport. Trends Cell Biol 2:41–48
Plutner H, Cox AD, Pind S, Khosravi-Far R, Bourne JR, Schwaninger R, Der CJ, Balch WE (1991) Rab1B regulates vesicular transport between the endoplasmic reticulum and successive Golgi compartments. J Cell Biol 115:31–43
Resh MD (1989) Specific and saturable binding of pp60 v-src to plasma membranes: evidence for a myristyl-src receptor. Cell 58:281–286
Rexach MF, Schekman RW (1991) Distinct biochemical requirements for the budding, targeting and fusion of ER-derived transport vesicles. J Cell Biol 114:219–230
Rothman JE (1987) Transport of vesicular stomatitis glycoprotein to trans Golgi membranes in a cell-free system. J Biol Chem 262:12502–12510
Rothman JE, Orci L (1990) Movement of proteins through the Golgi stack: a molecular dissection of vesicular transport. FASEB J 4:1460–1468
Ruohola H, Kabcenell AK, Ferro-Novick S (1988) Reconstruction of protein transport from the endosplasmic reticulum to the Golgi complex in yeast: the acceptor Golgi compartment is defective in the sec23 mutant. J Cell Biol 107:1465–1475
Salamero JE, Sztul E, Howell KE (1990) Exocytic transport vesicles generated in vitro from the trans Golgi network carry secretory and plasma membrane proteins. Proc Natl Acad Sci USA 87:7717–7721

Salminen A, Novick PJ (1987) A ras-like protein is required for a post-Golgi event in yeast secretion. Cell 49:527–538

Sasaki T, Kaibuchi K, Kabcenell AK, Novick PJ, Takai Y (1991) A mammalian inhibitory GDP/GTP exchange protein (GDI) for smg p25A is active on the yeast SEC4 protein. Mol Cell Biol 11:2909–2912

Schmitt HD, Puzicha M, Gallwitz D (1988) Study of a temperature sensitive mutant of the ras related YPT1 gene product in yeast suggests a role in the regulation of intracellular calcium. Cell 53:635–647

Segev N (1991) Mediation of the attachment or fusion step in vesicular transport by the GTP-binding Ypt1 protein. Science 252:1553–1556

Segev N, Mulholland J, Botstein D (1988) The yeast GTP-binding YPT1 protein and a mammalian counterpart are associated with the secretion machinery. Cell 52:915–924

Serafini T, Orci L, Amherst M, Brunner M, Kahn RA, Rothman JE (1991) ADP ribosylation factor is a subunit of the coat of Golgi-derived COP-coated vesicles; a novel role for a GTP binding protein. Cell 67:239–253

Stearns T, Willingham MC, Botstein D, Kahn RA (1990) ADP-ribosylation factor is functionally and physically associated with the Golgi complex. Proc Natl Acad Sci USA 87:1238–1242

Takai Y, Kaibuchi K, Kikuchi A, Kawata M (1992) Small GTP-binding proteins. Int Rev Cytol 133:187–230

Tooze SA, Huttner WB (1990) Cell-free protein sorting to the regulated and constitutive secretory pathways. Cell 60:837–847

Wagner P, Molenaar CMT, Rauh AJG, Brökel R, Schmitt HD, Gallwitz D (1987) Biochemical properties of the ras-related YPT protein in yeast: a mutational analysis. EMBO J 6:2373–2379

Wattenberg BW (1990) The molecular control of transport vesicle fusion. The New Biologist 2:505–511

Wessling-Resnick M, Braell WA (1990) Characterization of the mechanism of endocytic vesicle fusion in vitro. J Biol Chem 265:16751–16759

Woodman PG, Edwardson JM (1986) A cell-free assay for the insertion of a viral glycoprotein into the plasma membrane. J Cell Biol 108:843–853

Yamamoto T, Kaibuchi K, Mizuno T, Hiroyoshi M, Shirataki H, Takai Y (1990) Purification and characterization from bovine brain cytosol of proteins that regulate the GDP/GTP exchange reaction of smgp21s, ras p21-like GTP binding proteins. J Biol Chem 265:16626–16634

CHAPTER 30

Endocytic Function in Cell-Free Systems

M. WESSLING-RESNICK

A. Introduction

Endocytosis is a dynamic cellular process through which the vesicle-mediated uptake of macromolecules can occur. The internalization of extracellular material proceeds via two related pathways: *pinocytosis* of macromolecules in the bulk fluid phase and *receptor-mediated endocytosis* of ligand bound to specific cell surface receptors which are internalized by endocytic vesicles. Functionally, endocytosis enables the cell to harvest key nutrient factors, including iron from transferrin, cholesterol from low-density lipoprotein (LDL), and vitamin B_{12}. Furthermore, receptor-mediated endocytosis plays a vital role in the downregulation of receptors in response to target hormones. In fact, characterization of the major cellular events in receptor-mediated endocytosis has been afforded by examining the internalization of ligands recognized by receptors (for a recent review, see SMYTHE and WARREN 1991).

Recent development of in vitro systems to reconstruct events of the endocytic pathway has led to the identification of critical elements and characterization of the conditions required for this activity. Other approaches involve the isolation of mutants defective in the endocytic process (KREIGER et al. 1985; RIEZMAN et al. 1987). However, genetic criteria implicating that a protein is involved in the endocytic machinery must ultimately be supplemented by demonstrating its direct biochemical involvement in membrane movement. Thus, cell-free systems are essential to better understand the factors and conditions which support the endocytic process. This chapter will describe the experimental strategies that have been taken to re-create events of the endocytic pathway in vitro. Cell-free studies have revealed a role for GTPases in endocytosis and provide the best available means to define how these factors function in vesicle traffic.

B. Development of Cell Free Assays

Key stages of endocytosis have been reconstituted in vitro, including: (a) the formation, invagination, and budding of coated vesicles; (b) fusion of endocytic vesicles with endosomal structures; and (c) sorting, processing, and recycling of internalized material to other cellular compartments.

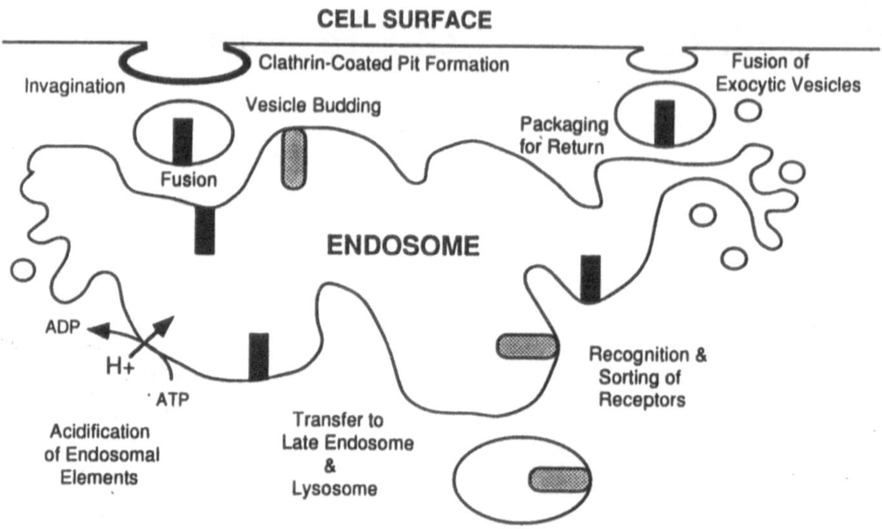

Fig.1. Endocytic functions that have been reconstituted in cell-free systems

Central to these steps is entry and exit from the endosomal compartment; thus, many cell-free studies have focused on the reconstitution of the fusion of vesicles with the endosome.

I. Endosomal Fusion

The basic element required to study the fusion between endocytic vesicles and endosomes is an assay for the colocalization of two different internalized probes. This is generally performed by allowing two populations of cells to endocytose the different probes and preparing vesicle fractions from each set. When the collected vesicle fractions fuse with one another or to a common endosomal compartment, the two probes are placed in the same proximal location and a signal is generated to indicate that vesicle fusion has occurred. Early experiments based on this strategy were performed by Davey and coworkers (1985) and Gruenberg and Howell (1986). In both of these cell-free systems, the fusion signal was generated within the fused vesicle by the action of enzyme on substrate, followed by neurominidase or lactoperoxidase activities, respectively. However, the interpretation of experimental results was complicated by the fact that conditions within the endosomal compartment could not be controlled with any certainty and could potentially result in variations of the generated signal unrelated to endocytic fusion. Furthermore, if the integrity of the vesicle populations had been breached in any way, it was possible that signal generation could continue long after specific fusion reactions colocalizing the probes were completed.

The search for alternative assay methods led BRAELL (1987) to develop a cell-free system that avoided these complications. This sensitive assay utilizes two fluid phase probes, avidin-linked ß-galactosidase and biotinylated IgG. Fusion reactions between endocytic vesicles containing the internalized markers can be detected by the avidin–biotin association reaction. To specify that the signal has been generated within an intact endosome, the assay mixture is supplemented with extravesicular biotin-insulin to scavenge any released avidin-ß-galactosidase. At the end of the in vitro reaction, the fusion signal that is produced can be assessed by lysing vesicles with detergent and captivating any avidin-ß-galactosidase: biotin-IgG complexes on microtiter wells coated with anti-IgG antibodies. The amount of avidin–biotin complex formed is measured with this modified enzyme-linked immunoadsorbent assay (ELISA) technique using a fluorogenic substrate to monitor ß-galactosidase activity.

A key feature of this assay is the critical choice of the avidin–biotin association reaction as a means of detecting the colocalization of endocytic markers (BRAELL 1987). Association between avidin and biotin is extremely rapid, and dissociation takes place on a very slow time scale. Moreover, the association of avidin with biotin is impervious to a wide range of pH and ionic strength conditions, both of which may vary with the cell's endocytic apparatus. This ensures that the signal produced by endocytic vesicle fusions will not be perturbed by environmental conditions within the endosome. Finally, this complex will remain intact even after lysis of the vesicles with detergents, a manipulation that is necessary in order to assay the amount of complex formed. Thus, this assay relies on formation of a stable complex, rather than enzyme–substrate interactions, and therefore provides a more accurate assessment of the extent of vesicle fusion.

This novel assay was eventually modified to include biotinylated transferrin as a second internalized probe (WESSLING-RESNICK and BRAELL 1990a). The resultant signal, a complex between avidin-ß-galactosidase and biotin-transferrin, can be assayed using the same ELISA method with anti-transferrin antibodies to capture the complex. Thus, vesicles that are involved in either fluid phase or receptor-mediated endocytosis can interact and fuse in a common endosomal compartment. By incorporating a third population of vesicles containing anti-transferrin IgG into the cell-free assay, it is possible to monitor multiple fusion events and obtain a comparative ratio of single to multiple fusion events. Results from studies using this cell-free system indicated that 40%–50% of the total in vitro fusion signal may arise from multiple vesicle fusions. This implies that the cell-free system itself is very efficient and that, overall, 40%–50% of the endocytic vesicles available for the assay will participate in fusion reactions (WESSLING-RESNICK and BRAELL 1990a).

Taking advantage of the ELISA-based format, GRUENBERG and coworkers (1989) developed a system to monitor the fusion of isolated endocytic vesicles. Vesicles containing is vesicular stomatitis virus (VSV) G-protein can be captivated using magnetic beads coated with antibodies that recognize the cytoplasmic tail of this membrane-spanning protein (Gruenberg and Howell 1986). In order to assess endosomal fusion, avidin is also internalized in the immuno-isolated vesicle population. Since the participatory vesicles are attached to a solid, magnetic support, they can be conveniently introduced and retrieved from assay mixtures. Cell-free fusion is detected in reactions with vesicles loaded with a biotin–horse radish peroxidase (HRP) conjugate. The avidin–biotin–HRP complex that results is captured on microtiter wells coated with anti-avidin antibody, and a signal based on HRP enzymatic activity is measured. Thus, in vitro systems utilizing the avidin–biotin association reaction to detect colocalization of endocytic probes can be adapted to study a wide range of endocytic events involving ligand-based (e.g., biotin-transferrin), fluid phase-associated (e.g., avidin-ß-galactosidase, biotin-HRP or biotin-IgG), and membrane-spanning (e.g., VSV G-protein) markers.

Other successful strategies that have been developed to monitor cell-free vesicle fusion utilize antibody–antigen recognition as a signaling event. Similar to detection based on avidin–biotin association, these cell-free systems also rely on the rapid, highly specific, and stoichiometric complex formation to monitor colocalization of endocytic markers. Diaz et al. (1988) described an in vitro assay to identify receptor-mediated endocytic events in macrophages using two internalized probes: a dinitrophenol (DNP)-ß-glucuronidase that is a ligand of the mannose receptor, and a mannose-derivitized anti-DNP IgG that could be internalized by the same receptor. Woodman and Warren (1988) utilized the formation of an immune complex between [^{125}I]transferrin and anti-transferrin IgG to detect fusion between vesicles involved in receptor-mediated endocytosis. The efficiencies of these cell-free reactions are difficult to assess due to problems in resolving the proportion of endocytosed probe contained within fusion-competent vesicles. To this end, Woodman and Warren (1991) purified functional, clathrin-coated endocytic vesicles containing [^{125}I]transferrin. These vesicles undergo in vitro fusion with acceptor endosomes containing anti-transferrin IgG in the cell-free assay, resulting in immunoprecipitation of about 50% of latent [^{125}I]transferrin. As with the cell-free systems based on detection using avidin–biotin association, the utility of antigen–antibody recognition is universally applicable and can be employed to study both receptor-mediated and fluid phase endocytosis. The requirements for a cell-free system based on this approach are also minimal: an appropriately derivitized or radiolabeled ligand and an antibody that recognizes the latter probe.

II. Early Endocytic Events: Formation, Invagination, and Budding of Coated Vesicles

Using either the avidin–biotin association reaction or antigen–antibody recognition as a basis for detecting endocytic events, cell-free systems have also been developed that recreate events in the evolution of endocytic vesicles: clathrin coat assembly, coated pit formation, and vesicle budding. Anderson and colleagues (Moore et al. 1987; Mahaffey et al. 1989) devised an in vitro method to study clathrin coat assembly using a fibroblast cell monolayer attached to a polylysine-coated surface. Sonication disrupts the external face of the cells, leaving an exposed inner surface on the substratum. Conditions have been identified which support the assembly of coated pits on the face of this membrane, as documented by morphological studies. It is also possible to quantify clathrin association with the membrane surface (Mahaffey et al. 1989): a mouse anti-clathrin heavy chain antibody is used to detect coat proteins newly assembled on the exposed membrane surface, measured by biotinylated horse anti-mouse IgG and [^{125}I]streptavidin binding. Clathrin-coated pit budding from plasma membrane has been monitored in this in vitro system by simply following the loss of [^{125}I]streptavidin binding upon formation of endocytic vesicles (Lin et al. 1991). However, the reconstitution of cell-free coated pit budding implies that a vesicle population, yet to be identified, could be collected to confirm the conversion of these coated pits into coated vesicles.

Other cell-free methods provide a more direct examination of the formation of coated vesicles. An in vitro system developed by Smythe et al. (1989) utilizes a broken cell preparation generated from A431 cells by a swell-scrape technique; this method has been employed to study other membrane traffic events in the so-called semi-intact cells (Balch 1989). It has been observed that the broken A431 cells will bud coated vesicles containing [^{125}I]transferrin, as followed by the subsequent loss in the ability of exogenously added anti-transferrin antibodies to capture the labeled probe. Moreover, this cell-free system also appears to support production of coated pits, albeit at a rate slower than that observed in intact cells (about 40-fold less vesicles form per minute). The efficiency of budding is also relatively low, with only about 20%–30% of the coated pits forming vesicles.

The cell-free assay described above is also problematic since the apparent gain of [^{125}I]transferrin latency simply defines the accessibility of antibody to the probe. In fact, by manipulating levels of adenosine triphosphate (ATP) and cytosol, it is possible to define intermediate structures, deeply invaginated pits, that most likely serve as precursors to coated vesicles (Smythe et al. 1989). Deeply invaginated pits could exclude antibodies and therefore provide a domain that sequesters [^{125}I]transferrin, without the need for scission or budding to create a completely encapsulated membrane vesicle to protect the probe from immune recognition. Stage-specific cell-free assays to discriminate between coated pit invagination and coated vesicle budding have been established by Schmid and Smythe (1991). To investigate the mechanistic details of coated vesicle biogenesis, [^{125}I]transferrin is biotinylated via a cleavable disulfide-linked bond. The internalization of biotin-S-S-[^{125}]transferrin from the plasma membrane of semi-intact cells can be followed using acquisition of antibody inaccessibility to detect invagination. Later events involving scission and coated vesicle budding can be assessed by monitoring the probe's resistance to a membrane-impermeable reducing agent which releases any exposed biotin and enables the determination of protected biotin-S-S-[^{125}I]transferrin by subsequent absorption to avidin-Sepharose. Therefore, by manipulating reaction conditions, it is possible to define singular properties and elements involved in specific stages of early endocytic events.

III. Late Endocytic Events: Sorting, Processing, and Recycling

Late endocytic events subsequent to endosomal fusion include the recognition, sorting, and processing of internalized material. Fluid phase markers, along with ligands released from receptors in the early endosome, are sorted and segregated to late endosomes for their degradative processing to lysosomes. Recycling receptors, on the other hand, are delivered from the endosome to the exocytic pathway for return to the cell surface. Other destinations can be specified by the nature of the endocytic probe. For example, transferrin is a unique ligand since it is also recycled to the cell surface with its receptor after the release of iron to the acidic endosome. The development of in vitro assays focused on late endocytic events appears to be limited only by the availability of appropriate endocytic markers since, once internalized, a given probe's characteristics will dictate its cellular disposition and consequently its fate in the cell-free system.

WESSLING-RESNICK and BRAELL (1990a) extended their cell-free studies to examine the sorting and segregation mechanism of the endocytic pathway based on detection of the colocalization of avidin-ß-galactosidase and biotin-transferrin. After entry into endosomal compartments, the two probes are directed to divergent pathways: the fluid phase marker is sorted for delivery to the lysosome, while the ligand remains receptor associated and is segregated for return to the cell surface. This event can be monitored by assaying the amount of endocytic marker available for complex formation. As the in vitro reaction progresses, both endocytic markers are cleared from early fusogenic and sorting compartments, resulting in an overall loss of avidin–biotin association when compared to unreactant control samples. The loss of signal, detected by avidin-ß-galactosidase–biotin-transferrin complex formation, reflects the sequestration and/or exit of the probes from the early endosomal compartment. The fluid phase marker, avidin-ß-galactosidase, is delivered to dense, prelysosomal, or late endosomal membrane compartments, while biotin-transferrin remains in light vesicle fractions when reaction products are fractionated on Percoll gradients. Examination of the requirements for the in vitro sorting and segregation reaction verified that early endocytic vesicles containing the probes must first participate in a fusion reaction. This step would be essential to place both the fluid phase marker and receptor-associated ligand into an endosomal domain that is functionally capable of recognizing and processing internalized material. Thus, endosomal activities subsequent to vesicle fusion also occur in the cell-free system, enabling the analysis of molecular parameters that dictate subcellular localization of endocytosed material after entry into the cell.

BOMSEL et al. (1990) also studied processing steps after endosomal delivery utilizing the biotin-HRP and avidin-coupled assay previously described, except that these fluid phase probes were allowed to be endocytosed from either the apical or basolateral surface of a polarized cell line. This remarkable cell-free system allows for the analysis of elements required for the meeting of these two endocytic routes, presumably in the late endosome. While apical–apical or basolateral–basolateral endosomal fusion occurs in vitro, fusion between apical and basolateral "early" endosomes does not. The colocalization of apically- and basolaterally derived material was found to result only if both probes are transferred to a late endosomal compartment. This rather unique cell-free system also requires the presence of microtubules and the microtubule-associated motor proteins, kinesin or dynein. In vitro studies using similar approaches can provide answers to key questions concerning specialized vesicle traffic events in polarized cells, including transcytotic transport.

Other functions associated with the late endosomal compartment have also been studied in vitro. The reconstitution of vesicle traffic interactions between late endosomes and dense lysosomes (MULLOCK et al. 1989) and the *trans* Golgi network (GODA and PFEFFER 1988) has been established. ROEDERER et al. (1990) investigated delivery of another receptor-associated ligand [^{125}I]EGF, to lysosomes by fractionating the products of cell-free reactions on Percoll gradients. Material contained in isolated early endosomes could eventually be localized within "heavy" vesicles corresponding to the density of lysosomes. Based on these observations, the authors suggest that maturation of a post-sorting endosomal compartment results in the biogenesis of a lysosome (MURPHY 1991), although MULLOCK et al. (1989) have found that, in vitro, the presence of dense lysosomes is required for cell-free traffic from endosomes to lysosomes. While the debate between various models concerning the nature and origins of endocytic compartments continues (see MURPHY 1991; GRIFFITHS and GRUENBERG 1991), it is clear that data obtained using cell-free systems can provide valuable information about organelle biogenesis and the interactions between compartments of the lysosomal pathway.

Not all material that is internalized by the cell will follow the degradative pathway. Some membrane elements, like the transferrin receptor as discussed above, will be recycled to the cell surface. The final stages of this pathway, including the fusion of exocytic vesicles with the plasma membrane, have also been reconstituted. PODBILEWICZ and MELLMAN (1990) developed a novel cell-free system to investigate transferrin recycling using semi-intact MDCK cells. Using a nitrocellulose strip method to remove the apical surface of a cell monolayer (SIMONS and

VIRTA 1987), these investigators were able to reconstitute the fusion of exocytic carrier vesicles with the basolateral membrane. Recycling vesicles containing endocytosed [^{125}I]transferrin fuse with the exposed membrane surface, followed by the release of ligand into the media. Thus, the elements of the *entire* intracellular vesicle traffic pathway of transferrin have been established in cell-free systems: a) early events in transferrin internalization, including coated pit invagination and coated pit budding (SCHMID and SMYTHE 1991); b) fusion of endocytic vesicles containing the ligand with endosome (WOODMAN and WARREN 1989; WESSLING-RESNICK and BRAELL 1990a,b); and c) late endocytic events involving the sorting transferrin and its receptor from the endosome (WESSLING-RESNICK and BRAELL 1990a) and recycling to the cell surface (PODBILEWICZ and MELLMAN 1990).

C. GTPases Implicated in Endocytic Traffic

The establishment of endocytic function in the cell-free systems described above provides the foundation to investigate both the conditions that support cellular vesicle traffic (e.g., energy requirements) and the enzymology of molecular elements that mediate this pathway (e.g., proteins that enable vesicle fusion). Several activities, cytosolic and membrane-associated, have been identified that support in vitro vesicle traffic (BRAELL 1987; DIAZ et al. 1988, 1989; COLOMBO et al. 1991). Currently, the most detailed information about endocytic function has been obtained from in vitro studies of endosomal fusion, and in all of the cell-free systems detailed in this review, a role for GTPases has been clearly revealed. However, the precise identity of these factors remains elusive.

I. Evidence Supporting a Functional Role for GTPases

Work by several investigators established that the presence of the nonhydrolyzable analog, GTPγS, will inhibit cell-free reactions of endosomal fusion (MAYORGA et al. 1989a; WESSLING-RESNICK and BRAELL 1990b; BOMSEL et al. 1990). MAYORGA et al. (1989a) found that μM levels of GTPγS inhibit fusion in a cytosol dependent manner. Although the presence of GTPγS inhibits the fusion between macrophage-derived endosomes at high concentrations of cytosol, the cell-free reaction is stimulated by GTPγS when cytosol levels are less than 0.5 mg/ml (MAYORGA et al. 1989b). This evidence suggests that perhaps more than one GTP-binding element is involved in endosomal vesicle traffic, but it does not exclude the idea that both stimulatory and inhibitory actions are exerted by the same factor.

WESSLING-RESNICK and BRAELL (1990b) extensively characterized the mechanism of action of the inhibitory GTPγS-sensitive factor. The order of inhibitory potency of other nonhydrolyzable GTP analogs has been determined: GTPγS $\gg$ GMP-PNP $>$ GMP-PCP $>$ GDPßS. In addition to

GTPγS, the in vitro endosomal fusion reaction could be blocked by ATPγS. The latter result was perhaps not surprising, since it had already been well-established that vesicle fusion required energy in the form of ATP (BRAELL 1987). However, the effects of the nonhydrolyzable nucleotide triphosphate analogs are not synergistic. Furthermore, while the addition of GTP lifts the inhibitory block imposed by GTPγS, the nucleotide has less of an effect on the inhibition of fusion by ATPγS. Conversely, ATP will alleviate ATPγS inhibition to a greater degree than GTPγS inhibition. These combined results indicate that the hydrolysis of GTP and ATP are both required for vesicle fusion, but at different points in the fusion mechanism.

Further studies explored the relationship between the GTPγS-sensitive stage of the reaction and a second step requiring a cytosolic protein that is inactivated by *N*-ethylmaleimide (NEM) (WESSLING-RESNICK and BRAELL 1990b). The GTPγS-sensitive step in endocytic vesicle fusion occurs at a mechanistic stage that is entirely distinct from the NEM-sensitive step and at a temporally earlier point in the fusion process. Taking advantage of the nature of the GTPγS effect, it is possible to accumulate a membrane reaction intermediate in the presence of the nonhydrolyzable analog. Moreover, subsequent incubation of this vesicular intermediate with cytosol and GTP restored its fusion competency (WESSLING-RESNICK and BRAELL 1990b). Thus, upon binding GTPγS, the inhibitory factor appears to cycle from a cytosolic form to exert its activity at the endosomal membrane surface. How the hydrolysis of GTP and/or guanine nucleotide exchange may be linked to endocytic vesicle fusion is still unknown.

II. Rab Proteins

The search for GTPases that are actively involved in vesicle traffic led to the identification of the Ras-related family of Rab proteins (ZAHRAOUI et al. 1989; CHAVRIER et al. 1990). Immunolocalization experiments defined that one of the Rab proteins, Rab5, is located on the cytoplasmic surface of the plasma membrane and on early endosomes (CHAVRIER et al. 1990). Subsequently, studies by VAN DER SLUIJS et al. (1991) demonstrated that a second GTPase, Rab4, is also associated with early endosomes.

Functional evidence for a role of Rab5 in endosomal vesicle traffic relied on cell-free endosomal fusion studies that utilized avidin and biotin-HRP as internalized markers (GORVEL et al. 1991). Fusion between endocytic vesicles is blocked by anti-Rab5 antibody, but not by antibodies against Rab2 or Rab7, which are localized to the Golgi and late endosomes, respectively. Cytosol that contains overexpressed Rab5 was found to stimulate the fusion reaction, while cytosol that contained a mutant form of Rab5 (Rab5-Ile^{133}) that is unable to bind GTP inhibited the in vitro assay. Although this in vitro evidence suggests that GTP binding is essential for Rab5 function and that molecules lacking the ability to bind guanine nucleotide will disrupt vesicle traffic mediated by wild-type Rab5, it has yet to be established whether or not Rab5 will inhibit vesicle fusion in the presence of GTPγS. Thus, it still is not clear if Rab5 is the GTPγS-sensitive inhibitory factor. This is an important point to be considered since the role of a stimulatory GTPase in vesicle traffic has been defined (MAYORGA et al. 1989b) and the action of the overexpressed cytosolic Rab5 is, in fact, to promote the in vitro reaction.

Despite uncertainties surrounding the exact function of Rab5, limited structural information is available. Like other Ras-related proteins, Rab5 has C-terminal cysteine residues that are modified by a geranylgeranyl (C_{20}) moiety (KINSELLA and MALTESE 1991). GORVEL et al. (1991) found that deletion of nine carboxy-terminal amino acids, including the sites of prenylation, promotes the cytosolic expression of Rab5 and that this mutant form has no effect on in vitro endocytic vesicle fusion. CHAVRIER et al. (1991) later demonstrated that a Rab5 chimeric construct containing the last 34 C-terminal amino acids of Rab7 would localize to a late endosomal fraction. Moreover, when a Rab2/Rab5 chimera was constructed such that the last 35 C-terminal residues of Rab5 replaced the native Rab2 sequence, the mutant protein localized to the early endosome instead of the Golgi network as expected for wild-type Rab2. Thus, the C-terminus region defines Rab protein subcellular localization and the integrity of this domain is critical for membrane association.

Little information has been obtained concerning the function of Rab4, a second member of the Rab family that is known to be cytolocalized to the early endosome. However, a potentially critical observation is that Rab4 can be phosphorylated by cdc2 protein kinase (p34) (BAILLY et al. 1991). The $p34^{cdc2}$ is a subunit of maturation promoting factor, a key regulator of the cell's mitotic cycle. It has been established that, during mitosis, intracellular vesicle traffic is arrested (reviewed by WARREN 1989) and that, furthermore, cytosol preparations enriched for the cdc2 protein kinase activity will inhibit cell-free endocytic vesicle fusion (TUOMIKOSKI et al. 1989). These observations may provide the first clues to a regulatory function for Rab family members in vesicle traffic events. BAILLY et al. (1991) found that not only is phosphorylation of Rab4 mitosis specific, but it also correlates with the distribution of cytosolic and membrane-bound forms of the GTPase. Since phosphorylated Rab4 is predominantly found in cytosol, release of the protein from the membrane surface could play an essential role in the arrest of endocytosis during mitosis.

III. Heterotrimeric G-Proteins

In retrospect, the first observations that suggested a role for heterotrimeric G-proteins in endocytic vesicle traffic were early findings that AlF_n^- inhibits cell-free endosomal fusion (MAYORGA et al. 1989a; WESSLING-RESNICK and BRAELL 1990b). WESSLING-RESNICK and BRAELL (1990b) observed that aluminum fluoride, together with GDP and Mg^{2+}, can suppress the in vitro reaction, compatible with the idea that AlF_n^- interacts with GTPases by mimicking the gamma phosphate of GTP when GDP is bound (HIGASHIJIMA et al. 1991). However, AlF_n^-, in combination with GDP and Mg^{2+}, only partially blocks cell-free endocytic vesicle fusion (WESSLING-RESNICK and BRAELL 1990b). Therefore, it remains to be determined whether the inhibitory effects are due to inhibition of GTP hydrolysis required for vesicle fusion or whether these effects are exerted through other regulatory elements. We now know that AlF_n^- does not act on small GTPases of the Ras superfamily and is selective for members of the heterotrimeric GTP-binding regulatory protein family (KAHN 1991). This information suggests that, in addition to the endosomal Rab proteins (Rab4 and Rab5), heterotrimeric G-proteins may also be involved in endocytic events.

Further evidence in support for a role of trimeric G-proteins in endosome fusion has been provided by cell-free studies on the effects of $\beta\gamma$ subunits in in vitro fusion assays (COLOMBO et al. 1992). These investigators found that the addition of $\beta\gamma$ blocked the stimulatory action of GTPγS observed at low concentrations of cytosol (<0.5 mg/ml). Since the activation of the α subunits of G-proteins is also blocked by excess-free $\beta\gamma$ subunits, one interpretation of this result is that the dissociation of an $\alpha\beta\gamma$ complex is required for vesicle fusion to proceed. This is consistent with the fact that the activation of α subunits by the binding of GTP (or GTPγS) promotes subunit dissociation and that if GTPγS is added prior to the addition of $\beta\gamma$, the latter produces less of an effect on endosomal fusion (COLOMBO et al. 1992).

Additional studies employing mastoparan, a wasp venom known to accelerate guanine nucleotide exchange of G_o and G_i protein α subunits, support G-protein involvement in endocytosis. Mastoparan will block both the stimulation of vesicle fusion by GTPγS at low cytosol levels and the inhibition observed at high cytosol concentrations (COLOMBO et al. 1992). One cautionary note in evaluating this data is that nonspecific effects can be exerted by mastoparan, particularly on membrane phospholipases (JOYCE-BRADY et al. 1991). Since the physical union of membrane bilayers requires that the integrity of the endocytic vesicles is maintained during the cell-free assay, it is important that these effects are considered. COLOMBO et al. (1992) verified that vesicle morphology is not perturbed under the conditions of this study ($10\,\mu M$ mastoparan), at least at the level of electron microscopic analysis.

The effects of mastoparan and $\beta\gamma$ subunits on endocytic vesicle fusion are difficult to rectify when their respective actions on the cell-free reaction are considered. Mastoparan, which should promote guanine nucleotide exchange, might be expected to exert effects opposite to those observed for $\beta\gamma$. COLOMBO et al. (1992) propose a model for GTPase action in vesicle traffic that includes dual G-proteins. In this hypothetical scheme, mastoparan's action promotes the activation of one trimeric G-protein and thus releases $\beta\gamma$ subunits that would act on a second (stimulatory) G-protein to consequently block its action. Although yet to be examined critically, this model could be readily tested since one would predict that addition of GTPγS before mastoparan treatment would reduce the block on the stimulation of vesicle fusion, just as a similar preincubation prevents the action of added $\beta\gamma$.

IV. ADP-Ribosylation Factors

The complexity of the role of GTPases in endocytic vesicle traffic is underscored by the recent finding that the addition of recombinant ARF1 to the cell-free system, at both low and high concentrations of cytosol, results in GTPγS-dependent inhibition of vesicle fusion (Lenhard et al. 1992). Adenosine diphosphate (ADP) ribosylation factor (ARF) is a myristoylated species and only the lipid-modified form of the protein is capable of exerting effects on the in vitro assay. Since GTPγS promotes the stable binding of ARF1 to membranes, Lenhard et al. (1992) suggest that ADP-ribosylation factors may regulate vesicle traffic by transient association with the endosomal membrane surface. Thus, ARF would be envisioned to translocate to the endosomal membrane upon binding GTP or GTPγS, and thereby function in vesicle fusion. This is reminiscent of the observation made by Wessling-Resnick and Braell (1990b) that a GTPγS-blocked intermediate can be isolated that recovers fusion competence upon incubation with cytosol and GTP. This reaction intermediate corresponds to a GTPγS-resistant species that accumulates prior to the NEM-sensitive stage of vesicle fusion. Similarly, Lenhard and coworkers (1992) have found an ARF-resistant intermediate during vesicle fusion that may indeed be the equivalent vesicular form. Thus, ADP-ribosylation factors must also be considered as candidate GTP-binding elements that function in endocytic vesicle traffic and may be the elusive GTPγS-sensitive inhibitory components first identified in vitro.

D. Future Perspectives

This brief overview of cell-free reconsitution systems demonstrates their utility in investigations of GTPase function in endocytic vesicle traffic. A number of proteins have been implicated to play a role in vesicle fusion: Rab5, Rab4, heterotrimeric G-proteins, as well as ARFs. Each of these individual elements may contribute to the complex pattern of endocytosis and further in vitro studies will be necessary to distinguish their interactions and precise activities in membrane traffic. Future areas of research include a focus on the early events of endocytosis, including coated pit assembly, invagination, and budding. Of particular interest is the role of microtubule-associated proteins such as dynamin (Shpetner and Vallee 1992). Dynamin is a GTPase that is the gene product of *shibire*, a *Drosophila* locus that has been implicated in the budding of coated vesicles (KESSEL et al. 1989). Fusion of transport vesicles with late endosomes is also microtubule dependent, in addition to being sensitive to GTPγS (BOMSEL et al. 1990). Other late endosomal events, including the sorting of internalized material, are GTPγS-sensitive (Wessling-Resnick, personal observations). Thus, additional functions for GTPases may be revealed in subcellular membrane traffic. Established cell-free systems are readily available to explore the enzymology of endocytic traffic and provide the necessary means to ultimately define the critical GTP-binding factors and their precise functional role in endocytosis.

References

Balch WE (1989) Biochemistry of interorganelle transport. J Biol Chem 264:16965–16968

Bailly E, McCaffrey M, Touchot N, Zahraoui A, Goud B, Bornens M (1991) Phosphorylation of two small GTP-binding proteins of the rab family be $p34^{cdc2}$. Nature 350:715–718

Bomsel M, Parton R, Kuznetsov SA, Schroer TA, Gruenberg J (1990) Microtubule- and motor-dependent fusion in vitro between apical and basolateral endocytic vesicles from MDCK cells. Cell 62:719–731

Braell WA (1987) Fusion between endocytic vesicles in a cell-free system. Proc Natl Acad Sci USA 84:1137–1141

Chavrier P, Gorvel JP, Stelzer E, Simons K, Gruenberg J, Zerial M (1991) Hypervariable C-terminal domain of rab proteins acts as a targeting signal. Nature 353:769–772

Chavrier P, Parton RG, Hauri HP, Simons K, Zerial M (1990) Localization of low molecular weight GTP binding proteins to exocytic and endocytic compartments. Cell 62:317–329

Colombo MI, Gonzalo S, Weidman P, Stahl P (1991) Characterization of trypsin-sensitive factors required for endosome–endosome fusion. J Biol Chem 266:23438–23445

Colombo MI, Mayorga LS, Casey PJ, Stahl PD (1992) Evidence of a role for heterotrimeric GTP-binding proteins in endosome fusion. Science 255:1695–1697

Davey J, Hurtley SI, Warren G (1985) Reconstitution of an endocytic fusion event in a cell-free system. Cell 43:643–652

Diaz R, Mayorga L, Stahl P (1988) In vitro fusion of endosomes following receptor-mediated endocytosis. J Biol Chem 263:6093–6100

Diaz R, Mayorga L, Weidman PJ, Rothman J, Stahl P (1989) Vesicle fusion following receptor-mediated endocytosis requires a protein active in Golgi transport. Nature 339:398–400

Goda Y, Pfeffer SR (1988) Selective recycling of the mannose-6-phosphate/IGF-II receptor to the trans-Golgi network in vitro. Cell 55:309–320

Gorvel JP, Chavrier P, Zerial M, Gruenberg J (1991) Rab5 controls early endosome fusion in vitro. Cell 64:915–925

Griffiths G, Gruenberg J (1991) The arguments for pre-existing early and late endosomes. Trends Cell Biol 1:5–9

Gruenberg J, Howell KE (1986) Reconstitution of vesicle fusions occurring in endocytosis with a cell-free system. EMBO J 5:3091–3101

Gruenberg J, Griffiths G, Howell KE (1989) Characterization of the early endosome and putative endocytic carrier vesicles in vivo and with an assay of vesicle fusion in vitro. J Cell Biol 108:1301–1316

Higashijima T, Graziano MP, Suga H, Kainosho M, Gilman AG (1991) ^{19}F and ^{31}P NMR spectroscopy of G protein α subunits. J Biol Chem 266:3396–3401
Joyce-Brady M, Rubins JB, Panchenko MP, Bernardo J, Steele MP, Kolm L, Simons ER, Dickey BF (1991) Mechanisms of mastoparan-stimulated surfactant secretion from isolated pulmonary alveolar type 2 cells. J Biol Chem 266:6859–6856
Kahn RA (1991) Fluoride is not an activator of the smaller (20–25 kDa) GTP-binding proteins. J Biol Chem 266:15595–15597
Kessel I, Holst BD, Roth TF (1989) Membranous intermediates in endocytosis are labile, as shown in a temperature-sensitive mutant. Proc Natl Acad Sci USA 4968–4972
Kinsella BT, Maltese WA (1991) Rab GTP-binding proteins implicated in vesicular transport are isoprenylated in vitro at cysteines within a novel carboxyl-terminal motif. J Biol Chem 266:8540–8544
Kreiger M, Kingsley D, Sege R, Hobbie L, Kozarsky K (1985) Genetic analysis of receptor-mediated endocytosis. Trends Biochem Sci 10:447–452
Lenhard JM, Kahn RA, Stahl PD (1992) Evidence for ADP-ribosylation factor (ARF) as a regulator of in vitro endosome-endosome fusion. J Biol Chem 267:13047–13052
Lin HC, Moore MS, Sanan DA, Anderson RGW (1991) Reconstitution of clathrin-coated pit budding from plasma membranes. J Cell Biol 114:881–891
Mahaffey DT, Moore MS, Brodsky FM, Anderson RGW (1989) Coat proteins isolated from clathrin coated vesicles can assemble into coated pits. J Cell Biol 108:1615–1624
Mayorga LS, Diaz R, Stahl PD (1989a) Regulatory role for GTP-binding proteins in endocytosis. Science 244:1475–1477
Mayorga LS, Diaz R, Colombo MI, Stahl PD (1989b) GTPγS stimulation of endosome fusion suggests a role for a GTP-binding protein in the priming of vesicles before fusion. Cell Regul 1:113–124
Moore MS, Mahaffey DT, Brodsky FM, Anderson RGW (1987) Assembly of clathrin-coated pits onto purified plasma membranes. Science 236:558–563
Mullock BM, Branch WJ, van Schaik M, Gilbert LK, Luzio JP (1989) Reconstitution of an endosome-lysosome interaction in a cell-free system. J Cell Biol 108:2093–2099
Murphy RF (1991) Maturation models for endosome and lysosome biogenesis. Trends Cell Biol 1:77–82
Podbilewicz B, Mellman I (1990) ATP and cytosol requirements for transferrin recycling in intact and disrupted MDCK cells. EMBO J 9:3477–3487
Riezman H, Chvatchko Y, Dulic V (1986) Endocytosis in yeast. Trends Biochem Sci 11:325–328
Roederer M, Barry JR, Wilson RB, Murphy RF (1990) Endosomes can undergo an ATP-dependent density increase in the absence of dense lysosomes. Eur J Cell Biol 51:229–234
Schmid SL, Smythe E (1991) Stage-specific assays for coated pit formation and coated vesicle budding in vitro. J Cell Biol 114:869–880
Shpetner HS, Vallee RB (1992) Dynamin is a GTPase stimulated to high levels of activity by microtubules. Nature 355:733–735
Simons K, Virta H (1987) Perforated MDCK cells support intracellular transport. EMBO J 6:2241–2247
Smythe E, Warren G (1991) The mechanisms of receptor-mediated endocytosis. Eur J Biochem 202:689–699
Smythe E, Pypaert M, Lucocq J, Warren G (1989) Formation of coated vesicles from coated pits in broken A431 cells. J Cell Biol 108:843–853
Tuomikoski T, Felix MA, Doree M, Gruenberg J (1989) Inhibition of endocytic vesicle fusion in vitro by the cell-cycle control protein kinase cdc2. Nature 342:942–945

Van der Sluijs P, Hull M, Zahraoui A, Tavitian A, Goud B, Mellman I (1991) The small GTP-binding protein rab4 is associated with early endosomes. Proc Natl Acad Sci USA 88:6313–6317

Warren G (1989) Mitosis and membranes. Nature 342:857–858

Wessling-Resnick M, Braell WA (1990a) The sorting and segregation mechanism of the endocytic pathway is functional in a cell-free system. J Biol Chem 265:690–699

Wessling-Resnick M, Braell WA (1990b) Characterization of the mechanism of endocytic vesicle fusion in vitro. J Biol Chem 265:16751–16759

Woodman PG, Warren G (1988) Fusion between vesicles from the pathway of receptor-mediated endocytosis in a cell-free system. Eur J Biochem 173:101–108

Woodman PG, Warren G (1991) Isolation of functional, coated, endocytic vesicles. J Cell Biol 112:1133–1141

Zahraoui A, Touchot N, Chardin P, Tavitian A (1989) The Human Rab genes encode a family of GTP-binding proteins related to yeast Ypt1 and Sec4 products involved in secretion. J Biol Chem 264:12394–12401

CHAPTER 31

Synaptic Vesicle Membrane Traffic and the Cycle of Rab3

G. Fischer von Mollard, T.C. Südhof, and R. Jahn

A. Membrane Traffic of Synaptic Vesicles in Neurons

Neurons are highly specialized cells which process and transmit information. Information is transferred between neurons via small molecules, the neurotransmitters. Neurotransmitters are stored in synaptic vesicles and are released by Ca^{2+}-dependent exocytosis upon activation of the presynaptic cell (for reviews see Smith and Augustine 1988; De Camilli and Jahn 1990; Südhof and Jahn 1991).

Exocytosis and recycling of synaptic vesicles is part of a highly amplified and specialized membrane traffic pathway in neurons. In a fully differentiated neuron, synaptic vesicle membranes originate from the *trans* Golgi network and are transported by fast axonal transport to the nerve terminal. It is unclear whether the membranes are transported by a precursor compartment or as mature synaptic vesicles. It is possible that newly synthesized membranes first undergo constitutive, i.e., Ca^{2+}-independent, exocytosis before being sorted into the synaptic vesicle pool (Regnier-Vigouroux et al. 1991).

In the resting state, synaptic vesicles, filled with neurotransmitter, are stored in high numbers in the cytoplasm. A fraction of the vesicles is firmly docked to specialized zones of the presynaptic plasma membrane. Upon arrival of an action potential, voltage-gated Ca^{2+} channels open and trigger fusion of the synaptic vesicle with the plasma membrane. The delay between Ca^{2+} entry and exocytosis is extremely short, in the range of 200–300 μs (for a review see Smith and Augustine 1988). This implies that the Ca^{2+} channels are probably integrated into the active zone. It further implies that all components of the fusion apparatus are preassembled at the synaptic vesicle – plasma membrane complex, in contrast to other intracellular fusion events (see, for example, Rothman and Orci 1990). After exocytosis, the synaptic vesicle membrane is retrieved by endocytosis, most likely via coated pits and coated vesicles (Heuser and Reese 1973; reviewed in Ceccarelli and Hurlbut 1980). Synaptic vesicles are then re-formed to undergo another round of exo-endocytic membrane cycling. This process can be repeated at least 20 times in the isolated nerve terminal (Ceccarelli and Hurlbut 1980). It is unclear whether this recycling involves passage through an endosomal compartment at each cycle or whether functional synaptic vesicles are directly generated by decoating of coated vesicles. However, the nerve terminal must possess a sorting compartment which decides whether individual membrane components are reused for the formation of synaptic vesicles or whether they are sorted out for retrograde axonal transport to the cell body. Retrograde transport probably involves multivesicular bodies that are present in axons and that have been shown to contain synaptic vesicle proteins (JANETZKO et al. 1989). These membranes are reminiscent of organelles mediating transport from early to late endosomes in other cells (GRIFFITHS and GRUENBERG 1991). Upon arrival in the cell body, synaptic vesicle membrane components are probably sorted to late endosomes and then directed either to the *trans* Golgi network for reuse or to lysosomes for degradation.

B. Rab3 Proteins: Structure, Posttranslational Modifications and Subcellular Localization

In recent years it became evident that a group of small, ras-related GTPases is involved in the control of intracellular membrane traffic. Originally discovered in yeast by genetic analysis (GALLWITZ et al. 1983; SALMINEN and NOVICK 1987; also see Chaps. 4, 27), it appears that each step of membrane traffic in eukaryotes requires specific GTPases. Many of these proteins are ubiquitous components of all cells since they are involved in fundamental trafficking pathways such as transport from the endoplasmic reticulum to the Golgi apparatus or from the Golgi apparatus to lysosomes (BALCH 1990; PFEFFER 1992). In addition, cell-type specific pathways such as synaptic vesicle exocytosis may require additional small GTPases specialized for the particular function.

Recently, we have found that a small GTPase, rab3A, is specifically localized on the synaptic vesicle membrane (FISCHER VON MOLLARD et al. 1990). Rab3A belongs to the subfamily of rab proteins. Three highly related rab3 cDNAs have been characterized from mammalian brain (MATSUI et al. 1988; ZAHRAOUI et al. 1988, 1989). The cDNA sequences of rab3A (also referred to smg p25A), rab3B (smg p25B), and rab3C (smg p25C) encode for proteins with predicted molecular masses of 25.0 kDa, 24.7 kDa, and 26.0 kDa, respectively. A comparison of the bovine amino acid sequences revealed 77% identity between rab3A and rab3B, 85% identity between rab3A and rab3C, and 80% identity between rab3B and rab3C. The rab3 forms are most divergent in the C-terminal domain. In rab5 and rab7, these domains have been shown to be responsible for organelle-specific localization (CHAVRIER et al. 1991). In addition to the mammalian species, rab3-related cDNA sequences have been reported from the electric fish *Discopyge ommata* (VOLKNANDT et al. 1991) and *Drosophila* (JOHNSTON et al. 1991) (76%–78% amino acid identity to all three rab3 forms), demonstrating a high degree of evolutionary conservation.

At the protein level, information is only available for rab3A (KIKUCHI et al. 1988; FISCHER VON MOLLARD et al. 1990). Similar to other small GTPases, rab3A does not contain a transmembrane domain (MATSUI et al. 1988; ZAHRAOUI et al. 1988). Nevertheless, approximately 70% of rab3A in brain is firmly attached to membranes. Binding of rab3A is mediated by posttranslational hydrophobic modifications that are dependent on its carboxy-terminal Cys-X-Cys sequence (JOHNSTON et al. 1991). Mass spectrometry has shown that both cysteine residues at the carboxy terminus in rab3A are geranylgeranylated and that the carboxy terminus is also carboxy-methylated (FARNSWORTH et al. 1991). Geranylgeranylation or farnesylation of carboxy-terminal cysteine residues appear to be general mechanisms used by small GTPases for membrane attachment. The carboxy termini of small GTPases fall into three classes of consensus sequences: (1) proteins ending in Cys-X-X-X which are either farnesylated or geranylgeranylated, depending on the C-terminal amino acid residue, respectively (CASEY et al. 1991; KINSELLA et al. 1991); (2) proteins ending in Cys-Cys; and (3) proteins ending in Cys-X-Cys (KHOSRAVI-FAR et al. 1991; KINSELLA and MALTESE 1992). Both cysteines in the two latter consensus sequences are probably geranylgeranylated. Enzymatic activities for the geranylgeranylation of rab3A and other rabs have been demonstrated (HORIUCHI et al. 1991). Recently, a geranylgeranyl transferase was purified from rat brain that geranylgeranylates the carboxy-terminal cysteines of rab3A. The enzyme contains at least three subunits and is different from previously characterized isoprenyl transferases. Surprisingly, the purified enzyme geranylgeranylated small GTPases containing Cys-Cys ends as effectively as those containing Cys-X-Cys ends, suggesting that the same enzyme recognizes both consensus sequences (SEABRA et al. 1992). Furthermore, some data suggest the presence of an additional hydrophobic moiety linked by a thioester bond (JOHNSTON et al. 1991). This suggests that rab3A may be palmitoylated, similar to ras (HANCOCK et al. 1989).

It remains to be clarified how these modifications are related to membrane binding. It was shown that rab3A is localized predominantly in the soluble fraction when the last three amino acids are deleted or when geranylgeranylation is inhibited by antagonists of the mevalonate pathway (JOHNSTON et al. 1991). However, it is unlikely that dissociation of rab3A from the membrane (see below) requires removal of the geranylgeranyl groups. Furthermore, the ester-linked modification appears to be present both in the soluble and the membrane-bound pool (FISCHER VON MOLLARD et al. 1990).

Northern blot analysis has shown that rab3A is predominantly expressed in neurons and neuroendocrine tissues (OLOFSSON et al. 1988; SANO et al. 1989). In the brain, rab3A expression displays considerable regional variation, being approximately 20 times lower in cerebellum and brain stem than in the mesencephalon (AYALA et al. 1989). No expression was detected in other tissues such as the exocrine gland (salivary gland) or the adrenal cortex. In addition, several laboratories have studied the tissue distribution of rab3A using a variety of monoclonal and polyclonal antibodies. Whereas all of these studies confirmed the presence of rab3A in neurons and neuroendocrine cells, TAKAI's group also reported rab3A-like immunoreactivity in exocrine glands (pancreas, submaxillary gland) (MIZOGUCHI et al. 1989) and more recently in stomach epithelia, including nonsecretory cells (MATSUDA et al. 1992). However, the monoclonal antibody used in these studies recognizes additional proteins and, in particular, was not tested for cross-reactivity with rab3B and rab3C (KIM et al. 1989). Antibody specificity appears to be a major concern for the study of these highly homologous proteins. Of the two monoclonal antibodies raised in our laboratory against recombinant rab3A, one recognizes rab3A, rab3B and rab3C, whereas the second recognizes only rab3A, but may react with one or more additional small GTPases not yet characterized (MATTEOLI et al. 1991; and unpublished observations). In our hands, the antibody specific for rab3A reacts only with neurons and neuroendocrine tissues, both in immunoblotting and immunocytochemistry experiments. In contrast, strong immunoreactivity was found in the exocrine pancreas using the monoclonal antibody specific for rab3A, rab3B and rab3C, which was selectively associated with zymogen granules (SCHNEFEL et al. 1992). This suggests that rab3B or a related GTPase but not rab3A is expressed in the exocrine pancreas.

In neurons, rab3A was shown to be a major constituent of the membrane of synaptic vesicles using both subcellular fractionation and immunocytochemistry. Furthermore, rab3A copurified with synaptic vesicles by immunoisolation using beads coated with synaptic vesicle protein-specific monoclonal antibodies (FISCHER VON MOLLARD et al. 1990). No immunoreactivity was found on other membranes by light microscopy immunocytochemistry, including that of the Golgi apparatus (MATTEOLI et al. 1991). In neuroendocrine cells, rab3A is highly enriched in a population of microvesicles which has a protein composition very similar to that of synaptic vesicles and is different from the respective secretory granules (FISCHER VON MOLLARD et al. 1990). It is controversial whether rab3A is also present on the membrane of secretory granules. Whereas in our study no significant amounts of rab3A were detected on chromaffin granules, DARCHEN et al. (1990) reported coenrichment of rab3A with chromaffin granules. Preliminary results from our laboratory suggest that at least some of these discrepancies may be explained by cross-reactivity of some of the antibodies with additional rab3A-related small GTPases which may be present in chromaffin cells.

C. The Cycle of rab3A in Nerve Terminals

What is the role of rab3A in synaptic vesicle cycling? We have recently shown that stimulation of exocytosis in isolated nerve terminals leads to the generation of a vesicle population devoid of rab3A. This was reversible, suggesting that rab3A undergoes a cycle parallel to the recycling of synaptic vesicle membranes (Fischer von Mollard et al. 1991). Such a cycle is congruous with the current concept of the action of small GTPases in membrane traffic (Bourne et al. 1990; also see Chap. 1). According to these models, small GTPases containing GTP are bound to the trafficking membrane. Upon contact with the target membrane, GTP is thought to be cleaved due to interaction with a specific GTPase activating protein (GAP). It has not yet been established which molecules are influenced by this step, i.e., how GTP cleavage relates to biological function. The GDP form is believed to be removed from the membrane via interaction with specific proteins (see below) and converted into the GTP form with the aid of GTP–GDP exchange proteins (GDS or GRF). The GTP form then reassociates with new membrane in a highly specific manner, completing the cycle.

Figure 1 shows our current model describing how rab3A cycling may be linked to the synaptic vesicle cycle. In the resting state, all vesicles loaded with transmitter, including those docked at active zones, have rab3A bound on their surface, probably in the GTP form. It is unclear whether rab3A functions in the docking of synaptic vesicles to the active zone, i.e., before

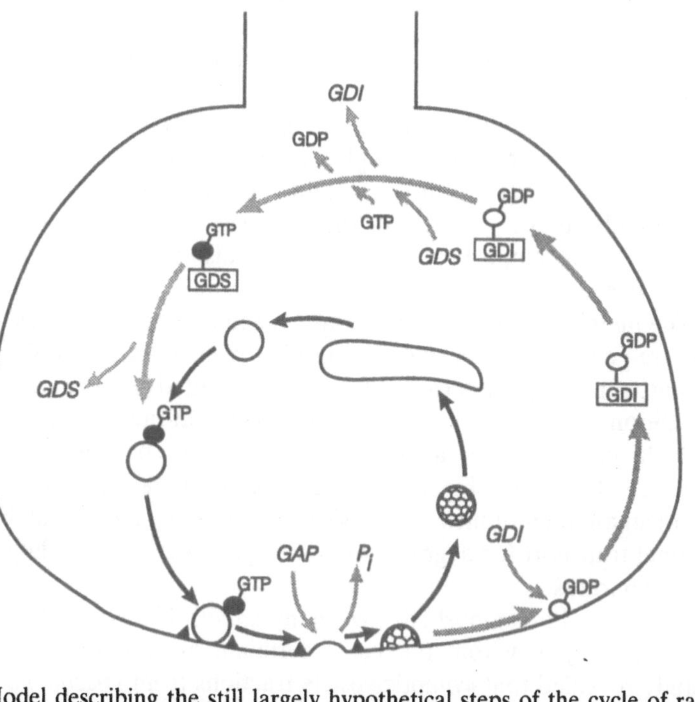

Fig. 1. Model describing the still largely hypothetical steps of the cycle of rab3A in nerve terminals. *GDI*, GDP dissociation inhibitor; *GDS*, guanine nucleotide dissociation stimulator; *GAP*, GTPase activating protein; P_i, inorganic phosphate

triggering exocytosis, or whether it is involved in exocytosis itself, as shown in the figure. In analogy to Ras, we assume that rab3A action involves GTP hydrolysis, which is induced by interaction of the GTP form with a Rab3A-specific GAP. Evidence for rab3A-specific GAP activity was recently reported in rat brain (BURSTEIN et al. 1991). Thus, it is conceivable that interaction with GAP and GTP hydrolysis occurs either during docking or after stimulatory Ca^{2+} influx.

After fusion of the vesicle with the plasma membrane, rab3A appears to be incorporated into the plasma membrane together with other vesicle proteins. This is supported by a localization of rab3A immunoreactivity at the plasmalemma after massive exocytosis elicited by α-latrotoxin in the absence of Ca^{2+}, similar to that of synapsin I and synaptophysin (MATTEOLI et al. 1991). Rab3A, now in the GDP form, then interacts with a GDP dissociation inhibitor (GDI) which allows its dissociation from the membrane, possibly complexed with GDI. GDI which has been purified and sequenced is a ubiquitous soluble protein that acts not only on rab3A, but also on other small GTPases such as SEC4p (MATSUI et al. 1990; ARAKI et al. 1990; SASAKI et al. 1990, 1991; see Chap. 39). Little information is available concerning GDP–GTP exchange and membrane reassociation of rab3A and the proteins involved in these processes. Recently, a factor stimulating nucleotide exchange of rab3A was enriched from rat brain cytosol, but its role in the rab3 cycle remains to be clarified (BURSTEIN and MACARA 1992).

The model predicts that rab3A is lacking from the endocytic limb of the synaptic vesicle cycle. The vesicle population devoid of rab3A which was isolated by us from stimulated synaptosomes (see above) may represent such endocytic compartments. Interestingly, we were unable to detect a concomitant increase of rab3A in the soluble fraction, indicating that GDI-mediated membrane dissociation of rab3A is slower than the first steps of endocytosis. Strong support for this model was also provided by our finding that rab3A is absent from clathrin-coated vesicles isolated from nerve terminals. The membrane protein composition of these coated vesicles revealed that they are generated from synaptic vesicles by endocytosis (MAYCOX et al. 1992). Thus, rab3A appears to be selectively excluded from coated pits.

In addition, rab3A labeling was not observed on membranes in the cell body which contain integral membrane proteins of synaptic vesicles (MATTEOLI et al. 1991). This suggests that rab3A is lacking from synaptic vesicle precursor membranes and vesicle membranes sorted out by retrograde axonal transport for degradation in the lysosomes or recycling through the Golgi apparatus.

It remains to be established how many small GTPases operate in the synaptic vesicle cycle within the nerve terminal in addition to rab3A. A recent study revealed that synaptic vesicle fractions from *Discopyge ommata* electric organ contain additional GTPases (NGSEE et al. 1990). Furthermore, the number of related genes isolated from the brain is still increasing. Thus, it is possible that the highly complex and specialized cycle of synaptic vesicles in the nerve terminal requires many more small GTPases for correct function.

References

Araki S, Kikuchi A, Hata Y, Isomura M, Takai Y (1990) Regulation of reversible binding of smgp25A, a ras p21 like GTP-binding protein, to synaptic plasma membranes and vesicles by its specific regulatory protein, GDP dissociation inhibitor (GDI). J Biol Chem 265:13007–13015

Ayala J, Olofsson B, Tavitian A, Prochiantz A (1989) Developmental and regional regulation of rab3: a new brain specific ras-like gene. J Neurosci Res 22:241–246

Balch WE (1990) Small GTP-binding proteins in vesicular transport. TIBS 15:473–477

Bourne HR, Sanders DA, McCormick F (1990) The GTPase superfamily: a conserved switch for diverse cell functions. Nature 348:125–132

Burstein ES, Macara IG (1992) Characterization of a guanine nucleotide-releasing factor and a GTPase-activating protein that are specific for the ras-related protein p25 rab3A. Proc Natl Acad Sci USA 89: 1154–1158

Burstein ES, Linko-Stentz K, Lu Z, Macara IG (1991) Regulation of the GTPase activity of the ras-like protein p25 rab3A evidence for a rab3A-specific GAP. J Biol Chem 266:2689–2692

Casey PJ, Thissen JA, Moolmaw JF (1991) Enzymatic modification of proteins with a geranylgeranyl isoprenoid. Proc Natl Acad Sci USA 88:8631–8635

Ceccarelli B, Hurlbut WP (1980) Vesicle hypothesis of the release of quanta of acetylcholine. Physiol Rev 60:396–441

Chavrier P, Gorvel JP, Stelzer E, Simons K, Gruenberg J, Zerial M (1991) Hypervariable C-terminal domain of rab proteins acts as a targeting signal. Nature 353:769–772

Darchen F, Zahraoui A, Hammel F, Monteils MP, Tavitian A, Scherman D (1990) Association of the GTP-binding protein Rab3A with bovine adrenal chromaffin granules. Proc Natl Acad Sci USA 87:5692–5696

De Camilli P, Jahn R (1990) Pathways to regulated exocytosis in neurons. Annu Rev Physiol 52:625–645

Farnsworth CC, Kawata M, Yoshida Y, Takai Y, Gelb MH, Glomset JA (1991) C terminus of the small GTP-binding protein smgp25A contains two geranylgeranylated cysteine residues and a methyl ester. Proc Natl Acad Sci USA 88:6196–6200

Fischer von Mollard G, Mignery G, Baumert M, Perkin MS, Hanson TJ, Burger PM, Jahn R, Südhof TC (1990) Rab3 is a small GTP-binding protein exclusively localized to synaptic vesicles. Proc Natl Acad Sci USA 87:1988–1992

Fischer von Mollard G, Südhof TC, Jahn R (1991) A small GTP-binding protein dissociates from synaptic vesicles during exocytosis. Nature 349:79–81

Gallwitz D, Donath C, Sander C (1983) A yeast gene encoding a protein homologous to the human c-has/bas proto-oncogene product. Nature 306:704–707

Griffiths G, Gruenberg J (1991) The arguments for pre-existing early and late endosomes. TICB 1:5–9

Hancock JF, Magee AI, Childs JE, Marshall CJ (1989) All ras proteins are polyisoprenylated but only some are palmitoylated. Cell 57:1167–1177

Heuser JE, Reese TS (1973) Evidence for recycling of synaptic vesicle membrane during transmitter release at the frog neuromuscular junction. J Cell Biol 57:315–344

Horiuchi H, Kawata M, Katayama M, Yoshida Y, Musha T, Ando S, Takai Y (1991) A novel prenyltransferase for a small GTP-binding protein having a C-terminal Cys-Ala-Cys structure. J Biol Chem 266:16981–16984
Jaentzko A, Zimmermann H, Volknandt W (1989) Intraneuronal distribution of a synaptic vesicle membrane protein: antibody binding sites at axonal membrane compartments and trans-Golgi network and accumulation at nodes of ranvier. Neuroscience 32:65–77
Johnston PA, Archer BT, Robinson K, Mignery GA, Jahn R, Südhof TC (1991) Rab3A attachment to the synaptic vesicle membrane mediated by a conserved polyisoprenylated carboxy-terminal sequence. Neuron 7:101–109
Khosravi-Far R, Lutz R, Cox AD, Conroy L, Bourne JR, Sinensky M, Balch WE, Buss JE, Der CJ (1991) Isoprenoid modification of rab proteins terminating on CC or CXC motifs. Proc Natl Acad Sci USA 88:6264–6268
Kikuchi A, Yamashita T, Kawata M, Yamamoto K, Ikeda K, Tanimoto T, Takai Y (1988) Purification and characterization of a novel GTP-binding protein with a molecular weight of 24000 from bovine brain membranes. J Biol Chem 263:2897–2904
Kim S, Kikuchi A, Mizoguchi A, Takai Y (1989) Intrasynaptosomal distribution of the ras, rho and smg-25A GTP-binding proteins in bovine brain. Mol Brain Res 6:167–176
Kinsella BT, Maltese WA (1992) rab GTP-binding proteins with three different carboxyl-terminal cysteine motifs are modified in vivo by 20-carbon isoprenoids. J Biol Chem 267:3940–3945
Kinsella, BT, Erdman RA, Maltese WA (1991) Posttranslational modification of Ha-ras p21 by farnesyl versus geranylgeranyl isoprenoids is determined by the COOH-terminal amino acid. Proc Natl Acad Sci USA 88:8934
Matsuda K, Sakamoto C, Nakano O, Konda Y, Matozaki T, Wada K, Kasuga M, Mizoguchi A, Kikuchi A, Takai Y (1992) Distribution of smg p25A and smg p21s, ras p21-like guanine nucleotide-binding proteins, in the rat stomach. Am J Physiol 262:G69–G73
Matsui Y, Kikuchi A, Kondo J, Hishida T, Teranishi Y, Takai Y (1988) Nucleotide and deduced amino acid sequences of a GTP-binding protein family with molecular weights of 25000 from bovine brain. J Biol Chem 263:11071–11074
Matsui Y, Kikuchi A, Araki S, Hata S, Kondo J, Teranishi Y, Takai Y (1990) Molecular cloning and characterization of a novel type of regulatory protein (GDI) for smg p25A, a ras p21-like GTP-binding protein. Mol Cell Biol 10:4116–4122
Matteoli M, Takei K, Cameron R, Hurlbut P, Johnston PA, Südhof TC, Jahn R, De Camilli P (1991) Association of rab3A with synaptic vesicles at late stages of the secretory pathway. J Cell Biol 115:625–633
Maycox PR, Link E, Reetz A, Morris SA, Jahn R (1992) Clathrin-coated vesicles in nervous tissue are involved primarily in synaptic vesicle recycling. J Cell Biol 118:1379–1388
Mizoguchi A, Kim S, Ueda T, Takai Y (1989) Tissue distribution of smgp25A, a ras p21-like GTP-binding protein, studied by use of a specific monoclonal antibody. Biochem Biophys Res Commun 162:1438–1445
Ngsee JK, Miller K, Wendland B, Scheller RH (1990) Multiple GTP-binding proteins from cholinergic synaptic vesicle. J Neurosci 10:317–322
Olofsson B, Chardin P, Touchot N, Zahraoui A, Tavitian A (1988) Expression or the ras-related ralIA, rhol2 and rab genes in adult mouse tissue. Oncogene 3:231–234
Pfeffer SR (1992) GTP-binding proteins in intracellular transport. TICB 2:41–46
Regnier-Vigouroux A, Tooze SA, Huttner WB (1991) Newly synthesized synaptophysin is transported to synaptic-like microvesicles via constitutive secretory vesicles and the plasma membrane. EMBO J 10:3589–3601
Rothman JE, Orci L (1990) Movement of proteins through the Golgi stack: a molecular dissection of vesicular transport. FASEB J 4:1460–1468

Salminen A, Novick PJ (1987) A ras-like protein is required for a post-Golgi event in yeast. Cell 49:527–538
Sano K, Kikuchi A, Matsui Y, Teranishi Y, Takai Y (1989) Tissue-specific expression of a novel GTP-binding protein (smg p25A) mRNA and its increase by nerve growth factor and cyclic AMP in rat pheochromocytoma PC-12 cells. Biochem Biophys Res Commun 158:377–385
Sasaki T, Kikuchi A, Araki S, Hata Y, Isomura M, Kuroda S, Takai Y (1990) Purification and characterization from bovine brain cytosol of a protein that inhibits the dissociation of GDP from and the subsequent binding of GTP to smg p25A, a ras p21 like GTP-binding protein. J Biol Chem 265:2333–2337
Sasaki T, Kaibuchi K, Kabcenell A, Novick PJ, Takai Y (1991) A mammalian inhibitory GDP/GTP exchange protein (GDP dissociation inhibitor) for smgp25A is active on the yeast sec4 protein. Mol Cell Biol 11:2909
Schnefel S, Zimmermann P, Pröfrock A, Jahn R, Aktories K, Hinsch KD, Haase W, Schulz I (1992) Multiple small and high molecular weight GTP-binding proteins in zymogen granule membranes of rat pancreatic acinar cells. Cell Physiol Biochem 2:77–89
Seabra MC, Goldstein JC, Südhof TC, Brown MS (1992) Rab Geranylgeranyl transferase: a multisubunit enzyme that prenylates GTP-binding terminating in Cys-X-Cys or Cys-Cys. J Biol Chem 267:14497–14503
Smith SJ, Augustine GJ (1988) Calcium ions, active zones and synaptic transmitter release. TINS 11:458–464
Südhof TC, Jahn R (1991) Proteins of synaptic vesicles involved in exocytosis and membrane recycling. Neuron 6:665–677
Volknandt W, Pevsner J, Elferink LA, Schilling J, Scheller RH (1991) A synaptic vesicle specific GTP-binding protein from ray electric organ. Mol Brain Res 11:283–290
Zahraoui A, Touchet N, Chardin P, Tavitian A (1988) Complete coding sequences of the ras related rab 3 and 4 cDNAs. Nucleic Acid Res 16:1204
Zahraoui A, Touchet N, Chardin P, Tavitian A (1989) The human rab genes encode a family of GTP-binding proteins related to yeast YPT1 and sec4 products involved in secretion. J Biol Chem 264:12394–12401

CHAPTER 32

Regulated Exocytosis and Interorganelle Vesicular Traffic: A Comparative Analysis

A. LUINI and M.A. DE MATTEIS

A. Introduction

There are two main pathways of vesicular protein traffic and secretion in eukaryotic cells. One, common to all cells, is constitutive. The other, termed regulated, is present in specialized cell types and serves to rapidly export selected proteins under controlled conditions (BURGESS and KELLY 1987). During transit towards the cell surface, all secretory proteins share a number of constitutive vesicular transport steps through the endoplasmic reticulum (ER) and the Golgi compartments. Their fates then diverge in the tract from the *trans* Golgi network (TGN) to the plasma membrane in two major respects. Proteins secreted by the regulated pathway are sorted, concentrated, and stored for variable periods of time in rather large specialized structures, the secretory granules. By contrast, constitutively secreted proteins are enclosed in small vesicles (50–60 nm in diameter), similar to those used in intra-Golgi transport, and flowing continuously towards the plasma membrane where they fuse. The second difference is that exocytosis of regulated granules occurs only upon triggering by an appropriate extracellular signal, whereas all other transport steps proceed continuously and, in fact, constitutively.

The concept of two different secretory pathways coexisting in the same cell has advanced our understanding of the mechanisms of secretion. At the same time, however, it may have retarded the appreciation of similarities between these transport processes, a problem compounded until recently by the relative lack of communication between the two areas of research, which have developed separately. However very recently, common mechanisms and components have been recognized between the constitutive and regulated exocytic pathways (SÖLLNER et al. 1993). The purpose of this chapter is to review and compare recent data with a focus on the role, organization, and type of GTPases involved in constitutive and regulated pathways and the control of these pathways by signal transduction systems, an aspect that has been extensively studied in regulated secretion and remarkably neglected in the case of constitutive traffic.

Table 1. Effects of GTPase activators/inactivators on constitutive and regulated membrane traffic

Agent	GTPase	Constitutive	Regulated
		Vesicle budding and formation	
GTPγS	all	Partial inhibition (1)	Partial inhibition (2)
GTP	all	Prevention of GTPγS inhibition (3)	Prevention of GTPγS inhibition (4)
Fluoroaluminate	heterotrimeric G-protein	Partial inhibition (5)	Inhibition (5)
βγ subunits	heterotrimeric G-protein	Stimulation (5)	Stimulation (5)
Pertussis toxin	heterotrimeric G-protein (G_i, G_o)	Partial prevention of GTPγS inhibition (5) Stimulation (6)	Partial prevention of GTPγS inhibition (5)
		Vesicle targeting and fusion	
GTPγS	all	No effect (7) stimulation (8) inhibition (9) depending on experimental conditions (see text)	Stimulation (10) or inhibition (11) depending on experimental conditions (see text)
GTP	all	Prevention of GTPγS effects (12)	Prevention of GTPγS inhibition (13)
GDPβS	all	Inhibition (14)	Inhibition (15)
Fluoroaluminate	heterotrimeric G-protein	See text (16)	Stimulation (17)
Mastoparan	heterotrimeric G-protein, rho-rac	Inhibition, reversal of GTPγS effects (18)	Stimulation (19)
Pertussis toxin	heterotrimeric G-protein (G_i, G_o)		Stimulation (20), prevention of mastoparan effect (21)
βγ subunit	heterotrimeric G-protein	Antagonism of GTPγS stimulation (22)	No effect (23)
α_n, α_i, α_{22}, α_{24}, α_{26} subunits	heterotrimeric G-protein		Inhibition, see text (24)

1) REXACH and SHECKMAN 1991; TOOZE and HUTTNER 1990; 2) TOOZE and HUTTNER 1990; BARR et al. 1991; 3) TOOZE et al. 1990; 4) TOOZE et al. 1990; 5) BARR et al. 1991; 6) STOW et al. 1991; 7) BECKERS and BALCH 1989; REXACH and SHECKMAN 1991; 8) MAYYORGA et al. 1989; COLOMBO et al. 1992; 9) MELANÇON et al. 1987; MAYORGA et al. 1989; WESSLING and BRAELL 1990; BOMAN et al. 1992; GODA and PFEFFER 1988; 10) BARROWMAN et al. 1986; MORGAN and BURGOYNE 1990; LUINI and DE MATTEIS 1988, 1990; ULLRICH et al. 1990; PADFIELD et al. 1991; KNIGHT and SCRUTTON 1986; DE MATTEIS et al. 1991; KOOPMAN and JACKSON 1990; HOWELL et al. 1987; 11) BITTNER et al. 1986; KNIGHT and BAKER 1985; DE MATTEIS et al. 1991; VAN DER MERWE et al. 1991; 12) BECKERS and BALCH 1989; MELANÇON et al. 1987; MAYORGA et al. 1989; 13) DE MATTEIS et al. 1991; VAN DER MERWE et al. 1991; 14) COLOMBO et al. 1992; 15) DE MATTEIS et al. 1991; BITTNER et al. 1986; BARROWMAN et al. 1986; 16) MELANÇON et al. 1987, BECKERS and BALCH 1989; WESSLING-RESNICK and BRAELL 1990; 17) NADIN et al. 1989; 18) COLOMBO et al. 1992; 19) NAKAMURA and UI 1985; ARIDOR et al. 1990, YOKOKAWA et al. 1989; 20) SONTAG et al. 1991; OHARA-IMAIZUMI et al. 1992; 21) SAITO et al, 1987; ARIDOR et al. 1990; 22) COLOMBO et al. 1992; 23) LINDAU and GOMPERTS 1991; 24) LINDAU and GOMPERTS 1991.

B. GTPases in Membrane Traffic: Experimental Approaches

Several approaches have been used to establish the involvement of GTPases in vesicular traffic. A common one has been based on the effects exerted by general GTPase activators such as GTPγS, mastoparan, and fluoraluminate in in vitro transport assays. This approach has provided important initial information about the existence and the function of GTPases in several membrane traffic steps. A problem connected with the use of these agents, however, has been that they have been employed in transport assays involving more than one molecular step (vesicle budding, targeting, fusion) and more than one GTPase. This has sometimes prevented a simple interpretation of the results. For example, GTPγS completely blocks constitutive transport in most assays, while it only retards vesicle formation from ER and TGN (Table 1), apparently suggesting that the targeting–fusion step is also blocked by the nucleotide. However, detailed kinetic and biochemical studies of ER-to-Golgi transport (BECKERS and BALCH 1989; REXACH and SCHEKMAN 1991) have indicated that GTP hydrolysis may actually be required for preparatory events such as vesicle uncoating MELANÇON et al. 1987), rather than for the final stages of vesicle targeting–fusion. Fusion can, in fact, occur in the presence of GTPγS, as has been shown for endosome fusion at low cytosol concentration (MAYORGA et al. 1989), for the fusion of regulated secretory granules with the plasma membrane (LINDAU and GOMPERTS 1991), and for the fusion occurring in ER-to-Golgi traffic in the presence of Brefeldin A (BFA) (ORCI et al. 1991).

More specific information about the role of individual GTPases has come from morphological, biochemical, and functional studies, in which selective tools such as antibodies, competitor peptides, antisense oligonucleotides, and overexpression of normal, or expression of mutated, GTPases have been employed. Finally, genetic evidence from yeast secretory mutants has allowed the identification and the functional analysis of many proteins, including GTPases, required for secretion (Table 2).

C. GTPases in Constitutive Transport

I. Vesicle Formation

In cell-free systems designed to specifically assay vesicle formation from the TGN and the ER (REXACH and SCHEKMAN 1991; TOOZE and HUTTNER 1990; SALAMERO et al. 1990; GRIMES and KELLY 1992), GTPγS and other nonhydrolyzable analogs of GTP have been shown to have a partial inhibitor effect that is competed by GTP (TOOZE et al. 1990). The inhibition resulted from retardation, not blocking, of the process (REXACH and SCHEKMAN 1991; TOOZE et al. 1990; BARR et al. 1991; GRIMES and KELLY

Table 2 Approaches to the study of individual GTPases in membrane traffic

Approach	GTPase	Effect
	Constitutive exocytosis and endocytosis	
Yeast sec mutants	YPT1	Defect in ER-Golgi transit GALLWITZ et al. 1983; SEGEV et al. 1988; SCHMITT et al, 1988; REXACH and SCHEKMAN 1991
	SEC4	Defect in vesicle transit to plasma membrane SALMINEN and NOVICK 1987
Drosophila mutants	dynamin-like protein	Defect in bud scission VAN DER BLIEK and MEYEROWITZ 1991
Gene overexpression	$G_{\alpha i3}$	Inhibition of constitutive secretion STOW et al. 1991
	rab5	Increases in early endosome size ZERIAL, see PFEFFER 1992
Expression of mutated genes	rab5	Fragmentation of early endosome ZERIAL, see PFEFFER 1992
Peptides		
rab effector domain	rab family	Inhibition of targeting/fusion of ER vesicles to the Golgi PLUTNER et al. 1990
N-terminal ARF1,4	ARF family	Prevention of GTPγS effects on endosome fusion LENHARD et al. 1992
Antibodies		
anti-YPT 1	YPT1	Block of transport between ER and Golgi REXACH and SCHEKMAN 1991;
anti-rab 1	rab1	Block of transport between ER and Golgi PLUTNER et al. 1991
anti-rab 5	rab5	Block of endosome fusion GORVEL et al. 1991
	Regulated exocytosis	
Antisense oligonucleotides		
rab3b specific	rab3b	Inhibition of the late component of the secretory response P.M. LLEDO 1992, personal communication
Peptides		
rab effector region	rab family	Stimulation of secretion PADIFIELD et al. 1992; LI et al. 1992
Antibodies		
anti-$G_{\alpha o}$	$G_{\alpha o}$	Stimulation of calcium-dependent exocytosis OHARA-LMAIZUMI et al. 1992

ER, endoplasmic reticulum; ARF, ADP-ribosylation factor.

1992). This conclusion may not apply to all stages of transport, since in an intra-Golgi transport assay GTPγS caused accumulation, rather than decrease, of Golgi-derived vesicles (MELANÇON et al. 1987); in these experiments, however, the steady state vesicle number, rather than the rate of formation, was measured. So an inhibitory effect of GTPγS on vesicle formation might have been obscured by a block in vesicle consumption.

GTPγS is probably a universal activator of GTPases. To distinguish between the involvement of ras-related small GTPases and that of heterotrimeric G-proteins in transport processes (in this case vesicle budding–formation), heterotrimeric G-protein activators such as fluoroaluminate (KAHN 1991) and mastoparan (HIGASHIJIMAN et al. 1988) or inhibitors such as $\beta\gamma$ subunits and pertussis toxin have been employed. Vesicle formation from TGN was inhibited by fluoroaluminate and enhanced by the addition of excess $\beta\gamma$ subunits (BARR et al. 1991). The GTPγS-induced block was partially relieved by pre-incubation with pertussis toxin, which inactivates the heterotrimeric G-proteins G_i and G_o (BARR et al. 1991). These results can be explained by the involvement of a heterotrimeric G-protein with inhibitory effects on vesicle budding, although the requirement of GTP hydrolysis may also suggest a role for a "cycling" GTPase (see reviews by BOURNE et al. 1991 and PFEFFER 1992, for definition of GTPase cycling in membrane traffic). In an unrelated set of experiments, STOW et al. (1991) showed that overexpression of $G_{\alpha i3}$, a heterotrimeric G-protein abundant in the Golgi (ERCOLANI et al. 1990), leads to inhibition of the constitutive secretion of proteoglycans, an effect that was prevented by pertussis toxin. Pertussis toxin also augmented proteoglycan release in control cells. These experiments did not distinguish whether transport was inhibited at the stage of vesicle formation or consumption. Nevertheless, it has been hypothesized that G_{i3} might be the heterotrimeric G-protein which mediates inhibition of vesicle formation from the TGN (BARR et al. 1992).

Another line of investigation has focused on the regulation of non-clathrin coat protein deposition on Golgi membranes, a process crucial for vesicle formation and forward membrane flow (MELANÇON et al. 1991). A family of small GTPases termed ARFs (adenosine diphosphate ribosylation factors) and one or more heterotrimeric G-proteins appear to be involved in this process. ARF has been found to be a major component of the vesicle coat and to be required for binding of another major coat component, the protein β-COP (β coat protein), to Golgi membranes. β-COP is a constituent of the large protein complex termed coatomer (MALHOTRA et al. 1989; SERAFINI et al. 1991; WATERS et al. 1991). Thus, ARF has been proposed to act as a trigger for coat assembly (SERAFINI et al. 1991). As noted above, another key player in coat assembly is likely to be a heterotrimeric G-protein. This was suggested by the fact that the adhesion of both ARF itself and β-COP to Golgi stacks is stimulated by the specific heterotrimeric G-protein activators fluoroaluminate (DONALDSON et al. 1991) and mastoparan (KTISTAKIS et al. 1992), and by the fact that pertussis toxin prevented the mastoparan effects (KTISTAKIS et al. 1992). As the activation of the heterotrimeric G-protein involved leads to coat association, a necessary event in vesicle formation, this heterotrimeric G-protein might have a stimulatory role in transport. If this is the case, vesicle formation might be under dual regulation by a stimulatory and an inhibitory heterotrimeric G-protein, similar to the familiar scheme originally proposed for the receptor-mediated control of adenylyl cyclase (GILMAN 1987).

In addition to ARFs and heterotrimeric G-proteins, a third class of GTPases is involved, at least in the case of vesicle budding, namely, the GTP-driven, dynamin-like motor proteins. The *Drosophila shibire* mutant lacks a protein 69% homologous to dynamin and is affected by temperature-sensitive paralysis due to the block of endocytosis and recycling of synaptic vesicles (VAN DER BLIEK and MEYEROWITZ 1991). The primary defect is the inability to release coated vesicles from deeply invaginated coated pits, suggesting that motor proteins are required for bud scission.

Finally, in yeast, a small GTPase having 35% homology with ARF, SAR1p, is involved in budding of vesicles from the ER through interaction with the SEC12 protein (D'ENFERT et al. 1991). No mammalian counterpart of SAR1 has so far been reported.

II. Vesicle Targeting and Fusion

The first identification of a GTPase involved in vesicle fusion was made in a yeast mutant accumulating secretory vesicles in the cytoplasm. The mutated gene (*SEC4*) was found to code for a ras-related small GTPase (SALMINEN and NOVICK 1987). Another yeast GTPase involved in vesicle fusion is YPT1, which is 48% identical to SEC4p and functions in ER-to-Golgi transit (SEGEV et al. 1988; SCHMITT et al. 1988). Subsequently, as a result of the search for the mammalian counterparts of the SEC4/YPT1 gene products, a large family of small GTPases, termed Rab, has been identified.

1. Rab Proteins

More than 20 Rab GTPases have been identified so far. Some of them have been shown to be localized in distinct subcellular organelles (GOUD and MCCAFFREY 1991; BALCH 1990; BOURNE et al. 1991). Rab2 is found on an intermediate compartment between ER and Golgi (CHAVRIER et al. 1990), Rab3a on neuronal synaptic vesicles and secretory granules (FISCHER VON MOLLARD et al. 1990; DARCHEN et al. 1990; MIZOGUCHI et al. 1990), Rab4 and 5 on early endosomes (VAN DER SLUIJS et al. 1991; CHAVRIER et al. 1990), and Rab7 on late endosomes (CHAVRIER et al. 1990). This organelle-selective location has suggested that Rab proteins might be involved in recognition between vesicles and their specific target membranes. Recently, however, other Rabs have been reported to have a more diffuse distribution. Rab1b is present on both ER and Golgi (PLUTNER et al. 1991), and Rab6 is found on medial and *trans* Golgi as well as on post Golgi vesicles (GOUD and McCAFFREY 1991).

Functional studies have supported an essential role of Rab proteins in vesicle targeting/fusion. Anti-Rab5 antibodies impaired endosome–endosome fusion, while elevated levels of native Rab5 protein stimulated it (GORVEL et al. 1991); in addition, high expression of nonfunctional Rab5 protein caused endosome fragmentation (see PFEFFER 1992). Anti-YPT1p antibodies inhibited fusion of ER-to-Golgi transport vesicles (REXACH and SCHEKMAN 1991) and the same effect was produced by peptides mimicking the common effector region of Rab proteins (PLUTNER et al. 1990). Antibodies against the mammalian YPT1 homolog Rab1b appeared to inhibit transport at a stage preceding fusion in ER-to-Golgi and intra-Golgi traffic, suggesting that this GTPase might be involved in more than one stage of transport (PLUTNER et al. 1991). Thus, while the available evidence indicates that Rab proteins play a role in the targeting/fusion process, the precise molecular action of these GTPases is still largely unclear.

2. ARF Proteins

The ARF protein family has been proposed to play a role in targeting/fusion, in addition to the previously mentioned role in vesicle formation (SERAFINI et al. 1991). The strongest support for this idea has come from studies of in vitro endosome–endosome fusion, where ARF-derived peptides were shown to abolish both the inhibitory and the stimulatory effect of GTPγS (LENHARD et al. 1992). ARF has been identified as the protein factor mediating GTPγS-induced inhibition of vesicle fusion in intra-Golgi transport in vitro (TAYLOR and MELANÇON 1991).

3. Heterotrimeric G-Proteins

There is little evidence that G-proteins have a role in targeting/fusion steps along the exocytic pathway. The inhibition of vesicle consumption by fluoroaluminate in intra-Golgi transport has been attributed to a block of the uncoating process, rather than to the involvement of a heterotrimeric G-protein specifically controlling targeting or fusion (BARR et al. 1992; MELANÇON et al. 1991). The inhibitory effect of $G_{\alpha i3}$ overexpression on proteoglycan secretion (STOW et al. 1991) is compatible with this G-protein having a role in targeting or fusion, but, as noted above, is more easily interpreted in terms of $G_{\alpha i3}$ being involved in vesicle formation.

Strong evidence that heterotrimeric G-proteins function in targeting–fusion has come from studies on early endosome–endosome fusion, where compounds specifically acting on heterotrimeric G-proteins (fluoroaluminate, mastoparan, $\beta\gamma$ subunits) have complex effects, indicating a role for one or more heterotrimeric G-proteins (COLOMBO et al. 1992) in this transport process.

D. GTPases in Regulated Exocytosis

I. Granule Formation

Less effort has been made so far to study of the role of GTPases in granule formation than to study other transport events. The available data suggest the involvement of an inhibitory heterotrimeric G-protein that retards granule formation from the TGN, in a manner similar to that observed for constitutive vesicles (TOOZE et al. 1990; BARR et al. 1991).

II. Granule Targeting and Fusion

The first evidence that a GTPase is involved in membrane traffic was provided in studies on regulated, calcium-dependent secretion by GOMPERTS (1983), who showed that guanine nucleotides have marked stimulatory effects on permeabilized mast cell degranulation. The independence of these effects on second messengers suggested that the involved GTPase (termed G_e, where e stands for exocytosis) directly controls the exocytotic machinery. Similar results were subsequently obtained in many other cell types (see review by GOMPERTS 1990). In most permeabilized secretory systems calcium is sufficient to induce exocytosis. The activation of a stimulatory G_e, however, may be required together with calcium in some cell types where the GTPase inactivator GDPβS was found to prevent calcium-triggered secretion (BITTNER et al. 1986; BARROWMAN et al. 1986; DE MATTEIS et al. 1991). An inhibitory G_e is also likely to exist, as GTPγS, in addition to having stimulatory effects, can also inhibit calcium-dependent secretion, under certain conditions. In two cases, GTP relieved this GTPγS inhibition (DE MATTEIS et al. 1991; VAN DER MERWE et al. 1991), suggesting the possible additional involvement of a "cycling" GTPase (BOURNE et al. 1991; PFEFFER 1992).

There is evidence that both heterotrimeric G-proteins and small GTPases may be involved in the action of stimulatory and inhibitory G_e. Morphological studies show that both G-proteins and small GTPases are localized on structures of the regulated pathway. Some small GTPases have been detected by ^{32}P-GTP binding assays in secretory granules from neutrophils (DEXTER et al. 1990; PHILIPS et al. 1991), chromaffin cells (BURGOYNE and MORGAN 1989), platelets (VAN DER MEULEN et al. 1991), and paramecium (PETERSON 1991). Two small GTPases have been immunologically identified: Rab3a on neuronal synaptic vesicles (FISCHER VON MOLLARD et al. 1990) and secretory granules in chromaffin cells (DARCHEN et al. 1990), and Rap1b/Rap2b on the specific, but not on azurophilic, granules in neutrophils (MARIDONNEAU-PARINI and DE GUNZBURG 1992). Rab3a was found to dissociate from synaptic vesicles in stimulated synaptosomes (FISCHER VON MOLLARD et al. 1991) and to redistribute on the cell surface in stimulated cultured neurons (MATTEOLI et al. 1991); Rap1b and Rap2b in neutrophils redistributed to the plasma membrane during stimulated exocytosis (MARIDONNEAU-PARINI and DE GUNZBURG 1992). Heterotrimeric G-proteins, too, have been localized on regulated secretory structures; G_o, G_{i1} and G_{i2} have been detected on chromaffin and neutrophils granules (TOUTANT et al. 1987; ROTROSEN et al. 1988).

An early piece of functional evidence that the inhibitory G_e protein can be heterotrimeric was obtained from studies on the effects of inhibitory hormones (somatostatin, epinephrine and galanin) on the secretion of adrenocorticotropic hormone (ACTH; LUINI and DE MATTEIS 1988, 1990) and insulin (ULLRICH and WOLLHEIM 1988, 1989; ULLRICH et al. 1990) which showed that the inhibitory hormones act through a pertussis toxin sensitive G-protein and independently of second messengers. Other experiments supporting the existence of a heterotrimeric inhibitory G_e showed that pertussis toxin stimulates calcium-dependent secretion in permeabilized chromaffin cells (SONTAG et al. 1991; OHARA-IMAIZUMI et al. 1992) and PC12 cells (AHNERT-HILGER et al. 1987). Furthermore, polyclonal antibodies against the α subunits of the pertussis toxin sensitive G_o have recently been shown to mimic the effects of pertussis toxin in decreasing the half maximal concentration of calcium required for triggering exocytosis in permeabilized chromaffin cells (OHARA-IMAIZUMI et al. 1992). The stimulatory G_e is also likely to be, or include, a heterotrimeric G-protein. This was suggested by results obtained with selective activators or inactivators of heterotrimeric G-proteins, such as mastoparan and pertussis toxin. Mastoparan and other polybasic compounds stimulated secretion in mast cells (NAKAMURA and UI 1985; ARIDOR et al. 1990; YOKOKAWA et al. 1989) and platelets (OZAKI et al. 1990; WHEELER-JONES et al. 1992), and this stimulation in mast cells was prevented by pertussis toxin (SAITO et al. 1987; MOUSLI et al. 1989). In melanotrophs dialyzed through a patch pipette with a medium containing chloride, pertussis toxin prevented the calcium-induced stimulation of exocytosis (R. ZOREC 1992, personal communication). GTPγS-activated G-proteins ($G_{n\alpha}$, G_{i}, G_{22}, G_{24}, G_{26}) introduced into mast cells via a patch pipette inhibited exocytosis. This was interpreted as being due to the ability of these G-proteins to bind to the effectors system of G_e without activating it, thereby preventing the action of the endogenous G_e (LINDAU and GOMPERTS 1991).

Only very recently functional evidence has been obtained in support of a role for small GTPases in regulated secretion. In prolactin-secreting pituitary cells, the injection of rab 3b-specific antisense oligonucleotides led to the disappearance of the late, but not of the early, component of the secretory response (P.M. LLEDO 1993). This suggested that rab3b may be involved in early events such as docking, but not in fusion of already docked secretory granules. Peptides mimicking the Rab effector region (a well-conserved sequence in several rab proteins) stimulated secretion in permeabilized pancreatic acinar cells (Padfield et al. 1992) and insulin-secreting β cells (Li et al. 1992), and induced massive degranulation in mast cells (Oberhauser et al. 1992). These data are in apparent contrast with the inhibitory effects of the same peptides in ER-to-Golgi traffic (Plutner et al. 1990). Perhaps the effector peptides can have agonist or antagonist effects, depending on the receptor protein they bind to. If this is the case, their stimulatory action suggests the involvement of a Rab protein (perhaps Rab3a) which would activate the targeting-fusion mechanism without participating directly in granule–plasma membrane recognition. Alternatively, one should admit the existence of another inhibitory Rab protein whose effect would be competed for by the effector peptides.

E. Regulation of the Secretory Pathways by Transduction Systems

I. Regulated Exocytosis

It is well known that GTPases control regulated exocytosis not only through G_e, but also by transducing the effect of ligand binding to membrane receptors into changes in second messenger levels, which in turn have profound effects on secretion. In particular, calcium, cyclic adenosine monophosphate (cAMP), and diacylglycerol have been studied extensively in secretory cells.

II. Constitutive Traffic

In contrast, little attention has been paid to the regulation of constitutive interorganelle traffic by receptors and second messengers. There is, however, some evidence in favor of the existence and physiological importance of such regulation (Table 3). Recently, the constitutive secretion of proteoglycans was found to be stimulated by the activation of IgE receptors and by PMA (phorbol 12-myristate 13-acetate) in rat basophilic leukemia (RBL) cells (de Matteis et al. 1993). This effect, but not the calcium-dependent release of serotonin through the regulated pathway, was inhibited by Brefeldin A (BFA). In pituitary prolactin-secreting cells a rapid increase in constitutive exocytotic membrane flow was observed in response to the secretagogue TRH (thyrotropin-releasing hormone; Tougard et al. 1983). Secretion of immunoglobulins and proteins from fibroblasts, but not from macrophages, was found to be inhibited by ethyleneglycol-bis-β-aminoethylether-N,N,N′,N′-tetraacetic acid (EGTA) in the presence of calcium ionophores (Tartakoff et al. 1978). More recently, in permeabilized baby hamster kidney (BHK) cells, TGN-plasma membrane traffic was shown to be inhibited by EGTA (de Curtis and Simons 1988). Calcium,

Table 3. Regulation of consitutive traffic by receptors and second messengers

	Level of regulation	Effects
Exocytosis	Receptor	
	IgE	Stimulation of BFA-sensitive release of GAG in RBL cells (De Matteis et al. 1993)
	TRH	Formation of small vesicles and their fusion with plasma membrane in PRL secreting cells (Tougard et al. 1983)
	Intracellular signals	
	Cytosolic calcium	Required for immunoglobulin secretion in plasmacells (Tartakoff et al. 1978)
		Required for protein secretion in fibroblasts (Tartakoff et al. 1978)
		Required for transport of Semliki Forest virus glycoproteins in BHK cells (de Curtis and Simons 1988)
	Kinases	PKC regulation of non-clathrin coat assembly and GAG release (De Matteis et al. 1993)
		Inhibition of ER-Golgi transit (Davidson et al. 1992)
Endocytosis	Receptor	
	EGF	Stimulation of fluid-phase endocytosis in A431 cells (Haigler et al. 1979)
	IgE	Stimulation of fluid-phase and adsorptive endocytosis in RBL cells (Furuichi et al. 1984; Pfeiffer et al. 1985)
	fMLP	Stimulation of fluid-phase endocytosis in neutrophils (Davis et al. 1982)
	Carbachol	Inhibition of fluid-phase endocytosis and endosome traffic in CHO cells transfected with muscarinic receptors (Haraguchi and Rodbell 1991)
	Isoproterenol	Stimulation of fluid-phase endocytosis in adipocytes (Haraguchi and Rodbell 1990)
	Intracellular signals	
	Cytosolic calcium	Inhibition of endocytosis in CHO cells transfected with muscarinic receptors (Haraguchi and Rodbell 1991)
	Kinases	Inhibition by of endosome fusion in vitro cdc2 Kinase (Tuomikoski et al. 1989)
		Inhibition of endosome fusion in vitro (Woodman et al. 1992)

GAG, Glycosaminoglycom cholius; RBL, Rat basophilic leukemia
BFA, Brefeldin A; PKC, protein kinase C; CHO, Chinese hamster ovary.

however, is not required for constitutive transport in other cases (Helms et al. 1990; Miller and Moore 1991). In lactating mammary cells, constitutive secretion of casein was inhibited by an autocrine inhibitory peptide (Rennison et al. 1992). Fluid phase and adsorptive endocytosis were increased by IgE receptor activation and by PMA (Pfeiffer et al. 1985;

FURUICHI et al. 1984), and fluid phase endocytosis was stimulated by epidermal growth factor, f-Met-Leu-Phe (fMLP), and isoproterenol in their target cells (HAIGLER et al. 1979; DAVIS et al. 1982; HARAGUCHI and RODBELL 1990). The stimulation of muscarinic receptors transfected into chinese hamster ovary (CHO) cells inhibited endosome trafficking (HARAGUCHI and RODBELL 1991).

The mechanisms underlying such regulations are unclear, but recent work in reconstituted transport systems might provide relevant clues. One such clue is that calcium is required, together with adenosine triphosphate (ATP) and cytosol, in different vesicular transport steps. In semi-intact cells calcium regulates targeting/fusion of ER-to-Golgi transport vesicles in a biphasic manner, with a peak of activity at 100 n*M*, the calcium intracellular resting concentration, and inhibition at higher or lower calcium levels (BECKERS and BALCH 1989). Additional evidence for a calcium role in ER-to-Golgi transport has come from studies on yeast YPT1 mutants. In these cells the secretory defect is corrected by the presence of elevated extracellular calcium concentrations (SCHIMTT et al. 1988), or by alterations of calcium homeostasis due to overexpression of the PMR1 gene, which encodes a calcium ATPase (RUDOLPH et al. 1989). So far, in discussing the physiological significance of calcium in constitutive traffic, emphasis has been palced on mitosis because calcium is elevated in mitotic cells and the calcium rise is associated with inhibition of vesicle traffic (WARREN 1985; BECKERS and BALCH 1989). Obviously, however, another condition that is often associated with wide cytosolic calcium oscillations is the hormonal stimulation of membrane receptors. Thus, calcium is a potential candidate for mediating receptor effects on vesicular transport.

Another regulatory mechanism involved in membrane traffic might be the phosphorylation of the transport machinery. The cdc2 mitotic kinase has been shown to be responsible for the endosome fusion block known to occur during mitosis (TUOMIKOSKI et al. 1989), but other kinases are likely to play a role in membrane traffic in interphase cells. In vitro ER-to-Golgi transport was inhibited by specific blockers of protein phosphatases and the block was reversed by kinase inhibitors (DAVIDSON et al. 1992). The kinase responsible for this effect has not been identified, but the involvement of PKC (protein kinase C), cAMP-dependent protein kinase, cdc2 mitotic kinase, and calcium-calmodulin-dependent kinase II has been shown to be unlikely (DAVIDSON et al. 1992). Similarly, and based on analogous evidence, it has been proposed that the hyperphosphorylation of an unidentified protein or proteins leads to inhibition of endosome–endosome fusion in an in vitro system from interphase cells (WOODMAN et al. 1992). These findings obviously raise questions concerning the identity of the involved kinases and their mechanism of action. Vesicle budding is likely to be rate limiting in constitutive secretory processes. It has been recently found that the membrane association–dissociation cycle of the coat proteins ARF and β-COP, a key event in budding, is markedly modulated by PKC (DE MATTEIS et al. 1993). Short treatments with the PKC activator PMA enhanced the GTP and Mg^{2+}-dependent reassociation rate of these proteins with Golgi compartments after dissociation in permeabilized RBL cells. In contrast, PKC downregulation was accompanied by inhibition of the coat protein association induced by GTP, Mg^{2+}, or fluoroaluminate. Similar effects were seen upon IgE receptor activation, which involves stimulation of PKC. In line with a modulatory role of PKC, the $\beta 2$ isoform of this enzyme has been found to be specifically associated with Golgi membranes (SAITO et al. 1989), and relatively high concentrations of the PKC activator diacylglycerol are present in this organelle (PAGANO 1988). The origin and significance of the Golgi diacylglycerol are unclear, but it is attractive to speculate that its presence might sustain a basal local activity of PKC.

Thus, there is evidence indicating that constitutive membrane traffic, both exocytic and endocytic, can be regulated by transduction systems. It would make physiological sense if the constitutive membrane traffic involved in functions such as antigen processing and presentation, production of blood proteins, and chronic hormone secretion were adjusted to the needs of the organism through receptors and second messengers, as in the case of regulated exocytosis.

F. Conclusions

GTPases of the Rab family appear to be required for the targeting–fusion step in both the regulated and the constitutive pathways. Rab and Rab accessory proteins such as GAP (GTPase activating protein) and GDI (GDP dissociation inhibitor) (BOURNE et al. 1991) might be involved in, and constitute part of, a targeting and fusion machinery common to many types of membrane traffic. Heterotrimeric G-proteins with stimulatory and inhibitory activity are found to function in both pathways. Altogether, the available evidence, albeit incomplete, suggests that in the regulated pathway a GTPase-mediated mechanism fundamentally similar to that at work in constitutive transport is integrated with another, second messenger-dependent control machinery, also regulated by G-proteins. A further similarity between constitutive and regulated traffic is that regulation by second messengers seems to function in both pathways, although at different molecular steps. While the control by second messengers on regulated exocytosis is absolute and is exerted at the targeting–fusion stage, it might be only modulatory and exerted mainly at the level of vesicle formation in constitutive traffic.

Acknowledgements. We would like to express our gratitude to all colleagues who made unpublished results known to us for the benefit of this review. We are also grateful to Ms. D. Spadano and Ms. C. Di Sebastiano for help with the bibliography. Work in the author's laboratory was supported in part by Fidia S.p.A, L'Agenzia per la Promozione e lo Sviluppo del Mezzogiorno (PS 3593/IND; 6168, PR2), and the National Research Council (Convenzione CNR-Consorzio Mario Negri Sud).

References

Ahnert-Hilger G, Bräutigam M, Gratzl M (1987) Ca^{2+}-stimulated catecholamine release from α-toxin-permeabilized PC12 cells: biochemical evidence for exocytosis and its modulation by protein kinase C and G proteins. Biochemistry 26:7842–7848

Aridor M, Traub LM, Sagi-Eisenberg R (1990) Exocytosis in mast cells by basic secretagogues: evidence for direct activation of GTP-binding proteins. J Cell Biol 111:909–917

Balch WE (1990) Small GTP-binding proteins in vesicular transport. Trends Biochem Sci 15:473–477

Barr FA, Leyte A, Mollner S, Pfeuffer T, Tooze SA, Huttner WB (1991) Trimeric G-proteins of the trans-Golgi network are involved in the formation of constitutive secretory vesicles and immature secretory granules. FEBS Lett 294:239–243

Barr FA, Leyte A, Huttner WB (1992) Trimeric G proteins and vesicle formation. Trends Cell Biol 2:91–94

Barrowman MM, Cockroft S, Gomperts BD (1986) Two roles for guanine nucleotides in the stimulus-secretion sequence of neutrophils. Nature 319:504–507

Beckers CJM, Balch WE (1989) Calcium and GTP: essential components in vesicular trafficking between the endoplasmic reticulum and Golgi apparatus. J Cell Biol 108:1245–1256

Bittner MA, Holz RW, Neubig RR (1986) Guanine nucleotides effects on cathecolamine secretion from digitonin-permeabilized adrenal chromaffin cells. J Biol Chem 261:10182–10188

Boman AL, Delannoy MR, Wilson KL (1992) GTP hydrolysis is required for vesicle fusion during nuclear envelope assembly in vitro. J Cell Biol 116:281–294

Bourne HR, Sanders DA, McCormick F (1991) The GTPase superfamily: conserved structure and molecular mechanism. Nature 349:117–127

Burgess TL, Kelly RB (1987) Constitutive and regulated secretion of proteins. Annu Rev Cell Biol 3:243–293

Burgoyne RD, Morgan A (1989) Low molecular mass GTP-binding proteins of adrenal chromaffin cells are present on the secretory granule. FEBS Lett 245:122–126

Chavrier P, Parton RG, Hauri HP, Simons K, Zerial M (1990) Localization of low molecular weight GTP binding proteins to exocytic and endocytic compartments. Cell 62:317–329

Colombo MI, Mayorga LS, Casey PJ, Stahl PD (1992) Evidence of a role for heterotrimeric GTP-binding proteins in endosome fusion. Science 255:1695–1697

Darchen F, Zarhoui A, Hammel F, Montelis MP, Tavitian A, Scherman D (1990) Association of the GTP-binding protein Rab3A with bovine adrenal chromaffin granules. Proc Natl Acad Sci USA 87:5692–5696

Davidson HW, McGowan CH, Balch WE (1992) Evidence for the regulation of exocytic transport by protein phosphorylation. J Cell Biol 116:1343–1355

Davis BH, Walter RJ, Pearson CB, Becker EL, Oliver JM (1982) Membrane activity and topography of f-Met-Leu-Phe-treated polymorphonuclear leukocytes. Acute and sustained responses to chemotactic peptide. Am J Pathol 108:206–216

de Curtis I, Simons K (1988) Dissection of Semliki Forest virus glycoprotein delivery from the trans-Golgi network to the cell surface in permeabilized BHK cells. Proc Natl Acad Sci USA 85:8052–8056

De Matteis MA, Di Tullio G, Buccione R, Luini A (1991) Characterization of calcium-triggered secretion in permeabilized rat basophilic leukemia cells. Possible role of vectorially acting G proteins. J Biol Chem 266:10452–10460

De Matteis MA, Santini G, Kahn RA, Di Tullio G, Luini A (1993) Receptor and protein kinasi C regulation of ARF binding to the Golgi complex. Nature 384:818–821

d'Enfert C, Wuestehube LJ, Lila T, Schekman R (1991) Sec12p-dependent membrane binding of the small GTP-binding protein Sar1p promotes formation of transport vesicles from the ER. J Cell Biol 114:663–670

Dexter D, Rubins JB Manning EC, Khachatrian L, Dickey BF (1990) Compartmentalization of low molecular mass GTP-binding proteins among neutrophil secretory granules. J Immunol 145:1845–1850

Donaldson JG, Kahn RA, Lippincott-Schwartz J, Klausner RD (1991) Binding of ARF and β-COP to Golgi membranes: possible regulation by a trimeric G protein. Science 254:1197–1199

Ercolani L, Stow JL, Boyle JF, Holtzman EJ, Lin H, Grove JR, Ausiello DA (1990) Membrane localization of the pertussis toxin-sensitive G-protein subunits αi-2 and αi-3 and expression of a metallothionein-αi-2 fusion gene in LLC-PK1 cells. Proc Natl Acad Sci USA 87:4637–4639

Fischer von Mollard G, Mignery G, Baumert M, Perin M, Hanson T, Jahn R, Sudhof T (1990) Rab 3 is a small GTP-binding protein exclusively localized to synaptic vesicles. Proc Natl Acad Sci USA 87:1988–1992

Fischer von Mollard G, Südhof TC, Jahn R (1991) A small GTP-binding protein dissociates from synaptic vesicles during exocytosis. Nature 349:79–81

Furuichi K, Rivera J, Isersky C (1984) The fate of IgE bound to rat basophilic leukemia cells. III. Relationship between antigen-induced endocytosis and serotonin release. J Immunol 133:1513–1520

Gallwitz D, Donath C, Sander C (1983) A yeast gene encoding a protein homologous to the human c-has/bas proto-oncogene product. Nature 306:704–707

Gilman AG (1987) G-proteins: transducers of receptor-generated signals. Annu Rev Biochem 56:615–649

Goda Y, Pfeffer SR (1988) Selective recycling of the mannose 6-phosphate/IGF-II receptor to the trans Golgi network in vitro. Cell 55:309–320

Gomperts BD (1983) Involvement of guanine-nucleotide binding protein in the gating of Ca^{2+} by receptors. Nature 306:64–66

Gomperts BD (1990) Ge: a GTP-binding protein mediating exocytosis. Annu Rev Physiol 52:591–606

Gorvel J-P, Chavrier P, Zerial M, Gruenberg J (1991) Rab5 controls early endosome fusion in vitro. Cell 64:915–925

Goud B, McCaffrey M (1991) Small GTP-binding proteins and their role in transport. Curr Opin Cell Biol 3:626–633

Grimes M, Kelly RB (1992) Intermediates in the constitutive and regulated secretory pathways released in vitro from semi-intact cells. J Cell Biol 11:539–549

Haigler HT, McKanna JA, Cohen S (1979) Rapid stimulation of pinocytosis in human carcinoma cells A-431 by epidermal growth factor. J Cell Biol 83:82–90

Haraguchi K, Rodbell M (1990) Isoproterenol stimulates shift of G proteins from plasma membrane to pinocytic vesicles in rat adipocytes: a possible means of signal dissemination. Proc Natl Acad Sci USA 87:1208–1212

Haraguchi K, Rodbell M (1991) Carbachol-activated muscarinic (M1 and M3) receptors transfected into chinese hamster ovary cells inhibit trafficking of endosomes. Proc Natl Acad Sci USA 88:5964–5968

Helms JB, Karrenbauer A, Wirtz WA, Rothman JE, Wieland FT (1990) Reconstitution of steps in the constitutive secretory pathway in permeabilized cells. J Biol Chem 265:20027–20032

Higashijiman T, Uzu S, Nakajima T, Ross EM (1988) Mastoparan, a peptide toxin from wasp venom, mimics receptors by activating. GTP-binding regulatory proteins (G proteins). J Biol Chem 263:6491–6494

Howell TW, Cockroft S, Gomperts BD (1987) Essential synergy between Ca^{2+} and guanine nucleotides in exocytotic secretion from permeabilized rat mas cells. J Cell Biol 105:191–197

Kahn RA (1991) Fluoride is not an activator of the smaller (20–25 kDa) GTP-binding proteins. J Biol Chem 266:15595–15597

Knight DE, Baker PF (1985) Guanine nucleotides and Ca-dependent exocytosis. FEBS Lett 189:345–349

Knight DE, Scrutton MC (1986) Effects of guanine nucleotides on the properties of 5-hyroxytryptamine secretion from electropermeabilised human platelets. Eur J Biochem 160:183–190

Koopmann WR Jr, Jackson RC (1990) Calcium- and guanine-nucleotide-dependent exocytosis in permeabilized rat mast cells. Biochem J 265:365–373

Ktistakis NT, Linder ME, Roth MG (1992) Action of brefeldin A blocked by activation of a pertussis-toxin-sensitive G protein. Nature 356:344–346

Lenhard JM, Kahn RA, Stahl PD (1992) Evidence for ADP-ribosylation factor (ARF) as a regulator of in vitro endosome-endosome fusion. J Biol Chem 267:13047–13052

Li GD, Regazzi R, Balch WE, Wollheim CB (1992) An effector domain peptide of GTP binding protein rab 3 stimulates insulin release from permeabilized HIT cells. Diabetologia 35a

Lindau M, Gomperts BD (1991) Techniques and concepts in exocytosis: focus on mast cells. Biochem Biophys Acta 1071:429–471

Lledo PM, Vernier P, Vincent JD, William TM, Zorec R (1993) Inhibition of RAB3 expression attenuates Ca^{2+}-dependent exocytosis in rat anterior pituitary cells. Nature 364:540–544

Luini A, De Matteis MA (1988) Dual regulation of ACTH secretion by guanine nucleotides in permeabilized AtT-20 cells. Cell Mol Neurobiol 8:129–138

Luini A, De Matteis MA (1990) Evidence that receptor-linked G protein inhibits exocytosis by a post-second messenger mechanism in AtT-20 cells. J Neurochem 54:30–38

Malhotra V, Serafini T, Orci L, Shepherd JC, Rothman JE (1989) Purification of a novel class of coated vesicles mediating biosynthetic protein transport through the Golgi stack. Cell 58:329–336

Maridonneau-Parini I, de Gunzburg J (1992) Association of rap1 and rap2 proteins with the specific granules of human neutrophils. J Biol Chem 267:6396–6402

Matteoli M, Takei K, Cameron R, Hurlbut P, Johnston PA, Sudhof TC, Jahn R, De Camilli P (1991) Association of rab3a with synaptic vesicles at late stages of the secretory pathway. J Cell Biol 115:625–633

Mayorga LS, Diaz R, Stahl PD (1989) Regulatory role for GTP-binding proteins in endocytosis. Science 244:1475–1477

Melançon P, Glick BJ, Malhotra V, Weidman PJ, Serafini T, Gleason ML, Orci L, Rothman JE (1987) Involvement of GTP-binding "G" proteins in transport through the Golgi stack. Cell 51:1053–1062

Melançon P, Franzusoff A, Howell KE (1991) Vesicle budding: insights from cell-free assays. Trends Cell Biol 1:165–171

Miller SG, Moore H-P (1991) Reconstitution of constitutive secretion using semi-intact cells: regulation by GTP but not calcium. J Cell Biol 112:39–54

Mizoguchi A, Kim S, Ueda T, Kikuchi A, Yorifuji H, Hirokawa N, Takai Y (1990) Localization and subcellular distribution of smg p25A, a ras p21-like GTP-binding protein, in rat brain. J Biol Chem 265:11872–11879

Morgan A, Burgoyne RD (1990) Stimulation of Ca^{2+}-independent catecholamine secretion from digitonin-permeabilized bovine adrenal chromaffin cells by guanine nucleotide analogues. Biochem J 269:521–526

Mousli M, Bronner C, Bueb J-L, Tschirhart E, Gies J-P, Landry Y (1989) Activation of rat peritoneal mast cells by substance P and mastoparan. J Pharmacol Exp Ther 250:329–335

Nadin CY, Rogers J, Tomlinson S, Edwardson JM (1989) A specific interaction in vitro between pancreatic zymogen granules and plasma membranes: stimulation by G protein activators but not by Ca^{2+}. J Cell Biol 109:2801–2808

Nakamura T, Ui M (1985) Simultaneous inhibitions of inositol phospholopid breakdown, arachidonic acid release, and histamine secretion in mast cells by islet-activating protein, pertussis toxin. A possible involvement of the toxin-specific substrate in the Ca^{2+}-mobilizing receptor-mediated biosignalling system. J Biol Chem 260:3584–3593

Oberhauser AF, Monck JR, Balch WE, Fernandez JM (1992) Exocytotic fusion is directly activated by rab 3 AL peptides. Nature 360:270–273

Ohara-Imaizumi M, Kameyama K, Kawae N, Takeda K, Muramatsu S, Kumakura K (1992) Regulatory role of the GTP-binding protein, G_o, in the mechanism of exocytosis in adrenal chromaffin cells. J Neurochem 58:2275–2284

Orci L, Tagaya M, Amherdt M, Perrelet A, Donaldson JG, Lippincott-Schwartz J, Klausner RD, Rothman JE (1991) Brefeldin A, a drug that blocks secretion, prevents the assembly of non-clathrin-coated buds on Golgi cisternae. Cell 64:1183–1195

Ozaki Y, Matsumoto Y, Yatomi Y, Higashihara M, Kariya T, Kume S (1990) Mastoparan, a wasp venom, activates platelets via pertussis toxin-sensitive GTP-binding proteins. Biochem Biophys Res Commun 170:779–785

Padfield PJ, Ding T-G, Jamieson JD (1991) Ca^{2+}-dependent amylase secretion from pancreatic acinar cells occurs without activation of phospholipase C linked G-proteins. Biochem Biophys Res Commum 174:536–541

Padfield PJ, Balch WE, Jamieson JP (1992) A synthetic peptide of the rab 3a effector domain stimulates amylase release from permeabilized pancreatic acini. Proc Natl Acad Sci USA 89:1656–1660

Pagano RE (1988) What is the fate of diacylglycerol produced at the Golgi apparatus? Trends Biochem Sci 13:202–205

Peterson JB (1991) Small GTP-binding proteins associated with secretory vesicles of Paramecium. J Protozool 38:495–501

Pfeffer SR (1992) GTP-binding proteins in intracellular transport. Trends Cell Biol 2:41–46

Pfeiffer JR, Seagrave JC, Davis BH, Deanin GG, Oliver JM (1985) Membrane and cytoskeletal changes associated with IgE-mediated serotonin release from rat basophilic leukemia cells. J Cell Biol 101:2145–2155

Philips MR, Abramson SB, Kolasinski SL, Haines KA, Weissmann G, Rosenfeld MG (1991) Low molecular weight GTP-binding proteins in human neutrophil granule membrane. J Biol Chem 266:1289–1298

Plutner H, Schwaninger R, Pind S, Balch WE (1990) Synthetic peptides of the Rab effector domain inhibit vesicular transport through the secretory pathway. EMBO J 9:2375–2383

Plutner H, Cox AD, Pind S, Khosravi-Far R, Bourne JR, Schwaninger R, Der CJ, Balch WE (1991) Rab1b regulates vesicular transport between the endoplasmic reticulum and successive Golgi compartments. J Cell Biol 115:31–43

Rennison ME, Kerr MA, Addey CPV, Wilde CJ, Burgoyne RD (1992) Inhibition of constitutive protein secretion from alctating mouse mammary epithelial cells by FIL (feedback inhibitor of lactation), a secreted milk protein (submitted)

Rexach MF, Schekman RW (1991) Distinct biochemical requirements for the budding, targeting, and fusion of ER-derived transport vesicles. J Cell Biol 114:219–229

Rotrosen D, Gallin JI, Spiegel AM, Malech HL (1988) Subcellular localization of $Gi\alpha$ in human neutrophils. J Biol Chem 263:10958–10964

Rudolph HK, Antebi A, Fink GR, Buckley CM, Dorman TE, LeVitre J, Davidow LS, Mao DT (1989) The yeast secretory pathway is perturbed by mutations in PMR1, a Ca^{2+} ATPase family. Cell 58:133–145

Saito H, Okajima F, Molski TFP, Sha'Afi RI, Ui M, Ishizaka T (1987) Effects of ADP-ribosylation of GTP-binding protein by pertussis toxin on immunoglobulin E-dependent and -independent histamine release from mast cells and basophils. J Immunol 138:3927–3934

Saito N, Kose A, Ito A, Hosoda K, Mori M, Hirata M, Ogita K, Kikkawa U, Ono Y, Igarashi K, Nishizuka Y (1989) Immunocytochemical localization of βII subspecies of protein kinase C in rat brain. Proc Natl Acad Sci USA 86:3409–3413

Salamero J, Sztul ES, Howell KE (1990) Exocytic transport vesicles generated in vitro from the trans-Golgi network carry secretory and plasma membrane proteins. Proc Natl Acad Sci USA 87:7717–7721

Salminen A, Novick PJ (1987) A ras-like protein is required for a post-Golgi event in yeast secretion. Cell 47:527–538

Schmitt HD, Puzicha M, Gallwitz D (1988) Study of a temperature-sensitive mutant of the ras-related YPT1 gene product in yeast suggests a role in the regulation of intracellular calcium. Cell 53:635–647

Segev N, Mulholland J, Botstein D (1988) The yeast GTP-binding YPT1 protein and a mammalian counterpart are associated with the secretion machinery. Cell 52:915–924

Serafini T, Orci L, Amherdt M, Brunner M, Kahn RA, Rothman JE (1991) ADP-ribosylation factor is a subunit of the coat of Golgi-derived COP-coated vesicles: a novel role for a GTP-binding protein. Cell 67:239–253

Söllner T, Whiteheart SW, Brunner M, Erdjument-Bromage H, Geromanos S, Tempst P, Rothman JE (1993) SNAP receptors implicated in vesicle targeting and fusion, Nature 362:318–323

Sontag J-M, Thierse D, Rouot B, Aunis D, Bader M-F (1991) A pertussis-toxin-sensitive protein controls exocytosis in chromaffin cells at a step distal to the generation of second messengers. Lack of correlation between insulin secretion and cyclic AMP levels. J Biol Chem 263:8615–8620

Stow JL, de Almeida B, Narula N, Holtzman EJ, Ercolani L, Ausiello DA (1991) A heterotrimeric G protein $G\alpha_{i\text{-}3}$, on Golgi membranes regulates the secretion of a heparan sulfate proteoglycan in LLC-PK_1, epithelial cells. J Cell Biol 114:1113–1124

Tartakoff A, Vassalli P, Detraz M (1978) Comparative studies of intracellular transport of secretory proteins. J Cell Biol 79:694–707

Taylor TC, Melançon P (1991) ADP-ribosylation factor (ARF) mediates the effect of GTPγS on a cell-free intra-Golgi transport assay. J Cell Biol 115:245a

Tooze SA, Huttner WB (1990) Cell-free protein sorting to the regulated and constitutive secretory pathways. Cell 60:837–847

Tooze SA, Weiss U, Huttner WB (1990) Requirement for GTP hydrolysis in the formation of secretory vesicles. Nature 347:207–208

Tougard C, Louvard D, Picart R, Tixier-Vidal A (1983) The rough endoplasmic reticulum and the Golgi apparatus visualized using specific antibodies in normal and tumoral prolactin cells in culture. J Cell Biol 96:1197–1207

Toutant M, Aunis D, Bockaert J, Homburger V, Rouot B (1987) Presence of three pertussis toxin substrates and $G_o\alpha$ immunoreactivity in both plasma and granule membranes of chromaffin cells. FEBS Lett 215:339–344

Tuomikoski T, Felix MA, Doree M, Gruenberg J (1989) Inhibition of endocytic vesicle fusion in vitro by the cell cycle control protein kinase cdc 2. Nature 342:942–945

Ullrich S, Wollheim CB (1988) GTP-dependent inhibition of insulin secretion by epinephrine in permeabilized RINm5F cells. Lack of correlation between insulin secretion and cyclic AMP levels. J Biol Chem 263:8615–8620

Ullrich S, Wollheim CB (1989) Galanin inhibits insulin secretion by direct interference with exocytosis. FEBS Lett 247:401–404

Ullrich S, Prentki MB, Wollheim CB (1990) Somatostatin inhibition of $Ca^{2(+)}$-induced insulin secretion in permeabilized HIT-T15 cells. Biochem J 270:273–276 van der Bliek AM, Meyerowitz EM (1991) Dynamin-like protein encoded by the Drosophila shibire gene associated with vesicular traffic. Nature 351:411–414

van der Merwe PA, Millar RP, Wakefield IK, Davidson JS (1991) Inhibition of luteinizing-hormone exocytosis by guanosine 5′-[γ-thio]triphosphate reveals involvement of a GTP-binding protein distal to second-messenger generation. Biochem J 275:399–405

van der Meulen J, Bhullar RP, Chancellor-Maddison K-A (1991) Association of a 24-kDA GTP-binding protein, G_n24, with human platelet α-granule membranes. FEBS Lett 291:122–126

van der Sluijs, Hull M, Zahraoui A, Tavitian A, Goud B, Mellman I (1991) The small GTP-binding protein rab4 is associated with early endosomes. Proc Natl Acad Sci USA 88:6313–6317

Warren G (1985) Membrane traffic and organelle division. Trends Biochem Sci 10:439–443

Waters MG, Serafini T, Rothman JE (1991) "Coatomer": a cytosolic protein complex containing subunits of non-clathrin coated Golgi transport vesicles. Nature 349:248–251

Wessling-Resnick M, Braell W (1990) Characterization of the mechanism of endocytic vesicle fusion in vitro. J Biol Chem 265:16751–16759

Wheeler-Jones CPD, Saermark T, Kakkar VV, Authi KS (1992) Mastoparan promotes exocytosis and increases intracellular cyclic AMP in human platelets. Evidence for the existence of a G_e-like mechanism of secretion. Biochem J 281:465–472

Woodman PG, Mundy DI, Cohen P, Warren G (1992) Cell-free fusion of endocytic vesicles is regulated by phosphorylation. J Cell Biol 116:331–338

Yokokawa N, Komatsu M, Takeda T, Aizawa T, Yamada T (1989) Mastoparan, a wasp venom, stimulates insulin release by pancreatic islets through pertussis toxin sensitive GTP-binding protein. Biochem Biophys Res Commun 158:712–716

CHAPTER 33

Regulated and Constitutive Secretion Studied In Vitro: Control by GTPases at Multiple Levels

H.-P.H. MOORE, L. CARNELL, R.A. CHAVEZ, Y.-T. CHEN, A. HWANG, S.G. MILLER, Y.-A. YOON, and H. YU

A. Introduction

All eukaryotic cells renew and expand their cell surface by a continuous supply of new plasma membrane components. These are provided by vesicles that bud from the Golgi complex and fuse with the plasma membrane. This process occurs continuously in the absence of any external signal and is thus named "constitutive secretion". In more differentiated cells, constitutive secretion alone is insufficient to meet other physiological needs. For example, in cells secreting chemical messengers as a means of communication, the special secretory products are often packaged in storage granules that undergo exocytosis only when the cells are appropriately stimulated by external signals. This process is termed "regulated secretion" (for reviews, see BURGESS and KELLY 1987; MOORE 1987; MILLER and MOORE 1990a). Many cells possess more than one type of constitutive or regulated secretory pathway. A well-known example is the polarized epithelium with two constitutive pathways for delivery of newly synthesized proteins to either the apical or the basolateral surface (for reviews, see SIMONS and FULLER 1985; RODRIGUEZ-BOULAN and NELSON 1989). The presence of multiple types of vesicles therefore offers the possibility of segregating secretion of different products both spatially and temporally.

In cells with more than one form of secretion, the pathways generally diverge at the level of the *trans* Golgi network (TGN; ORCI et al. 1987; TOOZE et al, 1987; TOOZE and HUTTNER 1990; SOSSIN et al. 1990), although sorting of constitutive proteins from the regulated secretory proteins may continue in immature secretory granules (CASTLE 1990; GRIMES and KELLY 1992). Proteins secreted via the constitutive and the regulated pathways traverse the same endoplasmic reticulum (ER) and Golgi compartments until they reach the TGN, where they are segregated into distinct sets of vesicles. It is now generally accepted that transfer between the early secretory compartments, and delivery from the TGN to the cell surface, is achieved by successive budding of transport vesicles carrying "cargos" and fusion of these vesicles with the next target compartment (reviewed in FARQUHAR 1985; ROTHMAN and ORCI 1990, 1992; MELLMAN and SIMONS 1992). At each step, a sorting event occurs in which cargo proteins are incorporated into transport vesicles whereas resident proteins are retained.

Sorting at the TGN is considerably more complex, since it must also include sorting among different secretory vesicles. In general, the problem of vesicular transport can be divided into two parts: sorting and transport machinery.

Recent genetic and biochemical studies have greatly advanced our knowledge of the mechanisms for sorting and the components involved in vesicular transport. An important realization is that many of the underlying principles are similar from step to step, and several common components are repeatedly used throughout the secretory pathway. Perhaps the most surprising finding is that components that were once considered as modulators of regulated secretion, i.e., GTPases, are now known to control membrane traffic through the constitutive pathway. Despite these similarities, unique components must also operate in each step to maintain the membrane selectivity of fusion and to differentiate secretion spatially and temporally. In this chapter, we will focus on the transport steps between the TGN and the cell surface in mammalian cells. We shall emphasize in particular the similarities and differences between constitutive and regulated secretion.

B. The Regulated Secretory Pathway: A General Mechanism for the Control of Cell–Cell Communication and Plasma Membrane Activities

Most cells have the capacity to secrete via a constitutive pathway. Unicellular organisms usually secrete their protein products constitutively. The budding yeast *Saccharomyces cerevisiae*, for example, appears to secrete all proteins by a single pathway (NOVICK and SCHEKMAN 1983; SCHEKMAN 1985); the same set of mutations affect transport of plasma membrane proteins, invertase, and also the pheromone alpha factor which is proteolytically processed by the *KEX2* gene product at pairs of basic residues (FULLER et al. 1989). Maturation and secretion of hormones in higher cells, by contrast, often occur in the regulated secretory pathway, even though a set of processing enzymes similar to Kex2p are involved (SMEEKENS and STEINER 1990; SMEEKENS et al. 1991; THOMAS et al. 1991). The regulated secretory pathway in animal cells has long been recognized as a secondary pathway used exclusively for the discharge of specialized secretory products (TARTAKOFF et al. 1978) and is often associated only with "professional" secretory cells such as endocrine, exocrine, and neuronal cells. Recent studies, however, indicate that this type of secretory pathway is much more common than originally thought. The pathway appears to serve as a general mechanism for regulating cell–cell communication, plasma membrane transporter activities, and formation of specialized membrane domains.

Table 1 summarizes the types of molecules transported to the cell surface in a regulated fashion in different cells. In general, these pathways

Table 1. Examples of secretion regulated at the level of exocytosis in mammalian cells

Tissue type	Molecules transported and packaged
Regulated secretory pathway – biosynthetic type	
Posterior pituitary	Oxytocin, arginine vasopressin, neurophysins
Hypothalamus	Corticotropin-releasing hormone; thyrotropin-releasing hormone; growth-hormone-releasing hormone; somatostatin; luteinizing hormone-releasing hormone; prolactin-releasing hormone; prolactin-inhibiting hormone
Anterior pituitary	Growth hormone, prolactin; β-lipotropin; adrenocorticotropic hormone; thyroid-stimulating hormone; follicle-stimulating hormone; luteinizing hormone; endopeptidase PC1/3; carboxypeptidase; peptidylgycine alpha-amidating monooxygenase; secretogranins
Brain	Enkephalins; substance P; etc.
Thyroid	Thyroglobulin
Pancreas, endocrine	Insulin; glucagon; somatostatin; pancreatic polypeptide; carboxypeptidase
Adrenal medulla	Norepinephrine; epinephrine; secretogranins; dopamine β-hydroxlyase; phenylethanolamine-*N*-methyltransferase
Kidney, endocrine	Renin
Heart, atria	Atrial natriuretic factor
Respiratory mucosa	Mucin
Pancreas, exocrine	Trypsinogen; chymotrypsinogen; proelastase; procarboxypeptidase A; prophospholipase A; α-amylase; esterase; deoxyribonuclease; ribonuclease; lipase
Gastrointestinal system	Cholecystokinin; gastrin; secretin; enteroglucagon; vasoactive intestinal peptide; gastric inhibitory peptide; motilin; substance P; bombesin; somatostatin; neurotensin; neuropeptide Y; peptide YY
Salivary gland	α-amylase; mucin
Blood vessels, endothelia	Von Willebrand factor; CD62
Platelets	Clotting factors; serotonin; ADP; ATP
Neutrophil, granulocytes	Gelatinase; CD11; CD18; etc.
Kidney, epithelia	Apical membrane proteins
Mast cells	Histamine
Regulated secretory pathway – recycling type:	
Central and peripheral nerves	Acetylcholine; serotonin; glycine; gamma-aminobutyric acid; etc.
Stomach, parietal cells	H,K-ATPase
Adipose	Glucose transporters GLUT4, GLUT1
Kidney, collecting duct epithelia	Water channels
Kidney, distal renal tubules	Ca^{2+} channels or regulators
Lung alveoli	Surfactant

can be classified into two categories based on the origin of their secretory vesicles. The "biosynthetic" type of regulated secretory pathway utilizes vesicles that are made in the TGN, and these vesicles usually contain peptides synthesized on membrane-bound ribosomes. In the "recycling" type of regulated secretory pathway, newly synthesized vesicle membrane components are first transported to the cell surface via the constitutive pathway (CUTLER and CRAMER 1990; REGNIER-VIGOUROUX et al. 1991). Assembly of vesicles then occurs during endocytosis and recycling from the endosomes (LINSTEDT and KELLY 1991). Molecules transported by the biosynthetic, regulated pathway include classical peptide hormones and their processing enzymes (GUEST et al. 1991; MILGRAM et al. 1992) in endocrine glands, and digestive enzymes in exocrine glands. The recycling regulated pathway, on the other hand, is involved in release of non-peptide neurotransmitters at synapses and endocrine cells. Cells capable of regulated secretion are also found in a variety of tissues. For example, the gastrointestinal system contains cells that secrete a variety of peptides; epithelia contain cells that secrete mucin; the heart atria contain cells secreting atrial natriuretic factor; and platelets, mast cells, neutrophils, and endothelial cells are triggered to secrete a number of factors during defense mechanisms. Even cells such as T-helper lymphocytes are now suspected to have a regulated secretory pathway (TAPLITS and HODES 1989). Thus, regulated secretion is not restricted to a few organs, but occurs in a wide range of tissues.

The function of regulated secretion is not limited to secreting soluble products. There is increasing evidence suggesting that a major role of this pathway is to regulate plasma membrane ion transporter activities. Gastric HCl secretion is regulated by insertion and retrieval of H,K-ATPase stored in membrane vesicles (for review, see FORTE et al. 1989), and the water permeability of the kidney collecting duct epithelium is controlled by vasopressin-induced recycling of water channels between an intracellular compartment and the plasma membrane (LENCER et al. 1990). Likewise, activation of calcium channels in renal epithelium may also involve agonist-induced exocytosis (BACSKAI and FRIEDMAN 1992). Adipocytes respond to insulin by translocating the glucose transporters GLUT1 and GLUT4 to the cell surface in a similar manner (SLOT et al. 1991; SMITH et al. 1991). In addition, a novel form of regulated secretion has been described for the generation of the apical plasma membrane during differentiation of epithelial cells (VEGA-SALAS et al. 1987). In this case, structures carrying apical membrane proteins, known as the vacuolar apical compartment (VAC), undergo exocytosis and contribute to the rapid formation of the apical surface (RODRIGUEZ-BOULAN and NELSON 1989). It is possible that most, if not all, cells possess some types of storage vesicles for communication and/or regulation of their membrane activities.

C. Controlling Passage Through the Regulated Secretory Pathway – Distinctions Between Constitutive and Regulated Secretion

I. Exocytosis

The most obvious distinction between the constitutive pathway and the regulated pathway is control of exocytosis. Exocytosis from the regulated pathway is modulated by extracellular signals, whose exact nature depends on the particular physiological process involved (see Table 1). The intracellular responses, however, are often elicited by a similar set of intracellular mediators. Studies using permeabilized cells have shown that exocytosis from the regulated pathway differs from the constitutive pathway in two respects: calcium and GTP. Regulated secretion in most cases is stimulated by micromolar concentration of free calcium ions. By comparison, constitutive secretion must be operational at resting levels of free Ca^{2+} (50–200 n*M*) since it is ongoing even in the absence of external signals. Such differences in the Ca^{2+} sensitivity may arise in one of two ways. First, a unique Ca^{2+}-responsive component may be specifically sorted to regulated secretory granules, but not constitutive secretory vesicles; this would make the regulated granules uniquely sensitive to calcium. Alternatively, the same Ca^{2+}-responsive component may be present on both types of vesicles, but modified in such a way that they differ in their affinities for Ca^{2+} – constitutive vesicles requiring physiological Ca^{2+} and regulated granules requiring higher levels for optimal secretory activities. Using a semi-intact Chinese hamster ovary (CHO) cell system to reconstitute constitutive secretion, we found that transport from the *trans* Golgi to the plasma membrane does not require physiological Ca^{2+} (MILLER and MOORE 1991). Thus, constitutive and regulated secretion appear to differ fundamentally with regard to their requirements for calcium (see Fig. 1). Recent studies have shown that calcium-binding proteins of the annexin family, along with other factors (MORGAN and BURGOYNE 1992), may mediate the responses of calcium during regulated secretion (ALI et al. 1989; BLACKWOOD and ERNST 1990; SARAFIAN et al. 1991), and also during endocytosis (LIN et al. 1992). The distinct requirements for calcium may be explained by the participation of these components in exocytosis of regulated, but not constitutive, vesicles.

Evidence for a second modulator of regulated secretion comes from studies using guanine nucleotides. Addition of nonhydrolyzable GTP analogs to semi-intact cells enhances regulated secretion from the biosynthetic pathway in a variety of cell types (reviewed in GOMPERTS 1990). These agents also induce regulated exocytosis from the recycling pathway, as has been shown for the translocation of GLUT4 to the plasma membrane

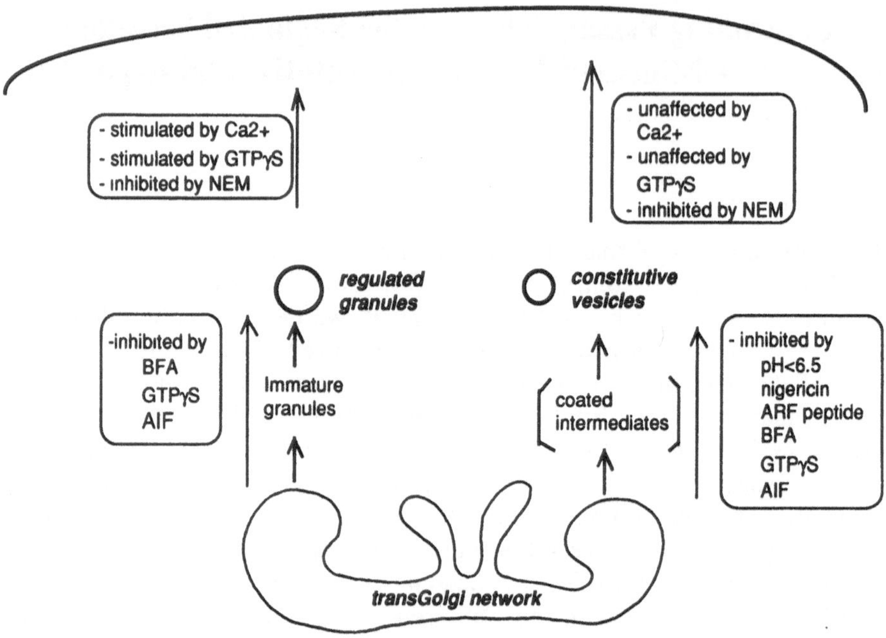

Fig. 1. Characteristics of transport between the *trans* Golgi network and the cell surface via *constitutive vesicles* and *regulated granules*. *BFA*, Brefeldin A; *ARF*, ADP-ribosylation factor; *NEM*, N-ethylmaleimide

in permeabilized adipocytes (BALDINI et al. 1991). These studies indicate that one or more GTPases may participate in late step(s) of regulated secretion. In contrast, GTPγS does not stimulate exocytosis of constitutive vesicles in semi-intact cells (see below), indicating that the exocytotic machinery for constitutive and regulated secretion must also differ in the GTPases involved.

As summarized in Table 1, many cell types possess a regulated secretory pathway. It remains to be determined whether all cells utilize a common set of calcium sensors and GTPases for regulated exocytosis, or each cell utilizes similar but distinct components.

II. Formation of Granules

Protein secretion by the regulated pathway is controlled at multiple points. In addition to granule exocytosis, the formation of secretory granules from the TGN is also likely to be regulated. Using sulfated glycosaminoglycan chains as a bulk-flow marker, we found that traffic out of the TGN via the constitutive and regulated pathways is differentially sensitive to the protein synthesis inhibitor cycloheximide. Constitutive secretion persists in the absence of new protein synthesis for several hours, whereas transport to the regulated secretory pathway is inhibited upon arrest of protein synthesis

(Brion et al. 1992). It is possible that the formation of storage granules is triggered by newly synthesized hormones and thus, in the absence of hormones in the Golgi complex, no new granules are formed. This would provide a simple way to regulate the number of secretory granules within the cell, i.e., by directly regulating the amount of hormone synthesized. If this model is indeed correct, then the question is how the hormones induce budding. The situation could be similar to budding of some enveloped viruses, in which binding of nucleocapsids to the membrane triggers budding of viral particles (Simons and Fuller 1987). Alternatively, binding of hormones to Golgi membrane could trigger a signal across the membrane to initiate vesicle coat assembly in a manner analogous to G-protein-mediated signal transduction across the plasma membrane. As will be discussed below, multiple GTPases are known to participate in intracellular transport steps. It will be of interest to determine if one of the trimeric G-proteins is involved in this signaling event.

III. Sorting of Contents

In addition to regulating formation of storage granules, cells also control the types of molecules entering this pathway. It is now generally accepted that sorting is accomplished by special structural features on proteins targeted to the regulated granules; proteins that are not actively segregated away are exported by a passive-flow process (for reviews, see Moore et al. 1989; Huttner and Tooze 1989; Kelly 1991). DNA transfection studies have shown that the sorting mechanism(s) in the different cell types mentioned in Table 1 appear to be quite similar; there are now numerous examples in which a given hormone can be recognized and targeted to the secretory granules of a different cell type upon transfection. Von Willebrand's factor provides an interesting case. This protein is normally targeted to a regulated secretory organelle, the Weibel-Palade bodies, in endothelial cells. When transfected into the pituitary cell line AtT-20, it induces the formation of a second regulated organelle resembling Weibel-Palade structures (Wagner et al. 1991). This suggests that the assembly of regulated secretory granules may be triggered by the transported protein itself, as mentioned earlier. Is there, then, a single universal mechanism for sorting proteins into regulated secretory granules from the biosynthetic pathway? The answer probably is, no. Several cells are known to contain granules with distinct contents (Fumagalli et al. 1985; Fisher et al. 1988), and von Willebrand's factor is sorted into a different population of secretory granules from the endogenous hormone ACTH in transfected AtT-20 cells (Wagner et al. 1991). These findings indicate that sorting of different molecules must be governed by different sets of biochemical interactions. This notion is supported by studies of the properties of the transported proteins in vitro. Several acidic proteins (including secretogranins, mast cell granule matrix, and sulfated proteoglycans) aggregate at low pH and/or high calcium in vitro, conditions that mimic the Golgi milieu (GERDES et al. 1989; FERNANDEZ et al. 1991; H.-P.H. MOORE, unpublished). This may provide a mechanism for sorting and concentrating this class of proteins at the TGN. Hormones, however, do not precipitate under the same conditions (H.-P.H. MOORE, unpublished), and mutations affecting the oligomeric states of insulin also do not abolish its ability to be sorted into regulated granules (QUINN et al. 1991). A similar finding was obtained with a mutant *Tetrahymena* cell line in which a mucocyst content protein fails to condense but is still sorted into dense core granules (TURKEWITZ et al. 1991).These proteins therefore must employ different sorting mechanisms.

D. Regulation of Traffic Through the Constitutive Pathway

The above discussion points out the differences between the constitutive secretory pathway and the regulated secretory pathway. While membrane traffic through the regulated secretory pathway is controlled at the levels of granule formation, sorting, and exocytosis, traffic through the constitutive pathway is not subject to these same types of regulation. This, however, does not mean that constitutive secretion is not regulated at all. In fact, recent evidence suggests that constitutive secretion is not only regulated, but the factors involved appear to be similar to those regulating exocytosis from storage granules. Thus, the nomenclature should be clarified. Regulated and constitutive secretion should not be viewed as a regulated vs. a nonregulated process, but should be both considered as regulated processes. The distinction is that regulated secretion is controlled *additionally* at other levels by extracellular signals as outlined above.

What regulates constitutive secretion? The first clue that constitutive secretion may be regulated by GTPases came from studies of the yeast gene encoding Sec4p. Temperature-sensitive mutants of this gene accumulate post-Golgi secretory vesicles at the restrictive temperature (NOVICK et al. 1981). Sequence analysis showed that it bears striking sequence homology to a family of small GTPases, including the proto-oncogene product Ras (SALMINEN and NOVICK 1987). This was quite unexpected, for GTPases were known to be associated with signal-transducing receptors that are involved in regulated secretion. Further support for the involvement of GTP in exocytosis of constitutive vesicles came from in vitro studies using reconstituted systems. In the system we have developed, CHO cells were perforated with the bacterial toxin streptolysin-O (SL-O). Transport from the TGN to the cell surface is reconstituted by the addition of an adenosine triphosphate (ATP) regenerating system and cytosol and can be monitored by insertion of a viral glycoprotein into the plasma membrane or by secretion of a bulk-flow marker – sulfated glycosaminoglycan chains (MILLER and MOORE 1992). Addition of micromolar concentrations of nonhydrolyzable GTP analogs to this system effectively blocks secretion, indicating that transport requires the hydrolysis of GTP. Nonhydrolyzable GTP analogs also block transport in earlier steps of the constitutive secretory pathway, indicating that this is a general phenomenon (MELANCON et al. 1987; BAKER et al. 1988; MAYORGA et al. 1988; RUOHOLA 1988; GRAVOTTA et al. 1990; MILLER and MOORE 1991).

Although constitutive secretion and regulated secretion are both regulated by GTP analogs, the effects are quite different. Whereas GTPγS effectively blocks constitutive secretion, it stimulates regulated exocytosis. There are two possible scenarios to explain this. One is that both effects are mediated by Sec4p-like small GTPases, but the two proteins involved have opposite states of activation by guanine nucleotides; for instance, exocytosis from the constitutive pathway may require the Sec4p-like protein in its GDP-bound state to allow exocytosis, whereas regulated exocytosis may require it to be in the GTP-bound state. An alternative hypothesis is that the different effects reflect regulation by distinct classes of GTPases. We prefer the latter hypothesis based on kinetic analysis of the GTPγS effects. Constitutive secretion is blocked only when GTPγS is added early in the transport reaction, during or soon after budding of vesicles from the TGN (see Fig. 1). Once the vesicles are formed, addition of GTPγS no longer exerts an inhibitory effect. This indicates that the exchange of GTP for GDP on Sec4p-like proteins must occur early during vesicle assembly, even though hydrolysis may take place later during vesicle fusion. If the Sec4p-like protein works in a similar fashion in regulated secretion, GTPγS should not affect Sec4p-like proteins when added to permeabilized systems measuring exocytosis of preformed granules. The stimulatory effects of GTPγS commonly observed in these systems are thus likely to be mediated by other types of GTPases such as G_e (see Chap. 42). This model also predicts that transport through the regulated secretory pathway should be inhibited by GTPγS if it is added during granule formation. Indeed, using a semi-intact system that reconstitutes transport via the regulated secretory pathway from the TGN to the cell surface, we found that GTPγS inhibits transport when added early, but stimulates secretion when added late after the transported molecules have entered mature granules. Thus, vesicular transport through the secretory pathway appears to involve more than one type of GTPase operating at several different levels.

E. GTPases and Intracellular Membrane Transport

During the last few years, several in vitro systems have been developed to examine the molecular components involved in Golgi-to-plasma-membrane transport (CURTIS and SIMONS 1988; TOOZE and HUTTNER 1990; SALAMERO et al. 1990; GRAVOTTA et al. 1990; MILLER and MOORE 1991). In this section, we shall summarize the findings from these systems. Some of the components discussed in this section are speculation based on analogy to other transport steps, and their participation in Golgi-to-cell-surface transport needs to be confirmed experimentally.

As we have discussed earlier, GTPases play a major role in vesicular transport. For the sake of our discussion, we shall classify them into four groups: SAR1, ARF, rab, and heterotrimeric G-proteins. Not all of these have been shown to operate between the TGN and the cell surface, although it is likely that components similar to these are involved in ER-to-Golgi, intra-Golgi, and Golgi-to-surface transport.

I. SAR1

The SAR1 gene of *S. cerevisiae* was identified as a multicopy suppressor of the yeast ER-to-Golgi transport mutant *sec12* (NAKANO and MURAMATSU 1989). It shares sequence homology with ras and other small GTPases, with the strongest homology to adenosine diphosphate (ADP) ribosylation factors (ARFs). Sar1p is bound to the ER membrane via a transmembrane protein, Sec12p (D'ENFERT et al. 1991), which interacts with the products of a group of genes (such as SEC23, SEC13, SEC16) controlling the formation of ER-to-Golgi transport vesicles (KAISER and SCHEKMAN 1990). Thus it is likely to participate in an early step to initiate vesicle budding. Although there is as yet no evidence for a Sar1 homolog in mammalian cells, a mammalian Sec23 homolog does exist; antibodies directed against yeast Sec23p recognize an 85-kDa protein from a variety of mammalian tissues (ORCI et al. 1991). The mammalian Sec23p is very similar to the yeast protein. Using polymerase chain reaction (PCR) amplification, we have isolated a 110 base pair cDNA fragment from human hepG2 cells. The fragment shares 88% sequence identity with the yeast SEC23 gene and thus probably encodes a bona fide mammalian homolog. Like the yeast protein, the mammalian sec23p exists in both cytosolic and membrane-bound forms. The protein purified from rat liver cytosol is complexed with a 110-kDa protein similar to that found in yeast (HWANG et al. 1991; HICKE and SCHEKMAN 1989). Addition of this protein complex to an yeast ER-to-Golgi transport assay inhibits transport, indicating that the mammalian components can interact with the transport machinery in yeast cells.

How does Sec23p/Sar1p control vesicle budding? A clue to the solution comes from localization studies in mammalian cells. Immunoelectron microscopic analyses of rat pancreatic cells with affinity-purified antibodies against Sec23p showed that the immunoreactivity was confined to the transitional zone between the ER and the *cis* Golgi complex; little immunoreactivity was found elsewhere (ORCI et al. 1991). Moreover, when cells were treated with energy poison to deplete the formation of vesicles from this region, the immunoreactivity remained localized to the fibrillar aggregates in the ER transitional zone that were previously reported by PALADE and coworkers (MERISKO et al. 1986). This raises the possibility that the Sec23p complex is normally anchored to this fibril network, and its function is to restrict vesicle budding spatially to transitional ER by activating Sar1p and subsequent coat assembly only in this region of the ER network. Consistent with this hypothesis, we found that 30% of the Sec23p in the particulate fraction is resistant to extraction by 1% octoxynol (Triton X-100). It will be of interest to analyze the molecular components of this cytoplasmic matrix.

At present, it is unknown whether budding of exocytic vesicles from the TGN is also controlled by similar complexes. Antibodies against Sec23p do not detect any immunoreactivity in the Golgi or the cytoplasm near TGN. This indicates that Sec23p participates specifically in ER-to-Golgi transport. This finding, of course, does not exclude the possibility that an immunologically distinct, Sec23p-like complex is involved in budding of exocytic vesicles. Future work will be necessary to determine if this is the case.

II. Trimeric G-proteins

Transport from the Golgi complex to the cell surface via both the constitutive pathway and the regulated pathway is inhibited by the addition of aluminum fluoride (see Fig. 1; L. Carnell and H.-P.H. Moore, unpublished; Barr et al. 1991). Since this agent is known to activate heterotrimeric G-proteins without affecting small GTPases (Kahn 1991), this result implies that trimeric G-proteins also control transport in some fashion. The heterotrimeric G-protein $G_{\alpha i3}$ is localized to the Golgi apparatus, and its overexpression in LLC-PK1 cells slows down secretion of a heparan sulfate proteoglycan (Stow et al. 1991). Pretreatment of cells with pertussis toxin reverses this effect. It remains to be established if this G-protein participates in TGN-to-cell-surface transport via the constitutive pathway. Whether the same G-protein regulates budding of regulated granules is also not known. Cells with regulated secretory function also contain other pertussis toxin-sensitive G-proteins, i.e., $G_{\alpha i1}$ and $G_{0\alpha}$ (reviewed in Kaziro et al. 1991). It will be important to determine which of these G-proteins is involved in constitutive and regulated transport from the TGN.

Although the determination of the exact role of trimeric G-proteins in TGN-to-surface transport requires further studies, it probably works by a similar mechanism as in intra-Golgi transport. Activation of trimeric G-proteins by GTPγS causes an increase in binding of coat proteins to the Golgi membrane, which is reversed by the addition of $\beta\gamma$ subunits of G-proteins (Donaldson et al. 1991) and inhibited by a peptide derived from ARF sequences (Donaldson et al. 1992). Likewise, aluminum fluoride causes an increase in the amount of ARF and coat proteins (see below) associated with coated vesicles generated during an in vitro Golgi transport reaction (Serafini et al. 1991). These studies are consistent with a model in which activation of a G-protein stimulates ARF and coat assembly, and persistent activation by AlF inhibits the uncoating step.

How does a trimeric G-protein activate coat assembly? Recent studies (DONALDSON et al. 1992) showed that stimulation of coat binding to Golgi membrane by AlF requires GTP. Since activation of trimeric G-proteins by AlF does not require GTP, this result suggests that activation of a G-protein does not directly stimulate binding of coat proteins. Instead, the effect is probably indirectly mediated by other factors. For instance, a trimeric G-protein may activate other small GTPases such as ARF, which is essential for coat binding (DONALDSON et al. 1992). It may also activate other cellular pathways required for ARF/coat binding. Recent studies (ZEUZEM et al. 1992) have shown that the binding of ARF to Golgi membranes requires a low intravesicular pH that is established by a vacuolar H^+-ATPase. Thus, regulation of ion transport may be a possible downstream effect of activation of a trimeric G-protein. It is interesting to note that trimeric G-proteins have been implicated in the regulation of intravesicular acidity; acidification of endosomes by the vacuolar H^+-ATPase is stimulated by GTPγS, and the effect is blocked by pertussis toxin (GURICH et al. 1991). Activation of heterotrimeric G-proteins may therefore regulate the assembly of coat proteins via changes in acidification of intracellular organelles.

The above model is consistent with in vitro transport studies. Transport between the *trans* Golgi and the cell surface requires acidification of the lumen of the *trans* Golgi by a vacuolar-type H^+-ATPase (MILLER and MOORE 1990b; S. MILLER and H.-P.H. MOORE, manuscript in preparation); it is blocked by protonophores, inhibitors of the vacuolar proton pump, or removal of chloride ions which limits the activity of the electrogenic pump. Pulse-chase experiments using nigericin to abolish the pH gradient across the TGN have demonstrated that the block to secretion occurs kinetically early in the reaction. Secretion of [^{35}S]-labeled glycosaminoglycan chains rapidly becomes insensitive to the addition of nigericin. Interestingly, this nigericin-insensitive population is also insensitive to inhibition by GTPγS, suggesting that the GTPγS-sensitive step occurs at the same point as, or earlier than, the nigericin-sensitive step (see Fig. 1). Taken together, these results are consistent with the hypothesis that inhibition of the vacuolar H^+-ATPase or neutralization of the lumenal pH blocks the assembly of coat proteins onto Golgi membranes, and thereby blocks vesicular transport at a kinetically early step.

III. The ADP-Ribosylation Factor Family

ADP-ribosylation factors (ARF), initially identified as cofactors required for in vitro cholera toxin-catalyzed ADP-ribosylation of the α subunit of the trimeric GTPase $G_{s\alpha}$ (KAHN 1986), are members of a growing gene family related to the ras-like small GTPases (see Chaps. 34, 35). Several reports have shown that ARF is involved in intra-Golgi transport; it is present on coated vesicles generated in intra-Golgi transport assay (SERAFINI et al.

1991), and depletion of two ARF-related factors from cytosol abolishes GTPγS-dependent inhibition of intra-Golgi transport (TAYLOR et al. 1992). Preliminary studies from our laboratory indicate that ARF is also involved in *trans* Golgi-to-cell-surface transport by the constitutive pathway, since a peptide corresponding to the N-terminal domain of ARF1P inhibits secretion of sulfated glycosaminoglycan chains (L. CARNELL, R. KAHN and H.-P.H. MOORE, unpublished).

How does ARF control vesicular traffic? Recent studies using isolated Golgi membranes showed that ARF is required for binding of non-clathrin coatomer proteins to the membranes (DONALDSON et al. 1992). While binding of ARF to the membrane is inhibited by Brefeldin A (BFA), subsequent binding of coat proteins to the membranes is not affected by BFA. This indicates that ARF participates in an early stage of coat assembly, and that BFA interferes with this step. Consistent with this model, we found that BFA blocks export from the TGN to the cell surface via the constitutive pathway; addition of BFA rapidly and reversibly inhibited cell surface transport of VSV G-protein that had been accumulated in the TGN by incubation of infected baby hamster kidney (BHK-21) cells at 20°C (MILLER et al. 1992). The block to secretion was not due to collapse of TGN to the ER, since VSV G-protein blocked in treated cells resided in compartments that were distinct from the ER/Golgi system. In addition to transport of membrane proteins, constitutive secretion of a soluble marker – sulfated glycosaminoglycan chains – from the TGN is also blocked by BFA. A similar effect was observed for export via the regulated secretory pathway; BFA potently inhibited secretion of sulfated secretogranin-II induced by K^+-depolarization from PC-12 cells. Inhibition was at the level of granule formation, since BFA had no effect on regulated secretion from preformed granules. Thus, budding of constitutive vesicles and regulated granules from the TGN appears to be governed by a common mechanism for coat assembly that is sensitive to BFA.

Although budding of constitutive vesicles and regulated granules from the TGN are both inhibited by BFA, the exact coat proteins involved may not be identical. Indeed, BFA not only affects the assembly of the non-clathrin coatomer protein, β-COP, but also the clathrin-associated coat protein, γ-adpatin, from the Golgi (WONG and BRODSKY 1992). The targets for BFA, however, appear to be different in these two cases, since in MDCK and PtK1 cells the assembly of γ-adaptin is sensitive to BFA even though the assembly of β-COP is BFA-resistant. Since β-COP is found on the same membrane as the vesicular stomatitis virus (VSV) G-protein blocked at 20°C, it is a likely candidate for the budding of constitutive vesicles from the TGN. Clathrin, on the other hand, is found on maturing regulated granule membranes (ORCI et al. 1985; TOOZE and TOOZE 1986) and thus is a good candidate for the generation of regulated granules. Whether this is indeed the case requires future functional tests.

IV. The rab Family

The rab proteins have been implicated in the control of vesicle targeting. Vesicular transport requires accurate delivery of cargos to the correct target compartment. Unlike the budding and fusion events, components involved in vesicle targeting must be specific to the individual step to prevent inappropriate delivery of cargos to the wrong membranes. H. Bourne proposed a model to explain the role of rab proteins in this process based on the function of the elongation factor, EF-Tu, in protein synthesis (Bourne 1988). According to this model, the GTPases act as a molecular switch that couples the hydrolysis of GTP to membrane fusion. Only when the vesicles interact with the correct target membranes will hydrolysis and fusion occur. The role of these proteins is thus to ensure the specificity of vesicle targeting. The model thus predicts the presence of a family of these small GTPases, each mediating targeting of a specific set of transport vesicles. Data accumulated thus far are consistent with such a prediction. Molecular cloning of Sec4p-related proteins in mammalian cells, also known as the "rab" proteins, indicated that this class of proteins belongs to a large family – approximately 20 or so of these proteins have been cloned (Zahraoui et al. 1989; Elferink et al. 1992; Chavrier et al. 1992; H. Yu, D. Leaf and H.P. Moore, manuscript in preparation), and many more remain to be discovered. Localization of some of these proteins indicates that they are specifically localized to different membranes – another prediction of the model.

In the budding yeast *S. cerevisiae*, consitutive secretion is mediated by the product of the *SEC4* gene (Salminen and Novick 1987; Goud et al. 1988). Thus far, the only functional *SEC4* homolog that has been isolated is the *YPT2* gene from *Schizosaccharomyces pombe* (Haubruck et al. 1990). Of the rabs known from mammalian and other sources, the two that share the closest homology with Sec4p and Ypt2p are rab8 and rab10 (Chavrier et al. 1990). Canine rab8 and rab10, also identified as the *ORA1* and *ORA2* genes from the electric fish *Omata* (Ngsee et al. 1991) and the MEL gene from human fibroblasts (Nimmo et al. 1991), show 48% and 49% amino acid sequence identity with the *S. cerevisiae* Sec4p, respectively. This degree of sequence homology is higher than for rab proteins involved in different steps of vesicular transport (for example, rat rab1 and rab2 are 35% identical). However, we found that neither rab8 nor rab10 could functionally complement a temperature-sensitive mutant of *sec4* (Y.T. Chen and H.P. Moore, unpublished). The lack of complementing activity means either rab8/10 performs a different function from Sec4p, or the post-Golgi pathways in mammalian cells are sufficiently different such that the rab protein or proteins involved in constitutive secretion are no longer conserved.

To analyze the intracellular location of rab8/10 and other members of this growing gene family, we have designed a mammalian expression vector with an influenza hemagglutinin epitope tag engineered at its extreme N terminus (the C terminus was not used to avoid interference with isoprenylation). To test whether the tag interferes with proper protein folding and function, we epitope-tagged the yeast Sec4p and tested its ability to rescue a mutant temperature-sensitive allele of yeast *sec4* cells. Yeast cells transformed with this altered *SEC4* gene can still grow at a restrictive temperature, indicating that the modified protein is functional. Moreover, when expressed in mammalian cells the Sec4 protein was localized to the cell surface. Thus, its localization signal is recognized by the mammalian secretory machinery. In transfected cells, rab8 is found on the cell surface, with the highest concentration in the ruffling regions of the cell periphery (Chen, Holcomb and Moore, submitted). On the other hand, rab10 is localized to the Golgi region as judged by co-staining with the Golgi markers, NBD-ceramide and β-COP. The cellular location of rab8 is consistent with a possible role in a late step of the constitutive secretory pathway. The staining in membrane ruffles may indicate that these regions are particularly active in exocytotic activities. In this regard, it is interesting to note that new membrane insertion and recycling often occurs in a polarized fashion. For example, newly synthesized membrane proteins are preferentially inserted at the leading edge or margin of cultured cells (Marcus 1962; Bergman et al. 1983). Recycling receptors such as transferrin receptor, LDL receptor, and fibronectin receptor are also inserted at the leading edge (Bretscher 1983, 1989). It will be of interest to determine if rab8 and/or rab10 indeed participate in constitutive transport between the *trans* Golgi and the cell surface in mammalian cells.

As discussed earlier, many cells contain pathways for regulated secretion, and they generally fall into two groups: the biosynthetic type and the recycling type. To date, only rab3A has been clearly localized to the synaptic vesicles in nerve terminals and neuroendocrine cells – a recycling type of regulated secretion (Fischer von Mollard et al. 1990). As shown in Table 1, the recycling type of regulated secretory vesicles is found in a wide variety of tissues.

rab3A, however, is restricted to a few tissues (i.e., neural and endocrine). This raises the possibility that regulated secretion in other tissues may be mediated by functionally similar, but distinct, rab proteins. Indeed, several rab proteins that are closely related to rab3A, i.e., rab3B, rab3C, and rab3D, have been identified; rab3D, for example, is enriched in adipocytes and may participate in the regulated insertion of the glucose transporter GLUT4 (Baldini et al. 1992). Many other *rab* genes also have closely related isoforms (>85% identify). There are at least three rab5's (Chavrier et al. 1992), four rab3's (Baldini et al. 1992), three rab1's (Zahraoui et al. 1989; Vielh et al. 1988; H. Yu, D. Leaf and H.-P.H. Moore, unpublished), and two rab11's (R. Chavez and H.-P.H. Moore, unpublished). Their functional significance is not yet clear.

Whether rab3 also participates in biosynthetic type of regulated secretion is not as clear. A peptide derived from the effector domain stimulates amylase secretion from permeabilized pancreatic cells, suggesting that rab3-like proteins may be involved (PADFIELD et al. 1992). Since the same peptide also inhibits ER-to-Golgi transport (PLUTNER et al. 1990), it may interact with effectors of other rab proteins. The protein rab3A is predominantly localized to the recycling type of microvesicles in chromaffin cells (FISCHER VON MOLLARD et al. 1990), although some has been also found on chromaffin granules (DARCHEN et al. 1990). Multiple GTPases have been detected by GTP-blotting and other techniques on exocytotic granules from a variety of tissues (e.g., RUBINS et al. 1992). It will be important to determine which of these GTPases functions in targeting of dense-core granules. We have cloned a number of rab proteins from the mouse anterior pituitary cell line, AtT-20 (YU et al. 1993). Their possible role(s) in granule targeting and membrane recycling are being investigated.

The rab proteins are thought to ensure the specificity of vesicle targeting by interacting with specific cellular targets. A possible factor associated with the target compartment is the GTPase activating protein (GAP). At present, it is unclear whether each rab has its own GAP, or a given GAP can act on more than one rab protein. We have begun to address this question by isolating the GAP involved in constitutive secretion. The effector domain of yeast Sec4p has the sequence FITTIGIDF (residues 32–40 on Ras). This region of the Ras protein is known to interact with the GAP protein. Several mammalian rabs have effector domains very similar to this sequence; rab8 and rab10 have the sequence FISTIGIDF, and *RAM* – a gene isolated from rat megakaryocytes – has a sequence encoding FITTVGIDF in this region (NAGATA et al. 1990). It is therefore possible that a mammalian GAP involved in exocytosis may interact with the yeast Sec4p. Using Sec4 protein purified from baculovirus-infected Sf9 cells, we have detected a GAP activity from crude rat liver extracts. Preliminary results show that the activity purifies as a 150-kDa and a 400-kDa complex. Interestingly, the GAP for rab3A also purifies as a large complex. It will be important to determine if rab proteins involved in constitutive and regulated secretion share the same GAP protein.

F. Conclusions

In summary, recent progress has uncovered a very large number of components involved in vesicular transport. Many of these turn out to be GTPases and their modulators. Collectively, they serve to control vesicular transport between cellular compartments. Figure 2 proposes a possible model in which sequential interactions of GTPases result in coat assembly, vesicle budding, uncoating, and targeting/fusion. It should be kept in mind that at the present time, many aspects of the model are hypothetical.

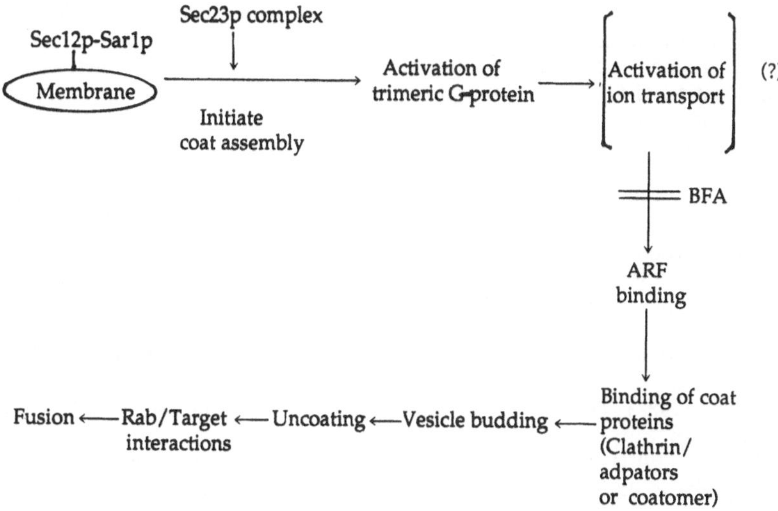

Fig. 2. A hypothetical model for the control of vesicular transport by GTPases. *ARF*, ADR-ribosylation factor; *BFA*, Brefeldin A

However, the model provides a useful conceptual framework for future experimentation. Several important questions remain. Does budding of exocytic vesicles from the TGN involve Sar1-like GTPases? If so, are they regulated by Sec23p-like proteins? Are coatomer and clathrin involved in the generation of constitutive and regulated granules from TGN, respectively, or are other coat proteins required? Do ARF proteins participate in the formation of regulated granules as well as constitutive vesicles, and if so, does each pathway utilize a different member of this family? What is the identity and downstream effector(s) of trimeric G-protein's regulatory vesicle traffic, and are they the same or different for constitutive and regulated vesicles? Finally, how is the rab protein in regulated secretion controlled differently such that fusion does not occur immediately upon correct targeting of vesicles to the membrane? Hopefully, answers to these questions will be forthcoming in the near future.

Acknowledgement. The authors thank members of the Moore laboratory for critical reading of the mansucript. This work was supported by grants (to H.P.M.) from National Institute of Health (GM35235), American Cancer Society (CD-497), and Lucille P. Markey Fund for Innovation.

References

Ali SM, Geisow MJ et al. (1989) A role for calpactin in calcium-dependent exocytosis in adrenal chromaffin cells. Nature 340:313–315

Bacskai B, Friedman P (1992) Activation of latent Ca^{2+} channels in renal epithelial cells by parathyroid hormone. Nature 347:388–391

Baker D, Hicke L et al. (1988) Reconstitution of SEC gene product-dependent intercompartmental protein transport. Cell 54:335–344

Baldini G, Hohman R et al. (1991) Insulin and nonhydrolyzable GTP analogs induce translocation of GLUT 4 to the plasma membrane in alpha-toxin-permeabilized rat adipose cells. J Biol Chem 266(7):4037–4040

Baldini G, Hohl T et al. (1992) Cloning of a Rab3 isotype predominantly expressed in adipocytes. Proc Natl Acad Sci USA 89(11):5049–5052

Barr FA, Leyte A et al. (1991) Trimeric Gk-proteins of the trans-Golgi network are involved in the formation of constitutive secretory vesicles and immature secretory granules. FEBS Lett 294:239–243

Bergman JE, Kupfer A et al. (1983) Membrane insertion at the leading edge of motile fibroblasts. Proc Natl Acad Sci USA 80:1367–1371

Blackwood RA, Ernst JD (1990) Characterization of Ca^{2+}-dependent phospholipid binding, vesicle aggregation and membrane fusion by annexins. Biochem J 266:195–200

Bourne HR (1988) Do GTPases direct membrane traffic in secretion? Cell 53(5):669–671

Bretscher MS (1983) Distribution of receptor for transferrin and low density lipoprotein on the surface of giant HeLa cells. Proc Natl Acad Sci USA 80:454–458

Bretscher MS (1989) Endocytosis and recycling of the fibronectin receptor in CHO cells. EMBO J 8(5):1341–1348

Brion C, Miller S et al. (1992) Regulated and constitutive secretion: differential effects of protein synthesis arrest on transport of glycosaminoglycan chains to the two secretory pathways. J Biol Chem 267:1477–1483

Burgess TL, Kelly RB (1987) Constitutive and regulated secretion. Annu Rev Cell Biol 3:243

Castle J (1990) Sorting and secretory pathways in exocrine cells. Am J Respir Cell Mol Biol 2(2):119–126

Chavrier P, Vingron M et al. (1990) Molecular cloning of YPT1/SEC4-related cDNAs from an epithelial cell line. Mol Cell Biol 10(12):6578–6585

Chavrier P, Simons K et al. (1992) The complexity of the Rab and Rho GTP-binding protein subfamilies revealed by a PCR cloning approach. Gene 112:261–264

Curtis Id, Simons K (1988) Dissection of Semliki forest virus glycoprotein delivery from the trans-Golgi network to the cell surface im permeabilized BHK cells. Proc Natl Acad Sci USA 85(21):8052–8056

Cutler D, Cramer L (1990) Sorting during transport to the surface of PC12 cells: divergence of synaptic vesicle and secretory granule proteins. J Cell Biol 110(3):721–730

d'Enfert C, Wuestehube LJ et al. (1991) Sec12p-dependent membrane binding of the small GTP-binding protein Sar1p promotes formation of transport vesicles from the ER. J Cell Biol 114(4):663–670

Darchen F, Zahraoui A et al. (1990) Association of the GTP-binding protein Rab3A with bovine adrenal chromaffin granules. Proc Natl Acad Sci USA 87(15):5692–5696

Donaldson JG, Kahn RA et al. (1991) Binding of ARF and beta-COP to Golgi membranes: possible regulation by a trimeric G protein. Science 254:1197–1199

Donaldson JG, Cassel D et al. (1992) ADP-ribosylation factor, a small GTP-binding protein, is required for binding of the coatomer protein β-COP to Golgi membranes. Proc Natl Acad Sci USA 89:6408–6412

Elferink LA, Anzai K et al. (1992) RAB15, a novel low molecular weight GTP-binding protein specifically expressed in rat brain. J Biol Chem 267(9):5768–5775

Farquhar MG (1985) Progress in unraveling pathways of Golgi traffic. Annu Rev Cell Biol 1:447–448

Fernandez JM, Villalon M et al. (1991) Reversible condensation of mast cell secretory products in vitro. Biophys J 59:1022–1027

Fisher JM, Sossin W et al. (1988) Multiple neuropeptides derived from a common precursor are differentially packaged and transported. Cell 54:813–822
Fischer von Mollard G, Mignery GA et al. (1990) rab3 is small GTP-binding protein exclusively localized to synaptic vesicles. Proc Natl Acad Sci USA 87(5):1988–1992
Forte J, Hanzel D et al. (1989) Pumps and pathways for gastric HCl secretion. Ann NY Acad Sci 574:145–158
Fuller R, Brake A et al. (1989) Intracellular targeting and structural conversation of a prohormone-processing endoprotease. Science 246(4929):482–486
Fumagalli G, Zanini B et al. (1985) In cow anterior pituitary, growth hormone and prolactin can be packed in separate granules of the same cell. J Cell Biol 100:2019–2024
Gerdes HH, Rosa P et al. (1989) The primary structure of human secretogranin II, a widespread tyrosine-sulfated secretory granule protein that exhibits low pH- and clacium-induced aggregation. J Biol Chem 264:12009–12015
Gomperts BD (1990) GE: a GTP-binding protein mediating exocytosis. Annu Rev Physiol 52:591–606
Goud B, Salminen A et al. (1988) A GTP-binding protein required for secretion rapidly associates with secretory vesicles and the plasma membrane in yeast. Cell 53:753–768
Gravotta D, Adesnik M et al. (1990) Transport of influenza HA from the trans-Golgi network to the apical surface of MDCK cells permeabilized in their basolateral plasma membranes: energy dependence and involvement of GTP-binding proteins. J Cell Biol 111:2893–2908
Grimes M, Kelley R (1992) Intermediates in the constitutive and regulated secretory pathways released in vitro from semi-intact cells. J Cell Biol 117(3): 539–549
Guest P, Ravazzola M et al. (1991) Molecular hetergeneity and cellular localization of carboxypeptidase H in the islets of Langerhans. Endocrinology 129(2):734–740
Gurich RW, Codina J et al. (1991) A potential role for guanine nucleotide-binding protein in the regulation of endosomal proton transport. J Clin Invest 87:1547–1552
Haubruck H, Engelke U et al. (1990) Structural and functional analysis of ypt2, an essential ras-related gene in the fission yeast Schizosaccharomyces pombe encoding a Sec4 protein homologue. EMBO J 9(6): 1957–1962
Hicke L, Schekman R (1989) Yeast Sec23p acts in the cytoplasm to promote protein transport from the ER to the Golgi complex in vivo and in vitro. EMBO J 8:1677–1684
Huttner W, Tooze S (1989) Biosynthetic protein transport in the secretory pathway. Opin Cell Biol 1(4):648–654
Hwang A, Yoshihisa T et al. (1991) Identification and isolation of homologues of yeast sec23 and sec21 proteins from mammalian tissues. J Cell Biol 115:62a
Kahn RA (1991) Fluoride is not an activator of the smaller (20–25 kDa) GTP-binding proteins. J Biol Chem 266(24):15595–15597
Kahn RA, Gilman AG (1986) The protein cofactor necessary for ADP-ribosylation of G_s by cholera toxin is itself a GTP-binding protein. J Biol Chem 261:7906–7911
Kaiser CA, Schekman R (1990) Distinct sets of SEC genes govern transport vesicle formation and fusion early in the secretory pathway. Cell 61(4):723–733
Kaziro Y, Itoh H et al. (1991) Structure and function of signal-transducing GTP-binding proteins. Annu Rev Biochem 60:349–400
Kelly R (1991) Secretory granule and synaptic vesicle formation. Opin Cell Biol 3(4):654–660
Lencer W, Verkman A et al. (1990) Endocytic vesicles from renal papilla which retrieve the vasopressin-sensitive water channel do not contain a functional H+ ATPase. J Cell Biol 111:379–389

Lin HC, Sudhof TC et al. (1992) Annexin VI is required for budding of Clathrin-Coated pits. Cell 70(2):283–291

Linstedt AD, Kelly RB (1991) Synaptophysin is sorted from endocytotic markers in neuroendocrine PC12 cells but not transfected fibroblasts. Neuron 7(2):d309–317

Marcus PI (1962) Dynamics of surface modification in myxovirus-infected cells. Cold Spring Harbor Sym Quant Biol 27:351–365

Mayorga LS, Diaz R et al. (1988) Plasma membrane-derived vesicles containing receptor-ligand complexes are fusogenic with early endosomes in a cell-free system. J Biol Chem 263(33):17213–17216

Melancon P, Glick BS et al. (1987) Involvement of GTP-binding "G" proteins in transport through the Golgi stack. Cell 51(6):1053–1062

Mellman I, Simons K (1992) The Golgi complex: in vitro veritas? Cell 68(5):829–840

Merisko EM, Fletcher M et al. (1986) The reorganization of the Golgi complex in anoxic pancreatic acinar cells. Pancreas (1):95–109

Milgram S, Johnson R et al. (1992) Expression of individual forms of peptidylglycine alpha-amidating monooxygenase in AtT-20 cells. endoproteolytic processing and routing to secretory granules. J Cell Biol 117(4):717–728

Miller SG, Moore H-PH (1990a) Regulated secretion. Curr Opin Cell Biol 2:642–647

Miller SG, Moore H-PH (1990b) Role of Golgi acidification in vesicular transport from the trans-Golgi to the plasma membrane in CHO cells. J Cell Biol 111(5, 2):325a

Miller SG, Moore H-PH (1991) Reconstitution of constitutive secretion using semi-intact cells: regulation by GTP but not calcium. J Cell Biol 112:39–54

Miller S, Moore H-PH (1992a) Movement from the trans-Golgi network to the cell surface in semi-intact cells. Methods Enzymol (in press)

Miller S, Carnell L et al. (1992b) Post-Golgi membrane traffic: brefeldin A inhibits export from the distal Golgi-Compartment to the cell surface but not recycling. J Cell Biol 118:267–284

Moore H-PH (1987) Factors controlling packaging of peptide hormones into secretory granules. Ann NY Acad Sci 493:50–61

Moore HPH, Brion C et al. (1989) Protein secretion by constitutive and regulated pathways. Soc Gen Physiol Ser 44:189–201

Morgan A, Burgoyne RD (1992) Exol and Exo2 proteins stimulate calcium-dependent exocytosis in permeabilized adrenal chromaffin cells. Nature 355:833–836

Nagata K-I, Satoh T et al. (1990) The ram: a novel low molecular weight GTP-binding protein from a rat megakaryocyte library. FEBS Lett 275:29–32

Nakano A, Muramatsu M (1989) A novel GTP-Binding protein, SAR1P, is involved in transport from the endoplasmic reticulum to the Golgi apparatus. J Cell Biol 109:N6:2677–2691

Ngsee JK, Elferink LA et al. (1991) A family of ras-like GTP-binding proteins expressed in electromotor neurons. J Biol Chem 266(4):2675–2680

Nimmo ER, Sanders PG et al. (1991) The MEL gene: a new member of the RAB/YPT class of RAS-related genes. Oncogene 6(8):1347–1351

Novick P, Schekman R (1983) Export of major cell surface proteins is blocked in yeast secretory mutants. J Cell Biol 96:541–547

Novick P, Ferro S, Schekman R (1981) Order of events in the yeast secretory pathway. Cell 25:461–469

Orci L, Ravazzola M et al. (1985) Clathrin-immunoreactive sites in the golgi apparatus are concentrated at the trans pole in polypeptide hormone-secreting cells. Proc Natl Acad Sci USA 82:5385–5389

Orci L, Ravazzola M et al. (1987) The trans-most cisternae of the Golgi complex: a compartment for sorting of secretory and plasma membrane proteins. Cell 51:1039–1051

Orci L, Ravazzola M et al. (1991) Mammalian Sec23p homologue is restricted to the endoplasmic reticulum transitional cytoplasm. Proc Natl Acad Sci USA 88:8611–8615

Padfield P, Balch W et al. (1992) A synthetic peptide of the rab3a effector domain stimulates amylase release from permeabilized pancreatic acini. Proc Natl Acad Sci USA 89(5):1656–1660

Plutner H, Schwaninger R et al. (1990) Synthetic peptides of the Rab effector domain inhibit vesicular transport through the secretory pathway. EMBO J 9:2375–2383

Quinn D, Orci L et al. (1991) Intracellular transport and sorting of mutant human proinsulins that fail to form hexamers. J Cell Biol 113:987–996

Regnier-Vigouroux A, Tooze SA et al. (1991) Newly synthesized synaptophysin is transported to synaptic-like microvesicles via constitutive secretory vesicles and the plasma membrane. EMBO J 10(12):3589–3601

Rodriguez-Boulan E, Nelson W (1989) Morphogenesis of the polarized epithelial cell phenotype. Science 245(4919):718–725

Rothman JE, Orci L (1990) Movement of proteins through the Golgi stack: a molecular dissection of vesicular transport. FASEB J 4:1460–1468

Rothman JE, Orci L (1992) Molecular dissection of the secretory pathway. Nature 355;409–415

Rubins J, Panchenko M et al. (1992) Identification of ras and ras-related low-molecular-mass GTP-binding proteins associated with rat lung lamellar bodies. Am J Respir Cell Mol Biol 6(3):253–259

Ruohola H, Kabcenell AK et al. (1988) Reconstitution of protein transport from the ER to the Golgi Complex in yeast: the acceptor Golgi compartment is defective in the sec23 mutant. J Cell Biol 107:1465–1476

Salamero J, Sztull ES et al. (1990) Exocytic transport vesicles generated in vitro grom the trans-Golgi network carry secretory and plasma membrane proteins. Proc Natl Acad Sci USA 87:7717–7721

Salminen A, Novick PJ (1987) A ras-like protein is required for a post-Golgi event in yeast secretion. Cell 49:527–538

Sarafian T, Pradel LA et al. (1991) The participation of annexcin II (calpactin I) in calcium-evoked exocytosis requires protein kinase C. J Cell Biol 114:1135–1147

Schekman R (1985) Protein localization and membrane traffic in yeast. Annu Rev Cell Biol 1:115–143

Serafini T, Orci L et al. (1991) ADP-ribosylation factor is a subunit of the coat of Golgo-derived COP-coated vesicles: a novel role for a GTP-binding protein. Cell 67:239–253

Simons K, Fuller SD (1985) Cell surface polarity in epithelia. Annu Rev Cell Biol 1:243–288

Simons K, Fuller S (1987) Biological organization: macromolecular interations at high resolution. Academic, New York, pp 139–150

Slot J, Geuze H et al. (1991) Immuno-localization of the insulin regulatable gucose transporter in brown adipose tissue of the rat. J Cell Biol 113(1):123–135

Smeekens S, Steiner D (1990) Identification of a human insulinoma cDNA encoding a novel mammalian protein structurally related to the yeast diabasic processing protease Kex2. J Biol Chem 265(6):2997–3000

Smeekens S, Avruch A et al. (1991) Identification of a cDNA encoding a second putative prohorome convertase related to PC2 in AtT20 cells and islets of langerhans. Proc Natl Acad Sci USA 88(2):340–344

Smith R, Charron M et al. (1991) Immunoelectron microscopic demonstration of insulin-stimulated translocation of glucose transporters to the plasma membrane of isolated rat adipocytes and masking of the carboxyl-terminal epitope of intracellular GLUT4. Proc Natl Acad Sci USA 88(15):6893–6897

Sossin WS, Fisher JM et al. (1990) Sorting within the regulated secretory pathway occurs in the trans-Golgi network. J Cell Biol 110:1–12

Stow J, Almeida J et al. (1991) A heterotrimerinc G protein, G alpha i-3, on Golgi membranes regulates the secretion of a heparan sulfate proteoglycan in LLC-PK1 epithelial cells. J Cell Biol 114(6):1113–1124

Taplits M, Hodes R (1989) The helper T lymphocyte as a secretory cell. Constitutive versus regulated secretion. Year Immunol 4:147–158

Tartakoff A, Vassalli P et al. (1978) Comparative studies of intracellular transport of secretory proteins. J Cell Biol 79:694–707

Taylor TC, Kahn RA et al. (1992) Two distinct members of the ADP-ribosylation factor family of GTP-binding proteins regulate cell-free intra-Golgi transport. Cell 70(1):69–79

Thomas L, Leduc R et al. (1991) Kex2-like endoproteases PC2 and PC3 accurately cleave a model prohormone in mammalian cells: evidence for a common core of neuroendocrine processing enzymes. Proc Natl Acad Sci USA 88(12):5297–5301

Tooze J, Tooze S (1986) Clathrin-coated vesicular transport of secretory proteins during the formation of ACTH-containing secretory granules in AtT-20 cells. J Cell Biol 103:839–850

Tooze SA, Huttner WB (1990) Cell-free protein sorting to the regulated and constitutive secretory pathways. Cell 60:837–847

Tooze J, Tooze SA et al. (1987) Sorting of progeny coronavirus from condensed secretory proteins at the exit from the trans-Golgi network of AtT-20 cells. J Cell Biol 105:1215–1216

Turkewitz A, Madeddu L et al. (1991) Maturation of dense core granules in wild type and mutant Tetrahymena thermophila. EMBO J 10(8):1979–1987

Vega-Salas DE, Salas P et al. (1987) Formation of the apical pole of epithelial (madin-darby canine kidney) cells: polarity of an apical protein is independent of tight junctions while segregation of a basolateral marker requires cell-ce. J Cell Biol 104:905–916

Vielh E, Touchot N et al. (1988) Nucleotide sequence of a rat cDNA: rab1B, encoding a rab1-YPT related protein. Nucleic Acids Res 17(4):1770

Wagner D, Saffaripour S et al. (1991) Induction of specific storage organelles by von Willebrand factor propolypeptide. Cell 64(2):403–413

Wong DH, Brodsky FM (1992). 100-kD proteins of Golgi-and trans-Golgi network-associated coated vesicles have related byt distinct membrane binding properties. J Cell Biol 117(6): 1171–1179

Yu H, Leaf DS et al. (1993) Gene cloning and characterization of a GTP-binding Rab protein from mouse pituitary AtT-20 cells. Gene (in press)

Zahraoui A, Touchot N et al. (1989) The human Rab genes encode a family of GTP-binding proteins related to yeast YPT1 and SEC4 products involved in secretion. J Biol Chem 264:12394–12401

Zeuzem S, Feick P et al. (1992) Intravesicular acidification correlates with binding of ADP-ribosylation factor to microsomal membranes. Proc Natl Acad Sci USA 89:6619–6623

CHAPTER 34

The Biology of ADP-Ribosylation Factors

R.A. KAHN

A. Introduction

Adenosine diphosphate (ADP) ribosylation factor (ARF) was originally defined as an activity required for the efficient activation of the purified, stimulatory, regulatory component of adenylyl cyclase, G_s, by cholera toxin (SCHLEIFER et al. 1982; KAHN 1989, 1991). The toxin catalyzes the covalent addition of the ADP-ribose moiety of nicotinamide adenine dinucleotide (NAD) onto the α subunit of G_s. The consequences of this modification include decreased intrinsic GTP hydrolysis by the G_s protein (CASSEL and SELINGER 1979) and its consequent activation and subunit dissociation (KAHN and GILMAN 1984a). Thus, while GTP alone is normally a very weak stimulator of membrane cyclase, after ADP-ribosylation GTP is nearly as potent as the slowly hydrolyzable analogs of GTP, e.g., GTPγS or Gpp(NH)p.

The original impetus for the purification and characterization of ARF derived from its role as a potential modulator of adenylyl cyclase activities. However, after purification of ARF was first achieved from rabbit liver (KAHN and GILMAN 1984b), it was possible to study the effects of added ARF on a number of in vitro assays related to G_s or adenylyl cyclase functions, e.g., efficiency of reconstituted purified β-adrenergic receptors: G-proteins coupling or G_s subunit interaction. To date, there have been no reports of any effect of any ARF protein on G_s or adenylyl cyclase functions or activities other than that of cofactor in the cholera toxin reaction. However, the early studies of ARF proteins led to the conclusion that ARF is ubiquitous in eukaryotic cells (KAHN et al. 1988) and is itself a GTP binding protein (KAHN and GILMAN 1986).

The ARF and related proteins are now recognized as one of the five families comprising the Ras superfamily of small GTPases (KAHN et al. 1992a). At least two distinct members of the ARF family are essential genes: ARFs in *Saccaromyces cerevisiae* (STEARNS et al. 1990a) and *arl* in *Drosphilia melanogaster* (TAMKUN et al. 1991). Substantial evidence now exists for the role(s) of ARF proteins in the protein secretory pathway, particularly at or near the Golgi complex. This review will focus on five aspects of ARF biology: (1) structural and functional information useful in defining the ARF family, (2) genetic analysis of yeast ARFs, (3) biochemical analysis of ARF proteins, (4) immunolocalization of ARF proteins, and (5) membrane and vesicle binding studies, which have recently been exploited to gain new insights into the processes regulated by ARF proteins.

B. The ARF Family of Small GTPases

I. Structural Definition

The ARF family of proteins is comprised of both ARF proteins, as defined functionally below, and a collection of structurally related ARF-like (ARL) proteins or cDNA-derived protein sequences which: (a) contain the consensus guanine nucleotide binding domains, (b) have predicted molecular masses of around 20 kDa, and (c) have greater homology to ARF than to members of other families, e.g., Ras, Rab, Rho, or TC4 (KAHN et al. 1992a). Members of the family generally share at least 30% sequence identity. Other structural features which may be useful in differentiating a member of the ARF family from one of the other families include an amino terminal extension of 14–20 amino acids relative to the other four families, a glycine at position 2 which has been shown to be the site of myristoylation for at least some ARF proteins (KAHN et al. 1988), the lack of a cysteine at or near the C terminus, and the well-conserved motif DXGGQ in the second of the consensus GTP-binding motifs, as opposed to the DTAGQ motif seen in the other four families at this location (SEWELL and KAHN 1988). This motif, and a few others, are also found in the α subunits of the G-proteins and contribute to the observation that the ARF proteins are actually as structurally related to the G-protein family as they are to the Ras family members (SEWELL and KAHN 1988).

II. Functional Definition

The bona fide ARF proteins have been shown to possess at least one of the following activities: activity in the in vitro ARF assay (WEISS et al. 1989; KAHN 1991; KAHN et al. 1991) or the ability to rescue the lethal double mutant *arf1⁻arf2⁻* when expressed in the yeast *S. cerevisiae* (KAHN et al. 1991). Included in this category are yeast ARF1 (SEWELL and KAHN 1988) and ARF2 (STEARNS et al. 1990a), and bovine (SEWELL and KAHN 1988) or human ARF1 (PENG et al. 1989; PRICE et al. 1988), ARF3 (BOBAK et al. 1989), and human ARF5p and ARF4 (KAHN et al. 1991; MONACO et al. 1990) CPS1p/ARF5p (TSUCHIYA et al. 1991). These proteins are at least 70% identical in primary sequence when aligned, with some (e.g., yARF1 vs yARF2, or hARF1 vs hARF3) sharing as much as 96% identity (see Fig. 1). Interspecies sequence conservation is also high; human ARF1 and bovine ARF1 are 100% identical.

The ARF-like group of proteins currently includes at least three members: SAR1p (NAKANO and MURAMATSU 1989) and CIN4p (BOTSTEIN et al.

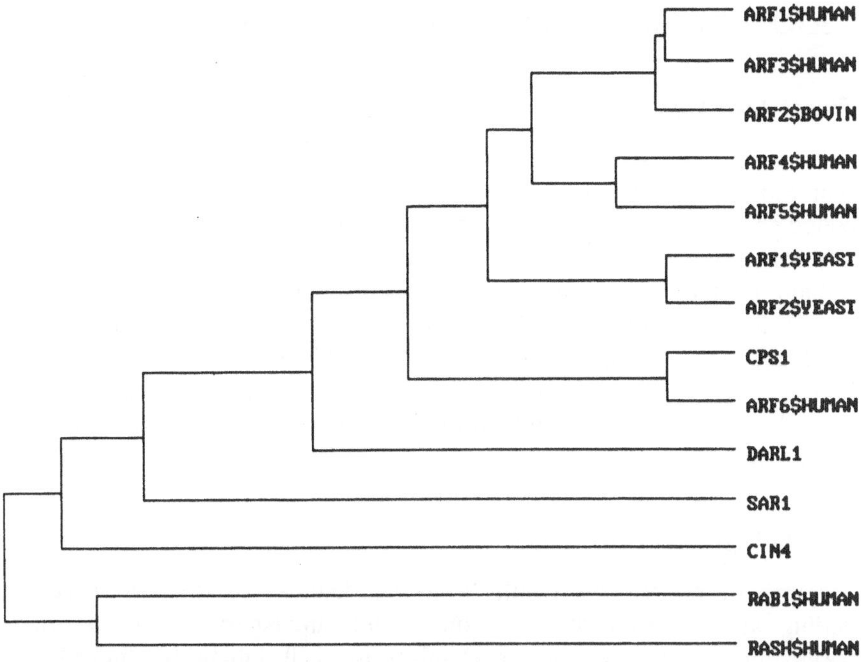

Fig. 1. Dendrogram of the members of the ARF family. Twelve members of the ARF family were aligned using the CLUSTAL program in PCGENE. Representatives of the Ras and Rab families were included for comparison. References are given in the text

1988) from yeast, and ARL1p (TAMKUN et al. 1991) from flies. These proteins are between 30% and 55% identical in sequence when aligned to any of the ARF or ARF-like proteins (see Fig. 1). One or more of these proteins may prove to have ARF activities and would later be classified as ARFs, but not all have been tested at this time.

The yeast genes *CIN4* and *SAR1* in single copy fail to complement the *arf*⁻ mutants and have been identified and cloned on the basis of functions which appear to be independent of ARF activities. The *D. melanogaster* ARL1p (TAMKUN et al. 1991) has been expressed in yeast and bacteria and found to be devoid of each of the ARF activities described above. Nevertheless, *arf*-like or *arl* is an essential gene in flies. Thus, ARF and related proteins are a large and functionally diverse family of small GTPases including at least a dozen distinct gene products.

[1] Although published under the name human ARF6, the predicted protein sequence of hARF6p and the previously reported CPS1p from chicken differ at only one position. In spite of the fact that CPS1 was reported as a pseudogene, in keeping with the recent suggestions for naming members of the Ras superfamily, the preferred name for this protein is human CPS1p.

C. ARF Functions in the Yeast, *Saccharomyces cerevisiae*

I. Yeast ARF Genes and Proteins

The yeast *S. cerevisiae* contains two ARF genes, *ARF1* and *ARF2*, located on chromosome IV approximately 25 cM apart (Stearns et al. 1990a). The encoded proteins are each 181 amino acids in length and are 96% identical. The purified proteins have identical specific activities in the ARF assay and bind guanine nucleotides at indistinguishable rates and with indistinguishable stoichiometries. ARF1p is expressed at significantly higher levels than ARF2p and accounts for 80%–90% of the ARF in a cell (Stearns et al. 1990a). The levels of the two ARF proteins appear to be independently regulated, as deletion of one gene causes no change in the level of expression of the other.

II. Phenotypes of *arf* Mutants

A number of phenotypes result from the deletion of the ARF1 gene, including an increased doubling time, and supersensitivity to fluoride (Stearns et al. 1990a). The mechanism for cell killing by fluoride is unknown. Deletion of ARF2 resulted in no discernible change in phenotype, while the loss of both ARF genes is lethal to cells. Thus, at least one ARF gene is required for cell viability in yeast.

It is likely that the two ARF genes are functionally indistinguishable and that the difference in the level of expression of the gene products accounts for the different phenotypes associated with the single gene disruptions (Stearns et al. 1990a). This conclusion was based on two observations: (1) a second copy of ARF2, on the chromosome or on a low copy plasmid, was capable of fully restoring $arf1^-$ cells to the wild type, and (2) one class of revertants of fluoride sensitivity of an *arf1* null mutation turned out to result from gene conversion in which most of the ARF2 coding sequence was inserted at the ARF1 locus to produce a fusion protein of mostly ARF2 under control of the ARF1 gene (Stearns et al. 1990a). An essential role for ARF in sporulation has recently been identified (Cavenagh and Kahn 1993). A single copy of ARF1 or ARF2 is sufficient to allow sporulation while a number of arf1 mutants do not (Cavenagh and Kahn 1993).

III. Evidence that ARF is Required in the Secretory Pathway

While the cloning of extragenic suppressors of ARF mutations may eventually prove to be a powerful approach to understanding ARF functions and pathways, the first clear indication of the physiological role of ARF1 in eukaryotic cells has already emerged from studies in yeast. A role for ARF in the protein secretory pathway was suggested from two results (STEARNS et al. 1990b). Crossing of *arf1*$^-$ strains to cells carrying different secretion gene (*SEC*) defects identified a subset of these genes (*sec7-1*, *sec21-1*, *ypt1-1*, and *bet2-1*) which were synthetically lethal. Analysis of the processing of the secreted protein invertase in arf1$^-$ cells revealed an impairment in the glycosylation pattern of this protein which is associated with a lesion at the ER or Golgi complex. A detailed molecular explanation of this defect in the secretory pathway is lacking, but together these results point to a role for ARF in this essential cellular function.

D. Biochemical Characterization of ARF Proteins

I. ARF Purified from Mammalian Sources is Heterogeneous

Purification of ARF from tissues – e.g., rabbit liver (KAHN and GILMAN 1984b) or bovine brain (KAHN and GILMAN 1986) – yielded a heterogeneous preparation as visualized by a doublet in sodium dodecyl sulfate (SDS) gels. The differently migrating species have been separated in active forms by both preparative SDS gels and by molecular sieve chromatography, and each has been shown to possess ARF activity (KAHN and GILMAN 1984b). However, even under these conditions it is not clear how many gene products are in each preparation. Similarly, ARF has been purified from both the soluble and particulate fractions of bovine brain, and the purified products are likely to be very similar in their protein content. It is likely that specific antibodies will be required to clearly determine the amounts of each individual ARF protein in preparations from mammalian sources.

The expression of ARF and ARF-like proteins in bacteria, engineered to express a specific gene product, has been used extensively to allow an abundant source of material for biochemical analysis of individual proteins. A simple purification scheme has been used to purify large amounts of several members of the ARF family, including yeast ARF1p and ARF2p, human ARF1p, ARF3p, ARF4p, and *Drosophila* ARL1p. The yeast CIN4 protein has also been expressed in active form in bacteria, but is unstable and thus has not yet been purified.

II. ARF Cofactor Activity

The cofactor activity of ARF in the cholera toxin catalyzed ADP-ribosylation of G_s has proven to be very useful as a sensitive, specific, and quantitative assay for ARF proteins (for recent reviews, see KAHN 1989, 1991). The ADP-ribosylation of G_s can be quantified either by the covalent incorporation of radioactivity from the substrate, α[^{32}P]NAD, into the α subunit of G_s, or by determination of the GTP-stimulated adenylyl cyclase activity after its reconstitution into cyc$^-$ membranes. In addition to allowing the first purification of an ARF protein (KAHN and GILMAN 1984b), this assay has been useful in providing a demonstration of the extent of functional conservation observed between ARF proteins expressed in yeast and man (KAHN et al. 1991). This assay also provided the first biochemical demonstration that the activity of a member of the Ras superfamily is regulated by GTP binding, as ARF·GDP is inactive in the ARF assay, but becomes activated upon binding GTP or a triphosphate analog (KAHN and GILMAN 1986). Less direct evidence suggests that the activated species of ARF may be binding G_s directly (KAHN and GILMAN 1986).

While useful as an ARF assay, its utility in determining the role of ARF proteins in cell physiology is clearly in question for a number of reasons. For example, no endogenous ADP-ribosyltransferase has been found to use G_s as a substrate, and the ARF assay is insensitive to myristoylation or specific mutations in ARF, each of which have dramatic effects on ARF activity in vivo. The physiological significance of ARF assays which utilize ADP-ribosylation of other substrates, e.g., agmatine, is even more suspect.

III. Guanine Nucleotide Binding

While only the ARF proteins have cofactor activity, each of the ARF preparations from animal sources as well as the recombinant ARF and ARF-like proteins has been shown to possess the capacity for high affinity guanine nucleotide binding. The most extensively characterized member of the ARF family is human ARF1p (WEISS et al. 1989).

The affinity of bovine brain ARF for guanine nucleotides was originally estimated by competition studies to be 20 nM, 90 nM, and 40 nM for GTPγS, GTP, and GDP (KAHN and GILMAN 1986). A more accurate, kinetic method was used to determine dissociation constants for hARF1p and resulted in values of 0.6–2.0 nM for GDP and 22–50 nM for GTPγS, depending on ionic strength (WEISS et al. 1989). The binding of guanine nucleotides to ARF proteins is very sensitive to a number of factors, including the concentration of phospholipids, detergents, salts, and magnesium.

ARF proteins are unique in having a near absolute dependence on phospholipids for nucleotide exchange (KAHN and GILMAN 1986). The requirement for phospholipids is intrinsic to the protein, as it is independent of the myristoylation status of the ARF protein (WEISS et al. 1989). Interestingly, an amino terminal deletion mutant of hARF1p, which lacks the first 17 amino acids, is fully capable of binding guanine nucleotides with high affinity, but no longer requires phospholipids or detergent to do so (KAHN et al. 1992b). The amino terminal domain of hARF1p is thought to form an α helix which is stabilized in aqueous solution by the presence of the myristate at the amino terminus (KAHN et al. 1992b). Thus, it is likely to be the interaction of the amino terminal domain with the phospholipid bilayer which promotes the release of the bound GDP. The N-terminal domain may also fold back to cover the nucleotide binding site, thereby acting as a physical barrier preventing GDP release. Structural studies are currently underway to determine the crystal and solution structures of hARF1p.

IV. GTPase Activity

Unlike other GTP binding proteins, purified mammalian ARF has no detectable ($<0.005\,min^{-1}$) GTPase activity (KAHN and GILMAN 1986; WEISS et al. 1989). Yet it is probably still safe to conclude that ARF proteins utilize intrinsic GTP hydrolysis as a mechanism for returning to a basal (GDP-bound) state in cells. Perhaps the best evidence in support of this conclusion comes from studies in yeast in which the dominant lethal mutant, [Q71L]yARF1p (analogous to the GTPase deficient [Q61L]*ras* p21 and [Q227L]$G_s\alpha$ mutants), was isolated from cells metabolically labeled with $^{32}P_i$ and found to have mostly GTP bound. Using the same protocol, the wild-type protein is recovered with only GDP bound to it. It is worth noting, however, that the purified recombinant yeast ARF proteins, unlike their mammalian counterparts, do have quite respectable intrinsic GTPase activities (0.2–$0.5\,min^{-1}$).

No GTPase activating protein (GAP) active on ARF proteins has been purified or characterized to date. The mechanism of activation or deactivation of ARF in cells remains purely speculative at this time. A role for nucleoside diphosphate kinase as an activator of ARF (RANDAZZO et al. 1991) has recently been shown to be completely artifactual (RANDAZZO et al. 1992).

V. The Role of Myristoylation

In contrast to the members of the other families of small GTPases which are modified at the C terminus with an isoprenyl group, most of the members of the ARF family are myristoylated at the amino terminus. This has been demonstrated directly for purified bovine brain ARF (KAHN et al. 1988). The presence of the consensus sequences for N-terminal myristoylation (TOWLER et al. 1987) on cDNA-derived ARF sequences suggests that this is likely to be true for most members of the ARF family. A more direct test of the ability of specific ARF proteins to be N-myristoylated can be performed by coexpression of an ARF or ARF-like protein and an N-myristoyltransferase in bacteria (DURONIO et al. 1990). This has been done for each of the ARF proteins (hARF1, hARF3, hARF4, yARF1, and yARF2) and each was found to incorporate [^{3}H]myristic acid.

The importance of N-myristoylation to proper ARF function can be demonstrated in yeast, in which the [G2A]yARF1 mutant (which cannot be myristoylated) was found to be completely inactive (ZHANG et al. 1992). When this mutant was purified from bacteria it was found to be indistinguishable from the wild-type protein with regard to guanine nucleotide binding, GTPase activity, and ARF cofactor activity (ZHANG et al. 1992). These results point to the importance of the proper processing of ARF for in vivo activity and also illustrate the limitation of the cofactor assay in defining the physiological role of ARF.

VI. Binding of ARF to Lipid Bilayers

The binding of an activating guanine nucleotide, GTP or GTPγS, to ARF is associated with a change in the intrinsic, tryptophan fluorescence (KAHN and GILMAN 1986), generally interpreted as a change in the conformation of the protein. This GTP-liganded conformation has also been shown to be the active species of ARF (KAHN and GILMAN 1986) and results in a dramatically enhanced affinity for phospholipid vesicles (KAHN and RULKA 1992) and biological membranes (REGAZZI et al. 1991). While the inactive, GDP-bound form of ARF migrates on sucrose density gradients as a monomer consistent with a protein of 21 kDa, the GTP- or GTPγS-bound protein migrates near the bottom of the gradient at the same location as the phospholipid vesicles (KAHN and RULKA 1992). The association of activated ARF with vesicles requires the myristoylated form of ARF, as the recombinant nonmyristoylated protein migrates on gradients as a monomer in both the GDP- or the GTP-bound forms (KAHN and RULKA 1992). As N-myristoylation in cells is currently not thought to be a regulatory process, it is likely that all the ARF in cells is properly myristoylated and the controlled association of ARF with intracellular membranes is regulated by the binding and hydrolysis of GTP.

VII. Evidence that ARF is Required at Several Steps in the Secretory and Endocytic Pathways

Guanine nucleotide binding proteins have been implicated as regulators of specific steps in the exocytic and endocytic pathways primarily on the basis of in vitro assays of specific steps in these processes that are inhibited by GTPγS (MELANCON et al. 1987; BECKERS and BALCH 1989). It is likely that ARF proteins are a dominant factor in the sensitivity of these assays to GTPγS. The best documented evidence supporting this assertion is found in TAYLOR et al. (1992), in which the factor responsible for the inhibition of an intra-Golgi transport assay by GTPγS was purified and shown to be an ARF protein. Addition of ARF to the endosome fusion (LENHARD et al. 1992) or ER-to-Golgi transport assays (BALCH et al. 1992) is also inhibitory in the presence of GTPγS. Thus, it seems that too much active ARF inhibits each of these assays. These results may be related to the toxicity observed in yeast when ARF is overexpressed.

While too much active ARF is clearly detrimental, it is not clear whether or not ARF is a *required* factor in any of these assays. It has proven technically difficult to deplete cytosol and membranes of ARF as it is so abundant, found in every eukaryotic cell, and present in both soluble and particulate fractions. However, a critical, functional domain of ARF has recently been identified (KAHN et al. 1992b) and synthetic peptides corresponding to this region of the protein have proven to be potent inhibitors of the ARF cofactor assay (Kahn et al. 1992b), intra-Golgi transport (KAHN et al. 1992b), Golgi-derived coated vesicle formation (KAHN et al. 1992b), ER-to-Golgi transport (BALCH et al. 1992), endosome–endosome fusion (LENHARD et al. 1992), and ARF and coat protein binding to Golgi membranes (DONALDSON et al. 1992). Thus, one or more ARF proteins may be required in a broad range of intracellular membrane budding or fusion processes. One or more of these in vitro assays are likely to prove useful in the further elucidation of the role of ARF in these pathways and in identifying other components in the ARF pathway.

E. Use of ARF Antibodies

I. Abundance of Different ARF Proteins is Quite Variable

Results of immunoblotting studies indicate that, in mammalian cells, ARF1 and/or ARF3 are by far the most abundant species of ARF expressed (there are currently no antibodies available capable of distinguishing between these two gene products as they are 96% identical). A comparison of the indirect immunofluorescence pattern seen in several cell lines (e.g., NIH 3T3, NRK, COS-7) between staining with the antipeptide antibody R-5 (STEARNS et al. 1990b) – which is known to recognize (at least) ARF1p, ARF3p, ARF4p, ARL1p, and even the two yeast ARF proteins – and the monoclonal 1D9 – which recognizes only mammalian ARF1 and ARF3 – reveals no discernible differences. We have been unable to detect the presence of any human ARF4 protein by immunoblotting of ten different mouse tissues using an antipeptide rabbit serum which can readily detect as little as 1 ng of purified recombinant hARF4p (ZHANG et al. 1992). Thus, while ARF1p and ARF3p represent as much as 1% of total cell protein in some cells and tissues, ARF4p appears to be expressed at less than 0.003% of cell protein in mouse tissues.

ARF proteins appear to be very well conserved among mammals, e.g., bovine ARF1p and human ARF1p are 100% identical. However, the extent to which the specificity of each of these antisera remains constant between species remains unproven. As described above, the yeast ARF1p is expressed at about tenfold higher levels than yeast ARF2p (STEARNS et al. 1990a).

II. Localization of ARF Proteins in Animal Cells

The subcellular distribution of ARF proteins has been determined by indirect immunofluorescence and more limited immunoelectron microscopy studies. ARF (ARF1 or ARF3) has been localized to the Golgi complex at the level of light microscopy and to the cytosolic surface of predominantly *cis* Golgi elements by electron microscopy (STEARNS et al. 1990b). The same antibodies have been used to demonstrate the presence of ARF proteins on Golgi-derived or non-clathrin-coated vesicles (SERAFINI et al. 1991). While the Golgi complex is the most easily identifiable structure observed after staining of cells with ARF antibodies, the ARF is clearly not limited to that structure. ARF is also seen as a diffuse, probably cytosolic protein, and as punctate staining essentially throughout the cytosol. It is not clear what compartment this represents, but the ER or the intermediate compartment are good candidates. These results are consistent with the reversible association of ARF with intracellular membranes which appears to be an important feature of its function in vivo.

Thus, the localization of ARF in mammalian cells to specific compartments along the secretory pathway is entirely consistent with observations made in yeast which indicate an essential role for ARF proteins in the protein secretory pathway and specifically at the ER and Golgi compartment. As newer antibodies are developed with greater specificity for specific gene products, it will be interesting to see if different ARF or ARF-like proteins are targeted to different organelles. Such data may help unravel the question as to why there are so many distinct, yet structurally conserved, ARF proteins.

F. ARF as a Regulator of Coat Protein Binding to Membranes

I. Brefeldin A Causes Rapid Release of ARF from Golgi Stacks

The fungal metabolite Brefeldin A (BFA) is a remarkable compound which, when added to animal cells in culture, results in a block in protein secretion (MISUMI et al. 1986; PELHAM 1991). The association of both ARF and the 110-kDa coat protein, β-COP, with the Golgi complex is reversed within minutes of addition of BFA (DONALDSON et al. 1991). This reversible association of ARF with Golgi membranes was shown in vitro to be an energy-requiring, GTPγS-stimulated, and BFA-reversible event (DONALDSON et al. 1991). The possible involvement of a heterotrimeric G-protein in the regulated binding of ARF to Golgi membranes (DONALDSON et al. 1991) is an intriguing but unproven hypothesis, particularly in light of the data from the ARF cofactor assay suggesting a possible direct ARF–G-protein interaction.

II. An in Vitro Assay for ARF as Regulator of Coat Protein Binding

The binding of ARF and β-COP is an ordered process, with ARF binding required prior to that of β-COP (DONALDSON et al. 1992). This conclusion is based on four observations: in the presence of GTPγS, (1) β-COP in a high molecular weight fraction from cytosol depleted of ARF does not bind to Golgi membranes, (2) purified myristoylated ARF will bind to Golgi membranes in the absence of cytosol, (3) addition of recombinant ARF to the β-COP fraction allows binding of both ARF and β-COP, and (4) a peptide inhibitor of ARF activities causes the release of both ARF and β-COP from Golgi membranes with the same dose dependence. The ability of ARF to serve as a regulator of β-COP, and presumably the entire coatomer, is largely dependent on its proper myristoylation and is reversible by previous addition of Brefeldin A. Thus, the regulation of coat protein binding to Golgi membranes may prove to be a useful in vitro assay of ARF function with far greater physiological relevance than that of the cofactor assay.

References

Alsip GR, Konkel DA (1986) A processed chicken pseudogene (CPS1) related to the ras oncogene superfamily. Nucleic Acids Res 14:2123–2138

Balch WE, Kahn RA, Schwaninger R (1992) ADP-ribosylation factor (ARF) is required for vesicular trafficking between the endoplasmic reticulum (ER) and the cis Golgi compartment. J Biol Chem 267:13053–13061

Beckers CJM, Balch WE (1989) Calcium and GTP: essential components in vesicular trafficking between the endoplasmic reticulum and golgi apparatus. J Cell Biol 108:1245–1256

Bobak DA, Nightingale MS, Murtagh JJ, Price SR, Moss J, Vaughan M (1989) Molecular cloning, characterization, and expression of human ADP-ribosylation factors: two guanine nucleotide-dependent activators of cholera toxin. Proc Natl Acad Sci USA 86:6101–6105

Botstein D, Segev N, Stearns T, Hoyt MA, Holden J, Kahn RA (1988) Diverse biological functions of small GTP-binding proteins in yeast. Cold Spring Harb Symp Quant Biol 53:629–636

Cassel D, Selinger Z (1979) Mechanism of adenylate cyclase activation by cholera toxin: inhibition of GTP hydrolysis at the regulatory site. Proc Natl Acad Sci USA 74:3307–3311

Cavenagh MM, Kahn RA (1993) Requirement for ARF in the sporulation process of S. cerevisiae (in preparation)

Donaldson JG, Kahn RA, Lippincott-Schwartz J, Klausner RD (1991) Binding of the BFA-sensitive proteins ARF and B-COP to Golgi membranes: involvement of a trimeric G protein. Science 254:1197–1199

Donaldson JG, Cassel D, Kahn RA, Klausner RD (1992) ADP-ribosylation factor, a small GTP binding protein, is required for binding of the coatomer protein β-COP to Golgi membranes. Proc Natl Acad Sci USA 89:6408–6412

Duronio RJ, Jackson-Machelski E, Heuckeroth RO, Olins PO, Devine CS, Yonemoto W, Slice LW, Taylor SS, Gordon JI (1990) Protein N-myristoylation in Escherichia coli: reconstitution of a eukaryotic protein modification in bacteria. Proc Natl Acad Sci USA 87:1506–1510

Kahn RA (1989) The ADP-ribosylation factor of adenylate cyclase: a 21 kDa GTP-binding protein. In: Birnbaumer L, Iyengar R (eds) G-Proteins. Academic, Orlando

Kahn RA (1991) Quantitation and purification of ADP-ribosylation factor. In: Johnson RA, Corbin JD (eds) Methods in enzymology, vol 195. Academic, Orlando

Kahn RA, Gilman AG (1984a) Effects of cholera toxin-dependent ADP-ribosylation on the regulatory component of adenylate cyclase. J Biol Chem 259:6235–6240

Kahn RA, Gilman AG (1984b) Purification of a protein cofactor required for ADP-ribosylation of the stimulatory regulatory component of adenylate cyclase by cholera toxin. J Biol Chem 259:6228–6234

Kahn RA, Gilman AG (1986) The protein cofactor necessary for ADP-ribosylation of Gs by cholera toxin is itself a GTP-binding protein. J Biol Chem 261:7906–7911

Kahn RA, Rulka C (1992) Association of a GTP-binding protein (ARF) with phospholipid vesicles is controlled by guanine nucleotide binding. J Biol Chem (submitted)

Kahn RA, Goddard C, Newkirk M (1988) Chemical and immunological characterization of the 21 kDa ADP-ribosylation factor (ARF) of adenylate cyclase. J Biol Chem 263:8282–8287

Kahn RA, Kern FG, Clark J, Gelmann EP, Rulka C (1991) Human ADP-ribosylation factors: a functionally conserved family of GTP-binding proteins. J Biol Chem 266:2606–2614

Kahn RA, Der CJ, Bokoch GM (1992a) The ras superfamily of GTP-binding proteins: guidelines on nomenclature. FASEB J 6:2512–2513

Kahn RA, Randazzo PA, Serafini T, Weiss O, Rulka C, Clark J, Amherdt M, Roller PP, Orci L, Rothman JE (1992b) The amino terminus of ADP ribosylation factor (ARF) is a critical determinant of ARF activities and is a potent and specific inhibitor of protein transport. J Biol Chem 267:13039–13046

Lenhard JM, Kahn RA, Stahl PD (1992) Evidence for ADP-ribosylation factor (ARF) as a regulator of in vitro endosome–endosome fusion. J Biol Chem 267:13047–13052

Melancon P, Glick BS, Malhotra V, Weidman PJ, Serafina T, Gleason ML, Orci L, Rothman JE (1987) Involvement of GTP-binding "G" proteins in transport through the Golgi stack. Cell 51:1053–1062

Misumi Y, Miki K, Takatsuki A, Tamura G, Ikehara Y (1986) Novel blockade by brefeldin A of intracellular transport of secretory proteins in cultured rat hepatocytes. J Biol Chem 261:11398–11403

Monaco L, Murtagh JJ, Newman KB, Tsai S, Moss J, Vaughan M (1990) Selective amplification of an mRNA and related pseudogene for a human ADP-ribosylation factor, a guanine nucleotide-dependent protein activator of cholera toxin. Proc Natl Acad Sci USA 87:2206–2210

Nakano A, Muramatsu M (1989) A novel GTP-binding protein, Sarlp, is involved in transport from the endoplasmic reticulum to the Golgi apparatus. J Cell Biol 109:2677–2691

Pelham HRB (1991) Multiple targets for Brefeldin A. Cell 67:449–451

Peng Z, Calver I, Clark J, Helman L, Kahn RA, Kung H (1989) Molecular cloning, sequence analysis and mRNA expression of human ADP-ribosylation factor. Biofactors 2:45–49

Price SR, Nightingale M, Tsai S, Williamson KC, Adamik R, Chen H, Moss J, Vaughan M (1988) Guanine nucleotide-binding proteins that enhance choleragen ADP-ribosyltransferase activity: nucleotide and deduced amino acid sequence of an ADP-ribosylation factor cDNA. Proc Natl Acad Sci USA 85:5488–5491

Randazzo PA, Northup JK, Kahn RA (1991) Activation of a small GTP binding protein by nucleoside diphosphate kinase. Science 254:850–853

Randazzo PA, Northup JK, Kahn RA (1992) Regulatory GTP binding proteins (ARF, Gt, and RAS) are not activated directly by nucleoside diphosphate kinase. J Biol Chem 267:18182–18189

Regazzi R, Ullrich S, Kahn RA, Wollheim CB (1991) Redistribution of ADP-ribosylation factor during stimulation of permeabilized cells with GTP analogues. Biochem J 275:639–644

Schleifer LS, Kahn RA, Hanski E, Northup JK, Sternweis PC, Gilman AG (1982) Requirements for cholera toxin-dependent ADP-ribosylation of the purified regulatory component of adenylate cyclase. J Biol Chem 257:20–23

Serafini T, Orci L, Amherdt M, Brunner M, Kahn RA, Rothman JE (1991) ADP-ribosylation factor is a subunit of the coat of Golgi-derived COP-coated vesicles: a novel role for a GTP binding protein. Cell 67:239–254
Sewell JL, Kahn RA (1988) Sequences of the bovine and yeast ADP-ribosylation factor and comparison to other GTP-binding proteins. Proc Natl Acad Sci USA 85:4620–4624
Stearns T, Hoyt MA, Botstein D, Kahn RA (1990a) ADP-ribosylation factor is an essential protein in yeast and is encoded by two genes. Mol Cell Biol 10:6690–6699
Stearns T, Willingham MC, Botstein D, Kahn RA (1990b) The ADP-ribosylation factor is functionally and physically associated with the Golgi complex. Proc Natl Acad Sci USA 87:1238–1242
Tamkun JW, Kahn RA, Kissinger M, Brizuela BJ, Rulka C, Scott MP, Kennison JA (1991) The arf-like gene encodes an essential GTP-binding protein in Drosophila. Proc Natl Acad Sci USA 88:3120–3124
Taylor TC, Kahn RA, Melancon P (1992) Two distinct members of the ADP-ribosylation factor family of GTP binding proteins regulate cell free intra-Golgi transport. Cell 70:69–79
Towler DA, Eubanks SR, Towery DS, Adams SP, Glaser L (1987) Amino-terminal processing of proteins by N-myristoylation. J Biol Chem 262:1030–1036
Tsuchiya M, Price SR, Tsai S-C, Moss J, Vaughan M (1991) Molecular identification of ADP-ribosylation factor mRNAs and their expression in mammalian cells. J Biol Chem 266:2772–2777
Weiss O, Holden J, Rulka C, Kahn RA (1989) Nucleotide binding and cofactor activities of purified bovine brain and bacterially expressed ADP-ribosylation factor (ARF). J Biol Chem 264:21066–21072
Zhang C-J, Clark J, Rulka C, Kahn RA (1993) Mutational analysis of yeast ARF1 (in preparation)

CHAPTER 35

Molecular Characterization of ADP-Ribosylation Factors

J. Moss and M. Vaughan

A. Introduction

Cholera toxin (CT) is largely responsible for the pathogenesis of cholera, a devastating diarrheal disease characterized by marked abnormalities in fluid and electrolyte flux (Carpenter 1980; Finkelstein 1973; Kelly 1986). The toxin, a product of *Vibrio cholerae*, is an oligomeric protein composed of one A subunit (CTA) of about 29 kDa and five B subunits of about 11.8 kDa each (Gill 1976, 1977). The five B subunits are responsible for binding the toxin to the cell surface, via interaction with the monosialoganglioside GM_1 (for review, see Moss and Vaughan 1988). The A subunit is synthesized as a single polypeptide chain and nicked by a *V. cholerae* protease to create two polypeptides of about 22 kDa (CTA1) and about 5 kDa (CTA2), linked by a single disulfide bond (Gill and Rappaport 1979). Reduction of nicked A subunit generates free CTA1, which exhibits ADP-ribosyltransferase activity (Moss et al. 1979).

The primary cellular substrate of CTA1 is $G_{s\alpha}$, the α subunit of the heterotrimeric guanine nucleotide-binding protein G_s (Johnson et al. 1978; Gill and Meren 1978; Cassel and Pfeuffer 1978; Northup et al. 1980). G_s is a regulatory complex composed of α, β and γ subunits responsible for the coupling of variety of receptors (e.g., β-adrenergic, glucagon) to their intracellular targets, which include the adenylyl cyclase catalytic unit and certain Ca^{2+} channels (Casey and Gilman 1988; Birnbaumer et al. 1990). $G_{s\alpha}$ is activated when dissociated from $\beta\gamma$ and containing bound GTP; inactivation results from hydrolysis of GTP to GDP by a GTPase activity intrinsic to $G_{s\alpha}$ (Casey and Gilman 1988; Birnbrumer et al. 1990; Moss and Vaughan 1988). CT-catalyzed ADP-ribosylation causes activation of $G_{s\alpha}$ by inhibiting the intrinsic GTPase activity, thereby preserving the GTP-bound state (Cassel and Selinger 1977; Navon and Fung 1984) and favoring the dissociation of α from $\beta\gamma$ (Kahn and Gilman 1984b). By ADP-ribosylation of $G_{s\alpha}$ and possibly other substrates, CT increases adenylyl cyclase activity and cyclic adenosine monophosphate (AMP) content, and exerts effects on other pathways, e.g., ion flux, eicosanoid synthesis (Peterson and Ochoa 1989).

It was observed that CT-catalyzed ADP-ribosylation and activation of adenylyl cyclase were enhanced by cellular factors, subsequently shown to be GTP, various phospholipids (detergents), and several small proteins, identified in a variety of tissues and species (Enomoto and Gill 1980; Bobak et al. 1990b; Schmidt et al. 1987; Pinkett and Anderson 1982; LeVine and Cuatrecasas 1981; Gill and Meren 1983; Enomoto and Asakawa 1982; Schleifer et al. 1982; Gill and Coburn 1987). A stimulatory protein of about 20kDa was purified from rabbit liver membranes and termed ADP-ribosylation factor (ARF) for its ability to stimulate the CT-catalyzed reaction (Kahn and Gilman 1984a). It was subsequently shown that ARF, isolated from bovine brain membranes, was a guanine nucleotide-binding protein (Kahn and Gilman 1986). ARFs purified from bovine brain cytosol had characteristics similar to those of the membrane species (Tsai et al. 1977, 1988).

B. Activation of Cholera Toxin by ADP-Ribosylation Factors

I. Mechanism of Activation of Cholera Toxin by ADP-Ribosylation Factors

To define the mechanism of action of ARFs, advantage was taken of the fact that CT, in addition to catalyzing the ADP-ribosylation of $G_{s\alpha}$, also ADP-ribosylates simple guanidine compounds (e.g., arginine, agmatine; Moss and Vaughan 1977), proteins unrelated to $G_{s\alpha}$, presumably secondary to its ability to modify an accessible arginine (Moss and Vaughan 1978), and its own CTA1 catalytic unit (Trepel et al. 1977; Moss et al. 1980). In the absence of an acceptor, the toxin hydrolyzes nicotinamide-adenine dinucleotide (NAD) to ADP-ribose, nicotinamide and H^+ (Moss et al. 1976). Bovine brain membrane ARF, mARF, and soluble ARF, sARF II, stimulated CT-catalyzed ADP-ribosylation of $G_{s\alpha}$, its CTA1 catalytic unit, and the simple guanidine compound, agmatine (Tsai et al. 1987, 1988; Bobak et al. 1990a,b; Noda et al. 1989, 1990). Stimulation was dependent on the presence of GTP or a nonhydrolyzable analog (e.g., GTPγS, Gpp(NH)p); GDP or GDPβS or adenine nucleotides were inactive (Tsai et al. 1987, 1988). Activation by ARF was associated with a decrease in K_m and relatively little change in V_{max}. sARF II enhanced the activity of a purified CTA1 fragment that was biotinylated on a cysteine residue near the carboxy terminus, indicating that the locus of ARF action was the toxin catalytic unit and that the carboxy terminus was not critical to ARF action (Noda et al. 1989).

Activation of CT by ARF plus GTP was modulated by phospholipids and detergents. In the presence of sodium dodecyl sulfate (SDS), ARF caused a substantial increase in maximal velocity and a significantly greater decrease in K_m than was observed in its absence (Noda et al. 1990). With or

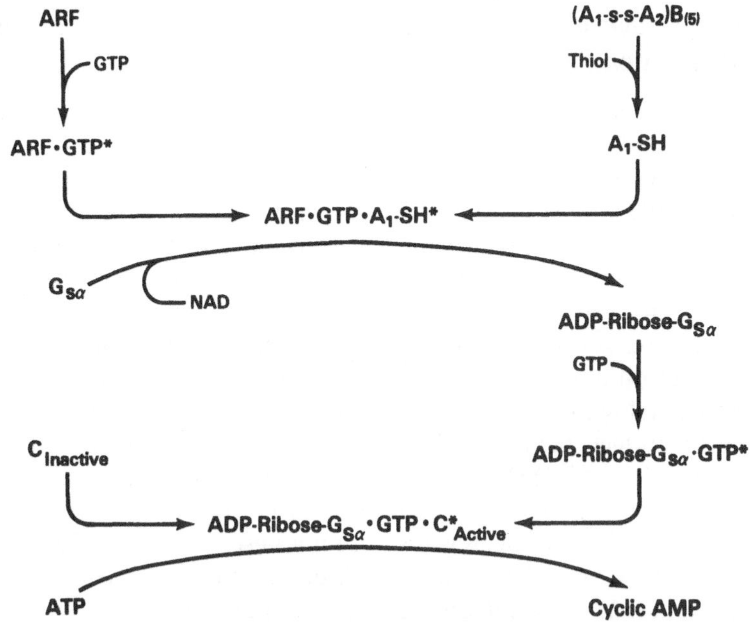

Fig. 1. Participation of a guanine nucleotide-binding protein cascade in the activation of adenylyl cyclase by cholera toxin. Cholera toxin is depicted schematically as an oligomer consisting of the A subunit, composed of protein fragments A1 and A2 linked through a single disulfide bond, joined noncovalently to the B pentamer. Activation of toxin by thiol results in the release of CTA1 from CTA2 and the B subunits. ADP-ribosylation factor (*ARF*), activated by bound GTP, in turn stimulates CTA1, enhancing its ability to ADP-ribosylate $G_{s\alpha}$, the stimulatory guanine nucleotide-binding protein of the adenylyl cyclase system. With GTP bound, ADP-ribosyl-$G_{s\alpha}$ activates the catalytic unit of adenylyl cyclase, increasing the conversion of ATP to cyclic AMP. Reprinted from TSAI et al. 1988

without SDS, the K_a for GTP was in the micromolar range. In contrast, activation of CT by ARF in the presence of dimyristoylphosphatidylcholine (DMPC)/cholate was associated with a nanomolar K_a for GTP. Consistent with these data, high affinity binding of guanine nucleotide to ARF was observed in the presence of DMPC/cholate but not SDS.

Since both $G_{s\alpha}$-dependent and $G_{s\alpha}$-independent reactions were enhanced by ARF, it appears that the stimulatory effect resulted from a direct allosteric activation of the toxin catalytic unit (TSAI et al. 1988; NODA et al. 1990) and not from interaction with $G_{s\alpha}$, as had been proposed (KAHN and GILMAN 1984a). Activation of adenylyl cyclase by CT appears in the in vitro experiments to operate through a guanine nucleotide-binding protein cascade (Fig. 1; TSAI et al. 1988). ARF with GTP bound activates CTA1, which in turn catalyzes the ADP-ribosylation of $G_{s\alpha}$. ADP-ribosyl-$G_{s\alpha}$ with bound GTP enhances the activity of the adenylyl cyclase catalytic unit.

II. Guanine Nucleotide-Dependent Association of Cholera Toxin with ADP-Ribosylation Factors

Based on its action, one would predict that CT, under appropriate conditions, would associate with ARF. Indeed, in the presence of GTPγS, but not GDPβS, sARF II formed a high molecular weight aggregate with CT (Tsai et al. 1991a). The GTPγS-dependent formation of the aggregate required the presence of submicellar concentrations of SDS (0.003%) and was not observed in the presence of DMPC/cholate (Tsai et al. 1991a). Only a small percentage of ARF was found in the complex, and the bulk of the ARF behaved on gel permeation chromatography as a monomeric species. The aggregate differed in substrate specificity from monomeric ARF and toxin, most notably in the enhanced auto-ADP-ribosylation of CTA1 (Tsai et al. 1991a). In fact, the markedly increased GTPγS-dependent auto-ADP-ribosylation observed in SDS is largely due to the formation of toxin–ARF aggregates (Tsai et al. 1987, 1988, 1991a). In the presence of GTPγS and DMPC/cholate, CT–ARF aggregates were not observed; an ARF aggregate was present, however, that a had a different effect on toxin substrate specificity from that of monomeric ARF (Tsai et al. 1991a).

III. Activation of *Escherichia coli* Heat-Labile Enterotoxin by ADP-Ribosylation Factor

Escherichia coli heat-labile enterotoxins (LT) appear to be responsible to a large extent for "traveller's diarrhea," a widespread disease similar to cholera, but less devastating in its clinical manifestations (Field 1979). These enterotoxins share numerous structural, immunological, and functional similarities with CT (Moss and Vaughan 1988). In addition, there are two other more different forms of LTs, known as LT-IIa and LT-IIb, that differ from LT-I, but have a mechanism of action similar to that of CT and LT (Pickett et al. 1986, 1987; Pickett and Holmes 1990; Holmes et al. 1986; Guth et al. 1986a,b). All of these toxins are composed of an A subunit associated with a B oligomer. Based on deduced amino acid sequences, CTA is similar in structure to LT-IhA, whereas the A subunits of LT-IIa and LT-IIb are different from each other and from CTA and LT-IhA. CTA is usually isolated as a proteolyzed species, requiring only reduction to display its latent ADP-ribosyltransferase activity (Moss et al. 1976). In LT-IhA, the A subunit is present primarily in an unnicked form. Activation required tryptic proteolysis and reduction by thiol to release the LT equivalents of CTA1 and CTA2 (Moss et al. 1981).

The structurally different A subunits of LT-Ih, LT-IIa, and LT-IIb possessed NAD–agmatine ADP-ribosyltransferase activity (Chang et al. 1987; Gill and Richardson 1980; Moss and Richardson 1978; Lee et al. 1991). In all instances, the activities were increased by ARF in the presence of GTP (Lee et al. 1991). As observed with CT, stimulation by ARF plus

GTP was further enhanced by SDS. The NAD–agmatine ADP-ribosyltransferase activities of LT-IIa and LT-IIb were less than 1% of those of CT and LT-Ih, whether assayed with or without ARF, GTP, and SDS (LEE et al. 1991). In contrast, the abilities of LT-IIa and LT-IIb to ADP-ribosylate $G_{s\alpha}$ were only slightly less than those of CT and LT-Ih. These data are consistent with the conclusion that LT-IIa and LT-IIb have a more restricted substrate specificity than do CT and LT-Ih. All four toxins appear to use the natural substrate $G_{s\alpha}$ with similar proficiency, whereas the model substrate, a simple guanidine compound, was used with considerably greater efficiency by CT and LT-Ih (LEE et al. 1991).

C. Structure of ADP-Ribosylation Factors

I. Deduced Amino Acid Sequences and Gene Structure of ADP-Ribosylation Factors

To determine the structures of ARFs, ARF-specific cDNA and oligonucleotide probes were used to isolate DNA from cDNA and genomic libraries and to generate cDNAs by polymerase chain reaction-based techniques (PRICE et al. 1988; SEWELL and KAHN 1988; BOBAK et al. 1989; MONACO et al. 1990; TSUCHIYA et al. 1991; TSAI et al. 1991c; LEE et al. 1992; MURTAGH et al. 1992a,b; SERVENTI et al. 1993; PENG et al. 1989; HAUN et al. 1992). At least six mammalian ARF genes have been identified. These fall into three classes based on size and deduced amino acid sequence of proteins, phylogenetic analysis, and gene structure (Fig. 2; Table 1; TSUCHIYA et al. 1991). ARFs 1–3 (class I) have 181 amino acids and differ from each other primarily near the amino and carboxy termini. ARFs 4 and 5 (class II) have 180 amino acids and differ from class I near the amino terminus and in the carboxy half of the protein. Class II ARFs are more similar to each other than they are to class I ARFs (Table 1). At present, ARF 6 is the only known member of class III. It has 175 amino acids and differs from other ARFs essentially throughout the coding region, although differences are most marked in the carboxy half of the protein (Fig. 2; Table 1; TSUCHIYA et al. 1991). All six ARF cDNAs when expressed in *E. coli* resulted in synthesis of proteins that stimulated the ADP-ribosyltransferase activity of CT (WELSH et al. 1991, 1992; WEISS et al. 1989; KAHN et al. 1991). All ARFs contain the consensus sequences of amino acids involved in guanine nucleotide-binding and GTP hydrolysis (phosphate-binding; PRICE et al. 1990).

The categorization of mammalian ARFs into three classes is borne out in the gene structures (TSAI et al. 1991c; SERVENTI et al. 1993; HAUN et al. 1992; LEE et al. 1992). In the human ARF 1 and 3 genes and the bovine ARF 2 gene, the positions of intron insertions in the coding region are identical. The intron insertion sites in the human ARF 4 and 5 genes are

```
                           10         20         30         40         50         60         70
hARF1          1   MGnifanLFk gLFGKKEMRI LMVGLDAAGK TTILYKLKLG EiVTTIPTIG FNVETVEYKN IsFTVWDVGG
bARF2          1   MGnvfekLFk sLFGKKEMRI LMVGLDAAGK TTILYKLKLG EiVTTIPTIG FNVETVEYKN IsFTVWDVGG
hARF3          1   MGnifgnLlk sLiGKKEMRI LMVGLDAAGK TTILYKLKLG EiVTTIPTIG FNVETVEYKN IsFTVWDVGG
hARF4          1   MGltiSsLFs rLFGKKqMRI LMVGLDAAGK TTILYKLKLG EiVTTIPTIG FNVETVEYKN IcFTVWDVGG
hARF5          1   MGltvSaLFs riFGKKqMRI LMVGLDAAGK TTILYKLKLG EiVTTIPTIG FNVETVEYKN IcFTVWDVGG
hARF6          1   MGkvlSk      iFGnKEMRI LMlGLDAAGK TTILYKLKLG qsVTTIPTvG FNVETVtYKN vkFnVWDVGG
yarf1          1   MGlfaSkLFs nLFGnKEMRI LMVGLDgAGK TTvLYKLKLG EviTTIPTIG FNVETVqYKN IsFTVWDVGG
yarf2          1   MGlyaSkLFs nLFGnKEMRI LMVGLDgAGK TTvLYKLKLG EviTTIPTIG FNVETVqYKN IsFTVWDVGG
gARF           1   MGqgaSkiFg kLFsKKEvRI LMVGLDAAGK TTILYKLmLG EvVTTvPTIG FNVETVEYKN InFTVWDVGG
CONSENSUS          MG                 K   RI LM GLD AGK TT LYKL LG    TT PT G FNVETV YKN   F VWDVGG

                           80         90        100        110        120        130          1
hARF1         71   QDkIRPLWRH YfQNTQGLIF VVDSNDRE  Rv nEAReELmRM LaEDELRDAV LLVFANKQDL PnAMnaaE
bARF2         71   QDkIRPLWRH YfQNTQGLIF VVDSNDRE  Rv nEAReELtRM LaEDELRDAV LLVFvNKQDL PnAMnaaE
hARF3         71   QDkIRPLWRH YfQNTQGLIF VVDSNDRE  Rv nEAReELmRM LaEDELRDAV LLVFANKQDL PnAMnaaE
hARF4         71   QDrIRPLWkH YfQNTQGLIF VVDSNDRE  Ri qEvadELqkM LlvDELRDAV LLlFANKQDL PnAMaısE
hARF5         71   QDkIRPLWRH YfQNTQGLIF VVDSNDRE  Rv qEsadELqkM LqEDELRDAV LLVFANKQDm PnAMpvsE
hARF6         67   QDkIRPLWRH YytgTQGLIF VVDcaDRd  Ri dEARqELhRi ındrEmRDAi iLiFANKQDL PdAMkphE
yarf1         71   QDrIRsLWRH YyrNTeGvIF VVDSNDRs  Ri gEARevmqRM LnEDELRnAa wLVFANKQDL PeAMsaaE
yarf2         71   QDrIRsLWRH YyrNTeGvIF ViDSNDRs  Ri gEARevmqRM LnEDELRnAV wLVFANKQDL PeAMsaaE
gARF          71   QDsIRPLWRH YyQNTdaLIy ViDSaDlEpkRi edARnELhtl LgEDELRDAa LLVFANKQDL PkAMsttd
CONSENSUS          QD IR LW H Y   T   I  V D  D    R                     E R A   L F NKQD  P AM

                          150        160        170              180
hARF1        141   dKLGLhSLRh RnWYIQATCA TSGdGLYEGL DWL  SNqL   rNqk           181
bARF2        141   dKLGLhSLRq RnWYIQATCA TSGdGLYEGL DWL  SNqL   KNqk           181
hARF3        141   dKLGLhSLRh RnWYIQATCA TSGdGLYEGL DWL  aNqL   KNkk           181
hARF4        141   dKLGLqSLRn RtWYvQATCA TqGtGLYEGL DWL  SNeLs  Kr             180
hARF5        141   dKLGLqhLRs RtWYvQATCA TqGtGLYdGL DWL  SheLs  Kr             180
hARF6        137   eKLGLtriRd RnWYvQpsCA TSGdGLYEGL tWL tSNy    Ks             175
yarf1        141   eKLGLhSiRn RpWfIQATCA TSGeGLYEGL eWL  SNsL   KNst           181
yarf2        141   eKLGLhSiRn RpWfIQsTCA TSGeGLYEGL eWL  SNnL   KNqs           181
gARF         143   erLGLqeLkk RdWYIQpTCA rSGdGLYqGL DWL  SdyıfdkKNkkkgkkr     191
CONSENSUS            LGL      R W  Q  CA   G GLY GL  WL
```

Fig. 2. Comparison of deduced amino acid sequences of mammalian, yeast, and *Giardia* ARFs. Deduced amino acid sequences of bovine ARF 2 (bARF2), human ARFs 3–6 (hARFs 3–6), yeast ARF 1 (yARF1), and *Giardia* ARF (gARF) are compared to that of human ARF 1 (hARF1). hARF1 and 3–6, human ARFs 1 and 3–6; bARF2, bovine ARF 2; yARF, yeast ARF; gARF, *Giardia* ARF. Sources of sequences are: hARF1, Bobak et al. 1989; bARF, Price et al. 1988; hARF3, Bobak et al. 1989; hARF4, Monaco et al. 1990; hARF5, Tsuchiya et al. 1991; hARF6 Tsuchiya et al. 1991; yARF, Sewell and Kahn 1988; gARF, Murtagh et al. 1992b. Amino acids identical in at least six of the sequences are indicated in capital letters. Data are from Tsuchiya et al. 1991, and Murtagh et al. 1992b

Table 1. Comparison of ARF coding region nucleotide and deduced amino acid sequences

	hARF1	bARF1	bARF2	hARF3	hARF4	hARF5	hARF6	gARF
hARF1	–	100	96	96	80	80	68	70
bARF1	91	–	96	96	80	80	68	70
bARF2	79	80	–	95	80	80	69	70
hARF3	84	84	80	–	79	79	68	69
hARF4	67	68	68	71	–	90	64	69
hARF5	75	73	71	73	77	–	64	69
hARF6	68	69	64	66	60	65	–	63
gARF	65	67	62	64	61	66	62	–

Relatedness of amino acid sequences, given as percent identity, is above the diagonal, relatedness of coding region nucleotide sequences below. Abbreviations and sources are given in the legend to Fig. 2. Data from Murtagh et al. (1992b)

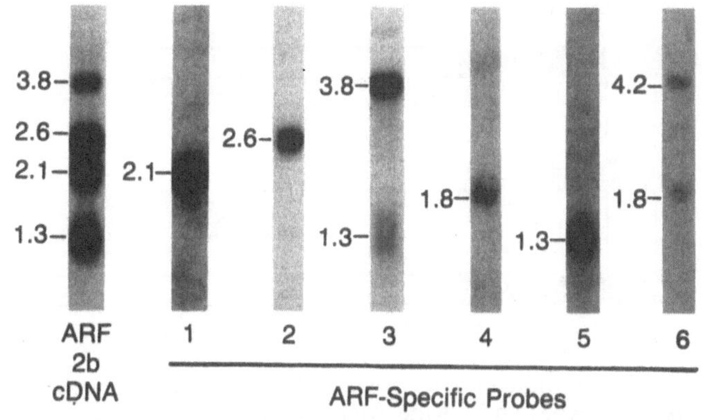

Fig. 3. Hybridization of rat brain poly(A)$^+$ RNA with ARF-specific cDNA and oligonucleotide probes. Rat brain poly(A)$^+$ RNA was hybridized with a bovine ratinal ARF 2 cDNA probe or with oligonucleotide probes specific for the individual ARFs. Reprinted from Moss et al. (1992)

identical, but differ from those in the class I ARF genes (HAUN et al. 1992). These data are consistent with the hypothesis that the ARFs in each class arose through a gene duplication event. In the class I ARFs, the consensus sequences for guanine nucleotide binding are distributed among the coding region exons (Fig. 4). The Gly-X-X-X-Gly-Lys sequence (where X is another amino acid) is encoded in exon 2, and the Asp-Val-Gly-Gly sequence in exon 3. The Asn-Lys-Gln-Asp sequence is divided between exons 4 and 5, with the intron located after the codon for Gln (TSAI et al. 1991c; SERVENTI et al. 1993; LEE et al. 1992).

II. Expression of ADP-Ribosylation Factors in Eukaryotic Species

Expression of the mammalian ARFs was assessed on northern blots using ARF-specific cDNA and oligonucleotide probes (Fig. 3). In poly(A)$^+$ RNA from rat brain, ARF 1 was represented by a 2.1 kb mRNA, ARF 2, 2.6 kb, ARF 3, 3.7 kb and 1.3 kb, ARF 4, 1.8 kb, ARF 5, 1.3 kb, and ARF 6, 4.2 kb and 1.8 kb (MOSS et al. 1992). The two ARF 3 mRNAs result from the use of alternative polyadenylation sites. It appears that in mammalian tissues there is considerable size conservation of ARF mRNAs (TSUCHIYA et al. 1989, 1991; MOSS et al. 1992; MONACO et al. 1990). For example, ARF 3 mRNAs of about 3.7 kb and 1.3 kb were found when probing human RNA (TSAI et al. 1991c).

ARFs are ubiquitous and highly conserved proteins in eukaryotic cells. Human ARF 1 is identical in deduced amino acid sequence to its bovine and rat counterparts (Fig. 2; BOBAK et al. 1989; SEWELL and KAHN 1988; PRICE et al., unpublished observations). Human ARF 6 differs from a chicken pseudogene in one amino acid, a threonine vs. a serine at position 158

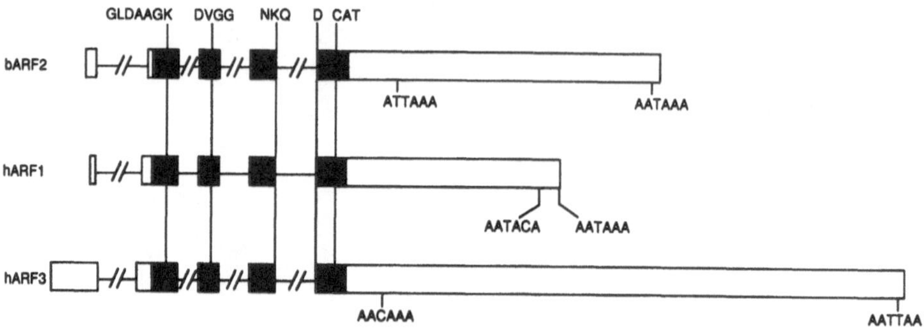

Fig. 4. Structures of human and bovine class I ARF genes. Human ARFs 1 (hARF1) and 3 (hARF3) and bovine ARF 2 (bARF2) genes were described by Lee et al. 1992; Tsai et al. 1991c and Serventi et al. 1993, respectively. Exons are *boxed*, *open boxes* noncoding and *filled boxes* coding. Above the bARF2 exons are consensus sequences of amino acids believed to be involved in guanine nucleotide-binding or GTP hydrolysis (phosphate-binding) (i.e., GLDAAGK, DVGG, NKQ/D, CAT). Nucleotide sequences below the noncoding 3′-untranslated regions are proposed polyadenylation signals in hARF3 and bARF2 and actual polyadenylation signals used in hARF1

(Alsip and Konkel 1986; Tsuchiya et al. 1991). The intestinal parasite *Giardia lamblia*, perhaps the most ancient surviving eukaryote, has ARFs that contain 191 amino acids, with more than 61% amino acid and coding region nucleotide identity to mammalian ARFs (Murtagh et al. 1992b; Table 1). ARFs have also been detected in *Saccharomyces cerevesiae*, *Candida albicans*, and *Drosophila melanogaster* (Sewell and Kahn 1988; Stearns et al. 1990a; Murtagh et al. 1992a; Denich et al. 1992). In these organisms, the conclusion that the genes encode ARFs was confirmed either by the ability of the gene to replace the two deleted ARF genes in a mutant *S. cerevesiae* strain (Stearns et al. 1990a) or by demonstrating that the protein synthesized by *E. coli* expressing an ARF cDNA stimulated the ADP-ribosyltransferase activity of CT (Lee et al., unpublished observations). In addition to at least one ARF gene, *D. melanogaster* contains an ARF-like gene that encodes a protein which, although very similar to ARFs in deduced amino acid sequence, does not have ARF activity (Tamkun et al. 1991).

D. Hormonal and Developmental Regulation of ADP-Ribosylation Factors

ARFs have been detected in numerous tissues and species using anti-ARF protein and anti-ARF peptide polyclonal antibodies (Kahn et al. 1988; Tsai et al. 1991b). Immunoreactive ARFs are present in a variety of eukaryotes. Rabbit polyclonal antibodies prepared against soluble bovine brain sARF II reacted with proteins of about 20 kDa on immunoblots from bovine, rat, frog, and chicken tissues (TSAI et al. 1991b; unpublished data). Using anti-peptide antibodies, immunoreactivity was found in *S. cerevesiae* and *Dictyostelium discoidium*, but not in *E. coli* (KAHN et al. 1988). The immunological data thus support the molecular biological findings of the ubiquitous expression of ARF in eukaryotes.

In rat and bovine brain, most of the ARF activity is soluble (TSAI et al. 1991b). On immunoblots of rat brain-soluble proteins, anti-bovine brain sARF II polyclonal antibodies reacted with two bands of about 20 kDa that are probably the equivalents of soluble ARFs, sARF I and sARF II, isolated from bovine brain (TSAI et al. 1988). Among rat tissues, the content of sARF I and sARF II was considerably higher in brain than in peripheral tissues (Tsai et al. 1988). On the second postnatal day, the amounts of sARF I and sARF II were similar. By the tenth postnatal day and thereafter, the amount of sARF II exceeded that of sARF I (TSAI et al. 1991b). The increase in immunoreactive ARF paralleled an increase in functional ARF, determined by the ability of soluble material, purified by gel permeation chromatography, to stimulate CT-catalyzed ADP-ribosylation of a simple guanidine compound (TSAI et al. 1991b).

Age also affected the levels of ARF mRNA in rat brain (TSAI et al. 1991b). Northern blots of rat brain poly(A)$^+$ RNA were hybridized with cDNA and oligonucleotide probes specific for ARFs 1–6 (Fig. 3). From the second to the 27th postnatal day, mRNAs corresponding to ARF 3 increased, whereas those for ARFs 2 and 4 decreased and those for ARFs 1, 5, and 6 did not change significantly (TSAI et al. 1991b). Based on amino acid sequences of peptides from sARF I and sARF II, they apparently correspond to the ARF 1 and 3 gene products, respectively (TSAI et al. 1991b; unpublished data). Thus, the increase in immunoreactive sARF II observed in rat brain represents an increase in AFR 3 that is associated with an increase in ARF 3 mRNA.

Expression of ARF 1 and ARF 3 is modulated by corticosterone (DUMAN et al. 1990). Chronic administration of the steroid to rats increased the amounts of ARF 1 and ARF 3 mRNA and the quantity of immunoreactive ARF in cerebral cortex (DUMAN et al. 1990). Similarly, bilateral adrenalectomy decreased the ARF 1 and ARF 3 mRNAs (DUMAN et al. 1990).

E. Physiological Roles for ADP-Ribosylation Factors

It is not known whether the different classes of mammalian ARFs, which were defined on the basis of structure, including deduced amino acid sequences, phylogenetic analysis, and gene structures, also have different functions. There is evidence that ARFs from at least two classes may be involved in the vesicular transport of proteins in the Golgi complex, as discussed below.

The evidence that ARFs play a role in toxin action or in ADP-ribosylation catalyzed by endogenous transferases in animal cells is much less substantial. It has been reported, however, that two ARFs purified from rat brain by the procedure described for purification of sARF I and sARF II from bovine brain (Tsai et al. 1988) had different effects on activities of four ADP-ribosyltransferases purified from rat brain (Matsuyama and Tsuyama 1991). The transferases, each of about 66 kDa, catalyzed the ADP-ribosylation of nonmuscle actin to the extent of about 1 mol per mole of G actin (not F actin) and inhibited polymerization. They also had similar pH optima and K_ms for NAD of about $20\,\mu M$. The enzymes differed from each other, however, in their abilities to use certain proteins as ADP-ribose acceptors as well as in the effects of four phospholipids and of the two purified ARF proteins on their activity. sARF I stimulated the activity of transferase II, inhibited transferase III, and had no effect on the other two. sARF II activated transferases I and IV, inhibited III, and had no effect on transferase II (Matsuyama and Tsuyama 1991). The ARF effects in these studies are, however, difficult to interpret as there was no comment on the apparent absence of GTP (or analog), which is necessary for ARF activation of CT. In the other experiments, sARF I and sARF II had no effect on activity of an NAD–agmatine ADP-ribosyltransferase purified from turkey erythrocytes (Moss et al., unpublished data).

ARF proteins in eukaryotes are components of the vesicular system that transports newly synthesized proteins from the endoplasmic reticulum to the Golgi complex and perhaps function also in endocytosis. There are two ARF genes in *S. cerevisiae* encoding proteins of 181 amino acids that are 96% identical. ARF 1 accounts for about 90% of total ARF, ARF 2, only about 10% (Stearns et al. 1990b). Deletion of both ARF genes was lethal, but rescue was effected by products of human ARF 1 and ARF 4 genes, consistent with the notion that human and yeast ARFs share at least some similar functions. No detectable alteration in phenotype resulted from disruption of the ARF 2 gene. On the other hand, disruption of ARF 1, perhaps due to the major reduction of total ARF, caused sensitivity to cold and to fluoride at concentrations that do not kill wild-type cells, slow growth, and defective secretion of invertase. Increasing expression of ARF 2 overcame the effects of disruption of the ARF 1 gene, as did spontaneous reversion that produced a functioning ARF 1 gene as a result of recombination with the ARF 2 gene (Stearns et al. 1990a). Overexpression of ARF, however, interfered with growth (Stearns et al. 1990a), an effect perhaps related to the inhibition of maturation of *Xenopus* oocytes that was produced by microinjection of ARF(s) (Bahnson et al. 1989).

Immunoreactive ARF was localized to the cytoplasmic surface of *cis* Golgi membranes in cultured NIH 3T3 cells (Stearns et al. 1990b). The antibodies used were raised by immunization with a synthetic peptide conjugated to keyhole limpet hemocyanin (KLH) via a cysteine added at the

N terminus and affinity-purified by binding to the peptide immobilized on CNBr-activated Sepharose beads (KAHN et al. 1988). The peptide sequence corresponded to that of a CNBr fragment (amino acids 23–36) of bovine ARF 1, which is identical in ARFs 1–5 and differs in ARF 6 only by one leucine for valine replacement. The same antibody was apparently used in studies of RINm5F cells, cultured rat insulinoma cells that have been used to study mechanisms of insulin release (REGAZZI et al. 1991). Incubation of permeabilized cells with GTPγS increased the amount of a 20-kDa [^{32}P]GTPase recovered with a particulate fraction. Gpp(NH)p (at the same concentration) was somewhat less effective; GTP, GDP, GDPβS, and ATP were ineffective. It was concluded that the increase in GTPase was entirely in a Golgi complex- and plasma membrane-enriched fraction. This protein was identified as ARF after ruling out *krev1*, *rab3*, and *rho* on the basis of electrophoretic mobility, immunoreactivity, and ADP-ribosylation catalyzed by C3 transferase from *Clostridium botulinum*. The evidence that the 20-kDa protein that bound [^{32}P]GTPγS on western blots was ARF appears still less than completely satisfying, although it seems quite clear that immunoreactive ARF became associated with membranes in the presence of GTPγS.

In homogenates of cultured PC12 pheochromocytoma cells, ARF that reacted with antibodies against class I ARFs was accumulated in a membrane fraction during incubation with GTPγS, and to a lesser degree with Gpp(NH)p (WALKER et al. 1992). ATPγS had a smaller effect, presumably via conversion to GTPγS catalyzed by nucleoside diphosphate kinase. GTP, GDP, GMP, and ATP were ineffective. Nucleotide effects on association of cytosolic ARF with phospholipid vesicles prepared from phosphatidylserine, phosphatidylcholine, or cardiolipin were similar (WALKER et al. 1992). Purified bovine brain ARF also associated with phospholipid vesicles in the presence of GTP, but not GDP (KAHN 1991). The observation that AlF^-_4 (with GDP) did not permit ARF binding was added to further evidence that, like other small (20- to 25-kDa) GTPases, ARF is not activated by AlF^-_4, and supports the view that AlF^-_4 could be useful for assessing participation of a G_α subunit (as opposed to a small GTPase) in cellular processes.

It is increasingly apparent that ARF proteins have an important role in the structure and function of the coated vesicles that serve in transport of proteins (and lipids) from *cis* to medial Golgi cisternae. ARF is a major component of these coated vesicles, which contain approximately three molecules of ARF per molecule of α-COP (SERAFINI et al. 1991). There are four relatively large (61- to 160-kDa) COP proteins associated with three smaller proteins in a complex referred to as a coatomer (WATERS et al. 1991). SERAFINI et al. (1991), based on their own studies and those of others, suggested that an ARF-specific nucleotide exchange protein in the donor Golgi membrane would convert ARF-GDP to ARF-GTP, which could then interact with the phospholipid membrane to initiate association and assembly of coatomers from the cytoplasm. The resultant COP-coated vesicle would remain intact until the ARF-GTP interacted with a specific GTPase activating protein (GAP) in the acceptor membrane. After GTP hydrolysis, the ARF-GDP would dissociate and coatomers would be released separately into the cytosol, permitting fusion of the then uncoated vesicle with the target membrane (Serafini et al. 1991).

In this scheme, the nonhydrolyzable GTPγS should block transport and cause accumulation of coated vesicles, as it does. The model does not, of course, necessarily require a specific nucleotide exchange protein or a specific GAP. One or the other could as well be replaced by a specific nucleoside diphosphate kinase, although an initial report of such an activity has subsequently been shown to be artifactual (see Chap. 34; Randazzo et al. 1991). The critical requirement is specificity, so that ARF-GTP is generated at the donor membrane to initiate vesicle budding and is converted to ARF-GDP at the proper acceptor membrane. It was suggested (Serafini et al. 1991) that the inhibitory factor described by Melançon et al. (1987) as a cytosolic factor required for GTPγS-induced vesicle accumulation and inhibition of vesicle transport in vitro is ARF. Perhaps this kind of competition of free ARF in cytosol with ARF in coated vesicles contributes to the deleterious effect of overexpression of ARF in *S. cerevisiae*. Serafini et al. (1991) noted that the 20-kDa protein in the Golgi-derived vesicles that bound [^{32}P]GTP on nitrocellulose blots was not proven to be the immunoreactive ARF also present on the blots. It is probable, however, that the model presented is basically correct and could, in principle, describe the functions of the several ARFs that have been identified, each operating with its own nucleotide exchange protein and GAP (or nucleoside diphosphate kinase) between specific donor and acceptor membranes.

It is possible also that certain ARFs function not in exocytic but in endocytic pathways, which have several shared or homologous components. These questions are difficult to answer at present, perhaps chiefly for technical reasons. Although it is possible to identify specific mRNAs for all six known mammalian ARFs, antibodies that can identify all six proteins are apparently not available. Polyclonal antibodies that can distinguish class I (1, 2, and 3) and class II ARFs (4 and 5) have been prepared and used successfully with western blots to demonstrate selective and independent interactions of ARFs 1, 3, and 5 with several subcellular fractions from rat brain (Tsai et al. 1992). Their application with fluorescence microscopy shows clearly that in many tissues and cells, class I and class II ARFs are differently distributed (Tsai et al. unpublished). For meaningful extension of these studies, however, antibodies with requisite specificity for individual ARFs will be required.

Acknowledgement. We thank Carol Kosh for expert secretarial assistance.

References

Alsip GR, Konkel DA (1986) A processed chicken pseudogene (CPS1) related to the *ras* oncogene superfamily. Nucleic Acids Res 14:2123–2138

Bahnson TD, Tsai S-C, Adamik R, Moss J, Vaughan M (1989) Microinjection of a 19-kDa guanine nucleotide-binding protein inhibits maturation of *Xenopus* oocytes. J Biol Chem 264:14824–14828

Birnbaumer L, Mattera R, Yatani A, Codina J, Van Dongen AMJ, Brown AM (1990) Recent advances in the understanding of multiple roles of G proteins in coupling of receptors to ionic channels and other effectors. In: Moss J, Vaughan M (eds) ADP-ribosylating toxins and G proteins: insights into signal transduction. American Society for Microbiology, Washington DC, p 225

Bobak DA, Nightingale MS, Murtagh JJ, Price SR, Moss J, Vaughan M (1989) Molecular cloning, characterization, and expression of human ADP-ribosylation factors: two guanine uncleotide dependent activators of cholera toxin. Proc Natl Acad Sci USA 86:6101–6105

Bobak DA, Bliziotes MM, Noda M, Tsai S-C, Adamik R, Moss J (1990a) Mechanism of activation of cholera toxin by ADP-ribosylation factor (ARF): both low- and high-affinity interactions of ARF with guanine nucleotides promote toxin activation. Biochemistry 29:855–861

Bobak DA, Tsai S-C, Moss J, Vaughan M (1990b) Enhancement of cholera toxin ADP-ribosyltransferase activity by guanine nucleotide-dependent ADP-ribosylation factors. In: Moss J, Vaughan M (eds) ADP-ribosylating toxins and G proteins: insights into signal transduction. American Society for Microbiology, Washington DC, p 439

Carpenter CCJ (1980) Clinical and pathophysiologic features of diarrhea caused by *Vibrio cholerae* and *Escherichia coli*. In: Field M, Fordtran JS, Schultz SG (eds) *Secretory diarrhea*. American Physiological Society, Bethesda MD, p 67

Casey PJ, Gilman AG (1988) G protein involvement in receptor-effector coupling. J Biol Chem 263:2577–2580

Cassel D, Pfeuffer T (1978) Mechanism of cholera toxin action: covalent modification of the guanyl nucleotide-binding protein of the adenylate cyclase system. Proc Natl Acad Sci USA 75:2669–2673

Cassel D, Selinger Z (1977) Mechanism of adenylate cyclase activation by cholera toxin: inhibition of GTP hydrolysis at the regulatory site. Proc Natl Acad Sci USA 74:3307–3311

Chang PP, Moss J, Twiddy EM, Holmes RK (1987) Type II heat-labile enterotoxin of *Escherichia coli* activates adenylate cyclase in human fibroblasts by ADP-ribosylation. Infact Immun 55:1854–1858

Denich KT, Malloy PJ, Feldman D (1992) Cloning and Characterization of the gene encoding the ADP-ribosylation factor in *Candida albicans*. Gene 110:123–128

Duman RS, Winston SM, Clark JA, Nestler EJ (1990) Corticosterone regulates the expression of ADP-ribosylation factor messenger RNA and protein in rat cerebral cortex. J Neurochemistry 55:1813–1816

Enomoto K, Asakawa T (1982) Partial purification and properties of a cytosolic protein factor required for the activation of rat liver adenylate cyclase by cholera toxin. Biomed Res 3:122–131

Enomoto K, Gill DM (1980) Cholera toxin activation of adenylate cyclase. Roles of nucleoside triphosphates and a macromolecular factor in the ADP-ribosylation of the GTP-dependent regulatory component. J Biol Chem 255:1252–1258

Field M (1979) Mechanisms of action of cholera and *Escherichia coli* enterotoxins. Am J Clin Nutr 32:189–196

Finkelstein RA (1973) Cholera. CRC Crit Rev Microbiol 2:553–623

Gill DM (1976) The arrangement of subunits in cholera toxin. Biochemistry 15:1242–1248

Gill DM (1977) Mechanism of action of cholera toxin. In: Greengard P, Robison GA (eds) Advances in cyclic nucleotide research. Raven, New York, p 85

Gill DM, Coburn J (1987) ADP-ribosylation by cholera toxin: functional analysis of a cellular system that stimulates the enzymatic activity of cholera toxin fragment A. Biochemistry 26:6364–6371

Gill DM, Meren R (1978) ADP-ribosylation of membrane proteins catalyzed by cholera toxin: basis of the activation of adenylate cyclase. Proc Natl Acad Sci USA 75:3050–3054

Gill DM, Meren R (1983) A second guanyl nucleotide-binding site associated with adenylate cyclase: distinct nucleotides activate adenylate cyclase and permit ADP-ribosylation by cholera toxin. J Biol Chem 258:11908–11914

Gill DM, Rappaport RS (1979) Origin of the enzymatically active A fragment of cholera toxin. J Infect Dis 139:674–680

Gill DM, Richardson SH (1980) Adenosine diphosphate-ribosylation of adenylate cyclase catalyzed by heat-labile enterotoxin of *Escherichia coli*: comparison with cholera toxin. J Infect Dis 141:64–70

Guth BEC, Pickett CL, Twiddy EM, Holmes RK, Gomes TAT, Lima AAM, Guerrant RL, Franco BDGM, Trabulsi LR (1986a) Production of type II heat-labile enterotoxin by *Escherichia coli* isolated from food and human feces. Infect Immun 54:587–589

Guth BEC, Twiddy EM, Trabulsi LR, Holmes RK (1986b) Variation in chemical properties and antigenic determinants among type II heat-labile enterotoxins of *Escherichia coli*. Infect Immun 54:529–536

Haun RS, Serventi IM, Tsai S-C, Lee C-M, Cavanaugh E, Stevens L, Moss J, Vaughan M (1992) Characterization of mammalian genes encoding ADP-ribosylation factors, 20 kDa guanine nucleotide-binding protein activators of cholera toxin. Clin Res 40:148A

Holmes RK, Twiddy EM, Pickett CL (1986) Purification and characterization of type II heat-labile enterotoxins of *Escherichia coli*. Infect Immun 53:464–473

Johnson GL, Kaslow HR, Bourne HR (1978) Genetic evidence that cholera toxin substrates are regulatory components of adenylate cyclase. J Biol Chem 253:7120–7123

Kahn RA (1991) Fluoride is not an activator of the smaller (20–25 kDa) GTP-binding proteins. J Biol Chem 266:15595–15597

Kahn RA, Gilman AG (1984a) Purification of a protein cofactor required for ADP-ribosylation of the stimulatory regulatory component of adenylate cyclase by cholera toxin. J Biol Chem 259:6228–6234

Kahn RA, Gilman AG (1984b) ADP-ribosylation of G_s promotes the dissociation of its α and β subunits. J Biol Chem 259:6235–6240

Kahn RA, Gilman AG (1986) The protein cofactor necessary for ADP-ribosylation of G_s by cholera toxin is itself a GTP-binding protein. J Biol Chem 261:7906–7911

Kahn RA, Goddard C, Newkirk M (1988) Chemical and immunological characterization of the 21-kDa ADP-ribosylation factor of adenylate cyclase. J Biol Chem 263:8282–8287

Kahn RA, Kern FG, Clark J, Gelmann EP, Rulka C (1991) Human ADP-ribosylation factors – a functionally conserved family of GTP-binding proteins. J Biol Chem 266:2606–2614

Kelly MT (1986) Cholera: a worldwide perspective. Pediatr Infect Dis 5:5101–5105

Lee C-M, Chang PP, Tsai S-C, Adamik R, Price SR, Kunz BC, Moss J, Twiddy EM, Holmes RK (1991) Activation of *Escherichia coli* heat-labile enterotoxins by native and recombinant adenosine diphosphate-ribosylation factors, 20 kD guanine nucleotide-binding proteins. J Clin Invest 87:1780–1786

Lee C-M, Haun RS, Tsai S-C, Moss J, Vaughan M (1992) Characterization of the human gene encoding ADP-ribosylation factor 1, a guanine nucleotide-binding activator of cholera toxin. J Biol Chem 267:9028–9034

LeVine H III, Cuatrecasas P (1981) Activation of pigeon erythrocyte adenylate cyclase by cholera toxin. Biochim Biophys Acta 672:248–261

Matsuyama S, Tsuyama S (1991) Mono-ADP-ribosylation in brain: purification and characterization of ADP-ribosyltransferases affecting actin from rat brain. J Neurochem 57:1380–1387

Melançon P, Glick BS, Malhotra V, Weidman PJ, Serafini T, Gleason ML, Orci L, Rothman JE (1987) Involvment of GTP-binding "G" proteins in transport through the Golgi stack. Cell 51:1053–1062

Monaco L, Murtagh JJ, Newman KB, Tsai S-C, Moss J, Vaughan M (1990) Selective amplification of an mRNA and related pseudogene for a human ADP-ribosylation factor, a guanine nucleotide-dependent protein activator of cholera toxin. Proc Natl Acad Sci USA 87:2206–2210

Moss J, Richardson SH (1978) Activation of adenylate cyclase by heat-labile *Escherichia coli* enterotoxin. Evidence for ADP-ribosyltransferase activity similar to that of choleragen. J Clin Invest 62:281–285

Moss J, Vaughan M (1977) Mechanism of action of choleragen: evidence for ADP-ribosyltransferase activity with arginine as an acceptor. J Biol Chem 252:2455–2457

Moss J, Vaughan M (1978) Isolation of an avian erythrocyte protein possessing ADP-ribosyltransferase activity and capable of activating adenylate cyclase. Proc Natl Acad Sci USA 75:3621–3624

Moss J, Vaughan M (1988) ADP-ribosylation of guanyl nucleotide-binding proteins by bacterial toxins. Adv Enzymol 61:303–379

Moss J, Manganiello VC, Vaughan M (1976) Hydrolysis of nicotinamide adenine dinucleotide by choleragen and its A protomer: possible role in the activation of adenylate cyclase. Proc Natl Acad Sci USA 73:4424–4427

Moss J, Stanley SJ, Lin MC (1979) NAD glycohydrolase and ADP-ribosyltransferase activities are intrinsic to the A_1 peptide of choleragen. J Biol Chem 254: 11993–11996

Moss J, Stanley SJ, Watkins PJ, Vaughan M (1980) ADP-ribosyltransferase activity of mono- and multi-(ADP-ribosylated) choleragen. J Biol Chem 255:7835–7837

Moss J, Osborne JC Jr, Fishman PH, Nakaya S, Robertson DC (1981) *Escherichia coli* heat-labile enterotoxin. Ganglioside specificity and ADP-ribosyltransferase activity. J Biol Chem 256:12861–12865

Moss J, Tsuchiya M, Price SR, Tsai S-C, Vaughan M (1992) Molecular characterization of the highly conserved, multigene family of human ADP-ribosylation factors, 20 kDa guanine nucleotide-binding protein activators of cholera toxin. 26th US Japan Cholera Conference 1990, Kyoto, Japan (to be published)

Murtagh JJ Jr, Lee F-JS, Lee C-M, Moss J, Vaughan M (1992a) Characterization of multigene families of guanine nucleotide-binding proteins: biotin-enhanced polymerase chain reaction strategies for rapid sequencing and cloning of ADP-ribosylation factors. Clin Res 40:223A

Murtagh JJ Jr, Mowatt MR, Lee C-M, Lee F-JS, Mishima K, Nash TE, Moss J, Vaughan M (1992b) Guanine nucleotide-binding proteins in the intestinal parasite *Giardia Lamblia*: isolation of a gene encoding a ~20 kDa ADP-ribosylation factor. J Biol Chem 267:9654–9662

Navon SE, Fung BKK (1984) Characterization of transducin from bovine retinal rod outer segments – mechanism and effects of cholera toxin-catalyzed ADP-ribosylation. J Biol Chem 259:6686–6693

Noda M, Tsai S-C, Adamik R, Bobak DA, Moss J, Vaughan M (1989) Activation of immobilized, biotinylated choleragen A1 protein by a 19-kilodalton guanine nucleotide-binding protein. Biochemistry 28:7936–7940

Noda M, Tsai S-C, Adamik R, Moss J, Vaughan M (1990) Mechanism of cholera toxin activation by a guanine nucleotide-dependent 19 kDa protein. Biochim Biophys Acta 1034:195–199

Northup JK, Sternweis PC, Smigel MD, Schleifer LS, Ross EM, Gilman AG (1980) Purification of the regulatory component of adenylate cyclase. Proc Natl Acad Sci USA 77:6516-6520

Peng Z, Calvert I, Clark J, Helman L, Kahn R, Kung H-F (1989) Molecular cloning, sequence analysis and mRNA expression of human ADP-ribosylation factor. Biofactors 2:45–49

Peterson JW, Ochoa LG (1989) Role of prostaglandins and cAMP in the secretory effects of cholera toxin. Science 245:857–859

Pickett CL, Holmes RK (1990) Nucleotide sequence of *Escherichia coli* heat-labile enterotoxins type IIa and IIb and comparisons to type I enterotoxin and cholera toxin. In: Sack RB, Zinnaka Y (eds) Advances in research in cholera and related diarrheas, vol 7. KTK Scientific Publishers, Tokyo, p 165

Pickett CL, Twiddy EM, Belisle BW, Holmes RK (1986) Cloning of genes that encode a new heat-labile enterotoxin of *Escherichia coli*. J Bacteriol 165:348–352

Pickett CL, Weinstein DL, Holmes RK (1987) Genetics of type IIa heat-labile enterotoxin of *Escherichia coli*: operon fusion, nucleotide sequence and hybridization studies. J Bacteriol 169:5180–5187

Pinkett MO, Anderson WB (1982) Plasma membrane-associated components that confer cholera toxin sensitivity to adenylate cyclase. Biochim Biophys Acta 714:337–343

Price SR, Nightingale MS, Tsai S-C, Williamson KC, Adamik R, Chen H-C, Moss J, Vaughan M (1988) Guanine nucleotide-binding proteins that enhance choleragen ADP-ribosyltransferase activity: nucleotide and deduced amino acid sequence of an ADP-ribosylation factor cDNA. Proc Natl Acad Sci USA 85:5488–5491

Price SR, Barber A, Moss J (1990) Structure-function relationships of guanosine nucleotide-binding proteins. In: Moss J, Vaughan M (eds) ADP-ribosylating toxins and G proteins: insights into signal transduction. American Society of Microbiology, Washington DC, p 397

Randazzo PA, Northup JK, Kahn RA (1991) Activation of a small GTP-binding protein by nucleoside diphosphate kinase. Science 254:850–853

Regazzi R, Ullrich S, Kahn RA, Wollheim CB (1991) Redistribution of ADP-ribosylation factor during stimulation of permeabilized cells with GTP analogues. Biochem J 275:639–644

Schleifer LS, Kahn RA, Hanski E, Northup JK, Sternweis PC, Gilman AG (1982) Requirements for cholera toxin-dependent ADP-ribosylation of the purified regulatory component of adenylate cyclase. J Biol Chem 257:20–23

Schmidt GJ, Huber LJ, Weiter JJ (1987) A-protein catalyzes the ADP ribosylation of G protein from cow rod outer segments. J Biol Chem 262:14333–14336

Serafini T, Orci L, Amherdt M, Brunner M, Kahn RA, Rothman JE (1991) ADP-ribosylation factor is a subunit of the coat of Golgi-derived COP-coated vesicles: a novel role for a GTP-binding protein. Cell 67:239–253

Serventi IM, Cavanaugh E, Moss J, Vaughan M (1993) Characterization of the gene for ADP-ribosylation factor (ARF) 2, A developmentally regulated, Selectively Expressed Member of the ARF family of ~20-kDa Guanine Nucleotide-binding Proteins. J Biol Chem 268:4863–4872

Sewell JL, Kahn RA (1988) Sequences of the bovine and yeast ADP-ribosylation factor and comparison to other GTP-binding proteins. Proc Natl Acid Sci USA 85:4620–4624

Stearns T, Kahn RA, Botstein D, Hoyt MA (1990a) ADP-ribosylation factor is an essential protein in *Saccharomyces cerevisiae* and is encoded by two genes. Mol Cell Biol 10:6690–6699

Stearns T, Willingham MC, Botstein D, Kahn RA (1990b) ADP-ribosylation factor is functionally and physically associated with the Golgi complex. Proc Natl Acad Sci USA 87:1238–1242

Tamkun JW, Kahn RA, Kissinger M, Brizuela BJ, Rulka C, Scott MP, Kennison JA (1991) The arf-like gene encodes an essential GTP-binding protein in *Drosophila*. Proc Natl Acad Sci USA 88:3120–3124

Trepel JB, Chuang D-M, Neff NH (1977) Transfer of ADP-ribose from NAD to choleragen: A subunit acts as catalyst and acceptor. Proc Natl Acad Sci USA 74:5440–5442

Tsai S-C, Noda M, Adamik R, Moss J, Vaughan M (1987) Enhancement of choleragen ADP-ribosyltransferase activities by guanyl nucleotides and a 19-kDa membrane protein. Proc Natl Acad Sci USA 84:5139–5142

Tsai S-C, Noda M, Adamik R, Chang P, Chen H-C, Moss J, Vaughan M (1988) Stimulation of choleragen enzymatic activities by GTP and two soluble proteins purified from bovine brain. J Biol Chem 263:1768–1772

Tsai S-C, Adamik R, Moss J, Vaughan M (1991a) Guanine nucleotide dependent formation of a complex between choleragen (cholera toxin) A subunit and bovine brain ADP-ribosylation factor. Biochemistry 30:3697–3703

Tsai S-C, Adamik R, Tsuchiya M, Chang PP, Moss J, Vaughan M (1991b) Differential expression during development of ADP-ribosylation factors, 20 kDa guanine nucleotide-binding protein activators of cholera toxin. J Biol Chem 266:8213–8219

Tsai S-C, Haun RS, Tsuchiya M, Moss J, Vaughan M (1991c) Isolation and characterization of the human gene for ADP-ribosylation factor 3, a 20-kDa guanine nucleotide-binding protein activator of cholera toxin. J Biol Chem 266:23053–23059

Tsai S-C, Adamik R, Haun RS, Moss J, Vaughan M (1992) Differential Interaction of ADP-ribosylation factors 1, 3, and 5 with rat brain Golgi membranes. Proc Natl Acad Sci USA 89:9272–9276

Tsuchiya M, Price SR, Nightingale MS, Moss J, Vaughan M (1989) Tissue and species distribution of mRNA encoding two ADP-ribosylation factors, 20-kDa guanine nucleotide binding proteins. Biochemistry 28:9668–9673

Tsuchiya M, Price SR, Tsai S-C, Moss J, Vaughan M (1991) Molecular identification of ADP-ribosylation factor mRNAs and their expression in mammalian cells. J Biol Chem 266:2772–2777

Walker MW, Bobak DA, Tsai S-C, Moss J, Vaughan M (1992) GTP but not GDP analogues promote association of ADP-ribosylation factors, 20 kDa protein activators of cholera toxin, with phospholipids and PC-12 cell membranes. J Biol Chem 267:3230–3235

Waters MG, Serafini T, Rothman JE (1991) "Coatomer": a cytosolic protein complex containing subunits of non-clathrin-coated Golgi transport vesicles. Nature 349:248–251

Weiss O, Holden J, Rulka C, Kahn RA (1989) Nucleotide binding and cofactor activities of purified bovine brain and bacterially expressed ADP-ribosylation factor. J Biol Chem 264:21066–21072

Welsh CF, Price SR, Tsuchiya M, Stanley SJ, Nightingale MS, Lee C-M, Moss J (1991) Molecular cloning, expression, and functional characterization of three classes of mammalian ADP-ribosylation factors. FASEB J 5:2621

Welsh CF, Moss J, Vaughan M (1992) ADP-ribosylation factor, a guanine nucleotide-binding protein activator of cholera toxin, is isolated in an activated state when expressed as a fusion protein in *Escherichia coli*. Clin Res 40:215A

C. rho and rho-Like Proteins

CHAPTER 36

rho and rho-Related Proteins

A.J. RIDLEY and A. HALL

A. Introduction

The Rho-like proteins form a distinct subgroup of the ras superfamily of small GTPases. They act as molecular switches having an inactive, GDP-bound and an active, GTP-bound conformation and have many structural and biochemical features in common with p21ras. A number of proteins that control this conformational switch – guanine nucleotide exchange proteins (GEPs) and GTPase activating proteins (GAPs) – have been identified, and some progress has been made in defining the biological roles of rho-related proteins. However, as for all small GTPases, their underlying biochemical mechanism of action is still unknown.

B. Sequence and Structure

Rho was first identified as a ras homologous sequence in a cDNA library from the marine snail Aplysia, and the clone was subsequently used to isolate three human (rhoA, rhoB and rhoC) and two *Saccharomyces cerevisiae* (RHO1 and RHO2) counterparts (MADAULE and AXEL 1985; MADAULE et al. 1987; YERAMIAN et al. 1987; CHARDIN et al. 1988). RhoA, B and C are about 85% identical to each other and like the three ras proteins, almost all the divergence is at the carboxy-terminal end of the proteins. RHO1 is the yeast homologue of mammalian rho and is essential for cell viability, whereas RHO2 is only 53% identical to either RHO1 or rho and is a non-essential gene. Five other mammalian proteins have been identified that belong to the rho subgroup. CDC42Hs and G25K differ from each other at only nine amino acids, while rac1 and rac2 differ at 15 amino acids (DIDSBURY et al. 1989; MUNEMITSU et al. 1990; SHINJO et al. 1990). The rac and CDC42 proteins are closer to each other (70%) than either is to rho (55%). Finally a cDNA, TC10, has been identified with about 55% identity to rho (DRIVAS et al. 1990). A yeast homologue of CDC24Hs/G25K has been identified, CDC42Sc, and its function has been examined in some detail genetically (see Chap. 37, this volume; JOHNSON and PRINGLE 1990).

Sequence alignment of rho and ras reveals just over 30% identity (Fig. 1). rho, rac, CDC42 and TC10, however, have at least 55% identity to each other and so they have been categorised as a distinct subfamily. There is

```
rasH          MTEY...V..A.GV..SA.T.QLIQNH.VDE.D.
rhoA        MAAIRKKLVIVGDGACGKTCLLIVFSKDQFPEVYVP
rac1          MQAI.C.V.....V.......SYTTNA..GE.I.
CDC42         MQTI.C.V.....V.......SYTTNK..SE...
TC10    MAHGPG.LML.C.V.....V......MSYAN.A...E...

rasH    .IEDS.RKQVVI..ETCL.DIL......E.SAM.DQYMRT
rhoA    TVFENYVADIEVDGKQVELALWDTAGQEDYDRLRPLSYPD
rac1    ...D..S.NVM....P.N.G...................Q
CDC42   ...D..AVTVMIG.EPYT.G.F.................Q
TC10    ...DH.AVSVT.G...YL.G.Y.................M

rasH    GEGF.CV.A.NNTK.F.D.HQYREQIKRVKDSDD..MV..
rhoA    TDVILMCFSIDSPDSLENIPEKWTPEVK HFCPNVPIILV
rac1    ...F.I...LV..A.F..VRA..Y...R .H...T.....
CDC42   ...F.V...VV..S.F..VK...V..IT .H..KT.FL..
TC10    ...F.I...VVN.A.FQ.VK.E.V..L. EYA....FL.I

rasH    ...C..AARTVE                SRQAQDLA.SYGIP
rhoA    GNKKDLRNDEHTRRELAKMKQEPVKPEEGRDMANRIGAFG
rac1    .T.L...D.KD.IEK.KEK.LT.ITYPQ.LA..KE...VK
CDC42   .TQI...D.PS.IEK...N..K.IT..TAEKL.RDLK.VK
TC10    .TQI...D.PK.LAR.ND..EK.ICV.Q.QKL.KE...CC

rasH    .I.T....RQ..EDA.YTLV.EIR.HKLR.LNPPDESGPG
rhoA    YMECSAKTKDGVREVFEMATRAALQARRGKKKSG
rac1    .L....L.QR.LKT..DE.I..V.CPPPV..RKRK
CDC42   .V....L.QK.LKN..DE.IL...EPPEPKKSRR
TC10    .V....L.QK.LKT..DE.II.I.TPKKHTV.KRIGSRCIN

rasH    CMSCKCVLS
rhoA         CLVL
rac1         CLLL
CDC42        CVLL
TC10        CCLIT
```

Fig. 1. Sequence comparisons of rhoA with rho-related proteins and with ras. The single letter amino acid code has been used and *dots* signify amino acid identity to rhoA

now growing biological and biochemical support for this classification. The identity between ras and the rho-like proteins is confined mainly to four regions conserved in all GTPases: the G1 region, $GX_4GK(S/T)$, residues 10–17 in ras; the G3 region, DX_2G, residues 57–60 of ras; the G4 region, (N/T)(K/Q)XD, residues 116–119 in ras; and the G5 region, EXSA(K/L), residues 143–147 in ras (Fig. 1; see BOURNE et al. 1991 and Chap. 1). Despite this, the almost identical spacing between these motifs suggest that rho proteins will adopt a similar overall three-dimensional structure to ras (see Chaps. 13, 14, this volume). Closer inspection of the sequences, however, reveals some potentially important differences. In particular, the

α_2 helix of ras (residue 66–74) is one of the few regions that undergo a conformational change when GDP is exchanged for GTP (MILBURN et al. 1990; SCHICHTING et al. 1990). All members of the rho subfamily have two proline residues in this region and it is unlikely that an α helix would be formed. Also, alignment of rho with ras requires the insertion of an extra 14 amino acids between residues 120 and 140 of ras (see Fig. 1). This insertion, though divergent in sequence, is present in all members of the rho subfamily.

C. Expression and Localisation

Expression of the rho, rac and CDC42 proteins appears to be ubiquitous, though some variations in levels have been observed. For example, rhoA has been shown to be expressed in a wide range of tissues at similar levels (OLOFSSAN et al. 1988). More detailed analysis in Rat-2 fibroblasts has revealed that although rhoA and rhoC are expressed constitutively, rhoB is almost undetectable in quiescent or density arrested cells, but 30 min after addition of growth factors its expression is induced (JÄHNER and HUNTER 1991). Whether rhoB is also an immediate early response gene in other cell types is not known. Rac1 is expressed in all cell types examined, whereas rac2 appears to be restricted to haemopoietic cells, and differentiation of HL60 to the macrophage lineage, for example, leads to a sevenfold increase in rac2 expression (DIDSBURY et al. 1989; SHIRSAT et al. 1990; MOLL et al. 1991). TC10 expression appears to have a much more restricted distribution (DRIVAS et al. 1990).

The carboxy termini of small GTPases encode post-translational modification and cellular localisation signals (see Chap. XXI by GIBBS). In particular ras and rho-like proteins have a C-terminal CaaX box (C, cysteine; a, an aliphatic amino acid; X, another amino acid) where an isoprenoid derivative is added to the cysteine residue, with the decision to add either C15 farnesyl or C20 geranylgeranyl isoprenoids being determined by the "X" residue. If X is serine or methionine, then C15 is added, as in ras, whereas if X is leucine, then C20 lipid is attached (SEABRA et al. 1991). rho, rac and CDC42 proteins end in leucine, and rhoA and G25K have been shown to have a C20 lipid attached at their C termini (KATAYAMA et al. 1991; YAMANE et al. 1991). In addition to the CaaX box, an additional motif close to the C terminus, a palmitoylation site or a polybasic region, has been shown to be essential for the localisation of ras to the plasma membranes (HANCOCK et al. 1989, 1991). rhoA/C, rac1/2 and CDC42/G25K have polybasic domains similar to Ki-ras, whilst rhoB and TC10 have potential palmitoylated sites similar to Ha- and N-ras.

Despite these similarities to ras, the cellular localisation of mammalian rho-like proteins has not been defined. Unlike ras, rho and rac have each been identified in both the cytosolic and the membrane fraction of cell homogenates (POLAKIS et al. 1989; KAWAHARA et al. 1990; HOSHIJIMA et al. 1990; ABO et al. 1991). It is possible that rho-like proteins cycle between the plasma membrane and a cytoplasmic location (HALL 1992a).

D. Upstream Regulation of rho-Like Proteins

I. Nucleotide Exchange

A variety of potential regulators of rho-like proteins have been identified. rho GDS (guanine nucleotide dissociation stimulator) and rho GDI (guanine nucleotide inhibitor) can stimulate or inhibit, respectively, the exchange of guanine nucleotide on rho-like proteins (FUKUMOTO et al. 1990; ISOMURA et al. 1990; HIRAOKA et al. 1992; and see Chap. 39). rho GDI can also solubilise rho-like proteins from membranes when they are in the GDP-bound form (ISOMURA et al. 1991), and this could provide a mechanism for the proposed cycling of rho-like proteins between membrane and cytosol. smg GDS was originally identified as a guanine nucleotide exchange factor for rap1, but was subsequently shown to act on Ki-ras and rho-like proteins (HIRAOKA et al. 1992, and Chap. 39, this volume). This provides a possible direct link between the regulation of the ras-like and rho-like subfamilies. More recently, the *dbl* oncogene has been shown to act as a GDS for CDC42 and perhaps rac (HART et al. 1991, and Chap. 37, this volume). Interestingly, this *dbl* exchange domain (DH domain) is also found in several other proteins, for example, bcr (the product of the breakpoint cluster region gene), and the oncogene *vav*, but whether these proteins can also act as exchange factors is unknown (ADAMS et al. 1992). Since *vav* and *dbl* are both oncogenes, it is tempting to speculate that activation of rho-like proteins could contribute to malignant transformation. However, the biological activities ascribed to rho-related proteins so far would not support this idea.

II. GTP Hydrolysis

A GTPase activating protein, rho GAP, has been identified in a variety of cells (GARRETT et al. 1989, 1991; MORII et al. 1991). Analysis of cDNA clones is not yet complete, though it is clear that the purified 29-kDa protein is a proteolytic fragment of a larger full-length rho GAP (A. HALL and C. LANCASTER, unpublished results). Sequence analysis of rho GAP has led to the identification of a family of related proteins. The ubiquitous *bcr* gene product and N-chimaerin, a brain specific cDNA, both encode GAP proteins with activity for rac and CDC42, but not rho (Fig. 2; DIEKMANN et al. 1991; A.H. and S. BRILL, unpublished results). More recently, a protein that interacts with ras GAP, p190, has been cloned and sequenced (SETTLEMAN et al. 1992, and Chap. 23). Its C-terminal domain is related to

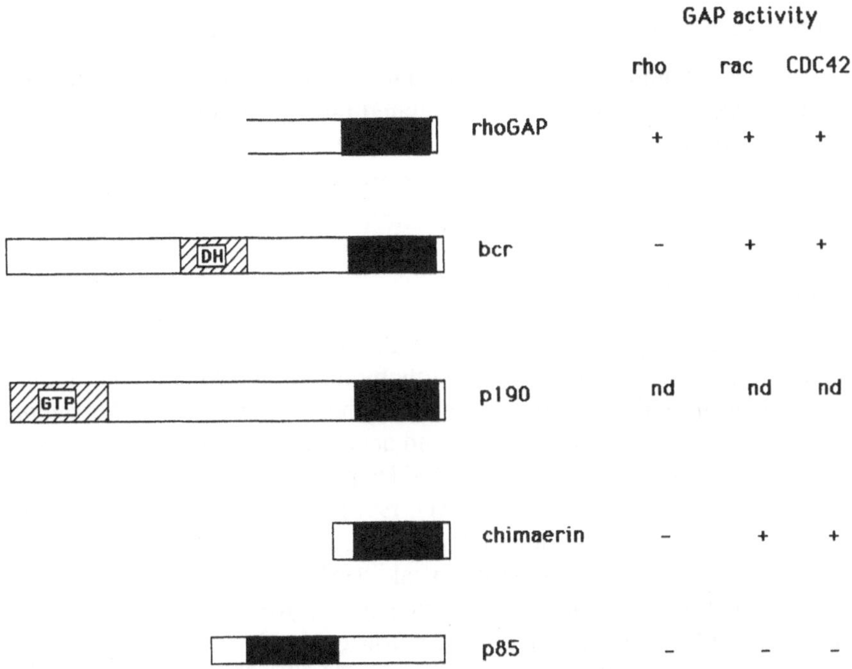

Fig. 2. Rho-GAP-related proteins. A full-length rho GAP cDNA clone has not yet been published. The *black box* represents the GAP (GTPase activating protein) domain, a segment of around 150 amino acids, with around 40% amino acid identity. The domain in p85 is significantly less related. Homology to dbl (*DH*) and GTP-binding consensus sequences are also shown

rho GAP and these domains of rhoGAP, chimaerin, bcr and p190 share about 40% amino acid identity to each other (see Fig. 2). The putative regulatory subunit of PI-3 kinase, p85, also has a domain with homology to rho GAP, though this is only around 25% homologous and it is not clear if this protein has any GAP activity (OTSU et al. 1991).

One further interesting feature of these rho-GAP-like proteins is their multidomain character (Fig. 2). Bcr, for example, in addition to its GAP domain has a dbl exchange-like domain, and its N terminus encodes a novel serine kinase activity (MARN and WITTE 1991). N-chimaerin has an N-terminal cysteine repeat sequence similar to that found in the regulatory domain of protein kinase C, and p190, in addition to the GAP domain, has a consensus sequence for binding GTP and encodes a potential transcription factor (HALL et al. 1990; SETTLEMEN et al. 1992). p190 also interacts with ras GAP. The significance of the multidomain character of these proteins and their relationship to rho (are they upstream regulators or downstream targets of rho?) are not at all clear (HALL 1992b). However, there must be an intricate network of interactions between signalling molecules at the plasma membrane with much cross-talk between ras and rho-like subfamilies.

E. Downstream Functions of rho-Like Proteins

The sequence homology and similar biochemical properties of rho-like proteins suggests that they may have related biological functions. Recently, a common link between the activities of several rho-like proteins has been identified – their involvement in processes regulating actin cytoskeletal organization.

I. Mammalian rho Proteins

Of the rho subfamily, rho proteins were the first to be studied functionally using a bacterial enzyme which covalently modifies the proteins and inhibits their biological activity. The C3 exoenzyme produced by *Clostridium botulinum* ADP-ribosylates rho on amino acid Asn41 (see Chap. 8). Since it has a very low activity on rac and CDC42Hs proteins, it is unlikely to affect these proteins in vivo (Ridley and Hall 1992). When C3 is introduced into a variety of cell types, actin microfilaments disassemble and the cells round up (Rubin et al. 1988; Chardin et al. 1989; Paterson et al. 1990), suggesting that rho proteins are involved in controlling actin cytoskeletal organization. Other effects of C3 are consistent with this hypothesis. For example, binucleate cells accumulate when NIH-3T3 fibroblasts are treated with C3, presumably due to inhibition of cell division, a process involving actin reorganisation (Rubin et al. 1988). In bovine neutrophils, C3 inhibits both spontaneous and chemoattractant-stimulated motility, which requires precisely controlled actin polymerisation and depolymerisation (Stasia et al. 1991). Finally, in PC12 cells C3 inhibits cell growth and stimulates the outgrowth of neurite-like processes (Rubin et al. 1988; Nishiki et al. 1990), a response resembling that of neuronal cells to cytochalasins. It has been suggested that disruption of filamentous actin can promote neurite outgrowth by allowing microtubules and organelles to invade a normally impenetrable cortical actin barrier (Goldberg and Burmeister 1989).

Another approach used to analyse the function of rho is the microinjection of cells with recombinant protein. rho induces the rapid assembly of stress fibres in fibroblasts (Paterson et al. 1990; Fig. 3). Stress fibres are one of the major actin cytoskeletal structures in many adherent cultured cells; they are thick bundles of actin filaments and associated proteins which traverse the cytoplasm and are attached to the plasma membrane at discrete structures known as focal adhesions. Here, integrin receptors are clustered and provide a link between extracellular matrix proteins and the cytoskeleton. A number of proteins are found associated with focal adhesions at the intracellular face of the plasma membrane, including vinculin and talin (Burridge et al. 1988). In addition to inducing stress fibre formation, microinjected rho proteins also stimulate focal adhesion assembly (Ridley and Hall 1992).

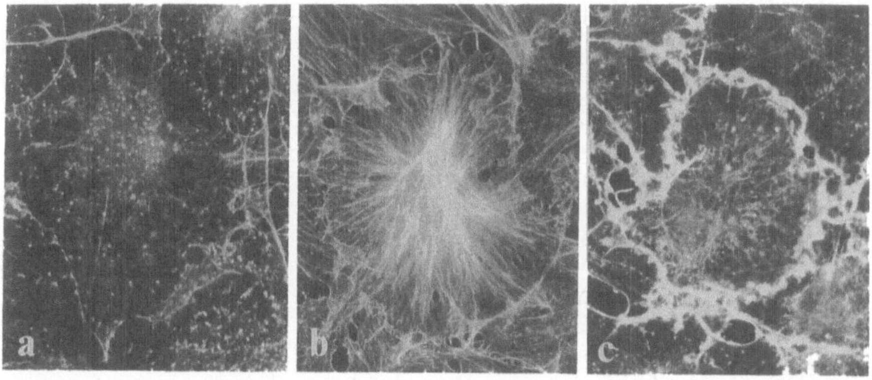

Fig. 3a–c. Actin reorganization induced by rho and rac proteins. The distribution of actin filaments in Swiss 3T3 fibroblasts revealed with rhodamine phalloidin in (**a**) uninjected cells, (**b**) 30 min after microinjection of Val-14 rhoA or (**c**) 15 min after microinjection of Val-12 rac1

The assembly of focal adhesions and stress fibres in fibroblasts is controlled by a number of extracellular factors. Serum-starved Swiss 3T3 cells, for example, contain very few stress fibres, but readdition of serum rapidly stimulates stress fibre and focal adhesion formation. The major component in serum responsible for this activity appears to be lysophosphatidic acid (LPA). Other growth factors, such as platelet-derived growth factor (PDGF), induce actin reorganisation, but in this case cell surface protrusions known as membrane ruffles are formed initially, followed later by new stress fibres. By microinjecting serum-starved Swiss 3T3 cells with C3 transferase to block rho activity and then adding growth factors, it has been shown that rho is essential for stress fibre and focal adhesion assembly, but not for the formation of membrane ruffles (Ridley and Hall 1992). These effects of LPA on actin reorganisation in Swiss 3T3 cells is of interest in light of responses to LPA observed with other cell types (Moolenaar et al. 1992). For example, LPA can induce platelet activation, which involves dramatic changes in cell shape and the actin cytoskeleton. In PC12 cells, LPA rapidly reverses morphological differentiation, inducing cell rounding and retraction of neurites, whereas as mentioned above C3 has an opposite effect, inducing neurite formation. It is likely, therefore, that all these responses to LPA are mediated by rho proteins.

By analogy to ras, growth factors and LPA are likely to activate a rho-dependent response by stimulating an increase in rho GTP in cells, either by stimulating nucleotide exchange or by inhibiting the activity of GAP proteins (Fig. 4). How rho GTP subsequently initiates focal adhesion and stress fibre assembly is not known, but one possibility is that it stimulates the formation of new nucleation sites for actin polymerisation at

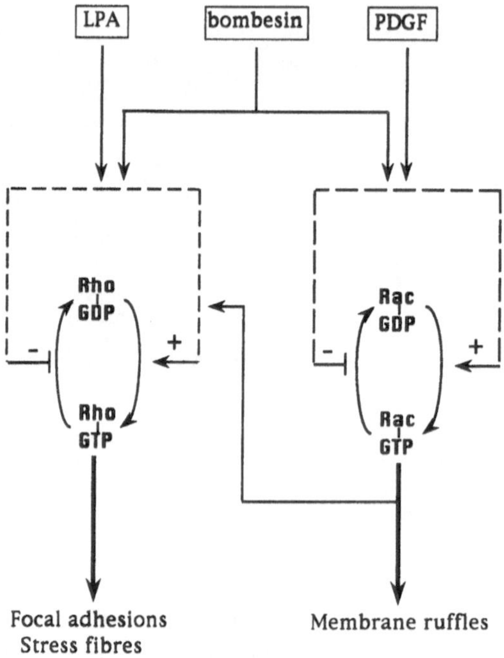

Fig. 4. A model for rho and rac function in fibroblasts. See text for description

the plasma membrane (RIDLEY and HALL 1992). These sites might include integrin receptors, and as new actin filaments polymerise and become cross-linked the nucleation sites would then cluster to form focal adhesions.

II. The rac Proteins

1. rac and the Actin Cytoskeleton

When recombinant rac1 protein is microinjected into confluent fibroblasts, it stimulates the accumulation of actin filaments in membrane ruffles (Fig. 3) (RIDLEY et al. 1992). This process is very rapid; an increase in actin at the plasma membrane is observed within 5 min of injection. Interestingly, in subconfluent fibroblasts the major phenotype observed is the accumulation of large intracellular vesicles by pinocytosis. This is probably a direct result of increased membrane ruffling, since other studies have shown that these two processes are closely linked (e.g., BAR-SAGL et al. 1987; KELLER 1990).

As described earlier, several extracellular factors stimulate membrane ruffling in fibroblasts. When N17rac1, a dominant negative mutant rac protein with a Ser to Asn mutation at codon 17, is injected into serum-starved Swiss 3T3 cells, membrane ruffling in response to PDGF, epidermal growth factor (EGF), insulin and bombesin is completely inhibited (RIDLEY et al. 1992; see FEIG, Chap. 19, this volume). These results indicate that endogenous rac function is required for the reorganisation of actin into membrane ruffles (Fig. 4).

In addition to stimulating membrane ruffling, several growth factors and also microinjected rac1 induce new stress fibre formation in serum-starved Swiss 3T3 cells, a process dependent on rho proteins (RIDLEY et al. 1992). However, growth factors can be divided into two classes according to both the time course of stress fibre formation and whether it is dependent on rac function. LPA (which does not induce membrane ruffling) and bombesin both stimulate stress fibre formation very rapidly, within 2 min of addition, and this is not inhibited by N17rac1. In contrast, with PDGF, EGF and insulin, stress fibre formation is a later response that is inhibited by N17rac1. These observations suggest that rac and rho have distinct but interconnecting roles in mediating growth-factor-induced actin organization, and a model for this is shown in Fig. 4. Growth factors such as PDGF initially activate rac, which stimulates membrane ruffling, and also leads to rho activation. rho subsequently induces stress fibre formation, accounting for the observed delay in this response. LPA and bombesin, on the other hand, can activate rho directly and independently of rac.

Growth factors presumably activate rac-dependent responses by stimulating an increase in rac GTP in cells, although this has not been directly demonstrated. This could be achieved either by stimulating proteins which increase exchange of GDP for GTP, or by inhibiting the activity of GAPs (see Sect. D above), and responses to different growth factors may be tailored through the involvement of distinct proteins. As yet, exchange factors which act preferentially on either rho or rac have not been identified, but GAP proteins show some specificity. For example, bcr is a GAP for rac but not rho, and could therefore be a target for growth factors such as PDGF, whereas rho GAP is active on both rac and rho, and might be involved in mediating responses to a growth factor such as bombesin (see Fig. 4).

Whatever the molecular mechanism whereby growth factor receptors activate rho or rac, it is clear that ras is not required (HALL 1990; RIDLEY et al. 1992). Inhibiting endogenous ras function blocks DNA synthesis induced by a variety of growth factors, but does not prevent either membrane ruffling or stress fibre formation. Val-12 H-ras proteins, however, do induce actin reorganisation when microinjected, and this response is blocked by N17rac1, implying that activated ras proteins stimulate signal transduction pathways, such as those acting through rac, not mediated by endogenous ras proteins (RIDLEY et al. 1992).

2. rac and Superoxide Production

Aside from its role in actin organisation, rac has been independently identified as an activator of superoxide production by the phagocyte-specific reduced nicotinamide-adenine-dinulcleotide phosphate (NADPH) oxidase.

This is part of a coordinated anti-bacterial defence mechanism, where bacteria are phagocytosed and toxic superoxide is released into the phagosome. Analysis of genetic lesions in humans with chronic granulomatous disease (CGD), in whom the NADPH oxidase is defective, has led to the identification of four components of this enzyme complex: the two subunits of the membrane-bound cytochrome b_{558} and two cytoplasmic components, p47 and p67 (Morel et al. 1991). Fractionation of cytosol from phagocytes revealed that a further component was required for stimulation of oxidase activity in a cell-free system. This component was purified and consisted of two proteins, rac1 and rhoGD1 (Abo et al. 1991). rac2 has also been independently purified as an activator of the NADPH oxidase (Knaus et al. 1991). Addition of purified recombinant rac1 protein bound to the non-hydrolysable GTP analogue GTPγS, but not to GDP, activated the NADPH oxidase in an in vitro reconstituted assay, but only in the presence of cytosolic fractions containing p47 and p67. The role of rho GD1 in this activation process is not clear.

Although the mechanism whereby the NADPH oxidase is activated in response to extracellular signals has not been established, it is known that p47 and p67 translocate to the plasma membrane (Morel et al. 1991). It has been postulated that rac acts to catalyse the formation of an active complex at the plasma membrane consisting of p47, p67 and cytochrome b (Abo et al. 1991). Another ras-related protein, rap1A, is also found in tight association with cytochrome b at the plasma membrane, and although it is not required for superoxide production in an in vitro assay, it may play a role in localization of the NAPDH oxidase in vivo (Bokoch et al. 1991; and see Chap. 25, this volume).

The two different functions described for rac could possibly be linked (Fig. 5), since extracellular factors such as *N*-formyl-methionyl-leucyl-phenylalanine (FMLP) which activate superoxide production in neutrophils also stimulate actin polymerisation and membrane ruffling leading to phagocytosis (Omann et al. 1987). rac might coordinate both these processes, perhaps by interacting with two different sets of proteins. If this is the case, then in cell types other than phagocytes, such as fibroblasts, rac may coordinate membrane ruffling with another, as yet uncharacterized, process which may bear some structural resemblance to superoxide production.

3. Other rho-Related Proteins

In *Saccharomyces cerevisiae*, the function of two members of the rho subfamily, RHO1 and CDC42Sc, has been studied. Deletion of *RHO1* is lethal, whereas expression of a constitutively active mutant inhibits sporulation (Madaule et al. 1987). *CDC42Sc* is also essential for viability, and by analysing temperature-sensitive mutants it has been shown that CDC42 is required for the assembly of components of the bud during cell

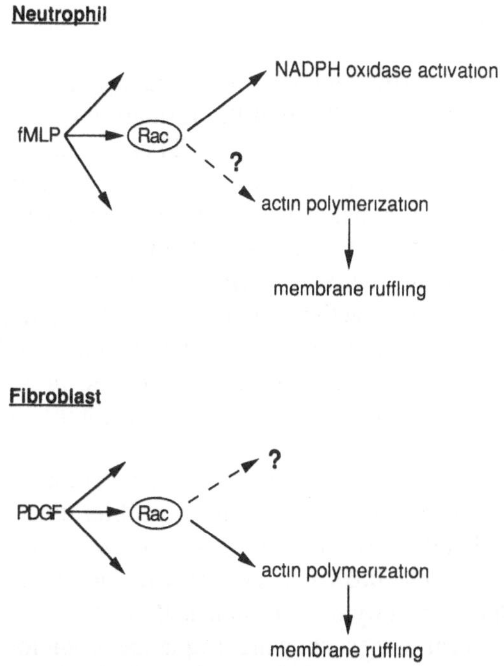

Fig. 5. Rac coordinates intracellular responses to extracellular factors. Rac proteins have been shown to be required for actin reorganisation to form membrane ruffles in *fibroblasts*, and for superoxide production in phagocytes. Neutrophil activation by agents such as *N*-formyl-methionyl-leucyl-phenylalanine (*FMLP*) leads to both membrane ruffling and superoxide production, and rac could function by coordinating the two processes. In fibroblasts, other components of the reduced nicotinamide-adenine-dinucleotide phosphate (*NADPH oxidase*) are not expressed, and therefore rac may coordinate membrane ruffling with another as yet undefined process

division. At the permissive temperature, these mutants are unable to bud, develop into greatly enlarged cells and display abnormalities in actin filament organisation similar to some actin mutants (Adams et al. 1990; see Chap. 37, this volume). A preliminary report has indicated that hyperactive RHO1 mutations can induce a similar phenotype (McCaffrey et al. 1991). Interestingly, RHO1 appears to localise predominantly on the Golgi and post-Golgi vesicles in close apposition to bud sites, and it has been speculated that RHO1 could be involved in modifying the structure of future plasma membrane, present in post-Golgi vesicles, to allow the attachment of actin microfilaments (McCaffrey et al. 1991). These results suggest that rho-related proteins in yeast, as well as in mammalian cells, are required for processes involving the formation of specific actin structures. Although no role for the mammalian proteins CDC42Hs or TC10 has yet been established, their close sequence similarity to rac and rho suggests that they may also function as regulators of actin organization.

F. Conclusions

The related actions of rho and rac in regulating specific types of actin reorganisation raise the possibility that the formation of other actin filament structures may also require rho-related GTPases. Such structures could include those associated with particular differentiated cell types, such as the microvilli of intestinal epithelial cells, or with specific stages of the cell cycle, such as the cleavage furrow formed at cell division. A possible molecular basis for the action of rho-related proteins is suggested by the role of rac in phagocyte NADPH oxidase activation. It appears that rac may control the assembly of cytoplasmic and membrane components to form an active complex at the plasma membrane. Other rho-related proteins could act by stimulating the formation of complexes between cytoplasmic and membrane proteins.

It is becoming apparent that growth factor receptors can activate both ras-dependent signalling pathways leading to proliferation and also distinct rho- and rac-mediated responses leading to actin reorganisation. As described above, possible links between growth factor receptors and rho-related proteins have recently been established on the molecular level, with the cloning of several genes showing sequence homology to the GAP domains of rho GAP and Bcr, including p190 and p85 (Fig. 2). In particular, the binding of p190 to ras GAP in growth-factor-stimulated cells may provide a direct connection between ras- and rho-related proteins. Functional studies in vivo and biochemical characterization of proteins such as p190 should soon converge to etablish a more complete picture of the molecular mechanisms underlying the activation and function of rho-related proteins.

References

Abo A, Pick E, Hall A, Totty N, Teahan CG, Segal AW (1991) Activation of the NADPH oxidase involves the small GTP-binding protein p21rac1. Nature 353:668–669

Adams AEM, Johnson DI, Longnecker RM, Sloat BF, Pringle JR (1990) *CDC42* and *CDC43*, two additional genes involved in budding and the establishment of cell polarity in the yeast *Saccharomyces cerevisiae.* J Cell Biol 111:131–142

Adams JM, Houston H, Allen J, Lints T, Harvey R (1992) The hematopoietically expressed vav proto-oncogene shares homology with the dbl GDP-GTP exchange factor, the bcr gene and a yeast gene (CDC24) involved in cytoskeletal organization. Oncogene 7:611–618

Bar-Sagi D, McCormick F, Milley RJ, Feramisco JR (1987) Inhibition of cell surface ruffling and fluid-phase pinocytosis by microinjection of anti-ras antibodies into living cells. J Cell Physiol 5:69–73

Bokoch GM, Quillam LA, Bohl BP, Jesaitis AJ, Quinn MT (1991) Inhibition of Rap1A binding to cytochrome b558 of NADPH oxidase by phosphorylation of Rap1A. Science 254:1794–1796

Bourne HR, Sanders DA, McCormick F (1991) The GTPase superfamily conserved structure and molecular mechanism. Nature 349:117–126

Burridge K, Fath K, Kelly T, Nuckolls G, Turner C (1988) Focal adhesions: transmembrane junctions between the extracellular matrix and the cytoskeleton. Annu Rev Cell Biol 4:487–525
Chardin P, Madaule P, Tavitian A (1988) Coding sequence of human rho cDNAs clone 6 and clone 9. Nucleic Acids Res 16:2717
Chardin P, Boquet P, Madaule P, Popoff MR, Rubin EJ, Gill DM (1989) The mammalian G protein rhoC is ADP-ribosylated by Clostridium botulinum exoenzyme C3 and affects actin microfilaments in Vero cells. EMBO J 8:1087–1092
Didsbury J, Weber RF, Bokoch GM, Evans T, Synderman R (1989) *rac*, a novel ras-related family of proteins that are botulinum toxin substrates. J Biol Chem 264:16378–16382
Diekmann D, Brill S, Garrett MD, Totty N, Hsuan J, Monfries C, Hall C, Lim L Hall A (1991) bcr encodes a GTPase-activating protein for $p21^{rac}$. Nature 351:400–402
Drivas GT, Shih A, Coutavas E, Rush MG, D'Eutsachio P (1990) Characterization of four novel ras-like genes expressed in a human teratocarcinoma cell line. Mol Cell Biol 10:1793–1798
Fukimoto Y, Kaibuchi K, Hori Y, Fujioka H, Araki S, Ueda T, Kikuchi A, Takai Y (1990) Molecular cloning and characterization of a novel type of regulatory protein (GD1) for the rho proteins. Oncogene 5:1321–1328
Garrett MD, Self AJ, Vanders C, Hall A (1989) Identification of distinct cytoplasmic targets for ras/R-ras and rho regulatory proteins. J Biol Chem 264:10–13
Garrett MD, Major GN, Totty N, Hall A (1991) Purification and N-terminal sequence of the $p21^{rho}$ GTPase-activating protein rhoGAP. Biochem J 276:833–836
Goldberg DJ, Burmeister DW (1989) Looking into growth cones. Trends Neurosci 12:503–506
Hall A (1990) The cellular function of small GTP-binding proteins. Science 249:635–640
Hall A (1992a) Ras-related GTPases and the cytoskeleton. Mol Biol Cell 3:(in press)
Hall A (1992b) Signal transduction and small GTPases: a tale of two GAPs. Cell 69:389–391
Hall C, Monfries C, Smith P, Lim HH, Kozma R, Ahmed S, Vanniasingham V, Leung T, Lim L (1990) Novel brain cDNA encoding at 34 KD protein, n-chimaerin. J Mol Biol 211:11–16
Hancock JF, Magee AI, Childs JM, Marshall CJ (1989) All ras proteins are polyisoprenylated but only some are palmitoylated. Cell 57:1167–1177
Hancock JF, Cadwallader K, Paterson HF, Marshall CJ (1991) A CAAX or a CAAL motif and a second signal are sufficient for plasma membrane targeting of ras proteins. EMBO J 10:4033–4039
Hart MJ, Eva A, Evans T, Aaronson SA, Cerione RA (1991) Catalysis of guanine nucleotide exchange on the CDC42Hs protein by the dbl oncogene product. Nature 354:311–314
Hiraoka K, Kaibuchi K, Ando S, Musha T, Takaishi K, Mizuno A, Menerd L, Tomhave E, Didsbury J, Sonyderman R, Takai Y (1992) Both stimulatory and inhibitory GDP/GTP exchange proteins, Smg GDS and rhoGD1, are active on multiple small GTP-binding proteins. Biochem Biophys Res Commun 182:921–930
Hoshijima M, Kondo J, Kikuchi A, Yamamoto K, Takai Y (1990) Purification and characterization from bovine brain membranes of a GTP-binding protein with an Mr of 21 000 ADP-ribosylated by an ADP-ribosyltransferase contaminated in botulinum toxin type C1-identification as the rhoA gene product. Mol Brain Res 7:9–16
Isomura M, Kaibuchi K, Yamamoto T, Kawamura S, Katayama M, Takai Y (1990) Partial characterization of GDP dissociation stimulation (GDS) for the rho proteins from brain cytosol. Biochem Biophys Res Commun 169:652–659

Isomura M, Kikuchi A, Ohga N, Takai Y (1991) Regulation of binding of rhoB p20 to membranes by its specific regulatory protein, GDP dissociation inhibitor. Oncogene 6:119–124

Jähner D, Hunter T (1991) The ras-related gene rhoB is an immediate-early gene inducible by v-fps, epidermal growth factor, and platelet-derived growth factor in rat fibroblasts. Mol Cell Biol 11:3682–3690

Johnson DI, Pringle JR (1990) Molecular characterization of CDC42, a *Saccharomyces cerevisiae* gene involved in the development of cell polarity. J Cell Biol 111:143–152

Katayama M, Kawata M, Yoshida Y, Horiuchi H, Yamamoto T, Matsura Y, Takai Y (1991) The post-translationally modified C-terminal structure of bovine aortic smooth muscle rhoA p21. J Biol Chem 266:12639–12645

Kawahara Y, Kawata M, Sunako M, Araki SI, Koide M, Tsuda T, Fukuzaki H, Takai Y (1990) Identification of a major GTP-binding protein in bovine aortic smooth muscle cytosol as the rhoA gene product. Biochem Biophys Res Commun 170:673–683

Keller HU (1990) Diacylglycerols and PMA are particularly effective stimulators of fluid pincytosis in human neutrophils. J Cell Physiol 145:465–471

Knaus UG, Heyworth PG, Evans T, Curnette JT, Bokoch GM (1991) Regulation of phagocyte oxygen radical production by the GTP-binding protein Rac2. Science 254:1512–1515

Madaule P, Axel R (1985) A novel ras-related gene family. Cell 42:31–40

Madaule P, Axel R, Myers AM (1987) Characterization of two members of the *rho* gene family from the yeast *Saccharomyces cerevisiae*. Proc Natl Acad Sci USA 84:779–783

Maru Y, Witte ON (1991) The bcr gene encodes a novel serine/threonine kinase activity within a single exon. Cell 67:459–468

McCaffrey M, Johnson JS, Goud B, Myers AM, Rossier J, Popoff MR, Madaule P, Boquet P (1991) The small GTP-binding protein Rho1p is localized on the Golgi apparatus and post-Golgi vesicles in *Saccharomyces cerevisiae*. J Cell Biol 115:309–319

Milburn MV, Tong L, DeVos AM, Brünger A, Yamaizumi Z, Nishimura S, Kim SH (1990) Molecular switch for signal transduction structural differences between active and inactive forms of ras proteins. Science 247:939–945

Moll J, Sansig G, Fattori E, van der Putten H (1991) The murine rac1 gene: cDNA cloning, tissue distribution and regulated expression of rac1 mRNA by disassembly of actin microfilaments. Oncogene 6:863–866

Moolenaar WH, Jalink K, van Corven EJ (1992) Lysophosphatidic acid: a bioactive phospholipid with growth factor-like properties. Rev Physiol Biochem Pharmacol 119:47–65

Morel F, Doussiere J, Vignais PV (1991) The superoxide-generating oxidase of phagocytic cells. Eur J Biochem 201:523–546

Morii N, Kawano K, Sekine A, Yamada T, Narumiya S (1991) Purification of GTPase activating protein, specific for the *rho* gene products. J Biol Chem 266:7646–7650

Munemitsu S, Innis MA, Clark R, McCormick F, Ullrich A, Polakis P (1990) Molecular cloning and expression of a G25K cDNA, the human homology of the yeast cell cycle gene CDC42. Mol Cell Biol 10:5977–5982

Nishiki T, Narumiya S, Morii N, Yamamoto M, Fujiwara M, Kamata Y, Sakaguichi G, Kozaki S (1990) ADP-ribosylation of the rho/rac proteins induces growth inhibition, neurite outgrowth and acetylcholine esterase in cultured PC-12 cells. Biochem Biophys Res Commun 167:265–272

Olofsson B, Chardin P, Touchot N, Zahraoui A, Tavitian A (1988) Expression of the ras-related ralA, rho12, and rab genes in adult mouse tissues. Oncogene 3:231–234

Omann GM, Allen RA, Bokoch GM, Painter RG, Traynore AE, Sklar LA (1987) Signal transduction and cytoskeletal activation in the neutrophil. Physiol Rev 67:285–322

Otsu M, Hiles I, Gout I, Fry MJ, Ruiz-Larrea F, Panayaton G, Thompson A, Dhand R, Hsuan J, Totty N, Smith AD, Morgan SJ, Courtneidge SA, Parker PJ, Waterfield MD (1991) Characterization of two 85 Kd proteins that associate with receptor tyrosine kinases, middle T, pp60c-src complexes and PI-3-kinase. Cell 65:91–104

Paterson HF, Self AJ, Garrett MD, Just I, Aktories K, Hall A (1990) Microinjection of recombinant $p21^{rho}$ induces rapid changes in cell morphology. J Cell Biol 111:1001–1007

Polakis PG, Weber RF, Nevins B, Didsbury JR, Evans T, Snyderdman R (1989) Identification of the ral and rac1 gene products, low molecular mass GTP-binding proteins from human platelets. J Biol Chem 264:16383–16389

Ridley AJ, Hall A (1992) The small GTP-binding protein Rho regulates the assembly of focal adhesions and actin stress fibres in response to growth factors. Cell 70:389–399

Ridley AJ, Paterson HF, Johnston CL, Diekmann D, Hall A (1992) The small GTP-binding protein Rac regulates growth factor-induced membrane ruffling. Cell 70:401–410

Rubin EJ, Gill DM, Boquet P, Popoff MR (1988) Functional modification of a 21-kilodalton G protein when ADP-ribosylated by exoenzyme C3 of Clostridium botulinum. Mol Cell Biol 8:418–426

Schichting I, Almo SC, Rapp G, Wilson K, Petratos K, Lentfer A, Wittinghofer A, Kabsch W, Pai EF, Goody RS (1990) Tissue-resolved X-ray study of the conformational change in Ha-ras p21 protein on GTP hydrolysis. Nature 345:309–315

Seabra MC, Reiss Y, Casey PJ, Brown MS, Goldstein JL (1991) Protein farnesyltransferase and geranylgeranyl transferase share a common α subunit. Cell 65:429–434

Settleman J, Narasimhan V, Forster LC, Weinberg RA (1992) Molecular cloning of cDNAs encoding the GAP associated protein p190: implications for a signalling pathway from Ras to the nucleus. Cell 69:539–550

Shinjo K, Koland JG, Hart MT, Narasimhan V, Johnson DI, Evan T, Cerione RA (1990) Molecular cloning of the gene for the human placental GTP-binding protein Gp (G25K): identification of this GTP-binding protein as the human homolog of the yeast cell-division cycle protein CDC42. Proc Natl Acad Sci USA 87:9853–9857

Shirsat NV, Pignolo RJ, Kreider BL, Rovera G (1990) A member of the *ras* gene superfamily is expressed specifically in T, B and myeloid hemopoietic cells. Oncogene 5:769–772

Stasia M-J, Jouan A, Bourmeyster N, Boquet P, Vignais PV (1991) ADP-ribosylation of a small size GTP-binding protein in bovine neutrophils by the C3 exoenzyme of Clostridium botulinum and effect on the cell motility. Biochem Biophys Res Commun 180:615–622

Yamane H, Farnsworth CC, Xiec H, Evans T, Howald WN, Gelb MH, Glomset JA, Clarke S, Fung BKK (1991) Membrane-binding domain of the small G protein G25K contains an S-(all-trans-geranylgeranyl) cysteine methyl ester at its carboxyl terminus. Proc Natl Acad Sci USA 88:286–290

Yeramian P, Chardin P, Madaule P, Tavitian A (1987) Nucleotide sequence of human rho cDNA clone 12. Nucleic Acids Res 15:1869

CHAPTER 37

The Mammalian Homolog of the Yeast Cell-Division-Cycle Protein, CDC42: Evidence for the Involvement of a Rho-Subtype GTPase in Cell Growth Regulation

M.J. HART, D. LEONARD, Y. ZHENG, K. SHINJO, T. EVANS, and R.A. CERIONE

A. Growth Factor-Coupled Signal Transduction

In recent years, through studies performed at the biochemical and molecular genetic level, the existence of an extended family of growth factor receptor tyrosine kinases has been documented. Members of this family include the epidermal growth factor (EGF) receptor, the neu/erbB-2 tyrosine kinase, the insulin receptor, the insulin-like (I) growth factor receptor, and the platelet-derived growth factor (PDGF) receptor (ULLRICH and SCHLESSINGER 1990; CADENA and GILL 1992). While a good deal of information has been generated regarding the general structural features of these receptors and how their tyrosine kinase activities become activated in response to growth factor binding, until recently, much less has been known about the sequence of events following receptor activation and, specifically, about the identity of specific phosphosubstrates for these receptors. However, through the development of antiphosphotyrosine antibodies, a number of potential participants in the signaling pathways of these growth factor receptors have been identified; these include the enzyme phospholipase C-γ (i.e., PLC-γ; WAHL et al. 1989; MARGOLIS et al. 1989; MEISENHELDER et al. 1989), the ras GTPase activating protein (ras GAP; MOLLOY et al. 1989; ELLIS et al. 1990), and the 85-kDa regulatory subunit of the phosphatidylinositol-3 (PI-3) kinase (CANTLEY et al. 1991; HU et al. 1992). In each of these cases, the interactions between the receptor tyrosine kinase and the enzyme/transducer occurs via the binding of a src homology region (SH-2 domain) on the enzyme/transducer to the autophosphorylated receptor (KOCH et al. 1991).

There appear to be a number of other cellular proteins which lack SH-2 domains but still are likely to be important participants in growth-factor-coupled signaling pathways. Some especially interesting examples are the GTPases, given their role as transducers in a number of hormone receptor signaling cascades. There have been a number of reports demonstrating that pertussis toxin treatment inhibits different growth factor-stimulated events (e.g., EGF-stimulated phosphatidylinositol lipid turnover in hepatocytes; JOHNSON and GARRISON 1987; YANG et al. 1991), which suggest the possibility of direct interactions between receptor tyrosine kinases and pertussis toxin-sensitive (heterotrimeric) GTPases.

Reconstitution approaches have proven invaluable (CERIONE 1991) in characterizing the functional interactions between hormone receptors (as well as the photoreceptor rhodopsin) and heterotrimeric GTPases. Thus, we first set out to establish reconstituted phospholipid vesicle systems containing the purified EGF receptor tyrosine kinases and different purified GTPases to determine whether analogies did in fact exist between hormone receptor/G-protein coupling and the involvement of G-proteins in the actions of the EGF receptor tyrosine kinase. It was through these reconstitution approaches that we discovered a 22-kDa GTPase which initially was felt to represent a potential transducer of EGF signaling (HART et al. 1990). It is the identification and biochemical characterization of this protein (now referred to as CDC42) which is the subject of this review.

B. Reconstitution of an Epidermal Growth Factor Stimulated Phosphorylation of a 22-kDa GTPase

Since all aspects of EGF action appear to require the tyrosine kinase activity of the receptor (cf. ULLRICH and SCHLESSINGER 1990), we examined the ability of different GTPases to serve as phosphosubstrates for the EGF receptor in reconstituted lipid vesicle systems. An initial examination of different heterotrimeric GTPases indicated that the EGF receptor could specifically phosphorylate the α subunits of the bovine brain inhibitory GTPases of the adenylyl cyclase system (i.e., α_i, α_o) and the α subunit of the related G_o protein, although in all cases the stoichiometries of phosphorylation were low, i.e., typically ≤ 0.1 mol ^{32}P incorporated per mol of α subunit. Likewise, the oncogenic ras proteins were very poor phosphosubstrates for the EGF receptor in these reconstituted systems. However, it was during the course of these studies that we discovered a 22-kDa protein, present in partially purified preparations of the bovine brain G_i and G_o proteins, which served as an excellent phosphosubstrate for the reconstituted EGF receptor. The phosphorylation of this 22-kDa protein was sharply attenuated by sub-micromolar concentrations of GTPγS (Fig. 1), which provided us with the first indication that this phosphosubstrate was a high affinity, guanine-nucleotide-binding protein. The phosphosubstrate was purified from bovine brain membranes by using two assays: [^{35}S]GTPγS binding activity and the guanine-nucleotide-sensitive, EGF-stimulated phosphorylation of this protein after its reconstitution into phospholipid vesicles (HART et al. 1990). The purification of this phosphosubstrate was achieved by a series of chromatographic steps which included DEAE-Sephacel, ultrogel AcA34, phenyl Sepharose, hydroxyapatite, and Mono-Q chromatographies. The stoichiometries of the EGF-stimulated phosphorylation of the purified

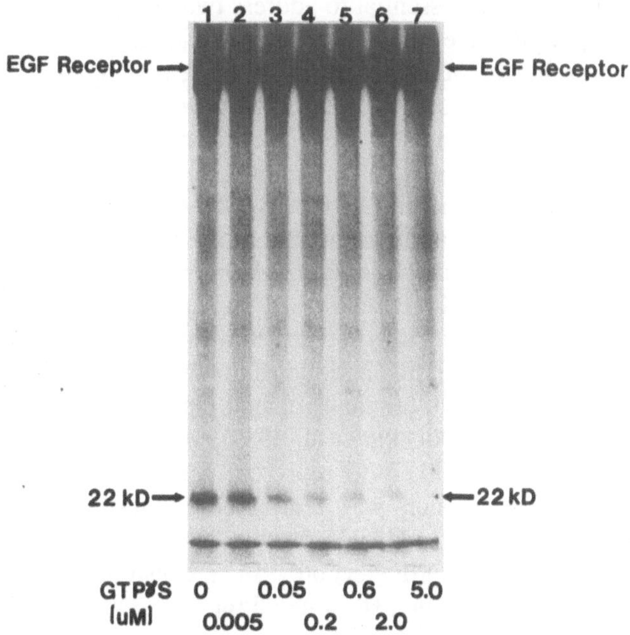

Fig. 1. The phosphorylation of a 22-kDa protein from bovine brain membranes by the epidermal growth factor (*EGF*) *receptor*. A fraction from the partially purified G_i and G_o proteins from bovine brain membranes was reconstituted with the purified human placental EGF receptor into phospholipid vesicles, as outlined in HART et al. (1990). The protein-lipid vesicles were added to phosphorylation incubations for 22 min at room temperature. *Lanes 1–7* represent lipid vesicles (containing the EGF receptor and the 22-kDa protein) that were preincubated with the indicated levels of GTPγS (plus 15 m*M* $MgCl_2$) for 20 min at room temperature prior to their addition to the phosphorylation incubations. These data were taken from HART et al. (1990)

22-kDa protein approached 2 mol of incorporated ^{32}P per mol GTPγ^{35}S binding activity; this represents the highest extent of phosphorylation that we have observed between any tyrosine kinase and any GTPase (as well as for any phosphosubstrate for the EGF receptor). The phosphorylation of the 22-kDa protein, which occurred entirely on tyrosine residues, was specifically catalyzed by the EGF receptor; neither the insulin receptor nor the src kinase was able to substitute for the EGF receptor. The abilities of both GTP (GTP analogs) and GDP to inhibit the EGF-stimulated phosphorylation of the 22-kDa protein coincided with the direct binding of these guanine nucleotides to the purified protein. However, while guanine nucleotide binding attenuated the phosphorylation once the 22-kDa protein was (tyrosine) phosphorylated, it bound [^{35}]GTPγS with a higher affinity than its non-phosphorylated counterpart.

At this point, it seemed essential to address two key questions regarding the 22-kDa phosphosubstrate; the first concerned the identity of this protein and/or its relationship to other members of the emerging superfamily of ras-related GTPases. The second question concerned the physiological relevance of the results obtained from the reconstituted EGF receptor system, namely what role (if any) does this GTPase play in cell growth regulation? Regarding the identity of the 22-kDa phosphosubstrate, it was relatively easy to eliminate various members of the ras family, including the ras proteins (based on the inability of the 22-kDa protein to cross-react with ras antibodies) and the rho GTPases (based on the inability of the 22-kDa phosphosubstrate to be ADP-ribosylated following exposure to botulinum exoenzyme C_3). Using two peptide antibodies that were specific for a human placental and platelet GTPase, originally named Gp (EVANS et al. 1986) and then later renamed G25K (POLAKIS et al. 1989), we demonstrated that the brain phosphosubstrate was likely to represent a form of the human Gp/G25K protein. This conclusion was reinforced by the finding that the purified human platelet Gp/G25K protein served as an excellent phosphosubstrate for the EGF receptor in reconstituted lipid vesicle systems, with the phosphorylation showing all the same characteristics (e.g., sharply attenuated by guanine nucleotides) as observed in studies with the brain 22-kDa protein.

Unfortunately, while determining the identity of the 22-kDa phosphosubstrate for the EGF receptor was relatively straightfoward, the question of the physiological role of the Gp/G25K protein and its possible involvement as a transducer of growth factor action still remains to be answered. Our initial attempts to demonstrate and/or characterize the tyrosine phosphorylation of the 22-kDa protein in vivo have not yielded conclusive information. Much of the difficulty has involved the development of conditions for maximal recognition of the tyrosine phosphorylated 22-kDa protein by antiphosphotyrosine antibodies and the development of highly sensitive anti-Gp/G25K antibodies. However, perhaps the chief difficulty stems from the findings that the 22-kDa protein is most susceptible to (EGF-stimulated) tyrosine phosphorylation in its guanine-nucleotide-depleted state and is least susceptible to phosphorylation when either GDP or GTP is bound. This, then, suggests that if tyrosine phosphorylation of the 22-kDa protein is occurring in vivo, it may happen within a narrow time period (i.e., while the protein is undergoing GDP–GTP exchange), which in turn suggests that the actions of the tyrosine kinase and the exchange factor need to be tightly correlated (this will be considered again below). An alternative approach that has been used to examine the role of proteins in cell growth involves gene transfer techniques. Specifically, the cDNA for the protein of interest is introduced into a fibroblast and then the influence of the expressed protein on the rate of cell growth is monitored. In order to use this approach to study the Gp/G25K protein, the molecular cloning of the cDNA of this protein was necessary.

```
                        10                  20                  30                  40
CDC42Hs   M Q T I K C - - V V V G D G A V G K T C L L I S Y T T N K F P S E Y V P T V F D N Y A V T V M I G G
CDC42Sc   M Q T L K C - - V V V G D G A V G K T C L L I S Y T T N Q F P A D Y V P T V F D N Y A V T V M I G D
rac1      M Q A I K C - - V V V G D G A V G K T C L L I S Y T T N A F P G E Y I P T V F D N Y S A N V M V D G
rac2      M Q A I K C - - V V V G D G A V G K T C L L I S Y T T N A F P G E Y I P T V F D N Y S A N V M V D S
rho12     M A A I R K K L V I V G D G A C G K T C L L I V F S K D Q F P E V Y V P T V F E N Y V A D I E V D G
Nras      M T E Y K L - - V V V G A G G V G K S A L T I Q L I Q N H F V D E Y D P T I E D S Y R K Q V V I D G

            50                  60                  70                  80                  90
CDC42Hs   E P Y T L G L F D T A G Q E D Y D R L R P L S Y P Q T D V F L V C F S V V S P S S F E N V K E K W V
CDC42Sc   E P Y T L G L F D T A G Q E D Y D R L R P L S Y P S T D V F L V C F S V I S P P S F E N V K E K W F
rac1      K P V N L G L W D T A G Q E D Y D R L R P L S Y P Q T D V F L I C F S L V S P A S F E N V R A K W Y
rac2      K P V N L G L W D T A G Q E D Y D R L R P L S Y P Q T D V F L I C F S L V S P A S Y E N V R A K W F
rho12     K Q V E L A L W D T A G Q E D Y D R L R P L S Y P D T D V I L M C F S I D S P D S L E N I P E K W T
Nras      E T C L L D I L D T A G Q E E Y S A M R D Q Y M R T G E G F L C V F A I N N S K S F A D I N L Y R E

            100                 110                 120                 130                 140
CDC42Hs   P E I T H H C P K - T P F L L V G T Q I D L R D D P S T I E K L A K N K Q K P I T P E T A E K L A R
CDC42Sc   P E V H H H C P G - V P C L V V G T Q I D L R D D K V I I E K L Q R Q R L R P I T S E Q G S R L A R
rac1      P E V R H H C P N - T P I I L V G T K L D L R D D K D T I E K L K E K K L T P I T Y P Q G L A M A K
rac2      P E V R H H C P S - T P I I L V G T K L D L R D D K D T I E K L K E K K L A P I T Y P Q G L A L A K
rho12     P E V K H F C P N - V P I I L V G N K K D L R N D E H T R R E L A K M K Q E P V K P E E G R D M A N
Nras      Q I K R V K D S D D V P M V L V G N K C D L P T R T V D T K Q A H E L A K S Y G I P F I E T S A K T

              150                 160                 170                 180                 190
CDC42Hs   D L K A V K Y V E C S A L T Q K G L K N V F D E A I L A A L E P P E P K K S R R C V L L
P1'                                             N V F D E A I L A A L E P P E T Q P K R K C C I F
CDC42Sc   E L K A V K Y V E C S A L T Q R G L K N V F D E A I V A A L E P P V I K K S K K C T I L
rac1      E I G A V K Y L E C S A L T Q R G L K T V F D E A I R A V L C P P P V K K R K R K C L L L
rac2      E I D S V K Y L E C S A L T Q R G L K T V F D E A I R A V L C P Q P T R Q Q K R A C S L L
rho12     R I G A F G Y M E C S A K T K D G V R E V F E M A T R A A L Q A R R G K K K S G C L V L
Nras      R Q G V E D A F Y T L V R E I R Q Y R L K K L N S S D D G T Q G C M G S P C V L M C K T L
```

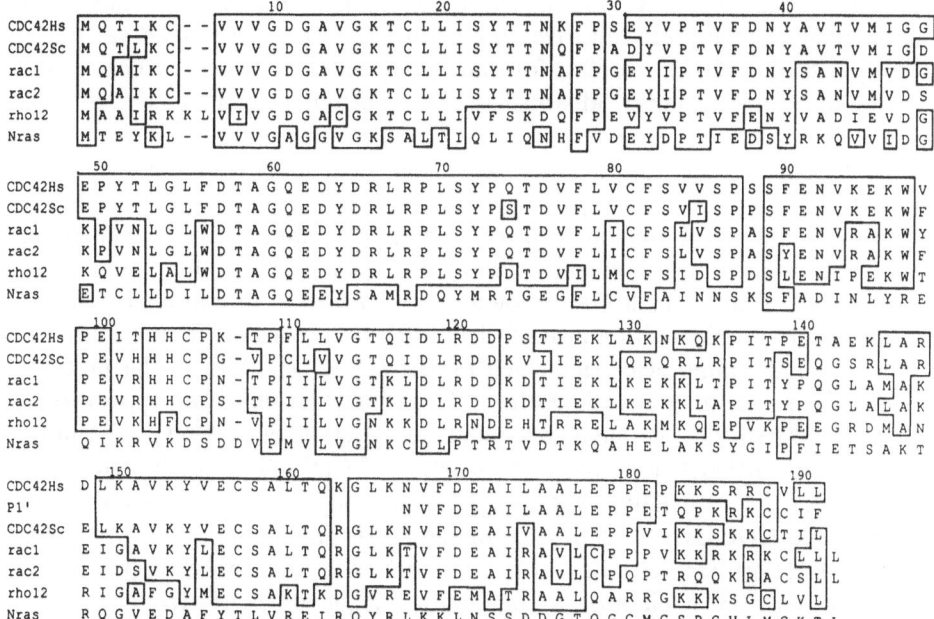

Fig. 2. Comparison of the amino acid sequence of the human placental CDC42Hs with the sequences of other small GTPases. The sequence *P1'* represents the carboxyl terminal of the human fetal brain GTPase which was cloned by MUNIMETSU et al. (1990). Residues identical to those of the CDC42Hs protein are *boxed*. These data were taken from SHINJO et al. (1990)

C. Molecular Cloning of the Human Gp/G25K Protein: Identification of this Protein as the Human Homolog of the Yeast Cell Division Cycle Protein CDC42Sc

Oligonucleotide probes based on two peptide sequences obtained for the human placental and platelet Gp/G25K protein were used to screen a human placental (λ gt11) cDNA library. A full-length cDNA clone was obtained which coded for the available Gp/G25K peptide sequences (Fig. 2) (SHINJO et al. 1990). The predicted amino acid sequence was approximately 30% identical to the ras GTPases and contained many of the residues in ras which, when mutated, result in oncogenic activation, including the glycine at position 12, an alanine at position 59, and a glutamine at position 61. The predicted sequence was 50% identical to the rho GTPases and approximately 70% identical to the rac1 and rac2 GTPases, which suggested that the Gp/G25K protein is a member of the rho subgroup of ras-related GTPases. Certainly the most interesting outcome of the molecular cloning was the realization that the amino acid sequence for Gp/G25K was 80% identical (and 90% similar) to that for a *Saccharomyces cerevisiae* cell division cycle protein, CDC42 (now designated CDC42Sc), first identified by

JOHNSON and PRINGLE (1990). The human cDNA was able to fully complement temperature-sensitive mutations of the yeast *CDC42* gene, whereas the cDNAs coding for rho, rac1, rac2, or ras were not. Thus, based on the sequence identity and the functional complementation, we felt that it would be more appropriate to name this protein after the yeast gene product, and so we now refer to the human form of the protein as CDC42Hs.

An alternative human form of CDC42 was cloned from a fetal brain cDNA library by MUNIMETSU and colleagues (1990); the amino acid sequence of the human brain protein is 95% identical to that of the human placental protein, with most of the differences existing in the carboxyl terminal 25 amino acids (indicated by P1′ in Fig. 2). Both of the proteins contain a cysteine four residues from the carboxyl terminus and appear to be isoprenylated (i.e., geranylgeranylated, see YAMANE et al. 1991). As will be discussed further below, the isoprenylation of CDC42Hs may be important both for its association with membranes and for its interactions with different CDC42Hs-regulatory proteins. It is interesting that while the CDC42Hs protein has many of the amino acid residues which appear to be essential for GTP binding, it contains the sequence TQID (residues 115–118) in place of the typical guanine nucleotide binding element NKXD found in ras and in most other GTPases. At the present time, the functional consequences of this sequence difference are not known.

D. Function of CDC42Sc in *Saccharomyces cerevisiae*

Given the ability of CDC42Hs to substitute for CDC42Sc in yeast, together with the high degree of sequence similarity between the yeast and human proteins, it seems likely that the structures of those proteins regulating the actions of, or interacting with, the CDC42 protein in yeast will be similar to those interacting with CDC42 in mammalian cells. The CDC42Sc protein appears to be essential for bud-site assembly in yeast and was first identified by screening for temperature-sensitive mutants which had phenotypes similar to mutants detected in the *CDC24* gene, i.e., another gene which is essential for normal budding in *S. cerevisiae* (ADAMS et al. 1990). A number of other gene products have been implicated in bud-site selection and assembly in *S. cerevisiae*, although the specific functions of many of these proteins are still not known (DRUBIN 1991). The CDC24 protein shares sequence homology with a mammalian protein, proto-dbl and its oncogenic counterpart dbl (RON et al. 1991); the latter has been shown to regulate GDP–GTP exchange on CDC42Hs (see below), thus suggesting that CDC24 may be an exchange factor for CDC42Sc. The CDC43 protein also has been implicated in bud-site assembly and has recently been shown to be responsible for catalyzing the isoprenylation of the CDC42Hs protein, i.e., the CDC43 protein appears to be the *S. cerevisiae* geranylgeranyltransferase (FINEGOLD et al. 1991). Two other gene products, BEM2 and BEM3, appear to be acting at a point downstream from CDC42Sc in the bud-site assembly pathway. Both of these proteins share homology with a family of GAPs (see below). A number of gene products, including Bud1 through Bud5, have been identified as being required for the correct positioning of the bud sites (CHANT and HERSKOWITZ 1991; CHANT et al. 1991). Bud1, which was previously called RSR-1 (BENDER and PRINGLE 1989), encodes the yeast homolog of rap1a, a ras-related GTPase. Bud5 is homologous to the *CDC25*-encoded GDP–GTP exchange protein (BROEK et al. 1987) and may promote guanine nucleotide exchange on the Bud1 protein. As more information becomes available regarding the interactions of these different yeast proteins in the formation of proper bud sites, important clues will be obtained regarding the physiological functions of CDC42 in mammalian cells and the types of mammalian regulatory proteins which work together with CDC42 in a common regulatory/signaling pathway.

E. Possible Involvement of CDC42Hs in Cell Growth Regulation

I. cDNa Transfection Studies

We have used gene transfer approaches to introduce the cDNAs for the wild-type CDC42Hs protein (with a glycine at position 12, i.e., CDC42Hs^{gly12}), and a GTPase-defective form of CDC42Hs (designated CDC42Hs^{val12}) into various fibroblasts (SHINJO et al., in preparation). Neither the expression of the wild-type CDC42Hs nor the CDC42Hs^{val12} species in NIH-3T3 cells, nor the expression of these proteins in NR6 fibroblasts (which lack detectable levels of the EGF receptor tyrosine kinase), resulted in a stimulation of the rate of growth of these cells or an enhanced focus formation. The expression of the CDC42Hs^{val12} protein in NR6 cells was accompanied by a multinucleate phenotype; it remains to be determined whether this phenotype is related to a similar phenotype observed in *S. cerevisiae* cells by JOHNSON and his colleagues upon the expression of GTPase-defective forms of CDC42Sc (ZIMAN et al. 1991) and to multinucleate phenotypes observed in NIH-3T3 cells upon the overexpression of the dbl oncogene product (a putative CDC42Hs-regulatory protein, see below).

While the expression of different forms of CDC42Hs in the normal NIH-3T3 fibroblasts (or in NR6 cells) did not have a significant influence on cell growth, some interesting results were obtained when CDC42Hs was overexpressed in NR6 fibroblasts overexpressing 20000–50000 EGF receptors per cell or in NR6 cells overexpressing 200000–500000 EGF receptor/erbB-2 chimeras per cell (these chimeras contained the extracellular domain of the EGF receptor and the membrane-spanning region and cytoplasmic domain of erbB-2). In these experiments, the overexpression of the wild-type CDC42Hs protein elicited a strong inhibition of the rate of cell growth (both in the absence and in the presence of EGF), while the expression of the CDC42Hs^{val12} species resulted in a marked, EGF-dependent inhibition of the rate of cell growth. In the NR6 cells overexpressing the EGF receptor/erbB-2 chimera, the EGF-dependent inhibition of growth (due to the expression of the CDC42Hs^{val12} species) was accompanied by striking changes in the morphology of the cells, i.e., they lost their adherence and "rounded up."

Although the results from these cDNA transfection studies reinforced the suggestion for some type of interplay between the EGF receptor tyrosine kinase (or related tyrosine kinases) and CDC42Hs (since the effects of CDC42Hs on cell growth were accentuated in cells expressing these tyrosine kinases), the molecular mechanisms underlying these effects are not clear. However, one possibility is that the complete GTP-binding/GTPase cycle of CDC42Hs is necessary for eliciting a normal growth signal and that the overexpression of either the wild-type CDC42Hs or GTPase-defective forms interrupts the cycle (i.e., due to the overexpressed GTPase "tying up" the regulatory proteins that are necessary for the normal stimulation of the cycle). This then prompted us to identify the different proteins responsible for regulating the GTP-binding/GTPase cycle of CDC42Hs.

II. CDC42Hs-Regulatory Proteins

1. CDC42Hs GTPase Activating Protein

The first CDC42Hs regulatory protein which we identified was a GAP purified from human platelets (HART et al. 1991a). CDC42Hs GAP activity was detected in both the cytosolic and solubilized membrane fractions, but the cytosolic GAP activity could not be recovered beyond a single (DEAE Sephacel) column step. The membrane-associated GAP activity was more stable and was purified, following its solubilization with Triton X-100, by a series of chromatography steps which included S-Sepharose, hydroxyapatite, Mono-Q, and Mono-S chromatographies. The purified GAP activity appeared to correspond to a protein with a molecular mass of approximately 25 kDa, i.e., similar in size to the rho GAPs purified from the cytosol of spleen (GARRETT et al. 1989) and bovine adrenal glands (MORII et al. 1991).

The CDC42Hs GAP was able to stimulate the GTPase activities of both the *Escherichia coli* expressed and the *Spodoptera frugiperda*-expressed CDC42Hs proteins (HART et al. 1991a), leading to the conclusion that geranylgeranylation, which does not occur in *E. coli*, is not necessary for CDC42Hs–GAP interactions. When the glycine at position 12 is changed to a valine, the *E. coli* expressed CDC42Hs has a greatly reduced GTPase activity and this activity is not stimulated further by the addition of the CDC42Hs GAP. The platelet CDC42Hs GAP is not able to stimulate the GTPase activities of either the H-ras or rap1a GTPases; however, it is capable of weakly stimulating the GTPase activity of the rhoA protein and also stimulates the GTPase activities of the rac1 and rac2 GTPases to extents that are comparable to the stimulation of the GTPase activity of CDC42Hs (M. HART, unpublished data).

The apparent similarities in the sizes of the CDC42Hs and rho GAPs, coupled with the finding that the CDC42Hs GAP can cross-react with the rho protein, suggests that these GAPs are likely to have very similar structures. A number of different proteins share homology with amino acid sequences first identified in the spleen rho GAP (Fig. 3). DIEKMANN et al. (1991) first reported that a 30 amino acid peptide derived from the spleen rho GAP was 50% identical to a region near the carboxyl terminus of the bcr protein. The bcr protein has received a great deal of attention because of its involvement in the development of certain leukemias following the fusion of a portion of this protein with the c-abl tyrosine kinase. DIEKMANN et al. demonstrated that the carboxl terminal half of bcr served as a GAP for the rac1 GTPase, but not for rho, more recently, we have shown that the *S. frugiperda*-expressed bcr protein serves as a GAP for CDC42Hs (HART et al. 1992). A second protein sharing homology with the rho GAP and bcr is chimerin, a brain protein with a molecular mass of approximately 30 kDa. The amino terminal half of this protein shares homology with the regulatory domains of members of the protein kinase C family, while the carboxyl terminal portion of chimerin contains GAP homology. Like bcr, chimerin was shown to serve as a GAP for the rac1 protein, but not for rho (DIEKMANN et al. 1991). A third member of the family is the ras-GAP-binding protein p190 (SETTLEMAN et al. 1992). This protein contains a number of interesting domains including a GTP-binding motif, a putative transcriptional regulatory domain, together with the GAP homology region. Recently, SETTLEMAN et al. (in press) have shown that p190 serves as a GAP for the rho, rac, and CDC42Hs proteins. The 85-kDa regulatory domain (p85) of the phosphatidylinositol-3 (PI-3) kinase also contains a region of homology with the above GAPs. However, thus far, we have not found p85 to serve as a GAP for any member of the rho subgroup (Y. ZHENG unpublished). Two other *S. cerevisiae* gene products, designated BEM2 and BEM3, which appear to function downstream from CDC42Hs in the bud-site assembly pathway, contain GAP homology regions, and we know now that only BEM3, and not BEM2, serves as a GAP for CDC42Hs and rac (ZHENG et al. 1993). A key question will be to determine whether all proteins (p85, BEM2) containing a GAP homology region are in fact GAPs and how the different regulatory domains of these proteins (e.g., bcr, p190) influence their GAP activities.

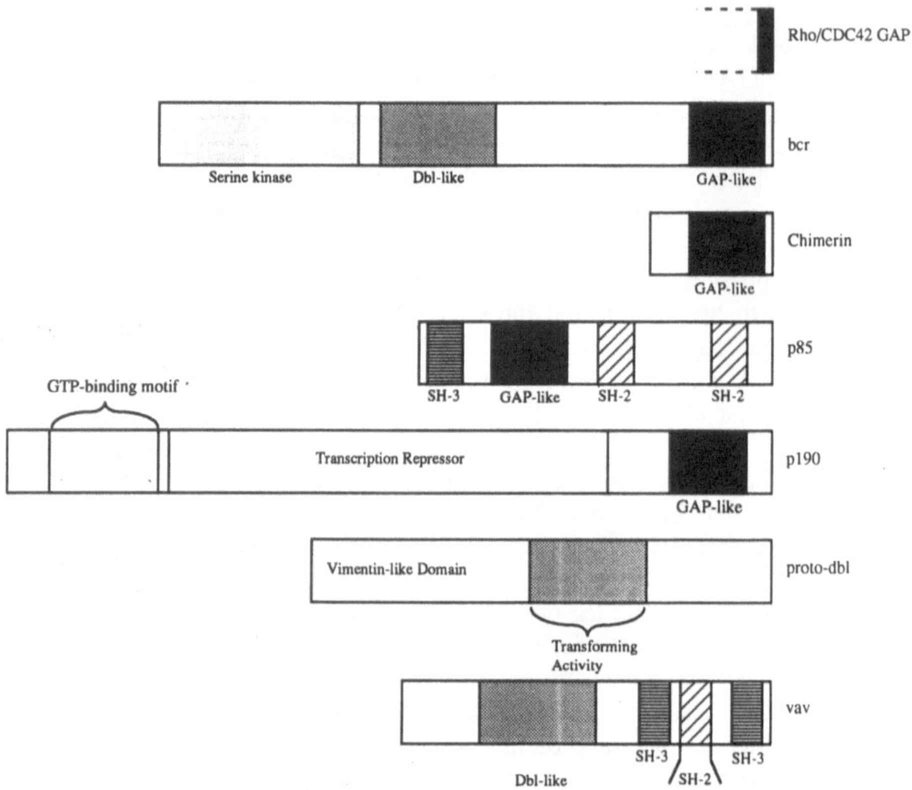

Fig. 3. Different members of the rho/CDC42 GAP and Dbl families. A 25-kDa, membrane-associated protein from human platelets, which serves as a GTPase activating protein (GAP) for CDC42Hs (HART et al. 1991a) appears to be similar in structure (based on the available amino terminal sequence) to a 29-kDa GAP (from spleen cytosol) for the rho GTPase (GARRETT et al. 1989). Based on the available (internal) sequence for the rho GAP (DIEKMANN et al. 1991), it appears that the rho/CDC42Hs GAPs share sequence homology with the bcr protein (M_r ~160 kDa), chimerin (M_r ~30 kDa), p85, and the p190 protein. Proto-dbl is a 115-kDa, cytoskeletal-associated protein which contains a region, in the carboxyl-terminal half of the molecule, that is capable of transforming NIH 3T3 cells and serving as an exchange factor (i.e., catalyzes the GDP–GTP activity) for CDC42Hs (HART et al. 1991b). The bcr and vav proteins also contain regions that are homologous to a region in dbl which appears to be essential for transformation

2. CDC42Hs Guanine Nucleotide Dissociation Stimulator

The fact that the bcr oncogene product and the ras-GAP-binding protein p190 serve as GAPs for CDC42Hs reinforces the notion that CDC42Hs may participate in a growth-regulatory pathway. Perhaps a stronger piece of evidence for such participation comes from the findings that the dbl oncogene product can influence the guanine nucleotide-binding activity of CDC42Hs (Hart et al. 1991b). The *dbl* oncogene was first identified by Eva and Aaronson (1985) by transfecting NIH 3T3 cells with the DNA from a human diffuse B cell lymphoma. *Proto-dbl* encodes for an approximately 115-kDa cytoskeletal-associated protein; oncogenic activation occurs as an outcome of a recombination event in which the 5′ terminal half of *proto-dbl* is replaced with some unidentified human sequences, yielding the approximately 66-kDa oncogenic dbl protein (designated p66-dbl). It is interesting that a region of p66-dbl which appears to be critical to its oncogenicity (i.e., between residues 498 and 735 of the proto-dbl sequence) is homologous to a 238 amino acid region from the center of the CDC24 molecule, a yeast cell division cycle protein which like CDC42 appears to be involved in bud-site assembly. Given this similarity, we examined whether p66-dbl might be capable of influencing the GTP-binding or GTPase activity of CDC42Hs. We first demonstrated that p66-dbl was not capable of acting as a GAP for CDC42Hs. However, it did strongly stimulate the dissociation of bound GDP from the CDC42Hs protein; specifically, in the absence of p66-dbl, the halftime for GDP dissociation (at Mg^{2+} levels $>1\,mM$) was at least 20 min, whereas in the presence of insect cell lysates expressing oncogenic dbl, the halftime for GDP dissociation was reduced to approximately 2 min (Fig. 4A). This effect was most pronounced for GDP, i.e., p66-dbl only weakly stimulated the dissociation of bound [^{35}S]GTPγS from CDC42Hs and showed no detectable stimulation of the dissociation of bound [γ^{32}P]GTP. Thus, p66-dbl was able to effectively stimulate the exchange of bound GDP for radiolabeled GTPγS on CDC42Hs (see Fig. 4B).

The stimulation of GDP dissociation by dbl appeared to be highly specific for the CDC42Hs protein, i.e., the insect cell lysates expressing p66-dbl showed only a weak ability to stimulate GDP dissociation from the rac1 protein and no ability to stimulate GDP dissociation from either the ras or rap1a proteins. The dbl-catalyzed dissociation of GDP can occur with the human platelet CDC42Hs, the *S. frugiperda*-expressed GTPase protein, and with the *E. coli*-expressed CDC42Hs. These results suggest that the geranylgeranylation of CDC42Hs is not necessary for the optimal coupling of this GTPase to its guanine nucleotide dissociation stimulator GDS. At this point, a key question concerns whether dbl itself functions as the GDS protein or whether other accessory proteins (which may be present in insect cells) are necessary to bind to dbl to elicit its full GDS activity. Proto-dbl appears to be present mainly in brain, adrenal glands, and gonads (Ron et al. 1988), whereas the CDC42Hs protein appears to be ubiquitously distributed. Thus, it is of interest to determine whether other dbl-related molecules might be capable of stimulating the guanine nucleotide dissociating activity of CDC42Hs in other cell types. There are in fact two interesting candidate proteins for such a role. One of these is the bcr protein, which we already have discussed (above) within the context of a rho GAP family (also

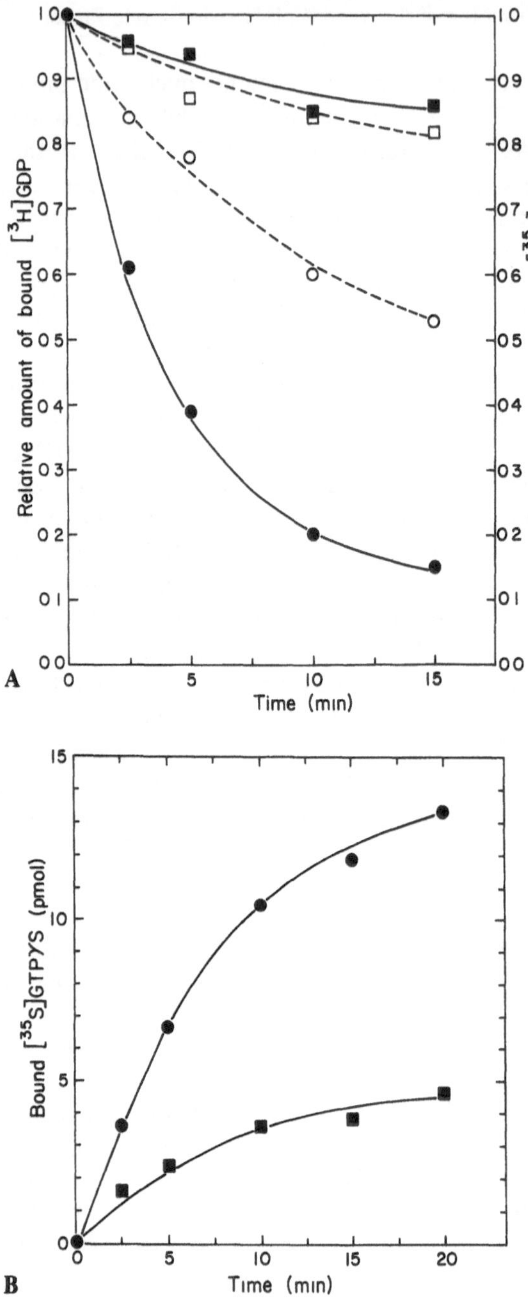

Fig. 4.A,B. Effects of *S. frugiperda* lysates expressing oncogenic dbl (M_r ~ 66 kDa) on [^{3}H]GTPγS binding to CDC42Hs. **A** Time course for the dissociation of bound [^{3}H]GDP or [^{35}S]GTPγS from platelet CDC42Hs. Samples from *S. frugiperda* control lysates (□,■) or from lysates expressing oncogenic dbl (○,●) were added to incubations containing approximately 5 μg of platelet CDC42Hs and [^{3}H]GDP (■,●) or [^{35}S]GTPγS (□,○) as described in Hart et al. (1991b). **B** Time course for the binding of [^{35}S]GTPγS to the platelet CDC42Hs protein. The CDC42Hs was preloaded with GDP and then added to incubations containing [^{35}S]GTPγS together with samples from the *S. frugiperda* control lysates (■) or from lysates expressing oncogenic dbl (●). These data were taken from Hart et al. (1992b)

see Fig. 3). The bcr protein contains a region within the center of the molecule (238 amino acids) which is homologous to the dbl and CDC24 proteins. A second protein which contains a dbl homology region is the vav oncogene product (KATZAU et al. 1989). This is particularly interesting because of our early reconstitution work, which indicated that CDC42Hs was most susceptible to tyrosine phosphorylation (by the reconstituted EGF receptor) when it was in a guanine-nucleotide-depleted state. This then suggested that maximal (in vivo) tyrosine phosphorylation of CDC42Hs would require the tight coordination of a GDS and a tyrosine kinase. The vav oncogene product (see Fig. 3) contains a SH-2 domain and recently has been shown to bind directly to tyrosine kinases such as the EGF receptor (BUSTELO et al. 1992; MARGOLIS et al. 1992). Thus, it is tempting to speculate that vav, or a vav-like molecule, might be capable of eliciting GDP dissociation from CDC42Hs (or a related rho-like molecule) and promoting its tyrosine phosphorylation by bringing a tyrosine kinase into the appropriate proximity of the GTPase.

3. CDC42Hs Guanine Nucleotide Dissociation Inhibitor

Recently, we identified a third regulatory activity for CDC42Hs. This activity, which represented the inhibition of GDP dissociation from CDC42Hs, was purified from the cytosol of bovine brain (LEONARD et al. 1992) and corresponded to a protein with an apparent molecular mass of 28 kDa. Although the amino terminus of the purified protein was blocked, it was possible to show that four cyanogen-bromide-generated fragments (10–20 amino acids each) were essentially identical to four regions within the carboxyl terminal portion of the rho guanine nucleotide dissociation inhibitor (GDI), first identified by UEDA et al. (1990) and then later cloned by FUKUMOTO et al. (1990). Thus, it is possible that a common GDI is involved in stabilizing the GDP-bound states of the different members of the rho subgroup.

The CDC42Hs GDI activity appears to be highly selective for the platelet CDC42Hs or for the insect cell-expressed GTPases and shows little or no capability of inhibiting GDP dissociation from the *E. coli*-expressed CDC42Hs. These results suggest that the isoprenylation of the carboxyl terminal cysteine of CDC42Hs is essential for proper coupling to the GDI. We have found that the GDI will effectively inhibit the dbl-catalyzed GDP dissociation (as well as the intrinsic dissociation of GDP which occurs at low Mg^{2+}; LEONARD et al. 1992); interestingly, the inhibition by the GDI cannot be overcome by the addition of increasing amounts of dbl-containing lysates to the assay solution. This then suggests that the GDI and the GDS do not bind in an identical manner to CDC42Hs.

In addition to stabilizing the GDP-bound state of CDC42Hs, the GDI also appears to influence the ability of CDC42Hs to associate with membranes. Specifically, the GDI causes a dose-dependent release of (endogenous) CDC42Hs from membranes prepared from human placenta and human epidermoid carcinoma (A431) cells (LEONARD et al. 1992). The mechanism by which the GDI mediates the release of CDC42Hs from membranes is not yet known, although it appears to occur equally well when CDC42Hs is in the GDP-bound state, a guanine nucleotide-depleted state, or even when the protein is in the GTP-bound state. The latter observation raised the question of whether the GDI might influence other activities of the GTP-bound CDC42Hs protein. Somewhat unexpectedly, we (HART et al. 1992) have recently determined that the GDI was capable of inhibiting both the intrinsic GTPase activity of CDC42Hs as well as GTPase activity stimulated by the platelet (25-kDa) GAP or by the bcr protein. To our knowledge, this represents the first identification of a GTPase inhibitory protein for a ras-like GTPase. At this point, we would suggest that in the case of CDC42Hs, the GDI stabilizes both the GDP-bound state as well as the GTP-bound state of the protein. This, together with the ability of the GDI to trigger the release of CDC42Hs from membranes, has caused us to speculate that the GDI may actually serve as a "shuttle molecule" for CDC42Hs. For example, it may be necessary for the CDC42Hs molecule to cycle between two intracellular compartments, such that, when the protein is in the GDP-bound state, it is able to bind to the membrane of one compartment (through its interaction with a GDS), whereas when the protein is in the GTP-bound state, it might bind to the membranes of the second compartment (through its interaction with a GAP or effector protein). The GDI might then mediate the movement of CDC42Hs between the two compartments by triggering the release of the GTPase from the membranes and insuring that the protein remains in the GDP-bound state (as it approaches the first compartment) or in the GTP-bound state as it approaches the second compartment.

The above speculation raises some important questions for the future, perhaps the foremost question being the specific role of CDC42Hs in cells and in particular whether CDC42Hs might function in some type of "trafficking" capacity, moving growth factor receptors or other growth regulatory molecules between different compartments in the cell? Immunofluorescence and subcellular localization studies should prove useful in addressing this possibility. It also will be interesting to determine just how many CDC42Hs regulatory proteins exist, i.e., how many types of GDS and GAP molecules are present, and if some of the GAPs also serve as downstream targets or effector molecules. Finally, it will be of interest to determine whether multiple forms of the GDI protein exist, if the GDI is capable of inhibiting the GTPase activities of other members of the rho subgroup, and if other types of GDI molecules function in a similar capacity for other members of the ras superfamily including ras itself?

References

Adams AEM, Johnson DI, Longnecker RM, Sloat BF, Pringle JR (1990) CDC42 and CDC43, two additional genes involved in budding and the establishment of cell polarity in the yeast Saccharomyces cerevisiae. J Cell Biol 111:131–142

Bender A, Pringle JR (1989) Multicopy suppression of the cdc24 budding defect in yeast by CDC42 and three newly identified genes including the ras-related gene RSR1. Proc Natl Acad Sci USA 86:9976–9980

Broek D, Toda T, Michaeli T, Levin L, Birchmeier C, Zoller M, Powers S, Wigler M (1987) The S. cerevisiae CDC25 gene product regulates the RAS/adenylate cyclase pathway. Cell 489:789

Bustelo XR, Ledbetter JA, Barbacid M (1992) Product of vav proto-oncogene defines a new class of tyrosine protein kinases. Nature 356:68–71

Cadena DL, Gill GN (1992) Receptor tyrosine kinases. FASEB J 6:2332–2337

Cantley LC, Auger KR, Carpenter C, Duckworth B, Graziani A, Kapeller R, Soltoff S (1991) Oncogenes and signal transduction. Cell 64:281–302

Cerione RA (1991) Reconstitution of receptor–GTP-binding protein interactions. Biochim Biophys Acta 1071:473–501 (Review in Biomembranes)

Chant J, Herskowitz I (1991) Genetic control of bud-site selection in yeast by a set of gene products that comprise a morphogenetic pathway. Cell 65:1203–1212

Chant J, Corrado K, Pringle JR, Herskowitz I (1991) The yeast BUD5 gene, which encodes a putative GDP–GTP exchange factor, is necessary for bud-site selection and interacts with a bud-formation gene BEM1. Cell 65:1213–1224

Diekmann D, Brill S, Garrett MD, Totly N, Hsuan J, Monfries C, Hall C, Lim L, Hall A (1991) Bcr encodes a GTPase-activating protein for $p21^{rac}$. Nature 351:400–402

Drubin DG (1991) Development of cell polarity in budding yeast. Cell 65:1093–1096

Ellis C, Moran M, McCormick F, Pawson T (1990) Phosphorylation of GAP and GAP-associated proteins by transforming and mitogenic tyrosine kinases. Nature 343:377–381

Eva A, Aaronson SA (1985) Isolation of a new human oncogene from a diffuse B-cell lymphoma. Nature 316:273–275

Evans T, Brown ML, Fraser ED, Northup JK (1986) Purification of the major GTP-binding proteins from human placental membranes. J Biol Chem 261:7052–7059

Finegold AA, Johnson DI, Farnsworth CC, Gelb MH, Judd SR, Glomset JA, Tamanoi F (1991) Geranylgeranyl: protein transferase of Saccharomyces cerevisiae is specific for Cys-Xaa-Xaa-Leu motif proteins and requires the CDC43 gene product, but not the DPR1 gene product. Proc Natl Acad Sci USA 88:4448–4452

Fukumoto Y, Kaibuchi K, Hori Y, Fujioka H, Araki S, Ueda T, Kikuchi A, Takai Y (1990) Molecular cloning and characterization of a novel type of regulatory protein (GDI) for the rho proteins, ras p21-like small GTP-binding proteins. Oncogene 5:1321–1328

Garrett MD, Self AJ, vanOers C, Hall A (1989) Identification of distinct cytoplasmic targets for ras/R-ras and rho regulatory proteins. J Biol Chem 264:10–13

Hart M, Polakis P, Evans T, Cerione RA (1990) Identification and characterization of a low molecular weight GTP binding protein which is a phospho-substrate for the epidermal growth factor receptor/tyrosine kinase. J Biol Chem 265:5990–6001

Hart MJ, Shinjo K, Hall A, Evans T, Cerione RA (1991a) Identification of the human platelet GTPase activating protein for the CDC42Hs protein. J Biol Chem 266:20840–20848

Hart MJ, Eva A, Evans T, Aaronson SA, Cerione RA (1991b) Catalysis of guanine nucleotide exchange on the CDC42Hs protein by the dbl oncogene product. Nature 354:311–314

Hart MJ, Maru Y, Leonard D, Witte O, Evans T, Cerione RA (1992) A GDP-Dissociation Inhibitor also serves as a GTPase-Inhibitor for the ras-like protein CDC42Hs. Science 58:812–815

Hu P, Margolis B, Skolnik EY, Lammers R, Ullrich A, Schlessinger J (1992) Interaction of phosphatidylinositol 3-kinase-associated p85 with epidermal growth factor and platelet-derived growth factor receptors. Mol Cell Biol 12:981–990

Johnson DI, Pringle JR (1990) Molecular characterization of CDC42, a Saccharomyces cerevisiae gene involved in the development of cell polarity. J Cell Biol 111:143–152

Johnson RM, Garrison JC (1987) Epidermal growth factor and angiotensin II stimulate formation of inositol 1,4,5- and inositol 1,3,4-trisphosphate in hepatocytes. J Biol Chem 262:17285–17293

Katzav S, Martin-Zanca D, Barbacid M (1989) vav, a novel human oncogene derived from a locus ubiquitously expressed in hematopoietic cells. EMBO J 8:2283–2290

Koch CA, Anderson D, Moran MF, Ellis C, Pawson T (1991) SH2 and SH3 domains: elements that control interaction of cytosolic signaling proteins. Science 252:668–674

Leonard D, Hart MJ, Platko JV, Eva A, Henzel W, Evans T, Cerione RA (1992) The identification and characterization of a GDP-dissociation inhibitor (GDI) for the CDC42Hs protein. J Biol Chem 267:22860–22868

Margolis B, Rhee SG, Felder S, Mervic M, Lyall R, Levitzki A, Ullrich A, Zilberstein A, Schlessinger J (1989) EGF induces tyrosine phosphorylation of phospholipase C-II: a potential mechanism of EGF receptor signaling. Cell 57:1101–1107

Margolis B, Hu P, Katzav S, Li W, Oliver JM, Ullrich A, Weiss A, Schlessinger J (1992) Tyrosine phosphorylation of vav proto-oncogene product containing SH2 domain and transcription factor motifs. Nature 356:71–74

Meisenhelder J, Suh P-G, Rhee SG, Hunter T (1989) Phospholipase C-γ is a substrate for the PDGF and EGF receptor protein-tyrosine kinases in vivo and in vitro. Cell 57:1109–1122

Molloy CJ, Bottaro DP, Fleming RP, Marshall MS, Gibbs JB, Aaronson SA (1989) PDGF induction of tyrosine phosphorylation of GTPase activating proteins. Nature 342:711–714

Morii N, Kawano K, Sekine A, Yamada T, Narumiya S (1991) Purification of GTPase-activating protein specific for rho gene products. J Biol Chem 266:7646–7650

Munimetsu S, Innis MA, Clark R, Ullrich A, Polakis P (1990) Molecular cloning and expression of G25K cDNA, the human homolog of the yeast cell cycle gene CDC42. Mol Cell Biol 10:5977–5982

Polakis PG, Snyderman R, Evans T (1989) Characterization of G25K, a GTP-binding protein containing a novel putative nucleotide binding domain. Biochem Biophys Res Commun 160:25–32

Ron D, Tronick SR, Aaronson SA, Eva A (1988) Molecular cloning and characterization of the human dbl proto-oncogene: evidence that its overexpression is sufficient to transform NIH/3T3 cells. EMBO J 7:2465–2473

Ron D, Zannini M, Lewis M, Wickner RB, Hunt LT, Grazīani G, Tronick SR, Aaronson SA, Eva A (1991) A region of proto-dbl essential for its transforming activity shows sequence similarity to a yeast cell-cycle gene, CDC24, at the human breakpoint cluster gene, bcr. New Biologist 3:372–379

Settleman J, Narasimhan V, Foster LC, Weinberg RA (1992) Molecular cloning of cDNAs encoding the GAP-associated protein p190: implications for a signaling pathway from ras to the nucleus. Cell 69:539–549

Settleman J, Albright CF, Foster LC, Weinberg RA (1992) Association between GTPase activators for rho and ras families. Nature 359:153–154

Shinjo K, Koland JG, Hart MJ, Narasimhan V, Johnson DI, Evans T, Cerione RA (1990) Molecular cloning of the gene for the human placental GTP-binding protein Gp (G25K): identification of this GTP-binding protein as the human homolog of the yeast cell-division-cycle protein CDC42. Proc Natl Acad Sci USA 87:9853–9857

Ueda T, Kikuchi A, Oiga N, Yamamoto J, Takai Y (1990) Purification and characterization from bovine brain cytosol of a novel regulatory protein inhibiting the dissociation of GDP from and the subsequent binding of GTP to rhoB p20, a ras p21-like GTP-binding protein. J Biol Chem 265:9373–9380

Ullrich A, Schlessinger J (1990) Signal transduction by receptors with tyrosine kinase activity. Cell 61:203–212

Wahl MI, Nishibe S, Suh PG, Rhee SG, Carpenter G (1989) Epidermal growth factor stimulates tyrosine phosphorylation of phospholipase C-II independently of receptor internalization and extracellular calcium. Proc Natl Acad Sci USA 86:1568–1572

Yamane HK, Farnsworth CC, Xie H, Evans T, Howald W, Gelb MH, Glomset JA, Clarke S, Fung BK-K (1991) The membrane binding domain of the small G-protein G25K contains an all-trans geranylgeranyl cysteinyl methyl ester at its carboxyl terminus. Proc Natl Acad Sci USA 88:286–290

Yang L, Baffy G, Rhee SG, Manning D, Hansen CA, Williamson JR (1991) Pertussis toxin-sensitive G_i protein involvement in epidermal growth factor-induced activation of phospholipase C-γ in rat hepatocytes. J Biol Chem 266:22451–22458

Zheng Y, Hart MJ, Shinjo K, Evans T, Bender A, Cerione RA (1993) Biochemical comparisons of the S. cerevisiae Bem2 and Bem3 proteins: Delineation of a limit Cdc42-GAP domain. J Biol Chem (in press)

Ziman M, O'Brien JM, Johnson DI (1991) Mutational analysis of CDC42Sc, a Saccharomyces cerevisiae gene that encodes a putative GTP-binding protein involved in the control of cell polarity. Mol Cell Biol 11:3537–3544

D. Regulation of and by Small GTPases

CHAPTER 38

Role of Rap1B and Its Phosphorylation in Cellular Function: A Working Model

D.L. ALTSCHULER, M. TORTI, and E.G. LAPETINA

A. Introduction: The Rap Family of Proteins

Rap proteins (*Ras proximate*) were discovered and identified within the last 4 years. This was accomplished through the use of four different approaches: (a) protein purification following GTP-binding activity, (b) low stringent hybridization using previously identified Ras probes, (c) screening of cDNA expression libraries by Ras antibodies, and (d) one of the members of the Rap family was identified as a Ras revertant clone.

GTPγS-binding activity led to the purification of Rap proteins from bovine brain membranes (KAWATA et al. 1988a), bovine aortic smooth muscle cell membranes (KAWATA et al. 1988b), and human platelet membranes (OHMORI et al. 1989) and cytosol (NAGATA et al. 1989). All of these proteins were later identified as members of the Rap1 family.

The first full-length sequence of a Rap protein was discovered after screening a Raji cell line cDNA library (PIZON et al. 1988a); using *Dras*-3 (a *Drosophila ras* allele) as a probe and low stringent hybridization conditions, two clones were isolated: Rap1 and Rap2. Rescreening of the library using the Rap1 clone as a probe resulted in the isolation of a closely related sequence, which was named Rap1b (PIZON et al. 1988b).

Interestingly, at about the same time, KITAYAMA and coworkers observed that when a K-Ras-transformed cell line was transfected with a human cDNA library, some cells reverted to a normal flat morphology. Sequence analysis of one of the genes responsible for this reversion, K-*Revl*, showed 100% homology with Rap1a (KITAYAMA et al. 1989).

Finally, the fourth member of the Rap family so far identified, Rap2b, was isolated by screening a human platelet cDNA expression library with a Ras monoclonal antibody (OHMSTEDE et al. 1990; FARRELL et al. 1990). Figure 1 shows the sequence alignment of the members of the Rap family of proteins in comparison with that of human K-ras. The close analysis of these sequences shows that Rap proteins share 50%–60% homology with Ras. The greatest homology is evident in the regions involved in basic functions, including the nucleotide-binding domain, the membrane localization domain, and most important, the putative effector domain. A striking difference between Ras and Rap proteins is that all Rap sequences have a threonine at position 61, instead of the glutamine found in Ras proteins.

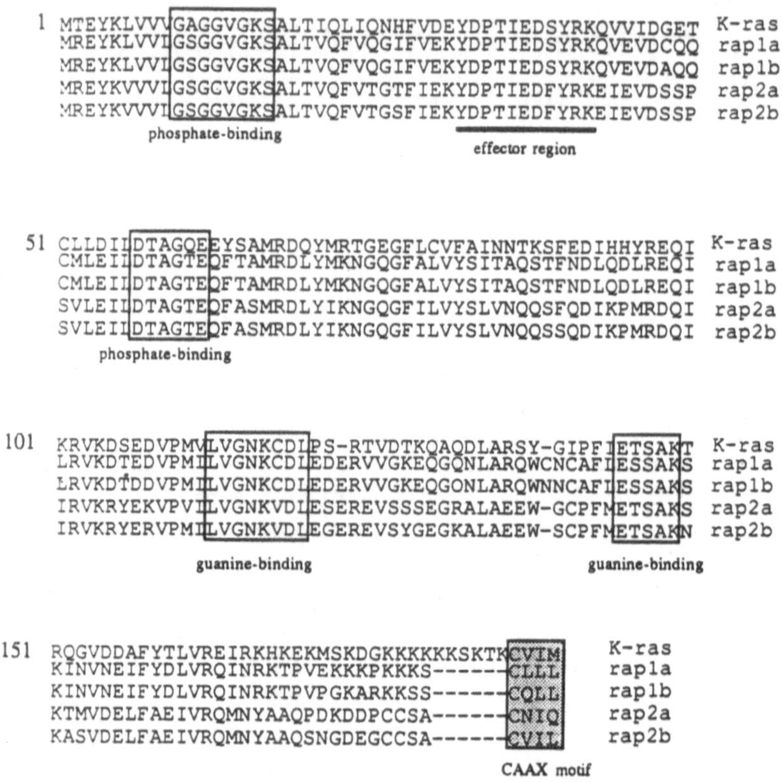

Fig. 1. Alignment of Rap proteins with K-Ras. Phosphate and guanine binding domains are indicated by *boxes*. The effector domain is shown by the *solid line* and the CAAX motif (membrane localization domain) is denoted by the *shaded box*.

This feature is discussed in Chap. 23, in relation to the interaction of the Rap proteins with their respectives GAP activities.

Although the function of Rap proteins is still unknown, it is now clear that the Rap1 proteins become phosphorylated after treatment with agonists that increase cAMP levels. Two cell systems have been studied in detail so far: neutrophils and platelets. Here we describe the current understanding of the regulatory function of Rap1b in platelets.

B. Phosphorylation of Rap1b

I. Structural Properties

1. cAMP-dependent Phosphorylation of Rap1b in Human Platelets

Several years ago we and other investigators showed that platelets possess distinct GTPases with molecular masses of between 21 and 31 kDa; these

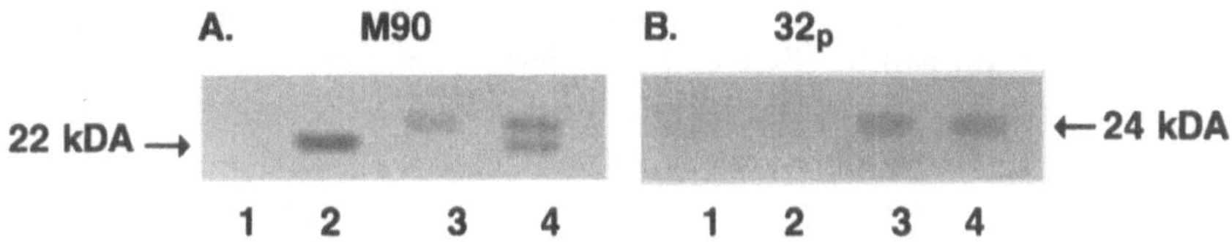

Fig. 2. Rap1b phosphorylation and translocation in human platelets. ^{32}P-labeled platelets were stimulated with iloprost for 10 min at 37°C. At the end of this incubation homogenates were prepared and particulate and cytosolic fractions were separated by centrifugation. Equivalent amount of protein were resolved on a 12.5% sodium dodecyl sulfate polycrylamide gel electrophoresis (SDS-PAGE) and electroblotted to nitrocellulose. *Panel A*, Western blot developed with the Ras monoclonal antibody M90. *Panel B*, The blot on panel A was exposed to X-ray film at −70°C with an intensifying screen for 16 h. *Lane 1*, Nonstimulated platelets; cytosolic fraction. *Lane 2*, Nonstimulated platelets; particulate fraction. *Lane 3*, Iloprost-stimulated platelets; cytosolic fraction. *Lane 4*, Iloprost-stimulated platelets; particulate fraction

were detected by binding of [α-^{32}P]GTP in western blots (LAPETINA and REEP 1987; BHULLAR and HASLAM 1987). Blotting with Y13-259, a monoclonal antibody that recognizes all the known ras-encoded p21 proteins, did not show any cross-reactivity with the small G-proteins present in platelets. Screening of platelet proteins with a set of monoclonal antibodies raised against Ha-Ras p21 showed that two of them did, in fact, recognize a 22-kDa protein present in a platelet particulate fraction, suggesting that this protein was a member of the Ras family (LAPETINA et al. 1989).

The increase of cAMP levels in platelets induced the phosphorylation of several proteins, the most prominent with molecular masses of 22–24 kDa and 50 kDa (FEINSTEIN et al. 1981). At the time, it was important to determine if the 22- to 24-kDa protein was the same protein that we detected with the Ras monoclonal antibody. In an effort to address this question, we looked for the effect of iloprost, a prostacyclin analog that increases platelet cAMP levels, on the proteins that are recognized by the ras antibody (LAPETINA et al. 1989). As shown in Fig. 2, in control platelets the antibody recognized a 22-kDa protein present in the particulate fraction. Treatment with iloprost induced the appearance of a 24-kDa protein in this fraction, which correlated with the decrease of the 22-kDa protein. Moreover, iloprost treatment induced the appearance of this 24-kDa protein in the cytosolic fraction. The mobility shift induced by iloprost was thought to indicate a covalent modification such as phosphorylation. Treatment with iloprost of ^{32}P-labeled platelets showed a perfect correlation between the 24-kDa immunoreactive band and the 24-kDa radiolabeled band.

The same results were observed in HEL cells (human erythroleukemia cells) (LAZAROWSKI et al. 1989). Moreover, in this cell line the protein kinase A (PKA) phosphorylation of the 22-kDa protein was correlated with inhibition of the phospholipase-C induced formation of inositol phosphates (LAZAROWSKI et al. 1990). The possibility of the involvement of this 22-kDa protein in phospholipase C regulation and cAMP-dependent inhibition of platelet activation will be discussed later in this chapter.

The first indication of the identity of this 22-kDa protein was advanced by KAWATA and coworkers. They showed that a 22-kDa G-protein from bovine brain membranes was stoichiometrically phosphorylated by PKA in a cell-free system. Sequence analysis of this protein, called smgp21, showed that it was identical to Rap1a (KAWATA et al. 1989). The same authors described that human platelets stimulated with PGE1 showed a phosphorylated band of 22–24 kDa that behaved like bovine brain smgp21 (Rap1a) phosphorylated in a cell-free system. The authors concluded, based on partial sequence analysis, that the 22- to 24-kDa protein from platelets was Rap1a. However, as mentioned before, Rap1a and Rap1b share more than 90% homology; therefore, only sequence analysis would allow the final determination of the identity of this 22-kDa protein from human platelets.

To unambiguously identify the protein undergoing phosphorylation, we isolated the phosphorylated protein and subsequently treated the labeled protein with proteases. The phosphopeptides were separated on high-pressure liquid chromatography (HPLC) and subjected to amino acid analysis, identifying the phosphoprotein as Rap1b (SIESS et al. 1990). To further confirm these results, synthetic peptides corresponding to the C-terminal regions of Rap1a and Rap1b were subjected to in vitro PKA phosphorylation. The results again confirmed that the 22-kDa protein phosphorylated in human platelets was Rap1b, and not Rap1a. Further studies confirmed that recombinant Rap1b (cloned from a platelet cDNA library) was a good substrate for PKA in vitro (SAHYOUN et al. 1991, and unpublished results).

2. Phosphorylation of Rap1b by a Neuronal Ca^{2+}/Calmodulin-dependent Protein Kinase, CaM Kinase Gr

Several small G-proteins were originally isolated from bovine brain preparations. Of particular interest was the presence of ras p21, smgp25/Rab3a, and smgp21/Rap1 in synaptic areas, with particular enrichment in synaptic vesicles (KIM et al. 1989, 1990).

Neuronal Ca^{2+}-signaling is partly mediated by two neuron-specific Ca^{2+}/calmodulin-dependent protein kinases, CaM kinase II and CaM kinase Gr. Because CaM kinase Gr also phosphorylates PKA-dependent sites in synapsin I, we decided to investigate the phosphorylation of Rap1b by these calmodulin-dependent kinases (SAHYOUN et al. 1991). Both CaM kinase Gr and PKA, but not CaM kinase II, stoichiometrically phosphorylated Rap1b in identical or contiguous serine residues, as judged by phosphopeptide mapping. The rate of phosphorylation of Rap1b by CaM kinase Gr was enhanced following autophosphorylation of the kinase, and the reaction was strictly dependent on Ca^{2+}/calmodulin.

Other small G-proteins used as substrate specificity controls, including Rab3A, Rap2b, and c-Ha-Ras, were not phosphorylated by CaM kinase Gr.

These experiments established a link between a neuronal Ca^{2+}/calmodulin-dependent kinase and a small G-protein signal transduction pathway. Further in vivo experiments will be necessary to demonstrate a physiological role for Rap1b in the regulation of critical neuronal functions.

3. Mutational Analysis of the Protein Kinase A-dependent Phosphorylation site of Rap1b

As discussed before, previous data from our laboratory and others (SIESS et al. 1990; HATA et al. 1991) showed that serine residues at the C-terminal portion of Rap1b are responsible for cAMP-dependent protein kinase-mediated phosphorylation. We used a mutational analysis to determine the precise site of phosphorylation in vitro and in intact cells. In vivo studies of a specific Rap1 protein have been difficult to conduct because of the high degree of homology (Rap1a and Rap1b differ only in nine out of 184 residues); thus, available antibodies are not able to discriminate between them. To overcome this problem we used a transient expression system in which all of the mutants were expressed as a fusion protein: the N terminus of the fusion protein consisted of an 18 amino acid sequence derived from the influenza virus hemagglutinin gene, which represents an epitope for a well-characterized monoclonal antibody. Using this system we were able to analyze the exogenous transfected protein without any cross-reactivity with the endogenous Rap1 proteins (ALTSCHULER and LAPETINA 1993). The C-terminal phosphopeptide sequence described below encompassed the sequence RKKSS (residues 176–180); both serine residues matched the consensus sequence for PKA phosphorylation (KENELLY and KREBS 1991). Single serine-to-alanine mutants as well as a double mutant were used to determine the exact site of phosphorylation. Both in vitro phosphorylation of the Rap1b immunoprecipitates and in vivo stimulation with forskolin of

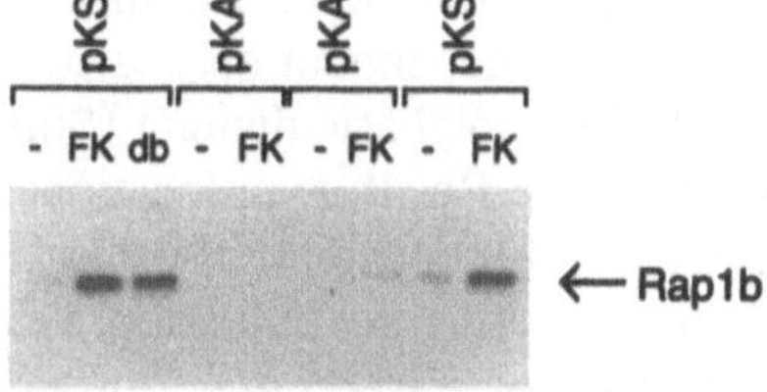

Fig. 3. Mutational analysis of the PKA-dependent phosphorylation of Rap1b: in vivo stimulation. COS-7 cells were transfected with the mutant constructs, using lipofectin reagent according to manufacturer's instructions. Forty eight hours after transfection cells were labeled with ^{32}P and stimulated for 15 min at 37°C with 50 μM forskolin (*F*), 0.5 m*M* dibutyryl cyclic AMP (*db*), or DMSO as control(−). Lysates were prepared and the Rap1b constructs were immunoprecipitated and resolved on a 12.5% SDS-PAGE. The gel was fixed, dryed, and exposed for 10 hs at −70°C with an intensifying screen. Plasmid nomenclature: three-letter code (pKSS, pKAA, pKAS, pKSA) corresponding to Rap1b amino acid sequence from position 178 to 180

transfected ^{32}P-labeled cells clearly showed that ser 179 is the preferred site of PKA phosphorylation (ALTSCHULER and LAPETINA 1993; a representative experiment of an in vivo stimulation is shown in Fig. 3).

Rap1b is known to be processed at the C terminus (KAWATA et al. 1990); this modification is thought to be necessary for the right subcellular localization. Interestingly, ser 179 resides just between the C terminus CAAX motif (Cys-Aliphatic-Aliphatic-X, where X is another amino acid) and a polybasic domain that is known to enhance the membrane-binding activity (HANCOCK et al. 1990). The interrelationship between these domains and the translocation induced by phosphorylation requires further investigation.

4. Phosphorylation-Dependent Activation of Rap1b: Role of Guanine Nucleotide Dissociation Stimulator

In vitro studies with phosphorylated and nonphosphorylated recombinant Rap1b did not give any clue to assessing a role for the phosphorylation event. The GDP- and GTP[γS]-bound forms were equally phosphorylated (HOSHIJIMA et al. 1988; KAWATA et al. 1989), and the phosphorylation did not affect its basal GDP/GTP exchange reaction, basal GTPase activity, or the GAP-stimulated GTPase activity (HATA et al. 1991).

Recent in vitro experiments suggest that phosphorylation may be involved in Rap1 activation: PKA-dependent phosphorylation made Rap1b more sensitive to the action of Rap-GDS (guanine nucleotide dissociation stimulator), resulting in the translocation of Rap1b from the membrane to the cytosol (HATA et al. 1991; ITOH et al. 1991).

GDS is a cytosolic factor that stimulates the GDP/GTP exchange reaction by stimulating the dissociation of GDP and subsequent binding of GTP (YAMAMOTO et al. 1990). In vitro studies show that GDS is able to inhibit the binding of Rap1b to membranes, and to induce the translocation of the protein to the cytosol (KAWAMURA et al. 1991). Interestingly, the C-terminal region of Rap1b (the membrane localization domain) interacts with GDS, and this interaction is dependent on posttranslational processing of Rap1b and the presence of its polybasic domain (SHIRATAKI et al. 1991).

These results suggest that the Rap1b translocational event that we first described as associated with PKA-dependent phosphorylation (LAPETINA et al. 1989) could, in fact, be reflecting an activation step. Even though the putative GTP-dependent effector target and subsequent steps in the pathway are still unknown, we are able to correlate these results with our recent data on the platelet model (see below).

The role of GDS in Rap1b activation will be discussed in greater detail in Chap. 39.

II. Physiological Properties: The Platelet Model

The function of platelets is regulated by a strict balance between stimulatory and inhibitory signals. Platelet activation induces a cascade of molecular events including shape change, secretion, adhesion, and aggregation, all of which lead, finally, to the formation of a localized hemostatic plug. Among the platelet activators, thrombin represents the most potent and physiologically relevant agonist. Activation of phospholipase C (a reaction coupled to the stimulation of the thrombin receptor) leads to the rapid formation of second messengers: 1,2-diacylglycerol to stimulate protein kinase C (PKC), and inositol 1,4,5-triphosphate that releases Ca^{2+} from intracellular stores.

The most important physiological platelet antagonists are prostaglandins (PGE_1, PGI_2), produced mainly in endothelial cells and then released into the bloodstream. Through specific membrane receptors, these prostaglandins activate adenyl cyclase, raising the levels of cAMP and activating the cAMP-dependent protein kinase (PKA). Pretreatment of platelets with these antagonists impairs the thrombin-induced hydrolysis of phosphoinositides.

Direct phosphorylation of phospholipase C by PKA has not been reported in platelets, suggesting that other phospholipase C regulators might mediate the cAMP-dependent inhibition.

All of these data stress the role of phospholipase C as a key element in platelet activation as well as a target for platelet antagonists.

1. Thrombin-induced Association of Rap1b with Ras-GTPase Activating Protein. Effect of Phosphorylation

The structural properties of Rap proteins indicate that, as with Ras-p21, they are regulatory GTPases that cycle between a GDP-bound inactive state and a GTP-bound active state. The balance between these two steps is regulated by factors that basically increase the dissociation of the bound GDP (exchange factors) or increase the intrinsic GTPase activity (GAP).

Ras-GAP was the first factor described: it is a 120-kDa cytosolic protein that associates with tyrosine kinase growth factors and which is tyrosine phosphorylated after growth factor stimulation, leading to its association with other tyrosine-phosphorylated proteins, p62 and p190 (Cantley et al. 1991). Even though the function of Ras-GAP as a negative regulator or a downstream effector is still unclear, it may represent a biochemical link between Ras and growth factors receptors. Rap1 proteins bind in vitro, with high affinity, with Ras-GAP in a GTP-dependent manner, although the intrinsic GTPase activity of Rap1 proteins is not stimulated (Frech et al. 1990). This suggests that Rap1 is able to compete with Ras-p21 for Ras-GAP, inhibiting in this way downstream signaling. Even though this might explain the reversion of Ras-transformed phenotype by Rap1a (Kitayama et al. 1989), and the inhibition of Ras-induced germinal vesicle breakdown in oocytes by Rap1b (Campa et al. 1991), the discovery of Rap-specific GAP proteins suggests that this is a much more complex system (Nice et al. 1992; Rubinfeld et al. 1991).

Platelets contain extremely low levels of Ras-p21 (LAPETINA et al. 1989), but these cells are rich in Rap1b. Ras-GAP and Rap-GAP activities are also present in platelets (BENCKE MARTI and LAPETINA 1992). So platelets represent a unique model in which to study a potential in vivo interaction between Rap1b and Ras-GAP, since Ras-p21 is practically nonexistent in these cells.

Using a specific Ras-GAP antiserum we showed that this protein is present in human platelets in considerable amounts, and that it is associated with p62 and p190 (TORTI and LAPETINA 1992). Stimulation of platelets with thrombin induced tyrosine phosphorylation of Ras-GAP. Immunoprecipitation of thrombin-stimulated human platelets with the Ras-GAP antiserum showed the presence of Rap1b in association with Ras-GAP (TORTI and LAPETINA 1992).

If the in vivo complex shows the same behavior as the in vitro association, we should expect the associated Rap1b to be in the GTP conformation, suggesting that thrombin induced the GDP/GTP exchange of Rap1b in human platelets. We are currently conducting transfection of the epitope-fusion constructs, described earlier, in thrombin-responsive cell lines with the hope of answering this question.

To study the role of phosphorylation in this complex formation, samples were treated with cAMP-elevating agents and then thrombin-dependent association was analyzed. Pretreatment with iloprost increased the amount of Rap1b associated with Ras-GAP in thrombin-induced platelets (unpublished results), in agreement with the postulated role for phosphorylation in the GDS-dependent activation of Rap1b.

2. Ras-GAP Associates with Phospholipase Cγ-1 in Human Platelets

As discussed before, it is possible that Ras-GAP plays a role in the transduction of signals from tyrosine kinases through p21-Ras. To understand this biochemical link, different laboratories started to look for stable complexes between GAP and GAP-associated proteins. Recent studies have shown that GAP associates with platelet-derived growth factor (PDGF) and epidermal growth factor (EGF) receptors (CANTLEY et al. 1991), with two phosphoproteins, p62 and p190 (SETTLEMAN et al. 1992; WONG et al. 1992), and, at least in fibroblasts, with c-src (BROTT et al. 1991). Specifically in human platelets, three src-related proteins – Fyn, Lyn and Yes (but not c-src itself) – were detected in association with Ras-GAP after thrombin stimulation (CICHOWSKI et al. 1992).

Using specific phospholipase Cγ1 (PLCγ-1) monoclonal antibodies, we recently found that Ras-GAP associates with PLCγ-1 in human platelets (TORTI and LAPETINA 1992): Ras-GAP immunoprecipitates showed the presence of PLCγ-1 by western blots and revealed PLC activity. This complex was formed in resting platelets, and thrombin stimulation did not modify either the amount of the enzyme in the complex or its catalytic activity, suggesting that active PLCγ-1 is constitutively associated with cytosolic Ras-GAP in resting human platelets. We estimated that PLCγ-1 represents 60% of the total PLC activity, and about half of it was associated with Ras-GAP.

In vivo phosphorylation studies revealed that only the PLCγ-1 associated with Ras-GAP was phosphorylated (serine/threonine residues), suggesting that this phosphorylation is a signal for the association, or that only the associated protein is a substrate for the corresponding protein kinase. The nature of this association is for the moment unknown: a direct interaction involving Ras-GAP SH2/SH3 domains and phosphoserine/threonine residues, as shown for other proteins (Pendergast et al. 1991), or an indirect interaction involving the PLCγ-1 SH_2 domain and phosphotyrosine residues from intermediates molecules (i.e., p62 and/or p190) might be possible.

III. A Working Model and Open Questions

In this section we will try to put together all the elements previously described and correlate them with the known properties of platelet activation (Fig. 4).

In resting platelets, Rap1b is associated with the particulate fraction, presumably in the GDP conformation; PLCγ-1 is constitutively complexed with Ras-GAP (and probably with other GAP-associated proteins) in the cytosolic fraction. After thrombin stimulation, a rapid activation of PLC is observed. About 60% of the total platelet PLC activity is PLCγ-1, and about half of this is associated with Ras-GAP in the cytosol. Thrombin induces the association of cytosolic Ras-GAP with Rap1b; as a consequence of this interaction the Ras-GAP–PLCγ-1 complex forms a ternary complex (Rap1b–Ras-GAP–PLCγ-1), allowing PLCγ-1 to be in close contact with its membrane-bound substrate. The existence of the putative ternary complex has not yet been demonstrated, and we cannot rule out the possibility that the Ras-GAP molecule that associates with Rap1b represents a subpopulation of PLCγ-1-free GAP. However, the fact that after thrombin stimulation Rap1b translocates to the triton insoluble fraction (Fischer et al. 1990), where PLC activity is detected (Grondin et al. 1991), strengthens the hypothesis of the colocalization of Ras-GAP–PLCγ-1 and Rap1b.

This proposed mechanism of PLC activation resolves, at least in part, the discrepancies between pertussis-sensitive and pertussis-insensitive pools of PLC activities previously observed (Crouch and Lapetina 1988): thrombin seems to activate independent pools of PLC through a G_i-like protein and activation of Rap1b. (Even when Gq was present in platelets, we could not detect appreciable amounts of PLCβ, suggesting that this isoenzyme is not significantly involved in platelet activation.)

The model predicts that thrombin induces activation of Rap1b (i.e., GTP/GDP exchange), as an explanation of the association with Ras-GAP, even though this has not yet been demonstrated.

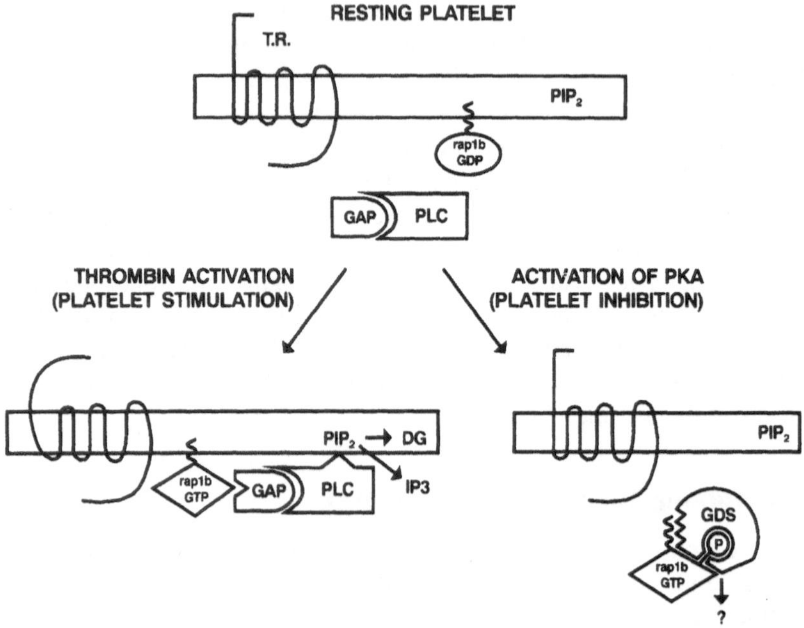

Fig. 4. A working model for the involvement of Rap1b in thrombin-induced phospholipase C (*PLC*) activation and protein kinase A (*PKA*)-dependent PLC inhibition. Three key elements are the basis for this model: (1) a cytosolic constitutive association between phospholipase Cγ-1 (PLC_γ-1) and Ras-GAP in resting platelets; (2) a thrombin-induced association of membrane-bound Rap1b with Ras-GAP; and (3) PKA-dependent phosphorylation and translocation of Rap1b, in association with *GDS* (GDP dissociation stimulator), from membrane to cytosol. According to this model Rap1b acts as a tag for signal transduction complexes, directing them to the right localization. The membrane-bound Rap1b colocalizes the PLC_γ-1 with its substrate; PKA-phosphorylation of Rap1b induces its translocation to the cytosol, inhibiting in this manner the access of PLC_γ-1 to its inositide substrate (PIP_2, phosphatidylinositol 4,5-bisphosphate). *T.R.*, thrombin receptor; *DG*, 1,2-diacylglycerol; *IP3*, inositol 1,4,5-trisphosphate; *GAP*, GTPase activating protein; *P*, phosphorylation site of Rap1b

According to this model, the mechanism by which antagonists that raise the levels of cAMP inhibit the thrombin-induced activation of PLC is easily envisaged: the PKA phosphorylation induces a GDS-dependent translocation of Rap1b to the cytosol, inhibiting the colocalization of PLC_γ-1 with its membrane-bound substrate. It is important to note here that by increasing cAMP levels, the activation of platelets is inhibited. However, this leads only to partial inhibition of platelet PLC (WATSON et al. 1984). If the inhibited PLC fraction corresponds to the PLC_γ-1 activity, as the model suggests, the importance of PLCγ-1 activity as a key element in platelet activation is strengthened.

It is intriguing to contemplate how two antagonistic events – PLC activation and PLC inhibition – might be undertaken upon Rap1b activation. It will be interesting to see whether the GDS-dependent activation and translocation is able to induce a protein modification (which could explain the observed mobility shift) and generate a new signal.

For reasons of simplicity, in the scheme of Fig. 4, Rap1b and the thrombin receptor are in the same plasma membrane fraction. The particulate fraction includes both the plasma membrane and internal membranes; thrombin might activate Rap1b in an indirect manner (i.e., inhibiting the specific Rap-GAP or stimulating an exchange factor). In fact, in fibroblasts, Rap1b localized by immunofluorescence in a perinuclear fraction (ALTSCHULER and LAPETINA 1992; BERANGER et al. 1991). Interestingly, we have observed an active Rap1b mutant (Gly12 to Val 12) in the perinuclear fraction and vesicle-like structures extending to the plasma membrane (unpublished results).

Other members of the Ras superfamily have been involved in vectorial transport of specific vesicles in the secretory pathway (CHAVRIER et al. 1990; BOURNE 1988). It is tempting to speculate that Rap proteins might have an analogous function in targeting "signal transduction vesicles" between different cellular compartments. Of course, these "transdusomes" are for the moment hypothetical, and their existence remains to be determined.

This speculative model (Fig. 4) describes a novel mechanism of PLC activation in human platelets and provides new observations on the mechanism of Rap function, in which signal transduction complexes are directed to the correct localization.

References

Altschuler D, Lapetina EG (1993) Mutational analysis of the PKA-dependent phosphorylation site of Rap1b. J Biol Chem 268:7527–7531

Bencke Marti K, Lapetina E (1992) Epinephrine suppresses rap1B-GAP-activated GTPase activity in human platelets. Proc Natl Acad Sci USA 89:2784–2788

Beranger F, Goud B, Tavitian A, de Gunzburg J (1991) Association of the Ras-antagonistic Rap1/Krev-1 proteins with the Golgi complex. Proc Natl Acad Sci USA 88:1606–1610

Bhullar R, Haslam R (1987) Detection of 23-27kDa GTP-binding proteins in platelets and other cells. Biochem J 245:617–620

Bourne H (1988) Do GTPases direct membrane traffic in secretion? Cell 53:669–671

Brott B, Decker S, Shafer J, Gibbs J, Jove R (1991) GTPase-activating protein interactions with the viral and cellular Src kinases. Proc Natl Acad Sci USA 88:755–759

Campa M, Chang K, Molina Y, Vedia L, Reep B, Lapetina EG (1991) Inhibition of Ras-induced germinal vesicle breakdown in Xenopus oocytes by Rap1-B. Biochem Biophys Res Commun 174:1–5

Cantley L, Auger K, Carpenter C, Duckworth B, Graziani A, Kapeller R, Soltoff S (1991) Oncogenes and signal transduction. Cell 64:281–302

Chavrier P, Parton R, Hauri H, Simons K, Zerial M (1990) Localization of low molecular weight GTP binding proteins to exocytic and endocytic compartments. Cell 62:317–329

Cichowski K, McCormick F, Brugge J (1992) P21^{ras} GAP association with Fyn, Lyn, and Yes in thrombin-activated platelets. J Biol Chem 267:5025–5028

Crouch M, Lapetina EG (1988) Role of Gi in control of thrombin receptor–phospholipase C coupling in human platelets. J Biol Chem 263:3363–3371

Farrell FX, Ohmstede CA, Reep BR, Lapetina EG (1990) cDNA sequence of a new ras-related gene (rap2b) isolated from human platelets with sequence homology to rap2. Nucleic Acids Res 18:14

Feinstein M, Rodan G, Cutler (1981) In: Gordon J (ed) Platelets in biology and pathology 2. Elsevier/North-Holland Biomed, Amsterdam, pp 437–442

Fischer T, Gatling M, Lacal J, White G (1990) Rap1B, a cAMP dependent substrate, associates with the platelet cytoskeleton. J Biol Chem 265:19405–19408

Frech M, John J, Pizon V, Chardin P, Tavitian A, Clark R, McCormick F, Wittinghoffer A (1990) Inhibition of GTPase activating protein stimulation of Ras-p21 GTPase by the Krev-1 gene product. Science 249:169–171

Grondin P, Plantavid M, Sultan C, Breton M, Mauco G, Chap H (1991) Interaction of $pp60^{c\text{-}src}$, phospholipase C, inositol-lipid, and diacylglycerol kinases with the cytoskeletons of thrombin-stimulated platelets. J Biol Chem 266:15705–15709

Hancock J, Paterson H, Marshall C (1990) A polybasic domain or palmitoylation is required in addition to the CAAX motif to localize $p21^{ras}$ to the plasma membrane. Cell 63:133–139

Hata Y, Kaibuchi K, Kawamura S, Hiroyoshi M, Shirataki H, Takai Y (1991) Enhancement of the actions of smg p21 GDP/GTP exchange protein by the protein kinase A-catalyzed phosphorylation of smg p21. J Biol Chem 266:6571–6577

Hoshijima M, Kikuchi A, Kawata M, Ohmori T, Hashimoto E, Yamamura H, Takai Y (1988) Phosphorylation by cyclic AMP-dependent protein kinase of a human platelet Mr 22,000 GTP-binding protein (smg p21) having the same putative effector domain as the *ras* gene products. Biochem Biophys Res Commun 157:851–860

Itoh T, Kaibuchi K, Sasaki T, Takai Y (1991) The smg GDS-induced activation of smg p21 is initiated by cyclic AMP-dependent protein kinase-catalyzed phosphorylation of smg p21. Biochem Biophys Res Commun 177:1319–1324

Kawamura S, Kaibuchi K, Hiroyoshi M, Hata Y, Takai Y (1991) Stoichiometric interaction of smg p21 with its GDP/GTP exchange protein and its novel action to regulate the translocation of smg p21 between membrane and cytoplasm. Biochem Biophys Res Commun 174:1095–1102

Kawata M, Matsui Y, Kondo J, Hishida T, Teranishi Y, Takai Y (1988a) A novel small molecular weight GTP-binding protein with the same putative effector domain as the ras proteins in bovine membranes. J Biol Chem 263:18965–18971

Kawata M, Kawahara Y, Araki S, Sunako M, Tsuda T, Fukusaki H, Mizoguchi A, Takai Y (1988b) Identification of a major GTP-binding protein in bovine aortic smooth muscle membranes as smg p21, a GTP-binding protein having the same effector domain as ras p21s. Biochem Biophys Res Commun 163:1418–1427

Kawata M, Kikuchi A, Hoshijima M, Yamamoto K, Hashimoto E, Yamamura H, Takai Y (1989) Phosphorylation of smg p21, a p21-like GTP-binding protein, by cyclic AMP-dependent protein kinase in a cell-free system and in response to protaglandin E1 in intact human platelets. J Biol Chem 264:15688–15695

Kawata M, Farnsworth C, Yoshida Y, Gelb M, Glomset J, Takai Y (1990) Posttranslationally processed structure of the human platelet protein smg p21: evidence for geranylgeranylation and carboxyl methylation of the C- terminal cysteine. Proc Natl Acad Sci USA 87:8960–8964

Kenelly P, Krebs E (1991) Consensus sequences as substrate specificity determinants for protein kinases and protein phosphatases. J Biol Chem 266:15555–15558

Kim S, Kikuchi A, Mizoguchi A, Takai Y (1989) Intrasynaptosomal distribution of the ras, rho and smg-25A GTP-binding proteins in bovine brain. Mol Brain Res 6:167–176

Kim S, Mizoguchi A, Kikuchi A, Takai Y (1990) Tissue and subcellular distribution of the smg-21/rap1/Krev-1 proteins which are partly distinct from those of c-ras p21s. Mol Cell Biol 10:2645–2652

Kitayama H, Sugimoto Y, Matsuzaki T, Ikawa Y, Noda M (1989) A ras-related gene with transformation suppressor activity. Cell 56:77–84

Lapetina EG, Reep B (1987) Specific binding of [α-^{32}P]GTP to cytosolic and membrane-bound proteins of human platelets correlates with the activation of phospholipase C. Proc Natl Acad Sci USA 84:2261–2265

Lapetina EG, Lacal J, Reep B, Molina Y, Vedia L (1989) A ras-related protein is phosphorylated and translocated by agonists that increase cAMP levels in human platelets. Proc Natl Acad Sci USA 86:3131–3134

Lazarowski E, Lacal J, Lapetina EG (1989) Agonist-induced phosphorylation of an immunologically ras-related protein in human erythroleukemia cells. Biochem Biophys Res Commun 161:972–978

Lazarowski E, Winegar D, Nolan R, Oberdisse E, Lapetina EG (1990) Effect of protein kinase A on inositide metabolism and rap1 G-protein in human erythroleukemia cells. J Biol Chem 265:13118–13123

Nagata K, Itoh H, Katada T, Takenaka K, Ui M, Kaziro Y, Nozawa Y (1989) Purification, identification, and characterization of two GTP-binding proteins with molecular weights of 25 000 and 21 000 in human platelet cytosol. J Biol Chem 264:17000–17005

Nice E, Fabri L, Hammacher A, Holden J, Simpson R, Burgess A (1992) The purification of a Rap1 GTPase-activating protein from bovine brain cytosol. J Biol Chem 267:1546–1553

Ohmori T, Kikuchi A, Yamamoto K, Kim S, Takai Y (1989) Small molecular weight GTP-binding proteins in human platelet membranes. J Biol Chem 264:1877–1881

Ohmstede C, Farrel F, Reep B, Clemetson K, Lapetina EG (1990) RAP2B: a RAS-related GTP-binding protein from platelets. Proc Natl Acad Sci USA 87:6527–6531

Pendergast A, Muller A, Havlik M, Maru Y, Witte O (1991) BCR sequences essential for transformation by the BCR-ABL oncogene bind to the ABL SH2 regulatory domain in a non-phosphotyrosine-dependent manner. Cell 66:161–171

Pizon V, Chardin P, Lerosey I, Oloffson B, Tavitian A (1988a) Human cDNAs rap1 and rap2 homologous to the Drosophila gene Dras3 encode proteins closely related to ras in the "effector" region. Oncogene 3:201–204

Pizon V, Lerosey I, Chardin P, Tavitian A (1988b) Nucleotide sequence of a human cDNA encoding a ras-related protein (rap1B). Nucleic Acids Res 16:7719

Rubinfeld B, Munemitsu S, Clark R, Conroy L, Watt K, Crosier W, McCorick F, Polakis P (1991) Molecular cloning of a GTPase activating protein specific for the Krev-1 protein p21^{rap1}. Cell 65:1033–1042

Sahyoun N, McDonald O, Farrel F, Lapetina EG (1991) Phosphorylation of a ras-related GTP-binding protein, Rap1-b, by a neuronal Ca^{2+}/calmodulin-dependent protein kinase, CaM kinase Gr. Proc Natl Acad Sci USA 88:2643–2647

Settleman J, Narasimhan V, Foster L, Weinberg R (1992) Molecular cloning of cDNAs encoding the GAP-associated protein p190: implications for a signalling pathway from ras to the nucleus. Cell 69:539–549

Shirataki H, Kaibuchi K, Hiroyoshi M, Isomura M, Araki S, Sasaki T, Takai Y (1991) Inhibition of the action of the stimulatory GDP/GTP exchange protein for smg p21 by the geranylgeranylated synthetic peptides designed from its C-terminal region. J Biol Chem 266:20672–20677

Siess W, Winegar D, Lapetina E (1990) Rap1-B is phosphorylated by protein kinase A in intact human platelets. Biochem Biophys Res Commun 170:944–950

Torti M, Lapetina EG (1992) The role of Rap1b and p21[ras] GTPase-activating protein in the regulation of phospholipase C-τ1 in human platelets. Proc Natl Acad Sci USA 89:7796–7800

Watson S, McConell R, Lapetina EG (1984) The rapid formation of inositol phosphates in human platelets by thrombin is inhibited by prostacyclin. J Biol Chem 259:13199–13203

Wong G, Muller O, Clark R, Conroy L, Moran M, Polakis P, McCormick F (1992) Molecular cloning and nucleic acid binding properties of the GAP-associated tyrosine phosphoprotein p62. Cell 69:551–558

Yamamoto T, Kaibuchi K, Mizunu, T, Hiroyoshi M, Shirataki H, Takai Y (1990) Purification and characterization from bovine brain cytosol of proteins that regulate the GDP/GTP exchange reaction of smg p21s ras p21-like GTP-binding proteins. J Biol Chem 265:16626–16634

CHAPTER 39

GDP/GTP Exchange Proteins for Small GTP-Binding Proteins

Y. Takai, K. Kaibuchi, A. Kikuchi, and T. Sasaki

A. Introduction

There is a superfamily of small GTP-binding proteins in which more than forty members are included (Bourne et al. 1991; Hall 1990; Takai et al. 1992). They exhibit GDP/GTP-binding and GTPase activities and have two interconvertible forms: GDP-bound inactive and GTP-bound active forms (Fig. 1). The GDP-bound form is converted to the GTP-bound form by the GDP/GTP exchange reaction which is regulated by GDP/GTP exchange proteins (GEPs). The GDP/GTP exchange reaction is initiated by the dissociation of GDP from the GDP-bound form of small GTPase followed by the association of GTP to the guanine-nucleotide-free form. The GTP-bound form is converted to the GDP-bound form by the GTPase reaction which is stimulated by the GTPase activating protein (GAP) (McCormick 1990). In our laboratory, we have isolated two types of GEP: a stimulatory type named guanine nucleotide dissociation stimulator (GDS) and an inhibitory type named guanine nucleotide dissociation inhibitor (GDI) (Takai et al. 1992). We have purified *smg* GDS, *rho* GDI and *smg* p25[1] GDI to homogeneity from mammalian tissues, cloned their cDNAs, and

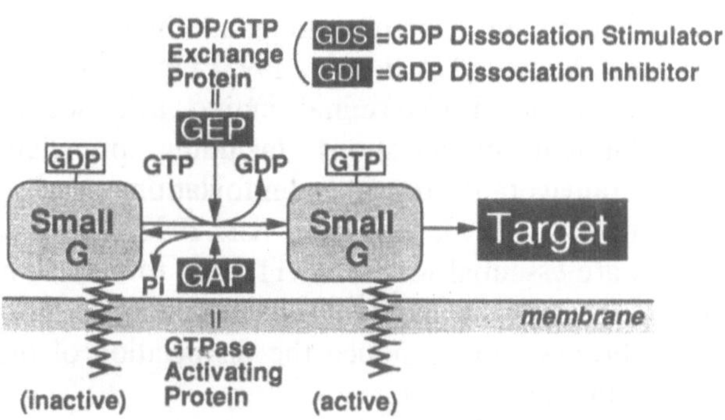

Fig. 1. Modes of activation and action of small GTPases (*small G*)

[1] *smg* p25 = Rab3

determined their primary structures, but we have only partially purified *rho* GDS. In this review article, we will describe these GEPs.

B. Physical Properties of GDP/GTP Exchange Protein

smg GDS, *rho* GDI, and *smg* p25 GDI are monomeric. *smg* GDS is composed of 558 amino acids with a M_r of about 61 kDa (KAIBUCHI et al. 1991; YAMAMOTO et al. 1990). *rho* GDI is composed of 204 amino acids with a M_r of about 23 kDa (FUKUMOTO et al. 1990; UEDA et al. 1990). *smg* p25 GDI is composed of 447 amino acids with a M_r of about 51 kDa (MATSUI et al. 1990; SASAKI et al. 1990). *rho* GDS shows a M_r of about 50 kDa as estimated by sucrose density gradient ultracentrifugation (ISOMURA et al. 1990). These GEPs are present in most mammalian tissues.

C. Two Actions of GDP/GTP Exchange Protein and Requirement of the Posttranslational Processing of Small GTPases for GDP/GTP Exchange Protein Actions

GDS stoichiometrically interacts with both the GDP-bound and GTP-bound forms of small GTPases and stimulates the dissociation of both nucleotides and thereby the subsequent association of both nucleotides in a cell-free system (YAMAMOTO et al. 1990). However, in intact cells, GDS may interact primarily with the GDP-bound form and stimulate the dissociation of GDP and the subsequent association of GTP, because the intracellular concentration of GTP is much higher than that of GDP. On the other hand, GDI forms a complex with only the GDP-bound form and inhibits the dissociation of GDP and the subsequent association of GTP (SASAKI et al. 1990; UEDA et al. 1990).

GEP has another function, namely, to regulate the reversible binding of small GTPases to membranes (TAKAI et al. 1992; Fig. 2A). Small GTPases have unique cDNA-predicted C-terminal amino acid sequences which undergo posttranslational modifications including prenylation (either farnesylation or geranylgeranylation), palmitoylation, methylation, and proteolysis (GLOMSET et al. 1990; RINE and KIM 1990; TAKAI et al. 1992). These modifications are essential for small GTPases to bind to membranes. Both GDS and GDI have the ability to block the association of small GTPases with membranes and to induce the dissociation of the prebound small GTPases from the membranes.

The posttranslational modifications are essential for small GTPases not only to bind to membranes but also to interact with GEP (Fig. 2B). GEP is active only on the posttranslationally processed form of small GTPases but is inactive on the posttranslationally unprocessed form (TAKAI et al. 1992).

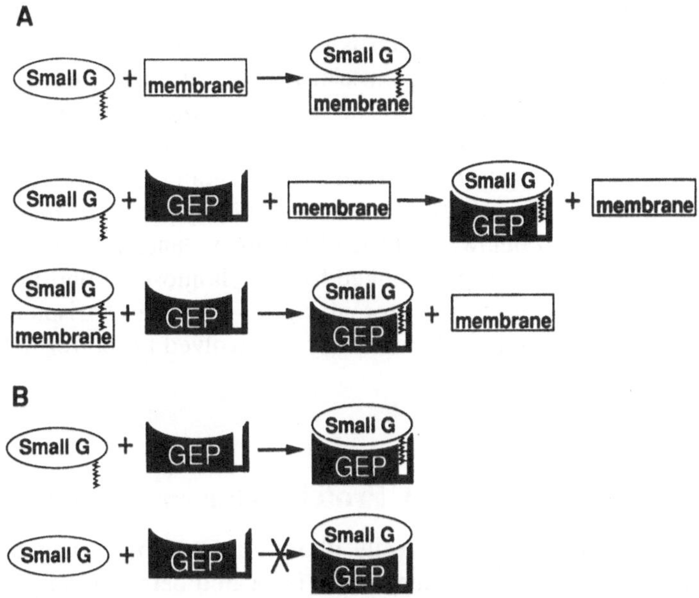

Fig. 2. A GDP/GTP exchange protein (*GEP*)-regulated reversible binding of small GTPases (*small G*) to membranes. **B** requirement of the posttranslational modifications of small GTPases for their interaction with GEP

D. Substrate Specificity of GDP/GTP Exchange Protein and Functional Cooperation Between Guanine Nucleotide Dissociation Stimulator and Guanine Nucleotide Dissociation Inhibitor

Each GEP was originally discovered as being specific for one small GTPase, but later found to be active on a group of small GTPases, i.e., *smg* GDS is active on at least Ki-*ras* p21, *smg* p21A[2], *smg* p21B[3], *rho*A p21, *rho*B p21, *rac*1 p21, and *rac*2 p21 (Hiraoka et al. 1992; Mizuno et al. 1991; Yamamoto et al. 1990); *rho* GDI is active on at least *rho*A p21, *rho*B p21, *rac*1 p21, and *rac*2 p21 (Hiraoka et al. 1992; Mizuno et al. 1992; Ueda et al. 1990); and *smg* p25 GDI is active on at least *smg* p25A[4], *rab*11 p24, and *SEC*4 p24 (Sasaki et al. 1990, 1991; Ueda et al. 1991). The precise substrate specificity of *rho* GDS has not been studied.

rho p21 and *rac* p21 are regulated by both GDS and GDI. In the simultaneous presence of equal amounts of *smg* GDS and *rho* GDI, the GDI activity is stronger than the GDS activity with *rho* p21 or *rac* p21 (Kikuchi et al. 1992; Mizuno et al. 1992). Similarly, the *rho* GDI activity is

---

[2] *smg* p21A = RapA
[3] *smg* p21B = RapB
[4] *smg* p25A = Rab3A

stronger than the *rho* GDS activity with *rho* p21 (Kuroda et al. 1992). In most tissues, the amount of *rho* GDI is much more than that of *smg* GDS and *rho* GDS. In resting smooth muscle, *rho* p21 is found mostly in the cytosol in complex with *rho* GDI, and not with *smg* GDS or *rho* GDS (Kuroda et al. 1992). Therefore, it is likely that some intracellular signal releases the inhibitory action of *rho* GDI, makes small GTPases sensitive to the *smg* GDS or *rho* GDS action, and consequently activates these small GTPases. As described below, we have obtained evidence that *rho* p21 and *rac* p21 are activated by receptors linked to phosphoinositide phospholipase C. Therefore, either diglyceride protein kinase C, inositol trisphosphate-Ca^{2+} or a still unidentified mechanism may be involved in the regulation of *rho* GDI activity.

E. Activation of *smg* p21 by Protein Kinases A and G

smg p21A and *smg* p21B are directly phosphorylated by protein kinases A and G at the same serine residue, which is located between the polybasic region and the geranylgeranylated cysteine in the C-terminal region (Takai et al. 1992). *smg* p21 is not activated by *smg* GDS in the presence of acidic phospholipids, such as phosphatidylserine, phosphatidic acid, and phosphoinositides. However, phosphorylated *smg* p21 is activated by *smg* GDS even in the presence of these phospholipids. Thus, *smg* p21 activation is initiated by protein kinase A- and protein kinase G-catalyzed phosphorylation of the small GTPases in a *smg*-GDS-dependent manner.

F. The Function of *smg* Guanine Nucleotide Dissociation Stimulator in Regulating Gene Expression and Cell Proliferation

The GTPγS-bound or the oncogenic form of *ras* p21 stimulates c-*fos* gene expression, DNA synthesis, and cell transformation in NIH/3T3 cells (Barbacid 1987; Takai et al. 1992). Coexpression of Ki-*ras* p21 with *smg* GDS also causes c-*fos* gene expression, DNA synthesis, and cell transformation in NIH/3T3 cells, indicating that *smg* GDS activates Ki-*ras* p21 in intact cells (Fujioka et al. 1992). K*rev*-1 p21 has been shown to inhibit the oncogenic Ki-*ras* p21-induced transformation of NIH/3T3 cells (Kitayama et al. 1989). Evidence is accumulating that *ras* p21 is a downstream molecule of the receptor-type tyrosine kinases, such as the platelet-derived growth factor (PDGF), epidermal growth factor (EGF), and nerve growth factor (NGF) receptors, and the nonreceptor-type tyrosine kinases, such as the *src*, *lck*, and *lyn* gene products (Fig. 3). It has also been shown that *ras* p21 activity is regulated by protein kinase C. It is

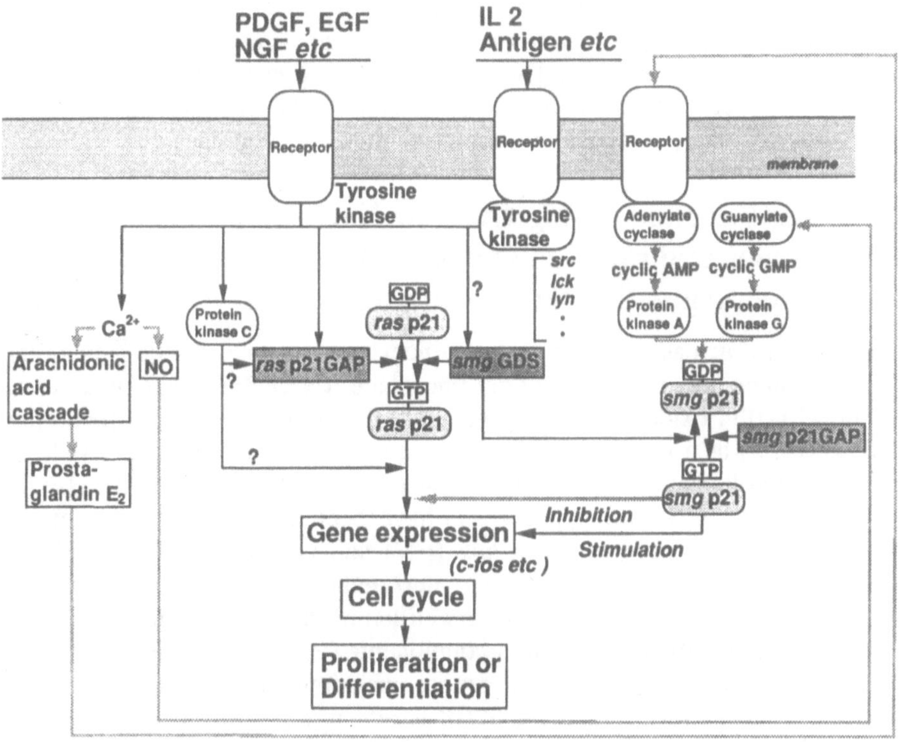

Fig. 3. Function of *smg* GDS to regulate gene expression and cell proliferation. *PDGF*, platelet-derived growth factor; *EGF*, epidermal growth factor; *NGF*, nerve growth factor; *IL*, interleukin

generally considered that *ras* p21 GAP is involved in the activation of *ras* p21 by these protein kinases, but it is very possible that *smg* GDS is also involved in this activation.

Cotransfection of *smg* p21 with *smg* GDS inhibits the expression of the c-*fos* gene induced by oncogenic Ki-*ras* p21 as well as by PDGF and a protein kinase C activating phorbol ester, but not that induced by dibutyryl cyclic AMP and oncogenic *raf* kinase in NIH/3T3 cells (Sakoda et al. 1992). The direct target protein for *ras* p21 is still unknown, but mitogen-activated protein (MAP) kinase activator, MAP kinase, and *raf* kinase may be the downstream molecules of *ras* p21 in that order. Our results, together with these earlier findings, suggest that *smg* p21 inhibits the signaling pathway between *ras* p21 and the MAP kinase activator in NIH/3T3 cells (Fig. 3).

In contrast to this antimitogenic action of *smg* p21 in NIH/3T3 cells, this small GTPase as well as *ras* p21 is mitogenic in Swiss 3T3 cells (Yoshida et al. 1992). Microinjection of the GTPγS-bound form of *smg* p21 into Swiss 3T3 cells causes DNA synthesis, but the GDP-bound form is inactive. Co-microinjection of the GDP-bound form of *smg* p21 with *smg* GDS is also active. It is well known that cyclic AMP is mitogenic in Swiss 3T3 cells. It is possible that at least a part of the mitogenic action of cyclic AMP is mediated through *smg* p21. The reason for the opposite actions of *smg* p21 in NIH/3T3 cells and Swiss 3T3 cells is unknown, but the target proteins for *ras* p21 and *smg* p21 may be different for these two cell lines.

G. The Function of *smg* Guanine Nucleotide Dissociation Stimulator and *rho* Guanine Nucleotide Dissociation Inhibitor Regulating Superoxide Generation

Phagocytes such as neutrophils and monocytes have a superoxide-generating nicotinamide-adenine-dinucleotide phosphate (NADPH) oxidase system. This NADPH oxidase system consists of membrane-associated cytochrome *b*558 (cyt *b*) as a terminal redox carrier and cytosolic components. At least three cytosolic components, named SOCI/NCF-3/σ1/Cl, SOCII/NCF-1/p47-*phox*, and SOCIII/NCF-2/p67-*phox*, have been identified (MOREL et al. 1991). The gene products p47-*phox* (p47) and p67-*phox* (p67) have been identified as the products causing autosomal recessive type of chronic granulomatous disease. Their primary structures have been determined, but neither the membrane-associated components (except cyt *b*) nor SOCI has been well characterized. Three groups including our own have recently revealed independently that the SOCI is *rac*1 p21 and *rac*2 p21 (ABO et al. 1991; KNAUS et al. 1991; MIZUNO et al. 1992). Moreover, we have shown that the NADPH oxidase activity is stimulated and inhibited through *rac* p21 by *smg* GDS and *rho* GDI, respectively. In the simultaneous presence of equal amounts of *smg* GDS and *rho* GDI, the stimulatory action of *smg* GDS is cancelled by the *rho* GDI action. The NADPH oxidase-catalyzed superoxide generation is stimulated by the receptors for immune complex, platelet-activating factor (PAF), and formyl-Met-Leu-Phe (fMLP) which are linked to the phosphoinositide phospholipase C. Therefore, as discussed above, this type of receptor releases the inhibitory action of *rho* GDI on *rac* p21 and makes it sensitive to the stimulatory action of *smg* GDS, resulting in its activation, which then stimulates the NADPH oxidase activity in cooperation with other factors (Fig. 4A).

H. The Function of *smg* Guanine Nucleotide Dissociation Stimulator, *rho*, Guanine Nucleotide Dissociation Stimulator, and *rho* Guanine Nucleotide Dissociation Inhibitor in Regulating the Actomyosin System

The actomyosin system, the contractile machinery of smooth muscle, is regulated by cytosolic Ca^{2+} (SOMLYO 1985). Elevated cytosolic Ca^{2+} activates calmodulin-dependent myosin light chain (MLC) kinase, which then phosphorylates and activates myosin ATPase, consequently inducing

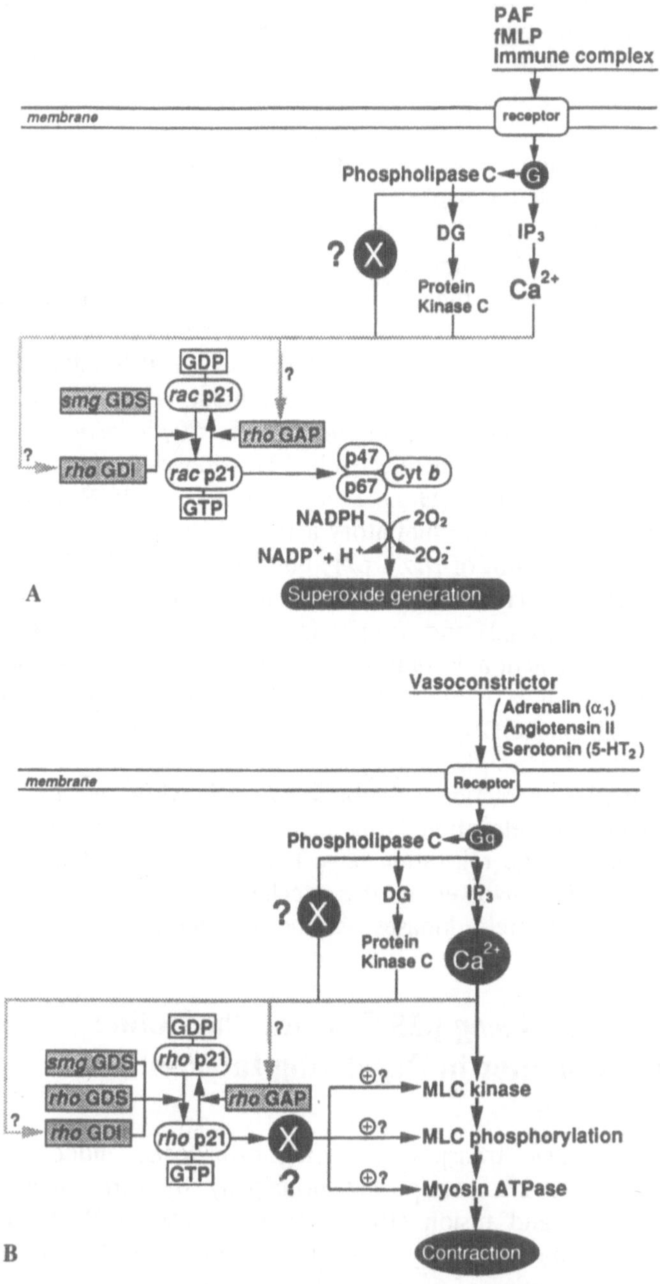

Fig. 4A,B. Function of *smg* GDS and *rho* GDI in regulating superoxide generation (A) and of *smg* GDS, *rho* GDS, and *rho* GDI in regulating smooth muscle contraction (B). *MLC*, myosin light chain; *GDS*, guanine nucleotide dissociation stimulator; *GDI*, guanine nucleotide dissociation inhibitor; *DG*, 1,2-diacylgycerol

smooth muscle contraction. However, the cytosolic Ca^{2+} level is not always parallel to the contraction level, and additional mechanisms to regulate the Ca^{2+} sensitivity of smooth muscle contraction have been proposed (BRADLEY and MORGAN 1987). One mechanism is that a GTPase may regulate the Ca^{2+} sensitivity on the basis of the observation that GTP or GTPγS lowers the Ca^{2+} concentrations necessary for the contraction. We have recently shown in collaboration with Dr. Saida's group (Bayer Yakuhin Ltd., Kobe, Japan) that *rho* p21 is responsible for this Ca^{2+} sensitization (HIRATA et al. 1992). In the rabbit mesenteric arterial smooth muscle skinned by saponin, Ca^{2+} induces contraction in a concentration-dependent manner. GTPγS lowers the Ca^{2+} concentrations required for this contraction and increases the Ca^{2+} sensitivity. GTPγS alone does not induce the contraction in the absence of Ca^{2+}. This GTPγS-enhanced Ca^{2+} sensitivity is completely abolished by treatment with C3 or EDIN, exoenzymes of *Clostridium botulinum* and *Staphylococcus aureus*, respectively. These exoenzymes selectively ADP-ribosylate *rho* p21 at Asn^{41} and inhibit *rho* p21 function (HALL 1990; SEKINE et al. 1989; SUGAI et al. 1992; TAKAI et al. 1992). The inhibitory action of these toxins is overcome by the GTPγS-bound form of *rho*A p21, but not by the GDP-bound form. *smg* p21B is inactive. The most abundant small GTPases in bovine aortic smooth muscle are *smg* p21 and *rho* p21 families. The Ca^{2+} sensitization of smooth muscle contraction is induced by the phospholipase C-linked receptors. Therefore, it is likely that *rho* p21 is regulated by this type of receptors through *smg* GDS, *rho* GDS, and *rho* GDI, as described above, and modulates Ca^{2+} action in smooth muscle contraction (Fig. 4B).

Smooth muscle contraction is inhibited by cyclic AMP- and cyclic GMP-elevating signals and relaxation is induced. Several mechanisms of these cyclic nucleotides on the relaxation have been reported, but it is possible that *smg* p21 is also involved in this mechanism, because it is directly phosphorylated by protein kinases A and G and is activated by this phosphorylation.

I. The Function of *smg* p25 Guanine Nucleotide Dissociation Inhibitor in Regulating Intracellular Vesicle Transport

In intracellular vesicle transport including exocytosis, endocytosis, and transcytosis, vesicles are transported principally by three mechanisms: budding, targeting, and fusion (ROTHMAN and ORCI 1992). Evidence is accumulating that the *rab* family is involved in these mechanisms (BALCH 1990). One mode of action of the small GTPase involved in this intracellular vesicle transport has been proposed, namely that a small GTPase is present in the GDP-bound form in the cytosol. This form is converted to the GTP-bound form which then binds to the vesicle. The vesicle is transported to the acceptor membrane and fuses with it. After this fusion, the GTP-bound form is converted to the GDP-bound form and is then translocated to the cytosol from the membrane. Thus, the cycling of a small GTPase between the cytosol and the membrane causes the transport of the vesicle to the membrane.

smg p25 GDI appears to play an important role in this small GTPase cycle because *smg* p25 GDI forms a complex only with the GDP-bound form, and regulates the association with and the dissociation from the membrane of the small GTPase. In collaboration with our group, Dr. Wollheim's group (University of Geneva, Switzerland) has recently revealed that the GDP-bound form of small GTPase is present in the cytosol in complex with *smg* p25 GDI in insulinoma cells (Regazzi et al. 1992). *smg* p25 GDI is active not only on *smg* p25A but also on *rab*11 p24 and *SEC4* p24 as described above. *smg* p25A is found only in cells with regulated secretion such as neurons, exocrine cells, and endocrine cells, whereas *rab*11 p24 is present not only in cells with regulated secretion, but also in cells with constitutive secretion such as hepatocytes and glia cells (Takai et al. 1992). *smg* p25 GDI is found in both types of cells. Therefore, it is likely that *smg* p25 GDI may control the cycling of a group of small GTPases between the cytosol and membrane for intracellular vesicle transport (Fig. 5).

J. Conclusions

We have focused here on the GEPs which we found in our laboratory, but several other GEPs have also been reported from other laboratories. The

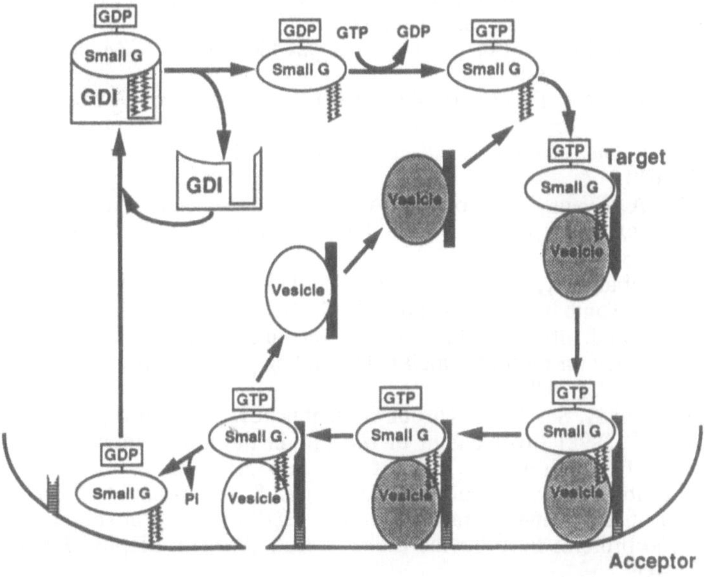

Fig. 5. Function of *smg* p25 GDI in regulating intracellular vesicle transport. *GDI*, guanine nucleotide dissociation inhibitor; *Small G*, small GTPase

CDC25 gene product as a GDS for the Ras protein in yeast was the first GEP reported for a small GTPase (TAKAI et al. 1992). Subsequently, several GEPs such as REP, GRF, and rGEF have been reported (TAKAI et al. 1992). Very recently, the *dbl* gene product, which was found as an oncogene for human lymphoma, has been shown to be a GDS for *CDC42* p25/G25K (HART et al. 1991). *RCC1*, which was found as a regulator for chromatin condensation, has been shown to be a GDS for *ran* p25/*SPI1* p25 (BISCHOFF and PONSTINGL 1991). Thus, GEPs regulate various cell functions through their respective small GTPases.

References

Abo A, Pick E, Hall A, Totty N, Teahan CG, Segal AW (1991) Activation of the NADPH oxidase involves the small GTP-binding protein p21^{rac1}. Nature 353:668–670

Balch WE (1990) Small GTP-binding proteins in vesicular transport. Trends Biochem Sci 15:473–477

Barbacid M (1987) *Ras* genes. Annu Rev Biochem 56:779–827

Bischoff FR, Ponstingl H (1991) Catalysis of guanine nucleotide exchange on Ran by the mitotic regulator RCC1. Nature 354:80–82

Bourne HR, Sanders DS, McCormick F (1991) The GTPase superfamily: conserved structure and molecular mechanism. Nature 349:117–127

Bradley AB, Morgan KG (1987) Alterations in cytoplasmic calcium sensitivity during porcine coronary artery contractions as detected by aequorin. J Physiol (Lond) 385:437–448

Fujioka H, Kaibuchi K, Kishi K, Yamamoto T, Kawamura M, Sakoda T, Mizuno T, Takai Y (1992) Transforming and c-*fos* promoter/enhancer-stimulating activities of a stimulatory GDP/GTP exchange protein for small GTP-binding proteins. J Biol Chem 267:926–930

Fukumoto Y, Kaibuchi K, Hori Y, Fujioka H, Araki S, Ueda T, Kikuchi A, Takai Y (1990) Molecular cloning and characterization of a novel type of regulatory protein (GDI) for the *rho* proteins, *ras* p21-like small GTP-binding proteins. Oncogene 5:1321–1328

Glomset JA, Gelb MH, Farnsworth CC (1990) Prenyl proteins in eukaryotic cells: a new type of membrane anchor. Trends Biochem Sci 15:139–142

Hall A (1990) The cellular functions of small GTP-binding proteins. Science 249:635–640

Hart MJ, Eva A, Evans T, Aaronson SA, Cerione RA (1991) Catalysis of guanine nucleotide exchange on the CDC42Hs protein by the *dbl* oncogene product. Nature 354:311–314

Hiraoka K, Kaibuchi K, Ando S, Musha T, Takaishi K, Mizuno T, Asada M, Ménard L, Tomhave E, Didsbury J, Snyderman R, Takai Y (1992) Both stimulatory and inhibitory GDP/GTP exchange proteins, *smg* GDS and *rho* GDI, are active on multiple small GTP-binding proteins. Biochem Biophys Res Commun 182:921–930

Hirata K, Kikuchi A, Sasaki T, Kuroda S, Kaibuchi K, Matsuura Y, Seki H, Saida K, Takai Y (1992) Involvement of *rho* p21 in the GTP-enhanced calcium ion sensitivity of smooth muscle. J Biol Chem 267:5719–5722

Isomura M, Kaibuchi K, Yamamoto T, Kawamura S, Katayama M, Takai Y (1990) Partial purification and characterization of GDP dissociation stimulator (GDS) for the *rho* proteins from bovine brain cytosol. Biochem Biophys Res Commun 169:652–659

Kaibuchi K, Mizuno T, Fujioka H, Yamamoto T, Kishi K, Fukumoto Y, Hori Y, Takai Y (1991) Molecular cloning of the cDNA for stimulatory GDP/GTP exchange protein for *smg* p21s (*ras* p21-like small GTP-binding proteins) and characterization of stimulatory GDP/GTP exchange protein. Mol Cell Biol 11:2873–2880

Kikuchi A, Kuroda S, Sasaki T, Kotani K, Hirata K, Katayama M, Takai Y (1992) Functional interactions of stimulatory and inhibitory GDP/GTP exchange proteins and their common substrate small GTP-binding protein. J Biol Chem 267:14611–14615

Kitayama H, Sugimoto Y, Matsuzaki T, Ikawa Y, Noda M (1989) A ras related gene with transformation suppressor activity. Cell 56:77–84

Knaus UG, Heyworth PG, Evans T, Curnutte JT, Bokoch GM (1991) Regulation of phagocyte oxygen radical production by the GTP-binding protein rac 2. Science 254:1512–1515

Kuroda S, Kikuchi A, Hirata K, Masuda T, Kishi K, Sasaki T, Takai Y (1992) Cooperative function of *rho* GDS and *rho* GDI to regulate *rho* p21 activation in smooth muscle. Biochem Biophys Res Commun 185:473–480

Matsui Y, Kikuchi A, Araki S, Hata Y, Kondo J, Teranishi Y, Takai Y (1990) Molecular cloning and characterization of a novel type of regulatory protein (GDI) for *smg* p25A, a *ras* p21-like GTP-binding protein. Mol Cell Biol 10:4116–4122

McCormick F (1990) The world according to GAP. Oncogene 5:1281–1282

Mizuno T, Kaibuchi K, Yamamoto T, Kawamura M, Sakoda T, Fujioka H, Matsuura Y, Takai Y (1991) A stimulatory GDP/GTP exchange protein for *smg* p21 is active on the post-translationally processed form of c-Ki-*ras* p21 and *rho*A p21. Proc Natl Acad Sci USA 88:6442–6446

Mizuno T, Kaibuchi K, Ando S, Musha T, Hiraoka K, Takaishi K, Asada M, Nunoi H, Matsuda I, Takai Y (1992) Regulation of the superoxide-generating NADPH oxidase by a small GTP-binding protein and its stimulatory and inhibitory GDP/GTP exchange proteins. J Biol Chem 267:10215–10218

Morel F, Doussiere J, Vignais PV (1991) The superoxide-generating oxidase of phagocytic cells: physiological, molecular and pathological aspects. Eur J Biochem 201:523–546

Regazzi R, Kikuchi A, Takai Y, Wollheim CB (1992) The small GTP-binding proteins in the cytosol of insulin-secreting cells are complexed to GDP-dissociation inhibitor proteins. J Biol Chem 267:17512–17519

Rine J, Kim SH (1990) A role for isoprenoid lipids in the localization and function of an oncoprotein. New Biologist 2:219–226

Rothman JE, Orci L (1992) Molecular dissection of the secretory pathway. Nature 355:409–415

Sakoda T, Kaibuchi K, Kishi K, Kishida S, Doi K, Hoshino M, Hattori S, Takai Y (1992) Inhibition by *smg/rap*1/*Krev*-1 p21s of the *ras* p21-stimulated c-*fos* promoter/enhancer in NIH/3T3 cells. Oncogene 7:1705–1711

Sasaki T, Kikuchi A, Araki S, Hata Y, Isomura M, Kuroda S, Takai Y (1990) Purification and characterization from bovine brain cytosol of a protein that inhibits the dissociation of GDP from and the subsequent binding of GTP to *smg* p25A, a *ras* p21-like GTP-binding protein. J Biol Chem 265:2333–2337

Sasaki T, Kaibuchi K, Kabcenell AK, Novick PJ, Takai Y (1991) A mammalian inhibitory GDP/GTP exchange protein (GDP dissociation inhibitor) for *smg* p25A is active on the yeast *SEC4* protein. Mol Cell Biol 11:2909–2912

Sekine A, Fujiwara M, Narumiya S (1989) Asparagine residue in the rho gene product is the modification site for botulinum ADP-ribosyltransferase. J Biol Chem 264:8602–8605

Somlyo AP (1985) Excitation-contraction coupling and the ultrastructure of smooth muscle. Circ Res 57:497–507

Sugai M, Hashimoto K, Kikuchi A, Inoue S, Okumura H, Matsumoto K, Goto Y, Ohgai H, Moriishi K, Syuto B, Yoshikawa K, Suginaka H, Takai Y (1992) Epidermal cell differentiation inhibitor (EDIN) ADP-ribosylates small GTP-binding proteins and induces hyperplasia in epidermis. J Biol Chem 267:2600–2604

Takai Y, Kaibuchi K, Kikuchi A, Kawata M (1992) Small GTP-binding proteins. Int Rev Cytol 133:187–222

Ueda T, Kikuchi A, Ohga N, Yamamoto J, Takai Y (1990) Purification and characterization from bovine brain cytosol of a novel regulatory protein inhibiting the dissociation of GDP from and the subsequent binding of GTP to *rho*B p20, a *ras* p21-like GTP-binding protein. J Biol Chem 265:9373–9380

Ueda T, Takeyama Y, Ohmori T, Ohyanagi H, Saitoh Y, Takai Y (1991) Purification and characterization from rat liver cytosol of a GDP dissociation inhibitor (GDI) for liver 24KG, a *ras* p21-like GTP-binding protein, with properties similar to those of *smg* p25A GDI. Biochemistry 30:909–917

Yamamoto T, Kaibuchi K, Mizuno T, Hiroyoshi M, Shirataki H, Takai Y (1990) Purification and characterization from bovine brain cytosol of proteins that regulate the GDP/GTP exchange reaction of *smg* p21s, *ras* p21-like GTP-binding proteins. J Biol Chem 265:16626–16634

Yoshida Y, Kawata M, Miura Y, Musha T, Sasaki T, Kikuchi A, Takai Y (1992) Microinjection of *smg*/*rap*1 p21 into Swiss 3T3 cells induces DNA synthesis and morphological changes. Mol Cell Biol 12:3407–3414

CHAPTER 40

GTP-Mediated Communication Between Intracellular Calcium Pools

D.L. GILL, T.K. GHOSH, A.D. SHORT, J. BIAN, and R.T. WALDRON

A. Intracellular Ca^{2+} Signaling Pools

I. Nature of Intracellular Ca^{2+} Pools

The release of intracellular Ca^{2+} is widely established as a major component of receptor-mediated cytosolic Ca^{2+} signals (GILL et al. 1989; BERRIDGE 1990; BERRIDGE and IRVINE 1989). Yet the nature and location of Ca^{2+} pools from which Ca^{2+} is released remain poorly characterized. There is good evidence that the endoplasmic reticulum (ER) is an important Ca^{2+}-sequestering site within cells and a probable major target for the intracellular mediator of Ca^{2+} release, inositol 1,4,5-trisphosphate ($InsP_3$) (GHOSH et al. 1989). However, the ER itself constitutes a large and heterogeneous organelle in most cells; obvious subdomains of the ER exist, including rough and smooth cisternae and the nuclear envelope. In addition, it is likely that microheterogeneity of ER results in subcompartments of ER that may have quite different properties, even though they are not morphologically distinguishable. For example, smooth cisternae are widely distributed in cells; some deep within the cell, others sometimes close to the plasma membrane. It is very possible that different Ca^{2+} handling properties are associated with ER at different locations. This does not necessarily imply that there are distinct Ca^{2+} transport mechanisms at such locations. Instead, differences may arise, for example, in proximity of ER cisternae to the plasma membrane, those closest receiving the most rapid exposure to the highest concentrations of $InsP_3$. Others believe that organelles perhaps closely associated with, but distinct from, ER may represent the major Ca^{2+}-signaling pools. Such organelles were termed "calciosomes" by MELDOLESI and colleagues (VOLPE et al. 1988). This view has been based on a number of observations including the immunocytochemical localization of Ca^{2+}-binding proteins, Ca^{2+} pumps, and $InsP_3$ receptors. However, as yet there does not seem to be overwhelming evidence for a clearly identifiable and discrete Ca^{2+}-signaling organelle (ROSSIER and PUTNEY 1991). Thus, whereas it may be reasonable to refer to the source of Ca^{2+} releasable by $InsP_3$ as "calciosomes", it is clear that we have no defined idea of what such organelles are or whether they are distinct from the ER.

The work described here is aimed at understanding the mechanisms that control Ca^{2+} within intracellular pools, to functionally distinguish between heterogeneous populations of Ca^{2+} pools, and to observe the process by which Ca^{2+} can be transferred between pools. Studies have identified and characterized both the Ca^{2+}-releasing (GILL et al. 1989; GHOSH et al. 1988, 1989, 1990; CHUEH et al. 1987; UEDA et al. 1986; CHUEH and GILL 1986) and Ca^{2+}-accumulating (GILL et al. 1984, 1989; GILL and CHUEH 1985; BIAN et al. 1991) properties of these organelles. The heterogeneity of the Ca^{2+}-handling properties of intracellular pools have become increasingly apparent (GHOSH et al. 1989; UEDA et al. 1986; CHUEH and GILL 1986; BIAN et al. 1991). Moreover, we have obtained intriguing results which we interprete as evidence that these distinct pools can undergo guanine nucleotide-dependent interactions and that Ca^{2+} can be transferred between different pools (GILL et al. 1986a, 1989a; GHOSH et al. 1989; CHUEH et al. 1987; UEDA et al. 1986; CHUEH and GILL 1986; MULLANEY et al. 1987, 1988). Using specific Ca^{2+}-pump blockers we have been able to gain insights into the functional significance of these pools, not only in their ability to generate Ca^{2+} signals, but also in their involvement in the control of cell growth (BIAN et al. 1991; GHOSH et al. 1991). More detailed descriptions of certain aspects of this work have appeared elsewhere (GILL et al. 1988a,b, 1989b, 1990).

II. Movements of Ca^{2+} Induced by Inositol Phosphates

The rapid and specific effects of $InsP_3$ on the release of Ca^{2+} from intracellular pools have been shown in a large number of studies (GILL et al. 1989a; BERRIDGE 1987; BERRIDGE and IRVINE 1984, 1989). Our work on permeabilized cells and microsomal vesicles has revealed that $InsP_3$ functions upon an internal organelle with Ca^{2+}-pumping properties similar to those of muscle sarcoplasmic reticulum (SR; GILL and CHUEH 1985) and quite distinct from the plasma membrane Ca^{2+} pump (GILL 1982; GILL et al. 1981, 1984, 1986). It is clear that $InsP_3$ releases only a fraction of Ca^{2+} within cells, indicating that multiple Ca^{2+}-pumping compartments exist, only some of which are sensitive to $InsP_3$ (CHUEH et al. 1987; UEDA et al. 1986; CHUEH and GILL 1986; GILL et al. 1986b). The close correlation between binding of $InsP_3$ and activation of Ca^{2+} channels in the ER membrane has come from studies on the potent and specific antagonism of action of $InsP_3$ by heparin (GHOSH et al. 1988). Heparin not only blocks, but also reverses, the action of $InsP_3$; the sustained action of $InsP_3$ is almost immediately reversed upon addition of heparin. Thus, $InsP_3$ appears to activate a channel for Ca^{2+} that remains open as long as $InsP_3$ is bound, but closes immediately upon $InsP_3$ dissociation from the receptor. Indeed observations by others on $InsP_3$-activated conductance provide more direct evidence for the existence of a direct $InsP_3$-receptor-activated Ca^{2+} channel (EHRLICH and WATRAS 1988).

$InsP_3$ induces a sustained release of Ca^{2+} from Ca^{2+} pump-loaded permeabilized cells (CHUEH et al. 1987; MULLANEY et al. 1987, 1988) indicating that channels can remain open for extended lengths of time. In many systems the actions of submaximal $InsP_3$ levels appear to be quantal in nature, the extent rather than rate of release being proportional to $InsP_3$ concentration (UEDA et al. 1986; MUALLEM et al. 1989). This has been interpreted to indicate that the $InsP_3$-sensitive Ca^{2+} pools themselves are heterogeneous with respect to sensitivity of $InsP_3$ (MUALLEM et al. 1989). Another suggestion is that the sensitivity of the pool, while more or less uniform, perhaps changes as Ca^{2+} is released (IRVINE 1990). It is possible the observations relate to fragmentation, vesiculation, or other modification of the pools under the conditions of Ca^{2+} flux measurement, which may occur as a result of physical damage to the ER and/or other organelles that make up Ca^{2+} pools. Since a definitive demonstration of quantal release in an intact system has yet to be made, it is well to keep open the possibility that the phenomenon may not occur under normal intracellular conditions. It is entirely possible that the mechanism of GTP-activated communication between Ca^{2+} pools described below may be part of the process by which a more homogeneous Ca^{2+}-signaling pool is maintained in intact cells.

III. Intracellular Ca^{2+} Channels

Much information on the molecular properties and arrangement of $InsP_3$ receptors has recently been gained (GILL 1989). The massive quantities of $InsP_3$-binding protein identified by SNYDER and coworkers in the cerebellum (WORLEY et al. 1987) indeed turned out to be the functional receptor. The same protein had been identified years earlier by several workers as a major substrate for cyclic adenosine monophosphate (AMP) dependent phosphorylation. Its cloning by MIKOSHIBA and coworkers (FURUICHI et al. 1989) was a remarkable revelation since the structure bore significant homology with the ryanodine receptor Ca^{2+} release channel from skeletal muscle (GILL 1989; FURUICHI et al. 1989; MIGNERY et al. 1989). A large N-terminal domain extending into the cytoplasm presumably contains the $InsP_3$-binding site; the transmembrane domain containing up to eight membrane-spanning α helices is situated near the C terminus (see Fig. 1). This latter region is largely homologous with the ryanodine receptor of muscle and is presumed to represent a similar Ca^{2+}-conducting channel. That the identified receptor is indeed the complete $InsP_3$-sensitive Ca^{2+} channel has been established from studies in which the purified protein from brain has been reconstituted in lipid bilayers (FURUICHI et al. 1989; FERRIS et al. 1989) and in which transfection of the $InsP_3$ receptor gene into cultured cells increases the levels of functional $InsP_3$ receptors (FURUICHI et al. 1989; NAKAGAWA et al. 1991). An interesting observation and a further indication of the similarities between the organization of $InsP_3$ and ryanodine receptors

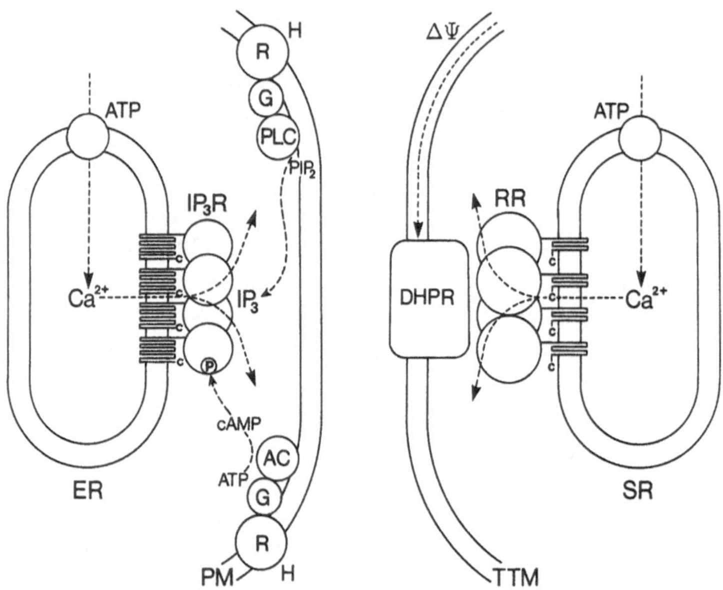

Fig. 1. Comparison of structure and function of the $InsP_3$ and ryanodine receptors. Comparisons are shown between the Ca^{2+}-signaling mechanisms of the $InsP_3$ receptor (*IP_3R*) and ryanodine receptor (*RR*) in nonmuscle tissue (*left*) and skeletal muscle (*right*), respectively. In nonmuscle, hormones (*H*) binding to receptors (*R*) coupled to G-proteins (*G*) activate phospholipase C (*PLC*) and the breakdown of phosphatidylinositol 4,5-bisphosphate (*PIP_2*) to release inositol 1,4,5-trisphosphate (*IP_3*) into the cytosol. The action of $InsP_3$ on Ca^{2+} release from endoplasmic reticulum (*ER*) is mediated via the $InsP_3R$, depicted here as being tetrameric. In addition, *cAMP*, produced by hormonal modification of adenylyl cyclase (*AC*), activates phosphorylation of the $InsP_3R$ and inhibits $InsP_3$-induced activation of Ca^{2+} release (GILL 1989). In skeletal muscle, the initial signal is an electrical depolarization (*$\Delta\Psi$*) of the T-tubule membrane (*TTM*), which is detected by the voltage-dependent dihydropyridine-sensitive receptor (*DHPR*). The signal is believed to be modulated via the RR (in close contact with the DHPR) into opening of a channel within the core of the RR tetrameric structure (COMERFORD and DAWSON 1988). In both cases, the end result is Ca^{2+} release into the cytosol

is that both receptors appear to exist in membranes as multimeric, probably tetrameric, complexes (FURUICHI et al. 1989) and that these may be important for functional Ca^{2+} conductance.

Immunocytochemical localization of the $InsP_3$ receptor has proven interesting, revealing its location within the ER and also the nuclear envelope (ROSS et al. 1989). Stacks of rough ER cisternae of the cerebellum can be heavily stained, although there are also areas of rough ER without staining; smooth ER is also clearly, though again heterogeneously, stained, and in certain cases smooth ER close to the plasma membrane is stained (FURUICHI et al. 1989; MIGNERY et al. 1989; ROSS et al. 1989). The functional homology between the $InsP_3$ receptor and the Ca^{2+}-releasing ryanodine receptor of fast muscle is intriguing (see Fig. 1). Based on the direct coupling between plasma membrane and sarcoplasmic reticulum, IRVINE (1990) postulated that $InsP_3$ receptors may also be modulated by interacting with a component in the plasma membrane. Indeed, it was hypothesized that signals that are triggered by emptying the internal pool and that activate Ca^{2+} entry (see below) may arise through this direct interaction. In addition, evidence that $InsP_4$ may be associated with Ca^{2+} entry (IRVINE 1991) was incorporated into the model such that the $InsP_4$ receptor might possibly be the plasma membrane component of this system (IRVINE 1990). Interestingly, our model of interaction between distinct membranes and modification of the $InsP_3$-sensitive pool (described in detail below) had suggested to us 2 years earlier that exactly this type of process might be occurring. Therefore, we fastidiously searched for any actions of $InsP_4$ on our systems, but our findings have been consistently negative. We are, therefore, in favor of such a model but at present skeptical of any direct involvement of $InsP_4$.

IV. Significance of Ca^{2+} Within the $InsP_3$-Sensitive Ca^{2+} Pool

Although $InsP_3$-induced emanation of Ca^{2+} from Ca^{2+}-accumulating organelles into the cytosol may be a major cellular signaling event, we should also consider the significance of intraorganelle Ca^{2+}. Thus, from a large body of recent evidence, the levels of Ca^{2+} remaining within such organelles appear to have important consequences for signaling and regulation within cells. Recently it has been shown that the emptying of such pools may be a direct trigger for the activation of entry of external Ca^{2+} ions across the plasma membrane (PUTNEY 1990), an important conclusion since the triggering of Ca^{2+} entry is fundamental to the generation of physiological Ca^{2+} signals. However, at present the nature of any signal generated from the ER as a result of pool emptying remains unclear (IRVINE 1990; PUTNEY 1990). Also, as described above, it has been proposed that the level of intraluminal Ca^{2+} may be a key regulator of the $InsP_3$ receptor itself, with recent evidence apparently directly supporting this view (Irvine 1990). Another recently considered aspect of the high Ca^{2+} content within the lumen of the ER is its potentially significant role in the control of essential functions of this organelle. Thus, ER is the major site for translation, translocation, and processing of membrane proteins and is intimately associated with the whole process of membrane biogenesis and trafficking. Recently, SAMBROOK (1990) described the possibility that intraluminal ER Ca^{2+} could have a major regulatory influence on these activities and that control of Ca^{2+} channel and/or pump activities in the ER membrane could be as important to these functions as the ER-derived Ca^{2+} signals that are generated in the cytosol. Our own studies on the modification of Ca^{2+} within the ER are correlated with some profound alterations of cell growth (GHOSH et al. 1991) and may reflect either an altered ability of the ER to generate specific Ca^{2+} signals or modification of key ER functions dependent on intraluminal Ca^{2+}. The observations that Ca^{2+} levels in different fractions of the ER may be differentially controlled and that interactions between ER subcompartments seem to be fundamentally associated with Ca^{2+} transfer between compartments both lend support to the view that Ca^{2+} regulation in ER is a key aspect of its functions. Studies supporting this view are presented below.

B. Ca^{2+} Movements Activated by Guanine Nucleotides

I. GTP-Induced Ca^{2+} Fluxes

During the last few years a number of studies have been directed towards understanding a process activated by guanine nucleotides that exerts profound influence on the movements of Ca^{2+} in both the $InsP_3$-sensitive and $InsP_3$-insensitive pools of Ca^{2+} in cells. The first description of a GTP-induced alteration of Ca^{2+} movements in cells was reported by DAWSON (1985), who observed that GTP appeared to alter the responsiveness of liver microsomes to $InsP_3$. Using permeabilized cells we observed a rather different effect, namely, that GTP alone induces a rapid and profound release of Ca^{2+} from internal Ca^{2+} pools without the addition of any $InsP_3$ (UEDA et al. 1986; GILL et al. 1986). Indeed, within a few seconds after its addition, GTP induces release of approximately 60%–70% of the total Ca^{2+} accumulated within either permeabilized N1E-115 neuroblastoma cells or DDT_1MF-2 smooth muscle cells (CHUEH et al. 1987; UEDA et al. 1986; CHUEH and GILL 1986; GILL et al. 1986). The effect of GTP has a K_m value of 0.7 μM and is highly specific, other nucleotides – even at millimolar concentrations – having no effect. With DDT_1MF-2 cells, 0.1 μM GTP induces release of Ca^{2+} even in the presence of 1 mM adenosine triphosphate (ATP), the latter routinely used to maintain Ca^{2+} pumping (GILL and CHUEH 1985). GDP also induces Ca^{2+} release, but only after a time lag of 30–60 s. Its effect is due to nucleoside diphosphokinase (NDPK) mediated conversion to GTP; thus, its action is blocked by adenosine diphosphate (ADP) which competes with GDP for the NDPK enzyme (GILL et al. 1986). Also, no effect of GDPβS, a poor substrate for γ-phosphorylation by NDPK, is observed (GILL et al. 1986). In the presence of ADP to prevent conversion, GDP competitively blocks the action of GTP on Ca^{2+} release with a K_i value of less than 3 μM. Significantly, the nonhydrolyzable GTP analogs, GTPγS and GppNHp are without effect on Ca^{2+} release. In fact, GTPγS blocks the action of GTP (K_i 1–10 μM) whereas GppNHp is almost ineffective in doing so up to 200 μM, indicating a considerable specificity of the GTP-binding site for the two analogs (a useful parameter to define nucleotide specificity). It seems most likely that the effect of GTP is dependent on terminal phosphate hydrolysis, since its action is blocked by GTPγS or GppNHp and is competitively inhibited by GDP. The latter indicates that both GDP and GTP interact at the same site.

II. Ca^{2+} Compartments Sensitive to GTP and $InsP_3$

It is apparent from the above that both the $InsP_3$- and GTP-induced Ca^{2+} release processes function on a similar intracellular Ca^{2+}-sequestering compartment. However, the size of the releasable pools of Ca^{2+} are quite distinct. For example, in the N1E-115 cell line, the pool of Ca^{2+} released by GTP is approximately twice the size of the $InsP_3$-releasable pool, as shown in Fig. 2. Using permeabilized N1E-115 cells, following maximal Ca^{2+} release by GTP, $InsP_3$ is ineffective in releasing further Ca^{2+} (Fig. 2B); however, following maximal release by $InsP_3$ (approximately 30% of

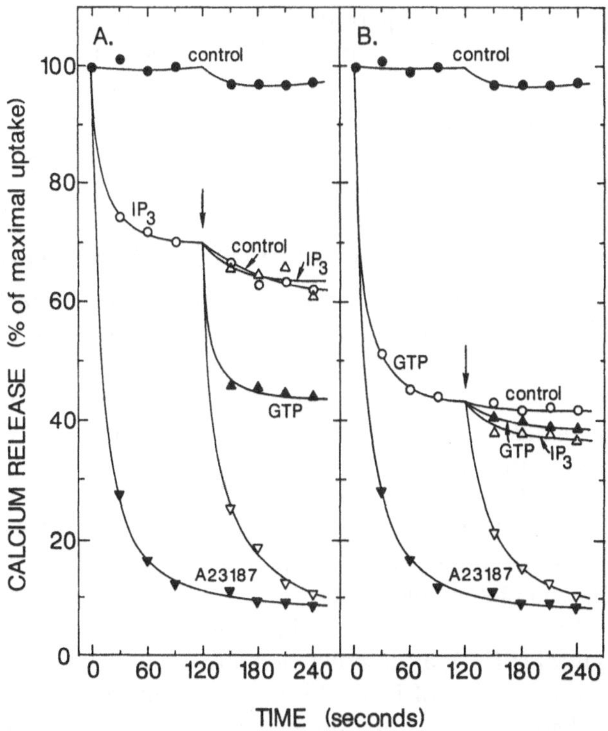

Fig. 2A,B. Relative sizes of the $InsP_3$- and GTP-sensitive Ca^{2+} pools. Effects of sequential addition of $InsP_3$ and GTP on Ca^{2+} release from permeabilized N1E-115 neuroblastoma cells. Ca^{2+} release was measured after loading for 5 min in the presence of $0.1\,\mu M$ free Ca^{2+}. **A** Immediately following uptake, release was observed after addition of either $10\,\mu M$ IP_3 (○), $5\,\mu M$ A23187 (▼), or buffer control (●); after 120 s of release in the presence of IP_3, measurement of release was continued after further additions of either $10\,\mu M$ IP_3 (△), $10\,\mu M$ GTP (▲), $5\,\mu M$ A23187 (▽), or buffer control (○). **B** Immediately follwoing uptake, release was observed after addition of either $10\,\mu M$ GTP (○), $5\,\mu M$ A23187 (▼), or buffer control (●); after 120 s of release in the presence of GTP, release was continued after further addition of either $10\,\mu M$ IP_3 (△), $10\,\mu M$ GTP (▲), $5\,\mu M$ A23187 (▽), or buffer control (○). In each case, samples of the Ca^{2+}-loaded permeabilized cell suspension were removed followed by rapid filtration and washing. Details of the experiment are given in Chueh et al. (1987)

accumulated Ca^{2+}), GTP does effect a further release of Ca^{2+} (Fig. 2A), in fact, down to the level GTP could induce when added alone (i.e., approximately 60% of accumulated Ca^{2+}). These results suggest that three compartments exist: one sensitive to both GTP and $InsP_3$, another releasing only in response to GTP, and a third not releasing in response to either agent. Thus, it is apparent that although the GTP-releasable pool differs from the $InsP_3$-releasable pool in that it is larger, at least a significant proportion of accumulated Ca^{2+} lies within a pool which can be released by either of the two agents. In other words, it appears that all of the Ca^{2+} within the $InsP_3$-sensitive Ca^{2+} pool is also releasable by the GTP-activated process, even if additional GTP-releasable Ca^{2+} also exists. This implies the probability of a proximal relationship between the $InsP_3$- and GTP-activated Ca^{2+} release processes, and permitted us to consider the existence of possible coupling events linking their modes of action. It should be noted that the GTP-activated Ca^{2+} pool, which includes that pool releasable by $InsP_3$, is exactly coincident with the pool of Ca^{2+} that is sensitive to the Ca^{2+} pump blocker thapsigargin and also that pool which can be released by certain sphingolipids, as described below.

III. Distinctions Between GTP- and $InsP_3$-Induced Ca^{2+} Transport

There are many clear parameters which reveal that GTP and $InsP_3$ activate Ca^{2+} movements by quite different processes (CHUEH and GILL 1986). First, the action of GTP is highly temperature dependent, its effect being absent at 4°C and reduced to a much slower rate at 25°C than at 37°C; in contrast, $InsP_3$ induces almost as rapid release at 4°C as at 37°C. Second, as mentioned above, the action of $InsP_3$ in permeabilized cells (normally conducted at $0.1\,\mu M$ free Ca^{2+}) is very sensitive to free $[Ca^{2+}]$, with half-maximal inhibition at $1\,\mu M$ and full inhibition at $10\,\mu M$ Ca^{2+}; GTP-activated Ca^{2+} release, on the other hand, is independent of free $[Ca^{2+}]$. Third, GTPγS, which completely blocks the action of GTP, has no effect on $InsP_3$-induced Ca^{2+} release. Fourth, the action of GTP is promoted by polyethylene glycol (PEG), as described in more detail below, whereas PEG has no effect on the action of $InsP_3$. Fifth, the specific action of heparin on $InsP_3$ (described above) is a further direct indication of separate mechanisms for $InsP_3$ and GTP, since heparin has no effect at all on the action of GTP (GHOSH et al. 1988). Overall, the relative temperature insensitivity and rapidity of $InsP_3$-induced release are consistent with its direct activation of a Ca^{2+} channel. In contrast, GTP appears to effect release via a temperature-sensitive process which most likely involves an enzymic hydrolysis of the terminal phosphate from GTP.

IV. Rationale for the Action of GTP

Observations of GTP actions on Ca^{2+} movements are intriguing. The major question arising from these results was what kind of Ca^{2+} transport process is being activated by GTP. It appeared to us that GTP might be inducing Ca^{2+} release by one of two mechanisms: (1) GTP-activated Ca^{2+} channels allowing efflux of Ca^{2+} from Ca^{2+}-pumping organelles, or (2) GTP-induced fusion between Ca^{2+}-pumping organelles, the open membrane structures so formed permitting Ca^{2+} release via an exocytotic-like event. To distinguish between these possibilities we undertook three different kinds of studies (CHUEH et al. 1987; GILL and CHUEH 1985; MULLANEY et al. 1987), which were designed to ascertain (a) whether the effect of GTP was reversible, (b) whether GTP-activated fusion could be directly visualized by electron microscopic analysis, and (c) whether GTP could induce release of oxalate-precipitated Ca^{2+}.

Considering the first question, we observed that the action of GTP could be reversed either by simple washing or by addition of GDP to the GTP-activated permeabilized cells or microsomal membrane vesicles. Washing of GTP-treated permeabilized cells with GTP-free medium appeared to completely reverse the action of GTP and restore uptake to its original state (CHUEH et al. 1987). GDP treatment after full GTP-activated release gave at least a partial restoration of uptake (GILL et al. 1986). These results would argue rather strongly against a simple membrane fusion mechanism, which would not be expected to be a reversible process. The second approach, electron microscopic visualization of microsomal vesicles after GTP treatment, indicated that GTP had no visible effect on their appearance. However, the addition of 1% PEG, a condition known to promote the action of GTP in releasing Ca^{2+} (UEDA et al. 1986; GILL et al. 1986), causes a very significant aggregation of vesicles although, again, there was no visible evidence of fusion. Thus, it was concluded that the action of GTP was enhanced under conditions of close membrane contact, another important clue to its possible mechanism of action. However, there was no visible evidence that GTP itself induces membrane fusion, a result which seemed to contrast with those of DAWSON and coworkers (DAWSON et al. 1986, 1987; COMERFORD and DAWSON 1988), as discussed below.

The third approach, analysis of oxalate-precipitated Ca^{2+}, gave a very unexpected result. Knowing that certain compartments within cells are permeable to Ca^{2+}-precipitating anions such as oxalate, we reasoned that Ca^{2+} precipitated within the Ca^{2+}-pumping compartment would not be rapidly released by a simple channel mechanism activated by GTP, whereas a gross fusion event might indeed release even the Ca^{2+}-oxalate precipitate. However, in the presence of even a low oxalate concentration (only modestly increasing the amount of Ca^{2+} accumulated within permeabilized cells), GTP induced a dramatic increase in Ca^{2+} accumulation (CHUEH et al. 1987; MULLANEY et al. 1987), as opposed to the rapid release of Ca^{2+} observed without oxalate. This result is shown in Fig. 3, where the effects of various levels of oxalate are compared; it can be seen that GTP-induced uptake is observed with oxalate concentrations as low as 2 m*M*. At first, we considered that this may indicate a completely separate GTP-activated

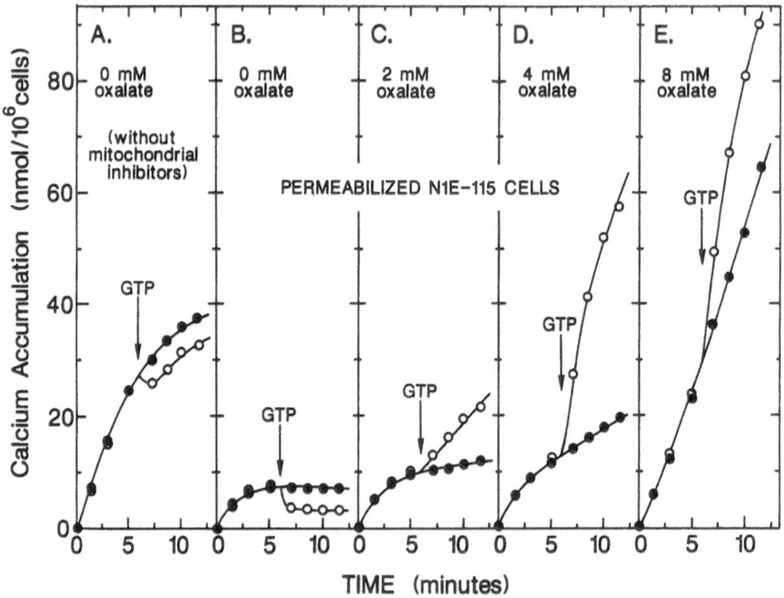

Fig. 3A–E. Oxalate-induced reversal of GTP-activated Ca^{2+} fluxes. GTP-activated movements of Ca^{2+} in permeabilized N1E-115 cells were measured with increasing oxalate concentrations as described in detail in MULLANEY et al. (1987). Experiments were undertaken either in the absence of mitochondrial inhibitors (**A**), or in the presence of 5 μM ruthenium red and 10 μM oligomycin. K-oxalate was either absent (**A** and **B**), or present from the beginning of uptake at a final concentration of 2 mM (**C**), 4 mM (**D**), or 8 mM (**E**). After 6 min of uptake 10 μM GTP (○) or control buffer (●) was added to the permeabilized cell suspensions

process that somehow results in stimulation of the Ca^{2+} pump. However, it soon became clear that each parameter defining the characteristics of the GTP-activated Ca^{2+} release process in the absence of oxalate was identical to that for GTP-stimulated uptake in the presence of oxalate (see Table 1). Thus, we concluded that the same mechanism mediates both processes. The only difference between the two effects is the dependence of the uptake process on Ca^{2+}-pumping activity, as revealed by its sensitivity to vanadate addition.

C. Interorganelle Translocation of Ca^{2+}

I. Model for GTP-Activated Ca^{2+} Translocation

The opposite effects of GTP described above in the presence and absence of oxalate, although at first perplexing, yielded the best clues to the possible mechanism of action of GTP. We considered the possibility of two pools of Ca^{2+}, only one of which is permeable to oxalate. There is good evidence to

Table 1. Summary of parameters of GTP-activated Ca^{2+} release and Ca^{2+} uptake in the absence and presence of oxalate, respectively

Parameter or condition	Calcium release (observed without oxalate)	Calcium uptake (observed with oxalate)
K_m for GTP	0.75 μM	0.9 μM
10 μM GDP	Delayed full effect	Delayed full effect
10 μM GDP (+ 1 mM ADP)	No effect	No effect
100 μM GDPβS	No effect	No effect
10 μM GTP + 100 μM GDP (+ 1 mM ADP)	GTP effect blocked	GTP effect blocked
10 μM GTP + 100 μM GDPβS	GTP effect blocked	GTP effect blocked
10 μM *GTPγ*S	Slight effect	Slight effect
10 μM GTP + 100 μM GTPγS	GTP effect blocked	GTP effect blocked
10 μM GppNHp	No effect	No effect
10 μM GTP + high 100 μM GppNHp	GTP effect not blocked	GTP effect not blocked
1–3% PEG	Stimulated	Stimulated
1 mM Vanadate	No effect	Blocked

Each of the parameters for GTP-activated Ca^{2+} uptake observed in the presence of oxalate refers to data presented in Chueh et al. (1987) and Mullaney et al. (1988). The observations relating to Ca^{2+} release (in the absence of oxalate) were published in Cheuh and Gill (1986), Gill et al. (1986) and Ueda et al. (1986). Explanation of the results is given in the text.

suppose that many cell membranes, including the plasma membrane, are impermeable to oxalate (Gill et al. 1984; Gill and Chueh 1985). In contrast, ER and SR appear to contain a nonspecific anion transporter (Martonosi 1982), the significance of which may be to assist in reducing charge buildup during rapid Ca^{2+} release events occurring during signal generation. The effect of GTP, which, as described above, is promoted under conditions inducing close contacts between membranes, may thus be to permit the "translocation" of Ca^{2+} between these compartments, perhaps by activating interorganelle junctional processes. In this event, communication of Ca^{2+} between oxalate-permeable and impermeable Ca^{2+}-pumping compartments would permit Ca^{2+} in the latter to have access to oxalate, and hence give rise to a net increase in total Ca^{2+} accumulation due to oxalate precipitation. In the absence of oxalate, no net effect would be observed. However, if the same GTP-activated Ca^{2+} conveyance occurs between intact and nonintact membrane structures (the latter being, for example, the plasma membrane), then the net effect would be release of Ca^{2+} into the medium. This relatively simple scheme is represented diagrammatically in Fig. 4 (top). Note that GTP-activated uptake in the presence of oxalate is dependent on Ca^{2+}-pumping activity, whereas release in the absence of oxalate can occur without simultaneous pumping activity,

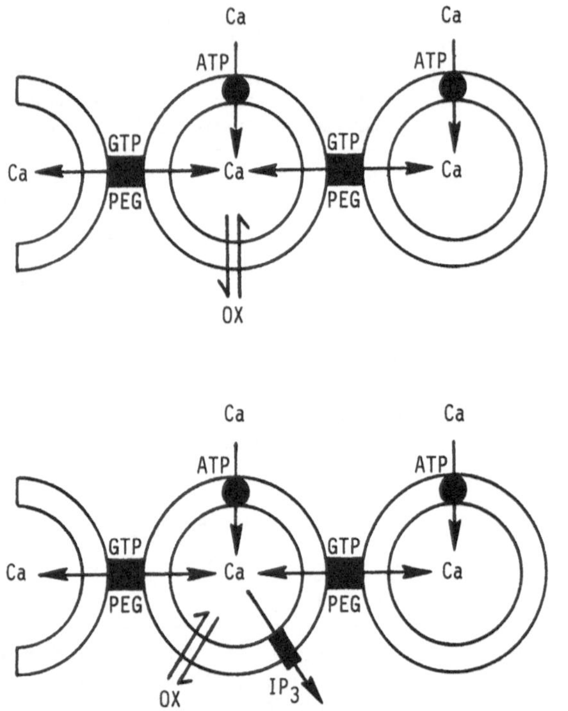

Fig. 4. Hypothetical models of GTP-activated Ca^{2+} translocation between pools explaining the two effects of GTP on Ca^{2+} movements in cells, i.e., GTP-mediated Ca^{2+} release and GTP-mediated Ca^{2+} uptake in the absence and presence of oxalate, respectively. Details of the evidence and an explanation of the proposed "conveyance" of Ca^{2+} between open and closed compartments via a transmembrane Ca^{2+} translocation process are given in the text and in MULLANEY et al. (1987, 1988). The model is depicted in the absence (*top*) and presence (*bottom*) of $InsP_3$. The model of GTP-activated Ca^{2+} translocation between oxalate-permeable and impermeable compartments (*top*) is extended to depict the action of $InsP_3$ (bottom) in releasing Ca^{2+} from the oxalate-permeable pool, preventing GTP-activated Ca^{2+} accumulation in the presence of oxalate and GTP (i.e., the result shown in Fig. 7)

in keeping with the observed effects of vanadate described above (see Table 1). Additional evidence in favor of this model comes from observations on the effect of adding GTP and oxalate at the same time (MULLANEY et al. 1988). Thus, GTP initially releases Ca^{2+}, but this changes with time to much increased accumulation, indicating a competition between GTP-activated uptake and release of Ca^{2+} within the same compartment. Thus, as the Ca^{2+} level approaches that of the threshold for Ca^{2+} precipitation with oxalate, the precipitated Ca^{2+} can no longer be released and remains trapped within the oxalate-permeable compartment. One major significance of such a model is that GTP may never induce Ca^{2+} release into the cytosol, but instead may only allow translocation of Ca^{2+} between Ca^{2+}-pumping compartments (and perhaps also the extracellular space) which have high

Ca^{2+} concentrations. Thus, the problem that cellular GTP levels (which are in excess of 100 μM) would collapse Ca^{2+} gradients within the cell would be obviated.

II. GTP-Activated Ca^{2+} Transfer into the $InsP_3$-Sensitive Ca^{2+} Pool

Another important question that remained to be answered was how and where $InsP_3$ might function within this scheme. Two significant observations indicated that the action of $InsP_3$ was closely linked with one of the two postulated Ca^{2+} pools. The first observation was that the $InsP_3$-sensitive Ca^{2+} pool is indeed an oxalate-permeable compartment. Thus, in permeabilized N1E-115 and DDT_1MF-2 cells, oxalate-enhanced Ca^{2+} accumulation is almost completely prevented by $InsP_3$ (MULLANEY et al. 1988). The second and most important observation is that $InsP_3$ blocks the GTP-mediated accumulation of Ca^{2+} into cells in the presence of oxalate. As described above, when GTP and oxalate are added at the beginning of Ca^{2+} uptake into permeabilized cells, the typical biphasic pattern of uptake is observed, that is, an initial GTP-activated release phase followed subsequently by a strong GTP-activated uptake phase. If $InsP_3$ is added along with GTP and oxalate at the beginning of uptake, the $InsP_3$ completely prevents the second phase of GTP-mediated Ca^{2+} uptake, as we described in detail (MULLANEY et al. 1988). Thus, combining both of these results we conclude that GTP activates Ca^{2+} entry into the same oxalate-permeable pool from which $InsP_3$ induces Ca^{2+} release. The scheme for Ca^{2+} transfer can thus be extended to include the actions of $InsP_3$, as shown in Fig. 4 (bottom). Such an effect of GTP may be very significant; thus, it is possible that the GTP-activated Ca^{2+} translocation process may control the size of the $InsP_3$-sensitive Ca^{2+} pool and possibly also the replenishment of this pool.

III. Isolation of $InsP_3$-Releasable and $InsP_3$-Recruitable Pools

The above description of the two Ca^{2+} pools remained an hypothesis until further cellular fractionation studies provided direct evidence to support their existence (GHOSH et al. 1989). Thus, starting from an ER-enriched microsomal membrane fraction derived from intact DDT_1MF-2 cells, two distinct classes of Ca^{2+}-pumping organelle were separated by density gradient centrifugation with properties which appeared to closely correspond to the two pools described above (GHOSH et al. 1989; ROSSIER and PUTNEY 1991). Fractionation studies (GHOSH et al. 1988) started with an 85000 g ER-enriched pellet (as determined by glucose-6 phosphatase and reduced nicotinamide-adenine-dinucleotide phosphate (NADPH) cytochrome c reductase activities), which, when centrifuged on a continuous sucrose gradient, separated into four discrete fractions with densities of 1.11, 1.14, 1.17, and >1.18, designated B1, B2, B3, and B4, respectively. All the fractions showed high ER marker enzyme activities as well as (Ca^{2+} + Mg^{2+}) ATPase and Ca^{2+}-pumping activites. The densest fractions (B3 and B4) were highly enriched in rough ER (RER; RNA contents >0.3 mg/mg protein), while the B1 and B2 fractions were enriched in smooth ER. The B1 fraction also contained significant plasma membrane marker activity (5′-nucleotidase and ouabain-sensitive ATPase).

Although all four fractions display similar Ca^{2+}-pumping activity, they markedly differ in Ca^{2+} release activated by $InsP_3$, as described in detail (GHOSH et al. 1989). Significantly, all four fractions contain the GTP-activated Ca^{2+} release process, which functions in most respects identically with that observed in permeabilized cells. However, only the denser (RER-containing) fractions release Ca^{2+} in response to $InsP_3$. In the B3 and B4 fractions, the effects of GTP and $InsP_3$ are approximately equal. Moreover, their effects are not additive, no further release being induced when both are added together. Thus, a single population of vesicles responds to both agents. Also, the levels of $InsP_3$ receptors determined from ^3H-$InsP_3$ binding studies (0.04, 0.07, 0.34, and 0.42 pmol/mg protein in the B1, B2, B3 and B4 fractions, respectively) closely correspond to the effectiveness of $InsP_3$ in inducing release. In additon, the effects of oxalate on Ca^{2+} uptake also strongly correlate with $InsP_3$ effectiveness. Thus, as shown in Fig. 5A, oxalate only slightly enhances Ca^{2+} uptake into the B1 fraction, whereas uptake into the B3 fraction (and B4 fraction, not shown) is enhanced manyfold (Fig. 5B). In fact, whereas $InsP_3$ does not alter uptake into the B1 fraction with or without oxalate, it almost completely reverses the oxalate-enhanced uptake into the B3 and B4 fractions (data not shown), as it does within permeabilized cells. Thus, the higher density fractions comprise RER-enriched, oxalate-permeable, $InsP_3$-sensitive, $InsP_3$ receptor-containing, GTP-sensitive, Ca^{2+}-pumping membrane vesicles; the lightest fractions contain smooth ER-enriched, GTP-responsive but $InsP_3$-unresponsive, relatively oxalate-impermeable, Ca^{2+}-pumping vesicles. One important property of all the "B" membrane fractions is that, whereas they all release Ca^{2+} in response to GTP, in the presence of oxalate they still only *release* Ca^{2+} wtih GTP in marked contrast to the GTP-activated *uptake* of Ca^{2+} observed with permeabilized cells in the presence of oxalate, as shown in Fig. 5. Thus, the effect of GTP on the B1 fraction (Fig. 5A) is independent of oxalate, consistent with release from the oxalate-impermeable pool. GTP completely blocks and reverses the large uptake component seen with oxalate in the B3 fraction (Fig. 5B), consistent with GTP-activated release from the oxalate-permeable pool. However, this is completely opposite to the effect in permeabilized cells, where GTP induces profound *uptake* of Ca^{2+} in the presence of oxalate (Fig. 5C). It is concluded that homogenization and fractionation of cells prevents the ability of GTP to translocate Ca^{2+} between intact Ca^{2+}-pumping organelles by destruction of the putative "junctions" between them (CHUEH et al. 1987; UEDA et al. 1986; CHUEH and GILL 1986).

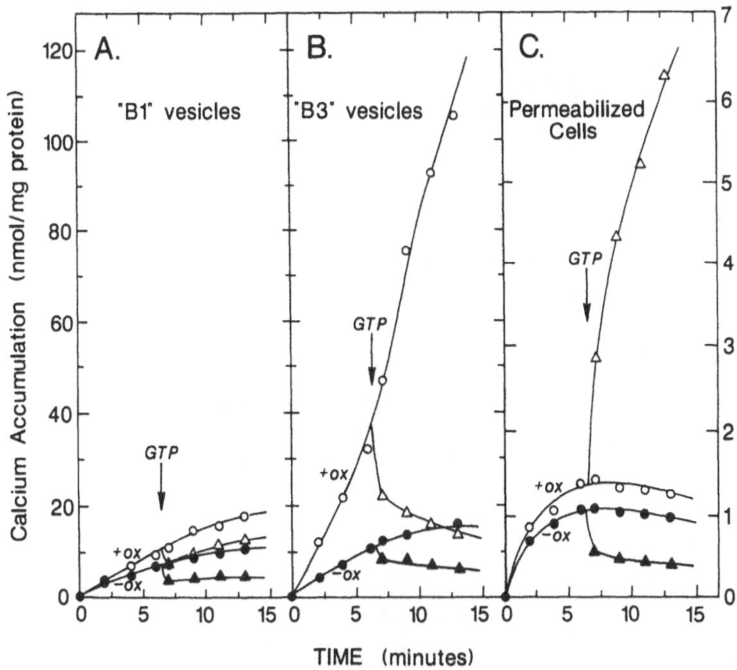

Fig. 5A–C. Difference in action of GTP on light and heavy membrane vesicles and permeabilized cells. The results show a comparison of the effects of GTP in the presence and absence of oxalate on B1 vesicles (**A**), B3 vesicles (**B**), and permeabilized cells (**C**). Oxalate was either absent (*closed symbols*) or present (*open symbols*) at the start of uptake, at 10 m*M* for vesicles (**A** and **B**) or 4 m*M* for cells (**C**). After 6.5 min, 10 μM GTP (▲,△) or control buffer (●,○) was added. Note, that GTP can induce uptake in permeabilized cells in the presence of oxalate, whereas GTP induces only Ca^{2+} release in vesicle preparations, with or without oxalate. See Ghosh et al. (1989) for experimental details

These studies provide evidence that the two predicted Ca^{2+} pools do exist within cells and that they can be physically separated. The first, which appears to fractionate with higher density, is termed the "$InsP_3$-releasable" pool, since it can be directly released via $InsP_3$ receptors. The second, which fractionates at lower density, corresponds to that pool which in permeabilized cells (and, we believe, in intact cells) can be transferred into the $InsP_3$-sensitive pool via the GTP-activated process, whereupon it becomes $InsP_3$ releasable; we have termed this second pool the "$InsP_3$-recruitable" pool.

IV. Functional Organization of Ca^{2+}-Regulatory Organelles

The above studies address two major questions concerning Ca^{2+} signaling: first, the identity and, second, the organization of the intracellular Ca^{2+} pools involved. Considering the first question, the evidence that ER is the

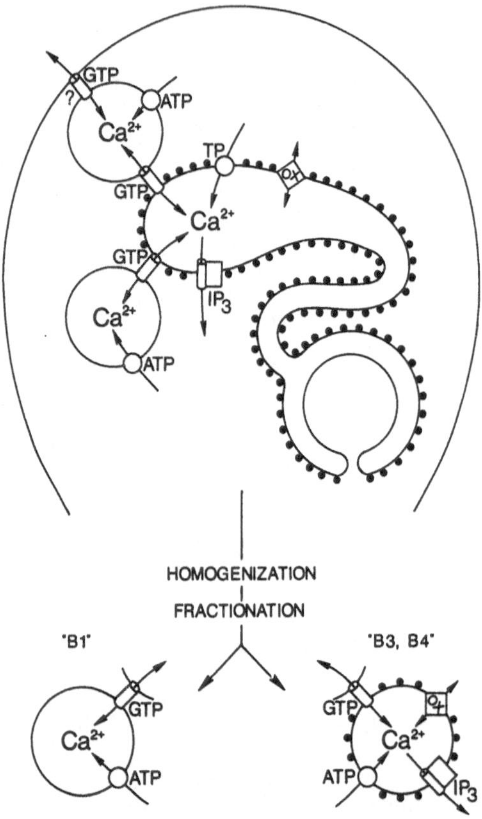

Fig. 6. Model depicting the organization of $InsP_3$-sensitive and -insensitive Ca^{2+}-pumping pools. The diagram describes the putative GTP-activated translocation process between the pools and the properties of the two types of vesicles derived from the pools by the fractionation procedure described in GHOSH et al. (1988, 1989). *GTP* is depicted as activating Ca^{2+} transfer between the pools via junctional connections existing at close appositions between membrane surfaces (CHEUH et al. 1987; MULLANEY et al. 1987, 1988); connections are proposed to exist either between intact compartments or between intact compartments and nonenclosed membrane surfaces, as described in Fig. 6. Dense, ribosome-rich membrane vesicles present in the *B3* and *B4* fractions have (*ATP* + Mg^{2+})-dependent Ca^{2+}-pumping activity, $InsP_3$-activated Ca^{2+} release, $InsP_3$-binding sites, mediated oxalate entry, and the GTP-activated Ca^{2+} transfer process (observed as Ca^{2+} release into the medium). These vesicles are considered to represent the continuous ribosome-rich compartment comprising the RER and nuclear envelope. The lighter, smooth membrane vesicles enriched in the B1 fraction (containing Ca^{2+} pump and GTP-activated Ca^{2+} release mechanisms) are of unknown cellular origin. This pool is depicted as communicating with RER and/or the plasma membrane. Involvement of the latter membrane is speculative and is based only on the observed plasma membrane content of the B1 fraction and previous assertions (MULLANEY et al. 1987, 1988)

site of both binding and action of $InsP_3$ is in good agreement with the studies of Ross et al. (1989), which immunocytochemically localized a purified $InsP_3$-binding protein, now established as having most properties of the functional $InsP_3$ receptor (FURUICHI et al. 1989; MIGNERY et al. 1989), on the RER and nuclear envelope of neuronal cells. A model combining these results and depicting the two possible Ca^{2+}-pumping pools is given in Fig. 6. It is important to note that the RER and nuclear envelope form a continuous ribosome-bearing organelle enclosing an extensive single compartment. The same Ca^{2+}-pump protein is expressed on both membranes (KAPRIELIAN and FAMBROUGH 1987), and both structures accumulate large deposits of precipitated Ca^{2+} in the presence of oxalate (HENKART 1980). These properties of Ca^{2+} pumping and oxalate permeability exactly coincide with those of the ribosome-rich membranes of the B3 and B4 fractions described here. In contrast, the identity of the second organelle involved is unclear. The B1 and B2 fractions are enriched in smooth ER; B1 also contains a significant plasma membrane content. The "calciosome" (a calsequestrin-containing, Ca^{2+}-pumping organelle, distinct from, but frequently closely associated with ER) described by VOLPE et al. (1988) is an intriguing candidate for this second compartment. However, unlike the conclusions of this latter study, which suggested that the calciosome is the site of action of $InsP_3$, our results indicate that the ER itself is a major site of functional $InsP_3$ receptors. Indeed, recent evidence on the action of the Ca^{2+}-pump blocker, thapsigargin, known to specifically inhibit only the internal Ca^{2+} pumps within ER or SR (THASTRUP et al. 1990), reveals that both the $InsP_3$-releasable and $InsP_3$-recruitable pools accumulate Ca^{2+} via Ca^{2+}-pumping activity that is highly sensitive to the Ca^{2+}-pump blocker (BIAN et al. 1991; GHOSH et al. 1991). Thus, it appears likely that all of the Ca^{2+} pool that is sensitive to the actions of GTP-mediated Ca^{2+} translocation is within a Ca^{2+}-pumping organelle which is functionally equivalent to ER or subcompartments thereof.

Considering the organization of the Ca^{2+} pools within cells, an arrangement as depicted in Fig. 6 is considered possible. In permeabilized cells, it is thought that the two pools and the connections between them remain intact and can be activated by GTP. Homogenization of cells is presumed to dissociate the interactions between the two intact compartments. The dissociated vesicles still retain the GTP-activated translocation mechanism; however, this results only in release to the medium, rather than transfer between pools. Even a crude homogenate prepared directly from permeabilized cells shows only release and not uptake in response to GTP, indicating that the putative connections between the two compartments are very labile. However, since the action of GTP appears to depend on the close association between membrane surfaces (CHUEH et al. 1987), it is possible that vesicles have not lost connections between membranes per se. Instead, the predominant connections may be between intact vesicles and nonintact membrane fragments (see Fig. 6). This could explain why in mixing experiments it has not been possible to reconstitute Ca^{2+} transfer between the two isolated fractions.

D. G-Proteins and Interorganelle Transfer of Ca^{2+}

I. Identification of Possible G-Protein Mediators of Ca^{2+} Transfer

Although the mechanistic basis for communication between these organelles is presently unknown, it appears to involve intermembrane junctions and possible later fusion between membranes. Although our own studies indicate that fusion may not account for the rapid and reversible actions of GTP, the studies of DAWSON and coworkers (DAWSON et al. 1987; COMERFORD and DAWSON 1988) suggest that fusion can occur in response to GTP. It is clear that this type of GTP-activated communication between membranes is highly reminiscent of the actions of a number of members of the class of small GTPases (BOURNE et al. 1990) also referred to by many as monomeric G-proteins. These proteins may be identified by their ability to retain GTP-binding activity even after blotting onto nitrocellulose, in contrast to the α subunits of heterotrimeric G-proteins and other GTPases (e.g., tubulin), which appear to lose the ability to bind GTP after immobilization. As shown in Fig. 7, after polyacrylamide gel electrophoresis (PAGE) separation, blotting onto nitrocellulose, and incubation in a reconstitution buffer, the membrane fractions from DDT_1MF-2 cells do show a clear array of at least nine (and probably considerably more) GTPases,varying in size from 16 to 26 kDa. No significant differences in this pattern were observed between the different fractions. This array of proteins appears to contain many of the known and now characterized families of monomeric G-proteins. For example, one of the monomeric G-proteins identified in DDT_1MF-2 cells, the rab1 protein, is the mammalian equivalent of the yeast Ypt1 protein. In preliminary studies in collaboration with Dr. Dieter GALLWITZ, a monoclonal antibody to the 15-amino-acid C terminus of the mammalian rab1 protein revealed a 23-kDa protein in the cell fractions by western blot analysis (not shown). Figure 7 shows a sample of bacterially expressed mammalian rab1 run along with the membrane fractions, where it clearly also binds GTP. An interesting finding was that an anti-rab1 antibody specifically blocks the action of GTP in inducing Ca^{2+} release from permeabilized DDT_1MF-2 cells. The effect was seen with the affinity-purified antibody, but not observed with control antibody preparations. However, at present we are not sure whether these results indicate that this is the protein mediating the GTP effect, or whether the steric effect of the binding of the large antibody protein to the surface of the membrane simply prevents close association between interacting membranes, hence preventing the GTP effect. In other words, it is not possible to distinguish yet whether the rab1 protein mediates the effect or

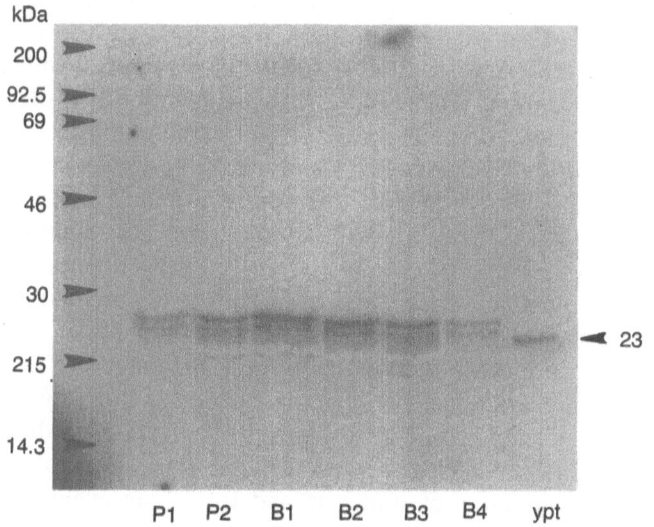

Fig. 7. [^{35}S]GTPγS-labeling of ER proteins after nitrocellulose immobilization and reconstitution. Proteins in the isolated ER fractions were electrophoresed on 12% polyacrylamide gels in 0.1% sodium dodecyl sulfate (SDS), blotted onto nitrocellulose, reconstituted for 30 min at 22°C in 50m*M* NaCl, 25 m*M* TRIS-Cl pH 7.4, 5 m*M* $MgCl_2$, 1 m*M* ethylene glycol bis(β-amino ethyl ether) N,N,N′,N′-tetra-acetic acid (EGTA), 1 m*M* ATP, and 0.25% Tween-20, and then placed in fresh buffer containing 2 n*M* [^{35}S]GTPγS at 1 μCi/ml. Blots were incubated 1 h at room temperature, rinsed four times in buffer, dried 3–4 h, and autoradiographed. *Lanes 1–6* each contain 20 μg total protein from fractions P1, P2, B1, B2, B3, and B4, respectively, and *lane 7* contains bacterially expressed mammalian ypt1 protein (known as rab1). Nine bands are identified with approximate molecular masses of 26.5, 25, 23.5, 22.5, 20, 19, 18, 17, and 16 kDa

exists with many other monomeric G-proteins close to the site of interaction between membranes.

A number of other studies were conducted in an attempt to shed light on the possible identity of the G-protein(s) mediating the actions of GTP on Ca^{2+} fluxes. Experiments were undertaken using the C3 botulinum toxin which effectively ADP-ribosylates members of the rho and rab families of monomeric G-proteins (RUBIN et al. 1988). The toxin clearly ADP-ribosylated the rho protein in the DDT_1MF-2 cell membrane fractions; however, it was ineffective in altering GTP-mediated Ca^{2+} movements in cells pretreated with toxin under the conditions where maximal labeling occurred. In addition, use of the *Pseudomonas aeruginosa* exoenzyme S, which ribosylates a much broader spectrum of monomeric G-proteins (COBURN et al. 1990), revealed no alteration of GTP-mediated Ca^{2+} movements under conditions where the exoenzyme S protein was effectively ribosylating many of the G-proteins present. In other experiments we tested the action of preventing isoprenylation of G-proteins, a posttranslational modification of several known monomeric G-proteins (including the rab family of proteins) which is believed to be important for the membrane attachment and action of these proteins (Kinsella and Maltese 1991). Pretreatment of cells with lovastatin can block this process and result in a much greater proportion of monomeric G-proteins in the cytosol as opposed to membrane-attached forms. However, such treatment of cells did not alter GTP-activated Ca^{2+} movements. This indicates either that isoprenylation of G-proteins is not important for this effect or that the experimental conditions were such that the effect of lovastatin was not complete. In other recent studies, we examined that effects of synthetic peptides of the rab protein effector domain which have been shown to inhibit rab-mediated transport through the secretory pathway (Plutner et al. 1990). However, no effects of these peptides were observed on GTP-mediated Ca^{2+} fluxes.

We also examined the actions of peptides derived from the sequence of another group of monomeric G-proteins, the ADP-ribosylation factor (ARF) proteins. It has been observed that the amino terminus of human ARF is a potent inhibitor of transport and trafficking events within cells believed to be mediated by ARF (Bourne 1988). Experiments in progress at the time of writing this manuscript indicate some very interesting results. The 17-amino-acid N-terminal ARF peptide, instead of blocking the action of GTP, actually appears to mimic the action of GTP on induction of Ca^{2+} release from permeabilized cells. It does so without any requirement for guanine nucleotides; the effect is also not blocked by GDP or GTPγS.

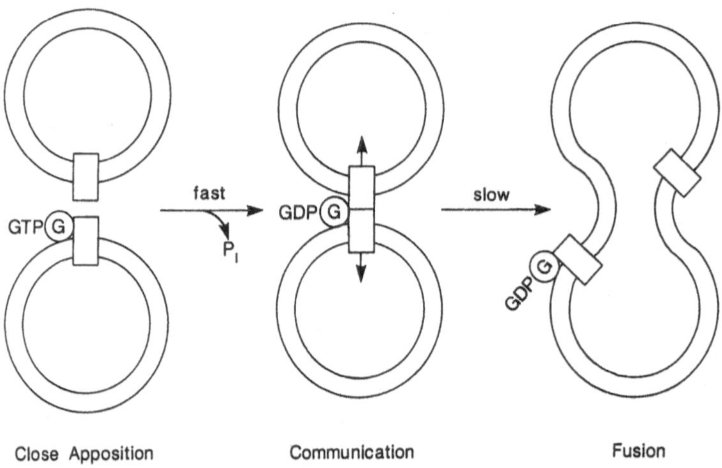

Fig. 8. Hypothetical scheme depicting GTP-activated communication between membrane surfaces. As described in the text and Ghosh et al. (1989) and Mullaney et al. (1987, 1988), *GTP* may initially activate a rapid and reversible interaction between membrane surfaces which involves formation of a communicating junctional process through which Ca^{2+} ions can flow. Thereafter, it is possible that a later irreversible *fusion* event occurs. It is likely that the initial *communication* process would be part of a series of recognition events that precede membrane fusion, as described by Bourne (1988) and Bourne et al. (1990). *G*, G-protein

The EC_{50} is between 5 and $10\,\mu M$, actually slightly below the IC_{50} for inhibition of trafficking (BOURNE 1988). The effect on Ca^{2+} release is very rapid, occurs even when the temperature is reduced to 4°C, and is independent of PEG. Peptide-induced Ca^{2+} release is specific to that total pool of Ca^{2+} activated by GTP; thus, the effects of the peptide do not seem to reflect a nonspecific emptying of all accumulated Ca^{2+}. GTP-mediated Ca^{2+} uptake in the presence of oxalate is blocked by the peptide as a result of induction of Ca^{2+} release. A truncated form of the peptide, devoid of the four N-terminal amino acids (i.e., a 13-amino-acid peptide), is without effect on either Ca^{2+} release or inhibition of trafficking (BOURNE 1988). Thus, it seems that the peptide mediates an effect similar to GTP on the same target organelles, but without any GTP requirement. Obviously this small fragment of the protein does not contain the GTP-recognition site; however, it may represent the portion of the molecule that directly interacts with some component of the membrane via the rest of the molecule. Although we do not know whether this reflects the function of this N-terminal peptide when present in the intact ARF protein, it is possible that this portion of the molecule may interact with the ER membrane surface, or perhaps some effector protein, and induce opening of a channel-like entity perhaps corresponding to the initial part of the diagram shown in Fig. 8 (described below).

II. Conclusions on the Role of G-Proteins

From the studies of others, particularly in yeast systems, it is clear that an array of monomeric GTPases (for example, ypt1 and sec4) are implicated in the membrane interaction events involving trafficking between ER, Golgi, secretory vesicles, and plasma membranes (BOURNE et al. 1990; BOURNE 1988). Similar proteins are believed to mediate equivalent events between the same organelles in mammalian cells (BOURNE et al. 1990; BOURNE 1988; BALCH 1989). In each case, membrane interactions appear to be mediated by a process that specifically requires GTP hydrolysis and is blocked by GTPγS, a specificity exactly coinciding with the process of GTP-activated Ca^{2+} transfer described by us. This blockade by nonhydrolyzable GTP analog serves as an important distinction between the actions of monomeric G-protein- vs. heterotrimeric G-protein-mediated events. BOURNE (1988; BOURNE et al. 1990) has pointed out that the monomeric G-proteins may act as GTP-hydrolyzing switches that control interaction events between membranes and that can go on to bring about membrane fusion.

The possibility that one or more monomeric G-proteins mediate the Ca^{2+} translocation event is an exciting one. The effect we observed of GTP on Ca^{2+} transfer may be a manifestation of an initial phase of the general phenomenon of G-protein-mediated membrane trafficking. As revealed by DAWSON and coworkers (DAWSON et al. 1987; COMERFORD and DAWSON 1988), GTP in the same type of system can lead to fusion between membranes, but probably on the scale of minutes as opposed to the very rapid (within 10–20 s) induction of Ca^{2+} transport that we observed with GTP. Also, the effect we observed is reversible (see above), as opposed to the irreversible fusion shown by DAWSON and coworkers (DAWSON et al. 1987; COMERFORD and DAWSON 1988). Combining our results with those of DAWSON and others, we have formulated the scheme shown in Fig. 8, whereby it is possible that GTP initially causes a rapid and reversible interaction between membrane surfaces which involves formation of a communicating junctional process (through which Ca^{2+} and probably other ions flow). Thereafter, it is feasible that a later irreversible fusion event occurs. Indeed it may be that the initial communication process is part of an important series of recognition events that precede membrane fusion. Thus, the G-protein-mediated switching process may be one of a series of events that allows activation of the communicating mechanism. The significance of this event may be as part of the recognition process itself. Thus, it might be important that a channel between the two interacting organelles be formed, permitting the conductance of ions (Ca^{2+} ions or perhaps protons), and that such a conductance event could be a signature of the identity, maturity, and/or contents of the two organelles that will eventually fuse. If the contents of the two organelles were such that the conduction of ions is deemed "correct," fusion might follow. In other words, the conduction process may be part of the mechanism that maintains "fidelity" of fusion, ensuring that only the correct organelles which have matured to contain the correct ionic content be permitted to fuse. We think that this type of event would be highly comparable to the action of GTPase synthetic factors, for example, elongation factor Tu, which enhances the fidelity of translation via a process termed "kinetic proofreading," in which the correctness of an incoming aminoacyl-tRNA is assessed before GTP hydrolysis is activated and the amino acid is permitted to be added to the growing peptide chain. Indeed, exactly this analogy between the actions of monomeric G-proteins and elongation factors was strongly emphasized by BOURNE (1988; BOURNE et al. 1990). There is certainly agreement from others that GTP may activate a "prefusion" event that functions as some kind of processing step occurring between membrane attachment and membrane fusion (BOURNE 1988; BALCH 1989). Such a prefusion event mediated by guanine nucleotides is not unprecedented; thus, the interaction between secretory membrane vesicles and the plasma membrane has long been known to involve a guanine nucleotide-mediated change in conductance (indicating ion transport) that lasts for a few seconds before a large capacitative change occurs which indicates actual fusion between membranes (FERNANDEZ et al. 1984). In addition, whatever the precise nature of the interactions between organelles the transfer of Ca^{2+} between compartments in the intact cell may have considerable significance in controlling the extent or replenishment of the $InsP_3$-activated Ca^{2+} pool and possibly the entry of external Ca^{2+}. This same mechanism may have a wide significance in controlling a broad range of membrane-trafficking events within cells.

References

Balch WE (1989) Biochemistry of interorganelle transport. A new frontier in enzymology emerges from versatile in vitro model systems. J Biol Chem 264:16965–16968

Berridge MJ (1987) Inositol trisphosphate and diacylglycerol: two interacting second messengers. Annu Rev Biochem 56:159–193

Berridge MJ (1990) Calcium oscillations. J Biol Chem 265:9583–9586

Berridge MJ, Irvine RF (1984) Inositol trisphosphate, a novel second messenger in cellular signal transduction. Nature 312:315–321

Berridge MJ, Irvine RF (1989) Inositol phosphates and cell signalling. Nature 341:197–205

Bian JH, Ghosh TK, Wang JC, Gill DL (1991) Identification of intracellular calcium pools. Selective modification by thapsigargin. J Biol Chem 266:8801–8806

Bourne HR (1988) Do GTPases direct membrane traffic in secretion. Cell 53:669–671

Bourne HR, Sanders DA, McCormick F (1990) The GTPase superfamily: a conserved switch for diverse cell functions. Nature 348:125–132

Chueh SH, Gill DL (1986) Inositol 1,4,5-trisphosphate and guanine nucleotides activate calcium release from endoplasmic reticulum via distinct mechanisms. J Biol Chem 261:13883–13886

Chueh SH, Mullaney JM, Ghosh TK, Zachary AL, Gill DL (1987) GTP- and inositol 1,4,5-trisphosphate-activated intracellular calcium movements in neuronal and smooth muscle cell lines. J Biol Chem 262:13857–13864

Coburn J, Wyatt RT, Iglewski BH, Gill DM (1989) Several GTP-binding proteins, including p21c-H-ras, are preferred substrates of Pseudomonas aeruginosa exoenzyme S. J Biol Chem 264:9004–9008

Comerford JG, Dawson AP (1988) The mechanism of action of GTP on Ca^{2+} efflux from rat liver microsomal vesicles. Measurement of vesicle fusion by fluorescence energy transfer. Biochem J 249:89–93

Dawson AP (1985) GTP enhances inositol trisphosphate-stimulated Ca^{2+} release from rat liver microsomes. FEBS Lett 185:147–150

Dawson AP, Comerford JG, Fulton DV (1986) The effect of GTP on inositol 1,4,5-trisphosphate-stimulated Ca^{2+} efflux from a rat liver microsomal fraction. Is a GTP-dependent protein phosphorylation involved? Biochem J 234:311–315

Dawson AP, Hills G, Comerford JG (1987) The mechanism of action of GTP on Ca^{2+} efflux from rat liver microsomal vesicles. Biochem J 244:87–92

Ehrlich BE, Watras J (1988) Inositol 1,4,5-trisphosphate activates a channel from smooth muscle sarcoplasmic reticulum. Nature 336:583–586

Fernandez JM, Neher E, Gomperts BD (1984) Capacitance measurements reveal stepwise fusion events in degranulating mast cells. Nature 312:453–455

Ferris CD, Huganir RL, Supattapone S, Snyder SH (1989) Purified inositol 1,4,5-trisphosphate receptor mediates calcium flux in reconstituted lipid vesides. Nature 342:87–89

Furuichi T, Yoshikawa S, Miyawaki A, Wada K, Maeda N, Mikoshiba K (1989) Primary structure and functional expression of the inositol 1,4,5-trisphosphate-binding protein p400. Nature 342:32–38

Ghosh TK, Eis PS, Mullaney JM, Ebert CL, Gill DL (1988) Competitive, reversible, and potent antagonism of inositol 1,4,5-trisphosphate-activated calcium release by heparin. J Biol Chem 263:11075

Ghosh TK, Mullaney JM, Tarazi FI, Gill DL (1989) GTP-activated communication between distinct inositol 1,4,5-trisphosphate-sensitive and -insensitive calcium pools. Nature 340:236–239

Ghosh TK, Bian J, Gill DL (1990) Intracellular calcium release is mediated by sphingosine derivatives generated in cells. Science 248:1653–1656

Ghosh TK, Bian J, Short AD, Rybak SL, Gill DL (1991) Persistent intracellular calcium pool depletion by thapsigargin and its influence on cell growth. J Biol Chem 266:24690–24697

Gill DL (1982) Sodium channel, sodium pump, and sodium–calcium exchange activities in synaptosomal plasma membrane vesicles. J Biol Chem 257:10986–10990
Gill DL (1989) Calcium signalling: receptor kinships revealed. Nature 342:16–18
Gill DL, Chueh SH (1985) An intracellular (ATP + Mg^{2+})-dependent calcium pump within the NIE-115 neuronal cell line. J Biol Chem 260:9289–9297
Gill DL, Grollman EF, Kohn LD (1981) Calcium transport mechanisms in membrane vesicles from guinea pig brain syncptosomes. J Biol Chem 256:184–192
Gill DL, Chueh SH, Whitlow CL (1984) Functional importance of the synaptic plasma membrane calcium pump and sodium–calcium exchanger. J Biol Chem 259:10807–10813
Gill DL, Chueh SH, Noel MW, Ueda T (1986a) Orientation of synaptic plasma membrane vesicles containing calcium pump and sodium–calcium exchange activities. Biochim Biophys Acta 856:165–173
Gill DL, Ueda T, Chueh SH, Noel MW (1986b) Ca^{2+} release from endoplasmic reticulum is mediated by a guanine nucleotide regulatory mechanism. Nature 320:461–464
Gill DL, Mullaney JM, Ghosh TK (1988a) Intracellular calcium translocation: mechanism of activation by guanine nucleotides and inositol phosphates. J Exp Biol 139:105–133
Gill DL, Chueh SH, Mullaney JM (1988b) In: Thorn NA, Treiman M, Petersen OH, Thaysen JH (eds) Molecular mechanisms in secretion. Munksgaard International, Copenhagen, pp 277–296 (Alfred Benzon Symposium 25)
Gill DL, Ghosh TK, Mullaney JM (1989a) Calcium signalling mechanisms in endoplasmic reticulum activated by inositol 1,4,5-trisphosphate and GTP. Cell Calcium 10:363–374
Gill DL, Mullaney JM, Ghosh TK, Chueh SH (1989b) In: Fiskum G (ed) Cell calcium metabolism: physiology, biochemistry, pharmacology, and clinical implications. Plenum, New York, pp 157–167
Gill DL, Mullaney JM, Ghosh TK, Tarazi FI (1990) In: Naccache PH (ed) G-proteins and calcium Mobilization. CRC Press, New York, pp 95–119
Henkart M (1980) Identification and function of intracellular calcium stores in axons and cell bodies of neurons. Fed Proc 39:2783–2789
Irvine RF (1990) "Quantal" Ca^{2+} release and the control of Ca^{2+} entry by inositol phosphates – a possible mechanism. FEBS Lett 263:5–9
Irvine RF (1991) Inositol tetrakisphosphate as a second messenger: confusions, contradictions, and a potential resolution. Bioessays 13:1–9
Kaprielian Z, Fambrough DM (1987) Expression of fast and slow isoforms of the Ca^{2+}-ATPase in developing chick skeletal muscle. Dev Biol 124:490–503
Kinsella BT, Maltese WA (1991) rab GTP-binding proteins implicated in vesicular transport are isoprenylated in vitro at cysteines within a novel carboxyl-terminal motif. J Biol Chem 266:8540–854
Martonosi AN (1982) In: Cheung WY (ed) Calcium in cell function, vol 3. Academic, New York, pp 37–102
Mignery GA, Südhof TC, Takei K, De Camilli P (1989) Putative receptor for 1,4,5-trisphosphate similar to ryanodine receptor. Nature 342:192–195
Muallem S, Pandol S, Beeker TG (1989) Hormone-evoked calcium release from intracellular stores is a quantal process. J Biol Chem 264:205–212
Mullaney JM, Chueh SH, Ghosh TK, Gill DL (1987) Intracellular calcium uptake activated by GTP. Evidence for a possible guanine nucleotide-induced transmembrane conveyance of intracellular calcium. J Biol Chem 262:13865–13872
Mullaney JM, Yu M, Ghosh TK, Gill DL (1988) Calcium entry into the inositol 1,4,5-trisphosphate-releasable calcium pool is mediated by a GTP-regulatory mechanism. Proc Natl Acad Sci USA 85:2499–2503

Nakagawa T, Okano H, Furuichi T, Aruga J, Mikoshiba K (1991) The subtypes of the mouse 1,4,5-trisphosphate receptor are expressed in a tissue-specific and developmentally specific manner. Proc Natl Acad Sci USA 88:6244–6248

Plutner H, Schwaninger R, Pind S, Balch WE (1990) Synthetic peptides of the Rab effector domain inhibit vesicular transport through the secretory pathway. EMBO J 9:2375–2383

Putney JW Jr (1990) Capacitative calcium entry revisited. Cell calicium 11:611–624

Ross CA, Meldolesi J, Milner TA, Tomohide S, Supattapone S, Snyder SH (1989) Inositol 1,4,5-trisphosphate receptor localized to endoplasmic reticulum in cerebellar Purkinje neurons. Nature 339:468–470

Rossier MF, Putney JW (1991) The identity of the calcium-storing inositol 1,4,5-trisphosphate-sensitive organelle in non-muscle cells: calciosome, endoplasmic reticulum . . . or both? Trends Neurosci 14:310–314

Rubin EJ, Gill DM, Boquet P, Popoff MR (1988) Functional modification of a 21-kilodalton G protein when ADP-ribosylated by exoenzyme C3 of Clostridium botulinum. Mol Cell Biol 8:418–426

Sambrook JF (1990) The involvement of calcium in transport of secretory proteins from the endoplasmic reticulum. Cell 61:197–199

Thastrup O, Cullen PJ, Drøbak BK, Hanley MR, Dawson AP (1990) Thapsigargin, a tumor promoter, discharges intracellular Ca^{2+} stores by specific inhibition of the endoplasmic reticulum Ca^{2+}-ATPase. Proc Natl Acad Sci USA 87:2466–2470

Ueda T, Chueh SH, Noel MW, Gill DL (1986) Influence of inositol 1,4,5-trisphosphate and guanine nucleotides on intracellular calcium release within the NIE-115 neuronal cell line. J Biol Chem 261:3184–3192

Volpe P, Krause K-H, Hashimoto S, Zorzato F, Pozzan T, Meldolesi J, Lew DP (1988) "Calciosome," a cytoplasmic organelle: the inositol 1,4,5-trisphosphate sensitive Ca^{2+} store of nonmuscle cells? Proc Natl Acad Sci USA 85:1091–1095

Worley PF, Baraban JM, Supattapone S, Wilson VS, Snyder SH (1987) Regulation by pH and calcium. J Biol Chem 262:12132–12136

CHAPTER 41

Coupling of ras to the T Cell Antigen Receptor

J. Downward

A. Introduction

The ras proto-oncogenes, Ha-, Ki- and N-*ras*, encode 21-kDa GTPases that are critical regulatory proteins in all eukaryotic cells. In most mammalian cell types, ras proteins are essential for cell growth and when constitutively activated by point mutation cause cellular transformation. In some cell types ras proteins have been implicated in differentiation pathways. Their activity is normally regulated by a cycle of binding GTP to give the biologically active form of the protein followed by hydrolysis of bound GTP to GDP. The GDP bound form of the protein is inactive: it is reactivated by exchange of bound GDP for free cytosolic GTP (see other chapters in this volume).

B. Receptors and Intracellular Signals that Regulate $p21^{ras}$

While ras proteins had long been postulated to couple extracellular signals to intracellular targets, it was not until 1990 that the first physiological stimulus to control the activity of $p21^{ras}$ was identified in studies showing that activation of T lymphocytes via the T cell antigen receptor (TCR) caused a very rapid stimulation of $p21^{ras}$ as measured by its conversion from the GDP- to the GTP-bound state in activated cells (Downward et al. 1990a). Since this initial finding a number of cell surface receptors have been identified in various cell systems which, upon triggering, cause stimulation of $p21^{ras}$.

I. Activation of $p21^{ras}$ in Cells Other than T Lymphocytes

Ras proteins can be regulated in fibroblasts by signals generated by triggering the receptors for platelet derived growth factor (PDGF; Gibbs et al. 1990; Satoh et al. 1990b), epidermal growth factor (EGF; Satoh et al. 1990a) and insulin (Burgering et al. 1991). In addition, fibroblasts which have been engineered to express the colony stimulating factor 1 (CSF-1) receptor also respond to this factor with a stimulation of the amount of GTP bound to $p21^{ras}$ (Heidaran et al. 1992). Common features of the receptors so far found to regulate $p21^{ras}$ in fibroblasts are that they all regulate cell growth in a positive manner and they all possess tyrosine kinase activity.

Table 1. Extracellular stimuli that activate p21[ras]

Stimulus	Cell type	GTP/GTP + GDP on p21[ras] (%)	
		Basal	Stimulated
Phorbol ester	T cell	5	85
Anti-TCR	T cell	5	50
Anti-CD2	T cell	5	50
IL-2	T cell	2	8
IL-3	B cell	3	16
IL-3	Mast cell	6	40
GM-CSF	Mast cell	3	16
Insulin	Fibroblast[a]	20	70
PDGF	Fibroblast	7	15
PDGF	Fibroblast[b]	0.5	1.0
EGF	Fibroblast[b]	0.5	2.0
NGF	PC12	2	18
EGF	PC12	2	18
EPO	HEL	5	23
SLF	Mast cell	3	15
IL5	Mast cell	3	12
CSF-1	Fibroblast[c]	4	12
TGFβ	Epithelial	5	80

TCR, T cell antigen receptor; IL, interleukin; GM-CSF, granulocyte/macrophage colony stimulating factor; PDGF, platelet derived growth factor; EGF, epidermal growth factor; NGF, nerve growth factor; EPO, erythropoietin; SLF, steel factor; TGF, transforming growth factor.
[a] Rat-1 cells overexpressing insulin receptor.
[b] Swiss 3T3 cells overexpressing normal c-Ha-*ras*.
[c] NIH 3T3 cells transfected with CSF-1 receptor (see text for references and details).

Ras is known to be important in regulating cell differentiation in neuronal cells. The best characterised system is the rat pheochromoytoma cell line PC12; this is induced to differentiate to a neuronal phenotype in response to nerve growth factor (NGF). This process has been known for some time to rely on endogenous ras proteins. It has recently been shown directly that NGF activates p21[ras] in PC12 cells (Muroya et al. 1992; Qiu and Green 1991). Interestingly, EGF also activates p21[ras] in these cells with a very similar time course; however, EGF does not promote differentiation, but instead cell growth in this system. Clearly events other than ras activation must also be important in determining the final cellular response to these two receptor tyrosine kinases.

In mast cells a number of stimuli have been directly demonstrated to activate p21[ras] including interleukin 3 (IL-3), granulocyte/macrophage colony stimulating factor (GM-CSF) and Steel factor (SLF), the ligand for the c-kit (Duronio et al. 1992). In B cells ras proteins can be activated by IL-3 (Satoh et al. 1991). In erythroleukemia cells ras is activated by erythropoietin (EPO; Torti et al. 1992). All of the above stimuli have in some way been linked to activation of tyrosine kinase activity (Satoh et al.

1992). One system that does not fit this pattern is the recently reported activation of ras in response to transforming growth factor β (TGFβ) (MULDER and MORRIS 1992).

II. Activation of p21^ras^ in T Lymphocytes

In T cells, it has been shown that $p21^{ras}$ proteins are regulated by receptors that function at different stages in cell cycle progression: the TCR and CD2 antigens that initiate $G_0 \rightarrow G_1$ transition (DOWNWARD et al. 1990a; GRAVES et al. 1991) and the IL-2 receptor that controls $G_1 \rightarrow S$ transition and ultimately controls T cell mitosis (GRAVES et al. 1992; SATOH et al. 1991). In T lymphocytes, phorbol esters and diacylglycerols that activate protein kinase C (PKC) can mimic TCR and CD2 antigen triggering and are potent stimulators of $p21^{ras}$. Since the TCR and CD2 antigen can activate PKC, it was accordingly postulated that PKC mediates all or part of the $p21^{ras}$ activation seen upon triggering these receptors (DOWNWARD et al. 1990a; GRAVES et al. 1991). However, this initial model for $p21^{ras}$ regulation was complicated by the discovery that activation of PKC by phorbol esters does not result in $p21^{ras}$ activation in cell types such as fibroblasts and mast cells, even though $p21^{ras}$ proteins can be activated via growth factor receptors in these cells. Furthermore, it was recognised that, even in the T cell, not all the receptors known to regulate $p21^{ras}$ can also activate PKC. For example the insulin, IL-3 and IL-2 receptors can stimulate $p21^{ras}$ in fibroblasts, mast cells and T cells, respectively, but these receptors probably do not activate PKC. There must, therefore, be an additional, non-PKC route for controlling $p21^{ras}$. The molecular details of this alternative $p21^{ras}$ control mechanism are not yet known. However, in fibroblasts and haemopoietic cells all the stimuli known to activate $p21^{ras}$ are activators of tyrosine kinases. A second route by which $p21^{ras}$ proteins can be stimulated could thus be mediated by tyrosine kinases.

In T lymphocytes there are indications that at least two mechanisms for $p21^{ras}$ regulation co-exist: one PKC mediated and one which is not, but involves tyrosine kinases (IZQUIERDO et al. 1992). Briefly, this second mechanism has been characterised in streptolysin O permeabilised peripheral blood lymphoblasts. In this system, activation of ras in response to PKC and TCR agonists can still be demonstrated (DOWNWARD et al. 1990a). However, it has now been shown that TCR agonists will still partially activate ras in this system under zero calcium conditions in which there is no activation of PKC (IZQUIERDO et al. 1992). Pseudosubstrate peptide inhibitors of PKC are capable of fully blocking the phorbol ester response, but not the TCR agonist response. The non-PKC-dependent component of the TCR ras activation is inhibitable by herbimycin A, an inhibitor of lymphocyte tyrosine kinases. The TCR and CD2 antigen can both stimulate PKC and regulate pathways of tyrosine phosphorylation and could work via both the PKC-dependent and PKC-independent mechanisms to activate p21^{ras}. These data do not exclude the possibility that the TCR and CD2 antigen could regulate ras via PKC, but do suggest that these receptors also use a PKC-"independent" mechanism. The IL-2 receptor, which does not activate PKC but does activate tyrosine kinases, must act on p21^{ras} proteins entirely through this second pathway or possibly even a third one: the IL-2 response does not work in permeabilised cells and so cannot be characterised at this level. However, it is known to be inhibitable in whole cells by herbimycin A (M. IZQUIERDO, D. CANTRELL and J. DOWNWARD, unpublished observations).

C. GTPase Activating Proteins Regulate p21^{ras} in T Lymphocytes

There are two possible ways in which increases in GTP levels on p21^{ras} proteins could be achieved: by stimulation of the rate of guanine nucleotide exchange onto p21^{ras} or by a decrease in p21^{ras} GTPase activity. The GTPase activity of p21^{ras} is controlled by GTPase activating proteins (GAPs). The best characterised of these is p120^{GAP}, which was the first such activity to be identified (TRAHEY and MCCORMICK 1987). Another related protein with GAP activity that has been recently identified is neurofibromin, the product of the *NF-1* gene: this is a candidate tumour suppressor gene, damage to which may give rise to the hereditary disease neurofibromatosis type 1. In T cells, the observed stimulation of p21^{ras} in TCR or phorbol ester stimulated cells correlated with a rapid decrease in the level of GAP activity measurable in cell extracts. It is not yet known whether p120^{GAP} or neurofibromin are inhibited upon T cell activation, since both are known to be present in these cells. A study using the detergent dodecylmaltoside, which inhibits the catalytic activity of the NF-1 protein, but not that of p120^{GAP}, suggests that both proteins are inhibited upon PKC activation of T cells (BOLLAG and MCCORMICK 1991). However, the interpretation of these results could be complicated if other p21^{ras} GAPs exist.

It has been shown that purified ras proteins exchange nucleotide only slowly with a turnover rate of the order of 1 h (HALL and SELF 1986). In contrast, kinetic studies of nucleotide exchange and hydrolysis on p21^{ras} proteins in permeabilised T lymphocytes have revealed that the rate at which guanine nucleotide exchanges on p21^{ras} is very rapid with a half-life of about 1 min (DOWNWARD et al. 1990a). A number of proteins have been described that will stimulate the exchange of nucleotide on p21^{ras} in vitro; these have been termed GDS (GDP dissociation stimulators; DOWNWARD et al. 1990b; HUANG et al. 1990; KAIBUCHI et al. 1991; MIZUNO et al. 1991; WEST et al. 1990; WOLFMAN and MACARA 1990). Recently a mammalian guanine nucleotide exchange factor for p21^{ras} has been cloned by a functional complementation assay in budding yeast, where the ras exchange factor CDC25 has been known for some time (MARTEGANI et al. 1992). The elevated exchange rate of nucleotides on $p21^{ras}$ in permeabilised T cells suggests that GDS proteins have high constitutive activity in T lymphocytes. Guanine nucleotide exchange on $p21^{ras}$, although rapid, is unchanged in both PKC and TCR activated cells. It appears, therefore, that the principal control mechanism for $p21^{ras}$ stimulation is mediated via inhibition of its GTPase activity: due to the high rate of nucleotide exchange on $p21^{ras}$ this is very rapidly translated into accumulation of GTP on $p21^{ras}$.

D. Mechanisms of Regulation of ras GTPase Activating Proteins in T Cells

From the data discussed above it seems likely that $p21^{ras}$ is controlled in the T cell by regulation of GAP-like proteins; however, it is still unclear how the GAPs themselves are controlled. The most straightforward mechanism for regulating GAPs in T lymphocytes would be by direct phosphorylation by PKC or by another kinase involved in the PKC-independent stimulation of $p21^{ras}$. However, available data show that $p120^{GAP}$ is not a substrate for PKC (J. DOWNWARD, unpublished data). Furthermore, although $p120^{GAP}$ can associate with activated tyrosine kinases via SH2 domains, it is phosphorylated on tyrosine residues at vanishingly low stoichiometry (about 0.1%; J. DOWNWARD, unpublished data). Similarly, preliminary analysis of the phosphorylation of neurofibromin in T cells reveals no obvious modulation. An alternative and perhaps more likely mechanism for the regulation of GAPs might involve the phosphorylation of proteins that regulate $p120^{GAP}$ or neurofibromin. For $p120^{GAP}$ two such proteins that stably associate with the GAP have been found: membrane-associated p62, which is phosphorylated on tyrosine at high stoichiometry in response to receptor stimulation and cytosolic p190, which has a very low level of phosphorylation on tyrosine and considerably more on serine and threonine (ELLIS et al. 1990). Association of $p120^{GAP}$ with p190 inhibits its GTPase activating ability, so if this interaction increased upon T cell activation, inhibition of GAP activity would result (MORAN et al. 1991). Neither of these two proteins seem to be greatly altered in their level of phosphorylation or of binding to $p120^{GAP}$ upon activation of the T cell. There is currently no data available regarding NF-1 associated proteins. Nevertheless, the role of GAP-associated proteins is an issue that needs to be addressed to further our understanding of the molecular mechanisms of $p21^{ras}$ regulation in T cells.

An alternative possibility for the control of GAP activity in T cells could be through the ras-related small GTPase $p21^{rap1}$. This protein is encoded by the *K-rev1* gene, which is known to be able to reverse $p21^{ras}$ transformation when expressed at high levels (KITAYAMA et al. 1989). GTP-bound $p21^{rap1}$ has been shown to competitively inhibit the interaction of $p21^{ras}$ with $p120^{GAP}$ in vitro (FRECH et al. 1990; HATA et al. 1990). As yet there is no evidence available that $p21^{rap1}$ is involved in the normal regulation of $p21^{ras}$ in T cells or any other cell type. Another mechanism whereby GAP activity could be regulated involves the production of mitogenic lipids. Lipids and lipid metabolites such as phosphatidic acid, polyphosphoinositides and arachidonic acid strongly inhibit the activity of the NF-1 protein and, to a lesser extent, that of $p120^{GAP}$ (BOLLAG and MCCORMICK 1991; TSAI et al. 1989). There are also reports of a GTPase inhibitory protein (GIP) that blocks the action of GAP proteins on $p21^{ras}$ and can be activated by diacylglycerols and the lipids that inhibit GAP proteins (TSAI et al. 1990). Studies to date suggest that neither these lipids nor GIP play a major role in GAP regulation in T cells. Firstly, phorbol ester and IL-2 treatment of T lymphocytes do not stimulate the production of these lipids within the time frame of $p21^{ras}$ activation, so they are unlikely to account for the PKC- or IL-2-mediated activation of $p21^{ras}$. Secondly, while lipids such as phosphatidic acid are produced in response to TCR stimulation, experiments in permeabilised cells indicate $p21^{ras}$ activation can occur efficiently under conditions in which cellular phospholipases are inactive (IZQUIERDO et al. 1992): the PKC-independent regulation of $p21^{ras}$ in T cells is therefore also unlikely to rely on lipid fluxes.

E. Function of $p21^{ras}$ in T Lymphocytes

A likely role for $p21^{ras}$ in T cells is to couple receptors such as the TCR to intracellular signalling pathways that ultimately control the expression of T cell activation induced genes such as those encoding lymphokines and their receptors. A distal gene regulation function for $p21^{ras}$ in T lymphocytes is compatible with observations in fibroblasts that $p21^{ras}$ can control nuclear transactivating factors such as signal recognition factor (SRF), c-jun and NFκB (BINETRUY et al. 1991; WASYLYK et al. 1988), all of which have been implicated in the regulation of gene expression in T lymphocytes. The details of the intracellular signalling routes that could link plasma membrane-associated $p21^{ras}$ to the nucleus are not known. In fibroblasts, $p21^{ras}$ has been described to modulate the function of serine/threonine kinases including PKC and c-raf (MORRIS et al. 1989; MORRISON et al. 1988). In T cells it is believed that PKC is an upstream regulator of $p21^{ras}$, and not an effector molecule. It is conceivable, however, that c-raf operates downstream of $p21^{ras}$, since receptors such as the TCR and IL-2 receptor that rapidly activate $p21^{ras}$ are known to have a subsequent regulatory effect on c-raf protein.

In T cells it has now been shown directly that activated $p21^{ras}$ is capable of stimulating gene expression (BALDARI et al. 1992; RAYTER et al. 1992). Mutationally activated ras proteins potentiate the ability of phorbol esters and TCR agonists to induce expression of the *IL-2* gene in the EL4 murine thymoma cell line. They also synergise strongly with a calcium ionophore.

Furthermore, it has been shown (RAYTER et al. 1992) that dominant negative mutants of ras (Asn17) prevent the activation of *IL-2* gene transcription by phorbol esters and by TCR agonists. The data is compatible with a system in which ras mediates a signal which is necessary, but not sufficient, for activation of gene transcription in response to the T cell receptor or protein kinase C. Thus, there must exist a ras-dependent PKC-mediated signal. Other pathways which must also exist are a calcium signal and a ras-independent PKC-mediated signal. It is not yet known whether the effects of TCR agonists in these experiments are mediated by PKC or by a PKC-independent tyrosine kinase mechanism.

It has been suggested that GAPs may be $p21^{ras}$ effectors, either alone or as a $p21^{ras}$/GAP complex (CALES et al. 1988; ADARI et al. 1988). This conclusion is based on the fact that $p21^{ras}$ interacts with these proteins via its effector regions (see Chap. 23 by McCormick). The ability of $p21^{ras}$ to interact in this way with $p120^{GAP}$ and neurofibromin, which have homology in their catalytic GAP-related domains but are otherwise structurally distinct, indicates that it is possible for $p21^{ras}$ proteins to interact with multiple effector molecules. One speculation in this context is that the receptor that induces $p21^{ras}$ activation might, by a separate regulatory event, be able to direct the interaction of $p21^{ras}$ with a different effector molecule. For example, the TCR and IL-2 receptor both activate $p21^{ras}$, but if the additional signals generated by these receptors controlled $p21^{ras}$ association with different proteins, then the $p21^{ras}$ effector complex generated by triggering the TCR or the IL-2 receptor would be different, as would the subsequent signalling pathways operating downstream of $p21^{ras}$. The only way to examine this hypothesis and gain insight regarding the role of $p21^{ras}$ in T cells activation is to identify these immediate proximal $p21^{ras}$ effector molecules. This is the ultimate goal and biggest challenge of studies of $p21^{ras}$ in the T lymphocyte system.

References

Adari H, Lowy DR, Willumsen BM, Der CJ, McCormick F (1988) Guanosine triphosphatase activating protein (GAP) interacts with the p21 ras effector binding domain. Science 240:518–521

Baldari CT, Macchia G, Telford JL (1992) Interleukin 2 promoter activation in T cells expressing activated Ha-ras. J Biol Chem 267:4289–4291

Binetruy B, Smeal T, Karin M (1991) Ha-ras augments c-jun activation and stimulates phosphorylation of its activation domains. Nature 251:122–127

Bollag G, McCormick F (1991) Differential regulation of rasGAP and neurofibromatosis gene product activities. Nature 351:576–579

Burgering BMT, Medema RH, Maassen JA, van de Wetering ML, van der Eb AJ, McCormick F, Bos JL (1991) Insulin stimulation of gene expression mediated by p21ras activation. EMBO J 10:1103–1109

Cales C, Hancock JF, Marshall CJ, Hall A (1988) The cytoplasmic protein GAP is implicated as the target for regulation by the ras gene product. Nature 332:548–551

Downward J, Graves JD, Warne PH, Rayter S, Cantrell DA (1990a) Stimulation of $p21^{ras}$ upon T-cell activation. Nature 346:719–723
Downward J, Riehl R, Wu L, Weinberg RA (1990b) Identification of a nucleotide exchange-promoting factor for p21ras. Proc Natl Acad Sci USA 87:5998–6002
Duronio V, Welham MJ, Abraham S, Dryden P, Scrader JW (1992) p21ras activation via hemopoietin receptors and c-kit requires tyrosine kinase activity but not tyrosine phosphorylation of GAP. Proc Natl Acad Sci USA 89:1587–1591
Ellis C, Moran M, McCormick F, Pawson T (1990) Phosphorylation of GAP-associated proteins by transforming and mitogenic tyrosine kinases. Nature 343:377–381
Frech M, John J, Pizon V, Chardin P, Tavitian A, Clark R, McCormick F, Wittinghofer A (1990) Inhibition of GTPase activating protein stimulation of ras-p21 GTPase by the Krev-1 gene product. Science 249:169–171
Gibbs JB, Marshall MS, Scolnick EM, Dixon RF, Vogel US (1990) Modulation of guanine nucleotides bound to ras in NIH 3T3 cells by oncogenes, growth factors and the GTPase activating protein (GAP). J Biol Chem 265:20437–20442
Graves JD, Downward J, Rayter S, Warne P, Tutt AL, Glennie M, Cantrell DA (1991) CD2 antigen mediated activation of the guanine nucleotide binding proteins p21ras in human T lymphocytes. J Immunol 146:3709–3712
Graves J, Downward J, Izquierdo-Pastor M, Rayter SI, Warne PH, Cantrell DA (1992) Interleukin 2 activates p21ras in normal human T lymphocytes. J Immunol 148:2417–2422
Hall A, Self AJ (1986) The effect of Mg^{2+} on the guanine nucleotide exchange rate of p21N-ras. J Biol Chem 261:10963–10965
Hata Y, Kikuchi A, Sasaki T, Schaber MD, Gibbs JB, Takai Y (1990) Inhibition of the ras p21 GTPase-activating protein-stimulated GTPase activity of c-Ha-ras p21 by smg p21 having the same putative effector domain as ras p21s. J Biol Chem 265:7104–7107
Heidaran MA, Molloy CJ, Pangelinan M, Choudhury GG, Wang L-M, Fleming TP, Sakaguchi AY, Pierce JH (1992) Activation of the colony-stimulating factor 1 receptor leads to the rapid tyrosine phosphorylation of GTPase-activating protein and activation of cellular p21ras. Oncogene 7:147–152
Huang YK, Kung H-F, Kamata T (1990) Purification of a factor capable of stimulating the guanine nucleotide exchange reaction of ras proteins and its effect on ras-related small molecular mass G proteins. Proc Natl Acad Sci USA 87:8008–8012
Izquierdo M, Downward J, Graves JD, Cantrell DA (1992) The role of protein kinase C in T cell antigen receptor regulation of p21ras: evidence that two ras regulatory pathways co-exist in T cells. Mol Cell Biol 12:(in press)
Kaibuchi K, Mizuno T, Fujioka H, Yamamoto T, Kishi K, Fukumoto Y, Hori Y, Takai Y (1991) Molecular cloning of the cDNA for stimulatory GDP/GTP exchange protein for smg p21s (ras p21-like small GTP bindng proteins) and characterisation of stimulatory GDP/GTP exchange protein. Mol Cell Biol 11:2873–2880
Kitayama H, Sugimoto Y, Matsuzaki T, Ikawa Y, Noda M (1989) A ras-related gene with transformation suppressor activity. Cell 56:77–84
Martegani E, Vanoni M, Zippel R, Coccetti P, Brambilla R, Ferrari C, Sturani E, Alberghina L (1992) Cloning by functional complementation of a mouse cDNA encoding a homologue of CDC25, a saccharomyces cerevisiae ras activator. EMBO J 6:2151–2157
Mizuno T, Kaibuchi K, Yamamoto T, Kawamura M, Sakodo T, Fujioka H, Matsuura Y, Takai Y (1991) A stimulatory GDP/GTP exchange protein for smg p21 is active on the post-translationally processed from of c-Ki-ras p21 and rhoA p21. Proc Natl Acad Sci USA 88:6442–6446

Moran MF, Polakis P, McCormick F, Pawson T, Ellis C (1991) Protein tyrosine kinases regulate the phosphorylation, protein interactions, subcellular distribution and activity of p21ras GTPase activating protein. Mol Cell Biol 11:1804–1812
Morris JD, Price B, Lloyd AC, Self AJ, Marshall CJ, Hall A (1989) Scrape-loading of Swiss 3T3 cells with ras protein rapidly activates protein kinase C in the absence of phosphoinositide hydrolysis. Oncogene 4:27–31
Morrison D, Kaplan D, Rapp U, Roberts T (1988) Activation of raf kinase in ras transformed cells. Proc Natl Acad Sci USA 85:8855–8859
Mulder KM, Morris SL (1992) Activation of p21ras by transforming growth factor β in epithelial cells. J Biol Chem 267:5029–5031
Muroya K, Hattori S, Nakamura S (1992) Nerve growth factor induces rapid accumulation of the GTP-bound form of p21ras in rat pheochromocytoma PC12 cells. Oncogene 7:277–281
Qiu M-S, Green SH (1991) NGF and EGF rapidly activate p21ras in PC12 cells by distinct, convergent pathways involving tyrosine phosphorylation. Neuron 7:937–946
Rayter SI, Woodrow M, Cantrell DA, Downward J (1992) p21ras mediates control of IL2 gene promoter function in T cell activation (submitted)
Satoh T, Endo M, Nakafuku M, Akiyama T, Yamamoto T, Kaziro Y (1990a) Accumulation of p21ras · GTP in response to stimulation with epidermal growth factor and oncogene products with tyrosine kinase activity. Proc Natl Acad Sci USA 87:7926–7929
Satoh T, Endo M, Nakafuku M, Nakamura S, Kaziro Y (1990b) Platelet-derived growth factor stimulates formation of active p21ras·GTP complex in Swiss mouse 3T3 cells. Proc Natl Acad Sci USA 87:5993–5997
Satoh T, Nakafuku M, Miyajima A, Kaziro Y (1991) Involvement of p21ras in signal transduction pathways from interleukin 2, interleukin 3 and granulocyte/macrophage colony stimulating factor, but not from interleukin 4. Proc Natl Acad Sci USA 88:3314–3318
Satoh T, Uehara Y, Kaziro Y (1992) Inhibition of interleukin 3 and granulocyte-macrophage colony stimulating factor stimulated increase of active ras · GTP by herbimycin A, a specific inhibitor of tyrosine kinases. J Biol Chem 267:2537–2541
Torti M, Marti KB, Altschuler D, Yamamoto EG, Lapetina E (1992) Erythropoietin induces p21ras activation and p120GAP tyrosine phosphorylation in human erythroleukemia cells. J Biol Chem 267:8293–8298
Trahey M, McCormick F (1987) A cytoplasmic protein stimulates normal N-ras p21 GTPase, but does not affect oncogenic mutants. Science 238:542–5
Tsai M-H, Yu C-L, Wei F-S, Stacey DW (1989) The effect of GTPase activating protein upon ras is inhibited by mitogenically responsive lipids. Science 243:522–526
Tsai M-H, Yu CL, Stacey DW (1990) A cytoplasmic protein inhibits the GTPase activity of H-ras in a phospholipid dependent manner. Science 250:982–985
Wasylyk C, Imler JL, Wasylyk C (1988) Transforming but not immortalizing oncogenes activate the transcription factor PEA1. EMBO J 7:2475–2483
West M, Kung H, Kamata T (1990) A novel membrane factor stimulates guanine nucleotide exchange reaction of ras proteins. FEBS Lett 259:245–248
Wolfman A, Macara I (1990) A cytosolic protein catalyzes the release of GDP from p21ras. Science 248:67–69

CHAPTER 42

GTPases as Regulators of Regulated Secretion

T.H.W. LILLIE and B.D. GOMPERTS

A. GTP: A *Sine Qua Non* for Exocytosis

This essay is presented as a polemic. Its object is to indicate that GTP plays the key role in the regulation of exocytosis from myeloid granulocytic cells at least, and in all probability from many other (regulated) secretory cells of diverse embryonic origins. As far as regulation of the exocytotic event is concerned, the role of Ca^{2+} is subsidiary. We provide a series of statements based on the determination of secretion from permeabilised cells treated under various conditions and use these to formulate an argument concerning the role and possible nature of the GTPase which mediates exocytosis. We have called this G_E. Its molecular identity remains undefined.

While we stress the paramountcy of GTP, we cannot escape a discussion of the role of Ca^{2+} in a consideration of the exocytotic process. An important place for Ca^{2+} in the signalling of regulated exocytotic processes has never been in doubt since the earliest descriptions of stimulus response coupling at the neuromuscular junction and the realisation that the stimulus is carried by secreted transmitters (HARVEY and MACINTOSH 1940). Ca^{2+}-binding proteins probably play a number of tasks at various stages of the exocytotic pathway. As an example, in the neuronally derived chromaffin cell, calpactin (annexin II) may act as a preliminary docking protein (ALI et al. 1989; NAKATA et al. 1990). However, although there are some hints relating to the identity of the Ca^{2+}-binding proteins which mediate exocytosis, they remain elusive and even in neuronal cells, in which an essential role for Ca^{2+} seems assured, their precise role is still unknown. For this reason we refer blandly to the Ca^{2+}-binding proteins which regulate exocytosis as C_E.

I. Ca^{2+}-Dependent Secretion in Myeloid Granulocytes

When stimulated by receptor-directed ligands, secretion from mast cells (FOREMAN and MONGAR 1972) and other granulocytes only occurs when Ca^{2+} is provided, and the binding of the ligand causes an elevation in the concentration of intracellular Ca^{2+} (WHITE et al. 1984; GOMPERTS and FEWTRELL 1985; MACGLASHAN and GOU 1991). Moreover, secretion can be elicited by application of Ca^{2+}-carrying ionophores (FOREMAN et al. 1973;

BENNETT et al. 1980; HENDERSON et al. 1983; DI VIRGILIO and GOMPERTS 1983; WRIGHT et al. 1977; ZABUCCHI and ROMEO 1976). Thus, there is little doubt that Ca^{2+}, by its interaction with C_E proteins, does mediate an essential signal at some point in the stimulus–secretion sequence. We shall propose that this may be unrelated to the immediacies of the membrane fusion process whereby secretory products are released, but that it regulates the exchange of guanine nucleotides on a GTPase, G_E, which has not yet been defined and which determines the commitment to the final step. The proposed sequence of activation is illustrated schematically in Fig. 1. The following paragraphs are intended to provide some substance to the conclusions outlined in Fig. 1.

B. Probing Exocytosis: Permeabilised Cells

In order to probe the post-receptor events in the stimulus–secretion sequence, little can be achieved with intact cells, because the plasma

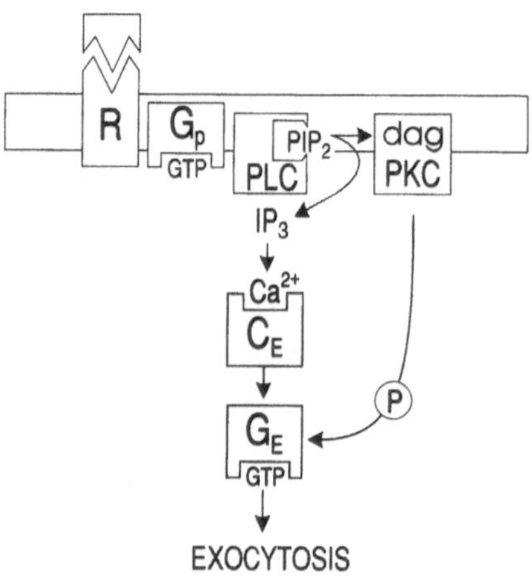

Fig. 1. The main regulatory components of secretion in myeloid cells. Two GTPases act in series to control stimulus–secretion coupling; G_p transduces external signals emanating from cell surface receptors and controls the activity of phospholipase C (*PLC*). Two products are generated. Inositol 1,4,5-trisphosphate (*IP_3*, which normally causes the release of Ca^{2+} in intact cells) leaks rapidly from permeabilised cells, but diglyceride (*dag*), the activator of protein kinase C, is retained. *Exocytosis* occurs from mast cells following pretreatment with metabolic inhibitors and permeabilisation in the presence of Ca^{2+} and GTPγS; the affinity for both is enhanced, due to a protein kinase C (*PKC*)-catalysed phosphorylation, when ATP is provided. Neither of the binding proteins for Ca^{2+} (C_E) or guanine nucleotide (G_E) has been identified, but we known that G_E is distinct from G_p. Mast cells permeabilised in zwitterionic electrolytes (glutamate) undergo secretion in response to GTPγS in the absence of Ca^{2+}; Ca^{2+} by itself is unable to elicit secretion, but spares the requirement for Mg^{2+} in GTPγS-induced secretion. These observations underlie the proposal that Ca^{2+} and guanine nucleotide act in series in the regulatory sequence leading to exocytosis (from LILLIE and GOMPERTS 1992a)

membrane constitutes a barrier rendering the exocytotic machinery inaccessible (LINDAU and GOMPERTS 1991). It is more or less imperative to work with permeabilised cells. In our work, we have mainly used the bacterial cytolysin streptolysin-O (SL-O) to permeabilise cells in suspension. This allows us to regulate the cytosol composition of multiple samples of cells with precision: it is possible to test in detail the effect of a wide range of conditions and concentrations of effectors (e.g. Ca^{2+}, buffered with appropriate chelators; activating and inhibitory nucleotides, etc.) and to do this in a flexible way (e.g. by varying the order of addition of the effectors, multiple timed sampling, etc.). A very powerful and complementary alternative is the use of the patch-clamp pipette applied in the whole cell mode to investigate single cells (LINDAU and GOMPERTS 1991). With regard to the general picture of control by Ca^{2+} and guanine nucleotides, the results which have been derived from the two approaches are in striking agreement (LINDAU and GOMPERTS 1991).

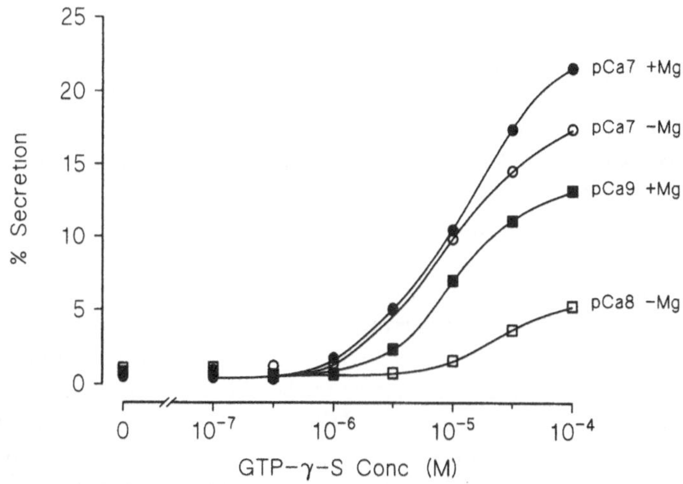

Fig. 2. GTPγS stimulates secretion at vanishingly low concentrations of Ca^{2+} in the absence of ATP. Mast cells, suspended in glutamate-based buffer and pretreated with metabolic inhibitors for 5 min to deplete intracellular ATP, were permeabilised by streptolysin-O at low (pCa7, ● ethylene glycol-bis-(2-aminoethyl)-tetra-acetic acid (EGTA); ○ (HEDTA)) or vanishingly low (■ 5 m*M* EGTA, □ 5 m*M* HEDTA) concentrations of Ca^{2+} and in the presence of various concentrations of GTPγS (as indicated). *Closed symbols* indicate Mg^{2+} (1 m*M*) and EGTA buffering, *open symbols* indicate suppression of Mg^{2+} to $<10^{-9}$ *M* with N′-(2-hydroxyethyl)-ethylenediamine-N,N,N′-triacetic acid (HEDTA) buffering. At the end of 10 min the cells were sedimented by centrifugation and the supernatant sampled for analysis of secreted hexosaminidase. Note: in order to ensure thorough depletion of Mg^{2+} in this experiment, we have used HEDTA (at pH 6.8, $\log K_{app(Mg^{2+})} = 4.27$) as an alternative to the conventional EGTA ($\log K_{app(Mg^{2+})} = 1.6$) in the buffering of Ca^{2+}. Since the affinity of HEDTA for Ca^{2+} (at pH 6.8, $\log K_{app(Ca^{2+})} = 5.13$) is less than that of EGTA ($\log K_{app(Ca^{2+})} = 6.19$), maximal Ca^{2+} suppression is probably not better than 10^{-8} *M*, and it is therefore just possible that this level is sufficient, in combination with the GTPγS, to elicit the very low level of exocytosis (~5%) that occurs (from LILLIE and GOMPERTS 1992a)

I. GTPγS-Induced, Ca^{2+}-Independent Exocytosis

The experiment illustrated in Fig. 2 shows that a non-hydrolysable guanine nucleotide (GTPγS) can induce a small extent of exocytosis (measured as secretion of hexosaminidase) from mast cells, permeabilised in isotonic buffered sodium glutamate, in the absence of ATP, and at vanishingly low concentrations of Ca^{2+} (5 m*M* ethylene glycol-bis-(2-aminoethyl)-tetra-acetic acid, ~10^{-9} *M* Ca^{2+}). In the absence of Ca^{2+}, secretion is strongly promoted by Mg^{2+} (2 m*M*), though we have been unable to confirm that its presence is absolutely essential. As the concentration of Ca^{2+} is elevated first to 10^{-7} *M* (i.e., up to about the normal cytosol resting level), the extent of secretion increases and the enhancement due to Mg^{2+} declines so that with 10^{-6} *M* Ca^{2+}, Mg^{2+} is without further effect. Higher concentrations of Mg^{2+} (in excess of 10 m*M*) are inhibitory to secretion in the absence of Ca^{2+}, but such inhibition can be overcome by addition of Ca^{2+} at 10^{-6} *M*, indicating that the two activating divalent cations interact with the system at entirely separate sites.

II. Ca^{2+}-Induced, GTP-Dependent Exocytosis

Unlike GTPγS, Ca^{2+} is unable by itself to induce secretion, though in the presence of ATP (and Mg^{2+}) a substantial extent (typically 60%) can be elicited. This is shown in Fig. 3. In addition, the figure illustrates the effect of adding GDP, in the range 50–1000 μM, on such Ca^{2+}-induced, ATP-dependent secretion. It can be seen that while high concentrations (>500 μM) of GDP are inhibitory, low concentrations have the effect of enhancing such Ca^{2+}-induced secretion. Clearly, GDP affects the system in two ways: (a) the inhibitory effect expressed at high concentrations is likely to be due to the occlusion of a guanine nucleotide binding site on a GTPase. An equivalent concentration of ADP is without effect; and (b) the enhancement by low concentrations of GDP of Ca^{2+}-induced secretion most likely arises as a result of conversion of GDP to GTP catalysed by nucleoside diphosphate kinase. GDP appears to be the limiting factor on Ca^{2+}-induced (ATP-dependent) secretion. Provision of low, non-inhibitory concentrations of GDP can permit a normal maximal extent (>80%) of exocytosis to be elicited. We are not in a position to speculate on whether the transphosphorylation of the guanine nucleotides occurs on G_E itself or at a closely proximal site on the membrane (for arguments concerning this point, see RANDAZZO et al. 1991; KIKKAWA et al. 1990, 1991; KIMURA and SHIMADA 1990; OTERO 1990).

In summary, we see that Ca^{2+} can only act to induce exocytosis in the presence of an activating guanine nucleotide; alone, it is without effect. In contrast, the non-hydrolysable guanine nucleotide is capable of inducing exocytosis in the effective absence of Ca^{2+}. This is strongly promoted by (and may even be dependent on) Mg^{2+}. ATP is not required. As the

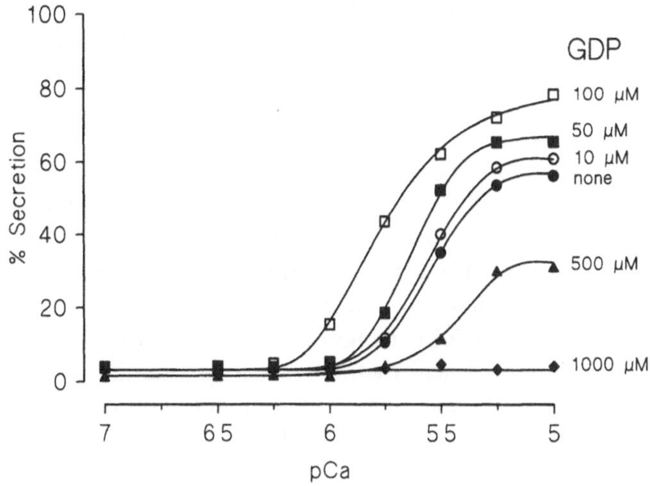

Fig. 3. ATP- and Ca^{2+}-dependent secretion in glutamate based buffer can be enhanced by low concentrations and inhibited by high concentrations of GDP. Mast cells, suspended in isotonic glutamate based buffers and pretreated with metabolic inhibitors, were incubated with SL-O and 1 m*M* Mg-ATP in the presence of Ca^{2+} (buffered as indicated) and in the presence of GDP at various concentrations (as indicated). At the end of 10 min the cells were sedimented by centrifugation and the supernatant sampled for analysis of secreted hexosaminidase (from LILLIE and GOMPERTS 1992b)

concentration of Ca^{2+} is elevated, even up to 10^{-7} *M*, the promotional effect of Mg^{2+} declines. These observations lead us to the conclusion that the role of Ca^{2+} in the induction of secretion from these cells resembles that of the orthodox receptor-directed ligands controlling G-protein-mediated processes such as the activation of adenylyl cyclase (RODBELL 1980). There is, however, one major difference, for while in terms of its function the calcium-binding protein C_E shares features with those cell surface receptors linked to G-proteins (enhancing the affinity for Mg^{2+} in the catalysis of guanine nucleotide exchange or activation; IYENGAR and BIRNBAUMER 1982), the site of ligand binding must be internally located.

III. One or Two Effectors?

For those familiar with earlier publications from our laboratory (HOWELL et al. 1987), a number of the above statements might appear to be in conflict with what we have previously written. In particular, our previous work has stressed an essential role for both effectors, Ca^{2+} and guanine nucleotide, tending to the conclusion that they operate in parallel to permit the final progression to membrane fusion (GOMPERTS 1990). The reason for this change in perception from a dual effector to a single effector mode of operation has been due to a difference in experimental approach and arises from the use of glutamate as the main anion in solutions used for cell permeabilisation (CHURCHER and GOMPERTS 1990). When work on permeabilised mast cells commenced, in about 1979 (initially using ATP in its tetrabasic form, ATP^{4-}, as a permeabilising agent; BENNETT et al. 1981), secretion was measured as the release of histamine. Although most other investigators in the field of exocytosis, following on the work of KNIGHT and BAKER (1982), preferred to use permeabilisation buffers formulated with glutamate as the predominant anion, we were compelled to use Cl^- (or other simple uni/univalent electrolytes) since it was not possible to measure secreted histamine against the overwhelming concentration of glutamate (125 m*M*, isotonic at pH 6.8). In more recent work we have turned to the measurement of released acid hydrolases (as hexosaminidase) and this has allowed us to work in the glutamate environment and, more importantly, to make a comparison between the two electrolyte systems (CHURCHER and GOMPERTS 1990; LILLIE and GOMPERTS 1992a,b). This is reviewed in the following paragraphs.

1. Chloride Suppresses and Glutamate Enhances Guanine Nucleotide Sensitivity of Exocytosis

Figure 4 illustrates the dependence on GTP and its non-hydrolysable analogue GTPγS and the effect of Mg^{2+}-deprivation for secretion from mast cells permeabilised in isotonic solutions of NaCl (a) and sodium glutamate (b). In these experiments, Ca^{2+} was buffered at pCa5. By comparing Fig. 4a and b, it is apparent that the sensitivity to both the stable and the native guanine nucleotides is substantially enhanced when Cl^- is substituted by glutamate (similar results have been obtained using either D- or L-glutamate, aspartate, GABA or glycine). Furthermore, withdrawal of Mg^{2+}, while having no effect on the secretion induced by the non-hydrolysable analogue, also enhances the sensitivity to GTP regardless of the nature of the solution anion. By substituting glutamate for Cl^-, or by removal of Mg^{2+}, the sensitivity of the secretory response to GTP shifts from concentrations in the millimolar into the micromolar range, approaching that of GTPγS (and becomes at least the equal of GppNHp; HOWELL et al. 1987). The most plausible explanation of these observations is that removal of Mg^{2+} permits the GTP to survive as a stable entity, resistant to hydrolysis. In addition, the glutamate environment also appears to confer some protection against hydrolysis even in the presence of Mg^{2+}. These experiments do not inform us whether the hydrolysis is a generalised activity of Mg^{2+}-dependent nucleotidases reducing the concentration of GTP within the confines of the permeabilised cell cytosol, or whether the conversion of GTP to GDP occurs on the G_E protein itself. We favour the latter explanation because we have found that ongoing GTP-induced secretion, initiated in the absence of Mg^{2+}, can be terminated within seconds of applying Mg^{2+}. It seems

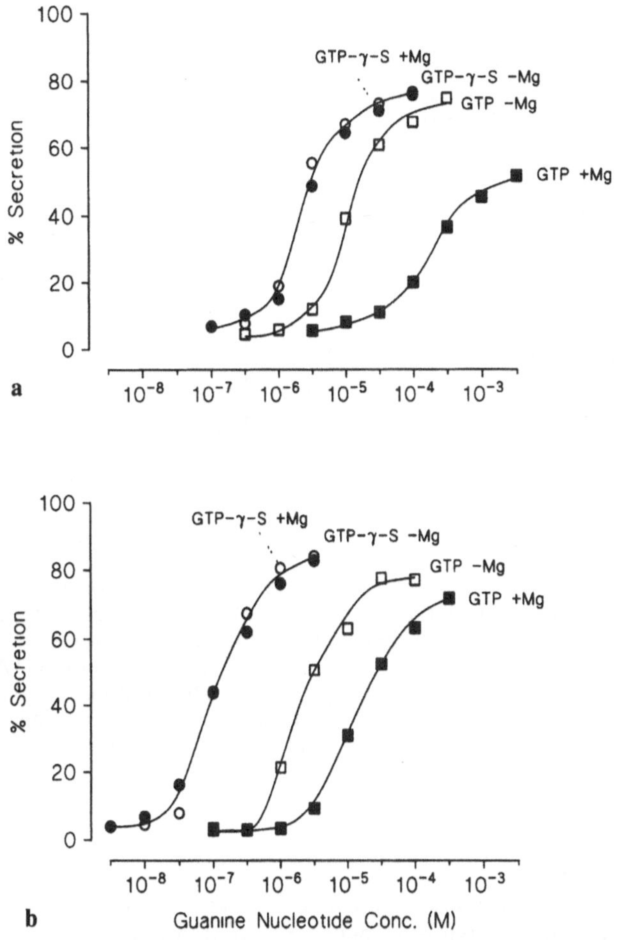

Fig. 4a,b. Exclusion of Mg^{2+} enhances the ability of GTP to induce exocytosis from permeabilised rat mast cells. Mast cells, suspended in buffered NaCl (**a**) or glutamate (**b**) and pretreated with metabolic inhibitors, were incubated with SL-O and Ca^{2+} (buffered at pCa5) and guanine nucleotides as indicated, in the presence and absence of Mg^{2+}. At the end of 10 min the cells were sedimented by centrifugation and the supernatant sampled for analysis of secreted hexosaminidase. Guanine nucleotide: ■,□ GTP; ●,○ GTPγS. *Closed symbols* indicate the presence of Mg^{2+} (1 m*M*, Ca^{2+} buffered by 5 m*M* EGTA); *open symbols*, Mg^{2+} absent (Ca^{2+} buffered by HEDTA 5 m*M*, $[Mg^{2+}] > 10^{-9}$) (from Lillie and Gomperts 1992b)

improbable that the concentration of GTP (applied as high as 1 m*M*) inside cells treated with SL-O – and therefore readily accessible to solutes having the dimensions of lactate dehydrogenase (LDH) – could drop so precipitously.

In summary, the affinity and the GTPase activity of G_E are sensitive to the electrolyte environment and the presence of Mg^{2+}. In comparison with chloride, an electrolyte solution formulated with isotonic glutamate

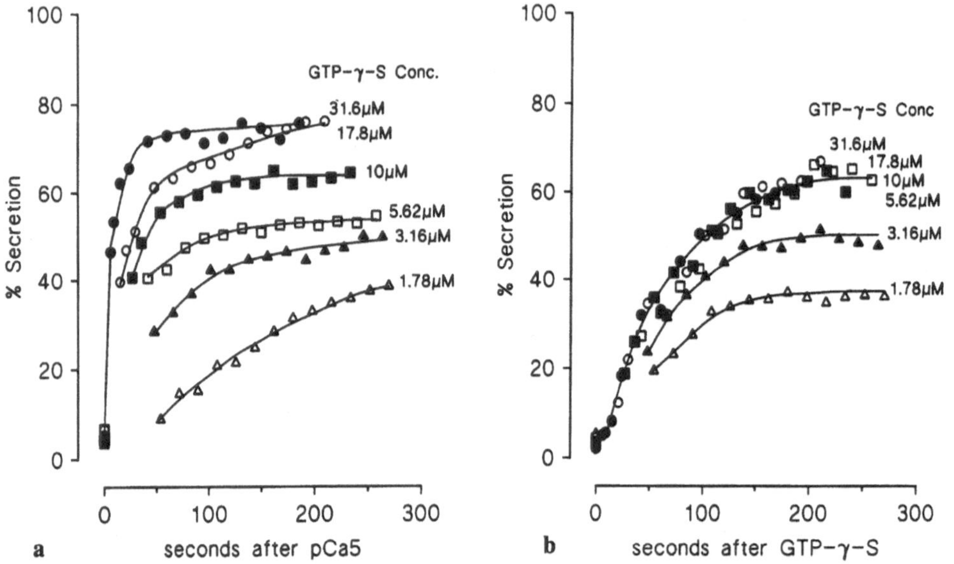

Fig. 5a,b. Time course of Ca^{2+}-triggered and GTPγS-triggered exocytosis from mast cells permeabilised by SL-O in the presence of Mg^{2+}. Mast cells, suspended in buffered NaCl and pretreated with metabolic inhibitors were permeabilised in the presence of **a** GTPγS at concentrations indicated, or **b** calcium buffer (EGTA, 5 m*M*, pCa5). One minute later exocytosis was triggered by addition of the complementary effector (**a** Ca^{2+}, pCa5; or **b** GTPγS at concentrations indicated) and timed samples were withdrawn and processed for measurement of secreted hexosaminidase. GTPγS concentration: ●, $10^{-4.5}$ *M*; ○, $10^{-4.75}$ *M*; ■, 10^{-5} *M*; □, $10^{-5.25}$ *M*; ▲, $10^{-5.5}$ *M*; △, $10^{-5.75}$ *M* (from LILLIE and GOMPERTS 1992a)

enhances the affinity for guanine nucleotides. The consequence is that in permeabilised cells it is possible to achieve a limited extent of secretion by applying GTPγS together with Mg^{2+} but in the absence of Ca^{2+}; GTP, because of its lability in the presence of Mg^{2+}, cannot serve as a single effector for exocytosis even in the glutamate environment, and for this reason GTP-induced secretion (as must occur in normal intact cells) necessitates the presence of Ca^{2+} which, through its interaction with C_E, enhances the affinity for the guanine nucleotide still further. Furthermore, Cl^-, although regarded as a relatively minor component in most cell types (where the main electronegative component is actually contributed by protein, not by glutamate!) is generally present in the range 10–20 m*M*. This is quite sufficient to suppress the affinity for GTPγS to the extent that secretion is substantially reduced (CHURCHER and GOMPERTS 1990).

IV. Kinetics of Exocytosis

All the experiments described above refer to completed events. As with intact mast cells stimulated through the IgE receptor, secretion from the

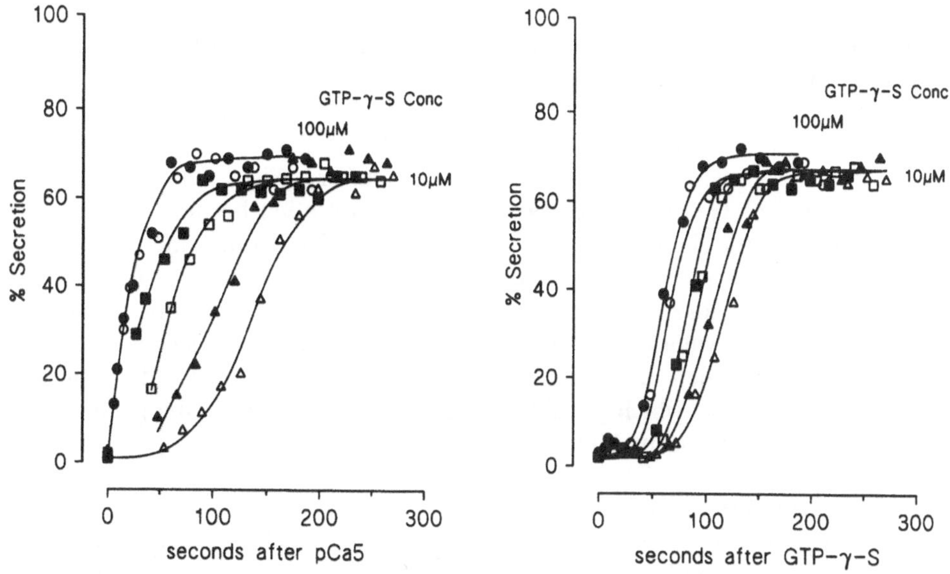

Fig. 6a,b. Time course of Ca^{2+}-triggered and GTPγS-triggered exocytosis from mast cells permeabilised by SL-O in the absence of Mg^{2+}. Mast cells, suspended in buffered NaCl and pretreated with metabolic inhibitors were permeabilised in the presence of **a**) GTPγS at concentrations indicated, or **b**) calcium buffer (HEDTA, 5 m*M*, pCa5). One minute later exocytosis was triggered by addition of the complementary effector (**a**: Ca^{2+}, pCa5; or **b**: GTPγS at concentrations indicated) and timed samples were withdrawn and processed for measurement of secreted hexosaminidase. GTPγS concentration: ●, $10^{-4}\,M$; ○, $10^{-4\,2}\,M$; ■, $10^{-4\,4}\,M$; □, $10^{-4\,6}\,M$; ▲, $10^{-4\,8}\,M$; △, $10^{-5}\,M$ (from Lillie and Gomperts 1992a)

permeabilised cells generally terminates within about 5 min of providing a stimulus. A rather different view of the mechanism of activation can be perceived by following the time course of secretion.

1. Mg^{2+} Permits Abrupt Onset

We have measured the rate of onset of exocytosis from permeabilised mast cells following the successive application of guanine nucleotides (GTP or GTPγS) and then Ca^{2+} (or in the reverse order) in the presence and absence of Mg^{2+}.[1] The experiments illustrated in Figs. 5 and 6 show that under

[1] The kinetic experiments have all been carried out on cells permeabilised in NaCl; there are two reasons for this: (1) in simple uni/univalent electrolytes both Ca^{2+} and a guanine nucleotide are necessary; we have exploited the dual effector system to explore the effects of equilibrating the permeabilised cells wth either Ca^{2+} or guanine nucleotide, and then applying the other as the trigger; and (2) cells permeabilised in glutamate do not remain viable for sufficiently long to permit sequential stimulation; for this reason it would be necessary to apply the stimulating effectors alongside the SL-O and the onset kinetics of exocytosis would necessarily be marred by artefacts due to permeabilisation.

certain conditions the onset of secretion is preceded by delays which are characteristic of the binding and activation reactions which must be satisfied before exocytosis can commence.

For cells permeabilised in the presence of Mg^{2+} and loaded for 1 min with GTPγS, exocytosis commences abruptly following provision of Ca^{2+}(pCa5, applied as 5 m*M* Ca-EGTA buffer; Fig. 5a). As the concentration of GTPγS is reduced, the rate of secretion becomes slower, but there is little indication of any substantial delay once the second effector, Ca^{2+}, is provided. All the progress curves in Fig. 5a appear to extrapolate back close to the origin of time and secretion. The situation is not very different when the order of addition of the two effectors is reversed (Fig. 5b). Here there is a constant delay of a few seconds' duration preceding the onset, which possibly reflects the time taken for the guanine nucleotide to diffuse to its binding site within the cells.

2. Mg^{2+} Deprivation Causes Onset Delays

Again, for Mg^{2+}-deprived cells initially loaded at sufficiently high concentrations of GTPγS, exocytosis commences abruptly when the Ca^{2+} trigger is provided (Fig. 6a). However, as the concentration of the guanine nucleotide is reduced, delays in onset become apparent. Since it is hardly plausible that the removal of Mg^{2+} could have slowed the time for diffusion beyond the few seconds revealed in the experiment of Fig. 5b, we are bound to conclude that these delays reflect subsequent events such as the time taken to bind and then to activate G_E. This becomes even more evident when GTPγS is applied as a trigger to Mg^{2+}-deprived cells permeabilised in the presence of Ca^{2+} (pCa5; Fig. 6b). Under these conditions of stimulation, delays are manifest at all concentrations of GTPγS.

The susceptibility of the onset kinetics of GTPγS-induced exocytosis to the presence or absence of Mg^{2+} stands in striking contrast to the insensitivity to Mg^{2+} of the concentration dependence of the completed secretory responses (see Fig. 4; Sect. BIII1). As can be seen (Fig. 6b) the delays are comprised of two components. One of these is dependent on the concentration of the guanine nucleotide and therefore most probably reflects the rate of binding. The other component is irreducible, but it is sensitive to the identity of the activating guanine nucleotide. Thus, the irreducible delay for GTPγS at about 50 s, as shown, is intermediate between that for GTP (about 20 s) and GppNHp (100 s). Such delays, insensitive to concentration but sensitive to the identity, are likely to reflect events which take place subsequent to binding of the activating guanine nucleotide.

V. GTPases Regulate and Modulate Exocytosis in Many Cells and Tissues

We have restricted our discussion of guanine nucleotide regulation of exocytosis to the mast cells and related myeloid granulocytes. However, it is apparent that guanine nucleotides are also instrumental in the late stages of the exocytotic pathway of a wide variety of other cells and tissues (LINDAU and GOMPERTS 1991). In most of these, non-hydrolysable analogues of GTP have been found to be capable of stimulating exocytosis independently of Ca^{2+}: platelets (ATHAYDE and SCRUTTON 1990; COORSSEN et al. 1990); bovine adrenal chromaffin cells (MORGAN and BURGOYNE 1990); parathyroid cells (OETTING et al. 1986); renal juxtaglomerular cells (NEWTON and KNIGHT 1991); pancreatic acinar cells (EDWARDSON et al. 1990; KITAGAWA et al. 1990). In others such as pituitary melanotrophs, GTP analogues modulate the sensitivity to Ca^{2+} by a mechanism not involving activation of protein kinase C (and thus independently of G_p-coupled phospholipase C; YAMAMOTO et al. 1987). In yet other secretory cells, GTP analogues have been found to mediate inhibitory mechanisms: pituitary gonadotrophs (VAN DER MERWE et al. 1991; DAVIDSON et al. 1991), sometimes linked to inhibitory receptors – insulin secreting cells (VALLAR et al. 1987; ULLRICH et al. 1990) and ACTH secreting cells (LUINI and DE MATTEIS 1988, 1990).

C. On the Nature of G_E

I. The Example of G_S

From the experimental findings summarised above we venture some suggestions about the nature of G_E as it is expressed in mast cells. We compare the control of secretion to the activation of adenylyl cyclase. The reason for this choice is twofold. Firstly, there are some obvious, even though possibly superficial, similarities between the two systems of control. These are summarized in Table 1. Secondly, however, it has to be said that the cyclase system of control by GTPases is one of only two (the other being transducin) against which any proper comparison, either positive or negative, can be made. Appropriate data are simply not available for other systems, neither those regulated by $\alpha\beta\gamma$ heterotrimeric G-proteins, nor for those regulated by the monomeric GTPases. In the latter class, characterization of the upstream regulators and (with one exception; ABO et al. 1991) downstream effectors is still at an early stage. From our data which relate to the cell physiology of secretion, not biochemistry, we cannot firmly place G_E into either the trimeric or monomeric category of GTPases. However, at the functional level similarities with the G_S system controlling adenylyl cyclase are plainly evident.

1. The Example of the Monomeric GTPases

In contrast to the heterotrimers, inhibition of the GTPase activity of many of the monomeric GTPases prevents the cycling of the protein between its active (GTP) and inactive (GDP) states, and freezes the processes that they direct,

Table 1. Cyclase and Exocytosis: activation by G_S and G_E

	Cyclase system	Exocytosis
Onset delays	For guanine nucleotide induced activation, onset delays expressed in the absence of Mg^{2+} (BIRNBAUMER et al. 1985)	Onset delays expressed in the absence of Mg^{2+}
	Duration of delays depend on the identity of the activating guanine nucleotide (BIRNBAUMER et al. 1985)	Duration of delays depend on the identity of the activating guanine nucleotide
	The rate of binding of guanine nucleotides to $\alpha\beta\gamma$ heterotrimeric GTPases is dependent on the presence of Mg^{2+} (IYENGAR and BIRNBAUMER 1981; IYENGAR 1981)	
GTPase activity	Maximal and persistent activation induced by hydrolysis resistant analogues of GTP	Maximal secretion induced by low concentrations of stable analogues of GTP
	Maximal and persistent activation following modification of active site by ADP-ribosylation catalysed by cholera toxin	Maximal secretion induced by GTP when Mg^{2+} is excluded, thus preventing hydrolysis

such as vesicle trafficking (WALWORTH et al. 1989; BOURNE 1988) or protein synthesis (TANAKA et al. 1977). Clearly, as we have shown, exocytosis proceeds unimpeded under conditions in which GTPases are maintained in the active (effectively GTP-bound) state. However, if the final reaction in the exocytotic pathway is a one-off event involving activation of a GTPase situated on the secretory vesicle, then inhibition of the GTPase function need not necessarily be an impediment. Furthermore, the finding that exocytosis is best activated under circumstances in which GTP hydrolysis cannot occur does not necessarily rule out a role for those monomeric GTPases whose biological activity is actually enhanced by inhibition of the GTPase activity (e.g. some mutant forms of the $p21^{ras}$ family). From this evidence it would appear that G_E requires Mg^{2+} for the binding of guanine nucleotide, its subsequent activation and for GTP hydrolysis. It shares these characteristics with the heterotrimeric GTPases.

D. Single Cell Analysis of GTPγS-Induced Exocytosis

Evidence from single-cell patch-clamp analysis of exocytosis is in good agreement with the data from permeabilised cells, but also points to a further level of complexity in the involvement of the GTPases. Individual events can be detected and measured, unsullied by the averaging which is an inevitable feature of standard secretion experiments involving hundreds or

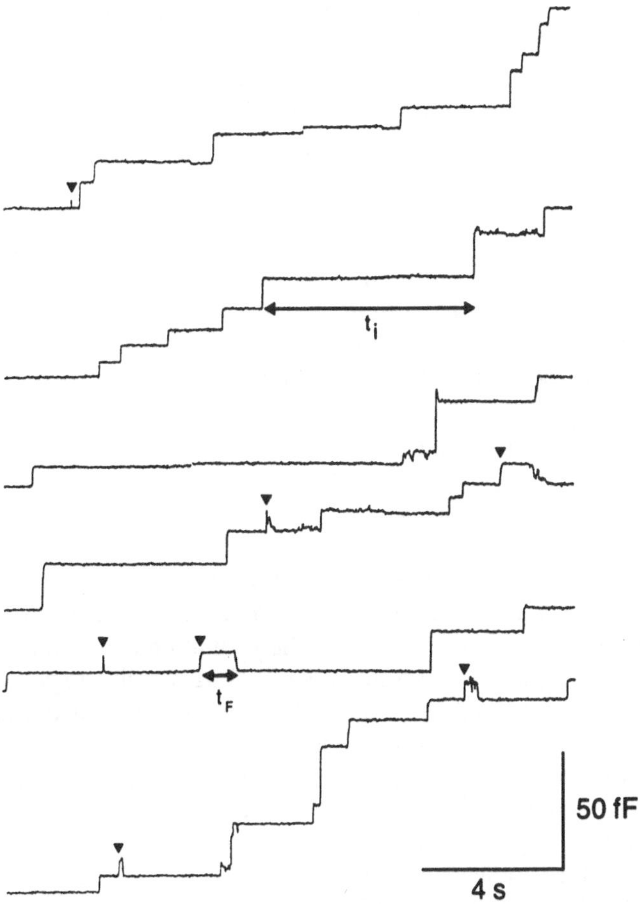

Fig. 7. Unitary exocytotic events revealed by high resolution capacitance measurements in a mouse mast cell. The record reveals step changes in cell membrane capacitance due to the fusion of individual granules with the plasma membrane. The 90 s of *trace* (starting at the *bottom*) is taken from a single record which was initiated 18 min previously by dialysing the cell interior with GTPγS (10 μM) at 13°. The *trace* reveals both irreversible fusion steps and reversible capacitance flicker events (marked with an *inverted triangle*); t_i indicates an interstep interval; t_F indicates duration of a flicker episode (from OBERHAUSER et al. 1992)

thousands of cells. By making measurements of membrane capacitance (proportional to membrane area), it is possible to monitor exocytosis at a level of resolution in space and time which permits the microscopic kinetic events controlling unit granule–plasma membrane fusions to be probed in detail.

An illustration of the degranulation of a single mast cell is shown in Fig. 7. Considered as a staircase, each riser constitutes the increase in capacitance due to the fusion of a single secretory granule, while the treads constitute the refractory intervals between the fusions. Frequency distribution analysis reveals that the duration of the refractory periods intervening between the irreversible fusion events is determined by the concentration of the activating guanine nucleotide (ALVAREZ DE TOLEDO and FERNANDEZ 1990). The higher the concentration (in the range 0.2–50 μM), the shorter is the time between individual fusion events, while the presence of Ca^{2+} serves to enhance the sensitivity to the guanine nucleotide. As shown, on occasion, capacitance increments of considerable magnitude occur which are then reversed (called "capacitance flicker"), indicating the existence of an intermediate stage (the construction of the fusion pore) which must be achieved before the final progression to stable membrane fusion and release of granule contents. The duration of such flicker episodes is very variable, extending from less than 14 ms (the limit for detection) up to several seconds.

The following scheme has been suggested (ALVAREZ DE TOLEDO and FERNANDEZ 1988):

$$U \rightleftharpoons F \rightarrow S$$

where those granules (F) that are neither unfused (U), nor stably fused (S) are classified together as a transition or flicker state. The exocytotic event clearly comprises two stages: the first is the formation of a reversible water-filled pore (U ~ F), and the second is the final irreversible expansion of the pore (F → S) which leads to release of the contained materials to the exterior (ZIMMERBERG et al. 1987; ALMERS 1990). The kinetics are explained if the transition from F to S is sensitive to the concentrations of GTPγS and Ca^{2+}, while the transition from F to U occurs at a constant rate. It is, however, evident from the secretion experiments related previously (see Sect. BIV2), which indicate delays preceding onset of secretion, that the guanine nucleotide is also required simply to enter the U state, which is a necessary preliminary to fusion. It is thus possible that fusion requires at least two interventions by guanine nucleotides so that a more complete representation of the sequence could be given by:

$$U \rightarrow U' \rightleftharpoons F \rightarrow S$$

with the transition U → U′ also dependent on the presence of guanine nucleotide.

E. Two GTPases in Regulated Exocytosis?

Recent experience of the most cell (and other myeloid granulocytes) is consistent with the idea that two GTPases might be involved in regulated exocytosis, as recently implied for vesicle traffic in the constitutive pathways of membrane traffic (BARR et al. 1991; STOW et al. 1991; DONALDSON et al. 1991). It has been well demonstrated that monomeric GTPases of the Rab and ARF classes regulate the assembly and disassembly of macromolecular complexes such as vesicle coats or docking and fusion proteins (ROTHMAN and ORCI 1992), but beyond this the overall rate of throughput appears to be under the control of heterotrimeric GTPases, which may provide the means by which cells are able to switch off secretion when entering mitosis (MELANÇON et al. 1987). It is possible that regulated exocytosis has a similar hierarchy of control. In this case the commitment of whether to undergo release is determined by a heterotrimeric GTPase. Kinetically, in its sensitivity to Mg^{2+} and its activation by stable analogues of GTP, this is the entity that we have called G_E. Beyond this point, further events sensitive to guanine nucleotides, such as the stability of the fusion pore, are probably again regulated by GTPases. These are likely to be monomeric, possibly Rab-like, ensuring accuracy and directionality by mechanisms not unlike those operating in the constitutive pathways of vesicle traffic (PADFIELD et al. 1992).

References

Abo A, Pick E, Hall A, Totty N, Teahan CG, Segal AW (1991) The small GTP-binding protein, p21rac 1, is involved in the activation of the phagocyte NADPH oxidase. Nature 353:668–670

Ali SM, Geisow MJ, Burgoyne RD (1989) A role for calpactin in calcium-dependent exocytosis in adrenal chromaffin cells. Nature 340:313–315

Almers W (1990) Exocytosis. Annu Rev Physiol 52:607–624

Alvarez de Toledo G, Fernandez JM (1988) The events leading to secretory granule fusion. In: Gunn RB, Parker JC (eds) Cell physiology of blood, vol 43. Rockefeller University Press, New York, pp 333–344 (Society of General Physiologists series)

Alvarez de Toledo G, Fernandez JM (1990) The effect of GTP-γ-S and Ca^{2+} on the kinetics of exocytosis of single secretory granules in peritoneal mast cells. Biophys J 57:495a

Athayde CM, Scrutton MC (1990) Guanine nucleotides and Ca^{2+}-dependent lysosomal secretion in electropermeabilised human platelets. Eur J Biochem 189:647–655

Barr FA, Leyte A, Mollner S, Pfeuffer T, Tooze SA, Huttner WB (1991) Trimeric G-proteins of the trans-Golgi network are involved in the formation of constitutive secretory vesicles and immature secretory granules. FEBS Lett 294:239–243

Bennett JP, Cockcroft S, Gomperts BD (1980) Ionomycin stimulates mast cell histamine secretion by forming a lipid soluble calcium complex. Nature 282:851–853

Bennett JP, Cockcroft S, Gomperts BD (1981) Rat mast cells permeabilised with ATP secrete histamine in response to calcium ions buffered in the micromolar range. J Physiol (Lond) 317:335–345

Birnbaumer L, Codina J, Mattera R, Cerione RA, Hildebrandt JD, Sunyer T, Rojas FJ, Caron MG, Lefkowitz RJ, Iyengar R (1985) Structural basis of adenylate cyclase stimulation and inhibition by distinct guanine nucleotide regulatory proteins. In: Cohen P, Houslay MD (eds) Molecular mechanisms of transmembrane signalling. Elsevier Science, Amsterdam, pp 131–182

Bourne HR (1988) Do GTPases direct membrane traffic in secretion? Cell 53:669–671

Churcher Y, Gomperts BD (1990) ATP dependent and ATP independent pathways of exocytosis revealed by interchanging glutamate and chloride as the major anion in permeabilised mast cells. Cell Regul 1:337–346

Coorssen JR, Davidson MML, Haslam RJ (1990) Factors affecting dense and α-granule secretion from electropermeabilized human platelets: Ca^{2+}-independent actions of phorbol ester and GTP-γ-S. Cell Regul 1:1027–1041
Davidson J, van der Merwe PA, Wakefield I, Millar RP (1991) Mechansisms of lutenizing hormone secretion: new insights from studies with permeabilized cells. Mol Cell Endocrinol 76:C33–33C38
Di Virgilio F, Gomperts BD (1983) Cytosol Mg^{2+} modulates Ca^{2+} ionophore induced secretion from rabbit neutrophils. FEBS Lett 163:315–318
Donaldson JG, Kahn RA, Lippincott-Schwartz J, Klausner RD (1991) Binding of ARF and β-COP to Golgi membranes: possible regulation by a trimeric G protein. Science 254:1197–1199
Edwardson JM, Vickery C, Christy LJ (1990) Rat pancreatic acini permeabilised with streptolysin O secrete amylase at Ca^{2+} concentrations in the micromolar range when provided with ATP and GTP-γ-S. Biochim Biophys Acta 1053:32–36
Foreman JC, Mongar JL (1972) The role of the alkaline earth ions in anaphylactic histamine secretion. J Physiol (Lond) 224:753–769
Foreman JC, Mongar JL, Gomperts BD (1973) Calcium ionophores and movement of calcium ions following the physiological stimulus to a secretory process. Nature 245:249–251
Gomperts BD (1990) G_E: a GTP-binding protein mediating exocytosis. Annu Rev Physiol 52:591–606
Gomperts BD, Fewtrell CMS (1985) The mast cell: a paradigm for receptor and exocytotic mechanisms. In: Cohen P, Houslay MD (eds) Molecular mechanisms of membrane signalling. Elsevier, Amsterdam, pp 377–409 (Molecular aspects of cellular regulation, vol 4)
Harvey AM, Maclntosh FC (1940) Calcium and synaptic transmission in a sympathetic ganglion. J Physiol (Lond) 97:408–416
Henderson WR, Chi EY, Jorg A, Klebanoff SJ (1983) Horse eosinophil degranulation induced by the ionophore A23187. Am J Pathol 111:341–349
Howell TW, Cockcroft S, Gomperts BD (1987) Essential synergy between Ca^{2+} and guanine nucleotides in exocytotic secretion from permeabilised mast cells. J Cell Biol 105:191–197
Iyengar R (1981) Hysteretic activation of adenylyl cyclases: Mg ion regulation of the activation of the regulatory component as analyzed by reconstitution. J Biol Chem 256:11042–11050
Iyengar R, Birnbaumer L (1981) Hysteretic activation of adenylyl cyclases: effect of Mg ion on the rate of activation by guanine nucleotides and fluoride. J Biol Chem 256:11036–11041
Iyengar R, Birnbaumer L (1982) Hormone receptor modulates the regulatory component of adenylyl cyclase by reducing its requirement for Mg^{2+} and enhancing its extent of activation by guanine nucleotides. Proc Natl Acad Sci USA 79:5179–5183
Kikkawa S, Takahashi K, Takahashi K, Shimada N, Ui M, Kimura N, Katada T (1990) Conversion of GDP into GTP by nucleoside diphosphate kinase on the GTP-binding proteins. J Biol Chem 265:21536–21540
Kikkawa S, Takahashi K, Takahashi K, Shimada N, Kimura N, Katada T (1991) Correction to conversion of GDP into GTP by nucleoside diphosphate kinase on the GTP-binding proteins. J Biol Chem 266:12795
Kimura N, Shimada N (1990) Evidence for complex formation between GTP binding protein (G_S) and membrane associated nucleoside diphosphate kinase. Biochem Biophys Res Commun 168:99–106
Kitagawa M, Williams JA, de Lisle RC (1990) Amylase release from streptolysin O-permeabilized pancreatic acini. Am J Physiol 259:G157–157G164
Knight DE, Baker PF (1982) Calcium-dependence of catecholamine release from bovine adrenal medullary cells after exposure to intense electric fields. J Membr Biol 68:107–140

Lillie THW, Gomperts BD (1992a) GTP, ATP, Ca^{2+} and Mg^{2+} as effectors and modulators of exocytosis in permeabilised rat mast cells. Philos Trans R Soc Lond [Biol] 336:25–34
Lillie THW, Gomperts BD (1992b) Guanine nucleotide is essential, Ca^{2+} is a modulator, in the exocytotic reaction of permeabilised rat mast cells. Biochem J 288:181–187
Lindau M, Gomperts BD (1991) Techniques and concepts in exocytosis: focus on mast cells. Biochim Biophys Acta 1071:429–471
Luini A, De Matteis MA (1988) Dual regulation of ACTH secretion by guanine nucleotides in permeabilized AtT-20 cells. Cell Mol Neurobiol 8:129–138
Luini A, De Matteis MA (1990) Evidence that receptor-linked G protein inhibits exocytosis by a post-second-messenger mechanism in AtT-20 cells. J Neurochem 54:30–38
MacGlashan D, Gou C (1991) Oscillations in free cytosolic calcium during IgE-mediates stimulation distinguish human basophils from human mast cells. J Immunol 147:2259–2269
Melancon P, Glick BS, Malhotra V, Weidman PJ, Serafini T, Gleason ML, Orci L, Rothman JE (1987) Involvement of GTP-binding "G" proteins in transport through the Golgi stack. Cell 51:1053–1062
Morgan A, Burgoyne RD (1990) Stimulation of Ca^{2+}-independent catecholamine secretion from digitoinin-permeabilized bovine adrenal chromaffin cells by guanine nucleotides: relationship to arachidonate release. Biochem J 269:521–526
Nakata T, Sobue K, Hirokawa W (1990) Conformational change and localization of calpactin I Complex involved in exocytosis as revealed by quick-freeze, deep-etch electron microscopy and immunocytochemistry. J Cell Biol 110:13–25
Newton JS, Knight DE (1992) Direct evidence of a role for Ca^{2+} and guanine nucleotide binding proteins in renin secretion from isolated. electropermeabilized, rat renal cortical cells. J Physiol (Lond) 446:585p
Oberhauser AF, Monck JR, Fernandez JM (1992) Events leading to the opening and closing of the exocytotic fusion pore have markedly different temperature dependencies: kinetic analysis of single fusion events in patch-clamped mouse mast cells. Biophys J 61:800–809
Oetting M, LeBoff M, Swiston L, Preston J, Brown E (1986) Guanine nucleotides are potent secretagogues in permeabilised parathyroid cells. FEBS Lett 208:99–104
Otero ADS (1990) Transphosphorylation and G protein activation. Biochem Pharmacol 39:1399–1404
Padfield PJ, Balch WE, Jamieson JD (1992) A synthetic peptide of the rab3a effector domain stimulates amylase release from permeabilized pancreatic acini. Proc Natl Acad Sci USA 89:1656–1660
Randazzo PA, Northup JK, Kahn RA (1991) Activation of a small GTP binding protein (ARF) by nucleoside diphosphate kinase catalysed phosphorylation of bound GDP. Science 254:850–853
Rodbell M (1980) The role of hormone receptors and GTP-regulatory proteins in membrane transduction. Nature 284:17–22
Rothman JE, Orci L (1992) Molecular dissection of the secretory pathway. Nature 355:409–415
Stow JL, de Almeida JB, Narula N, Holzman EJ, Ercolani L, Ausiello DA (1991) A heterotrimeric G protein, $G\alpha_{1\text{-}3}$, on Golgi membranes regulates the secretion of a heparan sulfate proteoglycan in LLC-PK_1 epithelial cells. J Cell Biol 114:1113–1124
Tanaka M, Iwasaki K, Kaziro Y (1977) Translocation reaction promoted by polypeptide chain elongation factor-2 from pig liver. J Biochem (Tokyo) 82:1035–1043
Ullrich S, Prentki M, Wollheim CB (1990) Somatostatin inhibition of Ca^{2+}-induced insulin secretion in permeabilized HIT-T15 cells. Biochem J 270: 273–276

Vallar L, Biden TJ, Wollheim CB (1987) Guanine nucleotides induce Ca^{2+} independent secretion from permeabilized RINm5F cells. J Biol Chem 262:5049–5056

van der Merwe PA, Millar RP, Wakefield IK, Davidson JS (1991) Inhibition of lutenizing hormone exocytosis by guanosine 5′-(gammathio)triphosphate reveals involvement of a GTP binding protein. Biochem J 275:399–405

Walworth NC, Goud B, Kabcenell AK, Novick PJ (1989) Mutational analysis of SEC4 suggests a cyclical mechanisms for the regulation of vesicular traffic. EMBO J 8:1685–1693

White JR, Ishizaka T, Ishizaka K, Sha'afi RI (1984) Direct demonstration of increased intracellular concentration of free calcium measured by quin-2 in stimulated rat mast cells. Proc Natl Acad Sci USA 81:3978–3982

Wright DG, Bralove DA, Gallin JI (1977) The differential mobilisation of human neutrophil granules. Effects of phorbol myristate and ionophore A23187. Am J Pathol 87:273–283

Yamamoto T, Furuki Y, Kebabian JW, Spatz M (1987) α-Melanocyte-stimulating hormone secretion from permeabilized intermediate lobe cells of rat pituitary gland. FEBS Lett 219:326–330

Zabucchi G, Romeo D (1976) The dissociation of exocytosis and respiratory stimulation in leucocytes by ionophores. Biochem J 156:209–213

Zimmerberg J, Curran M, Cohen FS, Brodwick M (1987) Simultaneous electrical and optical mesurements show that membrane fusion precedes secretory granule swelling during exocytosis of beige mouse mast cells. Proc Natl Acad Sci USA 84:1585–1589

CHAPTER 43

ADP-Ribosylation of Small GTPases by *Clostridium botulinum* Exoenzyme C3 and *Pseudomonas aeruginosa* Exoenzyme S

J. COBURN

A. Introduction

Bacterial toxins have served as important tools for the study of eukaryotic cell biology, especially signal transduction by the heterotrimeric G-proteins. Many of these toxins catalyze the ADP-ribosylation of specific proteins inside the host cell. In this reaction, the nicotinamide–ribose bond of nicotinamide-adenine dinucleotide (NAD) is split and the ADP-ribose moiety is transferred to the substrate protein. Most of these toxins, termed ADP-ribosyltransferases, modify G-proteins or other GTPases. The addition of the ADP-ribose group to the substrate protein alters the function of that protein dramatically. For example, ADP-ribosylation of $G_{s\alpha}$ by cholera toxin decreases GTP hydrolysis (CASSEL and SELINGER 1977), while the modification of $G_{i\alpha}$ by pertussis toxin alters its interaction with receptors (KUROSE et al. 1983; MURAYAMA and UI 1983). Diphtheria toxin and *Pseudomonas* exotoxin A both ADP-ribosylate the same residue in elongation factor 2 (EF-2), and inhibit interaction of EF-2 with the ribosome (COLLIER 1967; IGLEWSKI and KABAT 1975). In all cases, the actions of the toxins on the GTPases have drastic consequences for the eukaryotic cell, as normal signal transduction is disrupted. An additional characteristic of the bacterial ADP-ribosyltransferases is that they all have well-defined cell-binding or "B" components that are functionally distinct from the "A" domains, which are enzymically active after entry into the cytosol.

Recently, two bacterial ADP-ribosyltransferases that modify monomeric small GTPases have been identified. *Clostridium botulinum* exoenzyme C3 ADP-ribosylates the rho proteins, and *Pseudomonas aeruginosa* exoenzyme S modifies the *ras* gene products and several other ras-related proteins. The cellular functions of the small GTPases are generally not well understood, but exoenzyme C3 has already contributed to our understanding of the roles of the rho proteins and of the regulation of the cytoskeleton. Eventually exoenzyme S may be as useful in helping to understand the biology of several of the small GTPases as cholera and pertussis toxin have been in the study of the heterotrimeric G-proteins.

B. Small GTPases

The small GTPases comprise the ras superfamily and include the ras, rho, rap, rab, rac, and ral proteins. Relatively little is known about the cellular functions of the small GTPases, but they appear to be involved in the regulation of growth, vesicle trafficking, and the state of the cytoskeleton. Several lines of evidence suggest that the normal function of ras is the control of cellular growth (reviewed by BARBACID 1987; HALL 1990). The cellular functions of the more recently discovered ras-related gene products are also subjects of intense study. Studies with *C. botulinum* exoenzyme C3 suggest that the rho proteins may regulate actin microfilament polymerization (CHARDIN et al. 1989). The rab proteins are involved in vesicle trafficking (CHAVRIER et al. 1990; FISCHER VON MOLLARD et al. 1990, 1991; ZAHROUI et al. 1989). The rap gene family products may serve as antagonists to ras by competing for downstream effectors (HATA et al. 1990; KITAYAMA et al. 1989; PIZON et al. 1988) and may also have roles in the regulation of the cytoskeleton (FISCHER et al. 1990). The mammalian protein that activates cholera toxin, ARF (ADP-ribosylation factor; KAHN and GILMAN 1986), appears to be involved in protein trafficking (STEARNS et al. 1990). Studies on many of these mammalian proteins are complemented by genetic approaches taken toward understanding the cellular roles of their analogs in yeast and the developmental roles of the related proteins in *Drosophila*. The yeast proteins, however, may differ in their interactions and cellular roles from their counterparts in mammalian cells. For instance, the yeast Ras proteins regulate growth through control of adenylyl cyclase activity, whereas the mammalian adenylyl cyclase activity is controlled by the heterotrimeric GTPases G_s and G_i.

C. *Clostridium botulinum* Exoenzyme C3

C. botulinum is most famous for the production of botulinum neurotoxin, but certain strains also synthesize two ADP-ribosyltransferases. Toxin C2 modifies actin (AKTORIES et al. 1986) and will not be further discussed here. The recently discovered exoenzyme C3 ADP-ribosylates the rho proteins in eukaryotic cells (AKTORIES et al. 1989; BRAUN et al. 1989; CHARDIN et al. 1989; reviewed by AKTORIES et al. 1992) and Rho1p in yeast (MCCAFFREY et al. 1991). It has been reported that the rac proteins are also minor substrates for C3 (DIDSBURY et al. 1989). Early experiments indicated that ADP-ribosylation of rho does not alter its ability to bind or hydrolyze GTP (SEKINE et al. 1989), although a small (less than twofold) increase in GTPase activity upon ADP-ribosylation has recently been reported (MOHR et al. 1992). Nucleotide exchange rates and interaction with the rho-GAP are apparently unaffected (PATERSON et al. 1990), although the modification site is an asparagine residue within the putative effector domain (SEKINE et al. 1989). Thus, the precise effect of C3 action on rho remains unclear.

Despite the lack of an obvious effect at the biochemical level in vitro, ADP-ribosylation by C3 does alter the cellular functions of the rho proteins, as can be seen from the effects of C3 on intact cells (Rubin et al. 1988). C3 does not efficiently enter most cultured cells when added to the medium, but when C3 is artificially introduced into cells they undergo changes in morphology within a few hours (Rubin et al. 1988; Paterson et al. 1990); similar effects are seen when ADP-ribosylated rho is microinjected (Paterson et al. 1990). C3 disrupts the microfilament network without any apparent damage to actin itself (Chardin et al. 1989). These results may indicate that the cellular functions of the rho proteins include regulation of the actin microfilament network and possibly the intermediate filaments, but direct association of rho and cytoskeletal proteins has not been demonstrated. It is possible that rho affects other proteins involved in the regulation of the polymerization of actin, such as gelsolin, profilin, or the capping proteins. There is evidence of rho interaction with other, unidentified cellular proteins, as the rho proteins behave as considerably larger polypeptides on gel filtration under native conditions (E.J. Rubin, personal communication). Alternatively, the effects on the cytoskeleton may be far removed from the primary effect of the ADP-ribosylation of the rho proteins; the changes observed might be consequences of alteration of rho regulation of some other protein and reflect only the general health of the cell. In yeast, the rho homolog Rho1p appears to be associated with the late stages of the Golgi apparatus (McCaffrey et al. 1991). However, a conditional mutation in the gene encoding Rho1p prevents bud formation (McCaffrey et al. 1991) and results in cytoskeletal changes similar to those seen in actin mutants. It is likely, therefore, that the directed protein secretion required for bud formation requires interaction with the cytoskeleton and is disrupted in the mutant. Disruption of the Golgi is seen long before any effect on the cytoskeleton in Vero cells treated with a staphylococcal homolog of C3 (Sugai et al. 1992).

C3 is phage encoded and is secreted by *C. botulinum* strains C and D as a 211 amino acid polypeptide (Popoff et al. 1990, 1991). A very similar enzyme has also been identified in a strain of *C. limosum* (Aktories et al. 1992). In addition, a staphylococcal gene encoding a 212 residue enzyme with 35% homology to C3 at the amino acid level and similar enzymic activity has been cloned and characterized (Sugai et al. 1992). The C3 protein expressed in *Escherichia coli* shares the substrate specificity and enzymic activity of the native *C. botulinum* form. It has been reported that the enzymic activity of C3 is stimulated by a cytosolic protein from bovine brain (Ohtsuka et al. 1989), but other laboratories have not been able to repeat these results (E.J. Rubin and S.T. Dillon, personal communication). Another report from the same group describes a bovine brain enzyme that shares several properties with C3: it is activated by the cytosolic factor that activates C3, has a molecular mass similar to that of C3, and ADP-ribosylates the C3 substrates (Maehama et al. 1991). Both of these studies were performed with partially fractionated tissue homogenates, which raises the possibility that rho proteins or C3-like activities were present in different amounts in each fraction. The use of recombinant rho expressed in *E. coli* and more purified brain proteins, and (in the future) clones of the bovine genes would be helpful in assessing the roles of all of these factors in the ADP-ribosylation reaction.

The role of C3 in pathogenesis is unclear. C3 is not toxic when injected into mice or, in general, when added to the medium over tissue culture monolayers (RUBIN et al. 1988). It is likely that C3 represents the enzymically active or "A" component of an A–B toxin, and that an as yet unidentified cell-binding or "B" domain is secreted separately, as is the case for C2 toxin. C3 intoxication of host cells has not been demonstrated in vivo, but if it does occur C3 may cause a collapse of the Golgi apparatus and/or cytoskeleton that in some way contributes to the pathogenesis of the infection. There is evidence that a GTPase is involved in the regulation of the actin polymerization–depolymerization cycles required for neutrophil chemotaxis (BENGTSSON et al. 1990), so C3 might contribute to the pathogenesis of *C. botulinum* infections by paralyzing the chemotactic cells of the host defense system.

D. *Pseudomonas aeruginosa* Exoenzyme S

Exoenzyme S is an ADP-ribosyltransferase that is secreted by many strains of *P. aeruginosa* (IGLEWSKI et al. 1978) in addition to the better-known exotoxin A. Several experiments suggest that exoenzyme S plays a role in the pathogenesis of *P. aeruginosa* infections, but the evidence is indirect. Burned mice infected with *P. aeruginosa* display exoenzyme S activity in the serum before bacteria are detectable (BJORN et al. 1979). Transposon-insertion mutants of *P. aeruginosa* which no longer produce either form of exoenzyme S, but otherwise appear to be identical to the parental strains, have an LD_{50} in mice more than 5000 times that of the parental strain (NICAS and IGLEWSKI 1985; NICAS et al. 1985). Exoenzyme S^- mutants establish infection at the site of inoculation as well as the parental strain, but spread to other sites much less efficiently. When both strains are inoculated together, both disseminate efficiently, suggesting that exoenzyme S might be important in overcoming host defenses against spread at the site of infection, rather than for survival after dissemination (NICAS and IGLEWSKI 1985). The transposon insertion in this strain is in a region of DNA required for the production of exoenzyme S, rather than in the structural gene (FRANK and IGLEWSKI 1991), but the results obtained still support the idea that exoenzyme S does play a role in pathogenesis.

Exoenzyme S is secreted by *Pseudomonas* in two forms, with electrophoretic mobilities of 49 kDa and 53 kDa (for review see IGLEWSKI 1988). The two proteins share the same amino-terminal amino acid sequences (our unpublished observations), are immunologically cross-reactive, yield common peptides after digestion, and increase or decrease in parallel with different growth conditions. Transposon-insertion mutants deficient in exoenzyme S production do not secrete either form. The 49-kDa form is enzymically active, but the 53-kDa form is not, and neither one is toxic when injected into mice (NICAS and IGLEWSKI 1985). The two proteins may then represent only the A portion of an A–B toxin. The 53-kDa form may therefore be a precursor of the 49-kDa form, as activation by proteolysis appears to be a common feature of bacterial ADP-ribosyltransferases, including diphtheria toxin (DRAZIN et al. 1971), *Pseudomonas* exotoxin A (VASIL et al. 1977), and cholera toxin (MEKALANOS et al. 1979). Another possibility is that the two forms of exoenzyme S are products of distinct genes. If so, they would have to be encoded in the same transcription unit or regulon, given the transposon-insertion results. The cloning of the gene(s) encoding exoenzyme S will likely resolve these questions.

Exoenzyme S at first appeared to ADP-ribosylate many proteins, but a number of these can be accounted for by the abundant intermediate filament protein vimentin (COBURN et al. 1989a). Like the majority of bacterial ADP-ribosyltransferases, exoenzyme S preferentially ADP-ribosylates GTPases. These exoenzyme S substrates comprise a subset of the small GTPases and include the H-*ras*, K-*ras*, *rab3*, *rab4*, *ral*, and *rap1a* gene products (COBURN et al. 1989b; COBURN and GILL 1991). Rho, rab1, and ARF are not substrates for exoenzyme S. As is the case for cholera toxin, exoenzyme S activity is stimulated by a eukaryotic protein. However, while cholera toxin does display a low but detectable activity in the absence of ARF, exoenzyme S catalyzes the ADP-ribosylation of all substrates *only* in the presence of its activator protein, FAS (factor-activating exoenzyme S), which is distinct from ARF (COBURN et al. 1991). It is not known why bacterial enzymes that act inside the target cell should require eukaryotic proteins for activity, but this phenomenon is not unusual. In addition to ARF stimulation of cholera toxin and the requirement for FAS by exoenzyme S, the adenylyl cyclases of *Bordetella pertussis* and *Bacillus anthracis* both require the highly conserved eukaryotic protein calmodulin for activity (WOLFF et al. 1980; LEPPLA et al. 1985). As mentioned earlier, C3 may also be activated by a eukaryotic protein. It is possible that this requirement serves as insurance against ADP-ribosylation of bacterial proteins or against consumption of intracellular NAD or (in the case of the cyclases) ATP pools. It is possible that, like their mammalian counterparts, bacterial GTPases would be targets for ADP-ribosylation in the appropriate circumstances, especially in the cases of cholera toxin and exoenzyme S, which are less selective in substrate profile than many of the other enzymes. Bacterial GTPases have been less thoroughly studied that the eukaryotic GTPases, but at least two are essential for viability (reviewed by MARCH 1992).

ADP-ribosylation catalyzed by exoenzyme S does not significantly alter GTP binding, nucleotide exchange, or GTP hydrolysis by $p21^{ras}$ (COBURN and Gill 1991). Furthermore, after ADP-ribosylation no change is apparent in ras interaction with either guanine nucleotide releasing factor (GRF) or GTPase activating protein (GAP). ADP-ribosylation of $p21^{ras}$ by exoenzyme S does, however, cause a substantially larger shift in electrophoretic mobility than would be expected from the addition of the ADP-ribose group (Coburn et al. 1989b). This result suggests that a significant change in the conformation of the $p21^{ras}$ protein might occur upon ADP-ribosylation by exoenzyme S. If such a conformational change does occur, the ADP-ribosylation of $p21^{ras}$ might alter its interaction with cellular proteins aside from GRF and GAP. The existence of these other proteins is implied by numerous studies on the function of $p21^{ras}$, but not proven. If exoenzyme S does alter the functions of its substrates, it may prove useful as a tool to further explore the cellular roles of the proteins of the ras superfamily.

The rab proteins, including the exoenzyme S substrates rab3 and rab4, appear to regulate vesicle movement through the exocytic and endocytic pathways of eukaryotic cells (Chavrier et al. 1990; Goud et al. 1988; Melancon et al. 1987). Thus, exoenzyme S could interfere with secretion of microbicidal granules from cells of the host defense system by ADP-ribosylating the GTPases involved in vesicle trafficking and affecting their functions, or might disrupt the normal phagocytic functions of these cells. It has been previously suggested that exoenzyme S interferes with the host defense system at the site of infection (Nicas and Iglewski 1985; Iglewski 1988), and ADP-ribosylation of small GTPases in neutrophils and macrophages may be a particularly efficient way to disrupt their antimicrobial activities, allowing the spread of the exoenzyme S producing strains of *P. aeruginosa*. For example, GTPases appear to be involved in the secretion of neutrophil granules (Barrowman et al. 1986). These speculations are far from proven, but they are consistent with the evidence that exoenzyme S contributes to the efficiency of dissemination of *P. aeruginosa* in infection.

E. Conclusions

C3-catalyzed ADP-ribosylation of rho comprised the first suggestion of GTPase involvement in the regulation of the cytoskeleton. Exoenzyme C3 provides a relatively simple tool for probing the functions of rho proteins in mammalian cells, in which genetic approaches are difficult. C3 is therefore likely to prove valuable in the identification of physiological signals transduced by GTPases that influence the Golgi network and/or the state of the actin microfilaments in a variety of cell types.

Exoenzyme S is less specific than most other bacterial ADP-ribosyltransferases in substrate choice, but may provide an alternative strategy to understanding small GTPases that would complement genetic and immunochemical approaches. The functions of several GTPases in a variety of cellular compartments can be studied using exoenzyme S. This is particularly true of the rab proteins, which are closely related and unmerous, but do not all serve as substrates for ADP-ribosylation.

It is striking that so many bacterial toxins catalyze the ADP-ribosylation of eukaryotic GTPases (and even the C2 family, which ADP-ribosylates actin, modifies an ATP-binding protein). The basis for this selection is unknown, but the phenomenon has proven fortuitous for researchers interested in understanding the inner workings of eukaryotic cells. As more bacterial ADP-ribosyltransferases are identified and characterized, we can expect that other eukaryotic proteins and cellular processes will be open to study using these tools. In just the past few years, several clostridial toxins that modify actin have been identified, and exoenzymes C3 and S have been characterized. The staphylococcal protein that has some homology to C3 and appears to catalyze a reaction very similar to that catalyzed by C3 has been reported to stimulate the growth of, and to inhibit terminal differentiation of, cultured keratinocytes; the same results were obtained when C3 was used in parallel (SUGAI et al. 1992). Different systems, such as these two enzymes that complement each other, contribute to our knowledge of the structure and function of ADP-ribosyltransferases and to our understanding of the complex processes that occur in eukaryotic cell physiology.

Acknowledgements. I thank J. Leong and S. Dillon for critical review of the manuscript. This chapter was made possible by work done in the laboratory of D.M. Gill.

References

Aktories K, Barmann M, Ohishi I, Tsuyama S, Jakobs KH, Habermann E (1986) Botulinum C2 toxin ADP-ribosylates actin. Nature 322:390–392

Aktories K, Braun U, Rosener S, Just I, Hall A (1989) The *rho* gene product expressed in *E. coli* is a substrate of botulinum ADP-ribosyltransferase C3. Biochem Biophys Res Commun 158:209–213

Aktories K, Mohr C, Koch G (1992) *Clostridium botulinum* C3 ADP-ribosyltransferase. Curr Topics Microbiol Immunol 175:115–131

Barbacid M (1987) *ras* Genes Annu Rev Biochem 56:779–827

Barrowman MM, Cockcroft S, Gomperts BD (1986) Two roles for guanine nucleotides in the stimulus-secretion sequence of neutrophils. Nature 319:504–507

Bengtsson T, Sarndahl E, Stendahl O, Andersson T (1990) Involvement of GTP-binding proteins in actin polymerization in human neutrophils. Proc Natl Acad Sci USA 87:2921–2925

Bjorn MJ, Pavlovskis OR, Thompson MR, Iglewski BH (1979) Production of exoenzyme S during *Pseudomonas aeruginosa* infections in burned mice. Infect Immun 24:837–842

Braun U, Habermann B, Just I, Aktories K, Vandekerckhove J (1989) Purification of the 22 kDa protein substrate of botulinum ADP-ribosyltransferase C3 from porcine brain cytosol and its characterization as a GTP-binding protein highly homologous to the *rho* gene product. FEBS Lett 243:70–76

Cassel D, Selinger Z (1977) Mechanism of adenylate cyclase activation by cholera toxin: inhibition of GTP hydrolysis at the regulatory site. Proc Natl Acad Sci USA 74:3307–3311

Chardin P, Boquet P, Madaule P, Popoff MR, Rubin EJ, Gill DM (1989) The mammalian G protein rhoC is ADP-ribosylated by *Clostridium botulinum* exoenzyme C3 and affects actin microfilaments in Vero cells. EMBO J 8:1087–1092

Chavrier P, Parton RG, Hauri HP, Simons K, Zerial M (1990) Localization of low molecular weight GTP binding proteins to exocytic and endocytic compartments. Cell 62:317–329

Coburn J, Gill DM (1991) ADP-ribosylation of p21ras and related proteins by *Pseudomonas aeruginosa* exoenzyme S. Infect Immun 59:4259–4262

Coburn J, Dillon ST, Iglewski BH, Gill DM (1989a) Exoenzyme S of *Pseudomonas aeruginosa* ADP-ribosylates the intermediate filament protein vimentin. Infect Immun 57:996–998

Coburn J, Wyatt RT, Iglewski BH, Gill DM (1989b) Several GTP-binding proteins, including $p21^{c\text{-}H\text{-}ras}$, are preferred substrates of *Pseudomonas aeruginosa* exoenzyme S. J Biol Chem 264:9004–9008

Coburn J, Kane AV, Feig L, Gill DM (1991) *Pseudomonas aeruginosa* exoenzyme S requires a eukaryotic protein for ADP-ribosyltransferase activity. J Biol Chem 266:6438–6446

Collier RJ (1967) Effect of diphtheria toxin on protein synthesis: inactivation of one of the transfer factors. J Mol Biol 25:83–98

Didsbury J, Weber RF, Bokoch GM, Evans T, Snyderman R (1989) rac, a novel *ras*-related family of proteins that are botulinum toxin substrates. J Biol Chem 264:16378–16382

Drazin R, Kandel J, Collier RJ (1971) Structure and activity of diphtheria toxin: attack by trypsin at a specific site within the intact molecule. J Biol Chem 246:1504–1510

Fischer TH, Gatling MN, Lacal JC, White GC II (1990) rap 1B, a cAMP-dependent protein kinase substrate, associates with the cytoskeleton. J Biol Chem 265:19405–19408

Fischer von Mollard G, Migerny GA, Baumert M, Perin MS, Hanson TJ, Burger PM, Jahn R, Sudhof TC (1990) rab3 is a small GTP-binding protein exclusively localized to synaptic vesicles. Proc Natl Acad Sci USA 87:1988–1992

Fischer von Mollard G, Sudhof TC, Jahn R (1991) A small GTP-binding protein dissociates from synaptic vesicles during exocytosis. Nature 349:79–81

Frank DW, Iglewski BH (1991) Cloning and sequence analysis of a trans-regulatory locus required for exoenzyme S synthesis in *Pseudomonas aeruginosa*. J Bacteriol 173:6460–6468

Goud B, Salminen A, Walworth NC, Novick PJ (1988) A GTP-binding protein required for secretion rapidly associates with secretory vesicles and the plasma membrane in yeast. Cell 53:753–768

Hall A (1990) The cellular functions of small GTP-binding proteins. Science 249:635–640

Hata Y, Kikuchi A, Sasaki T, Schaber MD, Gibbs JB, Takai Y (1990) Inhibition of the *ras* p21 GTPase-activating protein-stimulated GTPase activity of c-Ha-*ras* p21 by *smg* p21 having the same effector domain as *ras* p21's. J Biol Chem 265:7104–7107

Iglewski BH (1988) Pseudomonas toxins. In: Hardegree MC, Tu AT (eds) Handbook of natural toxins, vol. 4. Dekker, New York

Iglewski BH, Kabat D (1975) NAD-dependent inhibition of protein synthesis by *Pseudomonas aeruginosa* toxin. Proc Natl Acad Sci USA 72:2284–2288

Iglewski BH, Sadoff J, Bjorn MJ, Maxwell ES (1978) *Pseudomonas aeruginosa* exoenzyme S: an adenosine diphophate ribosyl transferase distinct from toxin A. Proc Natl Acad Sci USA 75:3211–3215

Kahn RA, Gilman AG (1986) The protein cofactor necessary for ADP-ribosylation of Gs by cholera toxin is itself a GTP binding protein. J Biol Chem 261:7906–7911

Kitayama H, Sugimoto Y, Matsuzaki T, Ikawa Y, Noda M (1989) A *ras*-related gene with transformation suppressor activity. Cell 56:77–84

Kurose H, Katada T, Amano T, Ui M (1983) Specific uncoupling by islet-activating protein, pertussis toxin, of negative signal transduction via adrenergic, cholinergic, and opiate receptors in neuroblastoma X glioma hybrid cells. J Biol Chem 258:4870–4875

Leppla SH, Ivins BE, Ezzell JW Jr (1985) Anthrax toxin. In: Leive L (ed) Microbiology 1985. American Society for Microbiology, Washington

Maehama T, Takahashi K, Ohoka Y, Ohtsuka T, Ui M, Katada T (1991) Identification of a botulinum C3-like enzyme in bovine brain that catalyzes ADP-ribosylation of GTP-binding proteins. J Biol Chem 266:10062–10065

March PE (1992) Membrane-associated GTPases in bacteria. Mol Microbiol 6:1253–1257

McCaffrey M, Johnson JS, Goud B, Myers AM, Rossier J, Popoff MR, Madaule P, Boquet P (1991) The small GTP-binding protein Rho1p is localized on the Golgi apparatus and post-Golgi vesicles in *Saccharomyces cerevisiae*. J Cell Biol 115:309–319

Mekalanos JJ, Collier RJ, Romig WR (1979) Enzymic activity of cholera toxin: relationships to proteolytic processing, disulfide bond reduction, and subunit composition. J Biol Chem 254:5855–5861

Melançon P, Glick BS, Malhotra V, Weidman PJ, Serafini T, Gleason ML, Orci L, Rothman JE (1987) Involvement of GTP-binding "G" proteins in transport through the Golgi stack. Cell 51:1053–1062

Mohr C, Koch G, Just I, Aktories K (1992) ADP-ribosylation by *Clostridium botulinum* C3 exoenzyme increases steady-state GTPase activities of recombinant rhoA and rhoB proteins. FEBS Lett 297:95–99

Murayama T, Ui M (1983) Loss of the inhibitory function of the guanine nucleotide regulatory component of adenylate cyclase due to its ADP ribosylation by islet-activating protein, pertussis toxin, in adipocyte membranes. J Biol Chem 258:3319–3326

Nicas TI, Iglewski BH (1985) Contribution of exoenzyme S to the virulence of *Pseudomonas aeruginosa*. Antibiot Chemother 36:40–48

Nicas TI, Bradley J, Lochner JE, Iglewski BH (1985) The role of exoenzyme S in infections with *Pseudomonas aeruginosa*. J Infect Dis 152:716–721

Ohtsuka T, Nagata K, Iiri T, Nozawa Y, Ueno K, Ui M, Katada T (1989) Activator protein supporting the botulinum ADP-ribosyltransferase reaction. J Biol Chem 264:15000–15005

Paterson HF, Self AJ, Garret MD, Just I, Aktories K, Hall A (1990) Microinjection of recombinant $p21^{rho}$ induces rapid changes in cell morphology. J Cell Biol 111:1001–1007

Pizon V, Chardin P, Lerosey I, Oloffsson B, Tavitian A (1988) Human cDNAs rap1 and rap2 homologous to the *Drosophila* gene Dras3 encode proteins closely related to ras in the "effector" region. Oncogene 3:201–204

Popoff MR, Boquet P, Gill DM, Eklund MW (1990) DNA sequence of exoenzyme C3, an ADP-ribosyltransferase encoded by *Clostridium botulinum* C and D phages. Nucleic Acids Research 18:1291

Popoff MR, Hauser D, Boquet P, Eklund MW, Gill DM (1991) Characterization of the C3 gene of *Clostridium botulinum* types C and D and its expression in *Escherichia coli*. Infect Immun 59:3673–3679

Rubin EJ, Gill DM, Boquet P, Popoff MR (1988) Functional modification of a 21-kilodalton G protein when ADP-ribosylated by exoenzyme C3 of *Clostridium botulinum*. Mol Cell Biol 8:418–426

Sekine A, Fujiwara M, Narumiya S (1989) Asparagine residue in the *rho* gene product is the modification site for botulinum ADP-ribosyl transferase. J Biol Chem 264:8602–8605

Stearns T, Willingham MC, Botstein D, Kahn RA (1990) ADP-ribosylation factor is functionally and physically associated with the Golgi complex. Proc Natl Acad Sci USA 87:1238–1242

Sugai M, Chen C-H, Wu H (1992a) Bacterial ADP-ribosyltransferase with a substrate specificity of the rho protein disassembles the Golgi apparatus in Vero cells and mimics the action of brefeldin A. Proc Natl Acad Sci USA 89:8903–8907

Sugai M, Hashimoto K, Kikuchi A, Inoue S, Okumura H, Matsumoto K, Goto Y, Ohgai H, Moriishi K, Syuto B, Yoshikawa K, Suginaka H, Takai Y (1992b) Epidermal cell differentiation inhibitor ADP-ribosylates small GTP-binding proteins and induces hyperplasia of epidermis. J Biol Chem 267:2600–2604

Vasil ML, Kabat D, Iglewski BH (1977) Structure-activity relationships of an exotoxin of *Pseudomonas aeruginosa*. Infect Immun 16:353–361

Wolff J, Cook GH, Goldhammer AR, Berkowitz SA (1980) Calmodulin activates prokaryotic adenylate cyclase. Proc Natl Acad Sci USA 77:3841–3844

Zahraoui A, Touchot N, Chardin P, Tavitian A (1989) The human *rab* genes encode a family of GTP-binding proteins related to yeast YPT1 and SEC4 products involved in secretion. J Biol Chem 264:12394–12401

Subject Index

Springer-Verlag and the Environment

We at Springer-Verlag firmly believe that an international science publisher has a special obligation to the environment, and our corporate policies consistently reflect this conviction.

We also expect our business partners – paper mills, printers, packaging manufacturers, etc. – to commit themselves to using environmentally friendly materials and production processes.

The paper in this book is made from low- or no-chlorine pulp and is acid free, in conformance with international standards for paper permanency.